CHEMISTRY RESEARCH AND APPLICATIONS

MATHEMATICAL CHEMISTRY

CHEMISTRY RESEARCH AND APPLICATIONS

Additional books in this series can be found on Nova's website under the Series tab.

Additional E-books in this series can be found on Nova's website under the E-book tab.

CHEMISTRY RESEARCH AND APPLICATIONS

MATHEMATICAL CHEMISTRY

W.I. HONG
EDITOR

nova publishers
New York

For permission to use material from this book please contact us:
Telephone 631-231-7269; Fax 631-231-8175
Web Site: http://www.novapublishers.com

LIBRARY OF CONGRESS CATALOGING-IN-PUBLICATION DATA

Mathematical chemistry / [edited by] W.I. Hong.
p. cm.
Includes index.
ISBN 978-1-60876-894-3 (hardcover)
1. Chemistry--Mathematics. I. Hong, W. I. (Wei Inu), 1945-
QD39.3.M3M369 2009
540.15'1--dc22

2010001751

Published by Nova Science Publishers, Inc. † New York

CONTENTS

PREFACE

Mathematical chemistry is the area of research engaged in the novel applications of mathematics to chemistry; it concerns itself principally with the mathematical modeling of chemical phenomena. Mathematical chemistry has also sometimes been called computer chemistry, but should not be confused with computational chemistry.

Major areas of research in mathematical chemistry include chemical graph theory, which deals with topics such as the mathematical study of isomerism and the development of topological descriptors or indices which find application in quantitative structure-property relationships; chemical aspects of group theory, which finds applications in stereochemistry and quantum chemistry; and topological aspects of chemistry.

Low Rank Modification (LRM) is a new mathematical method by which one can express eigenvalues and eigenstates of a modified system in terms of the eigenvalues and eigenstates of some related parent system. In this respect LRM is similar to the perturbation theory which also expresses eigenvalues and eigenstates of the perturbed system in terms of the eigenvalues and eigenstates of the unperturbed system. However, there are profound differences. Perturbation theory relies on power series expansion, in most cases it produces only an approximate result, and it breaks down if the perturbation is to strong. On the other hand, LRM is exact regardless of how strong is the perturbation of the original system. Numerical efficiency of LRM depends essentially on the rank of the operators that represent modification of this system, and not on the strength of that modification. Hence LRM can successfully treat many systems where standard perturbation expansion fails. LRM applies to almost arbitrary eigenvalue problems. Moreover, it applies to standard as well to generalized eigenvalue equations. As long as modification of the parent system is represented by finite rank operators, it applies also to infinite-dimensional systems. In addition, it applies to time-independent as well as to time-dependent systems.

In Chapter 1 application of LRM to finite systems is illustrated with LRM treatment of the vibrational isotope effect in the harmonic approximation. Several examples of LRM treatment of infinite systems as well as of LRM treatment of time-dependent systems are also given.

Since ancient times, human beings have connected beauty with symmetry. In science, this correlation is found in many areas, from the fundamental laws of physics to the products of modern chemistry. However, it seems emphasize the deadness of perfect symmetry or the dissymmetry makes the phenomenon. As well known, most molecules will become less symmetrical as substitutions take place. In Chapter 2, the authors turn to a consideration of how to describe the effect of this kind of imperfect symmetry on molecular properties. Previously there are two general understandings on this issue: one is that the existence of the imperfect symmetry is denied, and there is not any symmetry here, i.e. imperfect symmetry means having none symmetry completely; another one is that the imperfection of the symmetry is completely neglected, and the system can be seen still having or almost having the previous symmetry. In fact, the previous symmetry has been damaged in local part of the system, however, the symmetry ought to be existence partially and may be called fuzzy symmetry. Recently, the authors have done some studies in this field and by using fuzzy mathematics theory the static and dynamic molecular fuzzy symmetry in relation to the point group symmetry, space periodic and some special transformations have been analyzed. The fuzzy symmetry characteristics of various molecules are studied in detail and the calculation methods for the relative parameters of molecule and molecular orbital are established. Some interesting results are obtained. It seems that there is a wonderful science virgin land waiting the theoretic chemists to open up.

Recent *Information-Theory* (IT) probes of the electronic structure and chemical bonds in molecules are surveyed. The electron redistribution accompanying the bond formation, from the (molecularly placed) free atoms $\{X^0\}$ of the "promolecule", exhibiting the ground-state densities $\{\rho_X{}^0\}$, to the bonded *Atoms-in-Molecules* (AIM) with densities $\{\rho_X\}$, is marked by the difference $\Delta\rho = \rho - \rho^0$ between the molecular ($\rho = \Sigma_X\rho_X$) and promolecular ($\rho^0 = \Sigma_X\rho_X{}^0$) electron densities, which mark the process final and initial stages, respectively. The entropy/information content of these distributions provides a basis for a novel IT perspective on the molecular electronic structure. The Schrödinger equations of quantum mechanics, which determine the equilibrium distribution of electrons and its time dependence for the fixed external potential due to the system nuclei, are shown to result from the constrained-extremum principles of the Fisher information, related to the system average kinetic energy. The densities of displacements in the Shannon entropy and missing information (cross entropy) relative to the promolecular reference are used as sensitive diagnostic tools for detecting the chemical bonds, and to monitor the promotion/hybridization changes the bonded atoms undergo in the molecular environment. When applied to a classical problem of density partition into AIM pieces the IT approach gives the familiar "stockholder" division scheme of Hirshfeld. The non-additive Fisher information density in the *Molecular Orbital* (MO) and *Atomic Orbital* (AO) resolutions respectively generate the *Electron Localization Function* (ELF) and *Contra-Gradience* (CG) probes for localizing electrons and for detecting the bonding regions in molecules, respectively. Representative examples of such an exploration of molecular ground-states will be presented. The *Orbital Communication Theory* (OCT) of the chemical bond uses the standard entropy/information descriptors of the Shannon theory of communication to characterize the AO scattering of electron probabilities, throughout the network of chemical bonds generated by the system occupied MO. The molecule is thus treated as an information system which propagates from the channel AO "inputs" to AO "outputs" the "signals" of electron allocations to these basis functions. The underlying conditional probabilities generated from the bond-projected superposition principle of quantum mechanics are shown to be proportional to the squares of the corresponding elements of the *first*-order density matrix in the AO representation, thus being related to Wiberg's quadratic indices of the chemical bond multiplicity. Such information propagation in molecules exhibits the communication "noise", due to electron delocalization via the system chemical bonds, which effectively lowers the information content in the output-signal distribution, compared to that contained in probabilities determining the channel input-signal, molecular or promolecular. The orbital information systems are used to generate the entropic measures of the chemical bond multiplicity and their covalent/ionic composition for both the molecule as a whole and its diatomic fragments. The average *conditional-entropy*, which measures the channel communication-noise due to electron delocalization, measures the IT-*covalency* in the molecule, while the complementary descriptor of the network *mutual-information* (information-capacity) reflects the electron localization effects and measures the system IT-*ionic* component. The illustrative examples of applying these IT tools for exploring the electronic and bonding structures of representative molecules are reported and discussed in Chapter 3. Other applications of the IT concepts to classical issues in quantum chemistry are also briefly summarized. The illustrative *Restricted-Hartree-Fock* (RHF) OCT results for diatomic fragments of selected molecules, which reproduce the Wiberg bond-order in diatomic molecules, are examined.

Polyhedra have attracted scientists' attentions due to their high-symmetric architectures since ancient times, and even served as common modes in natural world, for example, the polyhedral skeletons of molecular structures. Chemists, in the past few centuries, have strived to synthesize these polyhedral targets. Despite achievements that have been arrived, there is still a gap between the great varieties of polyhedral shapes observed in nature (such as viral protein capsids) and the relatively limited molecular polyhedra constructed by small organic molecules. Fortunately, DNA was shown to be an excellent material in molecular construction. The construction of polyhedral structures with DNA improves the development of synthetic chemistry, and meanwhile, produces some complicated molecules with novel topological and geometrical structures, such as the DNA tetrahedron, DNA cube, DNA octahedron, DNA dodecahedron, and DNA icosahedron, as well as the tetrahedral DNA cage and the hexahedral DNA box, and so on. Interests on these novel architectures have inspired mathematical considerations and modeling of a new theory of polyhedra, including structure construction, symmetry analysis, chirality detection and invariant computaion. In Chapter 4, the authors have summarized advances in the chemical syntheses of DNA polyhedra, which also have been envisaged and classified by underlying principles of geometry and symmetry. Moreover, this paper reviews their recent progresses in the theoretical investigation of a new mathematical theory of DNA polyhedra.

Chapter 5 deals with the design of high-gain observers for the unknown kinetics in two types of biochemical reactors: Batch (and Fed-Batch) Bioreactors, and Continuous Stirred Tank Bioreactors. The general dynamical state-space model of bioprocesses that take place inside these types of bioreactors is presented, and some particular bioprocesses are widely analyzed, such as wastewater biomethanation process, activated sludge bioprocesses, lipase production process. The models of the bioprocesses are obtained by using the classical modeling approach, which utilizes the mass balance of the components inside the bioreactor and a set of modeling rules. Also, a novel modeling procedure for bioprocesses is introduced, based on the pseudo Bond Graph approach, which uses the reaction schemes and the analysis of biochemical phenomena. The nonlinear bioprocesses are very complex and are characterized by the lack of on-line measurements. Therefore, in order to estimate the imprecisely known kinetic rates of these bioprocesses, some high-gain observers are designed. The estimation scheme does not require any model for the kinetic rates. The tuning of the proposed observers is reduced to the calibration of a single parameter. If the high-gain observers require on-line state estimates, these will be provided by asymptotic observers. In order to study the behavior and the performance of the proposed observers, extensive numerical simulations are performed.

In Chapter 6, some remarks, examples and further directions, concerning HOA as a tool to solve and provide analytical insight into solutions of dynamical systems occurring in, but not limited to Mathematical Chemistry, are posited. Among these more general considerations are the developement of Novel Exact Analytical solutions of Generalized Hamiltonian/Lagrangian Dynamical systems in the Quantum Phase Space Representation and Classical Connections; also the exact analytical solutions of attendant differential/difference equations. Though detailed treatment of these and many others shall appear elsewhere, the foundations of the HOA scheme will be sketched and some examples given in these Remarks, that will illustrate some of the numerous and sundry analyses that can be enhanced by HOA methods. For convenience, the authors begin with a recap of the HOA construction following from [ref. 6].

As explained in Chapter 7, there is an infinite series of 3-,4-connected, crystalline patterns that emerge from the elaboration of 1,4-cyclohexadiene-like, hydrocarbon building block generating fragments, onto the template of each and every possible tessellation. Probably, the archetypal crystalline pattern in this series is the so-called closoglitter lattice described below. Like all the infinite set of glitter series patterns, as just described, closoglitter is built from elaborating properly a quasi-closest packed tessellation thus with 1,4-cyclohexadiene-like moieties. And thus any arbitrary 2D tessellation can be elaborated like this, with 1,4-cyclohexadiene-like hydrocarbon fragments to produce novel 3D patterns thus. Furthermore it is true, that each and every one of these three-dimensional (3D) patterns, generated from two-dimensional (2D) tessellations, should have a set of "n" π-π* crystal orbitals (CO's), in a frontier manifold that determines the electronic properties of the lattice. Here "n" is the rotational order of the space group that the glitter series pattern lies in. And these π-π* CO's are able to be visualized at some symmetry point of the reciprocal space of the lattice, and are composed of pure p atomic orbitals in their symmetry adapted combinations, and they are beautiful to sketch on paper as they represent various shadings of the innumerable 2D tessellations thus.

In Chapter 8, several potential heuristic angles are explored with regard to the parametrization of the superconducting transition temperature, T(c), and its relationship to the chemical physics of bonding. Fruitful angles explored in this paper include the relationship of the gas of Cooper pairs in a superconductor to a van der Waals equation of state for such a fluid and how it may be controlled and exploited through the consideration of such variables as pressure, P, and volume, V, of the fluid of Cooper pairs resident in a superconductor. Other angles explored, in an attempt to get a heuristic handle on the superconducting transition temperature, T(c), for superconducting compositions across the Periodic Table, include the introduction of a Morse's anharmonic chemical bond potential for the Debeye frequency of the phonons as it emerges from the central result for T(c) in superconductors from the Bardeen-Cooper-Schreifer (BCS) formulation of superconductivity. Yet another angle explored in this heuristic reasoning, is the use of a Badger's relationship between chemical bond force constant, k, and equilibrium internuclear distance, $r_{(e)}$, to try to gain some understanding of the nature of the electron-phonon coupling mechanism in superconductors. A relationship between $r_{(e)}$ and $D_{(e)}$ emerges as a result of this line of thought, given as $r_{(e)} \cdot D_{(e)}$ = "critical temperature constant", from which examination of spectroscopic data across the Periodic Table may yield a compositional solution of the appropriate chemical bonding, with the assumption of the proper electron-phonon coupling constant, C, and density of states, N(0), that ultimately may yield more desirable superconductivity transition temperatures.

In Chapter 9, the 2D graphical representation of genetic sequence as proposed by Gates, Nandy and Leong & Morgenthaler have been enhanced by ensuring the number of overlaps is denoted by different line intensities. With the availability of this additional information, distribution of weak and strong H-bonds can be obtained from the intensity-enhanced Gates representation, while pyrimidine and purine distribution can be extracted based on intensity-enhanced Nandy representation. The intensity-enhanced Leong-Morgenthaler representation allows the distribution of keto and amino groups to be plotted. These intensity-enhanced graphical representations are herein illustrated for the first exon of human beta-globin.

Some physically interesting equations are naturally set on a semi-infinite interval and it would be desirable to have accurate and efficient numerical methods for such problems. The usual Laguerre polynomials have very poor resolution properties when compared with other types of orthogonal polynomials. However, the authors may obtain uniformly convergent Galerkin approximations using scaled Laguerre functions in regular unbounded domains, and construct stable and efficient numerical algorithms for their practical implementations. In Chapter 10, the authors introduce and illustrate some algorithmic ideas with several simple problems from mathematical chemistry.

Chapter 11 presents different approaches to the analysis of the stability of chemical and phase equilibrium in multicomponent heterogeneous systems. The traditional approach for the analysis of thermodynamic stability is presented. Le Chatelier – Braun principle is considered as alternative approach to the stability criteria. New variants of formulations of this statement are presented. The interconnection between non-conjugated thermodynamic parameters is also discussed on the base of Le Chatelier – Braun principle. Thermodynamic inequalities, both stability conditions and Le Chatelier – Braun principle, are considered for the case of phase transitions in multicomponent systems without chemical reactions. Thermodynamic inequalities are presented for the systems with chemical interactions. The general system of thermodynamic inequalities and chains of inequalities for non-reacting and reacting systems are considered at different external conditions. The possibility is discussed to describe different types of physicochemical processes on the base of the generalized concept of affinity. The illustration is given of the application of thermodynamic inequalities to various physicochemical systems and processes: e.g. the shifting of phase and chemical equilibrium in systems under different external conditions. Various transformations of the stability matrix are presented as variants of mathematical formulations of stability conditions. The possibility is discussed to use the stability criteria for the characterisation of the non-equilibrium process. Inequalities for the second derivatives of thermodynamic potentials give an opportunity to compare the changes of thermodynamic parameters in equilibrium and non-equilibrium processes. Processes in homogeneous and heterogeneous systems are discussed, including the processes accompanied by the generation of new phases. The interpretation is given for so-called jumps in the values of the second derivatives of thermodynamic potentials.

The Sanderson's geometric mean equalization principle for electronegativity has allowed expressions for chemical potential and molecular hardness to be derived within the framework of the Conceptual Density Functional Theory for obtaining molecular electronic properties of an entire molecule or supramolecule from the isolated fragments. Along with the determination of electronic properties, in former articles it has been demonstrated that Sanderson's scheme can be very useful as a method for rationalizing chemical reactions when both N (the number of electrons) and $\upsilon(\mathrm{r})$ (the external potential) change. The main problem that arises when using this approximation to estimate molecular properties is that the bonding potential and redistribution of the electron density among the fragments are not considered, and structural relaxation due to new specific through-bond and through-space interactions is not allowed within this scheme. Therefore this principle presents an important restriction to be used. In Chapter 12 a suitable example to expose this limitation is mathematically developed by means of a general deduced mathematical expression for a polymeric growing chain according to this principle. This reveals that a task for improving this scheme is pending.

Chapter 13 focuses on a recently developed generalized pseudospectral method for accurate, efficient treatment of certain central potentials of interest in various branches in quantum mechanics, usually having singularity. Essentially this allows *optimal, nonuniform* spatial discretization of the pertinent single-particle Schrödinger equation satisfying Dirichlet boundary condition leading to standard diagonalization of symmetric matrices. Its validity and feasibility have been demonstrated for a wide range of important potentials such as Hulthén, Yukawa, generalized spiked harmonic oscillators, Hellmann, Coulomb potentials without/with various perturbations (for instance, linear and quadratic) etc. Although initially designed for *singular* potentials, this has also been remarkably successful for various other cases such as power-law, logarithmic, harmonic potentials containing higher order perturbations, 3D rational potentials as well as confinement studies. Furthermore, a large number of low-, moderately high-, high-lying multiply excited Rydberg states such as singly, doubly excited He as well as triply excited *hollow 2l2l'2l'' (n ≥ 2) and 3l3l'3l'' doubly-hollow* resonances in manyelectron atoms have been treated by this approach within a Hohenberg-Kohn-Sham density functional theory with great success. This offers very high-quality results for *both ground and higher lying* states for arbitrary values of potential parameters (covering both *weak and strong* coupling) with equal ease and efficacy. In all cases, excellent agreement with literature results are observed; in many cases this surpasses the accuracy of all other existing results while in other occasions the authors' results are comparable to the best ones available in literature. This provides a simple general efficient route towards the understanding of a number of singular and other potentials of relevance in quantum mechanics including multiply excited Rydberg series of many-electron systems. This helps predicting many new states for the first time.

Simulations of complex stochastic reaction networks are often performed by Monte Carlo methods such as the Gillespie algorithm. In case that a stochastic reaction network contains certain fast reactions, simulations of the fast reactions by Monte Carlo methods are computationally inefficient and expensive, and researches have focused on reduction methods for fast reactions to reduce the computational intensity of the simulations. In Chapter 14, the authors first review reduction methods for multiple time scale stochastic reaction networks and then discuss possible future works on such methods.

In: Mathematical Chemistry
Editor: W.I. Hong, pp. 1-151

ISBN: 978-1-60876-894-3

Chapter 1

EXACT TREATMENT OF FINITE RANK MODIFICATIONS OF FINITE-DIMENSIONAL AND INFINITE-DIMENSIONAL QUANTUM SYSTEMS

Tomislav P. Živković***
Ruđer Bošković Institute, Zagreb, Croatia

Abstract

Low Rank Modification (LRM) is a new mathematical method by which one can express eigenvalues and eigenstates of a modified system in terms of the eigenvalues and eigenstates of some related parent system. In this respect LRM is similar to the perturbation theory which also expresses eigenvalues and eigenstates of the perturbed system in terms of the eigenvalues and eigenstates of the unperturbed system. However, there are profound differences. Perturbation theory relies on power series expansion, in most cases it produces only an approximate result, and it breaks down if the perturbation is to strong. On the other hand, LRM is exact regardless of how strong is the perturbation of the original system. Numerical efficiency of LRM depends essentially on the rank of the operators that represent modification of this system, and not on the strength of that modification. Hence LRM can successfully treat many systems where standard perturbation expansion fails. LRM applies to almost arbitrary eigenvalue problems. Moreover, it applies to standard as well to generalized eigenvalue equations. As long as modification of the parent system is represented by finite rank operators, it applies also to infinite-dimensional systems. In addition, it applies to time-independent as well as to time-dependent systems.

In this chapter application of LRM to finite systems is illustrated with LRM treatment of the vibrational isotope effect in the harmonic approximation. Several examples of LRM treatment of infinite systems as well as of LRM treatment of time-dependent systems are also given.

* E-mail address: zivkovic@irb.hr. (Corresponding author)

perturbation expansion [1,2]. The main idea of this method is simple: if one knows the solution of some parent system $\boldsymbol{A}$, one can find the solution of the "perturbed" system $\mathbf{C}$. It is assumed that the system $\mathbf{C}$ does not differ from the parent system very much, in which case "perturbation" is small. Next one expresses eigenvalues and eigenstates of the perturbed systems as a power series expansion. Each term in this expansion depends on the eigenvalues and eigenstates of the parent system. Hence in order to obtain the solution of the perturbed system, one has to know the solution of the parent system.

There are two problems with this approach. First, computation of each successive term in the perturbation expansion is in most cases much more time-consuming than the computation of the preceding term. One is hence usually restricted to the calculation of only few leading terms. Hence one obtains only an approximate and not an exact solution of the perturbed system. If the perturbation is not small, in order to obtain an acceptable solution one has to calculate so many terms that the entire approach becomes computationally unacceptable. Second, if the perturbation is large enough, perturbation expansion may diverge, in which case the solution cannot be obtained in this way. However, the problems which involve large perturbations are in many cases of prime interest. For example, if one knows the solution of some molecule $\boldsymbol{A}$, one would like to find the solution of the related "perturbed" molecule $\mathbf{C}$ which differs from the original molecule by a single heteroatom substitution or by few such substitutions. However, the replacement of an atom by a heteroatom is by no means a small perturbation. Hence such and similar problems cannot be treated efficiently within the standard perturbation theory.

A new approach, which avoids such problems, was developed some time ago [3-6]. This approach was originally named Low Rank perturbation (LRP). However, the term "perturbation" was misleading suggesting that this approach is a variant of the perturbation expansion, which is not the case. In order to avoid any further misunderstanding, this method was recently renamed Low Rank Modification (LRM) [7].

LRM is a general mathematical method by which one can express the eigenvalues and the eigenstates of the modified system $\mathbf{C}$ in terms of the eigenvalues and the eigenstates of the parent system $\boldsymbol{A}$. In this respect LRM is similar to the perturbation expansion. However, unlike perturbation expansion, LRM does not rely on any kind of power series, and it is exact however large "modification" of the original system. The only requirement on this method is that the operators which represent this modification should have a finite rank. Numerical complexity of LRM depends essentially on the rank ρ of those operators. This method is hence particularly suitable for those cases when this rank is relatively small. Substitution of an atom by a heteroatom in a large molecule, provided this substitution changes molecular geometry only in the immediate vicinity of this substitution, can be approximated as such a low rank modification. Similar example is a substitution of one or few atoms by isotopes of those atoms. Such substitutions have negligible effect on the molecular wavefunctions and on the corresponding energies. However, they have a significant influence on the molecular vibrations. Accordingly, vibrational isotope effect is another example of the modification of the original system which is not small but which can be represented as a low rank modification [6].

2. Description of Finite-Dimensional Systems

Consider a finite-dimensional time-independent quantum system $\boldsymbol{A}_n$ that contains n eigenvalues λ_i and n eigenstates $|\Phi_i\rangle$. Usually those eigenvalues and eigenstates satisfy a standard eigenvalue equation $\mathbf{H}_a|\Phi_i\rangle = \lambda_i|\Phi_i\rangle$ ($i = 1,\cdots,n$) where $\mathbf{H}_a$ is a Hermitean operator. However, in some cases instead of this simple eigenvalue equation one has to consider a generalized eigenvalue equation

$$\mathbf{H}_a|\Phi_i\rangle = \lambda_i\mathbf{S}_a|\Phi_i\rangle, \quad i = 1,\cdots,n, \tag{1a}$$

where $\mathbf{S}_a$ is positive-definite Hermitean operator. Eigenvalues λ_i of (1a) are real and the corresponding eigenstates $|\Phi_i\rangle$ span n-dimensional space X_n^a. Those eigenstates can be orthonormalized according to

$$\langle\Phi_i|\mathbf{S}_a|\Phi_j\rangle = \delta_{ij}, \quad i,j = 1,\cdots,n. \tag{1b}$$

If $\lambda_i \neq \lambda_j$ eigenstate $|\Phi_i\rangle$ is automatically orthogonal to the eigenstate $|\Phi_j\rangle$ according to (1b). However if $\lambda_i = \lambda_j$ one has to enforce this condition by some appropriate orthonormalization procedure such as Gram-Schmidt, QR, or alike [8,9].

Solution of a generalized eigenvalue equation of a type (1a) is for example required in the treatment of the molecular vibrations in the harmonic approximation. In the Cartesian coordinates this problem leads to the eigenvalue equation of a type $\mathbf{F}|\Phi_i\rangle = \lambda_i\mathbf{M}|\Phi_i\rangle$ where $\mathbf{F}$ and $\mathbf{M}$ are force field and mass operators, respectively [10-13]. Those operators are Hermitean and in addition operator $\mathbf{M}$ is positive definite. As another example, consider various methods for the calculation of molecular electronic structure [2]. Molecular systems are too complicated to be solved exactly. Hence one has to apply an approximate method. One of the most important approximate methods is a Linear Combination of Atomic Orbitals (LCAO). In this approach one expands each molecular wave-function in terms of the finite number of known atomic orbitals. This leads to a matrix eigenvalue equation of a type (1a). Matrix $\mathbf{S}_a$ in this expression contains overlaps between various atomic orbitals. Orbitals situated on the same atom are orthogonal to each other. Overlap between orbitals situated on two atoms that are far from each other in a molecule is usually so small that it can be neglected. However, overlap between orbitals situated on adjacent atoms is usually not negligible. There are many semi-empirical methods which further simplify LCAO approach [2]. Some of those methods are so extreme as to neglect all overlaps between mutually distinct orbitals [2]. This applies for example to the Hückel molecular orbital theory (HMO) which treats only π-electron conjugated systems. However, in the treatment of general molecular systems involving σ-orbitals more realistic semi-empirical methods are needed

[2]. Most of those methods require the solution of the eigenvalue equation of a general type (1a) where $\mathbf{S}_a$ is a nontrivial positive-definite operator. In order to treat such and similar systems, generalized eigenvalue equation of a type (1a) should be considered.

3. Internal and External Modifications of Finite-Dimensional Systems

One can modify parent system $\boldsymbol{A}_n$ described by the eigenvalue equation (1a) in two different ways. One can modify this system *internally* or *externally.*

In the case of an internal modification, modified system $\mathbf{C}_n$ has the same dimension as the parent system $\boldsymbol{A}_n$. In particular, eigenstates $|\Psi_s\rangle$ of $\mathbf{C}_n$ and eigenstates $|\Phi_i\rangle$ of $\boldsymbol{A}_n$ span the same space X_n^a. In the most general formulation, modification of the parent system is represented by two operators $\mathbf{V}$ and $\mathbf{P}$. Operator $\mathbf{V}$ modifies operator $\mathbf{H}_a$, while operator $\mathbf{P}$ modifies operator $\mathbf{S}_a$. Those operators are Hermitean, and they act in the space X_n^a. Both operators are scaled by the same coupling parameter β. This parameter is real and it is introduced for convenience, in order to have a clear insight into the effect of the change of the modification strength. If β is small, LRM expressions reduce to the well known expressions obtained within the formalism of the perturbation expansion. However if β is large, LRM produces qualitatively new features which cannot be obtained within a standard perturbation expansion approach.

In the case of an internal modification, modified system $\mathbf{C}_n$ is described by the generalized eigenvalue equation

$$\mathbf{H}_c|\Psi_s\rangle = \varepsilon_s \mathbf{S}_c|\Psi_s\rangle, \quad s = 1,\ldots,n, \tag{2a}$$

where

$$\mathbf{H}_c = \mathbf{H}_a + \beta\mathbf{V}, \quad \mathbf{S}_c = \mathbf{S}_a + \beta\mathbf{P}, \tag{2b}$$

In order to guarantee reality of the eigenvalues ε_s, operator $\mathbf{S}_c$ is required to be positive-definite in X_n^a. This requirement imposes some restrictions on the modification operator $\mathbf{P}$ and on the coupling parameter β. There are no restrictions on the modification operator $\mathbf{V}$, except that this operator should be Hermitean. In analogy to (1b), eigenstates $|\Psi_s\rangle$ of (2a) can be orthonormalized according to

$$\langle\Psi_s|\mathbf{S}_c|\Psi_p\rangle = \delta_{sp}, \quad s, p = 1,\ldots,n. \tag{2c}$$

The case of an internal modification is shown schematically in Figure 1.

Let me introduce the notion of *rank* and *range* to be used in this chapter. By definition, *range* of the operator $\mathbf{O}$ that acts in the space X is image of this operator [8,9]. In other words, this is the space spanned by all vectors of the type $\mathbf{O}|\phi\rangle$ where $|\phi\rangle \in X$. Dimension of this space is *rank* of $\mathbf{O}$ [8,9]. If operator $\mathbf{O}$ is Hermitean, rank of $\mathbf{O}$ equals the number (counting degeneracies) of nonzero eigenvalues of this operator.

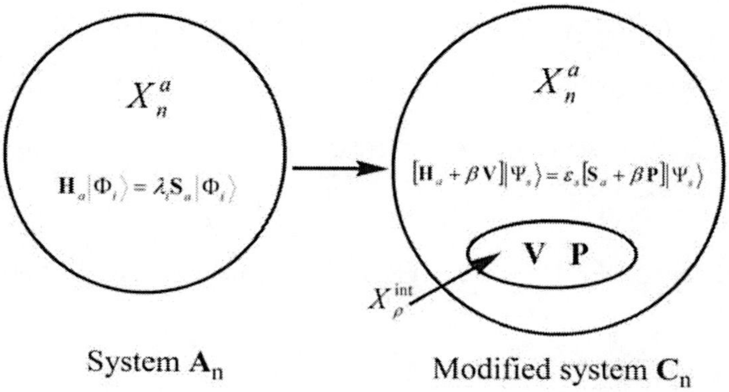

Figure 1. Internal modification of a finite-dimensional system. System $\mathbf{A}_n$ is modified internally by a generalized modification $(\mathbf{V},\mathbf{P})$. This modification effects interaction space X_ρ^{int}.

Above definition of rank and range of an arbitrary operator $\mathbf{O}$ can be generalized in a natural way to the notion of *rank* and *range* of the combined modification $(\mathbf{V},\mathbf{P})$. By definition, *range* of the combined modification $(\mathbf{V},\mathbf{P})$ is the space X_ρ^{int} spanned by all states of the type $\mathbf{V}|\phi\rangle$ and all states of the type $\mathbf{P}|\phi\rangle$ where $|\phi\rangle \in X_n^a$. Dimension of this space is *rank* of the combined modification and it is denoted by ρ. One easily finds $\rho \le \rho_h + \rho_s$ where ρ_h is the rank of the modification operator $\mathbf{V}$ while ρ_s is the rank of the modification operator $\mathbf{P}$. In particular, if $\mathbf{P} = 0$ one has $\rho = \rho_h$, while if $\mathbf{V} = 0$ one has $\rho = \rho_s$. One can also consider operator $\mathbf{O}(x) = \mathbf{V} + x\mathbf{P}$ which depends on a parameter x and on modification operators $\mathbf{V}$ and $\mathbf{P}$. One finds that for almost each x this operator satisfies $rank(\mathbf{O}(x)) = \rho$ and $range(\mathbf{O}(x)) = X_\rho^{\text{int}}$. One can have $rank(\mathbf{O}(x)) < \rho$ only for some isolated values of x. In this case range of the operator $\mathbf{O}(x)$ is a subspace of X_ρ^{int}. The space X_ρ^{int} which is the range of the combined modification is the *interaction space*. It is that part of the space X_n^a which is modified by the operators $\mathbf{V}$ and $\mathbf{P}$.

Let $\{|\mu\rangle\}$ be a base in X_ρ^{int} orthonormalized in a standard way

$$\langle \mu | \nu \rangle = \delta_{\mu\nu}, \quad \mu, \nu = 1, \ldots, \rho, \tag{3a}$$

This base can be extended to the orthonormalized base $\{|\alpha\rangle\}$ in the space X_n^a:

$$\langle \alpha | \beta \rangle = \delta_{\alpha\beta}, \quad \alpha, \beta = 1, \ldots, n, \tag{3b}$$

For the sake of reference, labels α and β will refer to the entire space X_n^a, while labels μ and ν will refer to the interaction space X_ρ^{int}, subspace of X_n^a. Base $\{|\mu\rangle\}$ considered as a set is a subset of the set $\{|\alpha\rangle\}$. In the base $\{|\alpha\rangle\}$ equations (1a) and (2a) are $n \times n$ matrix eigenvalue equations, while in the base $\{|\mu\rangle\}$ operators $\mathbf{V}$ and $\mathbf{P}$ are $\rho \times \rho$ matrices. Projection operator $\mathbf{I}_\rho^{\text{int}}$ on the space X_ρ^{int} can be expressed in terms of the vectors $|\mu\rangle$ as

$$\mathbf{I}_\rho^{\text{int}} = \sum_\mu^\rho |\mu\rangle\langle\mu|, \tag{3c}$$

while unit operator $\mathbf{I}_n^a$ in the space X_n^a can be expressed in terms of vectors $|\alpha\rangle$ as

$$\mathbf{I}_n^a = \sum_\alpha^n |\alpha\rangle\langle\alpha|. \tag{3d}$$

One finds that modification operators satisfy

$$\mathbf{V} = \mathbf{I}_\rho^{\text{int}} \mathbf{V} \mathbf{I}_\rho^{\text{int}}, \quad \mathbf{P} = \mathbf{I}_\rho^{\text{int}} \mathbf{P} \mathbf{I}_\rho^{\text{int}}. \tag{4}$$

As an example of the internal modification, consider replacement of one or several atoms in a large molecule with isotopes. To a very good approximation, isotopic substitutions do not change molecular geometry and they have negligible influence on the molecular electronic structure. However, such substitutions have significant influence on molecular vibrations. Given frequencies and normal modes of a parent molecule $\boldsymbol{A}_n$, of interest are frequencies and normal modes of various isotopomers $\mathbf{C}_n$ of this molecule. Similar examples of internal modifications are replacements of one or few selected atoms in a large molecule with heteroatoms, creation and/or destruction of chemical bonds, etc.

In the case of an external modification, system $\boldsymbol{A}_n$ interacts with another ρ-dimensional system $\boldsymbol{B}_\rho$ which is outside the original system $\boldsymbol{A}_n$. With the system $\mathbf{B}_\rho$ is associated ρ-

dimensional space X_ρ^b which is orthogonal to the space X_n^a. In analogy to (1a), system $\boldsymbol{B}_\rho$ alone is described by the generalized eigenvalue equation

$$\mathbf{H}_b|\Theta_r\rangle = E_r\mathbf{S}_b|\Theta_r\rangle, \quad r = 1,\cdots,\rho, \tag{5a}$$

where $\mathbf{H}_b$ and $\mathbf{S}_b$ are Hermitean operators that act in the space X_ρ^b and where $\mathbf{S}_b$ is in addition positive definite in this space. No other assumption about those operators is made. Hermiticity of those operators and positive definiteness of $\mathbf{S}_b$ ensures that the eigenvalues E_r of (5a) are real. In analogy to (1b), the corresponding eigenstates $|\Theta_r\rangle$ can be orthonormalized according to :

$$\langle\Theta_r|\mathbf{S}_b|\Theta_t\rangle = \delta_{rt}, \quad r,t = 1,\cdots,\rho. \tag{5b}$$

Since the states $|\Theta_r\rangle$ form a complete set in X_ρ^b, this implies

$$\sum_r^\rho |\Theta_r\rangle\langle\Theta_r|\mathbf{S}_b = \mathbf{I}_\rho^b. \tag{5c}$$

where $\mathbf{I}_\rho^b$ is a projection operator on the space X_ρ^b.

Inclusion of the interaction between initially noninteracting systems $\mathbf{B}_\rho$ and $\boldsymbol{A}_n$ creates a combined $(\rho+n)$-dimensional system $\mathbf{C}_{n+\rho}$. Each state in this combined system is contained in the $(\rho+n)$-dimensional space $X_{n+\rho}^c$, orthogonal sum of spaces X_ρ^b and X_n^a. Interaction between subsystems $\mathbf{B}_\rho$ and $\boldsymbol{A}_n$ of the combined system $\mathbf{C}_{n+\rho}$ is described by the Hermitean operators $\mathbf{V}$ and $\mathbf{P}$. In analogy to the case of an internal modification, those operators are scaled by the coupling parameter β. This parameter is again introduced for convenience, in order to have a clear insight into the differences between weak and strong modifications. In the case of an external modification, modification operators connect mutually distinct spaces X_ρ^b and X_n^a, and those operators vanish over the space X_ρ^b as well as over the space X_n^a. Combined system $\mathbf{C}_{n+\rho}$ that includes this interaction is described by the generalized eigenvalue equation

$$\mathbf{H}_c|\Psi_s\rangle = \varepsilon_s\mathbf{S}_c|\Psi_s\rangle, \quad s = 1,\ldots,n+\rho, \tag{6a}$$

where

$$\mathbf{H}_c = \mathbf{H}_a + \mathbf{H}_b + \beta \mathbf{V}, \quad \mathbf{S}_c = \mathbf{S}_a + \mathbf{S}_b + \beta \mathbf{P}, \tag{6b}$$

In order to guarantee reality of the eigenvalues ε_s, operator $\mathbf{S}_c$ is required to be positive-definite in $X^c_{n+\rho}$. In analogy to (2c), eigenstates $|\Psi_s\rangle$ of (6a) can be orthonormalized according to

$$\langle \Psi_s | \mathbf{S}_c | \Psi_p \rangle = \delta_{sp}, \quad s, p = 1,\dots,n+\rho. \tag{6c}$$

Let $\{|r\rangle\}$ be a base in X^b_ρ orthonormalized in a standard way

$$\langle r|t\rangle = \delta_{rt}, \quad r,t = 1,\dots,\rho, \tag{7a}$$

Let further $\{|\alpha\rangle\}$ be a base in X^a_n orthonormalized according to (3b). Projection operator $\mathbf{I}^b_\rho$ on the space X^b_ρ can be expressed in terms of the vectors $|r\rangle$ as

$$\mathbf{I}^b_\rho = \sum_r^\rho |r\rangle\langle r| \, . \tag{7b}$$

while projection operator $\mathbf{I}^a_n$ on the space X^a_n is given by (3d).

The case of an external modification is shown schematically in Figure 2. Note that the role of the mutually interacting systems $\mathbf{B}_\rho$ and $\mathbf{A}_n$ is symmetrical. One can consider system $\mathbf{B}_\rho$ to be a base system which is modified by the interaction with the system $\mathbf{A}_n$, or one can consider system $\mathbf{A}_n$ to be a base system which is modified by the interaction with the system $\mathbf{B}_\rho$. I will take this first point of view and consider system $\mathbf{B}_\rho$ to be a base system. All expressions to be derived are equally valid if the roles of systems $\mathbf{B}_\rho$ and $\mathbf{A}_n$ are permuted. Essentially what one has is the interaction of two systems that are both subsystems of the larger system $\mathbf{C}_{n+\rho}$. Each of those two systems can be considered to be modified by the interaction with another system.

The role of the space X^b_ρ in the case of an external modification is similar to the role of the interaction space X^{int}_ρ in the case of an internal modification. Note also that operator $\mathbf{I}^a_n$ is in the case of an internal modification a unit operator since X^a_n is the entire space considered, while in the case of an external modification $\mathbf{I}^a_n$ is a projection operator on the space X^a_n, subspace of the space $X^c_{n+\rho}$.

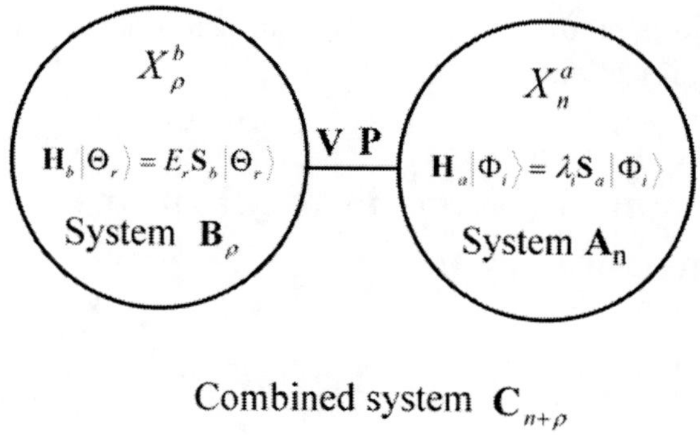

Figure 2. External modification of a finite-dimensional system. Finite systems $\mathbf{A}_n$ and $\mathbf{B}_\rho$ interact with generalized interaction $(\mathbf{V},\mathbf{P})$. This forms a combined system $\mathbf{C}_{n+\rho}$. From the point of view of the system $\mathbf{B}_\rho$, this system is modified externally by the interaction with the system $\mathbf{A}_n$.

Since in the case of an external modification operators $\mathbf{V}$ and $\mathbf{P}$ that describe interaction between systems $\mathbf{B}_\rho$ and $\boldsymbol{A}_n$ have non-vanishing matrix elements only between spaces X_ρ^b and X_n^a, one has

$$\mathbf{V} = \mathbf{I}_\rho^b\mathbf{V}\mathbf{I}_n^a + \mathbf{I}_n^a\mathbf{V}\mathbf{I}_\rho^b, \quad \mathbf{P} = \mathbf{I}_\rho^b\mathbf{P}\mathbf{I}_n^a + \mathbf{I}_n^a\mathbf{P}\mathbf{I}_\rho^b. \tag{8}$$

From the point of view of the base system $\mathbf{B}_\rho$, system $\boldsymbol{A}_n$ plays the role of the surrounding of $\mathbf{B}_\rho$. System $\mathbf{B}_\rho$ without the interaction with the system $\boldsymbol{A}_n$ is a closed system. This system alone is described by the eigenvalue equation (5a). System $\mathbf{B}_\rho$ subject to the interaction with its surrounding is an open system. If from the solution of the combined system $\mathbf{C}_{n+\rho}$ one extracts only those properties that refer to its subsystem $\mathbf{B}_\rho$, one obtains properties of an open system. Therefore expressions to be derived can be considered to describe an open quantum system.

A comment concerning a quantity ρ is in place here. In the case of an internal modification, ρ is defined as the rank of the combined modification $(\mathbf{V},\mathbf{P})$. The corresponding interaction space X_ρ^{int} is a range of this modification. In the case of an external modification, ρ is the dimension of the system $\mathbf{B}_\rho$ that interacts with the n-dimensional system $\boldsymbol{A}_n$. By the known theorem of algebra, rank of the $n\times m$ matrix is at most $\min(n,m)$ [8,9]. In matrix form operator $\boldsymbol{V}$ is represented by one (ρ,n) and one (n,ρ) sub-matrix. The same applies to the operator $\mathbf{P}$. Thus one finds that if $n\geq\rho$ (which is usually the case) rank of the combined modification $(\mathbf{V},\mathbf{P})$ is at most 2ρ. Hence from the more general point of view, external modifications as well as internal modifications are both low rank modifications.

4. LRM Treatment of External Modifications of Finite-Dimensional Systems

Consider an external modification of a ρ-dimensional base system $\mathbf{B}_\rho$. This system interacts with the n-dimensional parent system $\boldsymbol{A}_n$ through the generalized modification $(\mathbf{V},\mathbf{P})$. Combined system $\mathbf{C}_{n+\rho}$ is described by the generalized eigenvalue equation (6a). In analogy to the perturbation expansion, one assumes that eigenvalues λ_i and eigenstates $|\Phi_i\rangle$ of the system $\boldsymbol{A}_n$ are known. From those quantities one can construct Hermitean operator $\overline{\mathbf{\Omega}}(\varepsilon)$ which depends on a real parameter ε:

$$\overline{\mathbf{\Omega}}(\varepsilon)=\sum_{i(\lambda_i\neq\varepsilon)}^{n}\frac{|\Phi_i\rangle\langle\Phi_i|}{\varepsilon-\lambda_i}\,, \tag{9}$$

If ε equals some eigenvalue λ_i of the eigenvalue equation (1a) ($\varepsilon=\lambda_i$), the corresponding term or terms is excluded from the summation in (9).

Operator $\overline{\mathbf{\Omega}}(\varepsilon)$ is associated with the parent system $\boldsymbol{A}_n$. Since $\{|\Phi_i\rangle\}$ is a complete set in X_n^a, in the case $\varepsilon\notin\{\lambda_i\}$ one has $\overline{\mathbf{\Omega}}(\varepsilon)|\psi\rangle\neq 0$ for each nontrivial state $|\psi\rangle\in X_n^a$. Operator $\overline{\mathbf{\Omega}}(\varepsilon)$ is in this case nonsingular in X_n and hence $rank(\overline{\Omega}(\varepsilon))=n$. However, if $\varepsilon=\lambda_j\in\{\lambda_i\}$ one has $\overline{\mathbf{\Omega}}(\varepsilon)|\Phi_j\rangle=0$ and in this case $rank(\overline{\Omega}(\varepsilon))<n$.

With the base system $\mathbf{B}_\rho$ that interacts with the system $\boldsymbol{A}_n$ is associated Hermitean operator $\mathbf{\Omega}(\varepsilon)$

$$\mathbf{\Omega}(\varepsilon)=(\mathbf{V}-\varepsilon\,\mathbf{P})\overline{\mathbf{\Omega}}(\varepsilon)(\mathbf{V}-\varepsilon\,\mathbf{P})\,, \tag{10a}$$

This operator is a key quantity in the LRM formalism involving external modifications of finite dimensional quantum systems. Due to the expressions (8), operator $\mathbf{\Omega}(\varepsilon)$ has nonvanishing matrix elements only between the states contained in the space X_ρ^b. Over the space X_n^a this operator vanishes. In the base $\{|r\rangle\}$ of the space X_ρ^b this operator is a $\rho\times\rho$ matrix with matrix elements

$$\Omega_{rt}(\varepsilon) \equiv \langle r|\boldsymbol{\Omega}(\varepsilon)|t\rangle = \sum_{i(\lambda_i \neq \varepsilon)}^{n} \frac{\langle r|\mathbf{V}-\varepsilon\mathbf{P}|\Phi_i\rangle\langle\Phi_i|\mathbf{V}-\varepsilon\mathbf{P}|t\rangle}{\varepsilon-\lambda_i}, \quad r,t = 1,\ldots,\rho \qquad (10b)$$

In order to construct operator $\boldsymbol{\Omega}(\varepsilon)$ it is not necessary to have complete information about eigenstates $|\Phi_i\rangle$ of $\mathbf{A}_n$. It is sufficient to know only $2\rho n$ matrix elements $\langle r|\mathbf{V}|\Phi_i\rangle$ and $\langle r|\mathbf{P}|\Phi_i\rangle$ ($r = 1,\ldots,\rho$; $i = 1,\ldots,n$), as well as n eigenvalues λ_i of (1a). This makes a total of $(2\rho+1)n$ quantities. On the other hand, in order to specify all n eigenstates $|\Phi_i\rangle$ of $\mathbf{A}_n$ one has to know n^2 matrix elements $\langle\alpha|\Phi_i\rangle$ ($\alpha = 1,\ldots,n$; $i = 1,\ldots,n$). Hence if $\rho << n$ construction of the operator $\boldsymbol{\Omega}(\varepsilon)$ requires much less information than exact specification of all eigenstates $|\Phi_i\rangle$ of (1a).

For the sake of simplicity, operators that act in spaces X_n^a, X_ρ^b and $X_{n+\rho}^c$ will be identified with their representations in bases $\{|\alpha\rangle\}$, $\{|r\rangle\}$ and $\{|r\rangle,|\alpha\rangle\}$, respectively. For example, representation of the operator $\mathbf{I}_\rho^b$ in the base $\{|r\rangle\}$ is a unit matrix with matrix elements δ_{rt} ($r,t = 1,\ldots,\rho$). Considered as the operator in the space $X_{n+\rho}^c$, $\mathbf{I}_\rho^b$ is a projection operator on the subspace X_ρ^b of $X_{n+\rho}^c$. Strictly, this identification is not allowed. However, from the context it is usually clear whether $\mathbf{O}$ denotes an operator or a representation of this operator in the space X_n^a, X_ρ^b or $X_{n+\rho}^c$.

In a perturbation theory one treats in a different way degenerate and non-degenerate eigenvalues and eigenstates of the non-perturbed system [1,2,14]. In LRM one treats in a different way *cardinal* ($\varepsilon_s \notin \{\lambda_i\}$) and *singular* ($\varepsilon_s \in \{\lambda_i\}$) eigenvalues and eigenstates of the combined system [3-7]. Distinction between cardinal and singular eigenstates in LRM is not the same as the distinction between degenerate and non-degenerate eigenstates in the perturbation theory. For example, in the case of LRM cardinal solutions, it is not important whether the corresponding initial eigenvalue λ_i is degenerate or not, as long as modified eigenvalue ε_s differs from all initial eigenvalues.

In addition to LRM expressions that provide exact values for the modified eigenvalues ε_s and for the corresponding eigenstates $|\Psi_s\rangle$, qualitative distribution of modified eigenvalues ε_s relative to the parent eigenvalues λ_i is also of interest. Due to the interaction, modified eigenvalues ε_s of the combined system shift relative to the parent eigenvalues λ_i. One finds that maximum possible shift of those eigenvalues is not arbitrary, and that this shift is mainly determined by the dimension ρ of the space X_ρ^b. Let me first consider in qualitative terms how eigenvalues ε_s of the combined system $\mathbf{C}_{n+\rho}$ are related to the eigenvalues λ_i of the parent system $\boldsymbol{A}_n$. This is given by *interlacing rule* [4].

4.1. Interlacing of Modified Eigenvalues ε_s with Parent Eigenvalues λ_i

Eigenvalues ε_s of the combined system $\mathbf{C}_{n+\rho}$ and eigenvalues λ_i of the parent system $\boldsymbol{A}_n$ satisfy [4]:

Interlacing rule I (External Modifications)

Arrange eigenvalues λ_i ($i = 1,\cdots,n$) of $\boldsymbol{A}_n$ in a nondecreasing order and arrange also eigenvalues ε_s ($s = 1,\cdots,n+\rho$) of $\mathbf{C}_{n+\rho}$ in a non-decreasing order. These eigenvalues thus arranged are interlaced according to

$$\varepsilon_i \le \lambda_i \le \varepsilon_{i+\rho}, \; i = 1,\cdots,n. \tag{11}$$

In particular, condition $\lambda_1 \le \varepsilon_{1+\rho}$ implies that at most ρ modified eigenvalues $\varepsilon_1,\ldots,\varepsilon_\rho$ can be smaller than the first parent eigenvalue λ_1. Similarly, condition $\varepsilon_n \le \lambda_n$ implies that at most ρ modified eigenvalues $\varepsilon_{n+1},\ldots,\varepsilon_{n+\rho}$ can be larger than the last parent eigenvalue λ_n. Hence at least $n-\rho$ modified eigenvalues ε_s are confined to the interval $[\lambda_1, \lambda_n]$.

As a consequence of the above rule one has:
Corollary 1:

Let eigenvalue λ_j of $\boldsymbol{A}_n$ be η_j-degenerate. In this case at most $\eta_j + \rho$ modified eigenvalues ε_s of $\mathbf{C}_{n+\rho}$ can be equal to λ_j. In addition if $\eta_j > \rho$ at least $\eta_j - \rho$ modified eigenvalues ε_s of $\mathbf{C}_{n+\rho}$ are equal to λ_j.

This corollary directly follows from the above interlacing rule. Namely if λ_j is η_j-degenerate, one has

$$\lambda_{j-1} < \lambda_j = \lambda_{j+1} = \ldots = \lambda_{j+\eta_j-1} < \lambda_{j+\eta_j}.$$

Interlacing rule now implies $\varepsilon_{j-1} \le \lambda_{j-1} < \lambda_j$ and $\lambda_{j+\eta_j-1} < \lambda_{j+\eta_j} \le \varepsilon_{j+\eta_j+\rho}$. Hence only $\eta_j + \rho$ modified eigenvalues $\varepsilon_j, \varepsilon_{j+1},\ldots,\varepsilon_{j+\eta_j+\rho-1}$ can be equal to the parent eigenvalue λ_j. This proves first part of the above corollary. Similarly one derives second part of this corollary.

According to the corollary 1, in the case of an external modification, operators $\mathbf{V}$ and $\mathbf{P}$ can change (increase or decrease) degeneracy of each particular eigenvalue λ_j of $\boldsymbol{A}_n$ at most by ρ.

Let me now consider cardinal and singular solutions of the combined system $\mathbf{C}_{n+\rho}$ in more details.

4.2. Cardinal Solutions of the Combined System

Concerning cardinal solutions of $\mathbf{C}_{n+\rho}$ one has [4]:

Theorem I: (cardinal eigenvalues and eigenstates):
Let eigenstates $|\Phi_i\rangle$ of (1a) be orthonormalized according to (1b). Than

a) $\varepsilon_s \notin \{\lambda_i\}$ is a cardinal eigenvalue of the combined eigenvalue equation (6a) if and only if it satisfies

$$\left[\beta^2\mathbf{\Omega}(\varepsilon_s)+\mathbf{H}_b\right]|\theta_s\rangle=\varepsilon_s\mathbf{S}_b|\theta_s\rangle,\quad |\theta_s\rangle\in X_\rho^b, \tag{12a}$$

In the base $\{|r\rangle\}$ above expression is a $\rho\times\rho$ matrix equation. Each cardinal eigenvalue $\varepsilon_s \notin \{\lambda_i\}$ of (6a) is hence a root of the $\rho\times\rho$ determinant $h(\varepsilon)$

$$h(\varepsilon)\equiv\left|\beta^2\mathbf{\Omega}(\varepsilon)+\mathbf{H}_b-\varepsilon\mathbf{S}_b\right|=0. \tag{12b}$$

b) $|\Psi_s\rangle$ is an eigenstate of the eigenvalue equation (6a) that has cardinal eigenvalue $\varepsilon_s \notin \{\lambda_i\}$ if and only if it is a linear combination

$$|\Psi_s\rangle=\beta\sum_i^n\frac{\langle\Phi_i|\mathbf{V}-\varepsilon_s\mathbf{P}|\theta_s\rangle}{\varepsilon_s-\lambda_i}|\Phi_i\rangle+|\theta_s\rangle. \tag{13a}$$

where $|\theta_s\rangle\in X_\rho^b$ is an eigenstate of (12a) that corresponds to the eigenvalue $\varepsilon=\varepsilon_s$.

c) Degeneracy of cardinal eigenvalue $\varepsilon_s \notin \{\lambda_i\}$ of the combined system $\mathbf{C}_{n+\rho}$ equals nullity of the matrix $\beta^2\mathbf{\Omega}(\varepsilon_s)+\mathbf{H}_b-\varepsilon_s\mathbf{S}_b$. Since this is a $\rho\times\rho$ matrix, each cardinal eigenvalue $\varepsilon_s \notin \{\lambda_i\}$ can be at most ρ-degenerate.

Note that according to (13a) projection of the cardinal eigenstate (13a) on the space X_ρ^b equals $|\theta_s\rangle \in X_\rho^b$:

$$\mathbf{I}_\rho^b|\Psi_s\rangle = |\theta_s\rangle \in X_\rho^b. \tag{13b}$$

Cardinal eigenstates (13a) can be orthonormalised according to (6c). If $\varepsilon_s \neq \varepsilon_p$ eigenstates $|\Psi_s\rangle$ and $|\Psi_p\rangle$ automatically satisfy $\langle\Psi_s|\mathbf{S}_c|\Psi_p\rangle = 0$ [4]. However, if $\varepsilon_s = \varepsilon_p$ one has to enforce orthogonality of those eigenstates by some standard orthonormalization procedure [8,9].

There is a strong formal similarity between the LRM equation (12a) and standard equations that describe perturbation of the base system $\mathbf{B}_\rho$. Equation (12a) formally describes ρ-dimensional system $\mathbf{B}_\rho$ described by the eigenvalue equation (5a) that is perturbed by the perturbation $\beta^2\mathbf{\Omega}(\varepsilon_s)$. This perturbation depends on the eigenvalue ε_s, and LRM equation (12a) is hence an iterative eigenvalue equation. According to (13b), eigenstate $|\theta_s\rangle$ of this equation equals projection of the modified eigenstate $|\Psi_s\rangle$ on the space X_ρ^b. In conclusion, equation (12a) is a $\rho \times \rho$ eigenvalue equation, it acts entirely in the space X_ρ^b, and the solution to this equation describes system $\mathbf{B}_\rho$ subject to the interaction $(\mathbf{V},\mathbf{P})$ with the external system $\mathbf{A}_n$. In addition, according to the expression (13a), this equation also determines complete cardinal eigenstate $|\Psi_s\rangle$ of the combined system $\mathbf{C}_{n+\rho}$.

Unlike LRM equation (12a) which is a $\rho \times \rho$ eigenvalue equation, original eigenvalue equation (6a) that describes combined system $\mathbf{C}_{n+\rho}$ is a $(\rho+n)\times(\rho+n)$ eigenvalue equation. However large dimension n of the parent system $\mathbf{A}_n$, dimension of the corresponding LRM equation (12a) is still only ρ. Nevertheless, this LRM equation produces exact solutions of the $(\rho+n)\times(\rho+n)$ eigenvalue equation (6a). Hence if the system $\mathbf{A}_n$ is much larger than the system $\mathbf{B}_\rho$, LRM treatment will be numerically much faster than any other standard treatment of the eigenvalue equation (6a).

4.3. Singular Solutions of the Combined System

Concerning singular eigenvalues and eigenstates of the combined system $\mathbf{C}_{n+\rho}$ one has [4]

Theorem 2 (singular solutions): Let λ_j be a η_j-degenerate eigenvalue of the eigenvalue equation (1a) and let $X_{\eta_j}^{\lambda j}$ be the corresponding η_j-dimensional space spanned by η_j degenerate eigenstates $|\Phi_{jl}\rangle$ ($l = 1,\ldots,\eta_j$) of (1a). Let further $\mathbf{I}_{\eta_j}^{\lambda j}$ be the projection on the space $X_{\eta_j}^{\lambda j}$. Then:

Each singular eigenstate $|\Psi_s\rangle$ of the combined system $\mathbf{C}_{n+\rho}$ that has eigenvalue $\varepsilon_s = \lambda_j \in \{\lambda_i\}$ is a linear combination

$$|\Psi_s\rangle = \beta \sum_{i(\lambda_i \neq \varepsilon_s)}^{n} \frac{\langle \Phi_i | \mathbf{V} - \varepsilon_s \mathbf{P} | \theta_s \rangle}{\varepsilon_s - \lambda_i} |\Phi_i\rangle + |\theta_s\rangle + |\varphi_s\rangle \quad , \tag{14a}$$

where

$$|\theta_s\rangle \in X_\rho^b, \; |\varphi_s\rangle \in X_{\eta_j}^{\lambda j}. \tag{14b}$$

and where those two states satisfy

$$\left[\beta^2 \mathbf{\Omega}(\varepsilon_s) + \mathbf{H}_b - \varepsilon_s \mathbf{S}_b\right] |\theta_s\rangle + \beta(\mathbf{V} - \varepsilon_s \mathbf{P}) \mathbf{I}_{\eta_j}^{\lambda j} |\varphi_s\rangle = 0 \quad , \tag{15a}$$

$$\beta \mathbf{I}_{\eta_j}^{\lambda j} (\mathbf{V} - \varepsilon_s \mathbf{P}) |\theta_s\rangle = 0 \quad . \tag{15b}$$

In a matrix form, expressions (15) form a set of $(\rho + \eta_j)$ homogenous linear equations in $(\rho + \eta_j)$ unknowns, ρ unknown coefficients $C_r^{(s)}$ that determine the state $|\theta_s\rangle \in X_\rho^b$ and η_j unknown coefficients $D_l^{(s)}$ that determine the state $|\varphi_s\rangle \in X_{\eta_j}^{\lambda j}$:

$$|\theta_s\rangle = \sum_r^{\rho} C_r^{(s)} |r\rangle \quad , \quad |\varphi_s\rangle = \sum_l^{\eta_j} D_l^{(s)} |\Phi_{jl}\rangle \quad . \tag{16}$$

Combined system $\mathbf{C}_{n+\rho}$ has a singular eigenstate with the eigenvalue $\varepsilon_s = \lambda_j \in \{\lambda_i\}$ if and only if determinant of the system (15) vanishes. If λ_j is η_j-degenerate, this system may have at most $(\rho + \eta_j)$ modified eigenstates $|\Psi_s\rangle$ with the eigenvalue $\varepsilon_s = \lambda_j \in \{\lambda_i\}$. Further, since $|\Phi_i\rangle \in X_n^a$ and $|\varphi_s\rangle \in X_n^a$, X_ρ^b-component of the modified eigenstate (14a) equals

$$\mathbf{I}_\rho^b|\Psi_s\rangle = |\theta_s\rangle \in X_\rho^b, \tag{17a}$$

This expression is analogous to the expression (13b) that applies to cardinal eigenstates of the combined system. One also finds

$$\sum_l^{\eta_j} |\Phi_{jl}\rangle\langle\Phi_{jl}|\mathbf{S}_a|\Psi_s\rangle = |\varphi_s\rangle \in X_{\eta_j}^{\lambda_j}. \tag{17b}$$

It is convenient to distinguish *strongly singular* and *weakly singular* eigenstates. By definition, strongly singular eigenstate $|\Psi_s\rangle$ of the eigenvalue equation (6a) has no component in the space X_ρ^b, *i.e.* it satisfies $\mathbf{I}_\rho^b|\Psi_s\rangle = 0$. On the other hand, each weakly singular eigenstate $|\Psi_s\rangle$ satisfies $\mathbf{I}_\rho^b|\Psi_s\rangle \neq 0$. In addition, each weakly singular eigenstate should be orthogonal to all strongly singular eigenstates.

Theorem 2 implies:

Theorem 2a:

a). $|\Psi_s\rangle$ is *strongly singular* eigenstate of the eigenvalue equation (6a) with the eigenvalue $\varepsilon_s = \lambda_j$ if and only if this state satisfies:

$$|\Psi_s\rangle \equiv |\varphi_s\rangle \in X_{\eta_j}^{\lambda j}, \quad (\mathbf{V} - \varepsilon_s\mathbf{P})\mathbf{I}_{\eta_j}^{\lambda j}|\varphi_s\rangle = 0. \tag{18}$$

This state is hence a linear combination

$$|\Psi_s\rangle = \sum_l^{\eta_j} D_l^{(s)}|\Phi_{jl}\rangle, \tag{19a}$$

where the coefficients $D_l^{(s)}$ satisfy

$$\sum_l^{\eta_j} \langle r|\mathbf{V} - \varepsilon_s\mathbf{P}|\Phi_{jl}\rangle D_l^{(s)} = 0, \quad r = 1,\ldots,\rho. \tag{19b}$$

The set of all strongly singular eigenstates associated with the eigenvalue $\varepsilon_s = \lambda_j$ spans some r_j-dimensional space $X_{r_j}^{\lambda j-}$, subspace of the space $X_{\eta_j}^{\lambda j}$. Since the space $X_{\eta_j}^{\lambda j}$ is η_j-dimensional while the space X_ρ^b is ρ-dimensional, expression (19b) implies

$$\eta_j - \rho \le r_j \le \eta_j. \tag{20}$$

b). $|\Psi_s\rangle$ is weakly singular eigenstate of the generalized eigenvalue equation (6a) with the eigenvalue $\varepsilon_s = \lambda_j$ if and only if it is a linear combination (14a) where the states $|\theta_s\rangle$ and $|\varphi_s\rangle$ satisfy expressions (15) and where $|\theta_s\rangle \neq 0$. In addition, each weakly singular eigenstate should be orthogonal to all strongly singular eigenstates.

c). If $|\Psi_s\rangle$ is weakly singular eigenstate with the eigenvalue $\varepsilon_s = \lambda_j$, the state $|\varphi_s\rangle$ in the expression (14a) is uniquely determined by the state $|\theta_s\rangle \in X_\rho^b$ in this expression. Hence and since the space X_ρ^b is ρ-dimensional, one may have at most ρ weakly singular eigenstates with the eigenvalue $\varepsilon_s = \lambda_j$, however large degeneracy η_j of the parent eigenvalue λ_j.

According to the above Theorem, there is a substantial qualitative difference between strongly and weakly singular eigenstates of the combined system. If degeneracy η_j of the eigenvalue λ_j of the parent system $\mathbf{A}_n$ satisfies $\eta_j > \rho$, combined system $\mathbf{C}_{n+\rho}$ has at least $\eta_j - \rho$ strongly singular eigenstates with the eigenvalue $\varepsilon_s = \lambda_j$. If this degeneracy is very large, one may have very many strongly singular eigenstates with this eigenvalue. Those eigenstates do not depend on the coupling parameter β, and they exist however small this parameter. Further, each strongly singular eigenstate $|\Psi_s\rangle$ of the modified system $\mathbf{C}_{n+\rho}$ is at the same time an eigenstate of the parent system $\mathbf{A}_n$. Accordingly, modification operators $\mathbf{V}$ and $\mathbf{P}$ have no influence on strongly singular eigenstates. On the other hand, each weakly singular eigenstate $|\Psi_s\rangle$ depends on a coupling parameter β as well as on the modification operators $\mathbf{V}$ and $\mathbf{P}$. Unlike strongly singular eigenstates of the combined system, each weakly singular eigenstate of this system differs from all eigenstates of the parent system $\mathbf{A}_n$. In addition, modified system may have at most ρ weakly singular eigenstates with the eigenvalue $\varepsilon_s = \lambda_j$, however large degeneracy η_j of the eigenvalue λ_j. Since expressions (15) represent the set of $(\rho+\eta_j)$ homogenous linear equations in $(\rho+\eta_j)$ unknowns, and since each weakly singular eigenstate is required to satisfy $|\theta_s\rangle \neq 0$, and since in addition this state is required to be orthogonal to all strongly singular eigenstates, in the case of weakly singular eigenstates those expressions are highly overdetermined. Hence, unless there is some special reason, combined system contains no weakly singular eigenstate. This special reason is usually due to symmetry. One finds that combined system $\mathbf{C}_{n+\rho}$ may have some weakly singular eigenstates only if this system is more symmetrical than original systems $\mathbf{A}_n$ and $\mathbf{B}_\rho$.

Above two theorems produce all solutions of the combined system $\mathbf{C}_{n+\rho}$. All cardinal eigenvalues $\varepsilon_s \notin \{\lambda_i\}$ of the combined eigenvalue equation are roots of the function $h(\varepsilon)$. Once a particular root ε_s of $h(\varepsilon)$ is found, the corresponding eigenstate (or eigenstates) is given by eq. (13a) where the state $|\theta_s\rangle \in X_\rho^b$ satisfies (12a). Concerning remaining singular eigenvalues $\varepsilon_s \in \{\lambda_i\}$, each such eigenvalue coincides with some eigenvalue λ_j of a parent system, and all one has to do is to verify conditions (15). This verification is highly simplified by the expressions (19b) which produce all strongly singular eigenstates. If all modified solutions are required, in the numerical implementation of LRM one has first to calculate all cardinal solutions using expressions (12). If there are $(n+\rho)$ such solutions LRM calculation is completed, since combined system has exactly $(n+\rho)$ solutions. However if there are less than $(n+\rho)$ cardinal solutions, one has to calculate strongly singular solutions. This can be done using expressions (19). Only if the total number of all cardinal plus all strongly singular solutions is still less than $(n+\rho)$, one has to look for weakly singular solutions.

4.4. Numerical Considerations

In the computer implementation of the above method, most important and most time-consuming is the solution of the equation (12b) which produces all cardinal eigenvalues of the combined system. Operation count required to obtain singular solutions of the combined system is negligible with respect to the operation count required to obtain cardinal solutions of this system. In order to find a particular eigenvalue ε_s of the equation (12b), one has to recalculate in an iterative way function $h(\varepsilon)$ for many different values of ε. Each recalculation of this function requires calculation of matrix elements (10b) of the operator $\mathbf{\Omega}(\varepsilon)$. Operation count for the calculation of those matrix elements is of the order $O(\rho^2 n)$. Operation count for a single calculation of the function $h(\varepsilon)$ is dominated by the operation count for this operator. This operation count is hence also of the order $O(\rho^2 n)$. Though one has to recalculate this function many times, the number of those calculations is independent on ρ and on n [4]. Operation count for a single modified eigenvalue is hence still of the order $O(\rho^2 n)$. Accordingly, operation count for all eigenvalues is of the order $O(\rho^2 n^2)$. On the other hand, operation count of a standard diagonalization which produces all eigenvalues of the eigenvalue equation (6a) is of the order $O((n+\rho)^3) > O(n^3)$ [8,9]. Hence if all modified eigenvalues are needed, LRM will be numerically more efficient if $\rho << \sqrt{n}$. In addition, if a single or only few selected eigenvalues are required, LRM is comparatively even more efficient since it requires only $O(\rho^2 n)$ operations per eigenvalue, while all known numerical methods require substantially more than $O((n+\rho)^2) \approx O(n^2)$ operations per eigenvalue.

Once an eigenvalue $\varepsilon_s \notin \{\lambda_i\}$ is obtained, calculation of the corresponding eigenstate or eigenstates $|\Psi_s\rangle$ requires the solution of the expression (12a). Since ε_s is known, this calculation is of the order $O(\rho^3)$. Since $\rho << \sqrt{n} < n$ this operation count is negligible with respect to the operation count required to obtain the eigenvalue ε_s in the first place.

One finds that multiple recalculation of the function $h(\varepsilon)$ is more time-consuming in the case $\mathbf{P} \neq 0$ than in the case $\mathbf{P} = 0$ [4]. If $\mathbf{P} = 0$ one can devise much faster algorithm for this recalculation. This is due to the fact than in the case $\mathbf{P} \neq 0$ the unknown ε is in the nominator as well as in the denominator of the matrix elements (10b) of the operator $\mathbf{\Omega}(\varepsilon)$. One can express matrix $\mathbf{\Omega}(\varepsilon)$ in an equivalent form where this unknown is eliminated from the nominator of those matrix elements [4]:

$$\mathbf{\Omega}(\varepsilon) = \Omega'(\varepsilon) + \mathbf{K} + \varepsilon\,\mathbf{\Lambda} \quad , \tag{21a}$$

where

$$\Omega'(\varepsilon) = \sum_{i(\lambda_i \neq \varepsilon)}^{n} \frac{(\mathbf{V} - \lambda_i \mathbf{P})|\Phi_i\rangle\langle\Phi_i|(\mathbf{V} - \lambda_i \mathbf{P})}{\varepsilon - \lambda_i} \quad , \tag{21b}$$

$$\mathbf{K} = \sum_{i}^{n} \left[\lambda_i \mathbf{P}|\Phi_i\rangle\langle\Phi_i|\mathbf{P} - \mathbf{P}|\Phi_i\rangle\langle\Phi_i|\mathbf{V} - \mathbf{V}|\Phi_i\rangle\langle\Phi_i|\mathbf{P} \right] \quad , \tag{21c}$$

$$\mathbf{\Lambda} = \sum_{i}^{n} \mathbf{P}|\Phi_i\rangle\langle\Phi_i|\mathbf{P} \quad . \tag{21d}$$

Expressions (12) can be written in terms of the above operators as [4]

$$\left[\beta^2\left(\mathbf{\Omega}'(\varepsilon_s) + \mathbf{K}\right) + \mathbf{H}_b\right]|\theta_s\rangle = \varepsilon_s\left(\mathbf{S}_b - \beta^2\mathbf{\Lambda}\right)|\theta_s\rangle \; , \; |\theta_s\rangle \in X_\rho^b \; , \tag{22a}$$

$$h(\varepsilon) \equiv \left|\beta^2\left(\mathbf{\Omega}'(\varepsilon) + \mathbf{K}\right) + \mathbf{H}_b - \varepsilon\left(\mathbf{S}_b - \beta^2\mathbf{\Lambda}\right)\right| = 0 \quad . \tag{22b}$$

Expression (22b) seems to be more complicated than the equivalent expression (12b). However, it is numerically easier to obtain roots of the expression (22b) than roots of the expression (12b). This is due to the fact that according to (21b) operator $\mathbf{\Omega}'(\varepsilon)$ has unknown variable ε only in its denominator. Concerning operators $\mathbf{K}$ and $\mathbf{\Lambda}$, those operators do not depend on ε, and hence matrix elements of those operators can be calculated only once, i.e. there is no need to recalculate those matrix elements in each iteration cycle. Using expression (22b) instead of the expression (12b) one can substantially speed up LRM calculation of each particular eigenvalue of (12a) [4].

4.5. Normalization of Modified Eigenstates

Cardinal eigenstate $|\Psi_s\rangle$ as given by the expression (13a) is not normalized. This can be easily done once this expression is known. Eigenstate (13a) can be normalized according to

$$|\Psi_s\rangle \rightarrow |\Psi_s'\rangle = \frac{1}{\sqrt{N_s}}|\Psi_s\rangle \quad . \tag{23a}$$

where $|\Psi_s\rangle$ is given by (13a) while normalization constant N_s equals [4]

$$N_s \equiv \langle\Psi_s|\mathbf{S}_c|\Psi_s\rangle = \beta^2\sum_i^n \frac{\langle\theta_s|\mathbf{V}-\varepsilon_s\mathbf{P}|\Phi_i\rangle\langle\Phi_i|\mathbf{V}-\varepsilon_s\mathbf{P}|\theta_s\rangle}{(\varepsilon_s-\lambda_i)^2} + \langle\theta_s|\mathbf{S}_b|\theta_s\rangle +$$
$$+\beta^2\sum_i^n \frac{\langle\theta_s|\mathbf{P}|\Phi_i\rangle\langle\Phi_i|\mathbf{V}-\varepsilon_s\mathbf{P}|\theta_s\rangle + \langle\theta_s|\mathbf{V}-\varepsilon_s\mathbf{P}|\Phi_i\rangle\langle\Phi_i|\mathbf{P}|\theta_s\rangle}{\varepsilon_s-\lambda_i} \tag{23b}$$

This expression can be written in a compact form [4]

$$N_s = \beta^2\sum_i^n \frac{\langle\theta_s|\mathbf{V}-\lambda_i\mathbf{P}|\Phi_i\rangle\langle\Phi_i|\mathbf{V}-\lambda_i\mathbf{P}|\theta_s\rangle}{(\varepsilon_s-\lambda_i)^2} - \beta^2\langle\theta_s|\mathbf{\Lambda}|\theta_s\rangle + \langle\theta_s|\mathbf{S}_b|\theta_s\rangle \quad , \tag{23c}$$

In particular if $\mathbf{P}=0$ (the most important case) one has

$$N_s = \beta^2\sum_i^n \frac{\langle\theta_s|\mathbf{V}|\Phi_i\rangle\langle\Phi_i|\mathbf{V}|\theta_s\rangle}{(\varepsilon_s-\lambda_i)^2} + \langle\theta_s|\mathbf{S}_b|\theta_s\rangle \quad . \tag{23d}$$

It is remarkable that normalization constant N_s involves only calculations within the space X_ρ^b associated with a base system $\mathbf{B}_\rho$, and no calculation within the larger space X_n^a. In order to normalize cardinal eigenstate $|\Psi_s\rangle \in X_{n+\rho}^c$ it is not necessary to know X_n^a-component $|\Psi_s^a\rangle$ of this eigenstate. Eigenvalue equation (12a) produces the state $|\theta_s\rangle$ and the corresponding eigenvalue ε_s, and this is sufficient, in connection with known eigenvalues λ_i and known eigenstates $|\Phi_i\rangle$ of the parent system, to fix normalization constant N_s. Hence, though cardinal eigenstate $|\Psi_s\rangle$ extends over the entire space $X_{n+\rho}^c$, one can normalize this eigenstate using only its projection on the much smaller space X_ρ^b. The projection $\mathbf{I}_n^a|\Psi_s\rangle \in X_n^a$ of this eigenstate on the space X_n^a is not explicitly required for the calculation of the normalization constant N_s. This independence of the normalization constant N_s on the space X_n^a is an important result of the LRM approach.

Concerning normalization of singular eigenstates (14a), particularly simple is the normalization of strongly singular eigenstates (19a). As implied by this expression, in order to normalize a strongly singular eigenstate $|\Psi_s\rangle$, one has to know the state (19a) which is contained in the subspace $X_{\eta_j}^{\lambda j}$ of the space X_n^a. This differs from the normalization of cardinal eigenstate $|\Psi_s\rangle$ which does not require the knowledge of the X_n^a component of this eigenstate. However, as far as properties of the open system $\mathbf{B}_\rho$ are concerned, strongly singular eigenstates are not important since they have no component in the corresponding space X_ρ^b.

Above theorems apply to generalized eigenvalue equations (1a), (5a) and (6a). All the corresponding relations substantially simplify if those equations are not completely generalized. The most important special case is $\mathbf{P}=0$. This allows for the unperturbed eigenvalue equations (1a) and (5a) to be still of the most general type. Only the modified equation (6a) is slightly restricted with this requirement. This eigenvalue equation is still a generalized eigenvalue equation, though not of a most general type. If $\mathbf{P}=0$ operators $\mathbf{K}$ and $\mathbf{\Lambda}$ in the expressions (21,22) vanish, and hence one has $\mathbf{\Omega}(\varepsilon)=\mathbf{\Omega}'(\varepsilon)$. One has $\mathbf{P}=0$ in a special but highly important case when instead of generalized eigenvalue equations (1a), (5a) and (6a) one has standard eigenvalue equations $\mathbf{H}_a|\Phi_i\rangle=\lambda_i|\Phi_i\rangle$, $\mathbf{H}_b|\Theta_r\rangle=E_r|\Theta_r\rangle$ and $(\mathbf{H}_a+\mathbf{H}_b+\beta\mathbf{V})|\Psi_s\rangle=\varepsilon_s|\Psi_s\rangle$, respectively.

4.6. Numerical Example

In order to illustrate the above method, consider the interaction of a two-dimensional system $\mathbf{B}_2$ with a known three-dimensional system $\boldsymbol{A}_3$.

Let the system $\mathbf{B}_2$ be described by the generalized eigenvalue equation (5a) where in a base $\{|r\rangle\}$ one has

$$\mathbf{H}_b=\begin{bmatrix}1 & -0.5\\ -0.5 & 0\end{bmatrix},\quad \mathbf{S}_b=\begin{bmatrix}1 & 0.3\\ 0.3 & 1\end{bmatrix}. \tag{24}$$

Let the system $\boldsymbol{A}_3$ be described by the generalized eigenvalue equation (1a) where in a base $\{|\alpha\rangle\}$ one has

$$\mathbf{H}_a = \begin{bmatrix} 1 & 2 & 3 \\ 2 & 2 & 1 \\ 3 & 1 & 5 \end{bmatrix}, \quad \mathbf{S}_a = \begin{bmatrix} 1 & 0.5 & 0.2 \\ 0.5 & 1.2 & 0.5 \\ 0.2 & 0.5 & 1.3 \end{bmatrix}. \tag{25}$$

Let further the combined system $\mathbf{C}_5$ that includes the interaction between systems $\mathbf{B}_2$ and $\boldsymbol{A}_3$ be described by the generalized eigenvalue equation (6a) where in a base $\{|\alpha\rangle, |r\rangle\}$

$$\mathbf{H}_c = \begin{bmatrix} \mathbf{H}_a & \mathbf{V}^T \\ \mathbf{V} & \mathbf{H}_b \end{bmatrix}, \quad \mathbf{S}_c = \begin{bmatrix} \mathbf{S}_a & \mathbf{P}^T \\ \mathbf{P} & \mathbf{S}_b \end{bmatrix}, \tag{26a}$$

and where generalized interaction $(\mathbf{V}, \mathbf{P})$ is given by

$$\mathbf{V} = \begin{bmatrix} 1 & 2 & -1 \\ 1 & 2 & 3 \end{bmatrix}, \quad \mathbf{P} = \begin{bmatrix} 0.3 & -0.5 & 0.2 \\ 0.3 & 0.1 & -0.4 \end{bmatrix}. \tag{26b}$$

In the above expressions $\mathbf{V}^T$ is transpose of $\mathbf{V}$, $\mathbf{P}^T$ is transpose of $\mathbf{P}$, and parameter β is chosen to equal unity. One finds that combined system $\mathbf{C}_5$ has eigenvalues [4]

$$\varepsilon_1 = -2.32918, \varepsilon_2 = -1.50057, \varepsilon_3 = 0.06038, \varepsilon_4 = 4.35986, \varepsilon_5 = 60.42988, \tag{27a}$$

The corresponding normalized eigenstates are

$$|\Psi_1\rangle = \begin{pmatrix} 1.11912 \\ -0.49548 \\ -0.39826 \\ -0.51397 \\ 0.05630 \end{pmatrix}, \quad |\Psi_2\rangle = \begin{pmatrix} 0.35630 \\ 0.29638 \\ -0.02665 \\ -0.37694 \\ -0.73890 \end{pmatrix}, \quad |\Psi_3\rangle = \begin{pmatrix} 0.53374 \\ -0.49000 \\ 0.20859 \\ 0.40996 \\ -0.57247 \end{pmatrix},$$

$$|\Psi_4\rangle = \begin{pmatrix} 0.07052 \\ 0.63770 \\ 0.27950 \\ 0.44679 \\ 0.34263 \end{pmatrix}, \quad |\Psi_5\rangle = \begin{pmatrix} 1.17903 \\ -2.38652 \\ 1.73945 \\ -2.44302 \\ 1.36072 \end{pmatrix}. \tag{27b}$$

This can be verified by inserting those eigenvalues and eigenstates in the eigenvalue equation (6a) with matrices $\mathbf{H}_c$ and $\mathbf{S}_c$ as given by expression (26a). One can also verify that eigenstates (27b) are orthonormalized according to (6c), as required.

One can solve the same eigenvalue equation by the LRM approach. In LRM, in order to solve combined eigenvalue equation, one has to know the solution of the eigenvalue equation (1a) that describes the system $\boldsymbol{A}_3$. This system has three eigenvalues and three corresponding eigenstates. One finds

$$\lambda_1 = -1.66138, \ \lambda_2 = 1.89610, \ \lambda_3 = 5.20405, \tag{28a}$$

$$|\Phi_1\rangle = \begin{pmatrix} 0.99832 \\ -0.56029 \\ -0.32137 \end{pmatrix}, \ |\Phi_2\rangle = \begin{pmatrix} 0.23961 \\ 0.86544 \\ -0.26545 \end{pmatrix}, \ |\Phi_3\rangle = \begin{pmatrix} 0.45738 \\ -0.39004 \\ 0.86160 \end{pmatrix}. \tag{28b}$$

Eigenstates (28b) are orthonormalized according to (1b) where matrix $\mathbf{S}_a$ is given by the expression (25).

From (26b) and (28b) one derives matrix elements $\langle r|\mathbf{V}|\Phi_i\rangle \equiv \Delta H_{r,i}$ and $\langle r|\mathbf{P}|\Phi_i\rangle \equiv \Delta S_{r,i}$ [4]:

$$\Delta H_{1,1} = 0.19912, \ \Delta H_{2,1} = -1.08638, \ \Delta S_{1,1} = 0.51537, \ \Delta S_{2,1} = 0.37202,$$

$$\Delta H_{1,2} = 2.23594, \ \Delta H_{2,2} = 1.17415, \ \Delta S_{1,2} = -0.41393, \ \Delta S_{2,2} = 0.26461,$$

$$\Delta H_{1,3} = -1.18430, \ \Delta H_{2,3} = 2.26211, \ \Delta S_{1,3} = 0.50456. \tag{29}$$

In order to obtain cardinal eigenvalues of the combined system $\mathbf{C}_5$, one can use either expression (12b) or expression (22b). Let me use this latter expression. Using expressions (21), (28a) and (29) one obtains matrix elements $\mathbf{\Omega}'_{rt}(\varepsilon)$:,

$$\Omega'_{11}(\varepsilon) = \frac{1.11373}{\varepsilon - \lambda_1} + \frac{9.12516}{\varepsilon - \lambda_2} + \frac{14.51632}{\varepsilon - \lambda_3},$$

$$\Omega'_{22}(\varepsilon) = \frac{0.21932}{\varepsilon - \lambda_1} + \frac{0.45215}{\varepsilon - \lambda_2} + \frac{12.56382}{\varepsilon - \lambda_3},$$

$$\Omega'_{12}(\varepsilon) \equiv \Omega'_{21}(\varepsilon) = \frac{-0.49423}{\varepsilon - \lambda_1} + \frac{2.03125}{\varepsilon - \lambda_2} - \frac{13.50483}{\varepsilon - \lambda_3}, \tag{30a}$$

as well as matrices $\mathbf{K}$ and $\mathbf{\Lambda}$

$$\mathbf{K} = \begin{bmatrix} 4.04931 & -2.22630 \\ -2.22630 & 1.52069 \end{bmatrix}, \quad \mathbf{\Lambda} = \begin{bmatrix} 0.69151 & -0.04214 \\ -0.04214 & 0.26914 \end{bmatrix}. \tag{30b}$$

Inserting above expressions into (22b) one obtains [4]

$$h(\varepsilon) \equiv \begin{vmatrix} \Omega'_{11}(\varepsilon) + 5.04931 - 0.30849\varepsilon & \Omega'_{12}(\varepsilon) - 2.72630 - 0.34214\varepsilon \\ \Omega'_{21}(\varepsilon) - 2.72630 - 0.34214\varepsilon & \Omega'_{22}(\varepsilon) + 1.52069 - 0.73086\varepsilon \end{vmatrix} = 0 \tag{31}$$

Roots of this equation are cardinal eigenvalues of the combined system $\mathbf{C}_5$. One easily verifies that eigenvalues (27a) satisfy (31). In addition, one finds that there are no other solutions to (31). This shows that expression (12b) as well as equivalent expression (22b) produces correct cardinal eigenvalues of the combined system. Moreover, in this particular example the function $h(\varepsilon)$ has exactly five roots. Since the combined system is five-dimensional, those are all eigenvalues of this system, and there are no singular solutions.

Once a particular eigenvalue ε_s is found as a root of $h(\varepsilon)$, the corresponding cardinal eigenstate (13a) is determined by the state $|\theta_s\rangle$, nontrivial solution of (22a):

$$\begin{bmatrix} \Omega'_{11}(\varepsilon_s) + 5.04931 & \Omega'_{12}(\varepsilon_s) - 2.72630 \\ \Omega'_{21}(\varepsilon_s) - 2.72640 & \Omega'_{22}(\varepsilon_s) + 1.52069 \end{bmatrix} |\theta_s\rangle = \varepsilon_s \begin{bmatrix} 0.30849 & 0.34214 \\ 0.34214 & 0.73086 \end{bmatrix} |\theta_s\rangle. \tag{32}$$

Using eigenvalues ε_s which were obtained as roots of (31) one finds [4]:

$$|\theta_1\rangle = \begin{pmatrix} 1 \\ -0.10954 \end{pmatrix}, \quad |\theta_2\rangle = \begin{pmatrix} 1 \\ 1.96028 \end{pmatrix}, \quad |\theta_3\rangle = \begin{pmatrix} 1 \\ -1.39639 \end{pmatrix},$$

$$|\theta_4\rangle = \begin{pmatrix} 1 \\ 0.76687 \end{pmatrix}, \quad |\theta_5\rangle = \begin{pmatrix} 1 \\ -0.55698 \end{pmatrix}. \tag{33}$$

Inserting above states into the expression (13a) one obtains, up to the norm and phase, eigenstates (27b) of the combined system. If one normalizes those eigenstates according to (23), one reproduces eigenstates (27b) up to the phase. This shows that suggested method produces correct cardinal eigenstates of the combined system. Since above system has no singular eigenstates, LRM calculation is completed.

From a numerical point of view, above example is not very interesting. It can be solved more efficiently by many other methods, in particular by direct diagonalization. However, it illustrates main features of the LRM approach, its advantageous and possible drawbacks.

Eigenvalue equation (32) that describes the system $\mathbf{B}_2$ subject to the interaction (26b) with the known system $\boldsymbol{A}_3$ is a 2×2 eigenvalue equation. In this example, system $\boldsymbol{A}_3$ is relatively small three-dimensional system. However, dimension n of this system can be arbitrary large. All cardinal solutions of the combined system will be still the solution of the 2×2 eigenvalue equation that is similar to the eigenvalue equation (32). Thus if the dimension n of the system $\boldsymbol{A}_n$ is large enough, LRM approach will be numerically more efficient than any other known method.

5. Internal Modifications of Finite-Dimensional Systems

Consider now internal modifications of finite-dimensional systems (see Figure 1). Let again $\boldsymbol{A}_n$ be n-dimensional time-independent system described by the eigenvalue equation (1a), and let the corresponding eigenstates $|\Phi_i\rangle$ be orthonormalised according to (1b). Let further $\mathbf{C}_n$ be a modified system described by the eigenvalue equation (2a) where $\mathbf{V}$ and $\mathbf{P}$ are Hermitean operators which represent internal modification of the parent system $\boldsymbol{A}_n$ and where β is a coupling parameter. Let ρ be the rank of the combined modification $(\mathbf{V},\mathbf{P})$. As shown in section 3, range of this modification is ρ-dimensional interaction space X_ρ^{int}. This space is spanned by all vectors of a type $\mathbf{V}|\psi\rangle$ and all vectors of a type $\mathbf{P}|\psi\rangle$ where $|\psi\rangle \in X_n^a$.

Let $\{|\mu\rangle\}$ be a base in X_ρ^{int} orthonormalized according to (3a) and let $\{|\alpha\rangle\}$ be the base in X_n orthonormalized according to (3b). Projection operator $\mathbf{I}_\rho^{\text{int}}$ on the interaction space X_ρ^{int} can be expressed in terms of the states $|\mu\rangle$ according to (3c), while modification operators $\mathbf{V}$ and $\mathbf{P}$ satisfy (4).

In analogy to the case of external modifications, in the case of internal modifications one considers operator $\mathbf{\Omega}(\varepsilon)$ defined as

$$\mathbf{\Omega}(\varepsilon) = (\mathbf{V} - \varepsilon\,\mathbf{P})\overline{\mathbf{\Omega}}(\varepsilon)(\mathbf{V} - \varepsilon\,\mathbf{P}), \tag{34a}$$

In the base $\{|\mu\rangle\}$ of the interaction space operator $\mathbf{\Omega}(\varepsilon)$ is a $\rho\times\rho$ matrix with matrix elements

$$\Omega_{\mu\nu}(\varepsilon) \equiv \langle \mu | \mathbf{\Omega}(\varepsilon) | \nu \rangle = \sum_{i(\lambda_i \neq \varepsilon)}^{n} \frac{\langle \mu | \mathbf{V} - \varepsilon \mathbf{P} | \Phi_i \rangle \langle \Phi_i | \mathbf{V} - \varepsilon \mathbf{P} | \nu \rangle}{\varepsilon - \lambda_i}, \quad \mu, \nu = 1, \ldots, \rho. \tag{34b}$$

Expressions (34) are formally identical to the expressions (10). However, in the case of an internal modification operators $\mathbf{V}$ and $\mathbf{P}$ satisfy expressions (4), while in the case of an external modification those operators satisfy expressions (8).

In analogy to the interlacing rule that applies to external modifications, there is also an interlacing rule that applies to internal modifications.

5.1. Interlacing of Modified Eigenvalues ε_s with Eigenvalues λ_i

Eigenvalues ε_s of the modified system $\mathbf{C}_n$ and eigenvalues λ_i of the parent system $\mathbf{A}_n$ satisfy [7]:

Interlacing Rule 2 (Internal Modifications)

Let ranks of modification operators $\mathbf{V}$ and $\mathbf{P}$ be ρ_h and ρ_s, respectively. Let further r_h nonzero eigenvalues of $\mathbf{V}$ be positive, the remaining $\rho_h - r_h$ nonzero eigenvalues of $\mathbf{V}$ being negative. Arrange eigenvalues λ_i ($i = 1, \cdots, n$) of the parent system $\mathbf{A}_n$ in the non-decreasing order and arrange also eigenvalues ε_s ($s = 1, \cdots, n$) of the modified system $\mathbf{C}_n$ in the non-decreasing order. These eigenvalues thus arranged satisfy

$$\lambda_{s+r_h-\rho_s-\rho_h} \leq \varepsilon_s \leq \lambda_{s+r_h+\rho_s}. \tag{35}$$

5.2. Cardinal Solutions of the Modified System

Let me now consider solutions of the modified system $\mathbf{C}_n$ in more details. In analogy to the external modifications, in the case of internal modifications one can also have cardinal ($\varepsilon_s \notin \{\lambda_i\}$) and singular ($\varepsilon_s \in \{\lambda_i\}$) solutions of this system. Concerning cardinal solutions of $\mathbf{C}_n$, one has [7]:

Theorem 3 (cardinal solutions): Let the system $\mathbf{A}_n$ be described by the eigenvalue equation (1a) and let eigenstates $|\Phi_i\rangle$ of this system be orthonormalized according to (1b). Then:

a) $\varepsilon_s \notin \{\lambda_i\}$ is cardinal eigenvalue of the eigenvalue equation (2a) if and only if it satisfies

$$\left[\beta\mathbf{\Omega}(\varepsilon_s)-\mathbf{V}+\varepsilon_s\mathbf{P}\right]\left|\theta_s\right\rangle=0\ ,\ \left|\theta_s\right\rangle\in X_\rho^{\text{int}}\ , \tag{36a}$$

where operator $\mathbf{\Omega}(\varepsilon)$ is given by expressions (34).

In the base $\{|\mu\rangle\}$ expression (36a) is a $\rho\times\rho$ matrix equation. In particular, operator $\mathbf{\Omega}(\varepsilon)$ is in this base $\rho\times\rho$ matrix with matrix elements (34b). Each cardinal eigenvalue $\varepsilon_s\notin\{\lambda_i\}$ of (2a) is hence a root of the $\rho\times\rho$ determinant $g(\varepsilon)$:

$$g(\varepsilon)\equiv\left|\beta\mathbf{\Omega}(\varepsilon)-\mathbf{V}+\varepsilon\mathbf{P}\right|=0\ . \tag{36b}$$

b) $|\Psi_s\rangle$ is an eigenstate of the eigenvalue equation (2a) that has cardinal eigenvalue $\varepsilon_s\notin\{\lambda_i\}$ if and only if it is a linear combination

$$\left|\Psi_s\right\rangle=\beta\sum_i^n\frac{\left\langle\Phi_i\right|\mathbf{V}-\varepsilon_s\mathbf{P}\left|\theta_s\right\rangle}{\varepsilon_s-\lambda_i}\left|\Phi_i\right\rangle\ , \tag{37a}$$

where $|\theta_s\rangle\in X_\rho^{\text{int}}$ is an eigenstate of (36a) corresponding to the eigenvalue $\varepsilon=\varepsilon_s$.

c). Degeneracy of the cardinal eigenvalue $\varepsilon_s\notin\{\lambda_i\}$ of the modified system $\mathbf{C}_n$ equals number of linearly independent eigenstates $|\theta_s\rangle$ of (36a) that correspond to this eigenvalue. In other words, this degeneracy equals nullity of the matrix $\beta\mathbf{\Omega}(\varepsilon_s)-\mathbf{V}+\varepsilon_s\mathbf{P}$. Since this is a $\rho\times\rho$ matrix, each cardinal eigenvalue of a modified system can be at most ρ-degenerate.

d). X_ρ^{int}-component of the cardinal eigenstate (37a) equals eigenstate $|\theta_s\rangle$ of (36a):

$$\mathbf{I}_\rho^{\text{int}}\left|\Psi_s\right\rangle=\left|\theta_s\right\rangle\in X_\rho^{\text{int}}\ , \tag{37b}$$

This expression is analogous to the expression (13b) in the case of external modifications.

Eigenstates $|\Psi_s\rangle$ can be orthonormalized according to (2c). If $\varepsilon_s\neq\varepsilon_p$ eigenstates $|\Psi_s\rangle$ and $|\Psi_p\rangle$ automatically satisfy (2c). However, if $\varepsilon_s=\varepsilon_p$ one has to enforce this orthogonality by some standard orthonormalization procedure. In particular, using expressions (1b), (4) and (37), one finds that cardinal eigenstate (37a) can be normalized according to

$$\left|\Psi_s\right\rangle\rightarrow\left|\Psi_s^{'}\right\rangle=\frac{1}{\sqrt{N_s}}\left|\Psi_s\right\rangle\ , \tag{37c}$$

where normalization constant N_s equals

$$N_s \equiv \langle \Psi_s | \mathbf{S}_a + \beta \mathbf{P} | \Psi_s \rangle =$$
$$\beta^2 \sum_i^n \frac{\langle \theta_s | \mathbf{V} - \varepsilon_s \mathbf{P} | \Phi_i \rangle \langle \Phi_i | \mathbf{V} - \varepsilon_s \mathbf{P} | \theta_s \rangle}{(\varepsilon_s - \lambda_i)^2} + \beta \langle \theta_s | \mathbf{P} | \theta_s \rangle \tag{37d}$$

In the case of standard eigenvalue equations one has $\mathbf{P} = 0$ and above expression simplifies to

$$N_s = \beta^2 \sum_i^n \frac{\langle \theta_s | \mathbf{V} | \Phi_i \rangle \langle \Phi_i | \mathbf{V} | \theta_s \rangle}{(\varepsilon_s - \lambda_i)^2} \, . \tag{37e}$$

In analogy to the case of external modification, calculation of the normalization constant N_s requires only the knowledge of the X_ρ^{int}-component of the eigenstate $|\Psi_s\rangle$. The knowledge of the component of this eigenstate outside the interaction space is not required. Of course, as in the case of external modifications, this is due to the fact that X_ρ^{int}-component of the cardinal eigenstate $|\Psi_s\rangle$ uniquely determines complete eigenstate. Information about the complete cardinal eigenstate $|\Psi_s\rangle$ is hence implicitly contained in its X_ρ^{int}-component. Therefore this component alone should be sufficient to normalize this eigenstate. Nevertheless it is quite remarkable and highly convenient that this normalization can be done by the relatively simple expressions (23) and (37d).

5.3. Singular Solutions of the Modified System

Concerning singular solutions of $\mathbf{C}_n$, one has [7]:

Theorem 4 (singular solutions): Let λ_j be a η_j-degenerate eigenvalue of the eigenvalue equation (1a) that describes parent system $\boldsymbol{A}_n$ and let $X_{\eta_j}^{\lambda j}$ be the corresponding η_j-dimensional space spanned by η_j degenerate eigenstates $|\Phi_{jl}\rangle$ ($l = 1, \ldots, \eta_j$) of (1a). Let further $\mathbf{I}_{\eta_j}^{\lambda j}$ be the projection operator on the space $X_{\eta_j}^{\lambda j}$. Then:

Each singular eigenstate $|\Psi_s\rangle$ of the modified system $\mathbf{C}_n$ that has eigenvalue $\varepsilon_s = \lambda_j \in \{\lambda_i\}$ is a linear combination

$$\left|\Psi_s\right\rangle = \beta \sum_{i(\lambda_i \neq \varepsilon_s)}^{n} \frac{\left\langle \Phi_i \left| \mathbf{V} - \varepsilon_s \mathbf{P} \right| \theta_s \right\rangle}{\varepsilon_s - \lambda_i} \left|\Phi_i\right\rangle + \left|\varphi_s\right\rangle, \tag{38a}$$

where

$$\left|\theta_s\right\rangle \in X_\rho^{\text{int}}, \left|\varphi_s\right\rangle \in X_{\eta_j}^{\lambda j}. \tag{38b}$$

and where those two states satisfy

$$\left[\beta \mathbf{\Omega}(\varepsilon_s) - \mathbf{V} + \varepsilon_s \mathbf{P}\right]\left|\theta_s\right\rangle + \left(\mathbf{V} - \varepsilon_s \mathbf{P}\right)\mathbf{I}_{\eta_j}^{\lambda_j}\left|\varphi_s\right\rangle = 0, \tag{39a}$$

$$\beta \mathbf{I}_{\eta_j}^{\lambda j}\left(\mathbf{V} - \varepsilon_s \mathbf{P}\right)\left|\theta_s\right\rangle = 0. \tag{39b}$$

In addition, the state $\left|\theta_s\right\rangle$ is X_ρ^{int}-component of the eigenstate $\left|\Psi_s\right\rangle$, while the state $\left|\varphi_s\right\rangle$ is contained in the space $X_{\eta_j}^{\lambda_j}$:

$$\mathbf{I}_\rho^{\text{int}}\left|\Psi_s\right\rangle = \left|\theta_s\right\rangle \in X_\rho^{\text{int}}, \quad \left|\varphi_s\right\rangle = \sum_l^{\eta_j} \left|\Phi_{jl}\right\rangle \left\langle \Phi_{jl} \left| \mathbf{S}_a \right| \Psi_s \right\rangle \in X_{\eta j}^{\lambda j}. \tag{40}$$

In a matrix form expressions (39) form a set of $(\rho + \eta_j)$ homogenous linear equations in $(\rho + \eta_j)$ unknowns, ρ unknown coefficients $C_\nu^{(s)}$ and η_j unknown coefficients $D_l^{(s)}$ where

$$\left|\theta_s\right\rangle = \sum_\nu^\rho C_\nu^{(s)}\left|\nu\right\rangle, \quad \left|\varphi_s\right\rangle = \sum_l^{\eta_j} D_l^{(s)}\left|\Phi_{jl}\right\rangle. \tag{41}$$

Modified system $\mathbf{C}_n$ has a singular eigenstate with the eigenvalue $\varepsilon_s = \lambda_j \in \{\lambda_i\}$ if and only if determinant of the system (39) vanishes. Since this is a $(\rho + \eta_j) \times (\rho + \eta_j)$ system, one may have at most $(\rho + \eta_j)$ singular eigenstates with this eigenvalue.

One can again distinguish *strongly singular* and *weakly singular* eigenstates. In analogy to the case of external modifications, in the case of an internal modification strongly singular eigenstate $\left|\Psi_s\right\rangle$ of the eigenvalue equation (2a) has no component in the interaction space X_ρ^{int}, *i.e.* it satisfies $\mathbf{I}_\rho^{\text{int}}\left|\Psi_s\right\rangle = 0$ and hence $\left|\theta_s\right\rangle = 0$. On the other hand, each weakly singular eigenstate $\left|\Psi_s\right\rangle$ satisfies $\mathbf{I}_\rho^{\text{int}}\left|\Psi_s\right\rangle \neq 0$.

Theorem 4 implies:

Theorem 4a:

a). $\left|\Psi_s\right\rangle$ is *strongly singular* eigenstate of the eigenvalue equation (2a) with the eigenvalue $\varepsilon_s = \lambda_j$ if and only if this state satisfies

$$\left|\Psi_s\right\rangle \equiv \left|\varphi_s\right\rangle \in X_{\eta_j}^{\lambda j}, \quad (\mathbf{V} - \varepsilon_s \mathbf{P})\mathbf{I}_{\eta_j}^{\lambda j}\left|\varphi_s\right\rangle = 0 . \tag{42}$$

This state is hence a linear combination

$$\left|\Psi_s\right\rangle = \sum_l^{\eta_j} D_l^{(s)}\left|\Phi_{jl}\right\rangle , \tag{43a}$$

where the coefficients $D_l^{(s)}$ satisfy

$$\sum_l^{\eta_j}\left\langle\mu\right|\mathbf{V} - \varepsilon_s\mathbf{P}\left|\Phi_{jl}\right\rangle D_l^{(s)} = 0 , \quad \mu = 1,\ldots,\rho . \tag{43b}$$

b). $\left|\Psi_s\right\rangle$ is weakly singular eigenstate of the generalized eigenvalue equation (2b) with the eigenvalue $\varepsilon_s = \lambda_j$ if and only if it is a linear combination (38a) where the states $\left|\theta_s\right\rangle$ and $\left|\varphi_s\right\rangle$ satisfy expressions (39) and where $\left|\theta_s\right\rangle \neq 0$. In addition, each weakly singular eigenstate should be orthogonal to all strongly singular eigenstates.

According to the above expressions, the set of all strongly singular eigenstates associated with the eigenvalue $\varepsilon_s = \lambda_j$ spans some r_j-dimensional space $X_{rj}^{\lambda j-}$, subspace of the space $X_{\eta j}^{\lambda j}$. Expression (43b) implies

$$\eta_j - \rho \leq r_j \leq \eta_j . \tag{44}$$

Strongly singular eigenstates in the case of internal modifications have essentially the same properties as strongly singular eigenstates in the case of external modifications. In particular, expressions (42-44) are formally identical to the expressions (18-20). Thus according to (44), if degeneracy η_j of the eigenvalue λ_j of the parent system $\boldsymbol{A}_n$ satisfies $\eta_j > \rho$, modified system $\mathbf{C}_n$ has at least $\eta_j - \rho$ strongly singular eigenstates with the eigenvalue $\varepsilon_s = \lambda_j$. Those eigenstates do not depend on the coupling parameter β and each strongly singular eigenstate $\left|\Psi_s\right\rangle$ of the modified system $\mathbf{C}_n$ is at the same time an eigenstate of the parent system $\mathbf{A}_n$. Weakly singular eigenstates in the case of an internal modification have similar though not identical properties to the weakly singular eigenstates in the case of an external modification. In particular, expressions (39) differ from the corresponding expressions (15). However in analogy to the case of external modifications, each weakly singular eigenstate $|\Psi_s\rangle$ in the case of internal modifications also depends on a coupling parameter β as well as on the modification operators $\mathbf{V}$ and $\mathbf{P}$. Further, no weakly singular eigenstate of the modified system can be at the same time an eigenstate of the parent system. Finally, modified system may have at most ρ weakly singular eigenstates with the eigenvalue $\varepsilon_s = \lambda_j$.

Above two theorems produce all solutions of the modified system $\mathbf{C}_n$. All cardinal eigenvalues $\varepsilon_s \notin \{\lambda_i\}$ of the modified eigenvalue equation (2a) are roots of the function (36b). Once a particular root ε_s of this function is found, the corresponding eigenstate (or eigenstates) is given by eq. (37a) where the state $|\theta_s\rangle \in X_\rho^{\text{int}}$ satisfies (36a). Concerning remaining singular eigenvalues $\varepsilon_s \in \{\lambda_i\}$, each such eigenvalue coincides with some parent eigenvalue λ_j, and all one has to do is to verify conditions (39). This verification is simplified by the expressions (43) which produce all strongly singular eigenstates.

In analogy to the case of external modifications, in the case of internal modifications most important and most time-consuming is finding roots of the determinant (36b). Roots of this determinant are cardinal eigenvalues of the modified system. Operational count for a single cardinal eigenvalue is again of the order $O(\rho^2 n)$. Once an eigenvalue $\varepsilon_s \notin \{\lambda_i\}$ is obtained, LRM calculation of the corresponding eigenstate or eigenstates is of the order $O(\rho^3)$. Hence if $\rho << \sqrt{n}$, LRM will be more efficient than any other numerical method in finding eigenvalues and eigenstates of the modified system.

Note that in the case of an internal modification, it is not strictly required that one should use the interaction space X_ρ^{int} which is the range of the combined modification $(\mathbf{V},\mathbf{P})$. Instead of this space one can use any other κ-dimensional space $X_\kappa^{\text{int}} \subset X_n^a$ which contains the interaction space X_ρ^{int} as its subspace. Interaction space X_ρ^{int} is the smallest space with the property that modification operators $\mathbf{V}$ and $\mathbf{P}$ vanish in the orthogonal complement of this space. From the LRM point of view, the use of the interaction space X_ρ^{int} which is the smallest space with this property is most economical. However, in some cases it might be more practical to use some of those larger spaces instead of the interaction space.

In the original presentation of the LRM treatment of internal modifications described by eigenvalue equations of a general type (1a) and (2a), corresponding LRM expressions were not Hermitean [7]. Key expressions (36a) and (39) in this section are Hermitean. Those expressions can be easily derived from the corresponding expressions in ref. [7]. It is clearly advantageous to have Hermitean instead of non-Hermitean equations.

There are two special cases of particular interest when above LRM expressions substantially simplify. Those are cases $\mathbf{P} = 0$ and $\mathbf{V} = 0$. Consider first the case $\mathbf{P} = 0$.

5.4. Case $\mathbf{P} = 0$

If $\mathbf{P} = 0$ equation (2a) reduces to

$$(\mathbf{H}_a + \beta \mathbf{V})|\Psi_s\rangle = \varepsilon_s \mathbf{S}_a |\Psi_s\rangle, \quad s = 1,\ldots,n. \tag{45}$$

Define operator $\mathbf{\Omega}^0(\varepsilon)$:

$$\mathbf{\Omega}^0(\varepsilon) = \mathbf{I}_\rho^{\text{int}} \overline{\Omega}(\varepsilon) \mathbf{I}_\rho^{\text{int}}, \tag{46a}$$

This operator vanishes outside the interaction space X_ρ^{int}, and in the base $\{|\mu\rangle\}$ of this space it has matrix elements

$$\mathbf{\Omega}_{\mu\nu}^0(\varepsilon) = \sum_{i(\lambda_i \neq \varepsilon)}^{n} \frac{\langle\mu|\Phi_i\rangle\langle\Phi_i|\nu\rangle}{\varepsilon - \lambda_i}, \quad \mu,\nu = 1,\ldots,\rho. \tag{46b}$$

Theorem 3 now reduces to:

Theorem 3A (cardinal solutions): Let the system $\boldsymbol{A}_n$ be described by the eigenvalue equation (1a) and let eigenstates $|\Phi_i\rangle$ of this system be orthonormalized according to (1b). Let further modified system $\mathbf{C}_n$ be described by the eigenvalue equation (45). Then:

a) $\varepsilon_s \notin \{\lambda_i\}$ is a cardinal eigenvalue of the eigenvalue equation (45) if and only if it satisfies

$$\left[\beta \mathbf{\Omega}^0(\varepsilon_s) - \mathbf{V}^{-1}\right]|\theta_s\rangle = 0, \quad |\theta_k\rangle \in X_\rho^{\text{int}}, \tag{47a}$$

where $\mathbf{V}^{-1}$ is Hermitean operator, inverse of $\mathbf{V}$ in the interaction space X_ρ^{int}.

In the base $\{|\mu\rangle\}$ expression (47a) is a $\rho \times \rho$ matrix equation. Each cardinal eigenvalue $\varepsilon_s \notin \{\lambda_i\}$ of (45) is hence a root of a $\rho \times \rho$ determinant $f(\varepsilon)$:

$$f(\varepsilon) \equiv \left|\beta \mathbf{\Omega}^0(\varepsilon) - \mathbf{V}^{-1}\right| = 0. \tag{47b}$$

Note that since ρ is the rank of the modification operator $\mathbf{V}$, this operator is nonsingular in X_{ρ}^{int} and hence inverse $\mathbf{V}^{-1}$ exist.

b) Each normalized cardinal eigenstate $|\Psi_s\rangle$ of the generalized eigenvalue equation (45) that has eigenvalue $\varepsilon_s \notin \{\lambda_i\}$ is a linear combination

$$|\Psi_s\rangle = \frac{1}{\sqrt{N_s}} \sum_i^n \frac{\langle \Phi_i | \theta_k \rangle}{\varepsilon_s - \lambda_i} |\Phi_i\rangle \quad , \tag{48a}$$

where $|\theta_s\rangle \in X_{\rho}^{\text{int}}$ is an eigenstate of (47a) that corresponds to the eigenvalue ε_s, and where N_s is a normalization constant

$$N_s = \sum_i^n \frac{\langle \theta_s | \Phi_i \rangle \langle \Phi_i | \theta_s \rangle}{(\varepsilon_s - \lambda_i)^2} \quad . \tag{48b}$$

In addition, cardinal eigenstate (48a) is related to the state $|\theta_s\rangle \in X_{\rho}^{\text{int}}$ according to

$$|\theta_s\rangle = \sqrt{N_s}\,\mathbf{V}|\Psi_s\rangle \, . \tag{48c}$$

In a similar way can be treated singular solutions of (45).

Of special interest is the case $\rho = rank(\mathbf{V}) = 1$. In this case the space X_{ρ}^{int} reduces to a single state $|\tau\rangle$. Hence $\mathbf{V} = \Delta h_{\tau} |\tau\rangle\langle\tau|$, $|\theta_s\rangle = |\tau\rangle$, and expression (47b) reduces to

$$\beta \Omega_{\tau\tau}^{0}(\varepsilon) = \frac{1}{\Delta h_{\tau}} \, . \tag{49}$$

where $\Omega_{\tau\tau}^{0}(\varepsilon)$ is given by (46b).

Each root $\varepsilon = \varepsilon_s \notin \{\lambda_i\}$ of (49) is a cardinal eigenvalue of the modified system. The corresponding normalized eigenstate $|\Psi_s\rangle$ is nondegenerate and one has

$$|\Psi_s\rangle = \frac{1}{\sqrt{N_s}} \sum_i^n \frac{\langle \Phi_i | \tau \rangle}{\varepsilon_s - \lambda_i} |\Phi_i\rangle \quad , \tag{50a}$$

where

$$N_s = \sum_i^n \frac{\langle\tau|\Phi_i\rangle\langle\Phi_i|\tau\rangle}{(\varepsilon_s - \lambda_i)^2} \quad . \tag{50b}$$

Further, each normalized strongly singular eigenstate $|\Psi_s\rangle$ that has eigenvalue $\varepsilon_s = \lambda_j$ is a linear combination

$$|\Psi_s\rangle = \frac{1}{\sqrt{N_s}} \sum_l^{\eta_j} D_l^{(s)} |\Phi_{jl}\rangle \quad , \tag{51a}$$

where normalization constant N_s equals

$$N_s = \sum_l^{\eta_l} D_l^{(s)} D_l^{(s)*} \quad , \tag{51b}$$

and where the coefficients $D_l^{(s)}$ satisfy a single condition

$$\sum_l^{\eta_j} \langle\tau|\Phi_{jl}\rangle D_l^{(s)} = 0 \quad . \tag{51c}$$

In particular, if eigenvalue λ_j is degenerate ($\eta_j > 1$), modified system $\mathbf{C}_n$ has at least $\eta_j - 1$ strongly singular eigenstates with this eigenvalue. Concerning weakly singular eigenstates, one finds that weakly singular eigenstate $|\Psi_s\rangle$ associated with the eigenvalue $\varepsilon_s = \lambda_j$ exists if and only if all matrix elements $\langle\Phi_{jl}|\tau\rangle$ vanish

$$\langle\Phi_{jl}|\tau\rangle = 0 \quad , l = 1, \ldots, \eta_j , \tag{52a}$$

and if in addition this eigenvalue satisfies

$$\beta \Omega_{\tau\tau}^0(\varepsilon_s) = \frac{1}{\Delta h_\tau} \quad . \tag{52b}$$

If this is the case, the corresponding (normalized) weakly singular eigenstate is a linear combination

$$|\Psi_s\rangle = \frac{1}{\sqrt{N_s}} \sum_{i(\lambda_i \neq \varepsilon_s)} \frac{\langle\Phi_i|\tau\rangle}{\varepsilon_s - \lambda_i} |\Phi_i\rangle \quad , \ \varepsilon_s = \lambda_j . \tag{53a}$$

where

$$N_s = \sum_{i(\lambda_i \neq \varepsilon_s)}^{n} \frac{\langle\tau|\Phi_i\rangle\langle\Phi_i|\tau\rangle}{(\varepsilon_s - \lambda_i)^2} \quad . \tag{53b}$$

Those expressions are similar to the expressions (50) which apply to cardinal eigenstates, except that the term with $\lambda_i = \varepsilon_s$ is excluded from the summations in (53a) and (53b).

Concerning interlacing rule, in the case $\rho = 1$ and $\mathbf{P} = 0$ expression (35) reduces to

$$\lambda_1 \le \varepsilon_1 \le \lambda_2 \le \varepsilon_2 \le \ldots \le \varepsilon_n \,,\quad if\ \Delta h_\tau > 0 \,, \tag{54a}$$

$$\varepsilon_1 \le \lambda_1 \le \varepsilon_2 \le \lambda_2 \le \ldots \le \lambda_n \,,\quad if\ \Delta h_\tau < 0 \,. \tag{54b}$$

5.5. Case $\mathbf{V} = 0$

Another special case is $\mathbf{V} = 0$. In this case equation (2a) describing modified system $\mathbf{C}_n$ reduces to

$$\mathbf{H}_a|\Psi_s\rangle = \varepsilon_s(\mathbf{S}_a + \beta\mathbf{P})|\Psi_s\rangle \,,\quad s = 1,\ldots,n \,. \tag{55}$$

while **Theorem 3** reduces to:

Theorem 3B (cardinal solutions): Let the system $\mathbf{A}_n$ be described by the eigenvalue equation (1a) and let eigenstates $|\Phi_i\rangle$ of this system be orthonormalized according to (1b). Let further the system $\mathbf{C}_n$ be described by the eigenvalue equation (55). Then:

a) $\varepsilon_s \neq 0$ is a cardinal eigenvalue of the eigenvalue equation (55) if and only if it satisfies

$$\left[\beta\mathbf{\Omega}^0(\varepsilon_s) + \frac{\mathbf{P}^{-1}}{\varepsilon_s}\right]|\theta_s\rangle = 0 \,,\quad |\theta_s\rangle \in X_\rho^{\text{int}} \,, \tag{56a}$$

where $\mathbf{P}^{-1}$ is a Hermitean operator, inverse of $\mathbf{P}$ in the interaction space X_ρ^{int}.

In the base $\{|\mu\rangle\}$ expression (56a) is a $\rho \times \rho$ matrix equation. Each cardinal eigenvalue $\varepsilon_s \notin \{\lambda_i\}$ of (55) that differs from zero is hence a root of a $\rho \times \rho$ determinant

$$f(\varepsilon) \equiv \left| \beta \mathbf{\Omega}^0(\varepsilon) + \frac{\mathbf{P}^{-1}}{\varepsilon} \right| = 0 \quad . \tag{56b}$$

Note that if ρ is the rank of the modification operator $\mathbf{P}$, this operator is nonsingular in X_ρ^{int} and hence inverse $\mathbf{P}^{-1}$ exist.

b) Each eigenstate $|\Psi_s\rangle$ of the generalized eigenvalue equation (55) that has eigenvalue $\varepsilon_s \notin \{\lambda_i\}$ is a linear combination

$$|\Psi_s\rangle = \sum_i^n \frac{\langle \Phi_i | \theta_k \rangle}{\varepsilon_s - \lambda_i} |\Phi_i\rangle \quad , \tag{57a}$$

where $|\theta_s\rangle \in X_\rho^{\text{int}}$ is an eigenstate of (56a) that corresponds to the eigenvalue ε_s. In addition, one has

$$|\theta_s\rangle = -\varepsilon_s \beta \mathbf{P} |\Psi_s\rangle \quad . \tag{57b}$$

Above expressions produce all cardinal solutions of the combined system, except of the cardinal solution with the eigenvalue $\varepsilon_s = 0$ (provided such a solution exists). The possibility $\varepsilon_s = 0$ can be easily verified using general expressions (36). Inserting $\mathbf{V} = 0$ and $\varepsilon_s = 0$ into (36b) one derives condition $|\mathbf{\Omega}^0(0)| = 0$. Accordingly, cardinal eigenvalue $\varepsilon = \varepsilon_s = 0$ exist if and only if $0 \notin \{\lambda_i\}$ (no eigenvalue λ_i of $\mathbf{A}_n$ equals zero) and if in addition determinant of the matrix $\mathbf{\Omega}^0(0)$ vanishes. With the exception of this special case, above expressions produce all cardinal solutions of the modified system. In a similar way can be obtained expressions for the singular solutions of this system.

Concerning the interlacing rule, since $\mathbf{V} = 0$ expression (35) simplifies to $\lambda_{s-\rho} \le \varepsilon_s \le \lambda_{s+\rho}$ where $\rho = \rho_s$. This rule can be further tightened in some special but important cases. In particular, if no eigenvalue λ_i of (1a) is negative, one has

Interlacing rule 3 (case $\mathbf{V} = 0$): Let the rank of the operator $\mathbf{P}$ be ρ and let all eigenvalues λ_i of the system $\boldsymbol{A}_n$ be nonnegative. Let further r nonzero eigenvalues of the operator $\mathbf{P}$ be positive, the remaining $\rho - r$ nonzero eigenvalues being negative. Arrange eigenvalues λ_i of the parent system $\boldsymbol{A}_n$ in the non-decreasing order, and arrange also eigenvalues ε_s of the modified system $\mathbf{C}_n$ in the non-decreasing order. Those eigenvalues thus arranged satisfy:

$$\lambda_{s-r} \le \varepsilon_s \le \lambda_{s+\rho-r} . \tag{58}$$

Of special interest is the case $\rho = rank(\mathbf{P}) = 1$. In this case the space X_ρ^{int} reduces to a single state $|\tau\rangle$. Hence $\mathbf{P} = \Delta s_\tau |\tau\rangle\langle\tau|$, $|\theta_s\rangle = |\tau\rangle$, and expression (56b) reduces to

$$\beta\Omega_{\tau\tau}^0(\varepsilon) + \frac{1}{\varepsilon\,\Delta s_\tau} = 0 . \tag{59}$$

Each root $\varepsilon = \varepsilon_s \notin \{\lambda_i\}$ of (59) is a cardinal eigenvalue of the modified system. The corresponding eigenstate $|\Psi_s\rangle$ is nondegenerate and one has

$$|\Psi_s\rangle = \sum_i^n \frac{\langle\Phi_i|\tau\rangle}{\varepsilon_s - \lambda_i}|\Phi_i\rangle . \tag{60}$$

Further, each strongly singular eigenstate $|\Psi_s\rangle$ that has eigenvalue $\varepsilon_s = \lambda_j$ is a linear combination

$$|\Psi_s\rangle = \sum_l^{\eta_j} D_l^{(s)}|\Phi_{jl}\rangle , \tag{61a}$$

where the coefficients $D_l^{(s)}$ satisfy a single condition

$$\sum_l^{\eta_j}\langle\tau|\Phi_{jl}\rangle D_l^{(s)} = 0 . \tag{61b}$$

Concerning weakly singular eigenstates, one finds that weakly singular eigenstate $|\Psi_s\rangle$ associated with the eigenvalue $\varepsilon_s = \lambda_j$ exists if and only if all matrix elements $\langle\Phi_{jl}|\tau\rangle$ vanish

$$\langle\Phi_{jl}|\tau\rangle = 0 , \quad l = 1,\ldots,\eta_j , \tag{62a}$$

and if in addition this eigenvalue satisfies

$$\beta\Omega_{\tau\tau}^0(\varepsilon_s) + \frac{1}{\varepsilon_s\Delta s_\tau} = 0 . \tag{62b}$$

If this is the case, the corresponding weakly singular eigenstate is a linear combination

$$\left|\Psi_s\right\rangle = \sum_{i(\lambda_i \neq \varepsilon_s)} \frac{\left\langle\Phi_i\middle|\tau\right\rangle}{\varepsilon_s - \lambda_i}\left|\Phi_i\right\rangle, \quad \varepsilon_s = \lambda_j. \tag{63}$$

This expression is similar to the expression (60) which applies to cardinal eigenstates.

For simplicity in the above expressions normalization constants were not given. Those constants can be easily obtained observing the condition $\left\langle\Psi_s\middle|\mathbf{S}_a + \beta\mathbf{P}\middle|\Psi_s\right\rangle = 1$.

6. LRM Treatment of the Vibrational Isotope Effect in the Harmonic Approximation

Let me now show how one can apply LRM formalism to the concrete problem in quantum chemistry. In particular, I will analyze within the LRM formalism vibrational isotope effect in the harmonic approximation.

Molecular vibrations can be relatively well described within the harmonic approximation where atomic nuclei are represented by mathematical points with masses [10,11]. Displacements of nuclei from equilibrium positions can be described by Cartesian coordinates. If the molecule $\boldsymbol{A}_n$ contains N nuclei, there are $n = 3N$ such coordinates. Mathematical treatment of the molecular system $\boldsymbol{A}_n$ consisting of N masses connected by harmonic forces leads to the generalized eigenvalue equation

$$\mathbf{F}\left|\Phi_i\right\rangle = \lambda_i\mathbf{M}\left|\Phi_i\right\rangle, \quad i = 1,\ldots,n, \tag{64a}$$

where $\mathbf{F}$ and $\mathbf{M}$ are force field and mass operators, respectively. Eigenstates $\left|\Phi_i\right\rangle$ of (64a) describe molecular vibrations, while the corresponding eigenvalues λ_i are connected with the vibrational frequencies ν_i according to [10-13]

$$\lambda_i = 4\pi^2\nu_i^2. \tag{64b}$$

In view of the expression (1b), eigenstates $\left|\Phi_i\right\rangle$ should be orthonormalized according to

$$\left\langle\Phi_i\middle|\mathbf{M}\middle|\Phi_j\right\rangle = \delta_{ij}, \quad i, j = 1,\ldots,n. \tag{64c}$$

Expression (64a) is essentially a standard $(\mathbf{F},\mathbf{G})$ formalism [10-13]. In this formalism one usually uses internal coordinates which are suitable for the actual molecular geometry. In the LRM approach it is more appropriate to use Cartesian coordinates. In order to emphasize that Cartesian coordinates are used, the symbol $\mathbf{M}$ is retained.

If in the original molecule $\boldsymbol{A}_n$ some or all atoms are replaced by an isotope, one obtains modified molecule $\mathbf{C}_n$. To a very good approximation, force field is not changed by isotopic substitutions [10-13]. Within the harmonic approximation vibrations of the isotopomer $\mathbf{C}_n$ are hence described by the eigenvalue equation

$$\mathbf{F}|\Psi_s\rangle = \varepsilon_s(\mathbf{M}+\Delta\mathbf{M})|\Psi_s\rangle, \quad s = 1,\dots,n, \tag{65a}$$

where $\Delta\mathbf{M}$ is the operator of the isotope mass changes. Eigenstates $|\Psi_s\rangle$ of (65a) are vibrations of the isotopomer $\mathbf{C}_n$, while the corresponding eigenvalues ε_s are related to the vibrational frequencies ω_s of this isotopomer according to

$$\varepsilon_s = 4\pi^2\omega_s^2 . \tag{65b}$$

Apart from different notation, equation (64a) is identical to the equation (1a), while equation (65a) is identical to the equation (55) where $\beta = 1$. As required by the LRM formalism, operators $\mathbf{M}$ and $\mathbf{M}+\Delta\mathbf{M}$ are positive definite. According to the LRM treatment of those expressions, if one knows the solution of the eigenvalue equation (64a), one can obtain an exact solution of the eigenvalue equation (65a) where $\Delta\mathbf{M}$ is an arbitrary isotope mass change. In other words, if one knows frequencies and normal modes of a parent molecule $\boldsymbol{A}_n$, one can obtain frequencies and normal modes of all isotopomers $\mathbf{C}_n$ of this molecule. In the case of each particular isotopomer $\mathbf{C}_n$ all one needs to know are positions (x_μ, y_μ, z_μ) of isotopic substitutions and corresponding mass changes Δm_μ. It is remarkable that in the harmonic approximation normal modes and corresponding frequencies of a parent molecule $\boldsymbol{A}_n$ uniquely determine in this way all normal modes and vibrational frequencies of all possible isotopomers of this molecule.

One further point. In a standard mathematical treatment of molecular vibrations, equations (64a) and (65a) are expressed in a mass weighted coordinates where dependence on molecular masses is absorbed in the definition of related coordinates [10-13]. In the LRM approach explicit dependence on those masses is retained. Normal modes as defined in a standard treatment hence differ from normal modes as defined in the LRM. In a standard treatment normal modes $|\Phi_i^*\rangle$ of the parent molecule $\boldsymbol{A}_n$ are orthonormalized according to $\langle\Phi_i^*|\Phi_j^*\rangle = \delta_{ij}$, while in the LRM approach those normal modes are orthonormalized according to (1b) where $\mathbf{S}_a \equiv \mathbf{M}$. Connection between normal modes $|\Phi_i^*\rangle$ and $|\Phi_i\rangle$ is $|\Phi_i^*\rangle = \mathbf{M}^{1/2}|\Phi_i\rangle$. In similar way normal modes $|\Psi_s^*\rangle$ and $|\Psi_s\rangle$ of the modified molecule $\mathbf{C}_n$ are related to each other according to $|\Psi_s^*\rangle = (\mathbf{M}+\Delta\mathbf{M})^{1/2}|\Psi_s\rangle$.

If molecules $\mathbf{A}_n$ and $\mathbf{C}_n$ are planar, in-plane and out-of-plane vibrations separate and expressions (64a) and (65a) apply separately to in-plane and separately to out-of-plane vibrations. For the sake of simplicity I will consider here only the LRM treatment of the vibrational isotope effect of out-of-plane vibrations of planar molecules. LRM treatment of in-plane vibrations of planar molecules and of the vibrations of arbitrary (nonplanar) molecules is formally the same, though numerically more involved.

6.1. Out-Of-Plane Vibrations of Planar Molecules

Let $\mathbf{A}_n$ be a planar nonlinear molecule containing n atoms. Put the origin of the coordinate system in the centre of mass of this molecule and let *x*- , *y*- and *z*- axis coincide with principal axis of $\mathbf{A}_n$. Let further *z*-axis be perpendicular to the molecular plane, and let *x*- and *y*-axis lie in this plane. With this convention non-proper out-of-plane vibrations, translation $|T_z\rangle$ in the z -direction and rotations $|R_x\rangle$ and $|R_y\rangle$ around *x*- and *y*-axis of the parent molecule $\mathbf{A}_n$, are [15-17]

$$|T_z\rangle = \frac{1}{\sqrt{M}}\sum_{\alpha}^{n}|\alpha\rangle, \quad |R_x\rangle = \frac{1}{\sqrt{I_x}}\sum_{\alpha}^{n} y_\alpha|\alpha\rangle, \quad |R_y\rangle = \frac{-1}{\sqrt{I_y}}\sum_{\alpha}^{n} x_\alpha|\alpha\rangle. \tag{66}$$

where M is molecular mass of the molecule $\mathbf{A}_n$, while I_x and I_y are moments of inertia of this molecule around *x*- and *y*- axis, respectively. Further, $|\alpha\rangle$ is a unit displacement of the α *-th* atom in the *z*-direction perpendicular to the molecular plane, while (x_α, y_α) is the position of this atom in this plane. Unit displacements $|\alpha\rangle$ ($\alpha = 1,\ldots,n$) are orthonormalized according to (3b) and they span *n*-dimensional space X_n^a. If the origin of the coordinate system is in the centre of mass of molecule $\mathbf{A}_n$ and if *x*- and *y*-axis coincide with principal axis of this molecule, non-proper vibrations (66) are orthonormalized according to (1b), as required [15-17].

If one replaces $\rho \le n$ atoms in molecule $\mathbf{A}_n$ with various isotopes, one obtains an isotopomer $\mathbf{C}_n$. In the harmonic approximation out-of-plane vibrations of this isotopomer are described by the eigenvalue equation (65a). In the base $\{|\mu\rangle\}$ ($\mu = 1,\ldots,\rho$) operator $\Delta\mathbf{M}$ is diagonal and it satisfies $\Delta\mathbf{M}|\mu\rangle = \Delta m_\mu|\mu\rangle$ where the eigenvalue $\Delta m_\mu \ne 0$ is mass change at the substitution site μ. Matrix elements of the corresponding operator

$\mathbf{\Omega}^0(\varepsilon)$ are given by expression (46b). From this expression one can eliminate non-proper vibrations (66) to obtain [15-17].

$$\mathbf{\Omega}^0_{\mu\nu}(\varepsilon) = \frac{1}{\varepsilon}\left[\frac{1}{M} + \frac{x_\mu x_\nu}{I_y} + \frac{y_\mu y_\nu}{I_x}\right] + \sum_{i(\lambda_i \neq \varepsilon)}^{n-3} \frac{\langle \mu | \Phi_i \rangle \langle \Phi_i | \nu \rangle}{\varepsilon - \lambda_i}, \quad \mu,\nu = 1,\dots,\rho \tag{67a}$$

Since $\Delta\mathbf{M}$ is diagonal in the base $\{|\mu\rangle\}$, one has

$$\Delta\mathbf{M}^{-1}_{\mu\nu} = \frac{\delta_{\mu\nu}}{\Delta m_\mu}. \tag{67b}$$

In the expression (67a) $\langle \mu | \Phi_i \rangle$ is the amplitude of the normalized (out-of-plane) vibration $|\Phi_i\rangle$ of molecule $\boldsymbol{A}_n$ at the position (x_μ, y_μ) of the isotopic substitution, while λ_i is the corresponding eigenvalue. The summation in (67a) is performed over all proper vibrations $|\Phi_i\rangle$ that satisfy $\lambda_i \neq \varepsilon$.

According to **Theorem 3B**, each cardinal eigenvalue which differs from zero ($\varepsilon_s \neq 0$) of the isotopomer $\mathbf{C}_n$ satisfies

$$\left[\mathbf{\Omega}^0(\varepsilon_s) + \frac{\Delta\mathbf{M}^{-1}}{\varepsilon_s}\right]|\theta_s\rangle = 0, \quad |\theta_s\rangle \in X^{\text{int}}_\rho, \tag{68a}$$

It is hence a root of the $\rho \times \rho$ determinant

$$f(\varepsilon) \equiv \left| \mathbf{\Omega}^0(\varepsilon) + \frac{\Delta\mathbf{M}^{-1}}{\varepsilon} \right| = 0. \tag{68b}$$

This eigenvalue is related to the corresponding frequency ω_s according to (65b). Once ε_s and $|\theta_s\rangle$ are known, the corresponding eigenstate is given as a linear combination (57a). From this expression one can eliminate the contribution of non-proper vibrations to obtain

$$|\Psi_s\rangle = \frac{1}{\varepsilon_s}\left[T^{(s)}|T_z\rangle + R^{(s)}_x|R_x\rangle + R^{(s)}_y|R_y\rangle\right] + \sum_i^{n-3} \frac{\langle \Phi_i | \theta_s \rangle}{\varepsilon_s - \lambda_i}|\Phi_i\rangle, \tag{69a}$$

where $|\Phi_i\rangle$ are proper out-of-plane vibrations of the parent molecule $\boldsymbol{A}_n$, while $|T_z\rangle$, $|R_x\rangle$ and $|R_y\rangle$ are non-proper vibrations (66). Coefficients $T^{(s)}$, $R^{(s)}_x$ and $R^{(s)}_y$ are

$$T^{(s)} = \frac{1}{\sqrt{M}} \sum_{\nu}^{\rho} \langle \nu | \theta_s \rangle, \quad R_x^{(s)} = \frac{1}{\sqrt{I_x}} \sum_{\nu}^{\rho} y_\nu \langle \nu | \theta_s \rangle, \quad R_y^{(s)} = \frac{(-1)}{\sqrt{I_y}} \sum_{\nu}^{\rho} x_\nu \langle \nu | \theta_s \rangle, \tag{69b}$$

Due to (57b) and (65b), coefficients $\langle \nu | \theta_s \rangle$ are related to the cardinal eigenstate (69a) according to

$$\langle \nu | \theta_s \rangle = -4\pi^2 \omega_s^2 \Delta m_\nu \langle \nu | \Psi_s \rangle, \quad \nu = 1, \ldots, \rho. \tag{69c}$$

Vibrations $|\Psi_s\rangle$ can be orthonormalized according to (2c) where $\mathbf{S}_c = \mathbf{M} + \Delta\mathbf{M}$. If $\varepsilon_s \neq \varepsilon_p$ this expression is automatically satisfied. However, if $\varepsilon_s = \varepsilon_p$ one has to enforce this relation by some standard orthonormalization procedure. According to (69a), modified vibration $|\Psi_s\rangle$ is expressed as a linear combination $|\Psi_s\rangle = \sum_i c_i |\Phi_i\rangle$. In order to normalize this vibration, matrix elements $\langle \Phi_i | \mathbf{M} + \Delta\mathbf{M} | \Phi_j \rangle$ are required. Those matrix elements can be obtained using expression

$$\Delta\mathbf{M} = \sum_{\mu}^{\rho} |\mu\rangle \Delta m_\mu \langle \mu|, \tag{70a}$$

where Δm_μ is the change of the mass of atom μ by the isotope substitution. Since $|\Psi_s\rangle = \sum_i c_i |\Phi_i\rangle$ one has

$$\langle \Psi_s | \mathbf{M} + \Delta\mathbf{M} | \Psi_s \rangle = \sum_i c_i^* c_i + \sum_{i,j} c_i^* c_j \langle \Phi_i | \Delta\mathbf{M} | \Phi_j \rangle, \tag{70b}$$

where according to (70a) matrix elements $\langle \Phi_i | \Delta\mathbf{M} | \Phi_j \rangle$ equal

$$\langle \Phi_i | \Delta\mathbf{M} | \Phi_j \rangle = \sum_{\mu}^{\rho} \langle \Phi_i | \mu \rangle \Delta m_\mu \langle \mu | \Phi_i \rangle, \tag{70c}$$

In particular, in the case of a single isotopic substitution on the atomic site τ, expression (70c) reduces to

$$\langle \Phi_i | \Delta\mathbf{M} | \Phi_j \rangle = \langle \Phi_i | \tau \rangle \Delta m_\tau \langle \tau | \Phi_i \rangle. \tag{70d}$$

Using above expressions one can normalize each eigenstate $|\Psi_s\rangle$ of the modified system.

As emphasized in section 5.4., expression (68) does not produce cardinal solutions with the eigenvalue $\varepsilon_s = 0$. However if $\boldsymbol{A}_n$ is a rigid planar molecule, all proper frequencies ν_i of this molecule are positive, only three non-proper frequencies associated with non-proper vibrations (66) are zero. The same applies to the modified molecule $\mathbf{C}_n$. Since $\varepsilon_s = 4\pi^2\omega_s^2$, expression (68) produces all proper cardinal vibrations of the modified molecule $\mathbf{C}_n$ and no non-proper vibration of this molecule. Since in the analysis of the vibrational isotope effect one is interested only in the proper vibrations, this is exactly what is needed.

In a similar way can be treated singular vibrations of the isotopomer $\mathbf{C}_n$.

6.2. Interlacing Rule for out-of-Plane Vibrations of Planar Molecules

Vibrational spectra provide a large amount of information about a molecule. An efficient study and analysis of experimental frequencies requires correct assignment of those frequencies to various vibrational types and normal modes. Besides vibrational spectra of the original molecule, vibrational spectra of isotopically substituted molecules generate a huge amount of additional experimental data. Those data are important as a valuable source of information for the correct assignment and interpretation of experimental spectra. This assignment is highly facilitated by some simple rules which apply to vibrational frequencies of isotopically related molecules. One such rule is well known order rule [10-13]. If in a mechanical system that exhibits harmonic oscillations one increases the mass of one or few particles without changing the potential energy, the frequencies of normal vibrations do not increase. Those frequencies either decrease or remain the same. In particular, if frequencies ν_i of the parent molecule $\mathbf{A}_n$ are arranged in the increasing order, and if frequencies ω_s of the heavier isotopic molecule $\mathbf{C}_n$ are also arranged in the increasing order, then [12,13]

$$\omega_1 \le \nu_1, \omega_2 \le \nu_2, \ldots, \omega_s \le \nu_s, \ldots \tag{71}$$

Order rule involves all frequencies of such isotopic molecules $\mathbf{A}_n$ and $\mathbf{C}_n$ where all substituted isotopes in the molecule $\mathbf{C}_n$ are heavier than the corresponding isotopes in the molecule $\mathbf{A}_n$.

Since the order rule is derived within the harmonic approximation, anharmonicity effects may violate this rule in some cases. However, those effects are usually small and experimental frequencies almost always satisfy this rule.

Some additional types of connections between vibrational frequencies of isotopic molecules can be obtained by various other isotopic rules. Of such rules are known the product rule[18], the sum rule [19] and the complete isotopic rule [20,21]. For example, the complete isotopic rule concerns three isotopic molecules A, B and C. These molecules must be in a special relation: molecule A is a planar molecule in which two definite atoms form a symmetrically equivalent set. B is identical to A except that one of these two atoms is exchanged for an isotope, and C is identical to A and B except that both of these two atoms are exchanged for this isotope [20]. An example is the set of three molecules: ethylene, *d1*-ethylene and *d2-trans*-ethylene. From the known frequencies of the compounds A and C one can now calculate frequencies of the compound B. Unlike order rule which is more general, the complete isotopic rule requires a ruther special arangemet of three related molecules. All other isotopic rules have also rather special and narow application. From all those rules order rule (71) is most general and most powerful.

Within the LRM formalism one can derive vibrational interlacing rule [15-17]. According to the expressions (64b) and (65b) all eigenvalues λ_i of the parent molecule $\boldsymbol{A}_n$ as well as all eigenvalues ε_s of the modified molecule $\mathbf{C}_n$ are nonnegative. Hence to this case applies LRM interlacing rule 3 (expression (58)). Due to the expressions (64b) and (65b), the same rule applies also to the corresponding frequencies ν_i and ω_s. In the case of out-of-plane vibrations, if the modified molecule $\mathbf{C}_n$ differs from the original molecule $\boldsymbol{A}_n$ by ρ isotopic substitutions, modification operator $\Delta\mathbf{M}$ has rank ρ. In addition, if r out of ρ substituted isotopes are heavier in the molecule $\mathbf{C}_n$ and remaining $\rho - r$ isotopes heavier in the molecule $\boldsymbol{A}_n$, mass change operator $\Delta\mathbf{M}$ has r positive and $\rho - r$ negative eigenvalues. Hence expression (58) implies:

Vibrational interlacing rule (out-of-plane vibrations of planar molecules): Let $\boldsymbol{A}_n$ and $\mathbf{C}_n$ be two planar molecules and let molecule $\mathbf{C}_n$ differ from molecule $\boldsymbol{A}_n$ by ρ isotopic substitutions. Let further r substituted isotopes be heavier in a molecule $\mathbf{C}_n$, and let the remaining $\rho-r$ isotopes be heavier in a molecule $\boldsymbol{A}_n$. Arrange out-of-plane frequencies ν_i of the parent molecule $\boldsymbol{A}_n$ in the nondecreasing order and arrange also out-of-plane frequencies ω_s of the isotopomer $\mathbf{C}_n$ in the nondecreasing order. If the force field does not change upon isotopic substitutions and in the harmonic approximation, those frequencies satisfy:

$$\nu_{s-r} \leq \omega_s \leq \nu_{s+\rho-r}, \tag{72}$$

If all substituted isotopes are heavier in the isotopomer $\mathbf{C}_n$ one has $r = \rho$ and expression (72) reduces to $\nu_{s-\rho} \leq \omega_s \leq \nu_s$. In particular this implies $\omega_s \leq \nu_s$ ($s = 1,\ldots,n$). This is well known order rule (71) applied to out-of-plane frequencies: If all substituted isotopes are heavier in the isotopomer $\mathbf{C}_n$, modified frequencies ω_s do not increase [12,13]. Order rule is thus a special case of the interlacing rule. Interlacing rule is more general since it applies to arbitrary distribution of heavier and lighter isotopes in the isotopomer $\mathbf{C}_n$. It is also much stronger since it limits each frequency ω_s from both sides, while order rule limits each frequency ω_s only from one side.

Expression (72) applies to the vibrational isotope effect of out-of-plane vibrations of planar molecules. In the case of in-plane vibrations this expression is replaced with the expression $\nu_{s-2r} \le \omega_s \le \nu_{s+2\rho-2r}$, while in the case of arbitrary (non-planar) vibrations it is replaced with the expression $\nu_{s-3r} \le \omega_s \le \nu_{s+3\rho-3r}$ [17].

6.3. Single Isotopic Substitutions

LRM treatment of out-of-plane vibrations of planar molecules $\boldsymbol{A}$ and $\mathbf{C}_\tau$ which differ by a single isotopic substitution at atomic site τ is particularly simple. In the case of such a substitution one has $\rho = 1$ and expressions (67-69) simplify to

$$\Omega^0_{\tau\tau}(\varepsilon) = \frac{1}{\varepsilon}\left[\frac{1}{M} + \frac{x_\tau^2}{I_y} + \frac{y_\tau^2}{I_x}\right] + \sum_{i(\lambda_i \neq \varepsilon)}^{n-3} \frac{\langle \tau | \Phi_i \rangle \langle \Phi_i | \tau \rangle}{\varepsilon - \lambda_i}, \tag{73a}$$

$$f(\varepsilon) \equiv \Omega^0_{\tau\tau}(\varepsilon) + \frac{1}{\varepsilon\, \Delta m_\tau} = 0\,,\ \varepsilon \notin \{\lambda_i\}, \tag{73b}$$

$$|\Psi_s\rangle = \frac{1}{\varepsilon_s}\left[\frac{1}{\sqrt{M}}|T_z\rangle + \frac{y_\tau}{\sqrt{I_x}}|R_x\rangle - \frac{x_\tau}{\sqrt{I_y}}|R_y\rangle\right] + \sum_i^{n-3} \frac{\langle \Phi_i | \tau \rangle}{\varepsilon_s - \lambda_i} |\Phi_i\rangle. \tag{73c}$$

Each root $\varepsilon = \varepsilon_s \notin \{\lambda_i\}$ of (73b) is a cardinal out-of-plane eigenvalue of the isotopomer $\mathbf{C}_\tau$. Once $\varepsilon = \varepsilon_s$ is known, the corresponding vibration $|\Psi_s\rangle$ is given by (73c). Since $\rho = 1$, each cardinal vibration $|\Psi_s\rangle$ is nondegenerate.

Consider now singular solutions of $\mathbf{C}_\tau$. Each strongly singular vibration $|\Psi_s\rangle$ that corresponds to the eigenvalue $\varepsilon_s \equiv \lambda_j$ is a linear combination (61a) where coefficients $D_l^{(j)}$ satisfy a single condition (61b).

Concerning weakly singular vibrations with the eigenvalue $\varepsilon_s = \lambda_j$, expressions (62) imply

$$\Omega^0_{\tau\tau}(\varepsilon_s) + \frac{1}{\varepsilon_s \Delta m_\tau} = 0\,, \tag{74a}$$

$$\langle \Phi_{jl} | \tau \rangle = 0\,,\ l = 1, \ldots, \eta_j\,. \tag{74b}$$

Isotopomer $\mathbf{C}_\tau$ has weakly singular vibration $\left|\Psi_s\right\rangle$ with the eigenvalue $\varepsilon_s \equiv \lambda_j$ if and only if this eigenvalue satisfies (74a) and if in addition all amplitudes $\left\langle\Phi_{jl}\middle|\tau\right\rangle$ ($l=1,\ldots,\eta_j$) vanish.

In the case of a single isotopic substitution vibrational interlacing rule (72) also substantially simplifies. In particular, if molecule $\mathbf{C}_\tau$ is heavier than molecule $\boldsymbol{A}$ one has $\rho = r = 1$ and this rule reduces to

$$0 < \omega_1 \le \nu_1 \le \omega_2 \le \nu_2 \le \ldots \le \omega_{n-3} \le \nu_{n-3}\,. \tag{75}$$

Similar expression is obtained if molecule $\boldsymbol{A}$ is heavier than molecule $\mathbf{C}_\tau$.

According to the above expression, each out-of-plane frequency ω_s of the isotopomer $\mathbf{C}_\tau$ is bracketed between exactly two adjacent out-of-plane frequencies ν_i of a parent molecule $\boldsymbol{A}$, and vice-versa. In particular, a single isotopic substitution in a planar molecule can change (increase or decrease) degeneracy of each particular out-of-plane frequency ν_j at most by one. One similarly finds that a single isotopic substitution in a planar molecule can change (increase or decrease) degeneracy of each particular in-plane frequency ν_j at most by two, while a single isotopic substitution in a non-planar molecule can change (increase or decrease) degeneracy of each particular frequency ν_j at most by three [17].

Solution of the vibrational isotope effect in the case of the single isotopic substitution demonstrates efficiency of the LRM approach. In a standard approach one has to diagonalize $n \times n$ eigenvalue equation (65a). In the LRM approach one has to solve much simpler expression (73b) that produces all cardinal eigenvalues. Calculation of the remaining singular eigenvalues, if any, is numerically trivial.

6.4. Inversion Relations

Expressions (73) implies *inversion relation* [6,17,22]

$$\mathbf{Y} = \mathbf{G}^{-1}\mathbf{X}\,, \tag{76a}$$

where $\mathbf{G}$ is a square matrix with matrix elements G_{si} while $\mathbf{X}$ and $\mathbf{Y}$ are column vectors with components X_s and Y_i, respectively

$$G_{si} = \frac{1}{\varepsilon_s - \lambda_i}\,,\quad X_s = -\frac{1}{\varepsilon_s}\left[\frac{1}{M} + \frac{x_\tau^2}{I_y} + \frac{y_\tau^2}{I_x} + \frac{1}{\Delta m_\tau}\right], \tag{76b}$$

$$Y_i = \sum_{l}^{\eta_i} \langle \tau | \Phi_{il} \rangle \langle \Phi_{il} | \tau \rangle . \qquad (76c)$$

In choosing the set $\{\lambda_i\}$ of parent eigenvalues λ_i and the set $\{\varepsilon_s\}$ of modified eigenvalues ε_s that are required for the construction of the matrix $\mathbf{G}$ and vector $\mathbf{X}$, following rule applies [6,17,22]: The set $\{\lambda_i\}$ should contain all mutually distinct out-of-plane eigenvalues of a parent molecule $\boldsymbol{A}$ that satisfy $Y_i \neq 0$. By definition, each eigenvalue λ_i that satisfies $Y_i = 0$ is *passive* [17,22]. This eigenvalue has no influence on the vibrational isotope effect. Hence passive eigenvalues of a parent system $\boldsymbol{A}$ should not be considered. Concerning the set $\{\varepsilon_s\}$, this set should contain all mutually distinct cardinal out-of-plane eigenvalues of the isotopomer $\mathbf{C}_\tau$. One finds that both sets contain the same number of elements, and $\mathbf{G}$ is hence a square matrix. In addition, in almost all cases matrix $\mathbf{G}$ is nonsingular and hence invertible. Expression (76a) is hence well defined.

Inversion relation (76a) expresses squares of the amplitudes $\langle \tau | \Phi_i \rangle$ at the substitution site τ (or, if λ_i is degenerate, sum Y_i of such squares) in terms of active vibrations ν_i of the parent molecule $\boldsymbol{A}$, in terms of the cardinal vibrations $\omega_{s\tau}$ of the isotopomer $\boldsymbol{B}_\tau$, in terms of the position (x_τ, y_τ) of the isotopic substitution in molecule $\boldsymbol{A}$, in terms of mass change Δm_τ of this substitution, and finally in terms of the total mass M and moments of inertia I_x and I_y of the parent molecule $\boldsymbol{A}$. All those quantities are experimental quantities and they can be easily obtained from experimental data. Hence inversion relation provides squares of the vibrational amplitudes $\langle \tau | \Phi_i \rangle$ at the selected substitution site τ in terms of experimental data alone. Using several such substitution sites τ, one can derive in this way amplitude squares of all vibrations $|\Phi_i\rangle$ at all substitution sites τ thus selected. Only relative phases between amplitudes of those vibrations at different substitution sites remain to be determined. However, since all proper vibrations $|\Phi_i\rangle$ are orthogonal to non-proper vibrations $|\Phi_T\rangle$, $|\Phi_{Rx}\rangle$ and $|\Phi_{Ry}\rangle$, and since those proper vibrations must be mutually orthogonal, those orthogonality requirements in most cases uniquely determine all relative phases of amplitudes $\langle \tau | \Phi_i \rangle$ at various substitution sites τ [6,15-17].

Inversion relations that involve out-of-plane frequencies of planar molecule $\boldsymbol{A}_n$ and out-of-plane frequencies of several mono-substituted isotopomers $\mathbf{C}_{\tau s}$ of this molecule determine amplitudes $\langle \tau_1 | \Phi_i \rangle, \langle \tau_2 | \Phi_i \rangle, \ldots$, of the vibrations $|\Phi_i\rangle$ of $\boldsymbol{A}_n$ at all substitution sites τ thus selected. Once those amplitudes are known, one can derive out-of-plane frequencies and corresponding vibrations of all isotopomers $\mathbf{C}$ that contain substituted isotopes at thus selected substitution sites. If a parent molecule $\boldsymbol{A}_n$ has some symmetry (in addition to being planar) amplitudes $\langle\tau|\Phi_i\rangle$ can be partly or completely determined by this symmetry. This reduces the dimension of matrix $\mathbf{G}$ and vectors $\mathbf{Y}$ and $\mathbf{X}$ that are involved in the construction of the inversion relation (76a). Quantities $\mathbf{G}$, $\mathbf{X}$ and $\mathbf{Y}$ obtained in this way are accordingly modified, and they differ from their generic form (76).

In conclusion, inversion relation provides a systematic method for determining out-of-plane frequencies and normal modes of all isotopomers of a given planar molecule. All one needs are out of plane frequencies ν_i of a parent molecule $\mathbf{A}$, and several sets $\{\omega_s\}$ of out-of-plane frequencies of mono-substituted molecules $\mathbf{C}_\tau$.

6.5. Interlacing of Out-of-Plane Vibrational Frequencies of Thiophene and Deuterated Thiophenes

Let me illustrate the application of the LRM formalism to the vibrational isotope effect with few examples. Consider first vibrational interlacing rule. As an example of the application of this rule consider thiophene molecule [23]. Thiophene with the corresponding atom numbering is shown in Figure 3. There are 10 deuteration patterns of thiophene. The interlacing hierarchy of these isotopomers is shown in Figure 4. This hierarchy is an example of *partial ordering* [24]. One can make as many as 16 comparisons between various pairs of deuterated thiophenes that differ by a single isotopic substitution. Such pairs are in Figure 4. connected by an arrow. Frequencies of all such pairs should satisfy interlacing Rule 75.

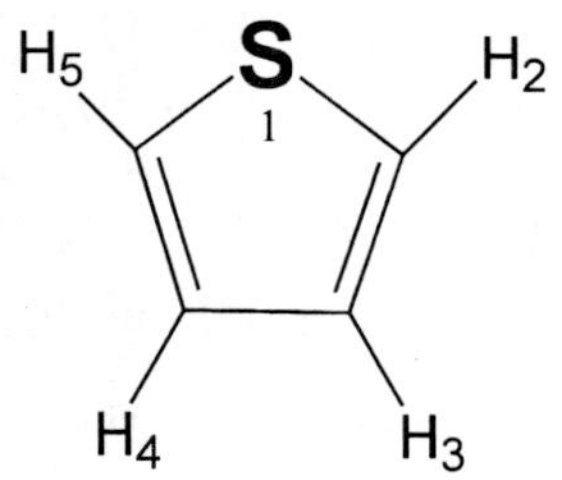

Figure 3. Thiophene.

Thiophene has 9 atoms and hence 9 out-of-plane degrees of freedom. One combination of the out-of-plane displacements describes the translation in the *z*-direction and other two combinations describe rotations around in-plane *x*- and *y*-axes, respectively. Hence each thiophene isotopomer has *9-3=6* proper out-of-plane vibrations.

Experimental out-of-plane frequencies [25] of thiophene (H,D)-isotopomers are given in Table 1. Using data from this Table one finds that thiophene and two d_1-thiophenes frequencies satisfy (all frequencies are in cm^{-1}).

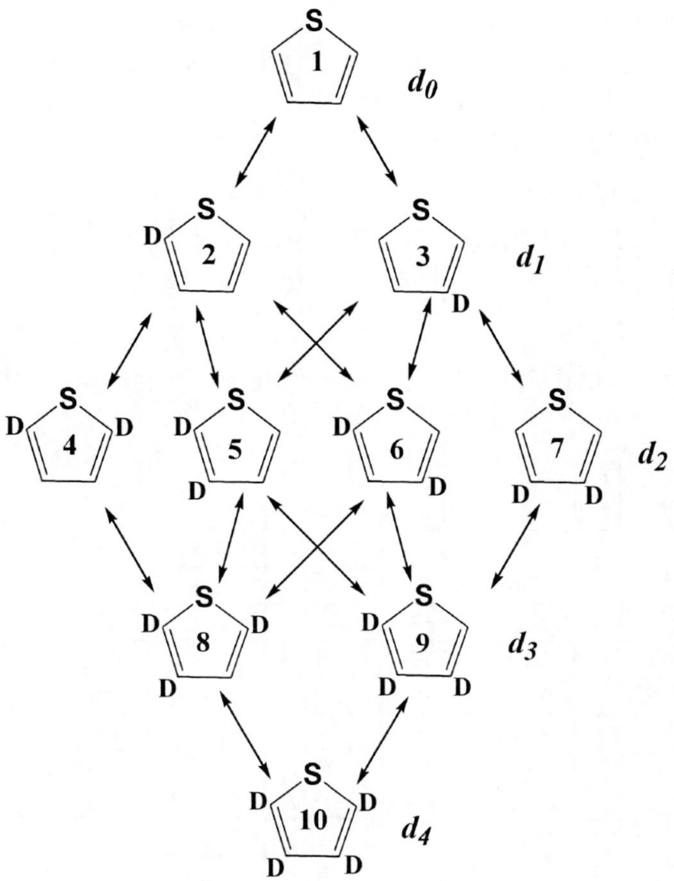

Figure 4. Interlacing hierarchy of deuterated thiophenes. Adopted from ref. 23.

$$\begin{pmatrix}433(2)\\446(3)\end{pmatrix}_1 < 452(1)_1 < \begin{pmatrix}533(2)\\516(3)\end{pmatrix}_2 < 565(1)_2 < \begin{pmatrix}567(2)\\645(3)\end{pmatrix}_3 < 683(1)_3 <$$
$$< \begin{pmatrix}701(2)\\685(3)\end{pmatrix}_4 < 712(1)_4 < \begin{pmatrix}839(2)\\800(3)\end{pmatrix}_5 < 867(1)_5 < \begin{pmatrix}896(2)\\882(3)\end{pmatrix}_6 < 898(1)_6 \tag{77a}$$

In this expression symbols (1), (2) and (3) denote thiophene and deuterated d_1-thiophenes 2-C_4H_3DS and 3-C_4H_3DS, respectively (see Figure 4). Subscripts 1 through 6 denote corresponding out-of-plane frequencies. As shown in this expression, all out-of-plane frequencies of those isotopomers satisfy interlacing rule 75.

Interlacing rule for isotopomer 2 (d_1-thiophene 2-C_4H_3DS) and three d_2-thiophenes is verified in the following expression

$$\begin{pmatrix} 419(4) \\ 427(5) \\ 427(6) \end{pmatrix}_1 < 433(2)_1 < \begin{pmatrix} 528(4) \\ 511(5) \\ 504(6) \end{pmatrix}_2 < 533(2)_2 < \begin{pmatrix} 566(4) \\ 553(5) \\ 561(6) \end{pmatrix}_3 < 567(2)_3 <$$

$$< \begin{pmatrix} 584(4) \\ 668(5) \\ 645(6) \end{pmatrix}_4 < 701(2)_4 < \begin{pmatrix} 819(4) \\ 744(5) \\ 786(6) \end{pmatrix}_5 < 839(2)_5 < \begin{pmatrix} 887(4) \\ 882(5) \\ 861(6) \end{pmatrix}_6 < 896(2)_6 \quad (77b)$$

Table 1. Experimental [25] out-of-plane vibrational frequencies of thiophene and deuterated thiophenes (all frequencies in cm^{-1}).

	1 d_0	**2** 2-d_1	**3** 3-d_1	**4** 2,5-d_2	**5** 2,3-d_2
1	452	433	446	419	427
2	565	533	516	528	511
3	683	567	645	566	553
4	712	701	685	584	668
5	867	839	800	819	744
6	898	896	882	887	882
	6 2,4-d_2	**7** 3,4-d_2	**8** 2,3,5-d_3	**9** 2,3,4-d_3	**10** d_4
1	427	442	416	427	414
2	504	488	504	486	484
3	561	587	531	531	531
4	645	671	579	614	531
5	786	775	711	738	684
6	861	811	860	788	750

Compound 4 is isotopomer 2,5-$C_4H_2D_2S$, compound 5 is isotopomer 2,3-$C_4H_2D_2S$, while compound 6 is isotopomer 2,4-$C_4H_2D_2S$ (see Figure 4). Again, the interlacing rule is satisfied for all out-of-plane frequencies considered.

Comparison of the out-of-plane frequencies of the isotopomer 3 (d_1-thiophene 3-C_4H_3DS) with out-of-plane frequencies of three d_2-thiophenes is shown in the following expression

$$\begin{pmatrix} 442(7) \\ 427(5) \\ 427(6) \end{pmatrix}_1 < 446(3)_1 < \begin{pmatrix} 488(7) \\ 511(5) \\ 504(6) \end{pmatrix}_2 < 516(3)_2 < \begin{pmatrix} 587(7) \\ 553(5) \\ 561(6) \end{pmatrix}_3 < 645(3)_3 \le$$

$$\le \begin{pmatrix} 671(7) \\ 668(5) \\ 645(6) \end{pmatrix}_4 < 685(3)_4 < \begin{pmatrix} 775(7) \\ 744(5) \\ 786(6) \end{pmatrix}_5 < 800(3)_5 < \begin{pmatrix} 811(7) \\ 882(5) \\ 861(6) \end{pmatrix}_6 \le 882(3)_6 \quad (77c)$$

Comparison of the out-of-plane frequencies of the isotopomer 8 (d_3-thiophene 2,3,5-$C_4H_1D_3S$) with out-of-plane frequencies of three d_2-thiophenes is shown in the following expression

$$416(8)_1 < \begin{pmatrix} 419(4) \\ 427(5) \\ 427(6) \end{pmatrix}_1 < 504(8)_2 < \begin{pmatrix} 528(4) \\ 511(5) \\ 504(6) \end{pmatrix}_2 < 531(8)_3 < \begin{pmatrix} 566(4) \\ 553(5) \\ 561(6) \end{pmatrix}_3 <$$

$$< 579(8)_4 < \begin{pmatrix} 584(4) \\ 668(5) \\ 645(6) \end{pmatrix}_4 < 711(8)_5 < \begin{pmatrix} 819(4) \\ 744(5) \\ 786(6) \end{pmatrix}_5 < 860(8)_6 < \begin{pmatrix} 887(4) \\ 882(5) \\ 861(6) \end{pmatrix}_6 \tag{77d}$$

Out-of-plane frequencies of the isotopomer 9 (d_3-thiophene 2,3,4-$C_4H_1D_3S$) are compared with out-of-plane frequencies of three d_2-thiophenes in the following expression

$$427(9)_1 \leq \begin{pmatrix} 442(7) \\ 427(5) \\ 427(6) \end{pmatrix}_1 < 486(9)_2 < \begin{pmatrix} 488(7) \\ 511(5) \\ 504(6) \end{pmatrix}_2 < 531(9)_3 < \begin{pmatrix} 587(7) \\ 553(5) \\ 561(6) \end{pmatrix}_3 <$$

$$< 614(9)_4 < \begin{pmatrix} 671(7) \\ 668(5) \\ 645(6) \end{pmatrix}_4 < 738(9)_5 < \begin{pmatrix} 775(7) \\ 744(5) \\ 786(6) \end{pmatrix}_5 < 788(9)_6 < \begin{pmatrix} 811(7) \\ 882(5) \\ 861(6) \end{pmatrix}_6 \tag{77e}$$

One can finally compare out-of-plane frequencies of d_4-thiophene and d_3-thiophenes 2,3,5-$C_4H_1D_3S$ and 2,3,4-$C_4H_1D_3S$:

$$414(10)_1 < \begin{pmatrix} 427(9) \\ 416(8) \end{pmatrix}_1 < 484(10)_2 < \begin{pmatrix} 486(9) \\ 504(8) \end{pmatrix}_2 < 531(10)_3 \leq \begin{pmatrix} 531(9) \\ 531(8) \end{pmatrix}_3 \leq$$

$$\leq 531(10)_4 < \begin{pmatrix} 614(9) \\ 579(8) \end{pmatrix}_4 < 684(10)_5 < \begin{pmatrix} 738(9) \\ 711(8) \end{pmatrix}_5 < 750(10)_6 < \begin{pmatrix} 788(9) \\ 860(8) \end{pmatrix}_6 \tag{77f}$$

As shown in expressions (77), all experimental out-of-plane frequencies of deuterated thiophenes satisfy the basic interlacing rule 75. This automatically guaranties that the generalized interlacing rule 72 is also satisfied. One finds that theoretical DFT scaled vibrational frequencies of thiophene isotopomers as calculated by A.A. El-Azhary and R.H. Hilal [26] also satisfy this rule [23].

6.6. Out-of-Plane Vibrations of Ethylene Isotopomers

As another example, consider deuterated ethylenes [15,6]. Ethylene molecule is shown in Figure 5. This molecule has three out-of-plane vibrations. Those vibrations are shown in Figure 6. while the corresponding experimental frequencies [12] are given in Table II. Following ref [12], frequency labels $\nu_4(a_{1u})$, $\nu_7(b_{1u})$ and $\nu_8(b_{2g})$ are used in order to denote those frequencies and the corresponding normal modes.

Concerning the interlacing rule, note that there are seven deuteration patterns of ethylene. The interlacing hierarchy of these isotopomers is shown in Figure 7. One can make as many as 8 comparisons between various pairs of deuterated ethylenes that differ by a single isotopic substitution. Frequencies of all such pairs should satisfy interlacing rule (75).

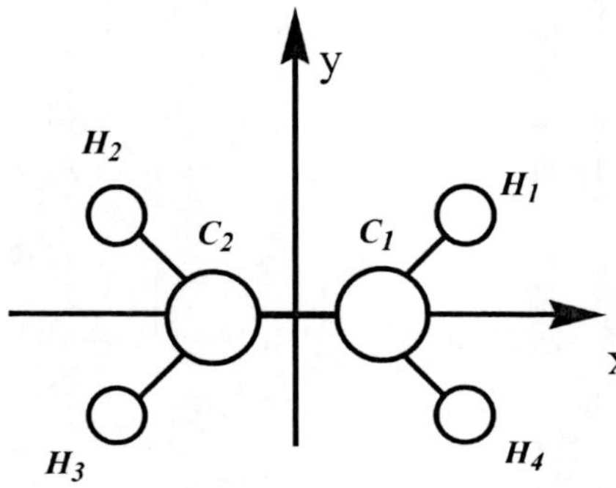

Figure 5. Ethylene.

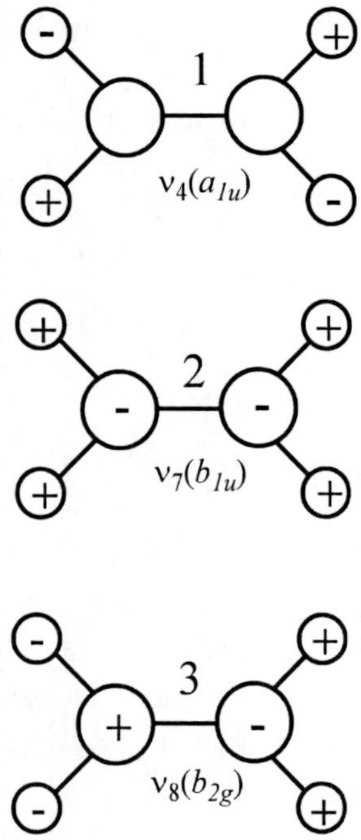

Figure 6. Out-of-plane vibrations of ethylene.

Table II. Experimental out-of-plane frequencies of ethylene. Data and frequency enumeration (ν_4, ν_7 and ν_8) are from Ref. [12].

	Symmetry	Vibrational type	$\nu(cm^{-1})$
ν_4	a_{1u}	$\chi^{CH_2}_{CH_2}$	(1024) (inactive)
ν_7	b_{1u}	ρ_{CH_2}	949.3 (IR, gas)
ν_8	b_{2g}	ρ_{CH_2}	943 (Raman, liquid)

Experimental out-of-plane frequencies of ethylene (H,D)-isotopomers [12] are shown in Table III. Using data in this Table and in Table II one can verify interlacing rule (75). In particular, ethylene and d_1-ethylene frequencies satisfy (experimental frequencies in cm^{-1}):

$$807(2)_1 < 943(1)_1 \leq 943(2)_2 < 943.3(1)_2 < 1001(2)_3 < 1024(1)_3, \tag{78a}$$

Similarly d_1-ethylene frequencies and three d_2-ethylenes frequencies satisfy

$$\begin{pmatrix} 763(3) \\ 725(4) \\ 751(5) \end{pmatrix}_1 < 807(2)_1 < \begin{pmatrix} 842(3) \\ 863(4) \\ 892(5) \end{pmatrix}_2 < 943(2)_2 \leq \begin{pmatrix} 978(3) \\ 987(4) \\ 943(5) \end{pmatrix}_3 < 1001(2)_3, \tag{78b}$$

while those three d_2-ethylene frequencies and d_3-ethylene frequencies satisfy

$$724(6)_1 < \begin{pmatrix} 763(3) \\ 725(4) \\ 751(5) \end{pmatrix}_1 < 765(6)_2 < \begin{pmatrix} 842(3) \\ 863(4) \\ 892(5) \end{pmatrix}_2 < 919(6)_3 < \begin{pmatrix} 978(3) \\ 987(4) \\ 943(5) \end{pmatrix}_3, \tag{78c}$$

Finally d_3-ethylene and d_4-ethylene out-of-plane frequencies satisfy

$$720(7)_1 < 724(6)_1 < 729(7)_2 < 765(6)_2 < 780(7)_3 < 919(6)_3. \tag{78d}$$

All those experimental frequencies satisfy interlacing rule (75).

For some deuterated ethylenes two experimental frequencies are available. One experimental frequency is obtained in a liquid phase, and another in a gas phase. Whereever possible gas experimental frequencies, as more reliable, should be considered. In a liquid phase there are additional interactions betwen molecules, and those interactions can shift original gas frequencies. Hence some liquid experimental frequencies may violate the interlacing rule. For example, in the comparation of monodeuterated (compound 2) and α-bideuterated (compound 5) ethylenes in the expession (78b) one has $943(2)_2 \leq 943(5)_3$ in agreement with the interlacing rule (75). However, in addition to the gas frequency $\nu = 943\ cm^{-1}$ of d_1-ethylene, there is another liquid frequency of d_1-ethylene: $\nu = 945\ cm^{-1}$ [12]. Clearly this frequency does not satisfy $945(2)_2 \leq 943(5)_3$.

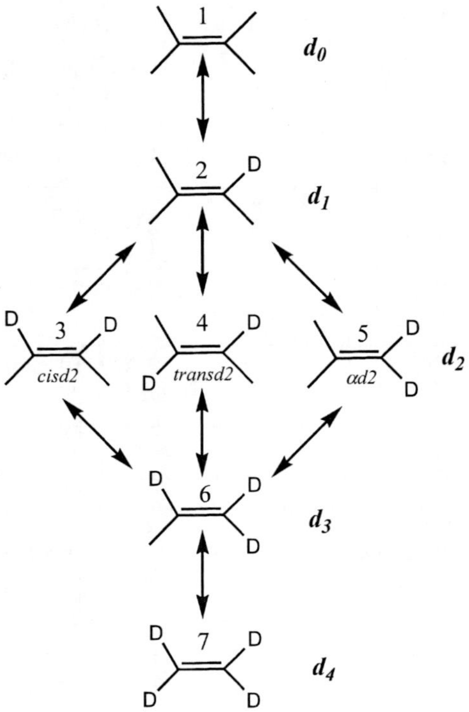

Figure 7. Interlacing hierarchy of deuterated ethylenes.

Table III. Theoretical (LRM) and experimental [12] out-of-plane frequencies (cm^{-1}) for six deuterated ethylenes. LRM frequencies were calculated using ethylene parameters (79a) and experimental ethylene frequencies from table II. *) liquid. Adopted from Ref. [15].

	2 C_2H_3D			3 $cis-C_2H_2D_2$			4 $trans-C_2H_2D_2$		
	Exp.	*LRM*	*Diff.*	*Exp.*	*LRM*	*Diff.*	*Exp.*	*LRM*	*Diff.*
ν_4	1001	*999.4*	*-1.6*	*978*	*980.2*	*2.2*	*987*	*987.7*	*0.7*
ν_7	*943*	*945.7*	*2.7*	*842*	*841.7*	*-0.3*	*725*	*720.7*	*-4.3*
ν_8	*807*	*804.8*	*-2.2*	*763*	*757.4*	*-5.6*	*863**	*864.7*	*1.7*
	5 $\alpha-C_2H_2D_2$			6 C_2HD_3			7 C_2D_4		
	Exp.	*LRM*	*Diff.*	*Exp.*	*LRM*	*Diff.*	*Exp.*	*LRM*	*Diff.*
ν_4	*892*	*886.9*	*-5.1*	*724*	*720.6*	*-3.4*	*(729)*	*724.4*	*-4.6*
ν_7	*751* *752*	*749.5*	*-1.5* *-2.5*	*765*	*762.2*	*-2.8*	*720*	*718.1*	*-1.9*
ν_8	*943*	*945.7*	*2.7*	*919*	*918.9*	*-0.1*	*780*	*778.6*	*-1.4*

Let me now consider in more quantitative terms LRM frequencies and normal modes of ethylene (H,D)-isotopomers. Ethylene parameters determined from the rotational structure of IR and Raman bands are [12]:

$$r_{CH} = 1.086\,A^0, \quad r_{CC} = 1.338\,A^0, \quad \alpha(HCH) = 117^0 30', \tag{79a}$$

In conjuncture with atomic masses as expressed in atomic units [27]

$$m_H = 1.007825, m_D = 2.014018, \; m_C = 12.0107\,. \tag{79b}$$

those parameters determine molecular mass, molecular geometry and moments of inertia of ethylene molecule. Atomic mass m_C in (79b) is taken to be atomic mass of a natural mixture of 98.93 % of ^{12}C (atomic mass 12.0000 u) and of 1.07 % of ^{13}C (atomic mass $13.00335\,u$ [27]).

All three ethylene out-of-plane normal modes are determined by the ethylene geometry and atomic masses of C and H atoms. From symmetry and data (79) one obtains amplitudes $\langle H_1|\Phi_i\rangle$ and $\langle C_1|\Phi_i\rangle$ at hydrogen atom 1 and carbon atom 1 [15]:

$$\langle H_1|\nu_4\rangle = 0.49806, \; \langle H_1|\nu_7\rangle = 0.46088, \langle H_1|\nu_8\rangle = 0.39756, \tag{80a}$$

$$\langle C_1|\nu_4\rangle = 0.00000, \; \langle C_1|\nu_7\rangle = -0.07735, \; \langle C_1|\nu_8\rangle = -0.12290. \tag{80b}$$

Carbon amplitudes (80b) are not needed in the calculation of deuterium isotope effects. However they are required for the calculation of carbon isotope effects.

Using hydrogen amplitudes (80a) and experimental ethylene out-of-plane frequencies from Table II, one can obtain out-of-plane frequencies and vibrations for all deuterated ethylenes. In using relations (68) one has to observe relative sign of the amplitudes $\langle\tau|\Phi_i\rangle$ on different deuterated sites. Those relative signs are needed in order to construct matrix elements (67a) of the operator $\Omega^0_{\mu\nu}(\varepsilon)$. Those relative signes are determined by symmetry alone, and they are obvious from Figure 6. In Table III. are given deuterated frequencies which were calculated in this way. The agreement between theoretical (LRM) and experimental frequencies is very good. A standard error for all LRM frequencies in Table III is only $\Delta = 2.92\ cm^{-1}$.

In conclusion, LRM frequencies are in excellent agreement with experimental frequencies. The agreement between LRM and experimental frequences seems to be better than expected from the harmonic approximation. Thus it is likely that unharmonicity effects partially cancel in the LRM approach.

LRM approach can be used in yet another way. Experimental deuterated ethylene frequencies can be used in order to improve ethylene out-of-plane frequencies. In particular, ethylene out of plane frequency $\nu_4 = 1024\ cm^{-1}$ is relatively unreliable. This vibration is of symmetry type a_{1u} and it is inactive both in Raman and in IR. Hence one has to obtain this frequency indirectly, which makes it very prone to errors. One can consider ethylene out of plane frequencies as adjustable parameters, and try to adjust these frequencies in such a way as to minimize standard error of LRM deuterated frequencies. If this is done one obtains $\nu_4 = 1028.2\ cm^{-1}$, $\nu_7 = 950.1\ cm^{-1}$ and $\nu_8 = 942.2\ cm^{-1}$ as optimal ethylene out of plane frequencies [15]. If the LRM calculation is done with these optimal ethylene frequencies, standard error for LRM frequencies of deuterated ethylenes decreases from $\Delta = 2.92\ cm^{-1}$ to $\Delta = 2.35\ cm^{-1}$. These optimal ethylene frequencies that minimize standard error differ from experimental values given in Table II by $\Delta\nu_4 = 4.2\ cm^{-1}$, $\Delta\nu_7 = 0.8\ cm^{-1}$ and $\Delta\nu_8 = -0.8\ cm^{-1}$. There is a negligible change of frequencies ν_7 and ν_8. This is in accord with the fact that these frequencies are experimentally quite reliable. Small discrepancies can be attributed to experimental error and/or to unharmonicity effects. However, frequency ν_4 which is inactive changes much more and the error $\Delta\nu_4 = 4.2\ cm^{-1}$ is very likely significant. This indicates that ethylene frequency ν_4 is probably underestimated, and that the true value of this frequency should be closer to $\nu_4 = 1028\ cm^{-1}$ than to the reported value of $\nu_4 = 1024\ cm^{-1}$ [15].

Once a particular modified frequency is obtained, one finds the corresponding normal mode using expressions (69) where $|\theta_s\rangle$ is the corresponding eigenstate of (68a). As an example, consider monodeuterated ethylene. In the case of a single isotope substitution, expressions (69) simplify to the expression (73c). Using LRM frequencies $\nu_4 = 999.4$, $\nu_7 = 945.7$ and $\nu_8 = 804.8$ from Table III one finds corresponding normal modes $|\Psi_4\rangle$, $|\Psi_7\rangle$ and $|\Psi_8\rangle$. Once those normal modes are known, they can be easily normalized. With the convention that x-axis points in the direction of the CC bond, normalized vibrations $|\Psi_4\rangle$, $|\Psi_7\rangle$ and $|\Psi_8\rangle$ thus obtained are [15]:

$$
\begin{aligned}
&|\Psi_4\rangle = \\
&0.01615|T_z\rangle + 0.04259|R_x\rangle + 0.02566|R_y\rangle - 0.85524|\Phi_4\rangle + 0.40313|\Phi_7\rangle + 0.30991|\Phi_8\rangle \\
&|\Psi_7\rangle = \\
&0.00204|T_z\rangle + 0.00539|R_x\rangle + 0.00325|R_y\rangle - 0.03125|\Phi_4\rangle - 0.65185|\Phi_7\rangle + 0.75760|\Phi_8\rangle \\
&|\Psi_8\rangle = \\
&0.08862|T_z\rangle + 0.23377|R_x\rangle + 0.14081|R_y\rangle - 0.37774|\Phi_4\rangle - 0.55284|\Phi_7\rangle - 0.50042|\Phi_8\rangle
\end{aligned} \tag{81}
$$

As required, these vibrations satisfy orthonormality relation (2c) where $\mathbf{S}_s = \mathbf{M} + \Delta\mathbf{M}$. This is an independent proof that LRM method is consistent and valid.

In the above example six deuterated ethylenes were considered. However, each ethylene hydrogen atom can be substituted either with a deuterium or with a tritium, in all possible combinations. There are 26 such isotopomers. To obtain LRM out-of-plane frequencies and normal modes of all these isotopomers no additional information (except of the mass of tritium atom) is required. To this list one can add various possible combinations with ^{12}C, ^{13}C and ^{14}C isotopes. This increases the number of all possible isotopomers to over two hundreds. Out-of-plane frequencies and the corresponding normal modes of all these isotopomers can be obtained by using only three ethylene out-of-plane frequencies. This relates a huge amount of experimental data to only three experimental quantities, ethylene out-of-plane frequencies ν_4, ν_7 and ν_8 from Table II.

6.7. Out-of-Plane Vibrations of Benzene Isotopomers

In the above example all amplitudes of the out-of-plane vibrations of a parent molecule **A** were determined by symmetry. Let me now consider another example where this is not the case. As such an example consider out-of-plane frequencies and normal modes of benzene (H,D)-isotopomers. Symmetry alone is not sufficient to fix all amplitudes of the benzene out-of-plane vibrations. Hence in order to fix those amplitudes one has to use either some independent calculation, or one can use inversion relations. I will demonstrate here how inversion relations can be utilized.

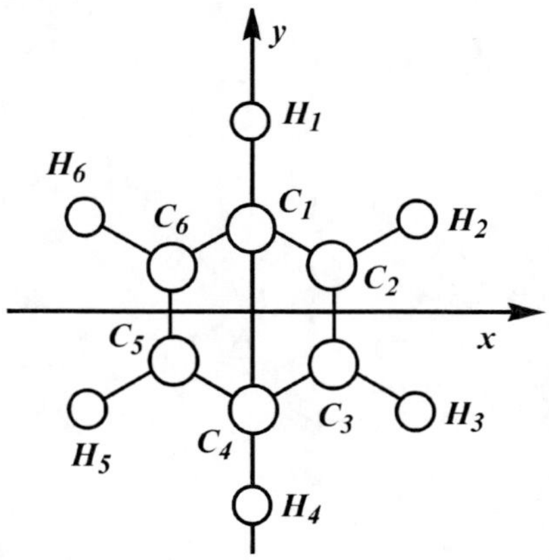

Figure 8. Benzene.

Benzene is a planar molecule with $n=12$ atoms (see Figure 8). Three vibrations are non-proper: translation in the z-direction perpendicular to the molecular plane, and rotations around two in-plane axes. Hence benzene has $n-3=9$ proper out-of plane vibrations. Equilibrium benzene geometry [12]

$$r_{CH} = 1.0897\,A^0\,,\; r_{CC} = 1.4000\,A^0\,. \tag{82}$$

in conjuncture with atomic masses (79b) determines molecular mass and moments of inertia of all (H,D)-benzene isotopomers. Those isotopomers are shown in Figure 9. Isotopomers connected by an arrow in this Figure differ from each other by a single isotopic substitution. In the harmonic approximation out-of-plane frequencies of all such isotopomer pairs should satisfy interlacing rule 75.

Benzene out-of-plane vibrations, arranged in the order of increasing frequency, are shown in Figure 10. In order to simplify presentation, notation in this Figure differs from the standard notation [13]. This standard notation is also shown in this Figure. For example, in a standard notation vibration $|\Phi_1\rangle$ is vibration $|\nu_{16b}\rangle$, vibration $|\Phi_2\rangle$ is vibration $|\nu_{16a}\rangle$, etc.

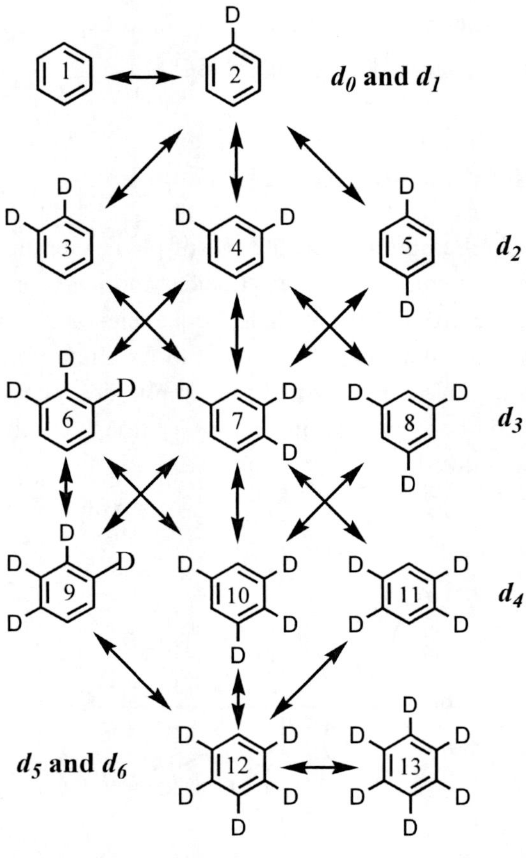

Figure 9. Interlacing hierarchy of deuterated benzenes. Adopted from ref. 22.

Symmetry constrains partially or complete determine all benzene out-of-plane vibrations. For example, symmetry alone completely determines all amplitudes of benzene vibrations $|\Phi_3\rangle \equiv |\nu_{11}\rangle$, $|\Phi_5\rangle \equiv |\nu_{10b}\rangle$ and $|\Phi_6\rangle \equiv |\nu_{10a}\rangle$. Using data (82) and (79b) one finds [22]

$$\langle H_1|\Phi_3\rangle = -0.39060,\ \langle C_1|\Phi_3\rangle = 0.03278. \tag{83a}$$

$$\langle H_1|\Phi_5\rangle = 0.51126,\ \langle H_2|\Phi_5\rangle = 0.25563,$$

$$\langle C_1|\Phi_5\rangle = -0.07629, \langle C_2|\Phi_5\rangle = -0.03815. \tag{83b}$$

$$\langle H_1|\Phi_6\rangle = 0, \langle H_2|\Phi_6\rangle = -0.44276,$$

$$\langle C_1|\Phi_6\rangle = 0, \langle C_2|\Phi_6\rangle = 0.06607. \tag{83c}$$

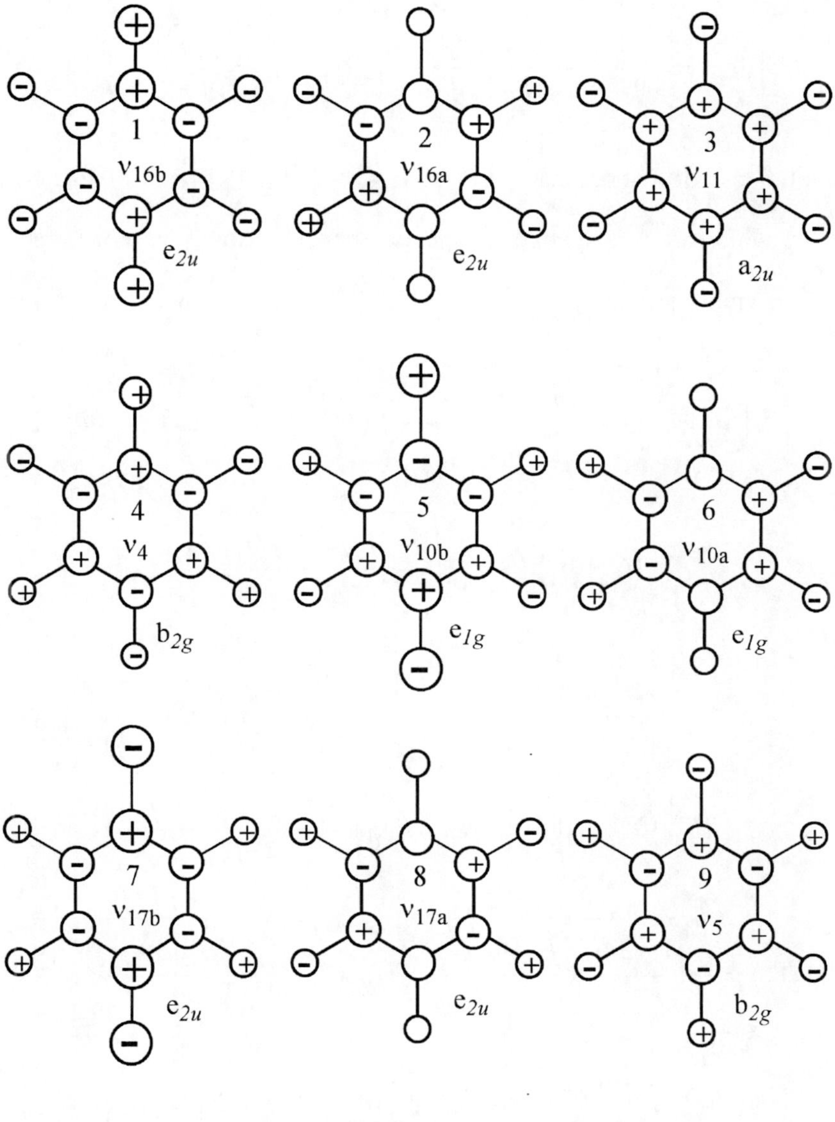

Figure 10. Out-of-plane vibrations of benzene. Adopted from ref. 22.

All remaining amplitudes of those vibrations are determined by the symmetry (see Figure 10).

All amplitudes of vibrations $|\Phi_4\rangle \equiv |\nu_4\rangle$ and $|\Phi_9\rangle \equiv |\nu_5\rangle$ which are of a symmetry type b_{2g} depend on only one amplitude. With the choice of the amplitude $\langle H_1|\Phi_4\rangle$ one finds [22]

$$\langle C_1|\Phi_4\rangle = \sqrt{\frac{1-6m_H\langle H_1|\Phi_4\rangle^2}{6m_C}}, \tag{84a}$$

$$\langle H_1|\Phi_9\rangle = -\sqrt{\frac{1-6m_H\langle H_1|\Phi_4\rangle^2}{6m_H}},\ \langle C_1|\Phi_9\rangle = \sqrt{\frac{m_H}{m_C}}\langle H_1|\Phi_4\rangle. \tag{84b}$$

All amplitudes of vibrations $|\Phi_1\rangle \equiv |\nu_{16b}\rangle$, $|\Phi_2\rangle \equiv |\nu_{16a}\rangle$, $|\Phi_7\rangle \equiv |\nu_{17b}\rangle$ and $|\Phi_8\rangle \equiv |\nu_{17a}\rangle$ which are of a symmetry type e_{2u} depend also on only one amplitude. With the choice of the amplitude $\langle H_1|\Phi_1\rangle$ one has [22]

$$\langle C_1|\Phi_1\rangle = \sqrt{\frac{1-3m_H\langle H_1|\Phi_1\rangle^2}{3m_C}},\ \langle H_2|\Phi_1\rangle = -0.5\langle H_1|\Phi_1\rangle,$$

$$\langle C_2|\Phi_1\rangle = -\frac{1}{2}\sqrt{\frac{1-3m_H\langle H_1|\Phi_1\rangle^2}{3m_C}}. \tag{85a}$$

$$\langle H_1|\Phi_2\rangle = 0,\ \langle H_2|\Phi_2\rangle = \frac{\sqrt{3}}{2}\langle H_1|\Phi_1\rangle,$$

$$\langle C_1|\Phi_2\rangle = 0,\ \langle C_2|\Phi_2\rangle = \sqrt{\frac{1-3m_H\langle H_1|\Phi_1\rangle^2}{4m_C}}. \tag{85b}$$

$$\langle H_1|\Phi_7\rangle = -\sqrt{\frac{1-3m_H\langle H_1|\Phi_1\rangle^2}{3m_H}},\ \langle H_2|\Phi_7\rangle = \frac{1}{2}\sqrt{\frac{1-3m_H\langle H_1|\Phi_1\rangle^2}{3m_H}},$$

$$\langle C_1|\Phi_7\rangle = \sqrt{\frac{m_H}{m_C}}\langle H_1|\Phi_1\rangle,\ \langle C_2|\Phi_7\rangle = -\frac{1}{2}\sqrt{\frac{m_H}{m_C}}\langle H_1|\Phi_1\rangle. \tag{85c}$$

$$\langle H_1|\Phi_8\rangle = 0,\ \langle H_2|\Phi_8\rangle = -\sqrt{\frac{1-3m_H\langle H_1|\Phi_1\rangle^2}{4m_H}},$$

$$\langle C_1|\Phi_8\rangle = 0,\ \langle C_2|\Phi_8\rangle = \sqrt{\frac{3m_H}{4m_C}}\langle H_1|\Phi_1\rangle. \tag{85d}$$

Amplitudes of all benzene out-of-plane vibrations are thus expressed in terms of only two amplitudes: amplitude $\langle H_1|\Phi_4\rangle$ and amplitude $\langle H_1|\Phi_1\rangle$. Those two amplitudes can be obtained from the frequencies of d_0- and d_1-benzene using inversion relation. Since d_0-benzene frequencies $\nu_1 = \nu_2$, $\nu_5 = \nu_6$ and $\nu_7 = \nu_8$ are doubly degenerate, d_1-benzene has three strongly singular vibrations with those frequencies. One can place deuterium at the benzene H_1 position. Since $\langle H_1|\Phi_2\rangle = \langle H_1|\Phi_6\rangle = \langle H_1|\Phi_8\rangle = 0$ (see Figure 10), with this choice d_0-benzene vibrations $|\Phi_2\rangle$, $|\Phi_6\rangle$ and $|\Phi_8\rangle$ are at the same time d_1-benzene strongly singular vibrations (those vibrations are also strongly singular vibrations of the para d_2-benzene). Remaining six out-of-plane d_1-benzene vibrations are cardinal. Since one has to fix only two amplitudes $\langle H_1|\Phi_4\rangle$ and $\langle H_1|\Phi_1\rangle$, inversion relation (75) involving 6×6 matrix $\mathbf{G}$ reduces to the inversion relation involving 2×2 matrix $\mathbf{G}$. One finds [22]

$$\mathbf{Y} = \mathbf{G}^{-1}\mathbf{X}, \tag{86a}$$

where 2×2 matrix $\mathbf{G}$ and 2-component column vectors $\mathbf{X}$ and $\mathbf{Y}$ have components

$$G_{i,1} = \frac{1}{\varepsilon_{si}-\lambda_1} - \frac{1}{\varepsilon_{si}-\lambda_7},\ G_{i,2} = \frac{1}{\varepsilon_{si}-\lambda_4} - \frac{1}{\varepsilon_{si}-\lambda_9},$$

$$Y_1 = \langle H_1|\Phi_1\rangle^2,\ Y_2 = \langle H_1|\Phi_4\rangle^2,\ X_i = g(\varepsilon_{si}), i=1,2. \tag{86b}$$

and where function $g(\varepsilon)$ is given by

$$g(\varepsilon) = -\frac{1}{\varepsilon}\left[\frac{1}{M} + \frac{y_{H1}^2}{I_x} + \frac{1}{\Delta m_\tau}\right] - \frac{m_C}{m_H M(\varepsilon-\lambda_3)} - \frac{m_C x_{C2}^2}{m_H I_y(\varepsilon-\lambda_5)} - \frac{1}{3m_H(\varepsilon-\lambda_7)} - \frac{1}{6m_H(\varepsilon-\lambda_9)} \tag{86c}$$

This function is determined by experimental data alone. $\Delta m_\tau = m_D - m_H = 1.006193$ is isotope mass change at the substitution site $\tau = 1$, y_{H1} is y-coordinate of the hydrogen atom H_1, x_{C2} is x-coordinate of the carbon atom C_2, etc. (see Figure 8). Matrix $\mathbf{G}$ in (86b) is determined by d_0-benzene frequencies ν_1, ν_4, ν_7 and ν_9 as well as by any two mutually distinct cardinal frequencies ω_{s1} and ω_{s2} of d_1-benzene. Since there are six such frequencies, one has 15 possible pairs ($\varepsilon_{s1} = 4\pi^2\omega_{s1}^2$, $\varepsilon_{s2} = 4\pi^2\omega_{s2}^2$) of cardinal eigenvalues ε_{si}. Each such pair determines amplitudes $\langle H_1|\Phi_1\rangle$ and $\langle H_1|\Phi_4\rangle$. Those two amplitudes determine all remaining amplitudes of all d_0-benzene out-of-plane vibrations. Once those amplitudes are known, expressions (67-69) produce out-of-plane normal modes and corresponding vibrational frequencies for all benzene isotopomers. If one considers all possible (H,D) and (^{12}C, ^{13}C) combinations, there are few hundreds of such isotopomers. If in addition one includes isotopomers that may contain ^{3}H and ^{14}C isotopes, one finds almost one hundred thousands of such isotopomers. LRM produces with a single stroke vibrational frequencies and normal modes for all those isotopomers!

LRM can be compared either with some reliable theoretical calculation, or directly with experimental results. Since there are no experimental frequencies for all (H,D)-benzene isotopomers, I will first compare LRM with theoretical out-of-plane vibrations of those isotopomers.

In order to calculate vibrational isotope effect one can use theoretical approaches such as MP2, HF or DFT theory [28-30]. Vibrational analysis by the DFT method, in particular the B3LYP density functional, is superior to alternant approaches [30]. Hence DFT calculation will be considered here. Wu and Cremer [30] derived all vibrational frequencies for all (H,D)-benzene isotopomers using DFT with the hybrid functional B3LYP [31-33] and Pople's 6-31G(d,p) basis set [34]. This calculation is done within the harmonic approximation and under standard assumption that force field does not change upon isotopic substitutions. LRM treatment of the vibrational isotope effect is based on the same assumptions. Hence if in the LRM approach as input data one uses DFT frequencies and DFT normal modes of d_0-benzene, as well as the same molecular geometry as in DFT, one should exactly reproduce DFT frequencies and normal modes for all isotopomers of this molecule.

In ref. [30] only benzene frequencies and not their normal modes are reported. However, amplitudes of those normal modes can be obtained using the inversion relation (86). As input to those relations one can use DFT d_0-benzene out-of-plane frequencies and a choice of any two DFT d_1-benzene out-of-plane cardinal frequencies.

Table IV. DFT out-of-plane frequencies of d_0- and d_1-benzene. B3LYP/6-31G(d,p) calculation[30]. All frequencies in cm^{-1}.

	1 C_6H_6 D_{6h}		2 C_6DH_5 C_{2v}	
1	e_{2u}	414	b_2	393
2	e_{2u}	414	a_2	414
3	a_{2u}	694	b_2	622
4	b_{2g}	718	b_2	714
5	e_{1g}	865	b_2	792
6	e_{1g}	865	a_2	865
7	e_{2u}	974	b_2	934
8	e_{2u}	974	a_2	974
9	b_{2g}	1013	b_2	1003

DFT d_0- and d_1-benzene out-of-plane frequencies are shown in Table IV. Using inversion relation (86) with d_0-benzene frequencies and d_1-benzene frequencies $\omega_1 = 393\ cm^{-1}$ and $\omega_3 = 622\ cm^{-1}$ from this Table, one finds [22]:

$$\langle H_1 | \Phi_1 \rangle = 0.30583,\ \langle H_1 | \Phi_4 \rangle = 0.17971, \tag{87a}$$

Expressions (84-85) now determine all amplitudes of out-of-plane benzene vibrations $|\Phi_1\rangle$, $|\Phi_2\rangle$, $|\Phi_4\rangle$, $|\Phi_7\rangle$, $|\Phi_8\rangle$ and $|\Phi_9\rangle$:

$$\langle H_2|\Phi_1\rangle = -0.15291,\ \langle C_1|\Phi_1\rangle = 0.14108,\ \langle C_2|\Phi_1\rangle = -0.07054,$$

$$\langle H_2|\Phi_2\rangle = 0.26485,\ \langle C_2|\Phi_2\rangle = 0.12218,\ \langle C_1|\Phi_4\rangle = 0.10567,$$

$$\langle H_1|\Phi_7\rangle = -0.48705,\ \langle H_2|\Phi_7\rangle = 0.24352,\ \langle C_1|\Phi_7\rangle = 0.08859,$$

$$\langle C_2|\Phi_7\rangle = -0.04429,\ \langle H_2|\Phi_8\rangle = -0.42180,\ \langle C_2|\Phi_8\rangle = 0.07672,$$

$$\langle H_1|\Phi_9\rangle = -0.36480,\ \langle C_1|\Phi_9\rangle = 0.05206. \qquad (87b)$$

Amplitudes (87) are derived from a particular choice $\omega_1 = 393\ cm^{-1}$ and $\omega_3 = 622\ cm^{-1}$ of two DFT d_1-frequencies and six mutually distinct d_0-benzene DFT frequencies. In an exact calculation, each particular choice of two DFT cardinal d_1-frequencies should produce the same numerical values for those amplitudes. However, DFT frequencies are in ref. [30] reported with the precision of $1\ cm^{-1}$. This introduces an initial uncertainty $\Delta\nu_i = \pm 0.5\ cm^{-1}$ which is reflected in the accuracy of LRM frequencies and corresponding vibrations of benzene isotopomers. Hence each choice of two DFT d_1-frequencies produces slightly different results.

6.8. Equality of LRM and DFT out-of-Plane Frequencies of Benzene (H,D)-Isotopomers

LRM and DFT out-of-plane frequencies of 12 deuterated (H,D)-isotopomers are compared in Table V. Two d_1-frequencies $\omega_1 = 393\,cm^{-1}$ and $\omega_3 = 622\,cm^{-1}$ in this Table serve as input in LRM, while all other LRM frequencies are obtained as the result of the LRM calculation. LRM calculation reproduces DFT frequencies with a standard deviation (on average) $\approx 0.38\,cm^{-1}$. This error is even smaller from the initial uncertainty ($\pm 0.5\ cm^{-1}$) of the input DFT d_0- and d_1-frequencies. LRM produces similar results with all other possible choices of two input DFT d_1-frequencies [22]. Such a high agreement between LRM and DFT frequencies proves that, given identical initial conditions, LRM exactly reproduces the corresponding DFT frequencies. The same applies to DFT normal modes. Since LRM is numerically much faster then DFT, this provides an efficient method to calculate vibrational isotope effect within the harmonic approximation.

Table V. DFT [30] and LRM [6] frequencies of out-of-plane vibrations of deuterated benzenes. All frequencies in cm^{-1}. LRM reproduces DFT frequencies within approximately $\pm 0.5 cm^{-1}$. Adopted from Refs. [6] and [22].

	2 C_6DH_5 Δ=0.26			3 1,2-$C_6D_2H_4$ Δ=0.34			4 1,3-$C_6D_2H_4$ Δ=0.23			5 1,4-$C_6D_2H_4$ Δ=0.40		
	C_{2v}	*DFT*	*LRM*	C_{2v}	*DFT*	*LRM*	C_{2v}	*DFT*	*LRM*	D_{2h}	*DFT*	*LRM*
1	a_2	393	*393.00*[a]	a_2	384	*383.89*	a_2	*386*	*386.02*	b_{1u}	*374*	*373.30*
2	b_2	414	*414.00*	b_2	*403*	*402.87*	b_2	*402*	*401.83*	a_u	*414*	*414.00*
3	b_2	622	*622.00*[a]	b_2	*591*	*590.81*	b_2	*582*	*581.71*	b_{1u}	*611*	*611.31*
4	a_2	714	*714.16*	a_2	*676*	*675.96*	b_2	*714*	*714.12*	b_{3g}	*648*	*647.61*
5	a_2	792	*792.46*	a_2	*782*	*782.58*	a_2	*719*	*719.48*	b_{3g}	*750*	*749.99*
6	b_2	865	*865.00*	b_2	*797*	*797.12*	b_2	*830*	*830.19*	b_{2g}	*865*	*865.00*
7	a_2	934	*933.84*	a_2	*900*	*900.25*	a_2	*934*	*933.73*	b_{1u}	*879*	*878.57*
8	b_2	974	*974.00*	b_2	*960*	*959.57*	b_2	*934*	*934.07*	a_u	*974*	*974.00*
9	a_2	1,003	*1,003.11*	a_2	*998*	*998.61*	b_2	*991*	*990.90*	b_{3g}	*978*	*978.16*

	6 1,2,3-$C_6D_3H_3$ Δ=0.28			7 1,2,4-$C_6D_3H_3$ Δ=0.53			8 1,3,5-$C_6D_3H_3$ Δ=0.36			9 1,2,3,4-$C_6D_4H_2$ Δ=0.36		
	C_{2v}	*DFT*	*LRM*	C_s	*DFT*	*LRM*	D_{3h}	*DFT*	*LRM*	C_{2h}	*DFT*	*LRM*
1	b_2	*383*	*383.05*	*a''*	370	*369.24*	*e''*	*386*	*386.02*	b_2	*368*	*367.78*
2	a_2	*386*	*386.02*	*a''*	*400*	*399.16*	*e''*	*386*	*386.02*	a_2	*384*	*383.70*
3	b_2	*558*	*558.10*	*a''*	*573*	*573.13*	a_2''	*545*	*545.16*	b_2	*544*	*544.23*
4	b_2	*670*	*669.72*	*a'*	*647*	*646.16*	a_2''	*714*	*714.11*	a_2	*640*	*639.65*
5	a_2	*719*	*719.48*	*a''*	*718*	*718.07*	*e''*	*719*	*719.48*	a_2	*698*	*698.13*
6	b_2	*790*	*790.16*	*a''*	*782*	*782.61*	*e''*	*719*	*719.48*	b_2	*746*	*745.73*
7	b_2	*840*	*840.53*	*a''*	*873*	*873.41*	*e''*	*934*	*933.73*	a_2	*822*	*822.60*
8	a_2	*934*	*933.73*	*a''*	*934*	*934.06*	*e''*	*934*	*933.73*	b_2	*873*	*873.61*
9	B_2	*991*	*990.88*	*a'*	*976*	*976.13*	a_2''	*934*	*934.74*	a_2	*976*	*976.07*

	10 1,2,3,5-$C_6D_4H_2$ Δ=0.38			11 1,2,4,5-$C_6D_4H_2$ Δ=0.62			12 C_6D_5H Δ=0.34			13 C_6D_6 Δ=0.41		
	C_{2v}	*DFT*	*LRM*	D_{2h}	*DFT*	*LRM*	C_{2v}	*DFT*	*LRM*	D_{6h}	*DFT*	*LRM*
1	b_2	*368*	*367.23*	a_u	361	*360.54*	a_2	*361*	*360.54*	e_{2u}	*361*	*360.54*
2	a_2	*386*	*386.02*	b_{1u}	*390*	*389.45*	b_2	*376*	*375.41*	e_{2u}	*361*	*360.54*
3	b_2	*537*	*537.14*	b_{1u}	*560*	*559.63*	b_2	*526*	*525.69*	a_{2u}	*509*	*509.54*
4	b_2	*646*	*645.81*	b_{3g}	*631*	*629.99*	a_2	*628*	*627.43*	b_{2g}	*615*	*614.31*
5	b_2	*718*	*717.73*	b_{2g}	*673*	*672.90*	a_2	*673*	*672.90*	e_{1g}	*673*	*672.90*
6	a_2	*719*	*719.48*	b_{3g}	*779*	*779.90*	b_2	*719*	*718.90*	e_{1g}	*673*	*672.90*
7	b_2	*810*	*810.33*	a_u	*791*	*791.16*	a_2	*791*	*791.16*	e_{2u}	*791*	*791.16*
8	a_2	*934*	*933.73*	b_{1u}	*931*	*931.37*	b_2	*825*	*825.14*	e_{2u}	*791*	*791.16*
9	b_2	*934*	*934.44*	b_{3g}	*936*	*936.87*	a_2	*934*	*934.11*	b_{2g}	*837*	*837.53*

[a] d_1-frequencies $\omega_1 = 393\,cm^{-1}$ and $\omega_3 = 622\,cm^{-1}$ serve as input in the LRM calculation.

6.9. Comparison of LRM, Frequency Scaled DFT and Experimental out-of-Plane Frequencies of Benzene (H,D)-Isotopomers

In order to improve the agreement of DFT frequencies with experimental frequencies, one usually applies scaling technique [30,35-39]. One can scale either theoretical force field [35-37], or one can scale theoretical frequencies[30,37-39]. In both approaches one uses one or several individual scaling factors which optimally reproduce frequencies of the parent molecule $\mathbf{A}$. It is assumed that those same factors should improve frequencies of all isotopomers $\mathbf{C}$ of $\mathbf{A}$.

Table VI. Scaled out-of-plane benzene frequencies [30]. Those frequencies equal experimental frequencies. All frequencies in cm^{-1}.

	C_6H_6 D_{6h}	
1	e_{2u}	398
2	e_{2u}	398
3	a_{2u}	673
4	b_{2g}	707
5	e_{1g}	846
6	e_{1g}	846
7	e_{2u}	967
8	e_{2u}	967
9	b_{2g}	990

There is a fundamental difference between force field scaling and frequency scaling. In a force field scaling, one replaces one force field with another (scaled) force field. The calculation done with a modified force field is still within the harmonic approximation. Hence one can apply LRM to the scaled force field DFT in the same way as to the initial unscaled DFT.

In a frequency scaling one uses several individual scaling factors in order to scale theoretical frequencies of a parent molecule. The same scaling factors are then applied to theoretical frequencies of all isotopomers of this molecule. Frequency scaling usually violates the assumption that force field does not change under isotopic substitutions and/or the harmonic approximation [22]. This is most drastically evident by the violation of the interlacing rule (75). Since this rule is based on the harmonic approximation and on the assumption that force field does not change upon isotopic substitutions, any violation of this rule implies that either of those two assumptions is not valid.

Scaled DFT frequencies which accurately reproduce all out-of-plane vibrational frequencies of benzene [30] are shown in Table VI. Since benzene has six mutually distinct out-of-plane frequencies, six scaling factors are required in order to reproduce exactly those six frequencies. Those scaling factors are used to obtain DFT scaled frequencies for all benzene (H,D) isotopomers. One finds that those scaled frequencies violate interlacing rule (75). For example, if one compares scaled out-of-plane frequencies of (H,D)-benzene isotopomers 7 and 11 as reported in reference [30] , one finds [22]

$$
\begin{aligned}
&347(11)_1 < 356(7)_1 < 375(11)_2 < 385(7)_2 < 582(11)_3 \,?\, 556(7)_3 < \\
&621(11)_4 < 637(7)_4 < 658(11)_5 < 702(7)_5 < 762(11)_6 < 765(7)_6 < \\
&785(11)_7 < 867(7)_7 < 915(11)_8 < 927(7)_8 \,?\, 924(11)_9 < 954(7)_9
\end{aligned} \tag{88}
$$

Those isotopomers differ by a single isotopic substitution (see Figure 9), and hence in the harmonic approximation the corresponding out-of-plane frequencies should satisfy interlacing rule (75). However, above frequencies violate this rule in two places. Particularly serious is the violation $582(11)_3 \,?\, 556(7)_3$ which amounts to $26\,cm^{-1}$. It is highly unlikely that violation as large as $26\,cm^{-1}$ is due to the anharmonicity effects. No force field can produce those scaled frequencies, provided harmonic approximation is valid and provided force field does not change under isotopic substitutions.

Since LRM is based on the harmonic approximation and on the assumption that force field does not change upon isotopic substitutions, it cannot reproduce frequency scaled frequencies. However, it should exactly reproduce force field scaled frequencies, provided as input d_0- benzene frequencies one uses d_0- benzene experimental frequencies. In addition to d_0- benzene frequencies one has to know the corresponding vibrational amplitudes. In the case of non-scaled frequencies shown in Tables IV and V, those amplitudes were derived using the inversion relation (86) with six theoretical DFT frequencies of d_0- benzene and two theoretical DFT frequencies of d_1- benzene. In a consistent approach based on experimental data alone, one should use d_0- benzene out-of-plane experimental frequencies and a choice of any two cardinal d_1-benzene out-of-plane experimental frequencies. In the absence of d_1-benzene experimental frequencies, one can combine d_0- benzene experimental frequencies with the amplitudes obtained in the above way from the DFT theoretical frequencies. This is a hybrid approach which combines experimental data (d_0- experimental frequencies) with theoretical data (DFT vibrational amplitudes $\langle H_1|\Phi_1\rangle$ and $\langle H_1|\Phi_4\rangle$).

Table VII. Comparison of DFT scaled frequencies [30] and LRM frequencies [6] with experimental frequencies [30]. All frequencies in cm^{-1}. Adopted from ref. [6]

	$C_6(DH)_3$			C_6D_6		
	DFT Δ=4.04	LRM Δ=2.78	exp.	DFT Δ=4.69	LRM Δ=3.78	exp.
1	371	371.33	368	347	*346.97*	*345*
2	371	371.33	368	347	*346.97*	*345*
3	528	530.01	531	494	*494.12*	*496*
4	703	701.67	697	606	*603.64*	*599*
5	703	705.81	708	658	*658.12*	*660*
6	703	705.81	708	658	*658.12*	*660*
7	913	913.15	917	785	*784.63*	*787*
8	927	923.65	924	785	*784.63*	*787*
9	927	923.65	924	818	*820.21*	*829*

LRM out-of-plane frequencies of (H,D)-benzene isotopomers obtained in this way differ from frequency scaled DFT frequencies on average by standard deviation $\Delta \approx 4.4\ cm^{-1}$

[22]. In view of the reported precision of the input DFT frequencies ($\pm 0.5\,cm^{-1}$), this difference is significant and it demonstrates that LRM differs qualitatively from frequency scaled DFT. Both approaches can be also compared with available experimental data [22]. This comparison is shown in Table VII. For reliable comparison, experimental data for isotopomer 8 ($C_6(DH)_3$) and isotopomer 13 (C_6D_6) which are reported in ref. [30] are used. In both cases above LRM hybrid approach is superior to frequency scaled DFT. Thus LRM out-of-plane frequencies for the isotopomer (8) ($C_6(DH)_3$) differ from the experimental frequencies by $\Delta_{LRM} = 2.78\,cm^{-1}$, while frequency scaled DFT out-of-plane frequencies differ from the same experimental frequencies by $\Delta_{DFT} = 4.04\,cm^{-1}$. In the case of benzene isotopomer C_6D_6 one finds $\Delta_{LRM} = 3.78\,cm^{-1}$ and $\Delta_{DFT} = 4.69\,cm^{-1}$[22]. Since LRM is consistent with force field scaling while it is not consistent with frequency scaling, this suggests that force field scaling is preferable to frequency scaling.

LRM can in the above way improve numerical efficiency of any theoretical method which is based on the harmonic approximation and on the assumption that force field does not change upon isotopic substitutions.

7. External Modification as a Special Case of Internal Modification

This completes the LRM treatment of finite-dimensional systems. One final point. I have mentained a strict separation of external and internal modifications. However, from a more general point of view each external modification can be considered as a special case of an internal modification. In an external modification system A_n interacts with another system B_ρ through the interaction represented by the modification operators $\mathbf{V}$ and $\mathbf{P}$. Emphasize is on the properties of the system B_ρ which is usually relatively small and which interacts with its "surrounding" represented by the system A_n. However, one can consider system $\mathbf{A}_{n+\rho}$ that consists of two non-interacting subsystems A_n and $\mathbf{B}_\rho$ as the initial parent system. From this point of view, interaction between subsystems A_n and B_ρ of $\mathbf{C}_{n+\rho}$ represents an internal modification. Parent system $\mathbf{A}_{n+\rho}$ is $(n+\rho)$-dimensional, and it is described by the eigenvalue equation

$$\mathbf{H}_a^{'}|\psi_s\rangle = x_s\mathbf{S}_a^{'}|\psi_s\rangle,\ s = 1,\cdots,n+\rho, \tag{89a}$$

where

$$\mathbf{H}_a^{'} = \mathbf{H}_a + \mathbf{H}_b,\ \mathbf{S}_a^{'} = \mathbf{S}_a + \mathbf{S}_b, \tag{89b}$$

and where

$$x_s = \begin{cases} E_s & if \quad s = 1,\ldots,\rho \\ \lambda_{s-\rho} & if \quad s = \rho+1,\ldots,\rho+n \end{cases}. \tag{89c}$$

Expressions (89) incorporate expressions (1) and (5). Each eigenstate $|\psi_s\rangle$ of (89a) is contained in the $(n+\rho)$-dimensional space $X^c_{n+\rho}$, orthogonal sum of spaces X^a_n and X^b_ρ. Modified system $\mathbf{B}_{n+\rho}$ is now created by introducing the interaction between the subsystems $\boldsymbol{A}_n$ and $\mathbf{B}_\rho$ of the system $\mathbf{A}_{n+\rho}$. This interaction is described by the Hermitean operators $\mathbf{V}$ and $\mathbf{P}$ scaled by the coupling parameter β. In the case of an external modification, those operators satisfy expressions (8) which are suitable for Theorems 1 and 2 that apply to external modifications. In order to use Theorems 3 and 4 that apply to internal modifications, expressions (8) should be transformed into expressions (4). This is nontrivial and it requires the construction of the interaction space X^{int}_ρ. In order to construct this space one has to consider eigenvalue equations $\mathbf{V}|h_s\rangle = h_s|h_s\rangle$ and $\mathbf{P}|s_s\rangle = s_s|s_s\rangle$. This can be done. However, it is more convenient to avoid this problem and to have two separate methods for external and internal modifications, as this was done here.

8. Infinite Dimensional Systems

In the previous sections finite-dimensional systems were considered. Let me now consider infinite-dimensional systems. In order to treat such systems one has to generalize finite-dimensional space X^a_n to an infinite-dimensional space X^a_∞. Concerning finite-dimensional space X^b_ρ in the case of external modifications and finite-dimensional interaction space X^{int}_ρ in the case of internal modifications, those spaces will not be generalized to infinite-dimensional spaces. Accordingly, one still considers finite rank modifications of the original infinite system.

Is there any need for such a generalization? Following general structure of LRM approach, the system $\boldsymbol{A}_\infty$ associated with the infinite-dimensional space X^a_∞ should be known, i.e. one should know eigenvalues and eigenstates of this system. This knowledge is required in order to obtain the solution of the corresponding modified system $\mathbf{C}_\infty$. Are there any infinite-dimensional systems with such a property? In particular, is there any infinite-dimensional system $\boldsymbol{A}_\infty$ with a known solution that interacts externally with a finite-dimensional system $\mathbf{B}_\rho$, and where the solution of the corresponding combined system $\mathbf{C}_\infty$ is of interest?

As one referee commented to a paper submitted to a journal, to consider such problems is ridiculous: what one needs, he claimed, is to consider perturbations of large unknown systems, and not to consider perturbations of large known systems!

However, this is a rather narrow point of view. Consider for example the interaction of an isolated molecule with the electromagnetic field. Of interest are effects such as shifts of molecular energy levels, broadening of those levels, emission and absorption of radiation, etc. All those effects are due to the interaction of the molecular system with the electromagnetic field. Isolated molecule can be approximated with a finite system X_ρ^b which contains a finite number of eigenvalues E_r and eigenstates $|\Theta_r\rangle$. Those eigenstates interact with one-photon states $|\Theta_t, \mathbf{k}\varpi\rangle$ where $|\mathbf{k}\varpi\rangle$ represents a state that contains a single photon with momentum $\mathbf{k}$ and polarization ϖ. States $|\Theta_t, \mathbf{k}\varpi\rangle$ interact with two-photon states $|\Theta_v, \mathbf{k}'\varpi', \mathbf{k}''\varpi''\rangle$, which in turn interact with three photon states, etc [1]. If there is no external electromagnetic field, one can to a very good approximation ignore all states containing multiple photons, and one can associate system $\mathbf{A}_\infty$ with the set of all one-photon states $|\Theta_r, \mathbf{k}\varpi\rangle$ with the corresponding eigenvalues. Those states span an infinite-dimensional space X_∞^a. The solution to this system is known, since the states $|\mathbf{k}\varpi\rangle$ are essentially plane waves, while $|\Theta_r\rangle$ are eigenstates of the isolated molecule which is approximated as a finite system. Hence one has formally the interaction of a finite system X_ρ^b with the known infinite system X_∞^a. With an appropriate modification, in the similar way can be treated the case when an external electromagnetic field is present.

As another example consider the interaction of the molecule situated in the surface of a solid with this solid. Molecule in isolation can be again approximated with a finite-dimensional system $\mathbf{B}_\rho$. Infinite-dimensional system $\mathbf{A}_\infty$ now represents a solid with a surface. The solution to this system consists of multiple eigenvalue bands $\lambda_s(\mathbf{k})$ with the corresponding eigenstates $|\Phi_s(\mathbf{k})\rangle$ ($s = 1,2,\ldots$) [1]. In addition, system $\mathbf{A}_\infty$ may contain some discrete eigenvalues that correspond to the surface states [40, 41]. One is usually interested in the properties of the molecule (system $\mathbf{B}_\rho$) subject to the interaction with the solid (system $\mathbf{A}_\infty$). Again one can assume that the solution to the system $\mathbf{A}_\infty$ is known. Usually one knows only an approximate solution to this system [40, 41]. Assuming this approximate solution is good enough, one has again the interaction of a finite system $\mathbf{B}_\rho$ with a known infinite system $\mathbf{A}_\infty$.

Those and similar problems are usually treated either within the formalism of the perturbation expansion approach (where $\mathbf{B}_\rho$ is treated as the unperturbed system), or using some approximate semi-classical model [1, 2]. If the interaction between the molecular system $\mathbf{B}_\rho$ and the infinite system $\mathbf{A}_\infty$ is relatively strong, convergence of the perturbation series may be very slow. In many cases of interest such as the interaction of the molecule with the surface of the solid, this interaction may be so strong that perturbation expansion diverges and the entire method fails. Concerning various semi-classical models, those models are only approximate, and they can never completely replace exact quantum treatment.

Such and similar problems can be efficiently treated within the LRM formalism of external modifications of quantum systems. Molecular system $\mathbf{B}_\rho$ is a base system which is modified by the interaction with an external infinite-dimensional system $\mathbf{A}_\infty$ with known eigenvalues and eigenstates. In the first example this external system is essentially a free electromagnetic field, while in the second example it is an infinite solid. In both cases emphasize is on the properties of the molecule, and not on the properties of the electromagnetic field or on the properties of a solid. Such LRM treatment of those systems provides a description of open quantum systems: molecular system $\mathbf{B}_\rho$ in isolation is a finite-dimensional closed system. Molecular system $\mathbf{B}_\rho$ subject to the interaction with its "surrounding" (either the electromagnetic field or a solid, or anything else) is an open quantum system. Unlike perturbation expansion which provides only an approximate solution to such systems, LRM provides an exact solution of the infinite combined system $\mathbf{C}_\infty$, regardless of how strong is the interaction between its subsystems $\mathbf{B}_\rho$ and $\mathbf{A}_\infty$. In general, LRM formalism of externally modified quantum systems can be applied to all those cases where one has some relatively small finite-dimensional system $\mathbf{B}_\rho$ in the interaction with its surrounding represented by a known infinite-dimensional system $\mathbf{A}_\infty$.

In the similar way can be treated internal modifications of infinite-dimensional systems. As an example of such a modification consider a perfect solid. This solid is an infinite quantum system and in LRM it represents a parent system $\boldsymbol{A}_\infty$. Though this system is infinite dimensional, since effective potential is periodic one can obtain more or less reliable solution to this system [40]. Consider now introduction of various impurities and/or defects in this solid, creation of a surface, etc. Most properties of interest are due exactly to such and similar modifications of a perfect solid. However, all such modifications break periodic symmetry and in addition those modifications are usually not small. Hence such effects cannot be efficiently treated within the standard formalism of the perturbation expansion. On the other hand, introduction of impurities and defects, as long as those modifications do not produce a global change of the crystal structure, can be very well approximated with operators that have finite rank. This is not the case with a creation of a crystal surface, since a surface of a real three-dimensional solid is two-dimensional, and hence such a modification should be represented with Hermitean operators that have infinite rank. However, one can still approximate a real surface with a small finite portion of this surface. Accordingly, the corresponding modification is approximated with finite rank operators. To the extent that this approximation is reliable, all problems associated with the creation of a surface, such as surface states, distortion of few unit cells close to the surface and alike can be also treated within the LRM formalism.

In conclusion, there are many important problems which can be treated within the LRM formalism and which involve either external or internal modifications of known infinite-dimensional systems. LRM can be also generalized to those cases where both types of modification, external and internal, are simultaneously present.

Development of LRM formalism is important for yet another reason. Consider for example LRM description of open quantum systems: finite system $\mathbf{B}_\rho$ in the interaction with its surrounding, an infinite system $\mathbf{A}_\infty$. From a point of view of quantum theory, each finite system $\mathbf{B}_\rho$ is an open quantum system, since there is always *some* interaction of this system with the surrounding. Of course, in most cases one does not know the solution of this surrounding, and hence one cannot predict what will be exact behavior of the system $\mathbf{B}_\rho$ subject to the interaction with its surrounding. However, from a broad theoretical point of view, the details of this interaction are not important. What is important is that each surrounding, however complicated, can be represented as *some* infinite-dimensional system $\mathbf{A}_\infty$. Since LRM formalism produces exact consequences of the interaction of a finite system $\mathbf{B}_\rho$ with its surrounding, however complex this surrounding may be, one can obtain a clear idea how this interaction may modify original system $\mathbf{B}_\rho$. All possibilities, however exotic, should be contained in this formalism.

9. Description of an Infinite-Dimensional System

Let me now generalize finite-dimensional system $\boldsymbol{A}_n$ described in section 2 to an arbitrary infinite-dimensional system $\boldsymbol{A}_\infty$. This generalization should be broad enough in order to include as many various infinite-dimensional systems as possible. Some pathological infinite-dimensional systems might be excluded, which is hopefully of no practical consequences.

Let $\boldsymbol{A}_\infty$ be such an infinite-dimensional quantum system. With this system is associated an infinite-dimensional space X_∞^a. In general, this system may contain several eigenvalue bands and/or several discrete eigenstates. Each of those eigenvalue bands may be one-parameter eigenvalue band (non-degenerate eigenvalue band) or it may be a many-parameter eigenvalue band (degenerate eigenvalue band).

Concerning discrete solutions of $\boldsymbol{A}_\infty$, in the most general formulation those solutions satisfy eigenvalue equation (1). For the sake of simplicity, instead of this generalized eigenvalue equation consider a standard eigenvalue equation

$$\mathbf{H}_a\left|\Phi_i\right\rangle = \lambda_i \mathbf{I}_\infty^a \left|\Phi_i\right\rangle,\ i = 1,\cdots,n, \tag{90a}$$

where $\mathbf{H}_a$ is Hermitean operator in X_∞^a and where $\mathbf{I}_\infty^a$ is a unit operator in this space. Since $\boldsymbol{A}_\infty$ is ∞-dimensional, one should include the possibility $n = \infty$.

Eigenstates $\left|\Phi_i\right\rangle$ of (90a) are orthonormalized in a standard way

$$\left\langle \Phi_i \middle| \Phi_j \right\rangle = \delta_{ij},\ i,j = 1,\cdots,n. \tag{90b}$$

In addition to the discrete eigenvalues and eigenstates described by the eigenvalue equation (90a), system $\mathbf{A}_\infty$ may contain κ eigenvalue bands where κ can assume any integer value, including the possibility $\kappa = \infty$. Eigenvalues and corresponding eigenstates of those eigenvalue bands satisfy

$$\mathbf{H}_a \left| \Phi_{\tau,m}(k,l) \right\rangle = \lambda_\tau(k) \left| \Phi_{\tau,m}(k,l) \right\rangle, \ k \in \left[k_{a\tau}, k_{b\tau} \right], \ \tau = 1,2,\ldots,\kappa, \tag{91a}$$

Index τ labels different eigenvalue bands. It is assumed that each eigenvalue $\lambda_\tau(k)$ is a continuous and monotonic function of a parameter *k* in the interval $\left[k_{a\tau}, k_{b\tau} \right]$ ($\tau = 1,\ldots,\kappa$). If this is not the case, one can always treat the corresponding eigenvalue band as if this band consists of several smaller eigenvalue bands. Parameters *l* and *m* are optional. Those parameters label possible degeneracy's inside eigenvalue band τ. Parameter *l* represents one or several continuous parameters and parameter *m* represents one or several discrete parameters. With this convention eigenstates $\left| \Phi_{\tau,m}(k,l) \right\rangle$ of (91a) can be orthonormalized according to

$$\left\langle \Phi_{\tau,m}(k,l) \middle| \Phi_{\tau',m'}(k',l') \right\rangle = \delta_{\tau,\tau'} \delta_{m,m'} \delta(k-k') \delta(l-l'), \tag{91b}$$

Without loss of generality one can assume that each function $\lambda_\tau(k)$ is monotonic increasing. In this case all eigenvalues of the eigenvalue band τ are confined to the interval $I_\tau = \left[a_\tau, b_\tau \right]$ where $a_\tau = \lambda_\tau(k_{a\tau})$ and $b_\tau = \lambda_\tau(k_{b\tau})$ are the smallest and the largest possible eigenvalue, respectively. One may have $b_\tau = \infty$. However, one may not have $a_\tau = -\infty$, since eigenvalues $\lambda_\tau(k)$ of $\mathbf{H}_a$ cannot assume arbitrary large negative values.

Since $\lambda_\tau(k)$ is monotonic, it has well defined inverse

$$\varepsilon = \lambda_\tau(k), \ k = \lambda_\tau^{-1}(\varepsilon), \ \varepsilon \in I_\tau \equiv \left[a_\tau, b_\tau \right]. \tag{91c}$$

Union of all eigenvalue intervals I_τ forms a range D of the continuous eigenvalues of the infinite system $\mathbf{A}_\infty$, $D = \bigcup_\tau I_\tau$. This range may consist of several disconnected intervals D_μ, where each D_μ is a union of one or several eigenvalue intervals I_τ. It is convenient to define a point set $\overline{D}$, a complement of D. Accordingly, $D \cup \overline{D} = R$ is the entire real axis.

In addition to orthonormality expressions (90b) and (91b), all discrete eigenstates $\left| \Phi_i \right\rangle$ are orthogonal to all continuous eigenstates $\left| \Phi_{\tau,m}(k,l) \right\rangle$

$$\left\langle \Phi_i \middle| \Phi_{\tau,m}(k,l) \right\rangle = 0. \tag{92}$$

Eigenvalue equations (90a) and (91a) can be written in a compact form

$$\mathbf{H}_a\left|\Phi(\ldots)\right\rangle = \lambda(\ldots)\left|\Phi(\ldots)\right\rangle, \tag{93a}$$

where

$$\mathbf{H}_a = \sum_i \lambda_i \left|\Phi_i\right\rangle\left\langle\Phi_i\right| + \sum_{\tau,m}\int \lambda_\tau(k)\left|\Phi_{\tau,m}(k,l)\right\rangle\left\langle\Phi_{\tau,m}(k,l)\right| dldk\,. \tag{93b}$$

and where dots (…) represent all continuous and/or discrete parameters. Eigenstates $\left|\Phi_i\right\rangle$ and $\left|\Phi_{\tau,m}(k,l)\right\rangle$ of $\boldsymbol{A}_\infty$ form a complete set in X^a_∞. Hence those eigenstates satisfy completeness relation

$$\sum_\tau\sum_m \int \left|\Phi_{\tau,m}(k,l)\right\rangle\left\langle\Phi_{\tau,m}(k,l)\right| dkdl + \sum_i \left|\Phi_i\right\rangle\left\langle\Phi_i\right| = \mathbf{I}^a_\infty\cdot \tag{94}$$

where $\mathbf{I}^a_\infty$ is a unit operator in X^a_∞.

Replacement of the generalized eigenvalue equation (1a) with the standard eigenvalue equation (93a) presents no restriction on generality. However, it may in some cases influence numerical efficiency of the derived LRM expressions. In this and in the following sections I would like to demonstrate that LRM can be successfully extended to infinite-dimensional systems. It is not my intension to present here a full scale and most efficient LRM treatment of such systems.

Let $\{\left|\alpha\right\rangle\}$ be a denumerable base in X^a_∞ orthonormalized according to:

$$\left\langle\alpha\middle|\beta\right\rangle = \delta_{\alpha\beta},\ \alpha,\beta = 1,\ldots,\infty\,. \tag{95a}$$

The base $\{\left|\alpha\right\rangle\}$ can be chosen in such a way that each scalar product $\left\langle\alpha\middle|\Psi_{\tau,m}(k,l)\right\rangle$ is a smooth function of parameters k and l for all admissible values of those parameters, with the possible exception of few isolated points. In most cases such a choice is a natural choice. For example, in the tight-binding approximation [2] each eigenstate $\left|\Phi_{\tau,m}(k,l)\right\rangle$ of an infinite solid is approximated as a linear combination of atomic orbitals. This eigenstate reflects periodic symmetry of a solid, and the states $\left|\alpha\right\rangle$ can be chosen to be atomic orbitals situated at various atoms of this solid. Those orbitals should be orthonormalized according to (95a) by some appropriate method. Alternatively those orbitals can be chosen to be Wannier functions [40]. One finds that scalar products $\left\langle\alpha\middle|\Psi_{\tau,m}(k,l)\right\rangle$ of such orbitals with the eigenstates $\left|\Phi_{\tau,m}(k,l)\right\rangle$ are smooth function of parameters k and l for almost all admissible values of those parameters, as required. In addition, each state $\left|\Psi\right\rangle \in X^a_\infty$ can be expressed as a linear combination of base orbitals $|\alpha\rangle$, and hence the set $\{|\alpha\rangle\}$ of all such orbitals spans the space X_∞^a.

A unit operator $\mathbf{I}_\infty^a$ in the space X_∞^a can be expressed in terms of the states $|\alpha\rangle$ as

$$\mathbf{I}_\infty^a = \sum_\alpha^\infty |\alpha\rangle\langle\alpha|. \tag{95b}$$

As shown in section 4, with each finite-dimensional system $\mathbf{A}_n$ is associated Hermitean operator $\overline{\mathbf{\Omega}}(\varepsilon)$ according to the expression (9). This operator contains complete information about discrete eigenvalues and corresponding eigenstates of the system $\mathbf{A}_n$. With the infinite-dimensional system $\boldsymbol{A}_\infty$ is associated in the analogous way Hermitean operator $\overline{\boldsymbol{\omega}}(\varepsilon)$. For each real ε this operator is uniquely determined by the eigenvalues and eigenstates of $\boldsymbol{A}_\infty$. Operator $\overline{\boldsymbol{\omega}}(\varepsilon)$ can be expressed as a sum of Hermitean operators $\overline{\boldsymbol{\omega}}_\tau(\varepsilon)$ that are associated with eigenvalue bands τ, and of the Hermitean operator $\overline{\mathbf{\Omega}}(\varepsilon)$ that is associated with discrete eigenvalues λ_i and corresponding eigenstates $|\Phi_i\rangle$ of $\boldsymbol{A}_\infty$:

$$\overline{\boldsymbol{\omega}}(\varepsilon) = \sum_\tau \overline{\boldsymbol{\omega}}_\tau(\varepsilon) + \overline{\mathbf{\Omega}}(\varepsilon), \tag{96a}$$

Operator $\overline{\mathbf{\Omega}}(\varepsilon)$ associated with discrete solutions of $\boldsymbol{A}_\infty$ is defined in the same way as in the case of a finite system $\mathbf{A}_n$

$$\overline{\mathbf{\Omega}}(\varepsilon) = \sum_{i(\lambda_i \neq \varepsilon)} \frac{|\Phi_i\rangle\langle\Phi_i|}{\varepsilon - \lambda_i}, \tag{96b}$$

Operator $\overline{\boldsymbol{\omega}}_\tau(\varepsilon)$ associated with the eigenvalue band τ is defined as [5]

$$\overline{\omega}_\tau(\varepsilon) = P\int \frac{\sum_m \int |\Phi_{\tau,m}(k,l)\rangle\langle\Phi_{\tau,m}(k,l)|dl}{\varepsilon - \lambda_\tau(k)} dk, \tag{96c}$$

where P denotes principal Cauchy integral value [14]. Definition (96c) generalizes definition (96b) to continuous eigenvalues and eigenstates.

In the base $\{|\alpha\rangle\}$ matrix elements of the operator $\overline{\boldsymbol{\omega}}_\tau(\varepsilon)$ are

$$\overline{\varpi}_{\alpha\beta}^{(\tau)}(\varepsilon) \equiv \langle \alpha | \overline{\boldsymbol{\varpi}}_{\tau}(\varepsilon) | \beta \rangle = P \int \frac{\sum_{m} \int \langle \alpha | \Phi_{\tau,m}(k,l) \rangle \langle \Phi_{\tau,m}(k,l) | \beta \rangle dl}{\varepsilon - \lambda_{\tau}(k)} dk ,$$

$$\alpha, \beta = 1, \ldots, \infty . \tag{96d}$$

If ε is outside the interval I_τ of the eigenvalue band τ ($\varepsilon \notin I_\tau$), expression (96d) is a standard integral. However, if $\varepsilon \in I_\tau$ in the point $\lambda_\tau(k) = \varepsilon \in I_\tau$ subintegral function in (96d) usually diverges. In this case one has to take principal integral value of this integral.

Above expressions generalize finite dimensional system $\mathbf{A}_n$ described in section 2 to an infinite-dimensional system $\boldsymbol{A}_\infty$. Most infinite-dimensional time-independent quantum system can be described in this way.

10. External Modifications of Infinite-Dimensional Systems

Consider finite dimensional system $\mathbf{B}_\rho$ described by the eigenvalue equation (5a) which interacts with the infinite dimensional system $\boldsymbol{A}_\infty$ described by the eigenvalue equation (93). With system $\mathbf{B}_\rho$ is associated ρ-dimensional space X_ρ^b while with the system $\boldsymbol{A}_\infty$ is associated ∞-dimensional space X_∞^a. Inclusion of the interaction between initially non-interacting systems $\mathbf{B}_\rho$ and $\boldsymbol{A}_\infty$ creates a combined ∞-dimensional system $\mathbf{C}_\infty$. Each state in this combined system is contained in the ∞-dimensional space X_∞^c, orthogonal sum of spaces X_ρ^b and X_∞^a. Interaction between subsystems $\mathbf{B}_\rho$ and $\boldsymbol{A}_\infty$ of $\mathbf{C}_\infty$ is described by the Hermitean operator $\mathbf{V}$ scaled by the coupling parameter β. This operator connects mutually distinct spaces X_ρ^b and X_∞^a, and it vanish over the space X_ρ^b as well as over the space X_n^a.

Combined system $\mathbf{C}_\infty$ is described by the generalized eigenvalue equation

$$\mathbf{H}_c |\Psi(\ldots)\rangle = \varepsilon(\ldots)\mathbf{S}_c |\Psi(\ldots)\rangle , \tag{97a}$$

where

$$\mathbf{H}_c = \mathbf{H}_a + \mathbf{H}_b + \beta \mathbf{V} , \ \mathbf{S}_c = \mathbf{I}_\infty^a + \mathbf{S}_b . \tag{97b}$$

In analogy to the system $\boldsymbol{A}_\infty$, combined system $\mathbf{C}_\infty$ can also contain continuous as well as discrete eigenvalues and eigenstates. This possibility is indicated by dots $(\ldots)$ in the expression (97a). Coupling parameter β is again introduced in order to make manifest dependence on the strength of the modification. If this parameter is small LRM expressions should reduce to the corresponding expressions obtained within the standard perturbation expansion approach. However if this parameter is large, qualitatively new results are obtained. In connection with the expressions (97), note that since $\mathbf{S}_b$ is positive-definite in X^b_ρ, operator $\mathbf{S}_c$ is automatically positive-definite in X^c_∞. Eigenvalues $\varepsilon(\ldots)$ of (97a) are hence guaranteed to be real.

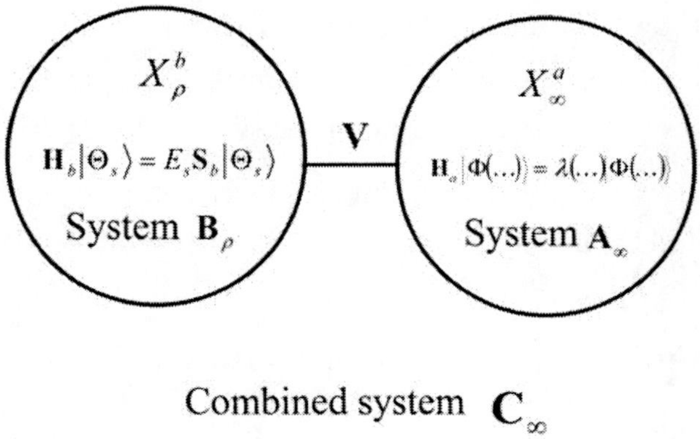

Figure 11. External modification of a finite-dimensional system by an infinite-dimensional system. Finite-dimensional system $\mathbf{B}_\rho$ is an open system which is modified externally by the interaction $\mathbf{V}$ with an infinite-dimensional system $\mathbf{A}_\infty$. Mutually interacting systems $\mathbf{B}_\rho$ and $\mathbf{A}_\infty$ form infinite-dimensional combined system $\mathbf{C}_\infty$.

Expression (97a) is still generalized eigenvalue equation, though not of a most general type. In order to avoid technical details which are not essential for the intended generalization of the LRM approach from the finite combined systems $\mathbf{C}_{n+\rho}$ to the infinite combined systems $\mathbf{C}_\infty$, instead of the modification of a general type $(\mathbf{V},\mathbf{P})$ expression (97a) contains a simpler modification where $\mathbf{P}=0$. This is in fact the most important case. Once this generalization is obtained, one can consider more general combined systems $\mathbf{C}_\infty$ that contain an arbitrary modification of a general type $(\mathbf{V},\mathbf{P})$.

In analogy to (8) one has

$$\mathbf{V} = \mathbf{I}^b_\rho \mathbf{V} \mathbf{I}^a_\infty + \mathbf{I}^a_\infty \mathbf{V} \mathbf{I}^b_\rho. \tag{98}$$

Since $\mathbf{P}=0$, there is no analogous expression for the operator $\mathbf{P}$.

From the point of view of the base system $\mathbf{B}_\rho$, system $\boldsymbol{A}_\infty$ plays the role of the surrounding of $\mathbf{B}_\rho$. System $\mathbf{B}_\rho$ without the interaction with the system $\boldsymbol{A}_\infty$ is a closed system. System $\mathbf{B}_\rho$ subject to the interaction with its surrounding represented by the system $\boldsymbol{A}_\infty$ is an open system.

The case of an external modification of the finite system $\mathbf{B}_\rho$ by the interaction with an infinite system $\boldsymbol{A}_\infty$ is shown schematically in Figure 11.

11. General Properties of The Solutions to the Infinite-Dimensional Combined System

Eigenvalue equation (97a) represents an infinite-dimensional eigenvalue problem. A general strategy in the LRM solution of this equation is to approximate infinite system $\boldsymbol{A}_\infty$ with a huge but finite system $\boldsymbol{A}_n$ containing n eigenvalues λ_i and n corresponding eigenstates. Infinite combined system $\mathbf{C}_\infty$ is accordingly approximated with a huge but finite $(n+\rho)$-dimensional combined system $\mathbf{C}_{n+\rho}$ that has $(n+\rho)$ eigenvalues ε_s and $(n+\rho)$ corresponding eigenstates. As shown in section 4, there are explicit LRM expressions for all eigenvalues and eigenstates of such finite systems. Given those expressions, one can derive their $n \to \infty$ limit. Provided the approximation of $\mathbf{C}_\infty$ with finite systems $\mathbf{C}_{n+\rho}$ is done in an appropriate way, this limit is well defined [5-7]. One thus derives the required expressions for the infinite-dimensional combined system $\mathbf{C}_\infty$.

Finite combined system $\mathbf{C}_{n+\rho}$ has two kinds of eigenvalues and eigenstates: If ε_s differs from all the eigenvalues λ_i of the parent system $\boldsymbol{A}_n$ ($\varepsilon_s \notin \{\lambda_j\}$) the corresponding solution is *cardinal*, otherwise it is *singular* [4]. In addition, one can have two kinds of singular eigenstates $|\Psi_s\rangle$ associated with the eigenvalue $\varepsilon_s \equiv \lambda_j \in \{\lambda_i\}$: If this eigenstate has no component in the space X_ρ^b ($\mathbf{I}_\rho^b|\Psi_s\rangle = 0$), it is *strongly singular*, otherwise ($\mathbf{I}_\rho^b|\Psi_s\rangle \neq 0$), it is *weakly singular*. As emphasized in section 4.3, there is an important difference between strongly and weakly singular eigenstates of finite systems: each strongly singular eigenstate $|\Psi_s\rangle$ of the combined system $\mathbf{C}_{n+\rho}$ is at the same time an eigenstate of the parent system $\boldsymbol{A}_n$. On the other hand, no weakly singular eigenstate $|\Psi_s\rangle$ of $\mathbf{C}_{n+\rho}$ is an eigenstate of $\boldsymbol{A}_n$. This distinction between cardinal, strongly singular and weakly singular eigenstates in the case of finite-dimensional systems generalizes to infinite-dimensional systems. In addition and in analogy to the infinite-dimensional system $\mathbf{A}_\infty$ which may contain discrete and continuous eigenvalues, combined system $\mathbf{C}_\infty$ may also contain discrete and continuous eigenvalues. In order to maintain clear distinction between the solutions of the combined system and the solutions of the parent system, each discrete eigenvalue ε_I of $\mathbf{C}_\infty$

is referred to as an *isolated* eigenvalue while each continuous eigenvalue ε of $\mathbf{C}_\infty$ is referred to as an *embedded* eigenvalue [5].

Eigenstates $|\Psi_I\rangle$ corresponding to isolated eigenvalues ε_I of $\mathbf{C}_\infty$ can be normalized to unity. This normalization is done in accord with the metrics induced by the operator $\mathbf{S}_c$. In particular one has [5]

$$\langle\Psi_I|\mathbf{S}_c|\Psi_J\rangle = \delta_{I,J}. \tag{99}$$

Concerning embedded solutions of $\mathbf{C}_\infty$, one finds that each $\varepsilon \in D$ is an embedded eigenvalue of the combined system [5]. Since embedded eigenvalues are part of a continuous band of eigenvalues, the corresponding embedded eigenstates are with respect to those continuous eigenvalues normalized to a δ-function. Those eigenstates are hence of a general type $|\Psi(\varepsilon,\ldots)\rangle$ where dots (...) denote additional discrete and/or continuous parameters, if any.

Embedded eigenvalues of the combined system $\mathbf{C}_\infty$ are confined to the same range D as continuous eigenvalues of the parent system $\mathbf{A}_\infty$. This is due to the fact that modification operator $\mathbf{V}$ has a finite rank. One finds that if the rank of this operator is infinite (the case not considered in this chapter), embedded eigenvalues of the combined system are not necessarily confined to the same range D as the continuous eigenvalues of the parent system.

One can generalize the notion of cardinal and singular solutions defined for finite modified systems $\mathbf{C}_{n+\rho}$ to all isolated as well as to all embedded solutions of the infinite system $\mathbf{C}_\infty$. By definition, each isolated eigenvalue ε_I of $\mathbf{C}_\infty$ is *cardinal* if it differs from all discrete eigenvalues λ_i of $\mathbf{A}_\infty$ ($\varepsilon_I \notin \{\lambda_j\}$). Otherwise it is *singular*. If the corresponding eigenstate $|\Psi_I\rangle$ has no component in the space X_ρ^b ($\mathbf{I}_\rho^b|\Psi_I\rangle = 0$), this eigenstate is *strongly singular*, otherwise ($\mathbf{I}_\rho^b|\Psi_I\rangle \neq 0$), it is *weakly singular*. This is a natural generalization of the notion of cardinal and singular solutions of finite combined systems to isolated cardinal and isolated singular solutions of infinite combined systems.

11.1. Fractional Shift

Concerning embedded solutions of $\mathbf{C}_\infty$, distinction between cardinal and singular solutions is more subtle. A key quantity in the treatment of embedded solutions is a *fractional shift* $x(\varepsilon)$ [5, 43-46]. In general, with each embedded eigenstate $|\Psi(\varepsilon,\ldots)\rangle$ of $\mathbf{C}_\infty$ is associated a fractional shift $x(\varepsilon,\ldots)$. Fractional shift has the following interpretation: Imagine infinite system $\boldsymbol{A}_\infty$ approximated with a huge but finite system $\boldsymbol{A}_n$ containing n eigenvalues λ_i. The corresponding combined system $\mathbf{C}_\infty$ is accordingly approximated with a huge but finite system $\mathbf{C}_{n+\rho}$ containing $n+\rho$ eigenvalues ε_s. In particular, each eigenvalue band τ of the infinite system $\mathbf{A}_\infty$ can be approximated with a huge but finite number of eigenstates with uniformly spaced eigenvalues λ_i inside the interval $I_\tau = [a_\tau, b_\tau]$. Eigenvalue band τ of the infinite modified system $\mathbf{C}_\infty$ is also approximated with huge but finite number of modified eigenstates with eigenvalues ε_s contained in the same interval I_τ. Consider quantities

$$x(\varepsilon_s) = \frac{\varepsilon_s - \lambda_{s-1}}{\lambda_s - \lambda_{s-1}}. \tag{100}$$

\where $\varepsilon_s \in I_\tau$ are eigenvalues of the combined system $\mathbf{C}_{n+\rho}$ while $\lambda_s, \lambda_{s-1} \in I_\tau$ are eigenvalues of the parent system $\mathbf{A}_n$. Each $x(\varepsilon_s)$ is a *fractional shift* of the modified eigenvalue ε_s relative to the initial eigenvalue λ_{s-1} [5, 43-46]. Fractional shift is thus defined as the ratio of two infinitesimal quantities: There is an infinitesimal shift $\Delta\varepsilon_s = \varepsilon_s - \lambda_{s-1}$ of the modified eigenvalue ε_s relative to the corresponding eigenvalue λ_{s-1}. Another infinitesimal quantity is the interval $\Delta\lambda_s = \lambda_s - \lambda_{s-1}$ between two adjacent eigenvalues $\lambda_s \in I_\tau$. As long as infinite systems $\mathbf{A}_\infty$ and $\mathbf{C}_\infty$ are approximated with finite systems $\mathbf{A}_n$ and $\mathbf{C}_{n+\rho}$, there is a clear distinction between cardinal and singular solutions. Due to the interlacing rule (11) fractional shift $x(\varepsilon_s)$ is confined to the interval $[1-\rho, 1]$, i.e. $x(\varepsilon_s) \in [1-\rho, 1]$. Further, if $|\Psi_s\rangle$ is a singular eigenstate of $\mathbf{C}_{n+\rho}$, fractional shift $x(\varepsilon_s)$ equals an integer to the order $O(n^{-1})$. In the limit $n \to \infty$ this is exactly an integer, and hence in an infinite system singular eigenstates are characterized by integer fractional shift. However, if $|\Psi_s\rangle$ is a cardinal eigenstate of $\mathbf{C}_{n+\rho}$, fractional shift $x(\varepsilon_s)$ may assume any non-integer value.

If the infinite system $\mathbf{C}_\infty$ is approximated with finite systems $\mathbf{C}_{n+\rho}$ in an appropriate way and if the base system $\mathbf{B}_\rho$ is one-dimensional ($\rho = 1$), in a limit $n \to \infty$ discrete quantities $x(\varepsilon_s)$ converge either to zero or to a single function $x(\varepsilon)$ of a continuous parameter $\varepsilon \in D$ [43-46]. However, if the base system $\mathbf{B}_\rho$ is ρ-dimensional where $\rho > 1$, in a limit $n \to \infty$ discrete quantities $x(\varepsilon_s)$ usually converge to several nontrivial functions $x_d(\varepsilon)$ [5]. In conclusion, in a limit $n \to \infty$ fractional shifts $x(\varepsilon_s)$ associated with cardinal eigenstates converge to one or several well defined continuous functions $x_d(\varepsilon)$, while fractional shifts associated with singular eigenstates converge to a trivial function $x(\varepsilon, \ldots) \equiv 0$. One finds that for each particular eigenvalue $\varepsilon \in D$ combined system $\mathbf{C}_\infty$

may contain at most $r(\varepsilon) \le \rho$ embedded cardinal eigenstates and $r(\varepsilon) \le \rho$ corresponding fractional shifts $x_d(\varepsilon)$. Those embedded eigenstates are hence of the type $|\Psi_d(\varepsilon)\rangle$ ($d = 1,\ldots,r(\varepsilon)$). On the other hand, for each particular eigenvalue $\varepsilon \in D$ combined system $\mathbf{C}_\infty$ may contain an infinite number of singular eigenstates [5]. Accordingly, embedded singular eigenstates of the combined system with the eigenvalue $\varepsilon \in D$ are of the form $|\Psi(\varepsilon,\ldots)\rangle$ where dots $(\ldots)$ indicate additional continuous and/or discrete parameters, if any. Further, as far as properties of embedded eigenstates are concerned, there is no difference between fractional shifts $x_d(\varepsilon)$ and $x_d(\varepsilon) \pm 1$ (see section 14.1) [43-46]. Fractional shift can be hence confined to the interval $[0,1)$. By definition, fractional shift confined to this interval is a *principal value* of a fractional shift [5]. Assuming fractional shift to be confined to its principal value, embedded cardinal eigenstates are characterized by fractional shift $x_d(\varepsilon) \ne 0$, while embedded singular eigenstates are characterized by fractional shift $x(\varepsilon) = 0$.

In conclusion and in view of the above picture of the combined system $\mathbf{C}_\infty$ as the $n \to \infty$ limit of finite combined systems $\mathbf{C}_{n+\rho}$, one can give a following interpretation to the notion of a fractional shift: Fractional shift $x(\varepsilon) = 0$ corresponds to the modified eigenvalue ε that (in the limit $n \to \infty$) coincides with some parent eigenvalue λ, while fractional shift $x(\varepsilon) \ne 0$ corresponds to the modified eigenvalue ε that does not coincide with any of the parent eigenvalues λ [5]. Accordingly, each embedded solution that satisfies $x(\varepsilon,\ldots) = 0$ is *singular* while each embedded solution that satisfies $x_d(\varepsilon) \ne 0$ is *cardinal.* This is a natural generalization of the notion of singular and cardinal solutions of the finite combined system $\mathbf{C}_{n+\rho}$ to embedded solutions of the infinite combined system $\mathbf{C}_\infty$.

In addition to embedded singular solutions which are characterized by fractional shift $x(\varepsilon,\ldots) = 0$, a special role is also played by embedded cardinal solutions with fractional shift $x_d(\varepsilon) = 0.5$. This fractional shift corresponds to the modified eigenvalue ε that is exactly in a middle between two adjacent infinitesimally close parent eigenvalues λ. Each solution that satisfies $x_d(\varepsilon) = 0.5$ is a *resonant* solution and the corresponding point $\varepsilon \in D$ a *resonant point.* This name is due to the resonant shape of the corresponding embedded cardinal eigenstates in the case of the weak coupling [5, 43-46] (see section 14.1.3).

11.2. Description of the Open System that Interacts with an Infinite System

Each isolated as well as each embedded eigenstate of the combined system can be written as a linear combination

$$|\Psi\rangle = |\Psi^b\rangle + |\Psi^a\rangle, \tag{101a}$$

where $\left|\Psi^b\right\rangle \in X_\rho^b$ and $\left|\Psi^a\right\rangle \in X_\infty^a$ are components of this eigenstate in spaces X_ρ^b and X_∞^a, respectively. Let $\mathbf{O}$ be an observable that describes some property of the open system $\mathbf{B}_\rho$ that interacts with the infinite system $\boldsymbol{A}_\infty$ and let $\left|\Psi_1\right\rangle$ and $\left|\Psi_2\right\rangle$ be two eigenstates of the combined system $\mathbf{C}_\infty$. Since operator $\mathbf{O}$ refers to the system $\boldsymbol{B}_\rho$, it has non-vanishing matrix elements only between the states contained in the space X_ρ^b. This implies

$$\left\langle\Psi_1\middle|\mathbf{O}\middle|\Psi_2\right\rangle = \left\langle\Psi_1^b\middle|\mathbf{O}\middle|\Psi_2^b\right\rangle. \tag{101b}$$

Hence, in order to obtain properties of the open system $\mathbf{B}_\rho$, it is sufficient to know X_ρ^b-components of (properly normalized) eigenstates of the combined system. The knowledge of X_∞^a-components of those eigenstates is not needed.

Consider now isolated and embedded eigenstates of the combined system from a point of view of the open system $\mathbf{B}_\rho$. In the case of a finite combined system $\mathbf{C}_{n+\rho}$, strongly singular eigenstates have no component in the space X_ρ^b. Hence and according to (101) those eigenstates do not contribute to the properties of the open system $\mathbf{B}_\rho$. Concerning weakly singular eigenstates of $\mathbf{C}_{n+\rho}$, those eigenstates do have non-vanishing X_ρ^b-component. However weakly singular eigenstates exist only if the combined system $\mathbf{C}_{n+\rho}$ has some symmetry which is not present in the initial non-interacting systems $\boldsymbol{B}_\rho$ and $\boldsymbol{A}_n$. One finds that the same applies to the infinite combined system $\mathbf{C}_\infty$. Accordingly, if one is interested in the open system $\boldsymbol{B}_\rho$ that interacts with the infinite system $\boldsymbol{A}_\infty$, most important are cardinal solutions of the combined system. In particular, those are X_ρ^b-components of embedded cardinal and X_ρ^b-components of the isolated cardinal eigenstates of the combined system. One finds that those components are described by two key eigenvalue equations. Those are *generic* eigenvalue equation and *fractional shift* eigenvalue equation. Both equations act in the space X_ρ^b and they can be both represented as $\rho \times \rho$ matrix eigenvalue equations. In addition to operators $\mathbf{H}_b$ and $\mathbf{S}_b$ that according to (5a) describe isolated system $\boldsymbol{B}_\rho$, those eigenvalue equations involve *characteristic operator* $\mathbf{f}(\varepsilon)$ and *derived operator* $\boldsymbol{\omega}(\varepsilon)$ [5, 43-46]. Characteristic operator $\mathbf{f}(\varepsilon)$ incorporates essential features of the interaction of the finite system $\boldsymbol{B}_\rho$ with the infinite system $\boldsymbol{A}_\infty$. This operator is basic. Derived operator $\boldsymbol{\omega}(\varepsilon)$ is uniquely determined by the corresponding characteristic operator

$\mathbf{f}(\varepsilon)$. This derived operator generalizes to infinite-dimensional systems operator $\mathbf{\Omega}(\varepsilon)$ defined in section 4.

12. Characteristic and Derived Operators

Key role in the LRM expressions that replace infinite-dimensional eigenvalue equation (97a) is played by the *characteristic* operator $\mathbf{f}(\varepsilon)$ which depends on the continuous parameter ε [5]. This operator acts in the space X_ρ^b and in this space it is represented as a $\rho \times \rho$ Hermitean matrix. Characteristic operator $\mathbf{f}(\varepsilon)$ is a sum of characteristic operators $\mathbf{f}_\nu(\varepsilon)$ associated with eigenvalue bands ν of the system $\boldsymbol{A}_\infty$ and of the characteristic operator $\mathbf{F}(\varepsilon)$ associated with discrete solutions of this system

$$\mathbf{f}(\varepsilon) = \sum_\nu \mathbf{f}_\nu(\varepsilon) + \mathbf{F}(\varepsilon), \tag{102a}$$

Operator $\mathbf{f}_\nu(\varepsilon)$ contains information about the interaction of the finite system $\boldsymbol{B}_\rho$ with the eigenvalue band ν of the infinite system $\boldsymbol{A}_\infty$. This operator is defined as

$$\mathbf{f}_\nu(\varepsilon) = \left. \frac{\sum_m \int \mathbf{V}|\Phi_{\nu,m}(k,l)\rangle\langle\Phi_{\nu,m}(k,l)|\mathbf{V}dl}{d\lambda_\nu(k)/dk} \right|_{k=\lambda_\nu^{-1}(\varepsilon)} \cdot \begin{cases} 1 & if \quad \varepsilon \in I_\nu \\ 0 & if \quad \varepsilon \notin I_\nu \end{cases} \quad \nu = 1,\ldots,\kappa, \tag{102b}$$

Above operator vanishes for each $\varepsilon \notin I_\nu$ and it is positive definite for almost each $\varepsilon \in I_\nu$. Only in some isolated points $\varepsilon_0 \in I_\nu$, if any, one may have $\mathbf{f}_\nu(\varepsilon_0) = 0$. Further, operator $\mathbf{f}_\nu(\varepsilon)$ vanishes over the entire space X_∞^a, and it can have non-vanishing matrix elements only over the finite space X_ρ^b. In the base $\{|r\rangle\}$ of this space operator $\mathbf{f}_\nu(\varepsilon)$ is a $\rho \times \rho$ matrix with matrix elements:

$$f_{rt}^{(\nu)}(\varepsilon) \equiv \langle r|\mathbf{f}_\nu(\varepsilon)|t\rangle = \left. \frac{\sum_m \int \langle r|\mathbf{V}|\Phi_{\nu,m}(k,l)\rangle\langle\Phi_{\nu,m}(k,l)|\mathbf{V}|t\rangle dl}{d\lambda_\nu(k)/dk} \right|_{k=\lambda_\nu^{-1}(\varepsilon)} \cdot \begin{cases} 1 & if \quad \varepsilon \in I_\nu \\ 0 & if \quad \varepsilon \notin I_\nu \end{cases}$$

$$\nu = 1,\ldots,\kappa, \; r,t = 1,\ldots,\rho. \tag{102c}$$

Each matrix element $f_{rt}^{(\nu)}(\varepsilon)$ is usually an analytic function of ε inside the interval I_ν. However, it is not analytic for each ε, since outside this interval $f_{rt}^{(\nu)}(\varepsilon)$ is identically zero. Since $\mathbf{f}_\nu(\varepsilon)$ is positive definite almost everywhere inside the interval I_ν and zero outside this interval, each eigenvalue $f_j^{(\nu)}(\varepsilon)$ ($j = 1,\ldots,\rho$) of this operator is nonnegative inside I_ν and zero outside I_ν. In particular, if ν is one-parameter eigenvalue band, operator $\mathbf{f}_\nu(\varepsilon)$ has rank one for each $\varepsilon \in I_\nu$ with a possible exception of few isolated points where this rank vanishes [5]. In this case each eigenvalue $f_j^{(\nu)}(\varepsilon)$ is identically zero ($f_j^{(\nu)}(\varepsilon) = 0$ for each ε) except for only one eigenvalue (say $f_1^{(\nu)}(\varepsilon)$ which is positive almost everywhere in the interval I_ν and zero outside this interval [5].

Operator $\mathbf{F}(\varepsilon)$ contains essential features of the interaction of the finite system $\boldsymbol{B}_\rho$ with discrete eigenstates of the infinite system $\boldsymbol{A}_\infty$. This operator is defined as

$$\mathbf{F}(\varepsilon) = \sum_i \mathbf{V}|\Phi_i\rangle\langle\Phi_i|\mathbf{V}\delta(\varepsilon - \lambda_i), \tag{102d}$$

Above operator vanishes for each $\varepsilon \notin \{\lambda_j\}$ and it is a δ-type operator in the points $\varepsilon \in \{\lambda_j\}$. This operator is also zero over the entire space X_∞^a, it is also positive definite, and in the base $\{r\}$ of the space X_ρ^b it is a $\rho \times \rho$ matrix with matrix elements $F_{r,t}(\varepsilon)$

$$F_{rt}(\varepsilon) \equiv \langle r|\mathbf{F}(\varepsilon)|t\rangle = \sum_i \langle r|\mathbf{V}|\Phi_i\rangle\langle\Phi_i|\mathbf{V}|t\rangle\delta(\varepsilon - \lambda_i),\ r,t = 1,\ldots,\rho. \tag{102e}$$

Characteristic operators $\mathbf{f}_\nu(\varepsilon)$ and $\mathbf{F}(\varepsilon)$ determine derived operators $\boldsymbol{\omega}_\nu(\varepsilon)$ and $\boldsymbol{\Omega}(\varepsilon)$, respectively. Unlike operator $\mathbf{f}_\nu(\varepsilon)$ that vanishes outside the interval I_ν, the corresponding derived operator $\boldsymbol{\omega}_\nu(\varepsilon)$ is nonzero in all points outside I_ν and in almost all points inside I_ν. This operator is expressed in terms of the characteristic operator $\mathbf{f}_\nu(\varepsilon)$ according to

$$\boldsymbol{\omega}_\nu(\varepsilon) = P\int \frac{\mathbf{f}_\nu(\lambda)}{\varepsilon - \lambda} d\lambda, \tag{103a}$$

where P denotes principal Cauchy integral value [14]. Matrix elements of this operator in the base $\{|r\rangle\}$ of the space X_ρ^b are

$$\omega_{rt}^{(\nu)}(\varepsilon) \equiv \langle r|\boldsymbol{\omega}_\nu(\varepsilon)|t\rangle = P\int \frac{f_{rt}^{(\nu)}(\lambda)}{\varepsilon - \lambda} d\lambda,\ r,t = 1,\ldots,\rho. \tag{103b}$$

If $\varepsilon \notin I_\nu$ above expression is a standard integral. However, if $\varepsilon \in I_\nu$ sub-integral function on the right-hand side of this expression usually diverges in a point $\lambda = \varepsilon$. In this case one has to take a principal Cauchy integral value of this expression.

In analogy to (103a), derived operator $\mathbf{\Omega}(\varepsilon)$ is expressed in terms of the characteristic operator $\mathbf{F}(\varepsilon)$ according to

$$\mathbf{\Omega}(\varepsilon) = P\int \frac{\mathbf{F}(\lambda)}{\varepsilon - \lambda} d\lambda \,, \tag{104a}$$

Hence and from (102d)

$$\Omega_{rt}(\varepsilon) \equiv \langle r|\mathbf{\Omega}(\varepsilon)|t\rangle = \sum_{i(\lambda_i \neq \varepsilon)} \frac{\langle r|\mathbf{V}|\Phi_i\rangle\langle\Phi_i|\mathbf{V}|t\rangle}{\varepsilon - \lambda_i}, \; r,t = 1,\ldots,\rho \,. \tag{104b}$$

This is identical to the expression (10b) (where $\mathbf{P} = 0$).

In analogy to (102a), derived operators $\boldsymbol{\omega}_\nu(\varepsilon)$ and $\mathbf{\Omega}(\varepsilon)$ combine into a global derived operator $\boldsymbol{\omega}(\varepsilon)$ according to

$$\boldsymbol{\omega}(\varepsilon) = \sum_\nu \boldsymbol{\omega}_\nu(\varepsilon) + \mathbf{\Omega}(\varepsilon) \,, \tag{105a}$$

Matrix elements $\omega_{rt}(\varepsilon)$ of $\boldsymbol{\omega}(\varepsilon)$ are hence

$$\omega_{rt}(\varepsilon) \equiv \langle r|\boldsymbol{\omega}(\varepsilon)|t\rangle = P\int \frac{f_{rt}(\lambda)}{\varepsilon - \lambda} d\lambda \,, \; r,t = 1,\ldots,\rho \,. \tag{105b}$$

One can express derived operators $\boldsymbol{\omega}_\nu(\varepsilon)$ and $\mathbf{\Omega}(\varepsilon)$ in terms of the operators $\overline{\boldsymbol{\omega}}_\nu(\varepsilon)$ and $\overline{\mathbf{\Omega}}(\varepsilon)$ that are associated with the system $\mathbf{A}_\infty$. Those operators were defined by the expressions (96). One has

$$\boldsymbol{\omega}_\nu(\varepsilon) = \mathbf{V}\overline{\boldsymbol{\omega}}_\nu(\varepsilon)\mathbf{V} \,, \; \mathbf{\Omega}(\varepsilon) = \mathbf{V}\overline{\mathbf{\Omega}}(\varepsilon)\mathbf{V} \,. \tag{106}$$

If the range D and the set $\{\lambda_j\}$ are bounded, for each state $|\Theta\rangle \in X_\rho^a$ derived operator $\boldsymbol{\omega}(\varepsilon)$ satisfies

$$\langle\Theta|\boldsymbol{\omega}(\pm\infty)|\Theta\rangle = 0 \,, \tag{107a}$$

This is true also if the range D and/or the set $\{\lambda_j\}$ is not bounded, provided in the limit $\lambda \to \infty$ matrix elements $f_{rt}(\lambda)$ satisfy some mild conditions. Further, since $\mathbf{f}(\lambda)$ is nonnegative, for each $\varepsilon \notin D$ and $\varepsilon \notin \{\lambda_j\}$ an arbitrary state $|\Theta\rangle \in X_\rho^b$ satisfies $d\langle\Theta|\boldsymbol{\omega}(\varepsilon)|\Theta\rangle / d\varepsilon \le 0$. This holds true also if $\varepsilon = \varepsilon_0 \in D$, provided the state $|\Theta\rangle$ satisfies $\mathbf{f}(\varepsilon_0)|\Theta\rangle = 0$. Since $\mathbf{f}(\varepsilon)|\Theta\rangle = 0$ whenever $\varepsilon \notin D$, this can be concisely written as

$$\frac{d}{d\varepsilon}\langle\Theta|\boldsymbol{\omega}(\varepsilon)|\Theta\rangle \le 0, \text{ if } \mathbf{f}(\varepsilon)|\Theta\rangle = 0 \quad \text{and } \varepsilon \notin \{\lambda_j\}. \tag{107b}$$

On the other hand, if $\varepsilon \in D$ and $\mathbf{f}(\varepsilon)|\Theta\rangle \neq 0$, one may have $d\langle\Theta|\boldsymbol{\omega}(\varepsilon)|\Theta\rangle / d\varepsilon \le 0$ as well as $d\langle\Theta|\boldsymbol{\omega}(\varepsilon)|\Theta\rangle / d\varepsilon > 0$. The condition $\mathbf{f}(\varepsilon)|\Theta\rangle = 0$ is hence essential. Concerning the condition $\varepsilon \notin \{\lambda_j\}$, in the point $\varepsilon \in \{\lambda_j\}$ above derivative is not well defined.

As a consequence of expressions (107), each eigenvalue $\omega_s(\varepsilon)$ of the derived operator $\boldsymbol{\omega}(\varepsilon)$ is a continuous decreasing function of ε for each ε that satisfies $\varepsilon \notin D$ and $\varepsilon \notin \{\lambda_j\}$. However, if $\varepsilon \in D$ eigenvalue $\omega_s(\varepsilon)$ may be decreasing as well as increasing function of ε, while in each point $\varepsilon = \lambda_i$ this eigenvalue usually diverges [5]. Similar conclusions apply to eigenvalues of operators $\boldsymbol{\omega}_\nu(\varepsilon)$ and $\boldsymbol{\Omega}(\varepsilon)$. In particular, each eigenvalue $\omega_s^{(\nu)}(\varepsilon)$ of the operator $\boldsymbol{\omega}_\nu(\varepsilon)$ is a continuous decreasing function for each real ε that satisfies $\varepsilon \notin I_\nu$. However if $\varepsilon \in I_\nu$ this eigenvalue may be decreasing as well as increasing function of ε.

Characteristic operator $\mathbf{f}(\varepsilon)$ contains all information necessary for the description of the open system $\boldsymbol{B}_\rho$ subject to the interaction with an infinite system $\boldsymbol{A}_\infty$. Since this operator vanishes over the infinite-dimensional space X_∞^a, in order to construct this operator it is not necessary to specify all the details of the infinite system $\boldsymbol{A}_\infty$. All what is needed is the knowledge of $\rho\cdot(\rho+1)/2$ functions (or more precisely distributions [1]) $f_{rt}(\varepsilon)$ ($r \le t$). There are an infinite number of possible systems $\boldsymbol{A}_\infty$ that generate the same functions $f_{rt}(\varepsilon)$. As far as the properties of the open system $\boldsymbol{B}_\rho$ that interacts with the infinite system $\boldsymbol{A}_\infty$ are concerned, huge amount of information about this system is redundant. Hence if the details of the system $\boldsymbol{A}_\infty$ and of the interaction of this system with the system $\mathbf{B}_\rho$ are not known, one can still model those functions in such a way as to satisfy some required properties and/or some known data of the combined system.

12.1. Calculation of the Derived Operators

In order to calculate matrix elements of the derived operator $\boldsymbol{\omega}(\varepsilon)$, one has to evaluate expressions of the types (103b) and (104b). There are no problems with the expression (104b). More problems present calculation of the integrals (103b). If $f_{rt}^{(\nu)}(\varepsilon) \equiv q(\varepsilon)$ is an analytic function in the interval I_ν, matrix element $\omega_{rt}^{(\nu)}(\varepsilon)$ can be usually expressed in a closed form as an infinite sum. In particular, if $q(\lambda)$ is analytic in this interval and if a and b are finite ($a \neq -\infty$ and $b \neq \infty$) one has [5, 46]

$$P\int_a^b \frac{q(\lambda)}{\varepsilon - \lambda} d\lambda = q(\varepsilon) \ln\left|\frac{a-\varepsilon}{b-\varepsilon}\right| - g(\varepsilon), \tag{108a}$$

where the function $g(\varepsilon)$ equals

$$g(\varepsilon) = \sum_{i=1}^{\infty} \frac{q^{(i)}(\varepsilon)}{i!\,i}\left[(b-\varepsilon)^i - (a-\varepsilon)^i\right]. \tag{108b}$$

and where ε can assume any real value. In the above expression $q^{(i)}(\varepsilon)$ is *i*-th derivative of the function $q(\varepsilon)$. If $q(\varepsilon)$ is polynomial, $g(\varepsilon)$ is also polynomial. More generally, if $q(\varepsilon)$ is an analytic function with the only singularity in infinity, infinite sum in (108b) is guaranteed to converge for each finite ε. Function $g(\varepsilon)$ defined with this sum is also analytic with the only singularity at infinity. Note however that principal integral value (108a) is not an analytic function of ε in the points $\varepsilon = a$ and $\varepsilon = b$ on the edgese of the interval I_ν where the function $\ln|(a-\varepsilon)/(b-\varepsilon)|$ is not analitic.

In addition to the integrals of the general type (108a), one can also consider integrals of a general type

$$P\int_a^b \frac{p(\lambda)}{(\varepsilon-\lambda)\sqrt{(\lambda-a)(b-\lambda)}} d\lambda, \tag{109a}$$

where $p(\lambda)$ is a polynomial. With a simple change of variables this integral can be transformed to the integral of a type

$$P\frac{1}{\pi}\int_{-1}^{1} \frac{q(\lambda)}{(\varepsilon-\lambda)\sqrt{1-\lambda^2}} d\lambda, \tag{109b}$$

where $q(\lambda)$ is polynomial. Since each polynomial can be written as a linear combination of Chebyshev polynomials $T_n(\lambda)$ of a first kind [47], integral (109b) can be written as a linear combination of integrals $I_n(\varepsilon)$ where

$$I_n(\varepsilon) \equiv P\frac{1}{\pi}\int_{-1}^{1}\frac{T_n(\lambda)}{(\lambda-\varepsilon)\sqrt{1-\lambda^2}}d\lambda\,,\; n=0,1,2,3,\cdots \tag{110}$$

One can obtain an exact solution of the integrals $I_n(\varepsilon)$ [43]

$$I_n(\varepsilon) = U_{n-1}(\varepsilon) + \frac{T_n(\varepsilon)}{\sqrt{x^2-1}}\Theta(\varepsilon)\,,\; n=0,1,2,3,\cdots \tag{111a}$$

where

$$\Theta(\varepsilon) = \begin{cases} 1 & \varepsilon < -1 \\ 0 & \varepsilon \in [-1,1] \\ -1 & \varepsilon > 1 \end{cases} \tag{111b}$$

and where $T_n(\lambda)$ and $U_n(\lambda)$ are Chebyshev polynomials of the first and of the second kind, respectively. Those polynomials are defined as [47]

$$T_n(\cos\theta) = \cos(n\theta), U_n(\cos\theta) = \frac{\sin(n+1)\theta}{\sin\theta},\; n=0,1,2,3,\cdots, \tag{112a}$$

and they satisfy recursive relations

$$T_{n+1}(x) = 2xT_n(x) - T_{n-1}(x),\; U_{n+1}(x) = 2xU_n(x) - U_{n-1}(x),\; n=1,2,3,\cdots, \tag{112b}$$

where

$$T_0(x) = U_0(x) = 1,\; T_1(x) = x,\; U_1(x) = 2x. \tag{112c}$$

First few Chebyshev polynomials are [47]

$$T_0(x)=1,\; T_1(x)=x,\; T_2(x)=2x^2-1,\; T_3(x)=4x^3-3x,$$
$$T_4(x)=8x^4-8x^2+1,\; \ldots \tag{113a}$$

$$U_0(x)=1,\; U_1(x)=2x,\; U_2(x)=4x^2-1,\; U_3(x)=8x^3-4x,$$
$$U_4(x)=16x^4-12x^2+1,\; \ldots \tag{113b}$$

Concerning the above definition of Chebyshev polynomials, two remarks are in place.

First, in view of the definition (112a), and since for real θ the value of $\cos\theta$ is confined to the interval $[-1,1]$, those polynomials are usually considered only for argument values $x \in [-1,1]$. However, since $T_n(x)$ and $U_n(x)$ are polynomials, they are well defined for each $x \in (-\infty, \infty)$. In particular, the expression (111) is valid for each $\varepsilon \in (-\infty, \infty)$. Note that this is not a standard way how those polynomials are treated.

Second, Chebyshev polynomial $T_n(x)$ and $U_n(x)$ are defined for $n = 0,1,2,3,\cdots$ [47]. In order for the expression (111) to be valid for each $n \geq 0$ including the case $n = 0$, one has to define polynomial $U_{-1}(x)$. One finds that the proper definition is

$$U_{-1}(x) = 0. \tag{113c}$$

This definition is in accord with expressions (112a). It is also in accord with iterative relations (112b).

Expressions (108) and (111) are needed in order to calculate matrix elements $\omega_{rt}(\varepsilon)$ of the derived operator $\boldsymbol{\omega}(\varepsilon)$. Those matrix elements are required in the LRM treatment of the eigenvalue equation (97) that describes infinite-dimensional combined system $\mathbf{C}_\infty$. Let me now consider LRM solution of this equation.

13. Isolated Solutions of the Infinite-Dimensional Combined System

Concerning isolated solutions of the infinite-dimensional combined system $\mathbf{C}_\infty$, most important are isolated cardinal solutions that satisfy $\varepsilon_I \notin \{\lambda_j\}$. Isolated cardinal eigenvalues and the corresponding eigenstates can be obtained as a solution of the *generic* eigenvalue equation [5]

$$\left[\beta^2 \boldsymbol{\omega}(\varepsilon_I) + \mathbf{H}_b\right]|\theta_I\rangle = \varepsilon_I \mathbf{S}_b |\theta_I\rangle, \; |\theta_I\rangle \in X_\rho^b, \tag{114a}$$

In the base $\{|r\rangle\}$ of the space X_ρ^b above expression is a $\rho \times \rho$ nonlinear eigenvalue equation. Each eigenvalue ε_I of this equation is hence a root of the $\rho \times \rho$ determinant $h(\varepsilon)$

$$h(\varepsilon) \equiv \left|\beta^2 \boldsymbol{\omega}(\varepsilon) + \mathbf{H}_b - \varepsilon \mathbf{S}_b\right| = 0. \tag{114b}$$

Expressions (114) generalize expressions (12) to infinite-dimensional systems. One finds that each solution $\varepsilon = \varepsilon_I \notin D$ of (114b) is an isolated eigenvalue of the infinite-dimensional combined system $\mathbf{C}_\infty$. Corresponding eigenstate $|\theta_I\rangle \in X_\rho^b$ of (114a) determines X_ρ^b-component $|\Psi_I^b\rangle \in X_\rho^b$ of the normalized isolated eigenstate $|\Psi_I\rangle \in X_\infty^c$ according to

$$\left|\Psi_I^b\right\rangle = \frac{1}{\sqrt{N_I}}\left|\theta_I\right\rangle, \qquad (115a)$$

where normalization constant N_I equals

$$N_I = \left\langle\theta_I\left|\mathbf{S}_b\right|\theta_I\right\rangle - \beta^2\frac{d}{d\varepsilon_I}\left\langle\theta_I\left|\boldsymbol{\omega}(\varepsilon)\right|\theta_I\right\rangle, \qquad (115b)$$

The corresponding X_∞^a-component $\left|\Psi_I^a\right\rangle \in X_\infty^a$ of this eigenstate equals [6]

$$\left|\Psi_I^a\right\rangle = \frac{\beta}{\sqrt{N_I}}\left[\sum_i \frac{\left\langle\Phi_i\left|\mathbf{V}\right|\theta_I\right\rangle}{\varepsilon_I-\lambda_i}\left|\Phi_i\right\rangle + \sum_\tau \int \frac{\sum_m \int\left\langle\Phi_{\tau,m}(k,l)\right|\mathbf{V}\left|\theta_I\right\rangle dl}{\varepsilon_I-\lambda_\tau(k)}\left|\Phi_{\tau,m}(k,l)\right\rangle dk\right], \qquad (115c)$$

Since $\varepsilon_I \notin D$, expression (103a) implies

$$\frac{d}{d\varepsilon_I}\left\langle\theta_I\left|\boldsymbol{\omega}(\varepsilon)\right|\theta_I\right\rangle = -\int\frac{\left\langle\theta_I\left|\mathbf{f}(\lambda)\right|\theta_I\right\rangle}{(\varepsilon_I-\lambda)^2}d\lambda. \qquad (115d)$$

In a point $\lambda = \varepsilon_I$ one has $\mathbf{f}(\lambda) = 0$. Expression (103a) is hence a standard integral, and there is no need to take principal integral value of this expression. As a consequence, one can take derivative of this expression in a standard way. This implies (115d). In particular one has $d\left\langle\theta_I\left|\boldsymbol{\omega}(\varepsilon)\right|\theta_I\right\rangle / d\varepsilon_I < 0$.

Concerning solutions $\varepsilon_I \in D$ of (114b), those solutions are *resonant points* [5, 43-46] and they are related to the embedded solutions of the combined system [5]. In general, in a resonant point expression (115d) is not valid, and one has to calculate this derivative more carefully. First one has to take principal integral value of the matrix element $\left\langle\theta_I\left|\boldsymbol{\omega}(\varepsilon)\right|\theta_I\right\rangle$, and then one has to take derivative of thus obtained expression. As a result, one may have $d\left\langle\theta_I\left|\boldsymbol{\omega}(\varepsilon)\right|\theta_I\right\rangle / d\varepsilon_I < 0$ as well as $d\left\langle\theta_I\left|\boldsymbol{\omega}(\varepsilon)\right|\theta_I\right\rangle / d\varepsilon_I \geq 0$. However, if in a point $\varepsilon = \varepsilon_I$ there is an eigenstate $\left|\theta_I\right\rangle$ of (114a) that in addition to (114a) satisfies

$$\mathbf{f}(\varepsilon_I)\left|\theta_I\right\rangle = 0. \qquad (116)$$

this point is an *anomal point*. In each anomal point combined system has an isolated solution. Eigenvalue of this isolated solution is eigenvalue ε_I of the generic eigenvalue equation (114a). In addition, components $\left|\Psi_I^b\right\rangle$ and $\left|\Psi_I^a\right\rangle$ of the corresponding isolated eigenstate $\left|\Psi_I\right\rangle$ are again given by expressions (115) where $\left|\theta_I\right\rangle$ satisfies (114a) and (116). In particular, since $|\theta_I\rangle$ satisfies (116) there is no divergence in the point $\lambda = \varepsilon_I$. Hence one again derives expression (115d).

In conclusion, eigenvalues and eigenstates of the generic eigenvalue equation determine all cardinal isolated eigenvalues $\varepsilon_I \notin \{\lambda_j\}$ of the combined system and all the corresponding eigenstates $|\Psi_I\rangle$. If $\varepsilon_I \notin D$, this solution is determined by the generic eigenvalue equation (114a). However, if $\varepsilon_I \in D$, the corresponding eigenstate $|\theta_I\rangle$ should satisfy additional condition (116).

In a similar way can be derived expressions for isolated singular solutions that satisfy $\varepsilon_I \in \{\lambda_j\}$. Those expressions are almost identical to the expressions (15) that apply to finite systems. However, those expressions are not that important. Strongly singular eigenstates have no component in the space X_ρ^b. Hence those eigenstates do not contribute to the properties of the open system $\mathbf{B}_\rho$. Concerning weakly singular eigenstates, those eigenstates exist only if the infinity combined $\mathbf{C}_\infty$ has some symmetry not present in the component subsystems $\mathbf{A}_\infty$ and $\mathbf{B}_\rho$. This is a rather exceptional situation and hence it can be usually neglected.

13.1. Resonant Points

As emphasized above, each eigenstate $|\theta_I\rangle$ of the generic eigenvalue equation (114a) that satisfies (116) generates an isolated cardinal eigenstate of the combined system $\mathbf{C}_\infty$. However if the eigenstate $|\theta_I\rangle$ of (114a) does not satisfy (116), this is not the case and the corresponding eigenvalue ε_I is a resonant point. Since $\mathbf{f}(\varepsilon) = 0$ for each $\varepsilon \notin D$, all resonant points are contained in the range D.

One can consider each eigenvalue ε_I of the generic eigenvalue equation (114a) as a continuous function of the coupling parameter β, $\varepsilon_I \equiv \varepsilon_I(\beta)$. From this point of view eigenvalue $\varepsilon_I(\beta)$ of (114a) may be connected with some eigenvalue E_r of the finite system $\mathbf{B}_\rho$, in which case one has $\varepsilon_I(0) = E_r$. It may be also connected with some discrete eigenvalues λ_i of the infinite system $\boldsymbol{A}_\infty$, in which case one has $\varepsilon_I(\beta) \to \lambda_i$ in a limit $\beta \to 0$. However, generic eigenvalue equation may contain some additional eigenvalues $\varepsilon_I(\beta)$ that are connected neither to some eigenvalue of $\mathbf{B}_\rho$, nor to some isolated eigenvalue λ_i of $\boldsymbol{A}_\infty$ [5]. In some cases such an additional eigenvalue $\varepsilon_I(\beta)$ may exist for each β however small, as long as $\beta \neq 0$. In some other cases such an additional eigenvalue

$\varepsilon_I(\beta)$ exists only if β is large enough, i.e. one must have $\beta > \beta_0$ where $\beta_0 \neq 0$ (see section 17.2. and Figure 22 for some examples).

In order to have a better insight into the dependence of the eigenvalue $\varepsilon_I(\beta)$ on the coupling parameter β, it is convenient to know the rate of change of $\varepsilon_I(\beta)$ with the change of this parameter. One finds [5]

$$\frac{\partial \varepsilon_I(\beta)}{\partial \beta} = \frac{2\beta \langle \theta_I | \boldsymbol{\omega}(\varepsilon_I) | \theta_I \rangle}{\langle \theta_I | \mathbf{S}_b | \theta_I \rangle - \beta^2 \dfrac{d}{d\varepsilon_I} \langle \theta_I | \boldsymbol{\omega}(\varepsilon) | \theta_I \rangle}, \quad \varepsilon_I \notin \{\lambda_j\}. \tag{117}$$

As shown in a previous section, if $\varepsilon_I(\beta)$ is an isolated eigenvalue one has $d\langle \theta_I | \boldsymbol{\omega}(\varepsilon) | \theta_I \rangle / d\varepsilon_I \leq 0$. The sign of the derivative $(\partial \varepsilon_I / \partial \beta)$ in this case equals the sign of the matrix element $\langle \theta_I | \boldsymbol{\omega}(\varepsilon_I) | \theta_I \rangle$. However, if $\varepsilon_I(\beta)$ is a resonant point and not an isolated eigenvalue of the combined system, one may have $d\langle \theta_I | \boldsymbol{\omega}(\varepsilon) | \theta_I \rangle / d\varepsilon_I \leq 0$ as well as $d\langle \theta_I | \boldsymbol{\omega}(\varepsilon) | \theta_I \rangle / d\varepsilon_I > 0$ [5]. In particular, in this case one may have $d\varepsilon_I(\beta)/d\beta = \infty$ (see Figure 22 for an example).

13.2. Probabilities Associated with Isolated Cardinal Eigenstates

Once component $|\Psi_I^b\rangle \in X_\rho^b$ of the isolated eigenstate $|\Psi_I\rangle \in X_\infty^c$ is known, one can find all related properties of the open system $\mathbf{B}_\rho$. In particular, in a metrics induced by the operator $\mathbf{S}_b$ probability amplitude to find isolated eigenstate $|\Psi_I\rangle$ in the local state $|\Theta_r\rangle$ equals $\langle \Theta_r | \mathbf{S}_b | \Psi_I \rangle$ [5]. Hence and according to (115), probability $w_{I,r}$ to find isolated eigenstate $|\Psi_I\rangle$ in the local state $|\Theta_r\rangle \in X_\rho^b$ equals

$$w_{I,r} \equiv \left| \langle \Theta_r | \mathbf{S}_b | \Psi_I \rangle \right|^2 = \frac{\langle \theta_I | \mathbf{S}_b | \Theta_r \rangle \langle \Theta_r | \mathbf{S}_b | \theta_I \rangle}{\langle \theta_I | \mathbf{S}_b | \theta_I \rangle - \beta^2 \dfrac{d}{d\varepsilon_I} \langle \theta_I | \boldsymbol{\omega}(\varepsilon) | \theta_I \rangle}, \tag{118a}$$

Since $\{|\Theta_r\rangle\}$ is a complete orthonormalized set in X_ρ^b, probability w_I to find isolated eigenstate $|\Psi_I\rangle$ in the open system $\mathbf{B}_\rho$ equals a sum $\sum_r w_{I,r}$. Hence and due to (5c)

$$w_I \equiv \sum_r w_{I,r} = \frac{\langle \theta_I | \mathbf{S}_b | \theta_I \rangle}{\langle \theta_I | \mathbf{S}_b | \theta_I \rangle - \beta^2 \frac{d}{d\varepsilon_I} \langle \theta_I | \boldsymbol{\omega}(\varepsilon) | \theta_I \rangle}, \tag{118b}$$

According to (115), this equals norm of the X_ρ^b-component of the isolated eigenstate $|\Psi_I\rangle$

$$w_I = \langle \Psi_I^b | \mathbf{S}_b | \Psi_I^b \rangle. \tag{118c}$$

All other properties associated with the open system $\mathbf{B}_\rho$ and with the isolated eigenstate $|\Psi_I\rangle$ can be also expressed in terms of the X_ρ^b-component $|\Psi_I^b\rangle$ of this eigenstate.

14. Embeded Solutions of the Combined System

Embedded solutions of the combined system depend on a continuous parameter ε and they exist for each $\varepsilon \in D$, while for $\varepsilon \notin D$ those solutions do not exist. With each embedded solution is associated a fractional shift $x(\varepsilon,\ldots)$ [5]. As explained in section 11.1, embedded singular solutions satisfy $x(\varepsilon,\ldots) = 0$ while embedded cardinal solutions satisfy $x_d(\varepsilon) \neq 0$. Most important are embedded cardinal solutions.

14.1. Embedded Cardinal Solutions

Embedded cardinal eigenstates of the combined system have non-vanishing fractional shift ($x_d(\varepsilon) \neq 0$) and non-vanishing X_ρ^b component. Hence those eigenstates contribute to the properties of the system $\mathbf{B}_\rho$ that interacts with the infinite system $\boldsymbol{A}_\infty$. Embedded cardinal eigenstates are obtained as a solution of the *fractional shift* eigenvalue equation [5]

$$\mathbf{h}(\varepsilon) | \varphi_d(\varepsilon) \rangle = X_d(\varepsilon) \mathbf{f}(\varepsilon) | \varphi_d(\varepsilon) \rangle, \; d = 1,2,\ldots,r(\varepsilon) \le \rho, \tag{119a}$$

where

$$\mathbf{h}(\varepsilon) = \beta^2 \boldsymbol{\omega}(\varepsilon) + \mathbf{H}_b - \varepsilon \mathbf{S}_b, \; \varepsilon \in D. \tag{119b}$$

Eigenvalues $X_d(\varepsilon)$ of this equation are related to the corresponding fractional shifts $x_d(\varepsilon)$ according to

$$X_d(\varepsilon) = -\pi\beta^2 \cot(\pi x_d(\varepsilon)), \; d = 1,2,\ldots,r(\varepsilon) \le \rho. \tag{119c}$$

Each (finite) eigenvalue $X_d(\varepsilon)$ of (119a) is hence associated with a nonzero fractional shift $x_d(\varepsilon) \neq 0$ and it corresponds to embedded cardinal solutions. Embedded singular solutions have fractional shift zero, and according to (119c) this fractional shift corresponds to the value $X_d(\varepsilon) = \pm\infty$. This value is not an eigenvalue of (119a). Hence fractional shift eigenvalue equation produces only cardinal embedded solutions, as required.

For each $\varepsilon \in D$ fractional shift eigenvalue equation is a generalized eigenvalue equation with a nontrivial operator $\mathbf{f}(\varepsilon)$ on the right hand side of this equation. In the base $\{|r\rangle\}$ of the space X_ρ^b this equation is represented as a $\rho \times \rho$ matrix eigenvalue equation. Given $\varepsilon = \varepsilon_0 \in D$, each eigenvalue $X = X_d(\varepsilon_0)$ of the fractional shift eigenvalue equation is a root of the $\rho \times \rho$ determinant $d(X)$

$$d(X) \equiv \left|\beta^2 \boldsymbol{\omega}(\varepsilon_0) + \mathbf{H}_b - \varepsilon_0 \mathbf{S}_b - X\mathbf{f}(\varepsilon_0)\right| = 0. \qquad (119d)$$

Fractional shift $x_d(\varepsilon)$ in conjuncture with the corresponding eigenstate $|\varphi_d(\varepsilon)\rangle$ of the fractional shift eigenvalue equation determines X_ρ^b-component $|\Psi_d^b(\varepsilon)\rangle$ of the normalized embedded eigenstate $|\Psi_d(\varepsilon)\rangle$ according to [5]

$$|\Psi_d^b(\varepsilon)\rangle = \frac{\sin(\pi x_d(\varepsilon))}{\pi\beta\sqrt{\langle\varphi_d(\varepsilon)|\mathbf{f}(\varepsilon)|\varphi_d(\varepsilon)\rangle}}|\varphi_d(\varepsilon)\rangle, \; d = 1,2,\ldots,r(\varepsilon) \le \rho, \qquad (120a)$$

Using (119c) one can eliminate fractional shift from this expression to obtain

$$|\Psi_d^b(\varepsilon)\rangle = \frac{\beta}{\sqrt{\pi^2\beta^4 + (X_d(\varepsilon))^2}\sqrt{\langle\varphi_d(\varepsilon)|\mathbf{f}(\varepsilon)|\varphi_d(\varepsilon)\rangle}}|\varphi_d(\varepsilon)\rangle. \qquad (120b)$$

Fractional shift eigenvalue equation thus produces X_ρ^b-components of cardinal eigenstates $|\Psi_d(\varepsilon)\rangle$ as well as the corresponding fractional shifts $x_d(\varepsilon)$. Since X_ρ^b-component of the cardinal eigenstate $|\Psi_d(\varepsilon)\rangle$ uniquely determines X_∞^a-component of this eigenstate, this equation determines complete cardinal eigenstate $|\Psi_d(\varepsilon)\rangle$, not only its X_ρ^b-component. However, from the point of view of the open system $\mathbf{B}_\rho$ of interest is only X_ρ^b-component of this eigenstate.

Index d in (120) is used to label multiple fractional shifts associated with degenerate cardinal eigenstates. Fractional shift $x_d(\varepsilon)$ is determined by the eigenvalue $X_d(\varepsilon)$ of the fractional shift eigenvalue equation according to (119c) and it is confined to the interval $[1-\rho,1]$ [5, 43-46].

$$1-\rho \leq x_d(\varepsilon) \leq 1. \tag{121}$$

Since for each integer m one has $\cot(\pi(x_d(\varepsilon)+m)) = \cot(\pi x_d(\varepsilon))$, expression (119c) determines fractional shift up to an additive integer constant. According to the expressions (120), X_ρ^b-component of the embedded cardinal eigenstate $|\Psi_d(\varepsilon)\rangle$ does not depend on this additive constant. One finds that neither X_∞^a-component of the cardinal eigenstate $|\Psi_d(\varepsilon)\rangle$ depends on this constant. Hence one can confine fractional shift $x_d(\varepsilon)$ to the interval $[0,1)$. This is a rational for the introduction of a notion of a *principal value* of a fractional shift as this was done in section 11.1. For all practical purposes one can assume that fractional shift is confined to its principal value. With this convention, fractional shift $x_d(\varepsilon) \neq 0$ corresponds to cardinal embedded eigenstates, while fractional shift $x(\varepsilon,\ldots) = 0$ corresponds to singular embedded eigenstates.

Fractional shift eigenvalue equation reduces to the generic eigenvalue equation in the case $X_d(\varepsilon) = 0$. In this case fractional shift equals $x_d(\varepsilon) = 0.5$. In particular, if $\varepsilon_I \equiv \varepsilon_I(\beta)$ is an eigenvalue of the generic eigenvalue equation (114a) and if $|\theta_I\rangle \equiv |\theta_I(\beta)\rangle$ is the corresponding eigenstate, than there is an eigenstate $|\varphi_d(\varepsilon)\rangle \equiv |\varphi_d(\varepsilon,\beta)\rangle$ of the fractional shift eigenvalue equation (119a) such that $X_d(\varepsilon_I(\beta),\beta) = 0$ and (up to the norm and phase) $|\theta_I(\beta)\rangle \equiv |\varphi_d(\varepsilon_I(\beta),\beta)\rangle$. Each solution to (114a) can be hence considered as a special case of solutions to (119a). Accordingly, one could obtain both, isolated as well as embedded solutions, using only fractional shift eigenvalue equation. However, it is important to have a clear separation between isolated eigenstates which are solutions of the generic eigenvalue equation (114) and embedded eigenstates which are solutions of the fractional shift eigenvalue equation (119). Hence both equations should be used.

14.1.1. Basic Properties of the Solutions to the Fractional Shift Eigenvalue Equation

In a base $\{|r\rangle\} \in X_\rho^b$ fractional shift equation is a $\rho \times \rho$ matrix eigenvalue equation. Standard $\rho \times \rho$ eigenvalue equation involving Hermitean matrix has ρ eigenvalues and ρ orthonormalized eigenstates. However, fractional shift equation is a generalized eigenvalue equation with Hermitean operator $\mathbf{f}(\varepsilon)$ on the right hand side of this equation. Hence for each $\varepsilon \in D$ the number $r(\varepsilon)$ of linearly independent eigenstates $|\varphi_d(\varepsilon)\rangle$ of this equation may be anything from $r(\varepsilon) = 0$ to $r(\varepsilon) = \rho$. As a consequence, one may have at most ρ

linearly independent cardinal eigenstates $\left|\Psi_d(\varepsilon)\right\rangle$ with each particular eigenvalue $\varepsilon \in D$. If $\mathbf{h}(\varepsilon)$ is regular, the number of linearly independent solutions to fractional shift equation equals rank of the operator $\mathbf{f}(\varepsilon)$, i.e. $r(\varepsilon) = rank(\mathbf{f}(\varepsilon))$. Since characteristic operator $\mathbf{f}(\varepsilon)$ is mainly determined by the continuous eigenstates of the infinite system $\boldsymbol{A}_\infty$, $r(\varepsilon)$ will mainly depend on the properties of those eigenstates. In particular, if the point $\varepsilon = \varepsilon_0 \in D$ is contained in a single eigenvalue band ν, one has $\mathbf{f}(\varepsilon_0) = \mathbf{f}_\nu(\varepsilon_0)$. If now eigenstates associated with this eigenvalue band are u-degenerate (i.e. if those eigenstates are of a type $\left|\Phi_{\nu,m}(k)\right\rangle$ where $m = 1,\cdots,u$) and if the modification operator $\mathbf{V}$ is nontrivial, one has almost certainly $rank(\mathbf{f}(\varepsilon_0)) = rank(\mathbf{f}_\nu(\varepsilon_0)) = \min(\rho,u)$. In particular if degeneracy u of those eigenstates is smaller than ρ, one has $rank(\mathbf{f}(\varepsilon_0)) = u$. On the other hand, if $u \geq \rho$ one has $rank(\mathbf{f}(\varepsilon_0)) = \rho$. Similarly, if eigenstates associated with this eigenvalue band are of a type $\left|\Phi_\nu(k,l)\right\rangle$ where l is a continuous parameter, those eigenstates are ∞-degenerate. In this case fractional shift equation has almost certainly ρ eigenstates $\left|\varphi_d(\varepsilon_0)\right\rangle$.

In general, eigenstates $\left|\varphi_d(\varepsilon)\right\rangle$ of the fractional shift eigenvalue equation are not orthogonal to each other. Hence usually $\left\langle\Psi_d^b(\varepsilon)\middle|\mathbf{S}_b\middle|\Psi_{d'}^b(\varepsilon)\right\rangle \neq 0$ ($d \neq d'$). Orthogonal to each other are complete embedded eigenstates $\left|\Psi_d(\varepsilon)\right\rangle$, not necessarily their X_ρ^b-components. However, those components satisfy [5]

$$\left\langle\Psi_d^b(\varepsilon)\middle|\mathbf{f}(\varepsilon)\middle|\Psi_{d'}^b(\varepsilon)\right\rangle = 0,\ d \neq d'. \tag{122}$$

Note further that all $r(\varepsilon)$ embedded eigenstates $\left|\Psi_d(\varepsilon)\right\rangle$ ($d = 1,\ldots,r(\varepsilon)$) have the same eigenvalue ε and hence each linear combination of those eigenstates is again an embedded eigenstate of the combined system with this eigenvalue. However, corresponding linear combination of the eigenstates $\left|\varphi_d(\varepsilon)\right\rangle$ of the fractional shift equation is usually not an eigenstate of this equation. Each eigenstate $\left|\varphi_d(\varepsilon)\right\rangle$ of (119a) has well defined fractional shift $x_d(\varepsilon)$. Hence, unless $x_d(\varepsilon) = x_{d'}(\varepsilon)$, linear combination $a\left|\varphi_d(\varepsilon)\right\rangle + b\left|\varphi_{d'}(\varepsilon)\right\rangle$ is not an eigenstate of (119a) and it has not well defined fractional shift.

14.1.2. Probabilities Associated with Embedded Cardinal Eigenstates

Once X_ρ^b-component $\left|\Psi_d^b(\varepsilon)\right\rangle$ of the embedded cardinal eigenstate $\left|\Psi_d(\varepsilon)\right\rangle$ is known, one can obtain all related properties. For example, probability amplitude to find local state $|\Theta_r\rangle$ in this eigenstate equals $\langle\Theta_r|\mathbf{S}_b|\Psi_d(\varepsilon)\rangle$ [5]. Hence and according to (120b) probability density $\rho_{d,r}(\varepsilon)$ to find this state in the embedded eigenstate $|\Psi_d(\varepsilon)\rangle$ equals

$$\rho_{d,r}(\varepsilon) \equiv \left|\langle\Theta_r|\mathbf{S}_b|\Psi_d(\varepsilon)\rangle\right|^2 = \frac{\beta^2\langle\varphi_d(\varepsilon)|\mathbf{S}_b|\Theta_r\rangle\langle\Theta_r|\mathbf{S}_b|\varphi_d(\varepsilon)\rangle}{\left[\pi^2\beta^4 + (X_d(\varepsilon))^2\right]\langle\varphi_d(\varepsilon)|\mathbf{f}(\varepsilon)|\varphi_d(\varepsilon)\rangle}, \tag{123}$$

One can also consider probability density $\sigma_r(\varepsilon)$ to find local state $|\Theta_r\rangle$ in any of the embedded eigenstates $|\Psi_d(\varepsilon)\rangle$ with the eigenvalue ε. One has

$$\sigma_r(\varepsilon) = \sum_d \rho_{d,r}(\varepsilon) = \sum_d \left|\langle\Theta_r|\mathbf{S}_b|\Psi_d(\varepsilon)\rangle\right|^2, \tag{124a}$$

Of interest is also total probability S_r to find local state $|\Theta_r\rangle$ in any of the embedded eigenstates of the combined system

$$S_r = \int \sigma_r(\varepsilon)d\varepsilon. \tag{124b}$$

In conjuncture with isolated eigenvalues ε_I and corresponding probabilities $w_{I,r}$, probability density $\sigma_r(\varepsilon)$ determines eigenvalue distribution (or spectral distribution) of the local state $|\Theta_r\rangle$. If one measures eigenvalue of the state $|\Theta_r\rangle$ in a closed system $\mathbf{B}_\rho$, one obtains with certainly $E = E_r$. However, if one measures eigenvalue of this state in the open system $\mathbf{B}_\rho$ that interacts with the infinite system $\mathbf{A}_\infty$, one finds each isolated eigenvalue ε_I with the probability $w_{I,r}$ and each embedded eigenvalue $\varepsilon \in D$ with probability density $\sigma_r(\varepsilon)$. Thus instead of a single eigenvalue $E = E_r$, one obtains a spectral distribution which depends on the details of the interaction of the system $\mathbf{B}_\rho$ with its surrounding (system $\mathbf{A}_\infty$).

In analogy to (118b), one can also consider probability density $\rho_d(\varepsilon)$ to find embedded eigenstate $|\Psi_d(\varepsilon)\rangle$ in the open system $\mathbf{B}_\rho$, i.e. to find this eigenstate in any of the states $|\Theta_r\rangle$. Due to (5c) one has

$$\rho_d(\varepsilon) \equiv \sum_r \rho_{d,r}(\varepsilon) = \frac{\beta^2\langle\varphi_d(\varepsilon)|\mathbf{S}_b|\varphi_d(\varepsilon)\rangle}{\left[\pi^2\beta^4 + X_d(\varepsilon)^2\right]\langle\varphi_d(\varepsilon)|\mathbf{f}(\varepsilon)|\varphi_d(\varepsilon)\rangle}, \tag{125a}$$

This probability density equals norm of the X_d^b-component $\left|\Psi_d^b(\varepsilon)\right\rangle$ of the embedded cardinal eigenstate $\left|\Psi_d(\varepsilon)\right\rangle$:

$$\rho_d(\varepsilon) = \left\langle \Psi_d^b(\varepsilon) \middle| \mathbf{S}_b \middle| \Psi_d^b(\varepsilon) \right\rangle. \tag{125b}$$

This expression is similar to the expression (118c) which gives probability w_I to find isolated eigenstate $\left|\Psi_I\right\rangle$ of the combined system in the subsystem $\mathbf{B}_\rho$.

14.1.3. Emergence of Resonance in the Case of the Weak Coupling

If the eigenvalue $X_d(\varepsilon)$ of the fractional shift equation is non-degenerate, one has [5]

$$\frac{\partial X_d(\varepsilon)}{\partial \varepsilon} = \frac{\beta^2 \frac{d}{d\varepsilon}\left\langle \varphi_d(\varepsilon) \middle| \boldsymbol{\omega}(\varepsilon) \middle| \varphi_d(\varepsilon) \right\rangle - X_d(\varepsilon)\frac{d}{d\varepsilon}\left\langle \varphi_d(\varepsilon) \middle| \mathbf{f}(\varepsilon) \middle| \varphi_d(\varepsilon) \right\rangle - \left\langle \varphi_d(\varepsilon) \middle| \mathbf{S}_b \middle| \varphi_d(\varepsilon) \right\rangle}{\left\langle \varphi_d(\varepsilon) \middle| \mathbf{f}(\varepsilon) \middle| \varphi_d(\varepsilon) \right\rangle}. \tag{126a}$$

where $\left|\varphi_d(\varepsilon)\right\rangle$ is eigenstate of the fractional shift eigenvalue equation corresponding to this eigenvalue.

Let the eigenvalue E_r of $\mathbf{B}_\rho$ satisfy $E_r \in D$ and let the corresponding eigenstate $\left|\Theta_r\right\rangle$ be non-degenerate. Let further E_r differ from all discrete eigenvalues λ_i of the infinite system $\mathbf{A}_\infty$. In this case there is one and only one isolated eigenvalue $\varepsilon_I = \varepsilon_I(\beta)$ of a generic equation that for $\beta = 0$ reduces to E_r: $\varepsilon_I(0) = E_r$. The corresponding eigenstate $\left|\theta_I\right\rangle \equiv \left|\theta_I(\beta)\right\rangle$ reduces (up to the norm and phase) to $\left|\Theta_r\right\rangle$: $\left|\theta_I(0)\right\rangle = \left|\Theta_r\right\rangle$. In addition, in a resonant point ε_I corresponding eigenvalue $X_I(\varepsilon)$ of a fractional shift equation vanishes: $X_I(\varepsilon_I) = 0$. Hence and according to (126a), in some small neighboorhood of the point $\varepsilon = \varepsilon_I$ eigenvalue $X_I(\varepsilon)$ can be approximated as

$$X_I(\varepsilon) \approx \frac{\beta^2 \frac{d}{d\varepsilon_I}\left\langle \theta_I \middle| \boldsymbol{\omega}(\varepsilon) \middle| \theta_I \right\rangle - \left\langle \theta_I \middle| \mathbf{S}_b \middle| \theta_I \right\rangle}{\left\langle \theta_I \middle| \mathbf{f}(\varepsilon_I) \middle| \theta_I \right\rangle}(\varepsilon - \varepsilon_I). \tag{126b}$$

This approximation is exact in the point $\varepsilon = \varepsilon_I$ and as $\left|\varepsilon - \varepsilon_I\right|$ increases it deteriorates. If $\mathbf{f}(E_r)\left|\Theta_r\right\rangle \neq 0$ and if β is sufficiently small, components $\left|\Psi_d^b(\varepsilon)\right\rangle \in X_\rho^b$ of embedded eigenstates $|\Psi_d(\varepsilon)\rangle$ can be in some small neighborhood $\Delta(\varepsilon_I)$ of the resonant point $\varepsilon = \varepsilon_I \equiv \varepsilon_I(\beta)$ approximated as [5]

$$|\Psi_d^b(\varepsilon)\rangle \approx$$

$$|\Psi_d^b(\varepsilon)\rangle^\circ = \frac{\beta\sqrt{\langle\theta_I|\mathbf{f}(\varepsilon_I)|\theta_I\rangle}}{\sqrt{\pi^2\beta^4\langle\theta_I|\mathbf{f}(\varepsilon_I)|\theta_I\rangle^2 + a_I^2(\varepsilon-\varepsilon_I)^2}}|\theta_I\rangle\delta_{I,d} \tag{127a}$$

where

$$a_I = \langle\theta_I|\mathbf{S}_b|\theta_I\rangle - \beta^2\frac{d}{d\varepsilon_I}\langle\theta_I|\boldsymbol{\omega}(\varepsilon)|\theta_I\rangle,\ \varepsilon\in\Delta(\varepsilon_I). \tag{127b}$$

In particular, if $\varepsilon\in\Delta(\varepsilon_I)$ and if $d \neq I$ one has $|\Psi_d^b(\varepsilon)\rangle \approx 0$. According to (127a), close to the point $\varepsilon = \varepsilon_I$ probability density $\rho_d(\varepsilon) = \langle\Psi_d^b(\varepsilon)|\mathbf{S}_b|\Psi_d^b(\varepsilon)\rangle$ to find embedded cardinal eigenstate $|\Psi_d(\varepsilon)\rangle$ in the open system $\mathbf{B}_\rho$ can be approximated as

$$\rho_d(\varepsilon) \approx \rho_d^0(\varepsilon) = \rho_I^0(\varepsilon)\delta_{d,I},\ \varepsilon\in\Delta(\varepsilon_I), \tag{128a}$$

where

$$\rho_I^0(\varepsilon) = \frac{\beta^2\langle\theta_I|\mathbf{f}(\varepsilon_I)|\theta_I\rangle\langle\theta_I|\mathbf{S}_b|\theta_I\rangle}{\pi^2\beta^4\langle\theta_I|\mathbf{f}(\varepsilon_I)|\theta_I\rangle^2 + a_I^2(\varepsilon-\varepsilon_I)^2}. \tag{128b}$$

and where a_I is given by (127b). Density $\rho_I^0(\varepsilon)$ represents a universal resonance curve [48] centered at the point $\varepsilon = \varepsilon_I(\beta)$ and with the width $\Delta\varepsilon_I(\beta)$

$$\Delta\varepsilon_I = \frac{2\pi\beta^2\langle\theta_I|\mathbf{f}(\varepsilon_I)|\theta_I\rangle}{\langle\theta_I|\mathbf{S}_b|\theta_I\rangle - \beta^2\frac{d}{d\varepsilon_I}\langle\theta_I|\boldsymbol{\omega}(\varepsilon)|\theta_I\rangle}, \tag{129a}$$

Height of this curve equals $\rho_I^0(\varepsilon_I)$ and it has the area $w_I^0 = \int\rho_I^0(\varepsilon)d\varepsilon$

$$\rho_I^0(\varepsilon_s) = \frac{\langle\theta_I|\mathbf{S}_b|\theta_I\rangle}{\pi^2\beta^2\langle\theta_I|\mathbf{f}(\varepsilon_I)|\theta_I\rangle}, \tag{129b}$$

$$w_I^0 = \frac{\langle \theta_I | \mathbf{S}_b | \theta_I \rangle}{\langle \theta_I | \mathbf{S}_b | \theta_I \rangle - \beta^2 \frac{d}{d\varepsilon_I} \langle \theta_I | \boldsymbol{\omega}(\varepsilon) | \theta_I \rangle}. \tag{129c}$$

Since β is small, this implies $w_I^0 \approx 1$.

The area w_I^0 is an approximation of a total probability to find any of the embedded cardinal eigenstates $|\Psi_d(\varepsilon)\rangle$ where $\varepsilon \in \Delta(\varepsilon_I)$ in the system $\mathbf{B}_\rho$. Expression (129c) is formally identical to the expression (118b) for the probability to find isolated eigenstate $|\Psi_I\rangle$ in a local system $\mathbf{B}_\rho$. Similar expression applies to the probability density $\sigma_I(\varepsilon)$ to find the state $|\Theta_r\rangle \in X_\rho^b$ in any of the embedded cardinal eigenstates with the eigenvalue $\varepsilon \in \Delta(\varepsilon_I)$. In an explicit notation, in order to emphasize dependence on the coupling parameter β, one should write $|\theta_I\rangle \equiv |\theta_I(\beta)\rangle$. Since $|\theta_I(0)\rangle = |\Theta_r\rangle$ and since β is small, one has $|\theta_I(\beta)\rangle \approx |\Theta_r\rangle$. Hence $\sigma_I(\varepsilon) \approx \rho_I(\varepsilon)$.

In conclusion, if $E_r \in D$ is non-degenerate and if $\mathbf{f}(E_r)|\Theta_r\rangle \neq 0$, for sufficiently small interaction parameter β there is an embedded eigenstate $|\Psi_I(\varepsilon)\rangle$ with the property that X_ρ^b-component $|\Psi_I^b(\varepsilon)\rangle$ of this eigenstate exhibits a strong resonance feature in a small neighborhood of the resonant point $\varepsilon = \varepsilon_I(\beta)$, while X_ρ^b-components of all other embedded eigenstates $|\Psi_d(\varepsilon)\rangle$ ($d \neq I$) are negligible in this neighborhood [5]. In particular, probability density $\sigma_I(\varepsilon)$ to find local state $|\Theta_r\rangle$ with the eigenvalue $\varepsilon \in D$ exhibits a strong resonance feature in the interval $\Delta(\varepsilon_I)$. The same applies to the probability density $\rho_I(\varepsilon)$, and if β is sufficiently small one has $\rho_I(\varepsilon) \approx \sigma_I(\varepsilon)$. This property is a rational for the name "resonant point" [5].

Similar results are obtained if $E_r \in D$ is degenerate. However, in this case there are several eigenstates $|\theta_I(\beta)\rangle$ of a generic eigenvalue equation that satisfy $|\theta_I(0)\rangle = |\Theta_r\rangle$ and one may have several closed spaced resonant structures corresponding to the degeneracy of E_r [5].

Emergence of resonance in the case of the weak coupling is usually derived within the formalism of the time-independent perturbation expansion approach [1, 2]. If a system of particles interacts weakly with the field (such as e.g. electromagnetic field), there are two main effects of such an interaction. First, due to this interaction each eigenvalue E_r of this system shifts to a new position $\varepsilon_I \equiv \varepsilon_I(\beta)$ where $\varepsilon_I(0) = E_r$. Second, this shifted eigenvalue is not sharp and it acquires the shape of the universal resonance curve [1]. As shown above, both effects are correctly described within the small β limit of the LRM approach.

14.1.4. Anomal Points and Isolated Solutions

As shown in a previous section, if β is sufficiently small the width $\Delta\varepsilon_I(\beta)$ of the universal resonance curve (128b), as calculated according to (129a), is small. This implies emergence of resonance in the case of small β. However, if $\langle\theta_I|\mathbf{f}(\varepsilon_I)|\theta_I\rangle$ is sufficiently small one may have small width $\Delta\varepsilon_I(\beta)$ of a resonant shape even for large β. Hence open system $\mathbf{B}_\rho$ may display a strong resonant feature even in the case of strong interaction with its surrounding. Such an extreme case is the case of anomal point $\beta=\beta_a$ where $\varepsilon_a=\varepsilon_I(\beta_a)$. In the anomal point there is at least one eigenstate $|\theta_I\rangle$ of the generic eigenvalue equation such that $\mathbf{h}(\varepsilon_a)|\theta_I\rangle=0$ and $\mathbf{f}(\varepsilon_a)|\theta_I\rangle=0$. This implies $\langle\theta_I|\mathbf{f}(\varepsilon_a)|\theta_I\rangle=0$ and hence formally $\Delta\varepsilon_I(\beta)=0$. Corresponding eigenvalue $X_d(\varepsilon_a)$ of a fractional shift eigenvalue equation can assume any value and fractional shift $x_I(\varepsilon_a)$ is hence not well defined [5]. This case can be analyzed as a $\beta\to\beta_a$ limit of the quantity $\langle\theta_I|\mathbf{f}(\varepsilon_I)|\theta_I\rangle$. Let $\varepsilon_I=\varepsilon_I(\beta)$ be nondegenerate in the point $\beta=\beta_a$. In the process $\beta\to\beta_a$ the width $\Delta\varepsilon_I(\beta)$ of resonance curve (128b) decreases, approximate probability (129c) improves, and in a limit $\Delta\varepsilon_I(\beta)\to 0$ it is exact. In this limit resonant shape at the anomal point $\varepsilon_a=\varepsilon_I(\beta_a)$ becomes infinitely narrow and infinitely high. Unlike in the case of small β, probability (129c) does not approximate to $w_I^0\approx 1$, and one has to use full expression (129c). In the anomal point this expression is exact and it is identical to the expression (118b). Formal similarity between expressions (129c) and (118b) is not accidental. One obtains qualitatively the same result in the case when $\varepsilon_I(\beta)$ is degenerate in the point $\beta=\beta_a$. Such infinitely narrow and infinitely high densities correspond to one or several isolated eigenstates [5]. One finds that those eigenstates are given by expressions (115) [5]. Thus all cardinal isolated eigenstates, those with eigenvalues $\varepsilon_I\notin D$ as well as those with eigenvalues $\varepsilon_I\in D$, are described in the same way.

14.2. Embedded Singular Solutions

Expressions (15) that apply to singular solutions of a finite system $\mathbf{C}_{n+\rho}$ imply analogous expressions for the embedded singular solutions of the infinite system $\mathbf{C}_\infty$:

$$\left[\beta^2\boldsymbol{\omega}(\varepsilon)+\mathbf{H}_b-\varepsilon\mathbf{S}_b\right]\left|\varphi(\varepsilon,\ldots)\right\rangle+\beta\mathbf{V}\mathbf{I}^a(\varepsilon)\left|\phi(\varepsilon,\ldots)\right\rangle=0, \tag{130a}$$

$$\mathbf{I}^a(\varepsilon)\mathbf{V}\left|\varphi(\varepsilon,\ldots)\right\rangle=0, \tag{130b}$$

where

$$\left|\varphi(\varepsilon,\ldots)\right\rangle\in X^b_\rho,\ \left|\phi(\varepsilon,\ldots)\right\rangle\in X^a(\varepsilon). \tag{130c}$$

and where $X^a(\varepsilon)\subset X^a_\infty$ is the space spanned by all those continuous eigenstates of the system $\mathbf{A}_\infty$ that have eigenvalue ε. Operator $\mathbf{I}^a(\varepsilon)$ is a projection operator on this space.

In analogy to a finite system $\mathbf{C}_{n+\rho}$, infinite system $\mathbf{C}_\infty$ may contain embedded strongly singular eigenstates which satisfy $\left|\varphi(\varepsilon,\ldots)\right\rangle=0$ and which have no X^b_ρ-component, as well as embedded weakly singular eigenstates which satisfy $\left|\varphi(\varepsilon,\ldots)\right\rangle\neq 0$ and which have a non-vanishing X^b_ρ-component. In the case of embedded strongly singular eigenstates expressions (130) reduce to

$$\mathbf{V}\mathbf{I}^a(\varepsilon)\left|\phi(\varepsilon,\ldots)\right\rangle=0, \tag{131a}$$

The corresponding normalized strongly singular eigenstate of the infinite-dimensional combined system $\mathbf{C}_\infty$ is

$$\left|\Psi(\varepsilon,\ldots)\right\rangle\equiv\left|\Psi^a(\varepsilon,\ldots)\right\rangle=\frac{1}{\sqrt{\left\langle\phi(\varepsilon,\ldots)\right|\left.\phi(\varepsilon,\ldots)\right\rangle}}\left|\phi(\varepsilon,\ldots)\right\rangle\in X^a(\varepsilon). \tag{131b}$$

Combined system may have an infinite number of such eigenstates associated with each $\varepsilon\in D$. For example, if the system $\boldsymbol{A}_\infty$ is described by the eigenvalue equation

$$\mathbf{H}_a\left|\Phi(k,l)\right\rangle=\lambda(k)\left|\Phi(k,l)\right\rangle,\ k\in\left[k_a,k_b\right], \tag{132a}$$

where l is a continuous parameter, each eigenvalue $\lambda(k)$ is ∞-degenerate. In this case the space $X^a(\varepsilon)$ is an ∞-dimensional space spanned by ∞ eigenstates $\left|\Phi(k,l)\right\rangle$ where $k=\lambda^{-1}(\varepsilon)$. Projection $\mathbf{I}^a(\varepsilon)$ on this space can be written as

$$\mathbf{I}^a(\varepsilon)=\int\left|\Phi(k,l)\right\rangle\left\langle\Phi(k,l)\right|dl\ ,\ k=\lambda^{-1}(\varepsilon), \tag{132b}$$

Each state $\left|\phi(\varepsilon,\ldots)\right\rangle\in X^a(\varepsilon)$ is hence a linear combination

$$\left|\phi(\varepsilon,\ldots)\right\rangle = \int c(l)\left|\Phi(k,l)\right\rangle dl\ ,\ k = \lambda^{-1}(\varepsilon)\ , \tag{133a}$$

where unknown function $c(l)$ satisfies

$$\int c(l)\left\langle s\right|\mathbf{V}\left|\Phi(k,l)\right\rangle dl = 0\ ,\ s = 1,\ldots,\rho\ ,\ k = \lambda^{-1}(\varepsilon)\ . \tag{133b}$$

This generalizes expressions (19). For each particular eigenvalue $\varepsilon \in D$ one has an infinite number of the solutions to (133b) and hence an infinite number of strongly singular eigenstates (131b) of the combined system. All those eigenstates of the combined system $\mathbf{C}_\infty$ are at the same time eigenstates of the parent system $\mathbf{A}_\infty$ with the same eigenvalue ε . Since those strongly singular eigenstates have no X_ρ^b-component, they do not contribute to the properties of the open system $\mathbf{B}_\rho$. In fact those eigenstates do not "see" the base system $\mathbf{B}_\rho$ since due to (133b) each such eigenstate $\left|\Psi(\varepsilon,\ldots)\right\rangle$ satisfies

$$\left\langle s\right|\mathbf{V}\left|\Psi(\varepsilon,\ldots)\right\rangle = 0\ ,\ s = 1,\ldots,\rho\ . \tag{133c}$$

Hence no strongly singular eigenstate $\left|\Psi(\varepsilon,\ldots)\right\rangle$ interacts with any state of the base system $\mathbf{B}_\rho$.

Concerning weakly singular eigenstates, one again finds that, unless the combined system $\mathbf{C}_\infty$ has some symmetry not contained in the systems $\mathbf{B}_\rho$ and $\mathbf{A}_\infty$, it contains no such eigenstates.

15. Completeness Relations

As emphasized above, non-vanishing X_ρ^b-component have only cardinal and weakly singular eigenstates of the combined system. Strongly singular eigenstates have no X_ρ^b-component. Hence to the properties of the open system $\mathbf{B}_\rho$ that interacts with the infinite system $\mathbf{A}_\infty$ can contribute only cardinal and weakly-singular eigenstates of the combined system. In addition, unless the combined system $\mathbf{C}_\infty$ has some symmetry not contained in its isolated subsystems $\mathbf{B}_\rho$ and $\boldsymbol{A}_\infty$, it has no weakly singular eigenstates. Hence in most cases isolated cardinal and embedded cardinal solutions are sufficient to provide complete description of the open system $\mathbf{B}_\rho$. In particular, if the combined system contains no weakly singular eigenstate, each state $|\Theta_r\rangle \in X_\rho^b$ must be found with certainty either in some isolated cardinal eigenstate of the combined system, or in some embedded cardinal eigenstate of this system. Hence and from (118) and (124)

$$\sum_I w_{I,r} + S_r = 1\,,\; r = 1,\ldots,\rho\,, \tag{134a}$$

where

$$w_{I,r} = \left|\langle\Theta_r|\mathbf{S}_b|\Psi_I\rangle\right|^2,\; S_r = \sum_d \int \left|\langle\Theta_r|\mathbf{S}_b|\Psi_d(\varepsilon)\rangle\right|^2 d\varepsilon\,, \tag{134b}$$

Quantity $w_{I,r}$ is a probability to find local state $|\Theta_r\rangle$ in the isolated cardinal eigenstate $|\Psi_I\rangle$ of the combined system, while S_r is a probability to find this state in any of the embedded cardinal eigenstate $|\Psi_d(\varepsilon)\rangle$ of this system. Those quantities are given by expressions (118a), (123) and (124). If the combined system contains no weakly singular eigenstates, expression (134a) should be satisfied. However, if combined system contains some weakly singular eigenstates, this expression should produce result which is less than one for at least one $r = 1,\ldots,\rho$. If this is the case, it indicates that combined system contains some weakly singular eigenstate which should be included in the calculation [5].

Summing expression (134a) over r one finds

$$\sum_I w_I + \sum_r S_r = \rho\,. \tag{134c}$$

where probabilities $w_I = \sum_r w_{I,r}$ are given by expression (118b).

Expressions (134) are *completeness relations*. Those expressions are satisfied for each value of the coupling parameter β, however large. Significance of those expressions is twofold. First, those expressions provide an efficient test for the validity of the LRM approach as applied to infinite-dimensional systems. Second, provided LRM is shown to be correct, those expressions present a simple test to verify whether combined system contains any weakly singular eigenstate [5].

16. One Dimensional Open Systems

A very important special case of open quantum systems is the case $\rho = 1$. In this case system $\mathbf{B}_1$ contains a single state $|\Theta\rangle$ with the eigenvalue E. This state interacts with an infinite-dimensional system $\mathbf{A}_\infty$. This situation is shown schematically in Figure 12. A large part of problems associated with the interaction of matter with radiation is exactly of this type: one considers for example a single molecular level $|\Theta\rangle$ that interacts with the radiation. Due to this interaction the corresponding eigenvalue is shifted to a new position and it is not sharp, but it acquires some natural width. Of interest are also various transition probabilities associated with this state, etc. Such and similar problems are usually treated within the perturbation expansion formalism [1, 2].

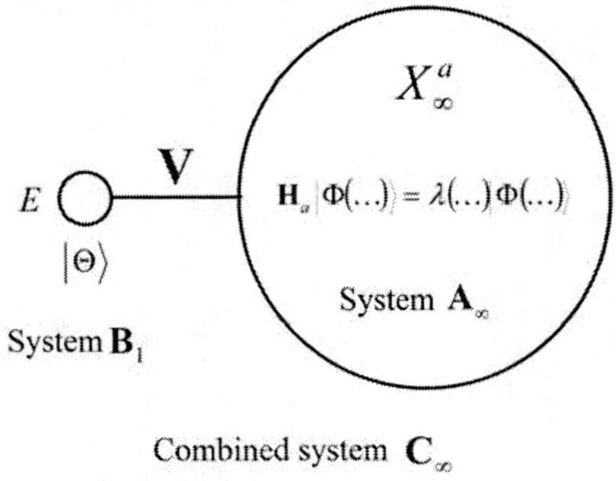

Figure 12. External modification of a single state $|\Theta\rangle$ by an infinite-dimensional system.

Consider now LRM treatment of such systems. Since the system $\mathbf{B}_1$ is one-dimensional, general LRM expressions substantially simplify. In particular, since $\rho = 1$ characteristic operator $\mathbf{f}(\varepsilon)$ reduces to a single matrix element and it becomes a characteristic function $f(\varepsilon)$. Similarly, derived operator $\boldsymbol{\omega}(\varepsilon)$ becomes derived function $\omega(\varepsilon)$. This derived function is expressed in terms of the characteristic function $f(\varepsilon)$ according to

$$\omega(\varepsilon) = P\int \frac{f(\lambda)}{\varepsilon - \lambda} d\lambda . \tag{135}$$

Concerning isolated cardinal solutions, generic equation (114) reduces to

$$\beta^2 \omega(\varepsilon) + E - \varepsilon = 0 , \tag{136a}$$

while condition (116) translates into

$$f(\varepsilon_I) = 0. \tag{136b}$$

Each root $\varepsilon = \varepsilon_I$ of (136a) that satisfies (136b) is an isolated eigenvalue of the combined system $\mathbf{C}_\infty$. According to (115), X_1^b-component of the corresponding eigenstate $|\Psi_I\rangle$ is

$$\left|\Psi_I^b\right\rangle = \frac{1}{\sqrt{N_I}}\left|\Theta\right\rangle. \tag{137a}$$

where

$$N_I = 1 - \beta^2 d\omega / d\varepsilon_I, \tag{137b}$$

and where $\left|\Theta\right\rangle$ is normalized ($\left\langle\Theta|\Theta\right\rangle = 1$). The corresponding X_∞^a-component $\left|\Psi_I^a\right\rangle$ of the isolated state $\left|\Psi_I\right\rangle$ is given by expression (115c) where $\left|\theta_I\right\rangle = \left|\Theta\right\rangle$.

According to (137), probability w_I to find isolated cardinal eigenstate $\left|\Psi_I\right\rangle$ in a state $\left|\Theta\right\rangle$ (system $\mathbf{B}_1$) equals

$$w_I \equiv \left|\left\langle\Theta|\Psi_I\right\rangle\right|^2 = \frac{1}{1-\beta^2\, d\omega(\varepsilon)/d\varepsilon_I}. \tag{137c}$$

Consider now embedded eigenstates. Since $\rho = 1$, for each $\varepsilon \in D$ one has only one fractional shift $x(\varepsilon)$. Fractional shift eigenvalue equation (119) hence reduces to

$$h(\varepsilon) \equiv \beta^2\omega(\varepsilon) + E - \varepsilon = X(\varepsilon) f(\varepsilon),\ \varepsilon \in D. \tag{138}$$

where $X(\varepsilon)$ is given by (119c). One finds

$$x(\varepsilon) = \frac{1}{\pi}\cot^{-1}\left(\frac{\varepsilon - E - \beta^2\omega(\varepsilon)}{\pi\beta^2 f(\varepsilon)}\right),\ \varepsilon \in D. \tag{139}$$

Further, expressions (120) reduce to

$$\left|\Psi^b(\varepsilon)\right\rangle = \frac{\sin(\pi x(\varepsilon))}{\pi\beta\sqrt{f(\varepsilon)}}\left|\Theta\right\rangle,\ \varepsilon \in D, \tag{140a}$$

$$\left|\Psi^b(\varepsilon)\right\rangle = \left[\frac{\beta^2 f(\varepsilon)}{\pi^2\beta^4 f^2(\varepsilon) + \left(\beta^2\omega(\varepsilon) + E - \varepsilon\right)^2}\right]^{1/2}\left|\Theta\right\rangle,\ \varepsilon \in D, \tag{140b}$$

Probability density $\rho(\varepsilon)$ to find embedded eigenstate $\left|\Psi(\varepsilon)\right\rangle$ in the state $\left|\Theta\right\rangle$ hence equals

$$\rho(\varepsilon) \equiv \left|\left\langle\Theta|\Psi(\varepsilon)\right\rangle\right|^2 = \frac{\beta^2 f(\varepsilon)}{\pi^2\beta^4 f^2(\varepsilon) + \left(\beta^2\omega(\varepsilon) + E - \varepsilon\right)^2},\ \varepsilon \in D. \tag{140c}$$

This probability density is the same as probability density to find the state $|\Theta\rangle$ with the (embedded) eigenvalue $\varepsilon \in D$. Since $\mathbf{B}_1$ is one-dimensional and since as a consequence for each $\varepsilon \in D$ there is only one embedded cardinal eigenstate $|\Psi(\varepsilon)\rangle$, one has $\rho(\varepsilon)=\sigma(\varepsilon)$ (see expressions (123-125).

Let $\varepsilon=\varepsilon_I \in D$ be a resonant point. By definition, $\varepsilon=\varepsilon_I$ is a root of (136a) that satisfies $f(\varepsilon_I)\neq 0$. According to (140c), in a resonant point one has

$$\rho(\varepsilon_I)\equiv\left|\langle\Theta|\Psi(\varepsilon_I)\rangle\right|^2=\frac{1}{\pi^2\beta^2 f(\varepsilon_I)},\quad \varepsilon_I \in D. \tag{140d}$$

If β is small, this density can be very large. In this case one can approximate density $\rho(\varepsilon)$ in a neighborhood of a resonant point $\varepsilon_I \in D$ according to

$$\rho(\varepsilon)\approx\rho_I^0(\varepsilon)=\frac{\beta^2 f(\varepsilon_I)}{\pi^2\beta^4 f^2(\varepsilon_I)+(\varepsilon-\varepsilon_I)^2},\quad \varepsilon\in D. \tag{141a}$$

This is a universal resonance curve centered at the point $\varepsilon=\varepsilon_I$, with the width

$$\Delta\varepsilon_I=2\pi\beta^2 f(\varepsilon_I), \tag{141b}$$

and with a unit area. If β is small the width $\Delta\varepsilon_I$ is also small. Approximation (141a) of the density $\rho(\varepsilon)$ in the neighborhood of the resonance point $\varepsilon=\varepsilon_I$ is hence reliable. The same expression can be obtained from the general discussion in section 14.1.3 and from the expression (128b). Since β is small and since X_ρ^b is one-dimensional, one can approximate quantity a_I with $a_I \approx 1$ and in addition one has $\langle\theta_I|f(\varepsilon_I)|\theta_I\rangle \to f(\varepsilon_I)$. Hence (128b) implies (141a). In addition, according to (136a), as a first approximation resonant point $\varepsilon=\varepsilon_I$ equals

$$\varepsilon_I\approx\varepsilon_I^0=E+\beta^2\omega(E), \tag{141c}$$

As a next approximation one has

$$\varepsilon_I\approx\varepsilon_I^1=E+\beta^2\omega(\varepsilon_I^0). \tag{141d}$$

which iteratively produces exact value of ε_I. Above LRM expressions provide a simple quantitative explanation of two main effects of the interaction of a single state with an infinite system: the shift of the original eigenvalue E to a new position ε_I, and the broadening of this eigenvalue to the resonance shape. For this to happen shifted eigenvalue ε_I (resonant point) must be contained inside the eigenvalue band D of the continuous eigenvalues of the infinite system $\mathbf{A}_\infty$ and the width $\Delta\varepsilon_I$ as calculated according to (141b) must be relatively small.

17. Examples

Let me illustrate LRM treatment of infinite-dimensional systems with some examples. In order to cover various aspects of open quantum systems, three different quantum systems in the interaction with various surroundings will be considered.

17.1. Example E1

Consider the system $\mathbf{B}_2$ that is in the base $\{|r\rangle\}$ characterized by matrices

$$\mathbf{H}_b = \begin{pmatrix} 0.5 & -0.25 \\ -0.25 & 0.6 \end{pmatrix}, \quad \mathbf{S}_b = \begin{pmatrix} 1.1 & 0.1 \\ 0.1 & 1 \end{pmatrix}, \tag{142a}$$

Eigenvalues and eigenstates of this system are

$$E_1 = 0.25474, \quad E_2 = 0.85536, \tag{142b}$$

$$|\Theta_1\rangle = \begin{pmatrix} 0.72621 \\ 0.57942 \end{pmatrix}, \quad |\Theta_2\rangle = \begin{pmatrix} 0.62454 \\ -0.82064 \end{pmatrix}. \tag{142c}$$

where $|\Theta_r\rangle$ are orthonormalized according to (5b).

Let the surrounding of the system $\mathbf{B}_2$ contain a single eigenvalue band in the interval $D \equiv I_1 = [a_1, b_1] = [-1, 1]$ and let this surrounding contain no isolated eigenvalues λ_i. Interaction of the system $\mathbf{B}_2$ with this eigenvalue band is described by the characteristic operator $\mathbf{f}_1(\varepsilon)$ which is identically zero outside the interval I_1 and which is nonnegative inside this interval. Let in the base $\{|r\rangle\}$ this operator be given by

$$\mathbf{f}_1(\varepsilon) = \begin{pmatrix} f_{11}^{(1)}(\varepsilon) & f_{12}^{(1)}(\varepsilon) \\ f_{21}^{(1)}(\varepsilon) & f_{22}^{(1)}(\varepsilon) \end{pmatrix} \equiv \begin{pmatrix} q_{11}^{(1)}(\varepsilon) & q_{12}^{(1)}(\varepsilon) \\ q_{21}^{(1)}(\varepsilon) & q_{22}^{(1)}(\varepsilon) \end{pmatrix} \cdot \begin{cases} 1 & if \quad \varepsilon \in I_1 \\ 0 & otherwise \end{cases} \tag{143a}$$

where $q_{rt}^{(1)}(\varepsilon)$ are polynomials

$$q_{11}^{(1)}(\varepsilon) = (\varepsilon^2 - 1)^2 (2\varepsilon^2 - 2\varepsilon + 1),$$

$$q_{22}^{(1)}(\varepsilon) = (\varepsilon^2 - 1)^2(\varepsilon^2 - 2\varepsilon + 2), \tag{143b}$$

$$q_{12}^{(1)}(\varepsilon) \equiv q_{21}^{(1)}(\varepsilon) = (\varepsilon^2 - 1)^3 .$$

Matrix elements $f_{rt}^{(1)}(\varepsilon)$ are identical to polynomials $q_{rt}^{(1)}(\varepsilon)$ inside the interval $I_1 = [-1,1]$, and they are identically zero outside this interval. Since $\boldsymbol{A}_\infty$ contains only this single eigenvalue band, one has $\mathbf{f}(\varepsilon) \equiv \mathbf{f}_1(\varepsilon)$ and hence $f_{rt}(\varepsilon) \equiv f_{rt}^{(1)}(\varepsilon)$. In Figure 13 are shown eigenvalues $f_1(\varepsilon)$ and $f_2(\varepsilon)$ of the characteristic matrix $\mathbf{f}(\varepsilon)$. Since polynomials $q_{rt}^{(1)}(\varepsilon)$ are zero on both end-points of the interval I_1, those eigenvalues are continuous at those end-points and they vanish outside the range $D \equiv I_1$. As required, eigenvalues $f_r(\varepsilon)$ are nonnegative. Rank of matrix $\mathbf{f}(\varepsilon)$ equals two for each interior point $\varepsilon \in D$, except for the critical point $\varepsilon = \varepsilon_c = 0.38197$ where $rank(\mathbf{f}(\varepsilon_c)) = 1$ and where eigenvalue $f_2(\varepsilon)$ of $\mathbf{f}(\varepsilon)$ vanishes: $f_2(\varepsilon_c) = 0$.

$$rank(\mathbf{f}(\varepsilon)) = \begin{cases} 2 \; if \; \varepsilon \in D \;\&\; \varepsilon \neq \varepsilon_c \\ 1 \; if \; \varepsilon = \varepsilon_c = 0.38197 \\ 0 \; if \; \varepsilon \in \overline{D} \end{cases} \tag{144}$$

Since for almost each $\varepsilon \in D$ one has $rank(\mathbf{f}(\varepsilon)) > 1$, this characteristic matrix corresponds to multi-parameter eigenvalue band.

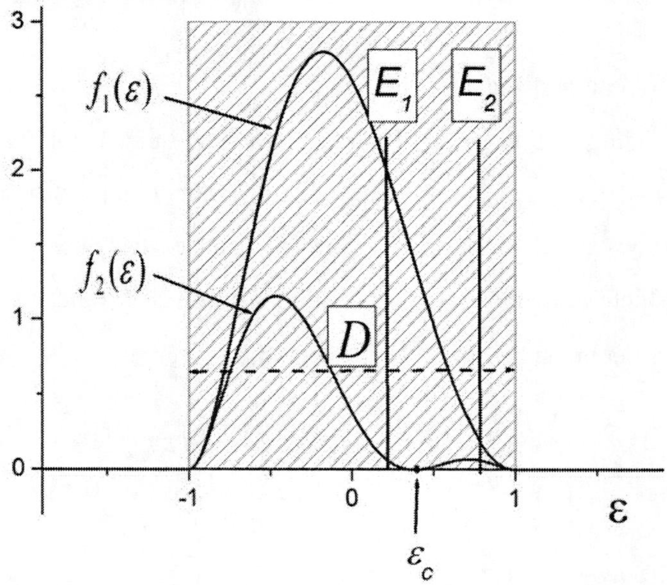

Figure 13. Example E1. Eigenvalues of the characteristic operator $\mathbf{f}(\varepsilon)$. Both eigenvalues E_r of the isolated base system $\mathbf{B}_2$ are inside the range D. Adopted from ref. [5].

Derived matrix $\boldsymbol{\omega}_1(\varepsilon)$ that corresponds to the characteristic matrix $\mathbf{f}_1(\varepsilon)$ is given by expressions (103). In particular, since matrix elements of $\mathbf{f}_1(\varepsilon)$ are polynomials inside the interval $D \equiv I_1$, one can use expressions (108) to obtain [5]:

$$\omega_{11}^{(1)}(\varepsilon) = q_{11}^{(1)}(\varepsilon)\ln\left|\frac{\varepsilon+1}{\varepsilon-1}\right| - g_{11}^{(1)}(\varepsilon),$$

$$\omega_{22}^{(1)}(\varepsilon) = q_{22}^{(1)}(\varepsilon)\ln\left|\frac{\varepsilon+1}{\varepsilon-1}\right| - g_{22}^{(1)}(\varepsilon), \tag{145a}$$

$$\omega_{12}^{(1)}(\varepsilon) \equiv \omega_{21}^{(1)}(\varepsilon) = q_{12}^{(1)}(\varepsilon)\ln\left|\frac{\varepsilon+1}{\varepsilon-1}\right| - g_{12}^{(1)}(\varepsilon),$$

where $q_{rt}^{(1)}(\varepsilon)$ are polynomials (143b), while $g_{rt}^{(1)}(\varepsilon)$ are polynomials:

$$g_{11}^{(1)}(\varepsilon) = 4\varepsilon^5 - 4\varepsilon^4 - \frac{14}{3}\varepsilon^3 + \frac{20}{3}\varepsilon^2 - \frac{6}{5}\varepsilon - \frac{32}{15},$$

$$g_{22}^{(1)}(\varepsilon) = 2\varepsilon^5 - 4\varepsilon^4 + \frac{2}{3}\varepsilon^3 + \frac{20}{3}\varepsilon^2 - \frac{28}{5}\varepsilon - \frac{32}{15}, \tag{145b}$$

$$g_{12}^{(1)}(\varepsilon) \equiv g_{21}^{(1)}(\varepsilon) = 2\varepsilon^5 - \frac{16}{3}\varepsilon^3 + \frac{22}{5}\varepsilon.$$

Since $\mathbf{f}(\varepsilon) \equiv \mathbf{f}_1(\varepsilon)$ one has $\boldsymbol{\omega}(\varepsilon) \equiv \boldsymbol{\omega}_1(\varepsilon)$. In Figure 14 are shown eigenvalues $\omega_1(\varepsilon)$ and $\omega_2(\varepsilon)$ of the derived matrix $\boldsymbol{\omega}(\varepsilon)$. Those eigenvalues are finite and continuous functions of ε for each real ε, including end-points $a_1 = -1$ and $b_1 = 1$ of the interval I_1. This follows from the property $\lim_{x\to 0}(x\ln(x)) = 0$ and from the fact that matrix elements of the operator $\mathbf{f}_1(\varepsilon)$ vanish at those end-points [5]. In accord with the expression (107b), both eigenvalues are decreasing functions of ε for each $\varepsilon \notin D = [-1,1]$. However, if $\varepsilon \in D$ those eigenvalues may be decreasing as well as increasing functions of ε.

Once operator $\boldsymbol{\omega}(\varepsilon)$ is known, one can write down generic eigenvalue equation (114). This equation produces all cardinal isolated solutions of the combined system and all resonant points. Solutions of the equation (114b) of a type $\varepsilon = \varepsilon(\beta)$ are shown in Figure 15. There are six such solutions: $\varepsilon_{R1}(\beta)$, $\varepsilon_{R2}(\beta)$, $\varepsilon_1(\beta)$, $\varepsilon_2(\beta)$, $\varepsilon_{L1}(\beta)$ and $\varepsilon_{L2}(\beta)$. Those solutions are eigenvalues of a generic eigenvalue equation considered as functions of the coupling parameter β. Eigenvalues inside the range D are resonant points, while eigenvalues outside this range are isolated eigenvalues of the combined system.

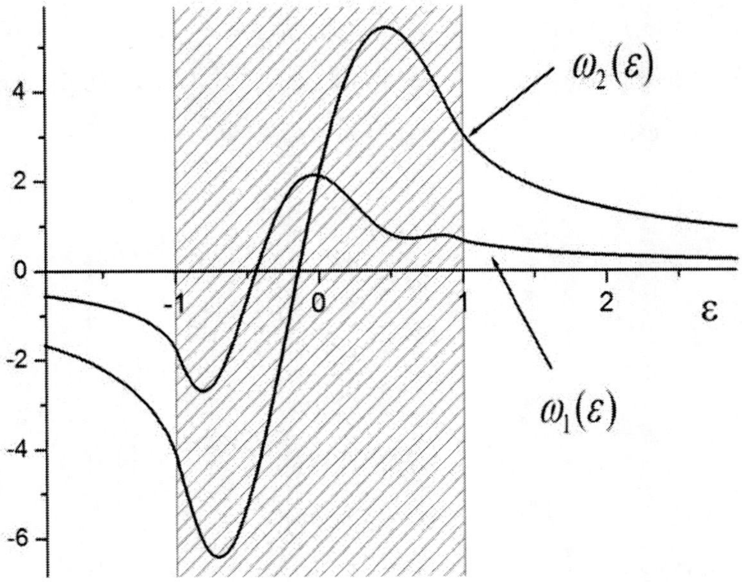

Figure 14. Example E1. Eigenvalues of the derived operator $\boldsymbol{\omega}(\varepsilon)$. Adopted from ref. [5].

If there is no interaction ($\beta = 0$) eigenvalues of the generic equation reduce to the eigenvalues E_r of the isolated base system $\mathbf{B}_2$. In particular, one has $\varepsilon_{R1}(0) = E_1$ and $\varepsilon_{R2}(0) = E_2$. Both eigenvalues are contained in the range D. Hence for small β one has $\varepsilon_{R1}(\beta), \varepsilon_{R2}(\beta) \in D$ and those eigenvalues are resonant points. However, if the coupling is as strong as $\beta > \beta_{R2}$ one has $\varepsilon_{R2}(\beta) \notin D$ and this eigenvalue becomes an isolated eigenvalue of the combined system. Also, if $\beta > \beta_{R1}$ one finds $\varepsilon_{R1}(\beta) \notin D$ and this eigenvalue becomes another isolated eigenvalue of the combined system. As $\beta > \beta_{R1}$ further increases, isolated eigenvalues $\varepsilon_{R1}(\beta)$ and $\varepsilon_{R2}(\beta)$ continue to increase. Hence for each $\beta > \beta_{R1}$ combined system has two isolated eigenvalues that satisfy $\varepsilon_I(\beta) > 1$. This is maximum number of extreme right isolated eigenvalues that combined system with $\rho = 2$ may have.

In the points β_{R1} and β_{R2} one has $\varepsilon = b_1 = 1$. Since $\mathbf{B}_2$ is a two-dimensional system, one can obtain a closed analytic expression for those points [5]. In fact one obtains a quadratic equation in the unknown $x = \beta^2$. Thus one finds, $\beta_{R2} = 0.21146$ and $\beta_{R1} = 1.12883$.

Consider now remaining four eigenvalues $\varepsilon_{L1}(\beta)$, $\varepsilon_{L2}(\beta)$, $\varepsilon_1(\beta)$ and $\varepsilon_2(\beta)$. Note that $\varepsilon_{L1}(\beta)$ and $\varepsilon_1(\beta)$ are two branches of one and the same analytic function $\beta = \beta_1(\varepsilon)$, while $\varepsilon_{L2}(\beta)$ and $\varepsilon_2(\beta)$ are two branches of another analytic function $\beta = \beta_2(\varepsilon)$. It is more convenient to analyze those functions in a form $\beta = \beta_1(\varepsilon)$ and $\beta = \beta_2(\varepsilon)$. Since $\mathbf{B}_2$ is a two-dimensional system, this can be done in a closed form [5].

One thus finds exact positions of critical points $(\beta_1,\varepsilon_1)=(0.67519,-0.76673)$ and $(\beta_2,\varepsilon_2)=(0.47058,-0.63563)$. Eigenvalues $\varepsilon_{L1}(\beta)$ and $\varepsilon_1(\beta)$ exist only if $\beta \geq \beta_1$ while eigenvalues $\varepsilon_{L2}(\beta)$ and $\varepsilon_2(\varepsilon)$ exist only if $\beta \geq \beta_2$. For each β (where defined) eigenvalues $\varepsilon_1(\beta)$ and $\varepsilon_2(\beta)$ are inside the range D and hence these eigenvalues are resonant points. Concerning eigenvalues $\varepsilon_{L1}(\beta)$ and $\varepsilon_{L2}(\beta)$, as β increases one has at first $\varepsilon_{L1}(\beta), \varepsilon_{L2}(\beta) \in D$ and those eigenvalues are resonant points. However, if the coupling is as strong as $\beta > \beta_{L2}$ one has $\varepsilon_{L2}(\beta) \notin D$ and this eigenvalue becomes an isolated eigenvalue of the combined system. Also, if $\beta > \beta_{L1}$ one finds $\varepsilon_{L1}(\beta) \notin D$ and this eigenvalue becomes another isolated eigenvalue of the combined system. As $\beta > \beta_{L1}$ further increases, isolated eigenvalues $\varepsilon_{L1}(\beta)$ and $\varepsilon_{L2}(\beta)$ continue to decrease. Hence for each $\beta > \beta_{L1}$ combined system has two isolated eigenvalues $\varepsilon_I(\beta)$ that satisfy $\varepsilon_I(\beta) < -1$. This is maximum number of left extreme isolated eigenvalues that combined system with $\rho = 2$ may have.

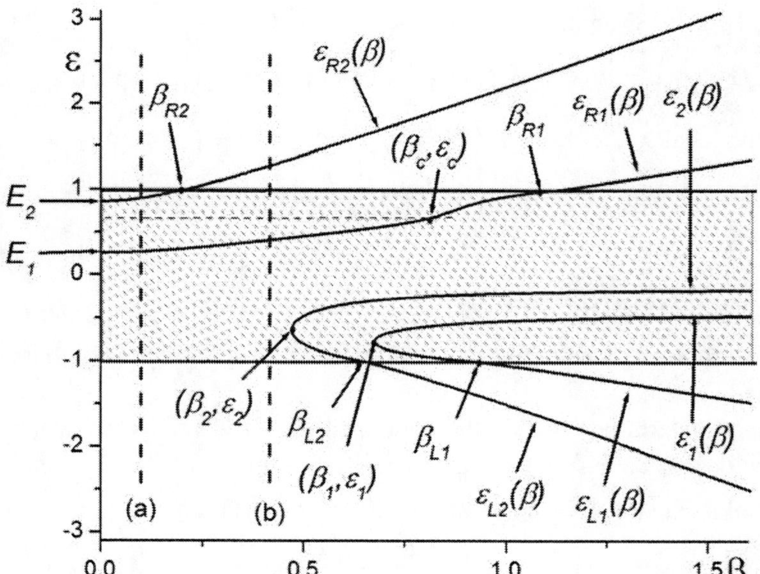

Figure 15. Example E1. Eigenvalues $\varepsilon_I(\beta)$ of the generic eigenvalue equation considered as functions of a coupling parameter β. Eigenvalues $\varepsilon_I(\beta)$ outside the range $D=[-1,1]$ are isolated eigenvalues of the combined system $\mathbf{C}_\infty$. Eigenvalues $\varepsilon_I(\beta)$ inside this range are resonant points. Eigenvalue distributions of local states $|\Theta_1\rangle$ and $|\Theta_2\rangle$ for the coupling $\beta = 0.1$ corresponding to the line (a) are shown in Figure 16. Eigenvalue distribution of a local state $|\Theta_1\rangle$ for the coupling $\beta = 0.4$ corresponding to the line (b) is shown in Figure 17. Adopted from ref. [5].

The position of the points β_{L1} and β_{L2} can be also obtained in the closed analytic form, and one finds $\beta_{L2} = 0.64717$ and $\beta_{L1} = 0.91845$ [5].

Figure 15 provides detailed interaction landscape of the open system $\mathbf{B}_2$ that interacts with its surrounding (system $\boldsymbol{A}_\infty$). Isolated eigenvalues $\varepsilon_{L1}(\beta)$ and $\varepsilon_{L2}(\beta)$ cannot be derived by any standard perturbation expansion method. Those eigenvalues do not exist in the point $\beta = 0$ and hence no perturbation expansion in this point can reproduce those eigenvalues. Note also that in a point (β_1, ε_1) one has $d\varepsilon_I(\beta)/d\beta = \infty$. Hence according to the expression (117), in this point one has $\langle \theta_I | \mathbf{S}_b | \theta_I \rangle = \beta^2 d\langle \theta_I | \boldsymbol{\omega}(\varepsilon) | \theta_I \rangle / d\varepsilon_I$. The same applies to the point (β_2, ε_2).

Once eigenvalues $\varepsilon_{R1}(\beta)$, $\varepsilon_{R2}(\beta)$, $\varepsilon_{L1}(\beta)$ and $\varepsilon_{L2}(\beta)$ are known, one can derive all corresponding isolated eigenstates. Those eigenstates can be also obtained in the closed analytic form [5]. They determine all related properties of the combined system. In particular, probabilities $w_{I,r}$ to find isolated eigenstate $|\Psi_I\rangle$ in a local state $|\Theta_r\rangle$ as well as probability $w_I = w_{I,1} + w_{I,2}$ to find this eigenstate in the base system $\mathbf{B}_2$ are given by expressions (118). For example, isolated eigenstate $|\Psi_{R1}(\beta)\rangle$ that has eigenvalue $\varepsilon_{R1}(\beta)$ exists only if $\beta > \beta_{R1} = 1.12883$. Hence probabilities $w_{R1,1}(\beta)$ and $w_{R1,2}(\beta)$ to find this eigenstate in the local states $|\Theta_1\rangle$ and $|\Theta_2\rangle$, respectively, equal zero if $\beta < \beta_{R1}$. The same of course applies for total probability $w_{R1}(\beta) = w_{R1,1}(\beta) + w_{R1,2}(\beta)$ to find this eigenstate in the system $\mathbf{B}_2$. At the point $\beta = \beta_{R1}$ those probabilities discontinuously jump from zero to $w_{R1}(\beta_{R1}+) = 0.44324$, $w_{R1,1}(\beta_{R1}+) = 0.43952$ and $w_{R1,2}(\beta_{R1}+) = 0.00372$, respectively [5]. Similarly one derives all other probabilities associated with isolated solutions of the combined system.

It remains to determine whether the combined system contains any anomal point where this system could have an isolated eigenstate in addition to isolated eigenstates contained outside eigenvalue band I_1. Each anomal point $\varepsilon = \varepsilon_c$ is an interior point of a range D, and in this point one must have a nontrivial state $|\theta_c\rangle$ that satisfies $\mathbf{f}(\varepsilon_c)|\theta_c\rangle = 0$ as well as $\mathbf{h}(\varepsilon_c)|\theta_c\rangle = 0$. First condition implies that in this point rank of a characteristic matrix $\mathbf{f}(\varepsilon)$ must be smaller than the dimension ρ of the space X_ρ^b. In the example considered and according to (144), only a point $\varepsilon = \varepsilon_c = 0.38197$ has such a property. Second condition implies that an anomal point must be a resonant point. The only candidate for the anomal point is hence the point (β_c, ε_c) on the intersection of the line $\varepsilon = \varepsilon_c$ and the function $\varepsilon = \varepsilon_{R1}(\beta)$ (see Figure 15). One finds $\beta_c = 0.32611$ [5]. One also finds that the state

$|\theta_c\rangle$ which satisfies $\mathbf{h}(\varepsilon_c)|\theta_c\rangle = 0$ does not satisfy $\mathbf{f}(\varepsilon_c)|\theta_c\rangle$. The point $\varepsilon = \varepsilon_c$ is hence not an anomal point. In conclusion, the system $\mathbf{C}_\infty$ contains no anomal points and hence no isolated eigenstates inside the range D.

Consider now embedded cardinal solutions of the combined system. Fractional shift equation (119a) produces all such solutions. In a 2×2 case the solution of this equation can be again obtained in the closed analytic form [5]. In particular, eigenvalues $X = X_{1,2}(\varepsilon)$ of a fractional shift eigenvalue equation are roots of the 2×2 determinant (119d). Those eigenvalues can be hence explicitly written as roots of a quadratic equation [5]. This determines eigenvalues $X_d(\varepsilon)$ of a fractional shift equation for each $\varepsilon \in D$. Once this eigenvalue is known, the corresponding X_ρ^b-component $|\Psi_d^b(\varepsilon)\rangle$ of the embedded eigenstate $|\Psi_d(\varepsilon)\rangle$ is given by the expressions (120) where $|\varphi_d(\varepsilon)\rangle$ can be obtained in a closed form as a solution to (119a). This determines all related properties of the open system $\mathbf{B}_\rho$.

As an example, in Figure 16 are shown eigenvalue distributions of local states $|\Theta_1\rangle$ and $|\Theta_2\rangle$ calculated in this way for the case $\beta = 0.1$. Those distributions correspond to the line (a) in Figure 15. For this value of β combined system has no isolated eigenstate and those distributions are hence completely determined by probability densities $\sigma_r(\varepsilon,0.1) = \rho_{1,r}(\varepsilon,0.1) + \rho_{2,r}(\varepsilon,0.1)$ to find local state $|\Theta_r\rangle$ with the eigenvalue $\varepsilon \in D$ (see expressions (123) and (124)). Since coupling parameter $\beta = 0.1$ is relatively small, interaction of the system $\mathbf{B}_2$ with its surrounding is relatively weak. Initial unperturbed eigenvalues $E_r \in D$ of $\mathbf{B}_2$ are hence only slightly shifted to the new positions $\varepsilon_r(0.1)$ (see Figure 16). Since $\varepsilon_r(0.1) \in D$, those shifted eigenvalues are not sharp and they acquire the shape of the universal resonance curves $\rho_r^0(\varepsilon,0.1)$ as given by expression (128b). The widths of those two resonant curves calculated according to (129a) are[1] $\Delta\varepsilon_1(0.1) = 0.00274$ and $\Delta\varepsilon_2(0.1) = 0.00286$, respectively [5]. Those widths are quite small and densities $\sigma_r(\varepsilon,0.1)$ are hence well approximated with the corresponding universal resonance curves $\rho_r^0(\varepsilon,0.1)$. In particular, density $\sigma_1(\varepsilon,0.1)$ has a prominent resonant shape centered at the point $\varepsilon = \varepsilon_{R1}(0.1) = 0.26749$, while density $\sigma_2(\varepsilon,0.1)$ has a prominent resonant shape centered at the point $\varepsilon = \varepsilon_{R2}(0.1) = 0.89403$. In Figure 16(b) is shown density $\sigma_1(\varepsilon,0.1)$ magnified. This density is almost identical to the universal resonance curve $\sigma_1^0(\varepsilon,0.1)$ [5]. On this scale there is virtually no difference between exact density $\sigma_1(\varepsilon,0.1)$ and approximate density $\sigma_1^0(\varepsilon,0.1)$. In Figure 16(c) is in the same way

[1] In ref. [5] half-width instead of width is given.

shown density $\sigma_2(\varepsilon,0.1)$ which is also almost identical to the corresponding approximate density $\sigma_2^0(\varepsilon,0.1)$. Further, one finds $S_1(0.1)=\int\sigma_1(\varepsilon,0.1)d\varepsilon=1$ and $S_2(0.1)=\int\sigma_2(\varepsilon,0.1)d\varepsilon=1$. Concerning component probabilities $C_{d,r}(0.1)=\int\rho_{d,r}(\varepsilon,0.1)d\varepsilon$, one has $C_{1,1}(0.1)=0.94312$ and $C_{1,2}(0.1)=0.05688$. Density $\sigma_1(\varepsilon,0.1)$ is hence essentially density of the state $|\Theta_1\rangle$ which interacts with its surrounding. Similarly one finds $C_{2,2}(0.1)=0.97613$ and $C_{2,1}(0.1)=0.02387$. Density $\sigma_2(\varepsilon,0.1)$ is hence essentially density of the state $|\Theta_2\rangle$ which interacts with its surrounding.

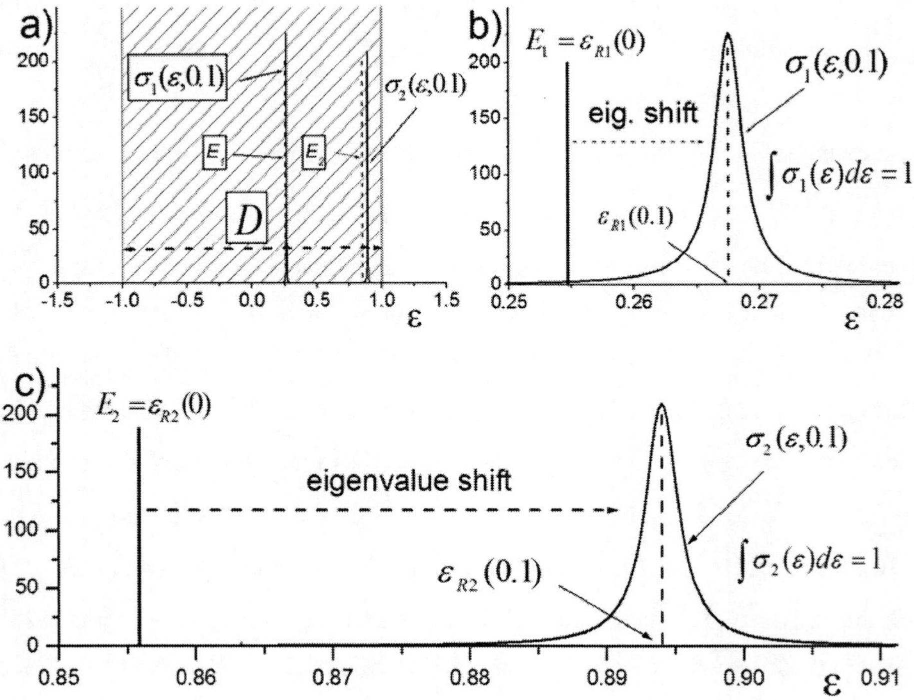

Figure 16. Eigenvalue distributions of local states $|\Theta_1\rangle$ and $|\Theta_2\rangle$ in the case $\beta=0.1$ (line (a) in Figure 15). a) Probability densities $\sigma_1(\varepsilon,0.1)$ and $\sigma_2(\varepsilon,0.1)$ are sharply localized inside the range D at positions $\varepsilon_{R1}(0.1)$ and $\varepsilon_{R2}(0.1)$, respectively. (b) Probability density $\sigma_1(\varepsilon,0.1)$ highly magnified. This probability density is almost identical to the universal resonance curve $\rho_1^0(\varepsilon,0.1)$ centered at the position $\varepsilon_{R1}(0.1)$. (c) Probability density $\sigma_2(\varepsilon,0.1)$ highly magnified. This probability density is almost identical to the universal resonance curve $\rho_2^0(\varepsilon,0.1)$ centered at the position $\varepsilon_{R2}(0.1)$. Adopted from ref. [5].

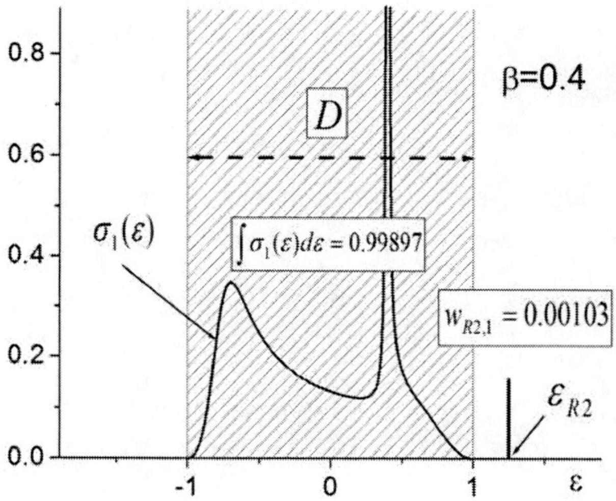

Figure 17. Eigenvalue distribution of local state $|\Theta_1\rangle$ in the case $\beta = 0.4$ (line (b) in Figure 15). In addition to embedded eigenstates, combined system contains an isolated eigenstate $|\Psi_{R2}(0.4)\rangle$ with the eigenvalue $\varepsilon_{R2}(0.4) = 1.24735$. Interaction is strong and this eigenvalue distribution has no resemblance to the universal resonance curve. Adopted from ref. [5].

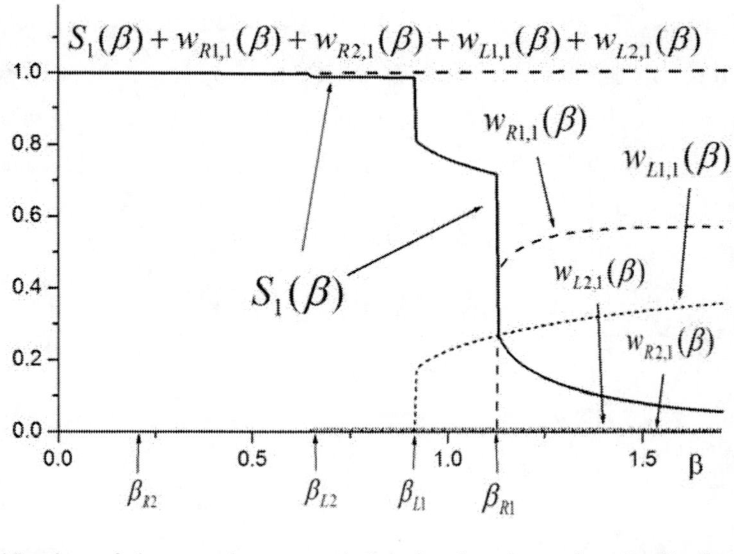

Figure 18. Verification of the completeness relation for the eigenvalue distribution of a local state $|\Theta_1\rangle$. Adopted from ref. [5].

As another example, in Figure 17 is shown a spectral distribution of a local state $|\Theta_1\rangle$ for the value $\beta = 0.4$. This value corresponds to the line (b) in Figure 15. This is relatively strong coupling and probability density $\sigma_1(\varepsilon, 0.4)$ does not resemble universal resonance curve centered at corresponding resonance point. For this value of β standard perturbation expansion fails. Fractional shift eigenvalue equation produces

$S_1(0.4) = \int \sigma_1(\varepsilon, 0.4)d\varepsilon = 0.99897$ [5]. Unlike in the case $\beta = 0.1$, this probability is less than one. In the case $\beta = 0.4$ in addition to embedded eigenstates combined system contains an isolated eigenstate $|\Psi_{R2}\rangle$ with the eigenvalue $\varepsilon_{R2}(0.4) = 1.24735$ (see Figure 15). Missing probability is due to this eigenstate. Using expressions (118) one finds $w_{R2,1}(0.4) = 0.00103$. As required by the completeness relation (134a), those probabilities satisfy $S_1(0.4) + w_{R2,1}(0.4) = 1$. In a similar way can be analyzed spectral distribution of another isolated eigenstate $|\Theta_2\rangle$ in the case $\beta = 0.4$ [5].

Completeness relations are verified in a systematic way in Figs. 18 and 19, respectively. In Figure 18 probability $S_1(\beta)$ to find local state $|\Theta_1\rangle$ in any of the embedded eigenstates of the combined system as well as probabilities $w_{R1,1}(\beta)$, $w_{R2,1}(\beta)$, $w_{L1,1}(\beta)$ and $w_{L2,1}(\beta)$ to find this state in the corresponding isolated eigenstates of the combined system are plotted as functions of a coupling parameter β. A sum of all those probabilities is also shown. As required by the completeness relation, this sum equals one for each value of β. In Figure 19 are in the same way plotted probabilities $S_2(\beta)$, $w_{R1,2}(\beta)$, $w_{R2,2}(\beta)$, $w_{L1,2}(\beta)$ and $w_{L2,2}(\beta)$ that correspond to the local state $|\Theta_2\rangle$. The sum of those probabilities also equals one for each value of β. Complete agreement of those probabilities with the requirements of the completeness relations (134) provides a strong verification of the suggested method. It also shows that in the example considered combined system $\mathbf{B}_\infty$ has no weakly singular eigenstates.

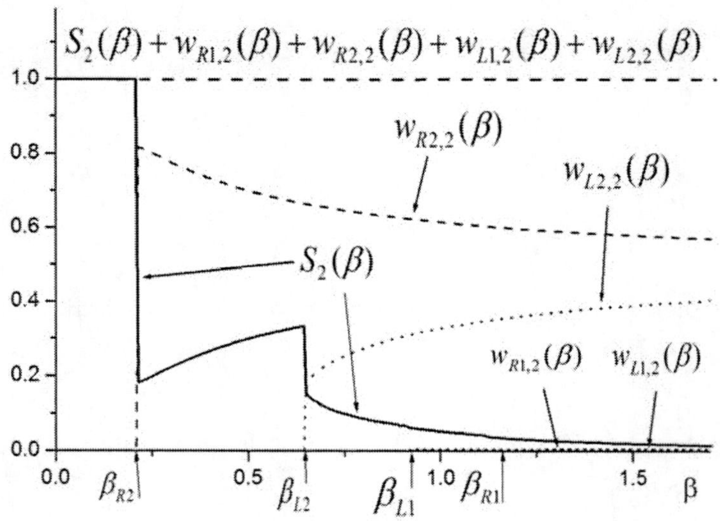

Figure 19. Verification of the completeness relation for the eigenvalue distribution of a local state $|\Theta_2\rangle$. Adopted from ref. [5].

17.2. Example E2

As another example consider the system $\mathbf{B}_2$ characterized by matrices

$$\mathbf{H}_b = \begin{pmatrix} 0.5 & -0.25 \\ -0.25 & 1.6 \end{pmatrix}, \ \mathbf{S}_b = \begin{pmatrix} 1.1 & 0.1 \\ 0.1 & 1 \end{pmatrix}, \tag{146a}$$

Those matrices are identical to matrices (142a) of the previous example, except for a single matrix element of the matrix $\mathbf{H}_b$. Eigenvalues and corresponding eigenstates of the system $\mathbf{B}_2$ are now

$$E_1 = 0.39164, \ E_2 = 1.72763, \tag{146b}$$

$$|\Theta_1\rangle = \begin{pmatrix} 0.91093 \\ 0.21799 \end{pmatrix}, \ |\Theta_2\rangle = \begin{pmatrix} 0.29604 \\ -0.98064 \end{pmatrix}. \tag{146c}$$

where $|\Theta_r\rangle$ are orthonormalized according to (5b).

Let the infinite system $\boldsymbol{A}_\infty$ that interacts with the system $\mathbf{B}_2$ contain two eigenvalue bands in the intervals $I_1 \equiv [a_1, b_1] = [-1,1]$ and $I_2 \equiv [a_2, b_2] = [2,3]$, respectively. Let further this system contain an isolated eigenstate $|\Phi_0\rangle$ with the eigenvalue $\lambda_0 = 1.2$. Assume that the characteristic operator $\mathbf{f}_1(\varepsilon)$ which describes the interaction of the system $\mathbf{B}_2$ with the first eigenvalue band is given by matrix elements (143), while the characteristic operator $\mathbf{f}_2(\varepsilon)$ which describes the interaction of this system with the second eigenvalue band is given by matrix elements

$$\mathbf{f}_2(\varepsilon) = \begin{pmatrix} f_{11}^{(2)}(\varepsilon) & f_{12}^{(2)}(\varepsilon) \\ f_{21}^{(2)}(\varepsilon) & f_{22}^{(2)}(\varepsilon) \end{pmatrix} \equiv \begin{pmatrix} q_{11}^{(2)}(\varepsilon) & q_{12}^{(2)}(\varepsilon) \\ q_{21}^{(2)}(\varepsilon) & q_{22}^{(2)}(\varepsilon) \end{pmatrix} \cdot \begin{cases} 1 & if \quad \varepsilon \in I_2 \\ 0 & otherwise \end{cases} \tag{147a}$$

where

$$\begin{aligned} q_{11}^{(2)}(\varepsilon) &= (\varepsilon - 2)^2 (2\varepsilon^2 - 14\varepsilon + 25), \\ q_{22}^{(2)}(\varepsilon) &= (\varepsilon - 2)^2 (\varepsilon^2 - 8\varepsilon + 17), \\ q_{12}^{(2)}(\varepsilon) &\equiv q_{21}^{(2)}(\varepsilon) = (\varepsilon - 2)^3 (\varepsilon - 4). \end{aligned} \tag{147b}$$

Let further the characteristic matrix $\mathbf{F}(x)$ which describes the interaction of the system $\mathbf{B}_2$ with the eigenstate $|\Phi_0\rangle$ be given by

$$\mathbf{F}(\varepsilon) = \begin{pmatrix} 0.5 & -0.3 \\ -0.3 & 0.2 \end{pmatrix} \cdot \delta(\varepsilon - \lambda_0). \tag{148}$$

In Figure 20 are shown eigenvalues $f_1(\varepsilon)$ and $f_2(\varepsilon)$ of the global characteristic matrix $\mathbf{f}(\varepsilon)=\mathbf{f}_1(\varepsilon)+\mathbf{f}_2(\varepsilon)+\mathbf{F}(\varepsilon)$. As required, those eigenvalues are nonnegative and they vanish outside the range D and outside the point $\varepsilon=\lambda_0$. Interaction of the system $\mathbf{B}_2$ with the system $\boldsymbol{A}_\infty$ described by this characteristic matrix is much more complex than in the previous example. There are several new features. First, the system $\boldsymbol{A}_\infty$ has two non-overlapping eigenvalue bands in the intervals I_1 and I_2, respectively. Second, characteristic matrix $\mathbf{f}(\varepsilon)$ is not continuous in the point $\varepsilon=b_2=3$ on the right edge of the interval I_2. Third, local eigenvalue $E_2\notin D$ is outside the range $D=I_1\cup I_2$, while in the previous example both eigenvalues of the isolated system $\mathbf{B}_2$ were contained inside the range D. Finally, the system $\boldsymbol{A}_\infty$ has an isolated eigenvalue at the point $\varepsilon=\lambda_0=1.2\notin D$ and in this point characteristic matrix $\mathbf{f}(\varepsilon)$ diverges. Rank of matrix $\mathbf{f}(\varepsilon)$ equals two for each interior point $\varepsilon\in D$, except for the critical point $\varepsilon_c=0.38197\in D$ where $rank(\mathbf{f}(\varepsilon_c))=1$. This is the same critical point as in the example ***E1.*** In addition, this rank equals two in the point $\varepsilon=\lambda_0$.

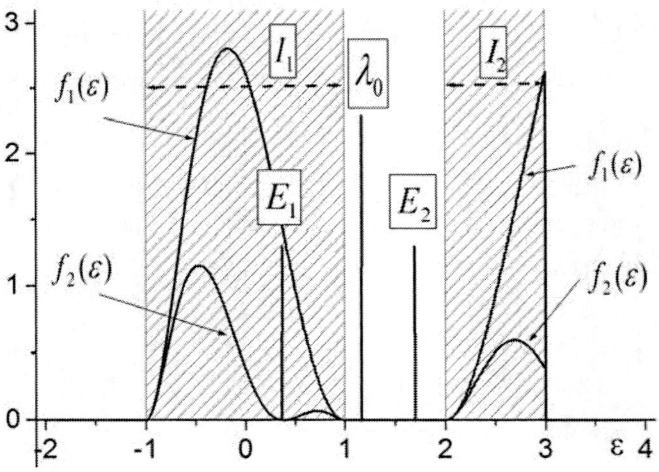

Figure 20. Example E2. Eigenvalues of the characteristic operator $\mathbf{f}(\varepsilon)$. Eigenvalue E_1 of the base system $\mathbf{B}_\rho$ is inside the range D, while another eigenvalue E_2 of this system is outside this range. Adopted from ref. [5].

Consider now global derived matrix $\boldsymbol{\omega}(\varepsilon)=\boldsymbol{\omega}_1(\varepsilon)+\boldsymbol{\omega}_2(\varepsilon)+\boldsymbol{\Omega}(\varepsilon)$. Matrix elements of the derived matrix $\boldsymbol{\omega}_1(\varepsilon)$ associated with the eigenvalue band $\nu=1$ are given by expressions (145). Concerning characteristic matrix $\mathbf{f}_2(\varepsilon)$, expression (108) implies:

$$\omega_{11}^{(2)}(\varepsilon)=q_{11}^{(2)}(\varepsilon)\ln\left|\frac{\varepsilon-2}{\varepsilon-3}\right|-g_{11}^{(2)}(\varepsilon),$$

$$\omega_{22}^{(2)}(\varepsilon) = q_{22}^{(2)}(\varepsilon)\ln\left|\frac{\varepsilon-2}{\varepsilon-3}\right| - g_{22}^{(2)}(\varepsilon), \tag{149a}$$

$$\omega_{12}^{(2)}(\varepsilon) \equiv \omega_{21}^{(2)} = q_{12}^{(2)}(\varepsilon)\ln\left|\frac{\varepsilon-2}{\varepsilon-3}\right| - g_{12}^{(2)}(\varepsilon),$$

where $q_{sp}^{(2)}(\varepsilon)$ are polynomials (147b), while $g_{sp}^{(2)}(\varepsilon)$ are polynomials:

$$g_{11}^{(2)}(\varepsilon) = 2\varepsilon^3 - 17\varepsilon^2 + \frac{140}{3}\varepsilon - \frac{121}{3},$$

$$g_{22}^{(2)}(\varepsilon) = \varepsilon^3 - \frac{19}{2}\varepsilon^2 + \frac{88}{3}\varepsilon - \frac{109}{4}, \tag{149b}$$

$$g_{12}^{(2)}(\varepsilon) \equiv g_{21}^{(2)}(\varepsilon) = \varepsilon^3 - \frac{15}{2}\varepsilon^2 + \frac{52}{3}\varepsilon - \frac{157}{12}.$$

Since $\mathbf{f}_2(\varepsilon)$ is nonzero in the point $\varepsilon = b_2 = 3$ on the right edge of the interval I_2, those matrix elements diverges in this point [5]. Finally, using (104) and (148) one finds derived matrix $\mathbf{\Omega}(\varepsilon)$

$$\mathbf{\Omega}(\varepsilon) = \begin{pmatrix} 0.5 & -0.3 \\ -0.3 & 0.2 \end{pmatrix} \cdot \frac{1}{\varepsilon - \lambda_0} \begin{cases} if\ \varepsilon \neq \lambda_0 \\ 0\ if\ \varepsilon = \lambda_0 \end{cases} \tag{150}$$

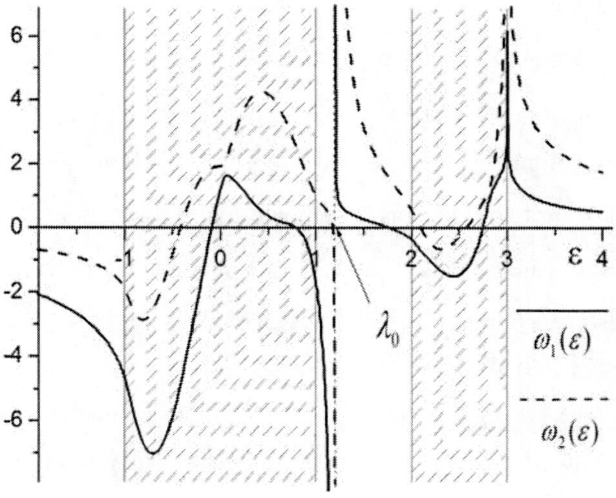

Figure 21. Example E2. Eigenvalues of the derived operator $\boldsymbol{\omega}(\varepsilon)$. Adopted from ref. [5].

In Figure 21 are shown eigenvalues $\omega_1(\varepsilon)$ and $\omega_2(\varepsilon)$ of the global derived matrix $\boldsymbol{\omega}(\varepsilon) = \boldsymbol{\omega}_1(\varepsilon) + \boldsymbol{\omega}_2(\varepsilon) + \mathbf{\Omega}(\varepsilon)$. In accord with the requirement (107b), those eigenvalues are decreasing functions of ε for each $\varepsilon \notin D = I_1 \cup I_2$ and $\varepsilon \neq \lambda_0$. However, in the interval I_1 as well as in the interval I_2 those eigenvalues are decreasing as well as increasing functions of ε. In the point $\varepsilon = \lambda_0$ those eigenvalues are not continuous and their $\varepsilon \to \lambda_0$ limits diverge. In addition, since characteristic matrix $\mathbf{f}(\varepsilon)$ is discontinuous in the point $\varepsilon = b_2 = 3$, both eigenvalues of $\boldsymbol{\omega}(\varepsilon)$ also diverge in the point $\varepsilon = 3$.

Consider now generic eigenvalue equation (114a). Solutions of the equation (114b) of a type $\varepsilon = \varepsilon(\beta)$ are shown in Figure 22. There are 12 such solutions: $\varepsilon_{R1}(\beta)$, $\varepsilon_{R2}(\beta)$, $\varepsilon_1(\beta)$, $\varepsilon_2(\beta)$, $\varepsilon_{L1}(\beta)$, $\varepsilon_{L2}(\beta)$, $\lambda_1(\beta)$, $\lambda_2(\beta)$, $E_1(\beta)$, $E_2(\beta)$, $d_1(\beta)$ and $d_2(\beta)$. If $\beta = 0$ eigenvalues of the generic equation coincide with local eigenvalues. In particular, one has $E_1(0) = E_1$ and $E_2(0) = E_2$. In addition, generic equation has two eigenvalues that in a $\beta \to 0$ limit coincide with isolated eigenvalue λ_0 of the $\boldsymbol{A}_\infty$ system. Those are eigenvalues $\lambda_1(\beta)$ and $\lambda_2(\beta)$:

$$\lim_{\beta \to 0} \lambda_1(\beta) = \lim_{\beta \to 0} \lambda_2(\beta) = \lambda_0 . \tag{151}$$

Consider eigenvalues $E_1(\beta)$ and $E_2(\beta)$ in more details. Since $E_1(0) \in D$, for small β eigenvalue $E_1(\beta)$ is contained in the range D. It is hence a resonant point. One finds that this is true for each β, however large [5]. On the other hand, since $E_2(0) \notin D$ for small β one has $E_2(\beta) \notin D$. This eigenvalue is hence an isolated eigenvalue of the combined system. However, if the coupling β is as strong as $\beta > \beta_{E2}$ one has $E_2(\beta) \in D$ and this eigenvalue becomes a resonant point. One finds $\beta_{E2} = 0.92203$ [5].

Consider now eigenvalues $\varepsilon_{L1}(\beta)$, $\varepsilon_{L2}(\beta)$, $\varepsilon_1(\beta)$ and $\varepsilon_2(\beta)$. Eigenvalues $\varepsilon_{L1}(\beta)$ and $\varepsilon_1(\beta)$ are two branches of one and the same analytic function $\beta = \beta_1(\varepsilon)$, while $\varepsilon_{L2}(\beta)$ and $\varepsilon_2(\beta)$ are two branches of another analytic function $\beta = \beta_2(\varepsilon)$. For each β (where defined) eigenvalues $\varepsilon_1(\beta)$ and $\varepsilon_2(\varepsilon)$ are inside the range D and hence these eigenvalues are resonant points. Concerning eigenvalues $\varepsilon_{L1}(\beta)$ and $\varepsilon_{L2}(\beta)$, for small β those eigenvalues are (where defined) resonant points. However, if the coupling is as strong as $\beta > \beta_{L2}$ eigenvalue $\varepsilon_{L2}(\beta)$ becomes an isolated eigenvalue of the combined system, while if $\beta > \beta_{L1}$ eigenvalue $\varepsilon_{L1}(\beta)$ becomes another isolated eigenvalue of the combined system. As $\beta > \beta_{L1}$ further increases, isolated eigenvalues $\varepsilon_{L1}(\beta)$ and $\varepsilon_{L2}(\beta)$ continue to decrease. Hence for each $\beta > \beta_{L1}$ combined system has two left isolated eigenvalues. In the points β_{L1} and β_{L2} one has $\varepsilon = a_1 = -1$. One finds $\beta_{L2} = 0.69110$ and $\beta_{L1} = 0.98637$.

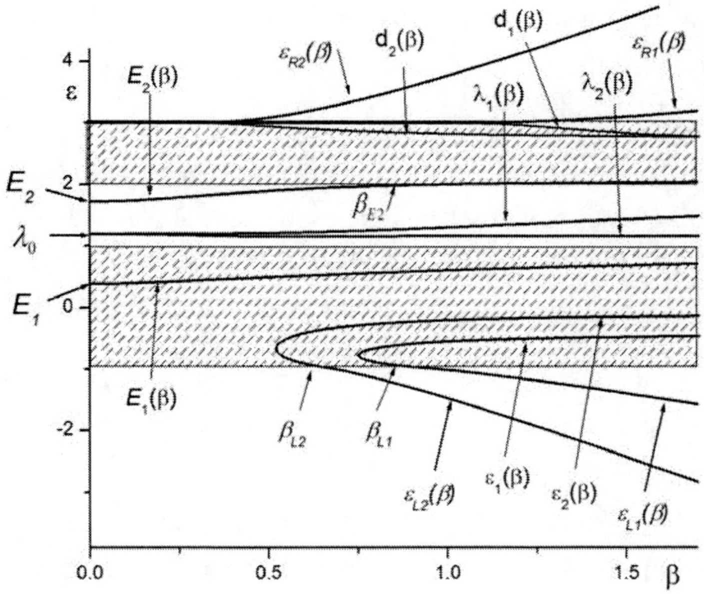

Figure 22. Example E2. Eigenvalues $\varepsilon_I(\beta)$ of the generic eigenvalue equation considered as functions of a coupling parameter β. Adopted from ref. [5].

Concerning eigenvalues $\varepsilon_{R1}(\beta)$, $\varepsilon_{R2}(\beta)$, $d_1(\beta)$ and $d_2(\beta)$, for each $\beta > 0$ eigenvalue $\varepsilon_{R1}(\beta)$ and $\varepsilon_{R2}(\beta)$ are isolated eigenvalues of the combined system, while eigenvalues $d_1(\beta)$ and $d_2(\beta)$ are resonant points of this system. In the limit $\beta \to 0$ all those four eigenvalues converge to the point $\varepsilon = 3$ where characteristic operator $\mathbf{f}(\varepsilon)$ is discontinuous and where derived operator $\boldsymbol{\omega}(\varepsilon)$ diverges. Due to this property, as $\beta > 0$ decreases those eigenvalues approach the value $\varepsilon = b_2 = 3$ asymptotically [5]. For example, if $\beta \approx 0.94$ one has $\varepsilon_{R1}(\beta) \approx 3.001$ and $d_1(\beta) \approx 2.999$, while if β decreases to $\beta = 0.70167$ one finds $\varepsilon_{R1}(\beta) \approx 3.000001$ and $d_1(\beta) \approx 2.999999$. Similarly, if $\beta \approx 0.278$ one has $\varepsilon_{R2}(\beta) \approx 3.001$ and $d_2(\beta) \approx 2.999$, while if β decreases to $\beta = 0.19069$ one finds $\varepsilon_{R2}(\beta) \approx 3.000001$ and $d_2(\beta) \approx 2.999999$. This is extremely close to the value $\varepsilon = 3$. As $\beta > 0$ further decreases it approaches this value even more. This asymptotic behavior is due to the fact that matrix elements of the characteristic operator $\mathbf{f}(\varepsilon)$ are discontinuous in the point $\varepsilon = b_2 = 3$ [5].

Consider finally eigenvalues $\lambda_1(\beta)$ and $\lambda_2(\beta)$. According to (151), in a limit $\beta \to 0$ those eigenvalues converge to the isolated eigenvalue $\lambda_0 = 1.2$ of the infinite system $\boldsymbol{A}_\infty$. Since $\lambda_0 \notin D$, for sufficiently small β those eigenvalues are isolated eigenvalues of the combined system. One finds that this is true for each β, however large [5].

In conclusion, for sufficiently big β combined system has 6 isolated eigenvalues, $\varepsilon_{R1}(\beta)$, $\varepsilon_{R2}(\beta)$, $\varepsilon_{L1}(\beta), \varepsilon_{L2}(\beta)$, $\lambda_1(\beta)$ and $\lambda_2(\beta)$. In addition, for each $\beta < \beta_{E2}$ there is isolated eigenvalue $E_2(\beta)$.

Once above eigenvalues are obtained as roots of (114b), one can derive X_2^b-components of the corresponding isolated eigenstates using expressions (114a) and (115). In analogy to the previous example, those eigenstates determine all related properties of the open system $\mathbf{B}_2$. In particular, probabilities $w_{L1,r}(\beta) = \left|\langle\Theta_r|\mathbf{S}|\Psi_{L1}\rangle\right|^2$ associated with the eigenvalue $\varepsilon_{L1}(\beta)$ are zero if $\beta < \beta_{L1} = 0.98637$. In a point $\beta = \beta_{L1}$ those probabilities discontinuously jump from zero to a finite nonzero values [5]. The same applies to the probabilities $w_{L2,r}(\beta) = \left|\langle\Theta_r|\mathbf{S}|\Psi_{L2}\rangle\right|^2$ associated with the eigenvalue $\varepsilon_{L2}(\beta)$. Those probabilities are zero as long as $\beta < \beta_{L2} = 0.69110$, while at this point those probabilities discontinuously jump from zero to finite nonzero values [5]. However, probabilities $w_{R1,r} \equiv \left|\langle\Theta_r|\mathbf{S}|\Psi_{R1}\rangle\right|^2$ and $w_{R2,r} = \left|\langle\Theta_r|\mathbf{S}|\Psi_{R2}\rangle\right|^2$ associated with discrete eigenvalues $\varepsilon_{R1}(\beta)$ and $\varepsilon_{R2}(\beta)$ have different behavior. For each $\beta \neq 0$ those probabilities are different from zero. However, if $\beta \neq 0$ is small enough, those probabilities are so extremely small that they are zero for all practical purposes. In the case of the probabilities $w_{R1,r}$ and $w_{R2,r}$ there is no discontinuous jump from zero to a finite nonzero value. This behavior is due to the fact that the characteristic operator $\mathbf{f}(\varepsilon)$ is discontinuous in the point $\varepsilon = 3$ [5].

Consider now cardinal embedded solutions of the combined system. As emphasized in the previous example, in a 2×2 case the solution of the fractional shift equation (119a) can be obtained in the closed form. In particular, eigenvalues $X = X_{1,2}(\varepsilon)$ of a fractional shift eigenvalue equation are roots of the 2×2 determinant (119d) and one obtains those eigenvalues in an explicit form as roots of a quadratic equation [5]. The same applies to the corresponding embedded eigenstates. One can thus obtain all related properties of the open system $\mathbf{B}_2$ in a closed analytic form [5].

In order not to be lost with detailed analyze of this example, let me consider only the final result. A powerful test for the LRM approach is given by completeness relations (134). Those relations are verified in a systematic way in Figures 23 and 24.

In Figure 23 probabilities $S_1(\beta)$, $w_{R1,1}(\beta)$, $w_{R2,1}(\beta)$, $w_{L1,1}(\beta), w_{L2,1}(\beta)$, $w_{\lambda 1,1}(\beta)$, $w_{\lambda 2,1}(\beta)$ and $w_{E2,1}(\beta)$ associated with a local state $|\Theta_1\rangle$ are plotted as functions of a coupling parameter β. A sum of all those probabilities is also shown. As required by the completeness relation, this sum equals one for each value of β. In Figure 24 are in the same way analyzed probabilities associated with the local state $|\Theta_2\rangle$. The complete agreement of LRM probabilities and probability densities with completeness relation (134) provides a strong demonstration that LRM expressions are correct. It also shows that in the example considered combined system has no weakly singular solution.

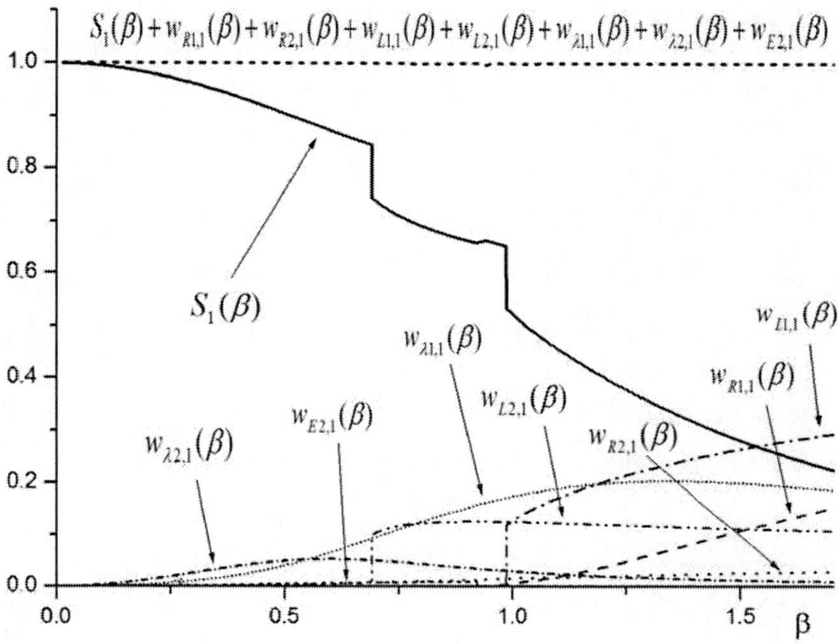

Figure 23. Example E2. Verification of the completeness relation for the eigenvalue distribution of a local state $|\Theta_1\rangle$. Adopted from ref. [5].

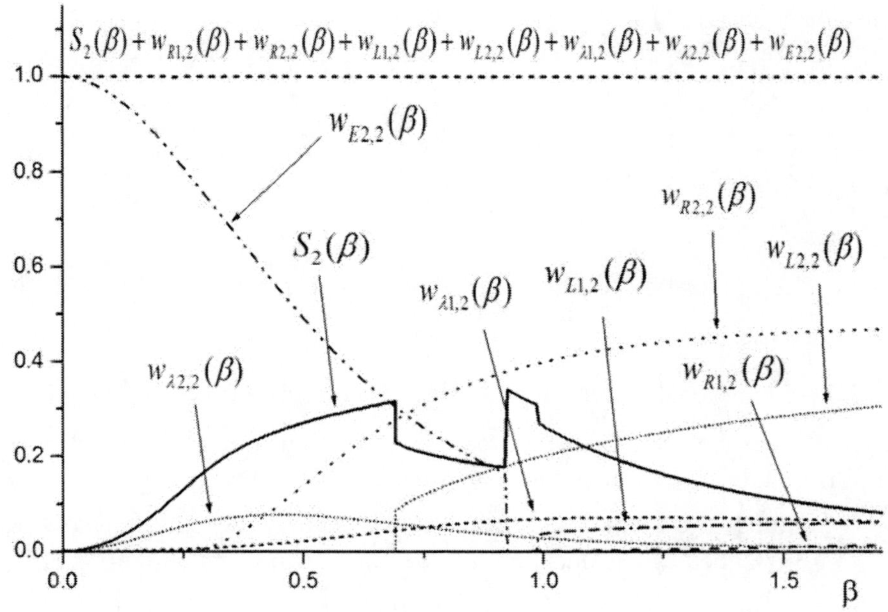

Figure 24. Example E2. Verification of the completeness relation for the eigenvalue distribution of a local state $|\Theta_2\rangle$. Adopted from ref. [5].

17.3. Example E3: Interaction of a Single State with one-Dimensional Solid in The Nearest-Neighbor Tight-Binding Approximation

Above two examples are rather artificial. Those examples are intended to illustrate various aspects of the LRM approach and to emphasize inner consistency of this approach. As a final example consider a very simple model of a molecule situated on the surface of a solid that interacts with this solid. Infinite system $\mathbf{A}_\infty$ is one-dimensional solid in the nearest-neighbor tight-binding approximation [2, 42]. With each site of this solid is associated a single state $|j\rangle$ ($j = 1,2,\cdots$). In this approximation all matrix elements $\langle i|\mathbf{H}|i\rangle$ between states on the same atomic site equal α, while matrix elements $\langle i|\mathbf{H}|j\rangle$ between states on the adjacent atomic sites equal γ. All remaining matrix elements are zero. This model is widely used in chemistry where it is known as a Hückel approximation [2]. Without loss of generality one can assume $\alpha = 0$ and $\gamma = 1$. The only effect of this assumption is the redefinition of zero eigenvalue and of eigenvalue rescaling. Eigenvalues λ_i and eigenstates $|\Phi_i\rangle$ of such one-dimensional solid containing n atoms are [2]:

$$\lambda_i = 2\cos\left(\frac{\pi}{n+1}i\right),\ |\Phi_i\rangle = \sqrt{\frac{2}{n+1}}\sum_{j=1}^{n}\sin\left(\frac{\pi}{n+1}ij\right)|j\rangle,\ i = 1,\cdots,n. \tag{152}$$

One can refer to such a solid as a Hückel chain. System $\mathbf{A}_\infty$ is an infinite Hückel chain, which is obtained in the limit $n \to \infty$. In this limit eigenvalues λ_i are replaced with a continuous function $\lambda(k) = 2\cos(k)$ of a parameter k ($0 < k < \pi$), and discrete eigenstates $|\Phi_i\rangle$ are replaced with continuous eigenstates $|\Phi(k)\rangle$:

$$\lambda(k) = 2\cos(k),\ |\Phi(k)\rangle = \sqrt{\frac{2}{\pi}}\sum_{j=1}^{\infty}\sin(jk)|j\rangle,\ 0 < k < \pi. \tag{153}$$

Relations (153) give all necessary information for the system $\mathbf{A}_\infty$. This system contains a single continuous eigenvalue band $\lambda(k)$ in the interval $[\lambda_a, \lambda_b] \equiv [-2,2]$.

As a finite-dimensional system $\mathbf{B}_\rho$ consider a molecule situated on the surface of this solid. In general, this molecule can be approximated as a finite dimensional system containing ρ eigenvalues and ρ corresponding eigenstates. If a particular eigenvalue E of this molecule is non-degenerate and if it is well separated from all other eigenvalues of this molecule, to a very good approximation one can consider in isolation the interaction of the corresponding molecular eigenstate $|\Theta\rangle$ with the infinite system $\mathbf{A}_\infty$. In this approximation system $\mathbf{B}_\rho$ is a one-dimensional system $\mathbf{B}_1$ that contains a single state $|\Theta\rangle$ with the eigenvalue E. An arbitrary interaction between the systems $\mathbf{B}_1$ and $\mathbf{A}_\infty$ can be written in the form $\beta\mathbf{V}$ ($\beta \geq 0$) where matrix element of the Hermitean operator $\mathbf{V}$ between the state $|\Theta\rangle$ and j-th state of the Hückel chain is $\langle\Theta|\mathbf{V}|j\rangle = \beta_j$, and where this operator is normalized according to $\langle\Theta|\mathbf{V}^2|\Theta\rangle = 1$. This normalization is equivalent to the condition $\sum_j \beta_j^2 = 1$. Hence and from (153) one obtains

$$\langle\Theta|\mathbf{V}|\Phi(k)\rangle = \sqrt{\frac{2}{\pi}}\sum_{j=1}^{\infty}\beta_j \sin(kj), \tag{154a}$$

$$\sum_j \beta_j^2 = 1. \tag{154b}$$

Relations (154) describe an arbitrary interaction of a single state $|\Theta\rangle$ with the infinite Hückel chain. In this general form the state $|\Theta\rangle$ is allowed to interact with each state $|j\rangle$ of the Hückel chain. Usually this interaction is confined to few states close to the surface of a solid. For the sake of simplicity assume that the state $|\Theta\rangle$ interacts only with the first state $|1\rangle$ of the infinite Hückel chain. In this case one has $\langle\Theta|\mathbf{V}|1\rangle = \beta_1 = 1$ and expressions (154) reduce to

$$\langle\Theta|\mathbf{V}|\Phi(k)\rangle = \sqrt{\frac{2}{\pi}}\sin(k),\quad 0 < k < \pi. \tag{155}$$

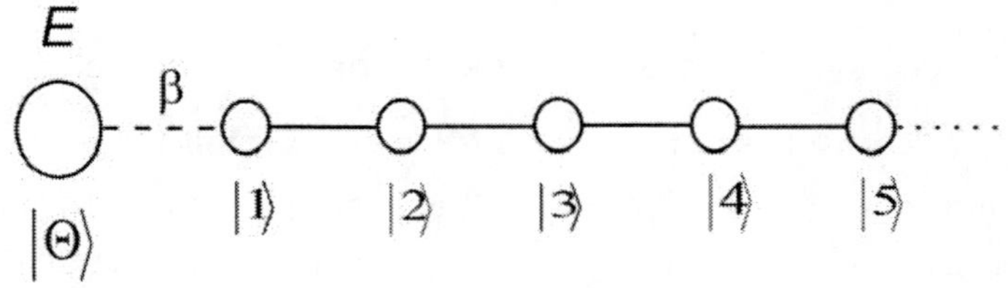

Figure 25. Example E3. Interaction of one-dimensional system $\mathbf{B}_1$ with infinite-dimensional system $\mathbf{A}_\infty$. System $\mathbf{B}_1$ contains a single eigenstate $|\Theta\rangle$ with the eigenvalue E. System $\mathbf{A}_\infty$ is an infinite one-dimensional solid (Hückel chain) in the nearest-neighbor tight-binding approximation. State $|\Theta\rangle$ interacts with a first atom of the one-dimensional solid. Coupling parameter is β. Adopted from ref. [43].

This situation is shown in Figure 25. One has one-dimensional solid approximated as an infinite Hückel chain. Surface of this solid is at the state $|1\rangle$, and with this surface interacts a single state $|\Theta\rangle$ that has eigenvalue E. Since the space X_1^b is one-dimensional, characteristic operator $\mathbf{f}(\varepsilon)$ reduces to a single function $f(\varepsilon)$, and derived operator $\boldsymbol{\omega}(\varepsilon)$ reduces to a single function $\omega(\varepsilon)$. Eigenvalue function $\lambda(k)$ as defined in (153) is non-increasing in the interval $[\lambda_a,\lambda_b]\equiv[-2,2]$, while we have assumed in the theoretical considerations that $\lambda(k)$ is non-decreasing. This can be easily corrected by a formal replacement of a parameter k with a parameter $k'=\pi-k$. As another solution, in the definition (102c) of the characteristic function $f(\varepsilon)$ one can take the absolute value of the derivative $d\lambda/dk$ in order to enforce the non-negativity of $f(\varepsilon)$.

Relations (102c), (153) and (155) imply [43]

$$f(\varepsilon)=\frac{\sin(k)}{\pi}\bigg|_{\varepsilon=2\cos(k)}=\frac{1}{\pi}\sqrt{1-\frac{\varepsilon^2}{4}}\,,\ \varepsilon\in[-2,2]. \tag{156}$$

Hence and according to (135b)

$$\omega(\varepsilon)=\frac{1}{\pi}P\int_{-2}^{2}\frac{\sqrt{1-\lambda^2/4}}{\varepsilon-\lambda}d\lambda\,,\ \varepsilon\in(-\infty,\infty). \tag{157}$$

One can integrate this expression to obtain [43]

$$\omega(\varepsilon)=\frac{1}{2}\begin{cases}\left(\varepsilon+\sqrt{\varepsilon^2-4}\right) & if \quad \varepsilon<-2\\ \varepsilon \quad if \quad \varepsilon\in D\equiv[-2,2] & \\ \left(\varepsilon-\sqrt{\varepsilon^2-4}\right) & if \quad \varepsilon>2\end{cases}\qquad \varepsilon\in(-\infty,\infty). \tag{158}$$

One has now all necessary information for the description of the combined system $\mathbf{C}_\infty$.

Let me first consider isolated eigenstates of this system. Since $\mathbf{B}_1$ is one-dimensional, those eigenstates are roots of the expression (136a) that satisfy (136b). Further, since the system $\mathbf{A}_\infty$ contains only one eigenvalue band in the interval $D=[-2,2]$ and since $\rho=1$, combined system $\mathbf{C}_\infty$ has at most one right ($\varepsilon_R>2$) and at most one left ($\varepsilon_L<-2$) isolated eigenvalue outside this eigenvalue band. Inserting (158) into the expression (136a), one finds that those two isolated eigenvalues satisfy

$$\frac{\beta^2}{2}\left(\varepsilon_R-\sqrt{\varepsilon_R^2-4}\right)+E-\varepsilon_R=0\,,\ \varepsilon_R>2\,, \tag{159a}$$

$$\frac{\beta^2}{2}\left(\varepsilon_L+\sqrt{\varepsilon_L^2-4}\right)+E-\varepsilon_L=0\,,\ \varepsilon_L<-2\,. \tag{159b}$$

In addition, one finds that resonant points $\varepsilon_0\in D$ satisfy

$$\frac{\beta^2}{2}\varepsilon_0 + E - \varepsilon_0 = 0,\ \varepsilon_0 \in [-2,2]. \tag{159c}$$

Those equations have a solution [43, 44]

$$\varepsilon_R = \frac{E(\beta^2-2)+\beta^2\sqrt{E^2+4(\beta^2-1)}}{2(\beta^2-1)} \text{ if } E > 2-\beta^2, \tag{160a}$$

$$\varepsilon_L = \frac{E(\beta^2-2)-\beta^2\sqrt{E^2+4(\beta^2-1)}}{2(\beta^2-1)} \text{ if } E < \beta^2-2. \tag{160b}$$

$$\varepsilon_0 = \frac{2E}{2-\beta^2}\begin{cases} \text{if } E \in [\beta^2-2, 2-\beta^2] \text{ and } \beta^2 < 2 \\ \text{or if } E \in [2-\beta^2, \beta^2-2] \text{ and } \beta^2 > 2 \end{cases} \tag{160c}$$

Eigenvalues ε_R and ε_L are right and left isolated eigenvalues of the combined system $\mathbf{C}_\infty$, while ε_0 is a resonant point of this system. According to the above expressions, isolated eigenvalue ε_R exist if and only if $E > 2-\beta^2$ and in this case it satisfies $\varepsilon_R > 2$, isolated eigenvalue ε_L exist if and only if $E < \beta^2 - 2$ and in this case it satisfies $\varepsilon_L < -2$, while resonant point ε_0 exist if $\beta^2 < 2$ and $E \in [\beta^2-2, 2-\beta^2]$ or if $\beta^2 > 2$ and $E \in [2-\beta^2, \beta^2-2]$. In addition, if the coupling β is as large as $\beta > \sqrt{2}$ combined system may contain both isolated eigenvalues, ε_R as well as ε_L.

Once the eigenvalue ε_I is known, one can find the corresponding isolated eigenstate $|\Psi_I\rangle$. Expressions (115) and (155) imply [43]

$$|\Psi_I\rangle = \frac{1}{\sqrt{N_I}}\left[|\Theta\rangle + \beta\sqrt{\frac{2}{\pi}}\int_0^\pi \frac{\sin(k)}{\varepsilon_I - 2\cos(k)}|\Phi(k)\rangle dk\right], \tag{161a}$$

where

$$N_I = 1-\beta^2\frac{d\omega(\varepsilon_I)}{d\varepsilon_I} = 1 + \frac{\beta^2}{2\sqrt{\varepsilon_I^2-4}}\begin{cases} \varepsilon_I + \sqrt{\varepsilon_I^2-4} & \text{if} \quad \varepsilon_I = \varepsilon_L < -2 \\ \varepsilon_I - \sqrt{\varepsilon_I^2-4} & \text{if} \quad \varepsilon_I = \varepsilon_R > 2 \end{cases} \tag{161b}$$

and where ε_I is given by (160).

Expressions (161) imply that the probability $w_R = \left|\langle\Theta|\Psi_R\rangle\right|^2$ to find right isolated eigenstate $|\Psi_R\rangle$ in a local state $|\Theta\rangle$ and the probability density $\rho_R(k) = \left|\langle\Phi(k)|\Psi_R\rangle\right|^2$ to find this eigenstate in the state $|\Phi(k)\rangle$ are [43]

$$w_R = \frac{1}{1 + \dfrac{\beta^2}{2}\dfrac{\varepsilon_R - \sqrt{\varepsilon_R^2 - 4}}{\sqrt{\varepsilon_R^2 - 4}}}, \quad \varepsilon_R > 2, \tag{162a}$$

$$\rho_R(k) = \frac{2\beta^2}{\pi\left[\varepsilon_R - 2\cos(k)\right]^2}\frac{\sin^2(k)}{1 + \dfrac{\beta^2}{2}\dfrac{\varepsilon_R - \sqrt{\varepsilon_R^2 - 4}}{\sqrt{\varepsilon_R^2 - 4}}}, \quad k \in [0, \pi]. \tag{162b}$$

In particular, if the coupling β between the state $|\Theta\rangle$ and the infinite Hückel chain is small, one has $w_R \approx 1$ and $\rho_R(k) \approx 0$. In this case modified eigenstate $|\Psi_R\rangle$ is essentially original eigenstate $|\Theta\rangle$ slightly perturbed by the interaction with this chain.

One similarly finds the probabilities $w_R^j = \left|\langle j|\Psi_R\rangle\right|^2$ to find eigenstate $|\Psi_R\rangle$ in the states $|j\rangle$ ($j = 1,2,\ldots$) of the infinite Hückel chain. One has

$$w_R^j = \frac{4\beta^2}{\pi^2}\frac{I_j^2(\varepsilon_R)}{1 + \dfrac{\beta^2}{2}\dfrac{\varepsilon_R - \sqrt{\varepsilon_R^2 - 4}}{\sqrt{\varepsilon_R^2 - 4}}}, \tag{163a}$$

where integrals $I_j(\varepsilon_R)$ ($j = 1,2,\ldots$) equal

$$I_j(\varepsilon_R) = \int_0^\pi \frac{\sin(k)\sin(jk)}{\varepsilon_R - 2\cos(k)}dk, \quad \varepsilon_R > 2. \tag{163b}$$

Due to the oscillatory factor $\sin(jk)$ in a sub-integral function of $I_j(\varepsilon_R)$, this integral decrease with the increase of j. Probabilities w_R^j hence decrease as j increases. Isolated state $|\Psi_R\rangle$ is hence mainly localized close to the surface of the infinite Hückel chain (position $j = 1$).

Similar expressions are obtained for the probability $w_L = \left|\langle\Theta|\Psi_L\rangle\right|^2$ and probability density $\rho_L(k) = \left|\langle\Phi(k)|\Psi_L\rangle\right|^2$.

Consider now embedded eigenstates of the combined system $\mathbf{C}_\infty$. Since $\mathbf{B}_1$ is a one-dimensional system, one can use expressions (138-140). In particular, probability density $\rho(\varepsilon) = \left|\langle\Theta|\Psi(\varepsilon)\rangle\right|^2$ to find modified eigenstate $|\Psi(\varepsilon)\rangle$ in a system $\mathbf{B}_1$ that contains a single state $|\Theta\rangle$ is given by expression (140c). Using expressions (156) and (158) for characteristic and derived functions, one finds

$$\rho(\varepsilon) \equiv \left|\langle\Theta|\Psi(\varepsilon)\rangle\right|^2 = \frac{\beta^2\sqrt{1-\varepsilon^2/4}}{\pi\beta^4\left(1-\varepsilon^2/4\right)+\pi\left(\beta^2\varepsilon/2+E-\varepsilon\right)^2}, \quad \varepsilon \in [-2,2]. \quad (164)$$

One can compare this probability density that is obtained for an infinite combined system $\mathbf{C}_\infty$ with probabilities $w_s^n = \left|\langle\Theta|\Psi_s\rangle\right|^2$ that are obtained in the case of the finite combined system $\mathbf{C}_{n+1}$. Since in the limit $n \to \infty$ probabilities w_s^n are replaced with $\rho(\varepsilon)d\varepsilon$, one should compare density $\rho(\varepsilon_s)$ with discrete probability w_s^n normalized per unit interval, i.e. one should compare $\rho(\varepsilon_s)$ with the ratio $w_s^n/\Delta\varepsilon_s$ where $\Delta\varepsilon_s = \varepsilon_s - \varepsilon_{s-1}$. Slightly better choice is to use the average of the intervals $\Delta\varepsilon_{s+1}$ and $\Delta\varepsilon_s$ on both sides of the eigenvalue ε_s instead of the interval $\Delta\varepsilon_s$ alone. From this comparison isolated eigenstates, if any, should be excluded. In addition, the smallest and the largest eigenvalue that in a limit $n \to \infty$ converge to some point inside the band $[\lambda_a, \lambda_b]$ should be separately normalized, since for the purpose of normalization only the intervals $\Delta\varepsilon_r$ that are inside the band $[\lambda_a, \lambda_b]$ can be utilized. Hence if $\mathbf{C}_\infty$ contains no isolated eigenvalue, extreme eigenvalues ε_1 and ε_{n+1} of $\mathbf{C}_{n+1}$ should be normalized according to $W_1^n = w_1^n/\Delta\varepsilon_2$ and $W_{n+1}^n = w_{n+1}^n/\Delta\varepsilon_{n+1}$, respectively [43].

In Figure 26 is compared in this way continuous probability density $\rho(\varepsilon)$ that corresponds to an infinite combined system $\mathbf{C}_\infty$ (expression (164) - dashed line), with normalized discrete probabilities W_s^{100} ($s = 1,\ldots,101$) that correspond to a finite combined system $\mathbf{C}_{n+1}$ with $n = 100$ (vertical columns). The case $E = 0.5$ and $\beta = 0.9$ is considered. The agreement between probabilities W_s^{100} and corresponding densities $\rho(\varepsilon_s)$ ($s = 1,\ldots,101$) is excellent. Standard deviation of normalized probabilities W_s^{100} from those densities is $\Delta \approx 0.00056$ [43]. If one excludes points W_1^b and W_{101}^b that are least reliable and that are close to the edge of the interval $[-2,2]$, standard deviation drops to $\Delta \approx 0.00004$. With further increase of n, results for a finite system $\mathbf{C}_{n+1}$ rapidly converge to the theoretical value $\rho(\varepsilon)$ as given by the expression (164).

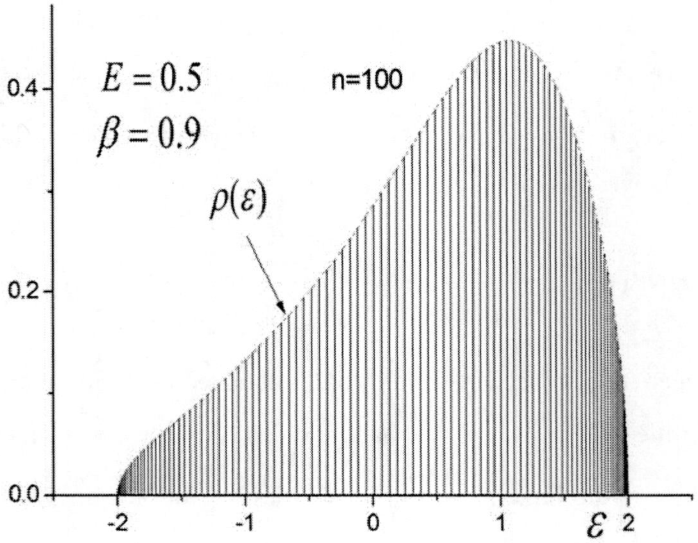

Figure 26. Example *E3*. Comparison of LRM density distribution $\rho(\varepsilon)$ of an infinite system $\mathbf{C}_\infty$ (dashed line) with normalized probabilities W_r^a of the corresponding finite combined system $\mathbf{C}_{n+1}$ (vertical columns) in the case $E = 0.5$ and $\beta = 0.9$. Finite combined system $\mathbf{C}_{n+1}$ contains a Hückel chain with $n = 100$ atoms. Adopted from ref. [43].

17.3.1. The Case of Weak Coupling

In the case when the coupling parameter β is small, above results reduce to the well known results which can be obtained within the formalism of the perturbation expansion approach [1, 2].

Consider first isolated solutions of the combined system in the case of small β. According to the expressions (160), if β is small isolated eigenvalue ε_I may exist only if the eigenvalue E of the base system is either outside the range $D=[-2,2]$, or if E is inside this range it must be very close either to the boundary point $a=-2$ or to the boundary point $b=2$. If this is the case, combined system has exactly one isolated eigenvalue which is close to the parent eigenvalue E and one corresponding eigenstate. For example, if $E > 2-\beta^2$ and if β is small combined system has one right isolated eigenstate with the eigenvalue $\varepsilon_R > 2$ as given by the expression (160a). This eigenvalue is close to the eigenvalue E. The corresponding eigenstate is given by expressions (161). Since β is small, this modified eigenstate is virtually identical to the (normalized) state $|\Theta\rangle$. Similarly, if $E < \beta^2 - 2$, combined system has exactly one left isolated eigenstate with the eigenvalue $\varepsilon_L < -2$. However, if the eigenvalue E is sufficiently deep inside the band $[-2,2]$ so that $\beta^2 - 2 < E < 2 - \beta^2$, combined system contains no isolated eigenstate.

Consider now embedded solutions. If β is small and if the eigenvalue E satisfies $\beta^2 - 2 < E < 2 - \beta^2$, combined system has the resonant point $\varepsilon_0 \in D$, as given by the expression (160c). In this case density (164) can be approximated with the universal resonance curve $\rho^0(\varepsilon)$. One finds

$$\rho(\varepsilon) \approx \rho^0(\varepsilon) = \frac{\beta^2\sqrt{1-\varepsilon_0^2/4}}{\pi\beta^4\left(1-\varepsilon_0^2/4\right)+\pi\left(\varepsilon-\varepsilon_0\right)^2}, \; \varepsilon \in [-2,2]. \tag{165}$$

where resonant point ε_o is given by (160c). The effect of the weak interaction of the eigenstate $|\Theta\rangle$ with an infinite Hückel chain is a) shift of the eigenvalue E to a new position ε_o and b) broadening of this eigenvalue to a universal resonance curve (165).

18. Time-Dependent Systems

In the previous sections time-independent systems were considered. Let me now consider time-dependent systems. LRM approach can be generalized to those systems in a rather straightforward way.

Consider an external modification of a finite-dimensional system $\mathbf{B}_\rho$ by the interaction $\mathbf{V}$ with an infinite-dimensional system $\mathbf{A}_\infty$. For the sake of simplicity, let the system $\mathbf{B}_\rho$ be described by the standard eigenvalue equation

$$\mathbf{H}_b|\Theta_r\rangle = E_r|\Theta_r\rangle, \; r = 1,\ldots,\rho. \tag{166a}$$

instead of by the generalized eigenvalue equation (5a). Orthonormalization (5b) in this case reduces to

$$\langle\Theta_r|\Theta_t\rangle = \delta_{rt}. \; r,t = 1,\ldots,\rho \tag{166b}$$

Let the system $\mathbf{A}_\infty$ be again described by the eigenvalue equations (93), and let the combined system be described by the eigenvalue equation

$$\mathbf{H}_c|\Psi(\ldots)\rangle = \varepsilon(\ldots)|\Psi(\ldots)\rangle, \tag{167a}$$

where

$$\mathbf{H}_c = \mathbf{H}_a + \mathbf{H}_b + \beta \mathbf{V} . \tag{167b}$$

This differs from the eigenvalue equation (97) considered in section 10. in that $\mathbf{S}_b \equiv \mathbf{I}_b$ is now a unit operator in the space X_ρ^b.

Equation (167a) is time-independent and it describes time-independent combined system. For example, (167a) may be a standard time-independent Schrödinger eigenvalue equation. The corresponding time-dependent equation is

$$i\hbar \frac{\partial}{\partial t}\left|\Psi(t)\right\rangle = \mathbf{H}_c \left|\Psi(t)\right\rangle . \tag{168}$$

General solution to this equation can be expressed in terms of the solutions to the time-independent equation (167)

$$\left|\Psi(t)\right\rangle = \sum_I c_I \left|\Psi_I\right\rangle \exp(-i\varepsilon_I t/\hbar) + \sum_{(\ldots)} \int c(\varepsilon,\ldots)\left|\Psi(\varepsilon,\ldots)\right\rangle \exp\left(-i\varepsilon t/\hbar\right) d\varepsilon \ldots \tag{169}$$

where first term contains summation over all isolated solutions of the eigenvalue equation (167), while second term contains summation and integration over all embedded solutions of this equation. Dots (...) represent any additional discrete and/or continuous parameters which are needed. Coefficients c_I and functions $c(\varepsilon,\ldots)$ are unknown quantities to be determined from the initial conditions.

Let at time $t = 0$ the combined system $\mathbf{C}_\infty$ be prepared in the system $\mathbf{B}_\rho$. In this case

$$\left|\Psi(0)\right\rangle = \left|\Theta\right\rangle \in X_\rho^b . \tag{170}$$

Since isolated and embedded eigenstates of the combined system $\mathbf{C}_\infty$ form a complete set in X_∞^c, each state $\left|\Psi\right\rangle \in X_\infty^c$ can be written as a linear combination of those eigenstates. In particular, each state $\left|\Theta\right\rangle \in X_\rho^b$ can be written in this way. Since strongly singular eigenstates of $\mathbf{C}_\infty$ have no X_ρ^a-component, the state $\left|\Theta\right\rangle$ contains no such eigenstates. As explained in previous sections, unless the system $\mathbf{C}_\infty$ has some symmetry not contained in the isolated systems $\mathbf{A}_\infty$ and $\mathbf{B}_\rho$, almost certainly it has no weakly singular eigenstates. In this case the state $\left|\Theta\right\rangle \in X_\rho^b$ is a linear combination of isolated cardinal and embedded cardinal eigenstates of the combined system. Hence

$$\left|\Theta\right\rangle = \sum_I \left|\Psi_I\right\rangle\left\langle\Psi_I\middle|\Theta\right\rangle + \sum_d \int \left|\Psi_d(\varepsilon)\right\rangle\left\langle\Psi_d(\varepsilon)\middle|\Theta\right\rangle d\varepsilon , \tag{171}$$

where the first term contains summation over all isolated cardinal eigenstates, while the second term contains summation and integration over all embedded cardinal eigenstates of the combined system. Using expressions (115) and (120b) one finds

$$|\Theta\rangle = \sum_I c_I(\Theta)|\Psi_I\rangle + \sum_d \int c_d(\varepsilon,\Theta)|\Psi_d(\varepsilon)\rangle d\varepsilon\,, \tag{172a}$$

where coefficients $c_I(\Theta)$ and $c_d(\varepsilon,\Theta)$ are given by

$$c_I(\Theta) \equiv \langle\Psi_I|\Theta\rangle = \frac{\langle\theta_I|\Theta\rangle}{\sqrt{\langle\theta_I|\theta_I\rangle - \beta^2\langle\theta_I|d\boldsymbol{\omega}/d\varepsilon_I|\theta_I\rangle}}\,, \tag{172b}$$

$$c_d(\varepsilon,\Theta) \equiv \langle\Psi_d(\varepsilon)|\Theta\rangle = \frac{\beta\langle\varphi_d(\varepsilon)|\Theta\rangle}{\sqrt{\pi^2\beta^4 + X_d(\varepsilon)^2}\sqrt{\langle\varphi_d(\varepsilon)|\mathbf{f}(\varepsilon)|\varphi_d(\varepsilon)\rangle}}\,. \tag{172c}$$

Hence

$$|\Psi(t)\rangle = \sum_I c_I(\Theta)|\Psi_I\rangle \exp(-\frac{i\varepsilon_I t}{\hbar}) + \sum_d \int c_d(\varepsilon,\Theta)|\Psi_d(\varepsilon)\rangle \exp\left(-\frac{i\varepsilon t}{\hbar}\right) d\varepsilon\,, \tag{173}$$

This is the required time-evolution of the state $|\Psi(t)\rangle$ which is at time $t = 0$ prepared in the system $\mathbf{B}_\rho$. Probability amplitude to find this state at time t in another state $|\Theta'\rangle \in X_\rho^b$ of this system equals

$$\langle\Theta'|\Psi(t)\rangle = \sum_I w_I(\Theta',\Theta)\exp(-\frac{i\varepsilon_I t}{\hbar}) + \sum_d \int \rho_d(\varepsilon,\Theta',\Theta)\exp\left(-\frac{i\varepsilon t}{\hbar}\right) d\varepsilon\,, \tag{174a}$$

where

$$w_I(\Theta',\Theta) \equiv \langle\Theta'|\Psi_I\rangle\langle\Psi_I|\Theta\rangle = \frac{\langle\Theta'|\theta_I\rangle\langle\theta_I|\Theta\rangle}{\langle\theta_I|\theta_I\rangle - \beta^2\langle\theta_I|d\boldsymbol{\omega}/d\varepsilon_I|\theta_I\rangle}\,, \tag{174b}$$

$$\rho_d(\varepsilon,\Theta',\Theta) \equiv \langle\Theta'|\Psi_d(\varepsilon)\rangle\langle\Psi_d(\varepsilon)|\Theta\rangle = \frac{\beta^2\langle\Theta'|\varphi_d(\varepsilon)\rangle\langle\varphi_d(\varepsilon)|\Theta\rangle}{[\pi^2\beta^4 + X_d(\varepsilon)^2]\langle\varphi_d(\varepsilon)|\mathbf{f}(\varepsilon)|\varphi_d(\varepsilon)\rangle}\,. \tag{174c}$$

Probability to find the state $|\Psi(t)\rangle$ which is at time $t = 0$ prepared in the state $|\Psi(0)\rangle \equiv |\Theta\rangle$ at some later time t in another state $|\Theta'\rangle$ equals square $|\langle\Theta'|\Psi(t)\rangle|^2$ of the amplitude (174a).

Probability amplitude (174a) for the transition between arbitrary states $|\Psi(t)\rangle$ and $|\Theta'\rangle \in X_\rho^b$ can be expressed in terms of the probability amplitudes $\langle\Theta_p|\Psi_r(t)\rangle$ where the state $|\Psi_r(t)\rangle$ is at time $t=0$ prepared in the eigenstate $|\Theta_r\rangle \in X_\rho^b$ of the closed system $\mathbf{B}_\rho$ ($|\Psi_r(0)\rangle = |\Theta_r\rangle$), while $|\Theta_p\rangle \in X_\rho^b$ is another eigenstate of this system. One has:

$$\langle\Theta_p|\Psi_r(t)\rangle = \sum_I w_{pr}^I \exp(-\frac{i\varepsilon_I t}{\hbar}) + \sum_d \int \rho_{pr}^d(\varepsilon) \exp\left(-\frac{i\varepsilon t}{\hbar}\right) d\varepsilon , \tag{175a}$$

where according to (172) matrix elements w_{rp}^I and $\rho_{rp}^d(\varepsilon)$ are

$$w_{rp}^I \equiv \langle\Theta_r|\Psi_I\rangle\langle\Psi_I|\Theta_p\rangle = \frac{\langle\Theta_r|\theta_I\rangle\langle\theta_I|\Theta_p\rangle}{\langle\theta_I|\theta_I\rangle - \beta^2\langle\theta_I|d\boldsymbol{\omega}/d\varepsilon_I|\theta_I\rangle}, \tag{175b}$$

$$\rho_{rp}^d(\varepsilon) \equiv \langle\Theta_r|\Psi_d(\varepsilon)\rangle\langle\Psi_d(\varepsilon)|\Theta_p\rangle = \frac{\beta^2\langle\Theta_r|\varphi_d(\varepsilon)\rangle\langle\varphi_d(\varepsilon)|\Theta_p\rangle}{[\pi^2\beta^4 + X_d(\varepsilon)^2]\langle\varphi_d(\varepsilon)|\mathbf{f}(\varepsilon)|\varphi_d(\varepsilon)\rangle}. \tag{175c}$$

Those matrix elements are determined by the solutions of the generic equation and fractional shift eigenvalue equation, respectively. Expressions (174) and (175) provide a complete description of the X_ρ^b-component $|\Psi^b(t)\rangle$ of each time-dependent state $|\Psi(t)\rangle$ which is at time $t=0$ prepared in the system $\mathbf{B}_\rho$. Note that diagonal matrix element $w_{rr}^I = w_{I,r} = |\langle\Theta_r|\Psi_I\rangle|^2$ is a probability to find the state $|\Theta_r\rangle$ in the isolated cardinal eigenstate $|\Psi_I\rangle$ of the combined system, while diagonal matrix element $\rho_{rr}^d(\varepsilon) = \rho_{d,r}(\varepsilon) = |\langle\Theta_r|\Psi_d(\varepsilon)\rangle|^2$ is a probability density to find this state in the embedded cardinal eigenstate $|\Psi_d(\varepsilon)\rangle$ of the combined system (see expressions (118a) and (123)).

Of particular interest is the probability $w^b(t)$ to find the state $|\Psi(t)\rangle$ at some later time in the system $\mathbf{B}_\rho$. This probability represents decay of the state $|\Psi(t)\rangle$. Due to the interaction with the infinite system $\mathbf{A}_\infty$, the state which is originally prepared in the system $\mathbf{B}_\rho$ after long enough time partially or completely decays to the system $\mathbf{A}_\infty$. Since the probability to find the state $|\Psi(t)\rangle$ in the eigenstate $|\Theta_r\rangle$ of the closed system $\mathbf{B}_\rho$ equals $|\langle\Theta_r|\Psi(t)\rangle|^2$, the probability $w^b(t)$ to find this state at time *t* anywhere in the system $\mathbf{B}_\rho$ equals

$$w^b(t) = \sum_r^{\rho} \langle \Psi(t) | \Theta_r \rangle \langle \Theta_r | \Psi(t) \rangle = \langle \Psi^b(t) | \Psi^b(t) \rangle . \tag{176}$$

where $|\Psi^b(t)\rangle$ is X_ρ^b-component of the state $|\Psi(t)\rangle$. Probability to find the state $|\Psi(t)\rangle$ which is at time $t = 0$ prepared in the system $\mathbf{B}_\rho$ ($|\Psi(0)\rangle \in X_\rho^b$) at some later time in the system $\mathbf{B}_\rho$ hence equals norm of the X_ρ^b-component $|\Psi^b(t)\rangle$ of this state.

Consider the $t \to \infty$ limit of the above probabilities. For sufficiently big times the second term on the right hand side of the expression (175a) becomes negligibly small. Hence

$$\lim_{t\to\infty} \langle \Theta_p | \Psi_r(t) \rangle = \sum_I w_{pr}^I \exp(-i\varepsilon_I t/\hbar), \tag{177a}$$

If all quantities w_{rp}^I are real, probability $w_{pr}^b(t)$ to find the state $|\Psi_r(t)\rangle$ which is at time $t = 0$ prepared in the state $|\Psi_r(0)\rangle = |\Theta_r\rangle \in X_\rho^b$ at large enough time in another state $|\Theta_p\rangle \in X_\rho^b$ of the same system $\mathbf{B}_\rho$ equals

$$w_{pr}^b(\infty) \equiv \left| \langle \Theta_p | \Psi_r(\infty) \rangle \right|^2 = \sum_I \left(w_{pr}^I \right)^2 + 2 \sum_{I<J} w_{pr}^I w_{pr}^J \cos\left(\frac{(\varepsilon_I - \varepsilon_J)t}{\hbar} \right). \tag{177b}$$

If the combined system contains no isolated eigenstate, after long enough time there is a complete decay of a state $|\Psi_r(t)\rangle$ and hence $w_{pr}^b(\infty) = 0$ for each *r* and each *p*. In this case the state $|\Psi_r(\infty)\rangle$ has no X_ρ^b-component and there is a complete decay of this state in the system $\mathbf{A}_\infty$. This is a usual case that can be also obtained within the formalism of the perturbation expansion [1, 2]. However, if the combined system contains only one isolated eigenstate $|\Psi_I\rangle$, one usually has $w_{pr}^b(\infty) = \left(w_{pr}^I \right)^2 \neq 0$. Therefore, at infinite time the state $|\Psi_r(\infty)\rangle$ will be found with a finite probability $\left(w_{pr}^I \right)^2$ in the state $|\Theta_p\rangle$. In particular, the state $|\Psi_r(\infty)\rangle$ will be found with finite probability $\sum_p \left(w_{pr}^I \right)^2$ in the system $\mathbf{B}_\rho$. The situation is more complex if the combined system contains several isolated eigenstates. In this case one has to use a general expression (177b). According to this expression, after long enough time the state $|\Psi_r(t)\rangle$ will only partially decay, and in a limit $t \to \infty$ there will be an oscillatory probability to find this state in any of the eigenstates $|\Theta_p\rangle$ of the isolated system $\mathbf{B}_\rho$. However, if the quantities $(\varepsilon_I - \varepsilon_J)$ ($\varepsilon_I \neq \varepsilon_J$) are large, those oscillations will be extremely fast. It is quite difficult to detect such fast oscillations experimentally.

Instead one should detect a time-average $\overline{w}_{pr}^{b}(\infty)$ of those oscillations. Time average of a second term in (177b) equals zero, and hence

$$\overline{w}_{pr}^{b}(\infty) = \sum_{I} \left(w_{pr}^{I}\right)^{2}. \tag{177c}$$

Similar results are obtained if the quantities w_{rp}^{I} are not real. Since an arbitrary state $|\Psi(t)\rangle$ that satisfies $|\Psi(0)\rangle \in X_{\rho}^{b}$ is a linear combination of the states $|\Psi_r(t)\rangle$ considered above, all obtained results generalize in a natural way to such an arbitrary state.

19. One Dimensional Open Systems – Time Dependent Case

Of particular interest is the case $\rho = 1$. In this case the system $\mathbf{B}_1$ contains a single state $|\Theta\rangle$ with the eigenvalue E. In section 16. time-independent one-dimensional open systems were considered. Let me now consider time-dependent one-dimensional open systems.

Since $\rho = 1$, for each $\varepsilon \in D$ there is only one embedded cardinal eigenstate $|\Psi(\varepsilon)\rangle$ of the combined system. Hence expression (173) reduces to

$$|\Psi(t)\rangle = \sum_{I} c_I |\Psi_I\rangle \exp(-i\varepsilon_I t/\hbar) + \int c(\varepsilon)|\Psi(\varepsilon)\rangle \exp(-i\varepsilon t/\hbar) d\varepsilon . \tag{178a}$$

where according to (137c) and (140c)

$$c_I \equiv \langle \Psi_I | \Theta \rangle = \sqrt{w_I} = \frac{1}{\sqrt{1 - \beta^2 \, d\omega(\varepsilon)/d\varepsilon_I}}, \tag{178b}$$

$$c(\varepsilon) \equiv \langle \Psi(\varepsilon) | \Theta \rangle = \sqrt{\rho(\varepsilon)} = \frac{\beta\sqrt{f(\varepsilon)}}{\sqrt{\pi^2 \beta^4 f^2(\varepsilon) + \left(\beta^2 \omega(\varepsilon) + E - \varepsilon\right)^2}} \tag{178c}$$

Consider now probability $w^b(t)$ to find the state $|\Psi(t)\rangle$ at time t in the initial state $|\Psi(0)\rangle = |\Theta\rangle$. This probability is a square of the amplitude $\langle \Theta | \Psi(t) \rangle$

$$w^b(t) = \left|\langle \Theta | \Psi(t) \rangle\right|^2, \tag{179a}$$

Above expressions imply

$$\langle\Theta|\Psi(t)\rangle=\sum_I w_I\exp(-i\varepsilon_I t/\hbar)+\int\rho(\varepsilon)\exp(-i\varepsilon t/\hbar)\,d\varepsilon\,. \tag{179b}$$

where density $\rho(\varepsilon)$ is given by (140c), eigenstates ε_I are roots of (136a), while probabilities w_I are given by (137c).

Consider now the $t\to\infty$ limit of the above amplitude. Since $\rho=1$ expressions (177) reduce to

$$\lim_{t\to\infty}\langle\Theta|\Psi(t)\rangle=\sum_I w_I\exp(-i\varepsilon_I t/\hbar), \tag{180a}$$

$$\lim_{t\to\infty} w^b(t)=\sum_I (w_I)^2+2\sum_{I<J} w_I w_J\cos\left(\frac{(\varepsilon_I-\varepsilon_J)t}{\hbar}\right), \tag{180b}$$

Since $\mathbf{B}_1$ is one dimensional, in each interval $(a,b)\subseteq\overline{D}$ that contains no eigenvalue λ_i of the parent system $\mathbf{A}_\infty$ there can be at most one isolated eigenstate. All quantities $(\varepsilon_I-\varepsilon_J)$ ($\varepsilon_I\neq\varepsilon_J$) are hence usually relatively large. Hence, If the system $\mathbf{A}_\infty$ contains several isolated eigenstates $|\Psi_I(t)\rangle$, almost certainly one has to use time-average expression (177c). One has

$$\overline{w}^b(\infty)=\sum_I (w_I)^2. \tag{180c}$$

19.1. Decay of a Local State in the Weak Coupling Limit

In the case of the weak coupling there are two qualitatively different cases, the case $E\in\overline{D}$ and the case $E\in D$. There are also small intermediate regions close to the boundaries between D and $\overline{D}$.

If $E\in\overline{D}$ is an interior point in $\overline{D}$ and if β is sufficiently small, there is one and only one isolated eigenvalue $\varepsilon_I\equiv\varepsilon_I(\beta)$ that satisfies $\varepsilon_I(0)=E$. For small β the corresponding eigenstate $|\Psi_I\rangle\equiv|\Psi_I(\beta)\rangle$ satisfies (up to the norm and phase) $|\Psi_I(\beta)\rangle\approx|\Theta\rangle$ and hence $w_I\approx 1$. In addition, one has $\rho(\varepsilon)\approx 0$. Relations (179) hence imply $w^b(t)\approx 1$. In conclusion, If $E\in\overline{D}$ is an interior point in $\overline{D}$ and if β is sufficiently small, the state $|\Theta\rangle$ is only slightly effected by the interaction with the system $\mathbf{A}_\infty$ and essentially it does not change with time, i.e. $|\Psi(t)\rangle\approx|\Theta\rangle$.

Another possibility is $E\in D$ where E is an interior point in D. If the coupling β is sufficiently small, resonant point $\varepsilon_I\equiv\varepsilon_I(\beta)$ that satisfies $\varepsilon_I(0)=E$ is close to E. In this case density $\rho(\varepsilon)$ is well approximated with the universal resonance curve $\rho_I^0(\varepsilon)$ given by the expression (141a). Hence and due to (179b) amplitude $\langle\Theta|\Psi(t)\rangle$ can be approximated as [44-46]

$$\langle\Theta|\Psi(t)\rangle \approx \int \rho_I^0(\varepsilon)\exp\left(-i\frac{\varepsilon t}{\hbar}\right)d\varepsilon = \exp\left(-i\frac{\varepsilon_I t}{\hbar}\right)\exp\left(-\frac{\pi\beta^2 f(\varepsilon_I)}{\hbar}t\right), \quad (181a)$$

where $\varepsilon_I \in D$ is a root of (136a). Probability $w^b(t)$ to find the state $|\Psi(t)\rangle$ at time t in the initial state $|\Theta\rangle$ hence equals

$$w^b(t) \approx w^{b0}(t) = \exp\left(-\frac{2\pi\beta^2 f(\varepsilon_I)}{\hbar}t\right). \quad (181b)$$

This expression describes well-known exponential decay of the state $|\Theta\rangle$.

If the coupling β is not small, instead of the above approximate expressions one has to use exact expressions (179)

19.2. Transitions of the State $|\Theta\rangle$ to The Eigenstates of the System $\mathbf{A}_\infty$

In the previous sections X_1^b-component of the time-dependent state $|\Psi(t)\rangle$ was analyzed. In LRM one can also obtain X_∞^a-component of this state [43-46]. This component determines probabilities to find the state $|\Psi(t)\rangle$ at time t in continuous eigenstates $|\Phi_{\tau,m}(k,l)\rangle \in X_\infty^a$ of the infinite system $\mathbf{A}_\infty$, as well as probabilities to find this state at time t in discrete eigenstates $|\Phi_i\rangle \in X_\infty^a$ of this system. Most important is the transition to the continuous eigenstates. Let $\rho_\tau^a(\varepsilon,t)$ be probability density to find the state $|\Psi(t)\rangle$ at time t in any of the eigenstates $|\Phi_{\tau,m}(k,l)\rangle \in X_\infty^a$ that have eigenvalue $\varepsilon = \lambda_\tau(k)$. Probability to find those eigenstates in the eigenvalue interval $d\varepsilon$ hence equals $\rho_\tau^a(\varepsilon,t)d\varepsilon$. This probability density is a square of the corresponding amplitude $u_\tau^a(\varepsilon,t)$:

$$\rho_\tau^a(\varepsilon,t) = \left|u_\tau^a(\varepsilon,t)\right|^2, \quad (182a)$$

If the system $\mathbf{A}_\infty$ contains several eigenvalue bands, the state $|\Psi(t)\rangle$ may decay via different channels $|\Psi(t)\rangle \to |\Phi_{\tau,m}(k,l)\rangle$ ($\tau = 1,\cdots,\kappa$). For example, one can consider the decay of some molecular state $|\Psi(t)\rangle$. This state may decay to each molecular state $|\Theta_f\rangle$ that has energy E_f lower than the energy E of the original state $|\Theta\rangle \equiv |\Psi(0)\rangle$. This decay is accompanied by the emission of a photon in the state $|\mathbf{k}\varpi\rangle$ where $\mathbf{k}$ denotes wavevector, while ϖ denotes photon polarization [1]. Hence one has the transition $|\Psi(t)\rangle \to |\Theta_f \mathbf{k}\varpi\rangle$. In the above notation $|\Phi_{\tau,m}(k,l)\rangle \equiv |\Theta_f \mathbf{k}\varpi\rangle$ is the state contained in the parent system $\mathbf{A}_\infty$. Transitions to various finale states $|\Theta_f\rangle$ are physically very different, they are usually well separated, and it is quite important to know relative probabilities of those transitions.

One finds that the amplitude $u_\tau^a(\varepsilon,t)$ can be expressed in terms of the probabilities w_I and in terms of the probability density $\rho(\varepsilon)$ according to [43, 45, 46]

$$u_\tau^a(\varepsilon,t) = \beta\sqrt{f_\tau(\varepsilon)}\left[\int \frac{\rho(\lambda)\left[e^{-i(\lambda-\varepsilon)t/\hbar}-1\right]}{\lambda-\varepsilon}d\lambda + \sum_I w_I \frac{\left[e^{-i(\varepsilon_I-\varepsilon)t/\hbar}-1\right]}{\varepsilon_I-\varepsilon}\right], \quad (182b)$$

Using (179b) this amplitude can be expressed in terms of the amplitude $\langle\Theta|\Psi(t)\rangle$ according to [45, 46]

$$u_\tau^a(\varepsilon,t) = -i\frac{\beta}{\hbar}\sqrt{f_\tau(\varepsilon)}\int_0^t \langle\Theta|\Psi(t)\rangle \exp\left(i\frac{\varepsilon t}{\hbar}\right)dt, \quad \tau = 1,\ldots,\kappa. \quad (182c)$$

This expression provides a connection between the amplitude $\langle\Theta|\Psi(t)\rangle$ that determines probability $w^b(t)$ to find the state $|\Psi(t)\rangle$ at time *t* in the system $\mathbf{B}_1$ (the state $|\Theta\rangle$) with the amplitudes $u_\tau^a(\varepsilon,t)$ that determine probability densities $\rho_\tau^a(\varepsilon,t)$ to find this state at time *t* in those eigenstates $|\Phi_{\tau,m}(k,l)\rangle$ of the infinite system $\mathbf{A}_\infty$ that satisfy $\varepsilon = \lambda_\tau(k)$.

At $t=0$ one has $u_\tau^a(\varepsilon,0)=0$ and hence $\rho_\tau^a(\varepsilon,0)=0$. Probability density to find the state $|\Psi(t)\rangle$ at initial time $t=0$ in any of the eigenstates $|\Phi_{\tau,m}(k,l)\rangle$ of the infinite system $\mathbf{A}_\infty$ equals zero. This is obvious and trivial, since at $t=0$ the state $|\Psi(t)\rangle$ has no X_∞^a-component. As t increases, one obtains non-vanishing probability densities $\rho_\nu^a(\lambda,t)$ to find this state in various continuous eigenstates $|\Phi_{\tau,m}(k,l)\rangle$ of the closed system $\mathbf{A}_\infty$. Above expressions imply

$$\frac{\rho_\nu^a(\varepsilon,t)}{\rho_\mu^a(\varepsilon,t)} = \frac{f_\nu(\varepsilon)}{f_\mu(\varepsilon)}. \quad (183)$$

For each eigenvalue ε the ratio of the probability densities to decay to different channels τ is independent on time.

If the combined system contains no isolated eigenstates, there is a well-defined $t \to \infty$ limit of the probability density $\rho_\tau^a(\varepsilon,t)$. Otherwise for large times this probability densities exhibit fast oscillations. This is similar and complementary to the analogous behavior of the probability $w^b(t)$ discussed above.

Probability $w_\tau^a(t)$ to find the state $|\Psi(t)\rangle$ at time t anywhere in the eigenvalue band τ equals

$$w_\tau^a(t) = \int \rho_\tau^a(\varepsilon,t)d\varepsilon \quad (184)$$

Unlike the ratio (183) that involves probability densities and that does not depend on time, the ratio $w_\mu^a(t)/w_\nu^a(t)$ of total probabilities to decay via two different channels μ and ν usually depends on time.

Above expressions provide probability amplitudes and corresponding probabilities to find the state $|\Psi(t)\rangle$ at time t in continuous eigenstates $|\Phi_{\tau,m}(k,l)\rangle \in X_\infty^a$ of the infinite system $\mathbf{A}_\infty$. In a similar way can be expressed probability amplitudes $u_j(t)$ to find this state at time t in the discrete eigenstates $|\Phi_j\rangle$ of this system. For simplicity those expressions are omitted here. If the system $\mathbf{A}_\infty$ contains no discrete eigenstates $|\Phi_i\rangle$, at each time $t \geq 0$ the state $|\Psi(t)\rangle$ must be found with certainty either in the original state $|\Theta\rangle \in X_1^b$ or in some state $|\Phi_{\tau,m}(k,l)\rangle \in X_\infty^a$. In this case

$$w^b(t) + \sum_\tau^K w_\tau^a(t) = 1 . \quad (185)$$

where probability $w^b(t)$ is given by the expressions (179), while probabilities $w_\tau^a(t)$ are given by the expressions (182) and (184). Relation (185) is a key completeness relation for time-dependent systems in the case when infinite system $\mathbf{A}_\infty$ contains no discrete eigenstates. This relation can be used as an efficient test for the validity of the LRM treatment of such systems.

19.3. Transition of the Local State to The System $\mathbf{A}_\infty$ in the Weak Coupling Limit

If the coupling parameter β is small, probability densities $\rho_\tau^a(\varepsilon,t)$ can be approximated as [44, 45]

$$\rho_\tau^a(\varepsilon,t) \approx \rho_\tau^{a0}(\varepsilon,t) = = \frac{f_\tau(\varepsilon_I)}{f(\varepsilon_I)} \rho_I^0(\varepsilon) \left[\exp\left(-\frac{2\pi\beta^2 f(\varepsilon_I)}{\hbar} t \right) - 2\exp\left(-\frac{\pi\beta^2 f(\varepsilon_I)}{\hbar} t \right) \cos((\varepsilon_I - \varepsilon)t/\hbar) + 1 \right]. \quad (186)$$

This expression is a product of three terms. First term is a ratio $f_\tau(\varepsilon_I)/f(\varepsilon_I)$ that determines relative probabilities of a transition via different channels τ. In particular, if $\varepsilon_I \notin I_\tau$ one has $f_\tau(\varepsilon_I) = 0$ and there is no transition to the channel τ. Second term is universal resonant curve (141a) that has a sharp peak in the resonant point $\varepsilon = \varepsilon_I$ and that satisfies $\int \rho_I^0(\varepsilon) d\varepsilon = 1$. Last term describes an exponential approach to the $t = \infty$ limit modified in an oscillatory way by the *cos()* term. This oscillatory modification is absent in the resonant point $\varepsilon = \varepsilon_I$ where density $\rho_I^0(\varepsilon)$ has a maximum. As one departs from this point, frequency of those oscillations increases, while probability $\rho_\tau^{a0}(\varepsilon,t)$ sharply decreases.

One finds that total probability $w_\tau^a(t)$ for the decay via channel τ can be approximated as [43, 45, 46]

$$w_\tau^a(t) \approx w_\tau^{a0}(t) = \int \rho_\tau^{a0}(\varepsilon,t) d\varepsilon = \frac{f_\tau(\varepsilon_I)}{f(\varepsilon_I)} \left[1 - \exp\left(-\frac{2\pi\beta^2 f(\varepsilon_I)}{\hbar} t \right) \right]. \quad (187)$$

Relations (181b) and (187) satisfy $w^{b0}(t) + \sum_\tau w_\tau^{a0}(t) = 1$ in accord with the completeness requirement (185). In addition one has

$$\frac{w_\nu^{a0}(t)}{w_\mu^{a0}(t)} = \frac{f_\nu(\varepsilon_I)}{f_\mu(\varepsilon_I)}. \quad (188)$$

which is similar to (183). Thus in the case of small β and if $\varepsilon_I(0) = E \in D$, ratios of the probabilities to decay via different channels are independent of time. Unlike exact expression (183) this result is only approximate, and it depends on the validity of the above approximations. One has also

$$w^a(t) \approx w^{a0}(t) \equiv \sum_\tau w_\tau^{a0}(t) = 1 - \exp\left(-\frac{2\pi\beta^2 f(\varepsilon_I)}{\hbar} t \right). \quad (189)$$

After long enough time the state $|\Theta\rangle$ completely decays into the system $\boldsymbol{A}_\infty$. Hence $w^{a0}(\infty) = 1$.

Above results are in accord with the well-known results obtained in a standard way within the formalism of the perturbation expansion approach [1, 2].

19.4. Example E4

Let me illustrate LRM treatment of time-dependent systems with a simple example. Consider the interaction of a single state with one-dimensional solid in the nearest-neighbor tight-binding approximation [42]. Time-independent properties of this system were discussed in section 17.3. Let me now consider time-dependent properties of this system.

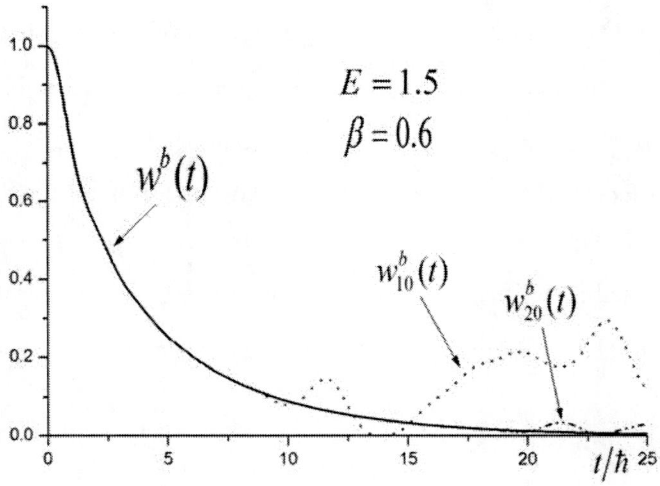

Figure 27. Example *E4*. Decay of a state $\left|\Psi(t)\right\rangle$ in the case $\beta = 0.6$ and $E = 1.5$. Probability $w^b(t)$ to find the state $\left|\Psi(t)\right\rangle$ at time t in the base system $\mathbf{B}_1$ (solid line) is compared with corresponding probabilities for finite combined systems $\mathbf{C}_{10+1}$ and $\mathbf{C}_{20+1}$ (other two lines). Since this interaction is relatively weak, combined system contains no isolated eigenstate. There is hence a complete decay of the eigenstate $\left|\Psi(t)\right\rangle$. Adopted from ref. [43]

Consider first probability $w^b(t)$ to find the state $\left|\Psi(t)\right\rangle$ which is at time $t = 0$ prepared in the state $\left|\Psi(0)\right\rangle = \left|\Theta\right\rangle$ at some later time in the same state $\left|\Theta\right\rangle$. This probability is given by the expressions (179). In order to verify those expressions one can compare probability $w^b(t)$ as given by those expressions with corresponding probabilities for a finite combined system $\mathbf{C}_{n+1}$. Those probabilities for a finite system $\mathbf{C}_{n+1}$ can be obtained in a standard way since they involve diagonalization of finite $(n+1)\times(n+1)$ matrices [43]. This comparison is shown in Figures 27 and 28. In Figure 27 combined system $\mathbf{C}_\infty$ with local eigenvalue $E = 1.5$ and with the coupling $\beta = 0.6$ is considered. In this figure probability $w^b(t)$ (solid line) is compared with corresponding probabilities $w_n^b(t)$ for finite combined systems $\mathbf{C}_{n+1}$. Those probabilities are given as functions of time t. Time is expressed in units $\hbar$. This is a natural time unit for a model considered. In the case $\beta = 0.6$

infinite system $\mathbf{C}_\infty$ contains no isolated eigenstate, and hence after long enough time the state $|\Psi(t)\rangle$ which is at time $t=0$ prepared in the state $|\Theta\rangle$ makes a complete decay to the infinite Hückel chain (system $\boldsymbol{A}_\infty$), hence $\lim_{t\to\infty} w^b(t)=0$. In order to illustrate the convergence of $w_n^b(t)$ to $w^b(t)$ as n increases, probability $w^b(t)$ is compared with two finite-chain probabilities $w_{10}^b(t)$ and $w_{20}^b(t)$. Those finite Hückel chains contain $n=10$ and $n=20$ atoms, respectively. For small times t both probabilities $w_n^b(t)$ follow theoretical curve $w^b(t)$ almost exactly. However, each curve $w_n^b(t)$ at some large enough time t' separates from $w^b(t)$. As n increases separation point t' also increases.

In Figure 28 the case $\beta=1.5$ is considered. In this case combined system $\mathbf{C}_\infty$ contains right isolated eigenstate $|\Psi_R\rangle$ and the state $|\Psi(t)\rangle$ only partially decays to the system $\boldsymbol{A}_\infty$. This decay exhibits damped oscillations, and for large times probability $w^b(t)$ converges to $\left(w_R^b\right)^2=0.36165$ [43]. This probability is compared with probabilities $w_{10}^b(t)$ and $w_{20}^b(t)$. For small times probabilities $w_n^b(t)$ again follow theoretical curve $w^b(t)$, while at some large enough time t' each curve $w_n^b(t)$ separates from $w^b(t)$. As n increases this point of separation again shifts towards higher values of t.

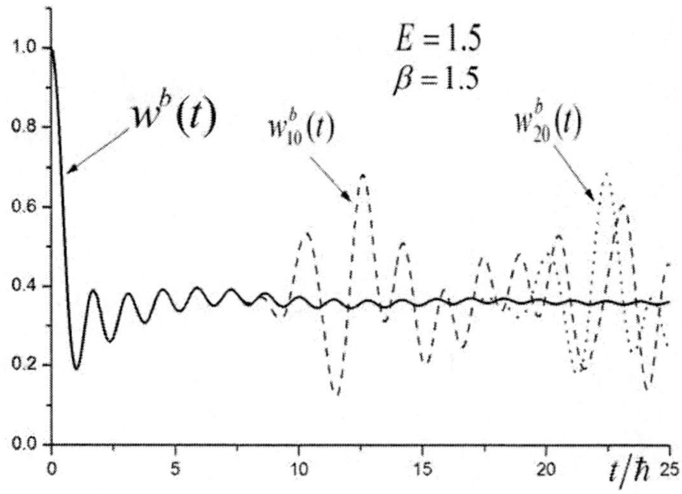

Figure 28. Example ***E4***. Decay of a state $|\Psi(t)\rangle$ in the case $\beta=1.5$ and $E=1.5$. Probability $w^b(t)$ (solid line) is compared with corresponding probabilities for finite combined systems $\mathbf{C}_{10+1}$ and $\mathbf{C}_{20+1}$ (other two lines). This interaction is relatively strong and combined system contains a single isolated eigenstate. Decay of the state $|\Psi(t)\rangle$ is not complete. Adopted from ref. [43]

Above property is general. Each probability $w_n^b(t)$ of a finite system $\mathbf{C}_{n+1}$ reproduces theoretical probability $w^b(t)$ of the corresponding infinite system $\mathbf{C}_\infty$ up to some point $t = t'$. If $t < t'$ probability $w_n^b(t)$ is virtually identical to the theoretical probability $w^b(t)$ for an infinite system $\mathbf{A}_\infty$. However, if $t > t'$ probability $w_n^b(t)$ deviates significantly from $w_n^b(t)$. As n increases the point $t = t'$ increases approximately linearly with n. In addition, the agreement between $w_n^b(t)$ and $w_n^b(t)$ for small values of t also improves [43, 45]. One finds that probabilities $w_n^b(t)$ converge to the probability $w^b(t)$ for an infinite system $\mathbf{C}_\infty$, i.e. one has $\lim_{n\to\infty} w_n^b(t) = w^b(t)$.

19.5. Example E5

In the previous example system $\mathbf{B}_1$ interacts with the infinite-dimensional system $\boldsymbol{A}_\infty$ that contains a single one-parameter eigenvalue band. In the example shown in Figures (29-32), system $\mathbf{B}_1$ that contains a single eigenstate $|\Theta\rangle$ with the eigenvalue $E = 1$ interacts with the infinite-dimensional system $\boldsymbol{A}_\infty$ that contains two eigenvalue bands $\tau = 1$ and $\tau = 2$ in the intervals $I_1 = [-2,2]$ and $I_2 = [-0.5,1.5]$, respectively. The details of this example are given elsewhere [45]. I will report here only main results.

In Figure 29 the case $\beta = 0.9$ is considered. In this Figure probability $w^b(t)$ to find the state $|\Psi(t)\rangle$ at time t in the initial state $|\Theta\rangle$ as well as probabilities $w_1^a(t)$ and $w_2^a(t)$ to find this state at time t in the two eigenvalue bands are shown. Since $\beta = 0.9$ is relatively weak coupling, combined system $\mathbf{C}_\infty$ contains no isolated eigenstate and hence in a limit $t \to \infty$ one has $w^b(\infty) = 0$. Probability $w^b(t)$ is calculated according to the expressions (179), while probabilities $w_\tau^a(t)$ are calculated according to the expressions (182) and (184). The sum $w^b(t) + w_1^a(t) + w_2^a(t)$ of those probabilities is also shown. As required by the completeness relation (185), for each time t this sum equals unity.

In Figure 30 the same system is considered, but this time coupling β is as strong as $\beta = 1.5$. In this case combined system $\mathbf{C}_\infty$ has a single isolate eigenstate $|\Psi_I\rangle$ with the eigenvalue $\varepsilon_R = 2.3186$ [45]. Hence in a limit $t \to \infty$ one has $w^b(\infty) = 0.2523 \neq 0$ [45]. As shown in this figure, completeness relation (185) is again satisfied.

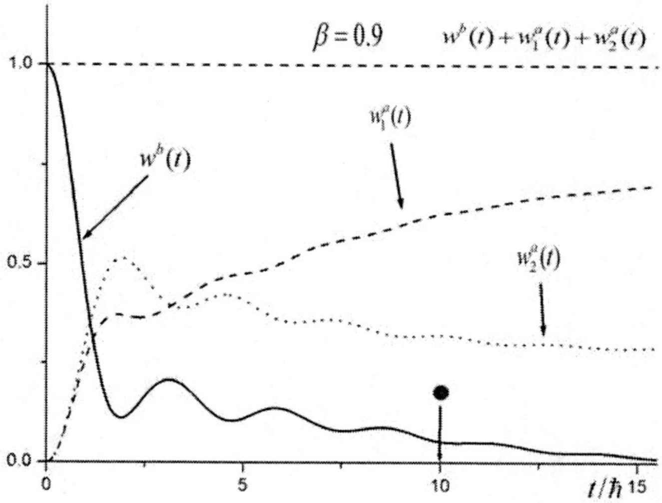

Figure 29. Example ***E5***. In this example the system $\mathbf{A}_\infty$ contains two eigenvalue bands. The case $\beta = 0.9$ is considered. Probabilities $w^b(t)$ (solid line) and probabilities $w_1^a(t)$ and $w_2^a(t)$ to find the state $|\Psi(t)\rangle$ at time t in the corresponding eigenvalue bands $\nu = 1$ and $\nu = 2$ (dashed and dotted line) are shown as functions of time t. In accord with the completeness relation (185), for each time t the sum of those probabilities equals unity. In the case $\beta = 0.9$ combined system contains no isolated eigenstate and hence $\lim_{t\to\infty} w^b(t) = 0$. Detailed probability densities $\rho_1^a(\varepsilon,t)$ and $\rho_2^a(\varepsilon,t)$ at time $t/\hbar = 10$ (position ●) are shown in Figures 31 and 32, respectively . Adopted from ref. [45].

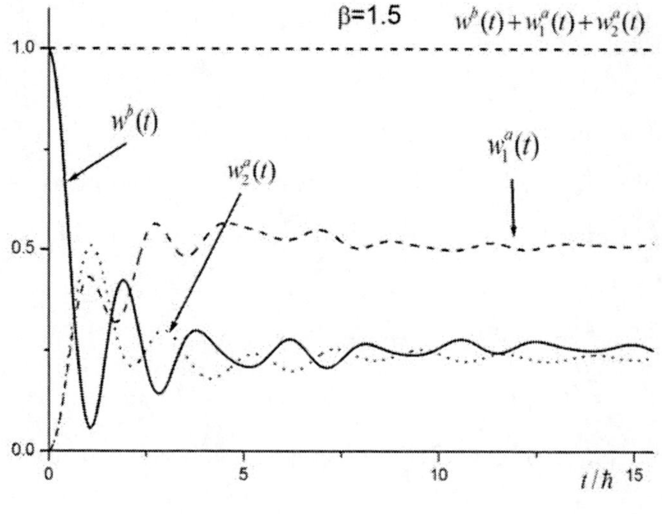

Figure 30. The same as in Figure 29, but for the coupling $\beta = 1.5$. In this case combined system contains an isolated eigenstate and hence $\lim_{t\to\infty} w^b(t) \neq 0$. In accord with the completeness relation, the sum of probabilities $w^b(t)$, $w_1^a(t)$ and $w_2^a(t)$ equals unity. Adopted from ref. [45].

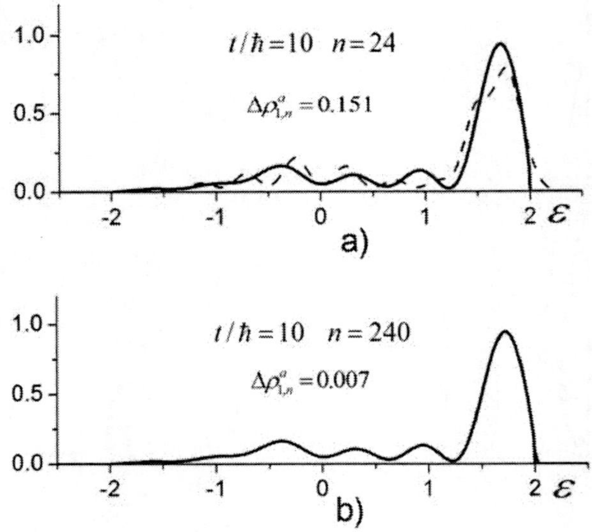

Figure 31. Probability density $\rho_1^a(\varepsilon,t)$ to find the state $|\Psi(t)\rangle$ at time $t/\hbar=10$ in the eigenvalue band $\nu=1$ and with the eigenvalue ε. This time corresponds to the position ● in Figure 29. a) Probability density $\rho_1^a(\varepsilon,t)$ (solid line) compared with the result for the corresponding finite system $\mathbf{A}_{24}$ (dashed line). Average standard deviation between those two curves is $\Delta\rho^a_{1,24}=0.151$. b) The same as a) but this time the corresponding finite system is $\mathbf{A}_{240}$. Average standard deviation drops to $\Delta\rho^a_{1,240}=0.007$. Adopted form ref. [45].

In Figures 31 and 32 are analyzed details of the probability densities $\rho_\tau^a(\varepsilon,t)$ ($\tau=1,2$). Those probability densities correspond to the case $\beta=0.9$ and to the time $t/\hbar=10$. This time is shown in Figure 29 with symbol ● . In Figure 31 probability density $\rho_1^a(\varepsilon,t)$ derived from the expressions (182) (solid lines) is compared with the corresponding results obtained in a standard way for finite systems $\mathbf{A}_n$. Systems $\mathbf{A}_n$ containing $n=24$ atoms (Figure 31a) and $n=240$ atoms (Figure 31b) are considered (dashed lines). Standard deviations $\Delta\rho^a_{1,n}$ between exact (combined system $\mathbf{C}_\infty$) and approximate (combined systems $\mathbf{C}_{n+1}$) results are also shown. Those standard deviations decrease from $\Delta\rho^a_{1,24}=0.151$ to $\Delta\rho^a_{1,240}=0.007$ with the increase of *n* from $n=24$ to $n=240$, respectively. In Figure 31b) there is no noticeable difference between exact ($n=\infty$) and approximate ($n=240$) calculation. In Figure 32 the same is shown for probability density $\rho_2^a(\varepsilon,t)$ associated with the eigenvalue band $\tau=2$. As shown in those Figures, as *n* increases approximate results obtained for finite systems converge to the exact LRM result.

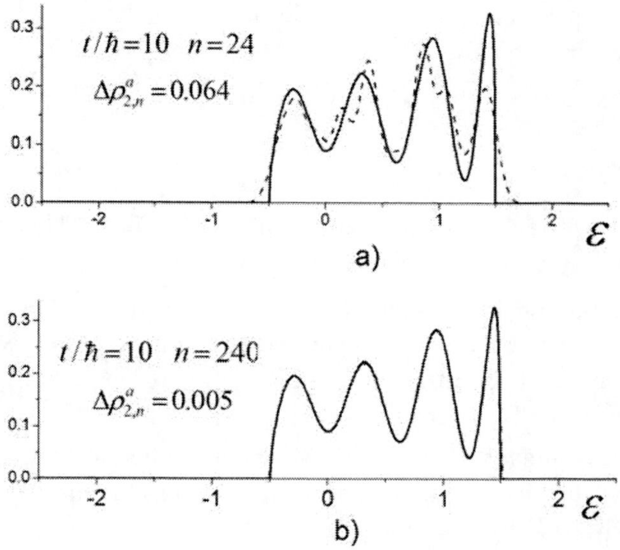

Figure 32. The same as Figure 31, but for the eigenvalue band $\nu = 2$. a) Comparison of the probability density $\rho_2^a(\varepsilon,t)$ with the result obtained for the approximate finite system $\mathbf{A}_{24}$. Average standard deviation between the exact and approximate calculation is $\Delta\rho_{2,24}^a = 0.064$. b) The same as a), but this time the corresponding finite system is $\mathbf{A}_{240}$. Average standard deviation drops to $\Delta\rho_{2,240}^a = 0.005$. Adopted form ref. [45].

Example ***E5*** demonstrated that LRM correctly predicts all the fine details of the time-dependent systems, however strong the coupling β between the base system $\mathbf{B}$ and the infinite system $\mathbf{A}$. The case $\beta = 0.9$ as well as the case $\beta = 1.5$ is an example of a rather strong coupling. As shown in Figure 29, even in the case $\beta = 0.9$ small coupling approximation (181b), which predicts exponential decay of the state $|\Psi(t)\rangle$, breaks down. In particular, as time t increases exact probability $w^b(t)$ at first decreases, but at some point it starts to increase. This behavior is contrary to the exponential decay approximation and it is not reproducible by the standard perturbation expansion. Deviation from this small coupling approximation is much more drastical in the case $\beta = 1.5$ (Figure 30).

Note that in a standard experimental set up one usually measures $t \to \infty$ limit of above probabilities and probability densities. Also standard perturbation formalism gives only those limit results. Of course, LRM provides information about those limit quantities as well. However, LRM provides in addition detailed information about probabilities $w_\tau^a(t)$ to find the state $|\Psi(t)\rangle$ at each particular time t in the eigenvalue band τ, as well as detailed probability densities $\rho_\tau^a(\varepsilon,t)$ to find this state in the eigenvalue band τ at each particular time t and with each particular eigenvalue ε. This additional information could be utilized, with an appropriate experimental set up, in order to enhance as much as possible transition probabilities into the appropriately selected final states.

20. Conclusion

In this chapter Low Rank Modification (LRM) is described. This approach is a general mathematical method by which one can express the eigenvalues and the eigenstates of the modified system $\mathbf{C}$ in terms of the eigenvalues and the eigenstates of the parent system $\mathbf{A}$. In this respect LRM is similar to the standard perturbation expansion which also expresses perturbed eigenvalues and eigenstates in terms of the unperturbed eigenvalues and eigenstates. However, unlike perturbation expansion, LRM is exact however large "modification" of the original system $\mathbf{A}$. LRM can be applied to finite as well as to infinite systems. It can be also applied to time-independent as well as to time-dependent systems. The only requirement on this method is that the operators which represent modification of the parent system $\mathbf{A}$ have finite rank. Numerical complexity of LRM depends essentially on the rank ρ of those operators.

LRM was originally developed for finite-dimensional time-independent systems [3]. Those systems are represented by generalized matrix eigenvalue equations that may contain arbitrary matrices of any type, not necessarily Hermitean [3]. In such a general formulation, left and right eigenstates may differ [3]. However, in physics and chemistry most important are Hermitean eigenvalue equations. Hence in this chapter are treated only such equations.

In LRM it is convenient to distinguish "internal" and "external" modifications of the parent system $\mathbf{A}$. In an internal modification modified system $\mathbf{C}$ has the same dimension as the parent system $\mathbf{A}$. Example of an internal modification of a finite-dimensional system is replacement of one or several atoms in a large molecule with various isotopes. Such a modification of the original system has a strong influence on molecular vibrations. In addition, provided relatively few isotopes are substituted, Such a modification is represented by operators that have relatively low rank. Hence vibrational isotope effect can be efficiently treated within the LRM formalism. In the similar way can be treated internal modifications of infinite-dimensional systems. As an example consider a perfect infinite-dimensional solid. Due to the periodicity of the effective potential one can obtain more or less reliable solution to this system [40]. However such perfect solid is an ideal approximation of a real solid, and most important and most interesting properties of real solids are due to various impurities and/or defects, existence of a surface which may generate surface states, etc. [40-42]. All such modifications of the original system break periodicity of this system and in addition those modifications are usually not small. Hence such modifications cannot be efficiently treated within the standard perturbation formalism. However, introduction of impurities and defects, as long as those modifications do not produce a global change of the crystal structure, can be very well approximated with operators that have finite rank. Though the creation of a crystal surface is not a low rank modification, it can still be approximated as such a modification. One can, for example, approximate a real two-dimensional crystal surface which is infinite-dimensional with a finite portion of this surface. This is of course an approximation. Nevertheless, to the extent this approximation is acceptable, all problems associated with the creation of a surface, such as surface states, distortion of few unit cells close to the surface and alike can be treated within the LRM formalism.

In an external modification parent system $\boldsymbol{A}$ interacts with an external finite-dimensional system $\boldsymbol{B}$ that is outside this parent system. Systems $\boldsymbol{A}$ and $\boldsymbol{B}$ that interact with each other form a combined system $\mathbf{C}$. An example of such a modification involving infinite-dimensional systems is the interaction of a molecule with a radiation. Molecule in isolation can be approximated with a finite-dimensional system $\boldsymbol{B}$. Radiation in isolation can be approximated with an infinite dimensional system $\mathbf{A}$. Reliable approximate solution to this system is known since this system essentially represents a free electromagnetic field. LRM provides (within the given approximations) an exact solution of the combined system $\mathbf{C}$. Within the LRM formalism one can describe time-independent effects such as shift of molecular eigenvalues due to the interaction with radiation as well as broadening of those levels due to this interaction. One can also describe time-dependent properties such as probabilities and probability densities for the transitions between various eigenstates of this molecule. Since the coupling with the radiation is very weak, all those effects can be efficiently described within the standard perturbation expansion [1, 2]. Nevertheless, LRM provides a new way of describing all such effects.

As another example of the external modification consider the interaction of a molecule situated on a surface of a solid with this solid. Unlike the interaction of a molecule with radiation, this interaction is usually so strong that standard perturbation expansion fails. Here LRM offers a possibility for completely new and innovative approach to such problems.

In conclusion, there are many important problems which can be treated within the LRM formalism and which involve either external or internal modifications of known finite and infinite-dimensional systems. LRM can be also generalized to those cases where both types of modification, external and internal, are simultaneously present. Even in the case of those systems that can be efficiently treated by some other known method (such as interaction of matter with radiation), LRM offers a new insight and a new way of thinking about such systems. As Sir William Brag remarked: "The most important thing in science is not so much to obtain new facts, as to discover new ways of thinking about them". LRM hopefully offers a new way of thinking about many old problems.

References

[1] Messiah, A. *Quantum Mechanics;* 3th ed; North-Holland: Amsterdam , 1965.

[2] Atkins, PW; Friedman, RS. *Molecular Quantum Mechanics;* 3th ed ; Oxford University Press: Oxford, 1997.

[3] Živković, TP. *Theor. Chim. Acta.,* 1989, 76, 331-351.

[4] Živković, TP. *J. Math. Chem.,* 2002, 32, 39-71.

[5] Živković, TP. *J. Math. Chem.,* 2009, 45, 627-701.

[6] Živković, TP. *J. Mol. Struct.*, 2009, vol 924-929, 477-492.

[7] Živković, TP. IJQC, 2009, 109, 2831-2859.

[8] Golub, GH; Van Loan, CF. *Matrix computations;* 3th ed; *The Johns Hopkins University Press: Baltimore*, 1996.

[9] Press, WH; Teukolsky, SA; Vetterling, WT; Flannery, BP. *Numerical Recipes in C, The Art of Scientific Computing;* 3th ed; Cambridge university press: *Cambridge*, 2007.

[10] Herzberg, G. *Molecular Spectra and Molecular Structure, II Infrared and Raman Spectra of Polyatomic Molecules*; Van Nostrand Company, New York, 1954.

[11] Bright Wilson, E; Decius, JC; Cross, PC. *Molecular Vibrations, The Theory of Infrared and Raman Vibrational Spectra;* Dover Publications, Inc.: New York, 1980.

[12] Sverdlov, LM; Kovner, MA; Krainov, EP. *Kolebatelnie spektri mnogoatomnih molekul;* Nauka: Moskva, 1970.

[13] Varsanyi, G. *Vibrational Spectra of Benzene derivatives*; Akademiai Kiado: Budapest, 1969.

[14] Mathews, J; Walker, RL. *Mathematical Methods of Physics*; Benjamin, W.A. Inc,Menlo Park: California, 1970.

[15] Živković, TP. *Croat. Chem. Acta.,* 1999, 72, 925-944.

[16] Živković, TP. *J. Math. Chem.*, 2000, 28, 267-285.

[17] Živković, TP. *J. Math. Chem.*, 2000, 28, 287-312.

[18] Teller, E; Angus, WR; Bailey, CR; Hale, JB; Ingold, CK; Leckie, AH; Raisin, CG; Thompson, JW; Wilson, CL. *J. Chem. Soc.*, 1936, 971-987.

[19] Decius, JC; Wilson, EB. *J. Chem. Phys.*, 1951, vol 19, 1409-1412.

[20] Brodersen, S; Langseth, AA. *J. Mol. Spectr.*, 1959, 3, 114-122.

[21] Brodersen, S; Iohannesen, T. *J. Mol. Spectr.*, 1967, 23, 293-301.

[22] Živković, TP. *J. Math. Chem.*, 2009, 45, 1060-1101.

[23] Tomić, A; Živković, TP. *Croat. Chem. Acta*, 2009, 82, 561-566.

[24] (a) Klein, DJ; Bytautas, L. *Commun. Math. Comput. Chem*. (MatCh), 2000, 42, 261-290. (b) Ivanciuc, T; Klein, DJ. *J. Chem. Inf. Comput. Sci*., 2004, 44, 610-617.

[25] Rico, M; Orza, JM; Morcillo, J. *Spectrochim. Acta,* 1965, 21, 689-719.

[26] El-Azhary, AA; Hilal, RH. *Spectrochimica Acta Part A,* 1997, 53, 1365-1373.

[27] www.periodictable.com.

[28] Moller, C; Plesset, MS. *Phys. Rev*., 1934, 46, 618-622.

[29] Roothaan, CCJ. *Rev. Mod. Phys*., 1951, 23, 69-89.

[30] Wu, A; Cremer, D. *J. Phys. Chem. A,* 2003, 107, 10272-10279.

[31] Becke, AD. *J. Chem. Phys*., 1993, 98, 5648-5652.

[32] Becke, AD. *Phys. Rev. A,* 1988, 38, 3098-3100.

[33] Lee, C; Yang, W; Parr, RP. *Phys. Rev. B*., 1988, 37, 785-789.

[34] Harihara, PC; Pople, JA. *Theor. Chim. Acta*, 1973, 28, 213-222.

[35] Rauhut, G; Pulay, P. *J. Phys. Chem*., 1995, 99, 3093-3100.

[36] Pulay, P; Fogarasi, G; Pongor, G; Boggs, JE; Vargha, A. *J. Am. Chem. Soc*., 1983, 105, 7037-7047.

[37] Baker, J; Jarzecki, A.A; Pulay, P. *J. Phys. Chem. A* 1998, 102, 1412-1424.

[38] Hrenar, T; Rauhut, G; Werner, H-J. *J. Phys. Chem. A* 2006, 110, 2060-2064.

[39] Scott, AP; Radon, L. *J. Phys. Chem*., 1996, 100, 16502-16513.

[40] Ashcroft, NW; Mermin, ND. *Solid State Physics*; Harcourt Brace: New York, 1976.

[41] Zangwill, A. *Physics at Surfaces*; Cambridge University Press: Cambridge, 1988.

[42] Kittel, G. *Introduction to Solid State Physics*, 5th ed; John Wiley & Sons. Inc. 1976.

[43] Živković, TP. *J. Math. Chem*., 2002, 32, 99-158.

[44] Živković, TP. *J. Math. Chem*., 2006, 39, 295-344.

[45] Živković, TP. *J. Math. Chem*., 2006, 39, 151-176.

[46] Živković, TP. *Croat. Chem. Acta.*, 2004, 77, 279-294.

[47] Abramowitz, M; Stegun, IA. *Handbook of mathematical functions*, National bureau of standards, Applied mathematical series, 55, 1964;

[48] Wichmann, EH. *Quantum Physics*; Berkeley Physics Course, vol 4; McGraw-Hill: New York, 1971.

In: Mathematical Chemistry
Editor: W.I. Hong, pp. 153-246

ISBN: 978-1-60876-894-3

Chapter 2

MOLECULAR SYMMETRY AND FUZZY SYMMETRY

Xuezhuang Zhao*
Department of Chemistry, Nankai University, Tianjin, P.R.China

Abstract

Since ancient times, human beings have connected beauty with symmetry. In science, this correlation is found in many areas, from the fundamental laws of physics to the products of modern chemistry. However, it seems emphasize the deadness of perfect symmetry or the dissymmetry makes the phenomenon. As well known, most molecules will become less symmetrical as substitutions take place. Here, we turn to a consideration of how to describe the effect of this kind of imperfect symmetry on molecular properties. Previously there are two general understandings on this issue: one is that the existence of the imperfect symmetry is denied, and there is not any symmetry here, i.e. imperfect symmetry means having none symmetry completely; another one is that the imperfection of the symmetry is completely neglected, and the system can be seen still having or almost having the previous symmetry. In fact, the previous symmetry has been damaged in local part of the system, however, the symmetry ought to be existence partially and may be called fuzzy symmetry. Recently, we have done some studies in this field and by using fuzzy mathematics theory the static and dynamic molecular fuzzy symmetry in relation to the point group symmetry, space periodic and some special transformations have been analyzed. The fuzzy symmetry characteristics of various molecules are studied in detail and the calculation methods for the relative parameters of molecule and molecular orbital are established. Some interesting results are obtained. It seems that there is a wonderful science virgin land waiting the theoretic chemists to open up.

1. Introduction

In Nature, there are many conservation rules to describe the invariants, of which certain symmetry properties are always included in the background. Seemingly, Nature is seeking the wonderful harmony of symmetry. In chemistry, accompanying the development of Man's knowledge toward the microcosmos, the inherent internal structure of molecules have been manifested and described by means of group theory with many successes. As the conservation

---

* E-mail address: zhaoxzh@nankai.edu.cn. (Corresponding author)

of orbital symmetry was established by R. B. Woodward and R. Hoffmann, more 40 years ago[1], symmetry was used to describe chemical reaction kinetics. These conservation rules and relative selection rules have been adopted by most chemists with many successful applications. However, István and Magdolna Hargittai[2] emphasize the deadness of perfect symmetry, citing Pierre Curie's "dissymmetry makes the phenomenon". As is well known, most molecules will become less symmetrical as substitutions take place. In such cases, although the higher symmetry ought to be damaged, it can be seen as still maintained partially. Here, we turn to a consideration of how to describe the effect of this kind of imperfect symmetry on molecular properties. Previously the general understandings on this issue are the following two extremism points: One is that the existence of the imperfect symmetry is denied, and there is no symmetry here, i.e. imperfect symmetry means having no symmetry. Another is that the imperfection of the symmetry is completely neglected, and the system may be seen still having or almost having the previous symmetry. Both viewpoints seem reasonable, but they are *ex parte*. In fact, the previous symmetry has been damaged in local part of the system. However, it ought to be existence partially. This kind of symmetry can be called fuzzy symmetry and analyzed by means of fuzzy mathematics theory. Even some persons say, "Are the concepts of chemistry all fuzzy?"[3]. However, the researches on the fuzzy characteristics of chemical symmetry are still poor. Recently, we have done some thing in connection with this field. Some interesting results are obtained, and going to appear. It may be expected that according to fuzzy symmetry, the corresponding selection rules ought to be shown not only the "yes or no" but also the "probability" for relative dynamic processes. To start with we will analyze some concepts, especially the ambiguous ones and using the some symmetry theorem in field theory[4], in relation to common molecular symmetry, and then introduce our work in connection with the fuzzy symmetry of molecule and molecular orbital (MO).

2. Generalized Parity

Last century, one of the most important achievements in theoretical chemistry was the conservation principle of orbital symmetry, i.e. Woodward-Hoffmann rules (W-H rule). Owing to the greatness of the accomplishment, some defects in connection with the W-H rule have been neglected. An important theorem (Noether's theorem[4]) reveals the inherent relation between the symmetry and the invariant in general. There are two different concepts (symmetry and invariant), and they are not distinguished in W-H rules. However, they ought to be distinguished. In fact, the molecular symmetry corresponds to the molecular symmetry group and relative invariant ought to the character of irreducible representation.[5] As for the point symmetry group, the character is the eigenvalue of a certain symmetric transformation. In field theory, the eigenvalue of inversion transformation is called the parity. And then, the eigenvalue of various point symmetry transformation is called various point-parity. Corresponding to space symmetry groups, there are various space-parity. Corresponding to other special symmetry groups, there is special parity. All these parities may be called the generalized parity.

2.1. Generalized Parity in the Environment Space with Point Group Symmetry

An orbital is a single electron wave function, $\Psi(\boldsymbol{r},t)$. The environment space is constructed by the field of other electrons and the nuclei or cores and it is not always homogeneity and isotropy. During a chemical process the environment space may vary, but it often may still keep certain point group symmetries throughout. The potential energy surface from the Born-Oppenheimer approximation [6,7] may comply with certain point group symmetries throughout the process. Certainly, there are some conservation rules which are responsive to various point group symmetries. Point group symmetry transformations ($\hat{G}$) are discrete and the corresponding invariance must be termed as generalized parities. In chemistry MO is the one electron wave function, and during chemical reaction, accompanying the movement of an electron in the environment space of the whole system a certain MO disappears while another MO appears.

Since all of the point-parity transformations ($\hat{G}$) are unitary, according to the Noether's theorem, a one-to-one correspondence can be established between the relative symmetry properties and the invariants [4]. For the common point group symmetry properties , i.e. the invariance under ($\hat{G}$), the relative transformations and their invariants as shown in Table 1.

By means of field theory, we may get the point-parities in spherical environment space as follows:

$$\text{the parity } p = p^0\,(-1)^l \quad (1a)$$

$$\text{the mirror-parity } m = m^0(-1)^{l+m} \quad (1b)$$

$$\text{the axis-parity } c = c^0 \exp(2\pi n' mi/n) \quad (1c)$$

where the p^0, m^0 and c^0 are the corresponding intrinsic point-parities. They may be called the intrinsic parity, intrinsic mirror-parity and intrinsic axis-parity, respectively. Supposing the spin quantum number (s) is included, so long as we substitute the m in eq(1a-c) for m+s. These results may be verified for atomic orbital(AO). As for the intrinsic point-parities of electron, because the number of electrons is invariable throughout the chemical processes, we need not consider them, which may always be cancelled. For convenience we can let all of the electronic intrinsic point-parities have equal unit value.

In fact, the point-parities are equal to the corresponding characters of the one-dimensional irreducible representations of the relative reaction system. According to Noether's theorem, we can conclude that so long as the symmetry of a given point group is maintained through the whole process of a chemical reaction, each point-parity corresponding to such symmetry must be invariant. That is the principle of "The conservation of point-parity" [5].

Considering that the character of the irreducible representation and point-parity are correlated, it is obvious that so long as the symmetry of a certain point group is retained throughout the whole process, the corresponding irreducible representation of this reaction must be invariable. Based on the MO approximation, the corresponding irreducible representation of the single electronic state, i.e. the orbital, must also be invariant. That is the essence of the Woodward-Hoffmann's rule [1].

Table 1. Some point group symmetry properties and their corresponding invariants.

No,	Symmetry properties Relative transformations	Operators Symbol	Invariants(eigenvalues) Point-parity	 Symbol
1	Point-parity transformation	$\hat{G}$	point-parity	g
2	Parity (space inversion) transformation	$\hat{P}$	P-parity	P
3	Axis-parity transformation (rotation by an angle 2πn'/n about an axis)	$\hat{C}$	axis-parity	c
4	Mirror-parity transformation (reflection in a mirror)	$\hat{M}$	mirror-parity	m

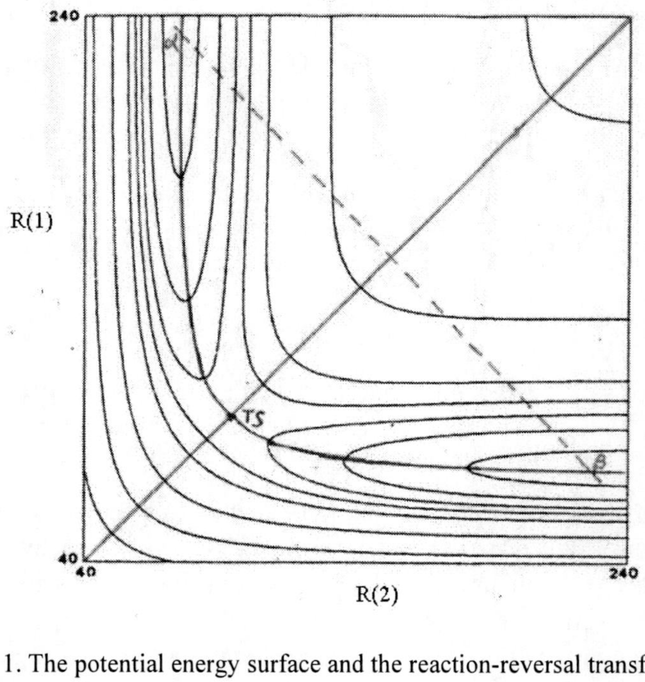

Figure 1. The potential energy surface and the reaction-reversal transformation.

According Noether's theorem [4], if the system is invariable under the certain unitary transformation group, there ought to be a conservation rule for a certain mechanical observable of the system. Here, the relative symmetry group must be the unitary group. Otherwise the Noether's theorem can be not set up. On the other hand, no matter what the continuous or the discrete symmetry group, the Noether's theorem may be true. But the relative observable invariants are different. Corresponding to the discrete symmetry the invariants are discrete generalized parity. And corresponding to the continuous symmetry, such as the time-space homogeneity and space isotropy, they are the continuous invariants, such as the energy, momentum, angular momentum and *etc*. By the way, the terminology 'continuous symmetry' had been introduced in some papers on the symmetric fuzzy subset theory, but the meaning seems not same as above 'continuous symmetry' in relation to the well-defined group.

2.2. Generalized Parity in the Intrinsic Reaction Space

For the sigmatropic reactions and some others, there is no any point group symmetry. For these reaction systems, there are some special symmetries other than the point group symmetry[5,8]. These symmetries are connected with the so-called reaction-reversal transformation, which ought to be a generalized parity transformation in the intrinsic reaction space.

For the interaction system of a chemical reaction, the configurations of the relative position of all the atomic cores or nuclei can be described with a multi-dimensional (*m*-dimensional) configuration space which is the frame of the potential energy surface. The reaction coordinate refers to a curve in this *m*-dimensional space. Across the center of this curve, there is an (m-1)-dimensional super-plane orthogonalized to the reaction coordinate curve. The operation of reflection at the (*m*-1)-dimensional super-plane is called the reaction-reversal transformation, and denoted $\hat{R}$. For a complex wave function, the transformation of Hermitian conjugation should be included in $\hat{R}$.

For example, there is the configuration space with the coordinates R(1) and R(2) (where $m = 2$), and blue curve is the so-called intrinsic reaction coordinate (IRC) and we may call such coordinate space the intrinsic reaction space as shown in Figure 1. The red diagonal line is so-call dividing line.

Where intersection point is in the center of IRC (blue) curve, and dividing (red) line is the (*m*-1)-dimensional super-plane (it is a straight line here) orthogonalized to the IRC curve, across the point TS. Chemists have an interest to the case where the center of the reaction coordinate is the transition state of the relevant reaction.

If $\Psi(\alpha)$ is a complex wave function corresponding to the point α, it will be transformed into the state function $\Psi^*(\beta)$ under the reaction-reversal transformation:

$$\hat{R}\,\Psi(\alpha) = \Psi^*(\beta) \tag{2a}$$

For real wave functions $\Psi = \Psi^*$, the $\hat{R}$ corresponds merely to the transition along the reaction coordinate from α to β, and

$$\hat{R}\,\Psi(\alpha) = \Psi(\beta) \tag{2b}$$

It may be noticed that by carrying out the corresponding operation $\hat{R}$, the molecular system of the chemical reaction may be transformed into a form which may be distinguished from the original one. That means:

$$\hat{R}\,\Psi(\alpha) \neq c\,\Psi(\alpha) \text{ nor } c\,\Psi^*(\alpha) \tag{3a}$$

where c is an arbitrary constant. However many reaction molecular systems, after carrying out the reaction-reversal transformation $\hat{R}$ and a certain point group symmetry operation $\hat{G}$ successively, will be transformed back into the original form, *i.e.*

$$\hat{R}\,\hat{G}\,\Psi(\alpha) = c\,\Psi(\alpha) \text{ or } c\,\Psi^*(\alpha) \qquad (3b)$$

Obviously, there is a new symmetry in such reaction systems. Such a kind of symmetry operator $\hat{R}\,\hat{G} = \hat{G}\,\hat{R}$ can be called the union transformation of point group and reaction-reversal, in which $\hat{G}$ usually is the transformation of $\hat{P}$, $\hat{M}$ or $\hat{C}_2$.

For example, considering the H_3-system as shown in Figure 2, the electronic configuration of H_3 is $(BMO)^2(NBMO)^1$, where BMO and NBMO are the bonding molecular orbital and non-bonding one, respectively. It can be seen that by performing the corresponding symmetry operation, $\hat{R}\,\hat{P} = \hat{P}\,\hat{R}$, the BMO (occupied by a pair of electrons) and NBMO (occupied by only one electron) are transformed. Both are the eigenstates of the symmetry transformation $\hat{R}\,\hat{P}$. However the eigenvalues, i.e. the invariants, r_p of BMO and NBMO are different in relation to $\hat{R}\,\hat{P}$. They are +1 and -1, respectively. Thus: $\hat{R}\,\hat{P}\,\Psi(BMO)=\Psi(BMO)$, and $\hat{R}\,\hat{P}\,\Psi(NBMO) = -\Psi(NBMO)$.

We must notice that such a system is provided with the symmetry of $\hat{R}\,\hat{P}$ but without that of the singular $\hat{R}$ nor $\hat{P}$. Notice that for this example (H_3) under the $\hat{P}$ both the geometry configuration and orbital phase are inverse through the symmetry center, while under the $\hat{R}$ there is only translation of the middle hydrogen atom, but the orbital phase is unaltered. Evidently, both $\hat{R}$ and $\hat{P}$ are the generalized parity transformations, and they are action on the intrinsic reaction space and orbital environment space, respectively.

Of course, the H_3-system is not the only system which possesses the union symmetry of point-group and reaction-reversal.

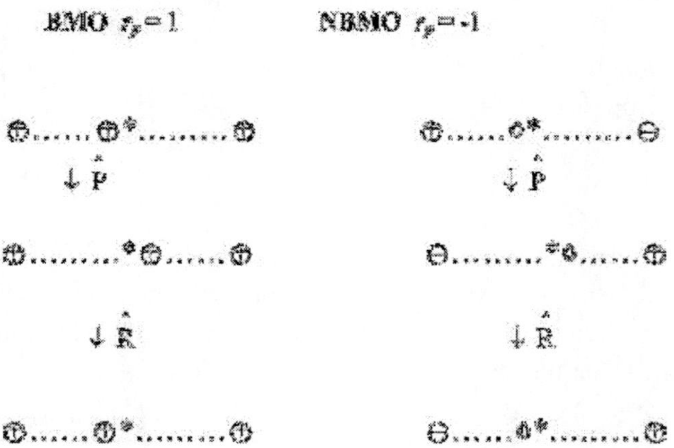

(The * is the inversion center, the ⓪ is the node, the ⊕ and ⊖ are the positive lobe and negative one of the orbital, respectively.)

Figure 2. The $\hat{P}\,\hat{R}$-symmetry of H_3-system: $(BMO)^2(NBMO)^1$.

2.3. Generalized Parity in the Environment Space with Space Periodic Symmetry

Similar to the point group transformation, if a system is invariant under certain space group transformation, according to Noether's theorem, from this symmetry property there follows the conservation of an observable quantity of this system [5,9]. We call this quantity the space-parity, which is the eigenvalue of the space group transformation. Obviously, the space-parity is the generalized parity corresponding to the discrete space group transformation. For example, corresponding to the translation transformation, $\hat{T}(na)$, the eigenvalue (wave-parity) of the eigenstate functional Φ_k should be exp(k*nai*) according to the Bolch's theorem[10], under the translation of the integral number (*n*) folds of the periodic length of potential field, *a* , so we get:

$$\hat{T}(na)\,\Phi_k = \exp(kna i)\,\Phi_k \tag{4}$$

where **k** is the wave vector. As mentioned above, we can be sure that so long as the symmetry of a certain translation transformation $\hat{T}(na)$ is maintained through the whole process, the wave-parity $\boldsymbol{\tau}^*(na) = \exp(kna i)$ must be invariant during this process. If *a* approaches zero, in such a case, the k will be the ordinary real number and means the momentum ought to be invariant. The conservation of wave-parity will then become the conservation of momentum. This situation is similar to what happens when the axis-parity conservation becomes angular momentum conservation in an isotopic environment space. In the finite periodic potential energy field, the conservation principle of wave-parity may illustrate the electron transition process of semi-conductor from the occupied band to the conduct band [5,11].

In addition to translation, the space symmetry transformations include the screw rotation and glide reflection. These space transformations can be analyzed in a similar manner. They may be also considered as the combination of translation and the corresponding point transformation. The corresponding space-parities may be obtained easily.

As for the environment time-space with time periodic symmetry, the relative system may be analyzed similarly with the space periodic symmetry [9]. For the environment time-space, if the time-space periodic length is very small but not null, we can get the formula similar to the uncertainty principle [12]. According to the relative result, it may be promote us, that there is a probable experimental definition for a time-space transformed from a periodic symmetry to homogeneity.

3. Molecular Fuzzy Symmetry Characteristics

As stated above, the molecular symmetry corresponds to the molecular symmetry group, *i.e.* the molecular symmetry is a property of molecular system: the molecular system ought to be unchanged under the symmetry transformation (the element of common symmetry transformation group) operation, in other words, the molecular invariable property under the symmetry transformation operation with the membership function to be one. On the ether hands, that it is "the deadness of perfect symmetry [2]". How we describe the imperfect symmetry? The fuzzy subset theory is one of the selectable ways. Similar as the common molecular symmetry, the molecular fuzzy symmetry corresponds to the molecular fuzzy symmetry subset, *i.e.* the molecular fuzzy symmetry is a property of molecular system, too: the molecule deviates somewhat from the original state under the symmetry transformation (the element of symmetry transformation fuzzy group or subset), in other words, it is the molecular invariable property under the symmetry transformation operation with the membership function less one. We may call it the fuzzy symmetry, reasonable. As the membership function less but near one, the fuzzy symmetry is the same as the approximate symmetry. However as membership function near zero, the fuzzy symmetry is not similar as the approximate symmetry. Although the terminology, 'fuzzy symmetry' may be disputed, still we use it.

In the area of theoretical chemistry, the molecular fuzzy symmetry subset is an interesting topic where some important results [13-23] have been obtained. Recently, we have done and going to analysis in connection with this field, more systematically [24-33]. Here, we turn to introduce that how to describe the molecular imperfect symmetry by means of the fuzzy set theory. Our purpose is to construct a generalized method of describing the fuzzy symmetry of various molecules and MOs using fuzzy set theory. These concepts ought to be useful to investigate the selection rules of spectrum and reaction process, not only"allowed or forbidden" but also the "probability". As mentioned above section, there are two different important concepts – symmetry and invariant. In connection with the fuzzy symmetry, there are two corresponding observable quanta – membership function and representation component, respectively.

3.1. The Membership Function of Molecular Fuzzy Symmetry

Corresponding to the molecular symmetry group, the fundamental question in designing a physically meaningful fuzzy set model is related to the choice of the membership function. In order to get the appropriate membership function, we label the atom of a molecule with its atomic number (Z_J) as a criterion (Y_J). If any atom J of a molecule M moves to the position of the atom GJ after the operation $\hat{G}$ of point group G is carried out, then the membership function of molecular skeleton in regards to operation $\hat{G}$ on the molecule M concerning the point group G is taken as follows,

$$\mu_Z(\hat{G}/G;M)=[\Sigma_J(Y_J \wedge Y_{GJ})]/[\Sigma_J(Y_J)]=[\Sigma_J(Z_J \wedge Z_{GJ})]/[\Sigma_J(Z_J)] \qquad (5)$$

Where, the atomic numbers Z_J and Z_{GJ} of atoms J and GJ represent the criteria, respectively. Here the operation $x \wedge y$ is introduced, and the result of this operation is equal to the value of the smaller one of x and y. Z in equation (5) is often designated by a general form of atomic criterion, Y. Here the atomic criteria(Y_J and Y_{GJ}) is the atomic numbers (Z_J and Z_{GJ}). We can also select other criteria for different objects. If the molecule M belongs to a common point group G, in the internal configuration space, then $Y_J=Y_{GJ}$ holds for all the atoms, consequently, the membership function of each operation of the point group G is one. While if the molecule M belongs to a fuzzy point group $\underset{\sim}{G}$ of the corresponding point group

G, then $Y_J = Y_{GJ}$ doesn't hold for some of the operations, and the values of the membership functions $\mu_Z(\hat{G} / \underset{\sim}{G};M)$ of these operations are smaller than one.

It is well known that, chemically, the most active part in a molecule or a collection of molecules is the electrons, which move in the space (or say the field) of the framework of the nuclei. Therefore, the symmetry of the electronic state shouldn't be neglected. Chemically, the state of the movement of electrons in a molecule is described with molecular orbital(MO). Obviously, it is unsuitable to describe the contribution of each atom to the symmetry of a certain MO with the membership function defined with the nuclear charge. It seems that the electron population on atoms in molecular orbital is more suitable criterion (Y_J) for the characterization of the membership function.

Table 2. The membership functions of the inversion for ten hetero-nuclear diatomic molecules.

molecule	μ	molecule	μ
HI	0.037	ClF	0.692
HBr	0.056	BN	0.833
HCl	0.111	CO	0.857
HF	0.200	CN	0.923
HLi	0.500	NO	0.933

First, we consider the diatomic molecules. For a homo-nuclear diatomic molecule, according to the internal configuration (one dimensional) space, there are only two symmetry operations, the identity and the inversion considered. Therefore, it belongs to the point group $C_i = \{\hat{E}, \hat{P}\}$. Since the two atoms of a homo-nuclear diatomic molecule (denoted as A_2) have the equal nuclear charge, Z_A, then

$$. \mu_Z(\hat{E} / C_i;A_2) = \mu_Z(\hat{P} / C_i;A_2) = 1 \tag{6}$$

For a hetero-nuclear diatomic molecule (denoted as AB), there is only one symmetry operation, the identity, in above point group. Hence, it doesn't belong to the point group C_i, but it can be classified to the fuzzy point group, $\underset{\sim}{C_i}$. For simplicity, we assume that $Z_A < Z_B$, then,

$$Z_A = Z_{EA} = Z_{PB} < Z_B = Z_{PA} = Z_{EB}$$

Further, for the whole framework of a hetero-nuclear diatomic molecule, the corresponding membership functions satisfy,

$$\mu_Z(\hat{E} / \underset{\sim}{C_i} ; AB) = \mu_{ZE} = 1 ; \tag{7a}$$

$$\mu_Z(\hat{P} / \underset{\sim}{C_i} ; AB) = \mu_{ZP} = 2Z_A/(Z_A+Z_B) < 1 \tag{7b}$$

According to the idea of Zadeh [34], the fuzzy point group to which a hetero-nuclear diatomic molecule belongs can be expressed as:

$$\underset{\sim}{C_i} = \frac{1}{\hat{E}} + \frac{\mu_P}{\hat{P}} \tag{8}$$

Where $\mu_P = \mu_{ZP}$ denoted the membership function in relation to inversion symmetry (7b).

In Table 2 lists the values of the membership functions of the inversion for ten hetero-nuclear diatomic molecules calculated according to equation (7b). We may denote their fuzzy point group according to eq(8) easily, *e. g.* the fuzzy point group to which ClF molecular skeleton belongs ought to be as:

$$\underset{\sim}{C_i} = \frac{1}{\hat{E}} + \frac{0.692}{\hat{P}} \tag{8a}$$

Since a group is a kind of algebra, it contains two essential constituents, a set of elements and well-defined combining operation. The set of elements of a fuzzy point group is called fuzzy set and operation, the group operation. Certainly, the set of elements of a fuzzy point group satisfy the rules concerning the combining operation. In addition, since the membership functions of all the elements (the symmetry operations) of the fuzzy point group take the values between 0 and 1, they should satisfy the following equations [21,34],

$$\mu(x) = \mu(x^{-1}) \tag{9a}$$

$$\mu(xy) \geq \mu(x) \wedge \mu(y) \tag{9b}$$

$$\mu(e) = 1 \tag{9c}$$

Where x and y are elements of the fuzzy set, and e is the identity. The equation (9a) indicates that the membership function of x equals that of its inverse, x^{-1}. The equation (9b) shows that the membership function of the combined element (xy) of x and y is no less than the smaller one of the two membership functions of the two elements, x and y. The equation (9c) implies that membership function of the identity is the greatest among those of all the elements of a fuzzy point group and it always takes the value of 1 for the fuzzy point groups. Obviously, the fuzzy point group $\underset{\sim}{C_i}$ to which the hetero-nuclear diatomic molecules belong satisfy all the rules. However, these rules are not always true for some polyatomic molecules.

As is well known, if a molecule belongs to a certain point group G, then any MO of this molecule belongs to an irreducible representation of the point group G. However if a molecule belongs to a certain fuzzy point group $\underset{\sim}{G}$, the membership functions of molecule and MOs may be different.

Based on the nuclear charge we have discussed the membership functions of operations on a molecular skeleton which belongs to a certain fuzzy point group. Comparatively, we can get the membership functions of operations on a MO with certain fuzzy symmetry by switching attention to the electron population in this MO. Thus, according to LCAO-MO theory, the ρth MO Ψ_ρ can be written as:

$$\Psi_\rho = \sum_J \sum_i a_\rho (J,i)\ \phi(J,i) \tag{10}$$

Where $\phi(J,i)$ is the ith AO of the Jth atom, and $a_\rho(J,i)$ is the corresponding linear combination coefficient of it.

According to simple MO theory,

$$NE_{J\rho} = \sum_i NE_{J\rho\iota} = \sum_i a_\rho^*(J,i)\ a_\rho (J,i) \tag{11a}$$

In which, $NE_{J\rho}$ and $NE_{J\rho\iota}$ can characterize the contribution of electrons of MO-Ψ_ρ to atom J and i-AO, respectively. They are the criteria ($Y_{J\rho}$ and $Y_{J\rho\iota}$) of J-atom and i-AO for MO-Ψ_ρ. According to the normalized condition,

$$\sum_J \sum_i a_\rho^*(J,i)\ a_\rho(J,i) = \sum_J\sum_i NE_{J\rho i} = \sum_J NE_{J\rho} = 1 \tag{11b}$$

Subsequently, replacing Z in equation (5) with NE we get the membership function for MO-Ψ_ρ. The membership function of operation $\hat{G}$ of a certain fuzzy point group $\underset{\sim}{G}$ for a MO such as Ψ_ρ can be written as,

$$\mu(\hat{G}/\underset{\sim}{G};\Psi_\rho) = [\textstyle\sum_J(NE_{J\rho} \wedge NE_{GJ\rho})] / [\sum_J(NE_{J\rho})]$$

$$= [\textstyle\sum_J((\sum_i a^*_\rho(J,i)\ a_\rho(J,i)) \wedge (\sum_i a^*_\rho(GJ,i)\ a_\rho(GJ,i))]$$

$$/[\textstyle\sum_J(\sum_i a^*_\rho(J,i)\ a_\rho(J,i))] \tag{12}$$

where $\sum_J$ indicates the sum over all the atoms in the molecule M. If the molecular orbital Ψ_ρ possesses the symmetry of a common point group G, then it belongs to a certain irreducible representation of G, and we can characterize the contribution of the molecular orbital Ψ_ρ to the electron population on atom J with $a^*_\rho(J,i)a_\rho(J,i)$. Further, by using a suitable normalized factor, the equation can be normalized as follows,

$$\mu(\hat{G}/\underset{\sim}{G};\Psi_\rho) = \textstyle\sum_J[(\sum_i a^*_\rho(J,i)\ a_\rho(J,i)) \wedge (\sum_i a^*_\rho(GJ,i)\ a_\rho(GJ,i))] \tag{13a}$$

Here, we first sum up all the electronic charge of all the AOs for atom J, then consider the effect of this total electronic charge on the fuzzy symmetry. While the more proper method is to consider the effect of each AO on the fuzzy symmetry, then get the sum, thus we can get equation (13b):

$$\mu(\hat{G}/\underset{\sim}{G};\Psi_\rho) = \textstyle\sum_J\sum_i [(a^*_\rho(J,i)\ a_\rho(J,i)) \wedge (a^*_\rho(GJ,i)\ a_\rho(GJ,i))]$$

$$= \textstyle\sum_J\sum_i [a^2_\rho(J,i) \wedge a^2_\rho(GJ,i)] \tag{13b}$$

The last equation in (13b) holds for real number $a_\rho(J,i)$. The fuzzy point group to which a MO of hetero-nuclear diatomic molecules belongs can be expressed as eq(8), too. However, the μ_P ought to get from (13).

3.2. The Fuzzy Representation and Representation Component of Molecular Orbital

It is well known that MOs of a homo-nuclear diatomic molecule, which belongs to the C_i point group, can be classified to A_g and A_u irreducible representations. The subscripts g (symmetrical) and u (anti-symmetrical) imply that the parities of the corresponding MOs are +1 and –1, respectively. For a hetero-nuclear diatomic molecule, it should belong to $\underset{\sim}{C}_i$ fuzzy point group, and the MOs of it should belong to main fuzzy irreducible representations $\underset{\sim}{A}_g$ and $\underset{\sim}{A}_u$. Similarly, the subscripts g and u imply that the fuzzy space parities of the corresponding MOs are +1 and –1, respectively. The so-called fuzzy parity means that only the significant phases of the LCAO for the MOs are considered and the combination coefficients are neglected. The difference in coefficients can be reflected by the difference in the values of the membership functions. The above discussion can also be applied to general linear molecules. The foregoing discussed $\underset{\sim}{A}_g$ or $\underset{\sim}{A}_u$ main fuzzy representation is the linear combination of the X_g-component, in relation to Ag-irreducible representation, and the X_u=(1-X_g)- component, in relation to A_u-irreducible representation. For example, in the ρ-MO, the coefficient of *i*-AO on atom J is $a_\rho(J,i)$, which can be divided into two parts: symmetrical (g) and anti-symmetrical (u),

$$a_\rho(J,i) = a_\rho(g;J,i) + a_\rho(u;J,i) \tag{14}$$

After operation $\hat{G}$, the corresponding normalized LCAO coefficient is:

$$a_\rho(GJ,i) = a_\rho(g;GJ,i) + a_\rho(u;GJ,i) = a_\rho(g;J,i) - a_\rho(u;J,i) \tag{14a}$$

Consequently, the $a_\rho(g;J,i)$ and $a_\rho(u;J,i)$ can be obtained using $a_\rho(J,i)$ and $a_\rho(GJ,i)$. It can be proved that, these coefficients satisfy the following equality,

$$a_\rho^2(J,i) + a_\rho^2(GJ,i) = a_\rho^2(g;J,i) + a_\rho^2(u;J,i) + a_\rho^2(g;GJ,i) + a_\rho^2(u;GJ,i) \tag{14b}$$

The components of symmetrical part and anti-symmetrical part in ρ-MO are

$$X_g = \sum_J \sum_i a_\rho^2(g;J,i) \tag{15a}$$

and

$$X_u = 1 - X_g = \sum_J \sum_i a_\rho^2(u;J,i) \tag{15b}$$

respectively. The relative fuzzy representations may be denoted as:

$$\Gamma = X_g A_g + X_u A_u. \tag{15c}$$

4. The Fuzzy Symmetry Characteristics of Linear Small Molecule

For a linear molecule, according to the internal configuration (one-dimensional) space, there may be only two symmetry point groups, $C_i = \{ \hat{E}, \hat{P} \}$ and $C_1 = \{ \hat{E} \}$. According to the fuzzy symmetry, we may consider the linear molecule belong to the fuzzy point group $\underset{\sim}{C_i}$, eq(8). Taking the ClF molecule as an example, at the AM1 and B3LYP/6-31G* levels, the results at both levels are similar, which are shown in Figure 3. In this Figure, the MOs areσ_{ss},σ_{ss}*,σ_{pp}, π_{pp}, π_{pp}, π_{pp}*, π_{pp}* andσ_{pp}* in the order number of MO from 7 to 14, respectively, that is the energy order (from low to high). It may be shown that four of the eight MOs have fuzzy representation with X_g>0.5(main fuzzy symmetrical) and the other with X_u=1-X_g>0.5 (main fuzzy anti-symmetrical). Though the membership function for the whole molecular skeleton of FCl is about 0.69, those for the MOs are very different, varying from small to great. Among them, theσ_{pp}-MO mainly forms σ-bond, its membership function is relatively greater.

4.1. The Fuzzy Symmetry Characteristics of Diatomic Molecule

To start with, we analyzed the simplest linear molecule, diatomic one. According to the conservation rule of the molecule orbital (W-H rule), both symmetry and invariant are important and different concepts [1,5]. Viewed from the chemical symmetric theory, they may denote either the point group to which the molecular skeleton belongs and the irreducible representation to which the MO belongs. In essence, the W-H rule means that as the molecular skeleton can always maintain the symmetry in relation to a certain group, then the MOs of corresponding molecule would be unchanged under the irreducible representations of such group. As to the fuzzy symmetric theory, we need to examine the fuzzy point group and fuzzy representation. For this molecular system, equation (8) and (15) are related to the above concepts. In these equations, membership function and representation component are involved. To understand their relationship, we consider the simplest model, a diatomic molecular AB, and suppose that the MO (ψ) is combined by two AOs (ϕ_A and ϕ_B), one from each atom, *i.e.*,

$$\psi = a_A\phi_A + a_B\phi_B = (a_g + a_u)\,\phi_A + (\,a_g - a_u)\,\phi_B \tag{16}$$

where a_A and a_B are the normalized LCAO-MO (ψ) coefficients of ϕ_A and ϕ_B, respectively. Set $a_A^2 > a_B^2$ and $a_A > 0$, these coefficients may be decomposed to two parts: the symmetric part a_g and the anti-symmetric part a_u, according to a certain $\hat{G}$ two-fold point symmetric transformation. From eq. (16), we have:

$$a_g = (a_A + a_B)/2 \tag{16-1}$$

$$a_u = (a_A - a_B)/2 \tag{16-2}$$

According to the definition of the membership function for MO ψ under the symmetric transformation $\hat{G}$, we have,

$$\mu = (a_A^2 \wedge a_B^2) + (a_B^2 \wedge a_A^2) = 2\, a_B^2 \tag{17}$$

Meanwhile for transformation $\hat{G}$, the symmetric and anti-symmetric components are:

$$Xg = 2\, a_g^2 \ \& \ Xu = 2\, a_u^2 \tag{18}$$

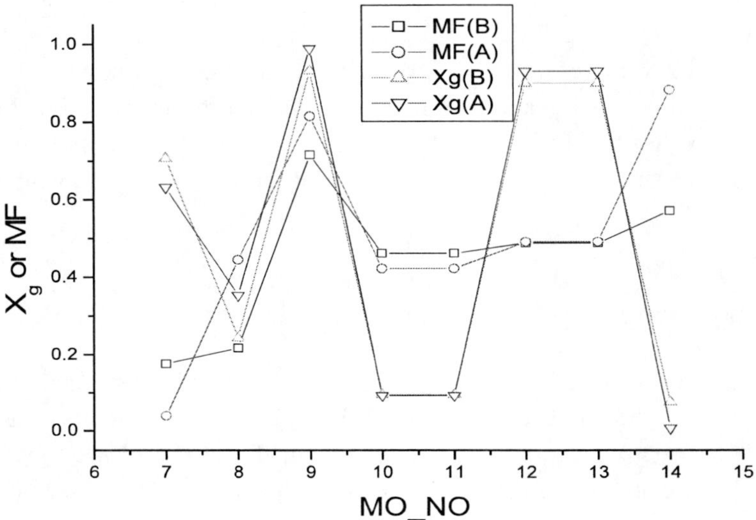

Figure 3. The membership functions (MF) and the symmetric compositions (X_g) vs some MOs of FCl about the space inversion, at AM1 (A) or B3LYP/6-31G* (B) level.

For a certain μ value, from the eq. (17), we can obtain a pair of a_B, *i.e.*, $a_B = \pm(\mu/2)^{0.5}$. Using the MO normalized condition and eq. (18), we may find a_A (>0) and Xg and Xu. Since a certain μ value, a_B has two values, the solution of Xg (or Xu) should also have two values. Therefore, for the simplest model, we obtain the relationship diagram of μ vs. Xg as shown in Figure 4, where the black curve responds to the Xg>0.5, indicating main symmetric representation, whereas the red curve responds to Xg<0.5, indicating main anti-symmetric representation. When Xg equals 0.5, the membership function equals zero, and a_g equals a_u, a_B equals 0, and a_A equals 1. In such case, the MO is only the AO ϕ_A. When Xg equals 0 (pure anti-symmetric representation) or 1 (pure symmetric representation), the membership function will be one. As shown in Figure 4, when the MO approaches to pure representation, the small change of the representation component may cause the large variation of the membership function. However, when the symmetry is poor, the membership function would be less, and its change slightly. For more complex MO, the relationship between the membership function vs. representation component will be also more complex, while the curve in Figure 4 will the boundary. Even a certain diatomic molecule, owing to there is more one AO included by each atom, we may get relative membership function and symmetric representation component by means of Section 3, but the membership function vs. symmetric representation component diagrams will be not same as Figure 4.

4.1.1. The Non-hydride Diatomic Molecule

Now we discuss the irreducible representations of some hetero-nuclear diatomic molecules. Different from MOs of homo-nuclear diatomic molecules, generally, each MO of these diatomic molecules is neither pure symmetrical (g) nor pure anti-symmetrical (u) component, but the result of the overlap, or called mix, of these two components. The components are obtained by means of eq(15). According to the idea of Zadeh [34], the fuzzy point group to which a MO of diatomic molecules belongs can be expressed as eq(8). The membership function in relation to the inversion symmetry may be get from eq(13). In Figure 5, there are the irreducible representation symmetric components X_g and membership functions μ of MOs in some diatomic molecules (hetero-nuclear BN, CO, FCl and homo-nuclear F_2). The MOs are σ_{SS}, σ_{SS}*, π_{PP}, π_{pp}, σ_{pp}, π_{pp}*, π_{pp}* and σ_{pp}* in the order number of MO (horizontal scale) from 1 to 8, respectively. These order and its energy order (from low to high) are consistent for CO and BN, but not for FCl and F_2. Have a care that the degenerate MOs, such as MOs 3 and 4 or 6 and 7, possess with not only the same energy but also the same fuzzy symmetry and fuzzy irreducible representation. All these MOs are composed by the LCAO of sp^3-AOs in the valance-shell of the two atoms.

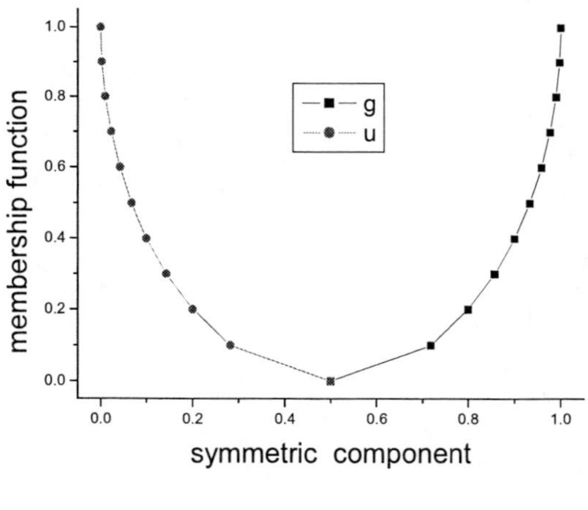

Figure 4. In connection to the two-fold transformation, $\hat{G}$, the membership function vs. the symmetric representation component diagram for the simplest MO.

As F_2 shown in Figure 5, for the MOs of a homo-nuclear diatomic molecule, the relative membership functions ought to be always equal to one, and the relative symmetric components must be either one or zero. For the MOs of hetero-nuclear diatomic molecule in the same figure, both the relative membership functions and the symmetric components would be between the one and zero. We can see that the symmetrical components X_g of the MOs is either near by one or by zero. Interestingly, as shown in Figure 5B, the trend of X_g of MOs is very similar for the different molecules. On the other hand, as shown in Figure 5A, the membership functions vary also between zero and one, but the trend is somewhat differ for the various molecules. Generally, as the X_g or X_u vary a litter from 1, the corresponding membership functions may vary more. While as both the X_g and X_u near by 0.5, such as MO-1 and MO-2 of FCl, the membership function will be quite small.

Generally, the effect of X_g and X_u to the membership functions is complex. If we draw the diagram of the relative membership functions vs the symmetric components for the MOs in relation to Figure 5, as shown in Figure 6, the differ from the simplest MOs (Figure 4) seems not so much. However, the difference will be more obvious for more complex MO.

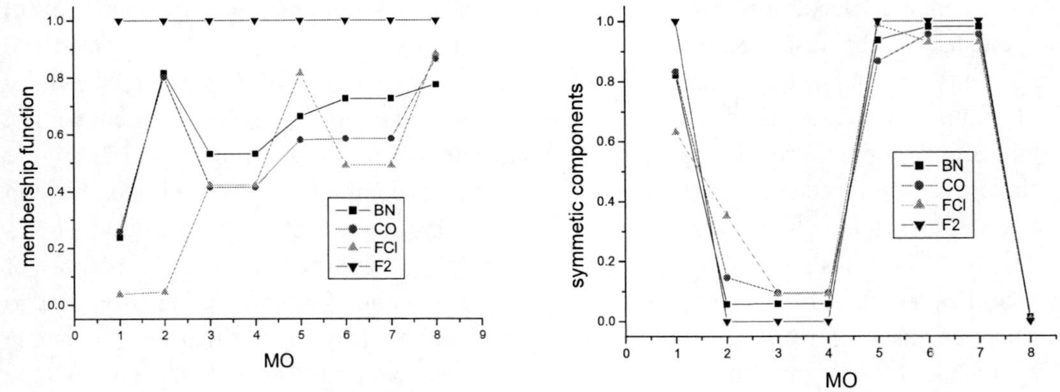

(a) (b)

Figure 5. The membership function μ (A) and the symmetric composition X_g (B) of MOs for several diatomic molecules, at AM1 level.

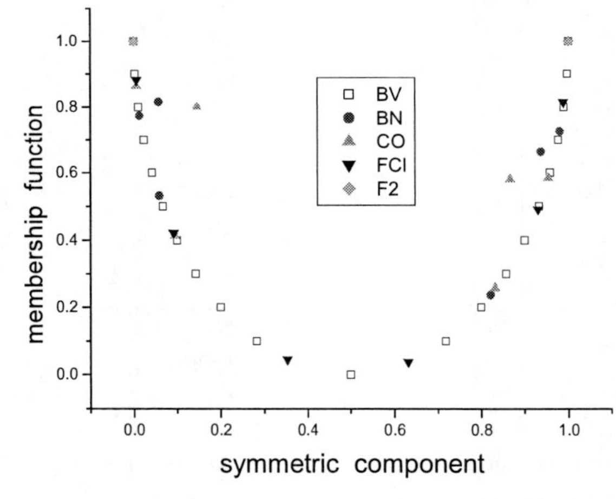

Figure 6. In connection to the inversion transformation ($\hat{P}$), the membership function vs. symmetric representation component diagram for the MOs of some diatomic molecules. Where the BV (□) denote the simplest MOs as Figure 4.

Now we analyze the fuzzy symmetry of hydride diatomic molecules. Here, we take four hydrogen halide (HX; X=F, Cl, Br, I) molecules as prototypical ones. In Figure 7 the AM1 calculated results for these molecules have been shown. The MOs are σ_{SS}, σ_{SSP}, π_{pp}, π_{pp} and σ_{SSP}* in the order number of MO (horizontal scale) from 1 to 5, respectively. In fact, the MOs-3 and -4 are the p_y-AOs and p_z-AOs of X. Where the subscript SSP denotes the relative

MO ought to be the linear combination of the 1s-AO of H atom and the sp-hybrid AO of halide atom.

4.1.2. The Hydride Diatomic Molecule

Comparing with the non-hydride diatomic molecules, the membership functions of the hydride diatomic molecule skeletons is much smaller (Table 2). This indicates that the fuzzy symmetry of inversion for each of these molecules is not obvious, especially for molecule including the larger halogen atom such as iodine and bromine. However, the membership functions for some MOs, such as the occupied MOs 2 and vacant MOs 5, are larger, that is to say the fuzzy symmetry of space inversion for these MOs is quite obvious. To be clear at a glance, we show the pictures of the MOs 2 in Figure 8. One can see from this Figure 8 that the fuzzy symmetry of space inversion for MOs 2 is obvious. The membership function of MO 2 of HX increases while that of the molecule HX skeleton decreases in the order of HF, HCl, HBr and HI. This indicates that although some symmetries of some molecules are not obvious, those of some MOs of these molecules may be obvious and significant, which is why some chemists are more interested in MOs than in molecule, itself.

It can be seen from Figure 7B that, for MOs 3 and 4, the symmetrical and the anti-symmetrical components account are the same, i.e. $X_g = X_u = 0.5$, while the corresponding membership functions (Figure 7B) are all zero. The underlying reason for this result is that the MOs 3 and 4 are formed from the p_y and p_z AOs of X atom (X = F, Cl, Br, I) only, and not related to H atom.

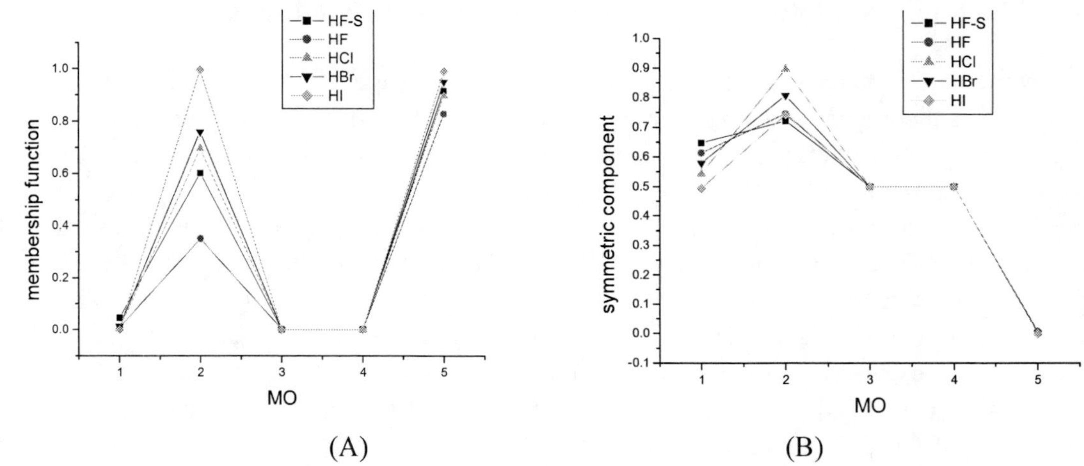

Figure 7. The membership functions (A) and symmetric components (B) of space inversion for the MOs of HX (X=F, Cl, Br, I) , at AM1 level. The HF-S denotes for the MOs of HF at HF/STO-3G level.

MOs 1, 2 and 5 are more significant from the viewpoint of bonding. However, the three MOs play different roles for bonding: MO-1 is mainly composed of s-AO of X atom, thus the X_g is nearly equal to X_u, i.e. $X_g \approx X_u \approx 0.5$, and the membership function is very small. This indicates that MO-1 contributes little to the bonding. MO-2 is the σ -BMO with $X_g > 0.5 > X_u$, indicating that the symmetrical component is crucial. While, MO-5 is the vacant σ^*-ABMO with $X_g < 0.5 < X_u$, indicating that the anti-symmetrical component is the majority. On the other hand, we note that the membership functions of MOs 2 and 5 increase in the order of HF, HCl, HBr and HI, which can be more clearly seen in Figure 7A. This implies that the fuzzy symmetry of space inversion for these MOs 2 and 5 becomes obvious in this order.

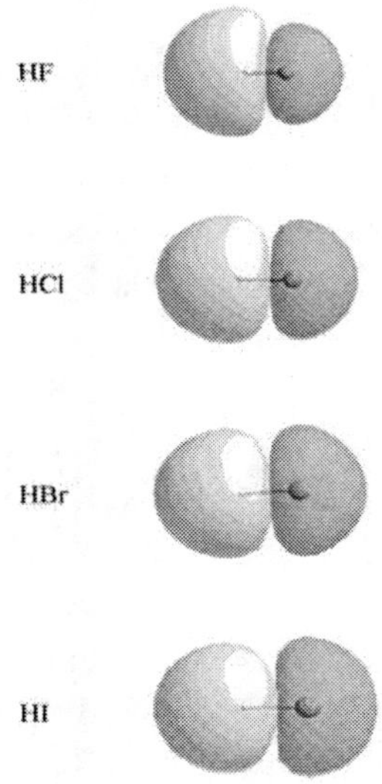

Figure 8. The MOs 2 of hydrogen halide HX (X=F, Cl, Br, I).

4.2. The Fuzzy Symmetry Characteristics of Tri-Atomic Linear Molecule

Now the prototypical tri-atomic molecules under investigation are the XCN molecules, where X is hydrogen or halogen atom (correspondingly, XCN is hydrogen cyanide or cyanogens halide). We are still interested in the fuzzy symmetry of space inversion. For the linear tri-atomic molecules, the center of inversion can be set in two kinds of proper ways: one is set at carbon atom, and the other is set between atoms C and N. If the center of inversion is set at carbon atom, the atoms X, C, and N will move to the place of N, C and X in the original fashion of the molecule, respectively, under the operation of inversion. While if the center of inversion is set between the C and N, then the X, C and N atoms will move to the place of a pseudo-atom, N and C, respectively. Corresponding to the two kinds of centers of inversion, the related membership functions for these linear tri-atomic molecules (skeleton) are listed in Table 3. It can be seen from Table 3 that except for HCN, no matter the center of inversion is set at atom C or between atoms C and N, the trend of the membership functions μ is the same for the rest molecules XCN (X=F, Cl, Br, I), both decreasing in the order of FCN, ClCN, BrCN and ICN. It also can be seen from Table 3 that the membership functions for molecules XCN (X=F, Cl, Br, I) obtained with the center of inversion set at the C atoms are greater than those as the center is set between the atoms C and N. In contrast, the membership functions for HCN obtained with the center of inversion set at the C atoms are smaller than those as the center is set between the atoms C and N. Therefore, for molecules XCN (X=F, Cl, Br, I) it is proper to set the center of inversion at carbon atom, while for HCN it is proper to set the center of inversion between atoms C and N.

Now we turn to discuss the fuzzy symmetries of MOs of these linear tri-atomic molecules. For HCN, 9 MOs are formed from the combination of the 1s-AO of H and the outermost shell sp^3-AOs of C and N atoms. However, the character of fuzzy symmetry of MOs-1 and -2, which are the lowest in energy among the 9 MOs of each molecule, are neglected in Figure 9. Where MOs-4,5 and MOs-6,7 are two pairs of degenerate π-MOs, and all of others are σ-MOs. There are 3 occupied MOs (from MO-3 to MO-5) and 4 vacant MOs (from MO-6 to M-9) among these MOs.

As shown in Figure 9, for the MOs at AM1 and HF/STO-3G the results are similar. The π-MOs are the LCAO only with the atoms C and N, as using the center of inversion between the point at CN bond, then the membership functions (μ_{CN}) will be bigger than the membership functions (μ_C) as using the C-atomic position, similar to the skeleton, meanwhile these π-MOs are near the pure representation, i.e. the symmetric components Xg(CN) near either 0 or1. Owing to the role of hydrogen AO, above appearance will be not shown obviously for the σ-MOs.

For molecules XCN (X=F, Cl, Br, I), we set the center of inversion at carbon atom. There are 12 MOs formed from the combination of the outermost shell sp^3-AOs of X, C and N atoms. However, the character of fuzzy symmetry of MOs -1 and -2, which are the lowest in energy among the 12 MOs of each molecule, and they are neglected in Figure 10. The MOs are in the order number of MOs (horizontal scale) from 3 to 12, respectively. These order and its energy order (from low to high) are consistent for ClCN, BrCN and ICN, but not for FCN. There are six occupied MOs (from MO-3 to MO-8) and four vacant MOs (from MO-9 to M-12) among the 10 MOs of XCN. The MO pairs, i.e. MOs-4 & 5, MOs-7 & 8, and MOs-10 & 11 are degenerate π-MOs, and the rest four are σ-MOs. Where the σ-MO energy of MO-9 is more than that of MO-10 or 11 for FCN, but it is less than MO-10 or 11 for other XCN. For XCN (X=Cl, Br, and I), Figure 10 indicates that the X_g of their MOs for the three molecules behavior similarly in trend, only for the individual MO of FCN there is somewhat differ. This may result from the similar construction of these MOs.

Table 3. Membership functions of inversion for linear tri-atomic molecules XCN (X=H, F, Cl, Br, I) as the center of inversion is at atom C and between atoms C and N, respectively (these membership functions, calculated with the atomic number on atoms as the criterion).

XCN	the center of inversion is at atom C	the center of inversion is between atoms C and N
	μ_C	μ_{CN}
HCN	0.5714	0.8571
FCN	0.9091	0.5455
ClCN	0.6667	0.4000
BrCN	0.4167	0.2500
ICN	0.3030	0.1818

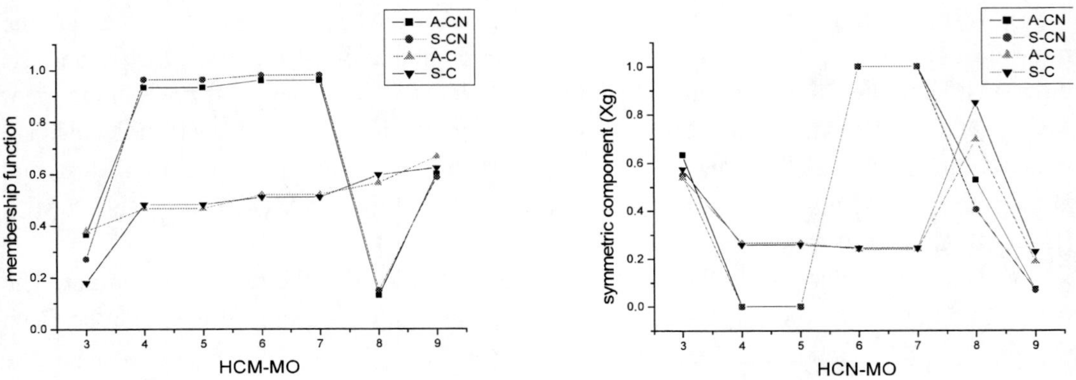

(A) (B)

Figure 9. The membership functions (A) and symmetric components (B) vs MOs of HCN molecule. Where A-C and A-CN denote that the membership functions (A) and symmetric components (B) of MOs for HCN molecules that are calculated at AM1 level using the C-atomic position and between the point at CN bond as the center of inversion, respectively. As for S-C and S-CN, they are denoted the same meaning as A-C and A-CN, except at HF/STO-3G level. For various calculation levels, these relationships are similar.

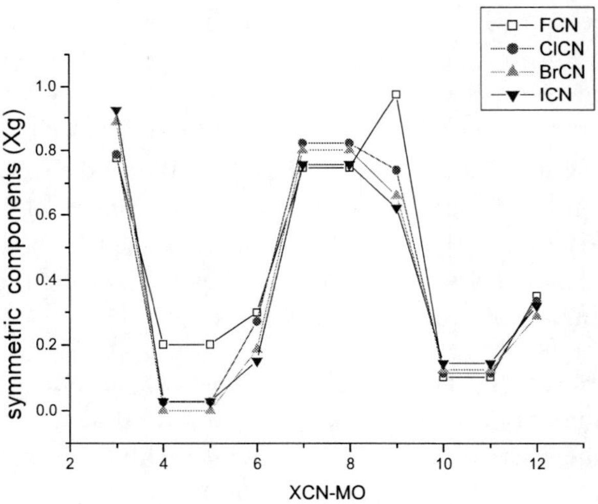

Figure 10. The symmetric components X_g of MOs for XCN(X=F, Cl, Br, I) molecules, at AM1 level and using the C-atomic position as the center of inversion.

5. The Fuzzy Symmetry Characteristics of Planar Molecule

For a planar molecular skeleton, according to the internal configuration (two-dimensional) space, the possible symmetry transformations ought to be the identity or two-fold, sometime only one higher(>2)-fold rotation may be included for each molecule. The two-fold transformations may be the inversion, reflection and two-fold rotation. The reflection about the molecular plane may often be neglected for molecular skeleton and two-dimensional crystallography, but it is important for some MO symmetry. As for the possible only one higher(>2)-fold rotation axis, it must be orthogonal with the molecular plane. If a plane molecular fuzzy point symmetry belongs to the fuzzy point group or set, $\underset{\sim}{G}$, corresponding to the point group, $G = \{\hat{G}_j \,; j = 0,1,2,\ldots\ldots,n\text{-}1\}$, according to Zadeh[34], we can get,

$$\underset{\sim}{G} = \sum_{j=0}^{n-1} \frac{\mu_j}{\hat{G}_j} \tag{17}$$

In which the μ_j denotes the membership function of the symmetry transformations, $\hat{G}_j$, and it can be obtained by eq(5) for molecular skeleton or by eq(13) for MO, similarly. As the irreducible representations included in G may be the $\Gamma^{(1)}$, $\Gamma^{(2)}$, the fuzzy representation of relative MO may be shown as:

$$\underset{\sim}{\Gamma} = X_1\Gamma^{(1)} + X_2\Gamma^{(2)} + X_3\Gamma^{(3)} + \tag{18}$$

The relative representation components X_1, X_2, X_3,⋯⋯, may be get by using various methods. The prototypical planer molecules we have chosen to study are the ethylene tetra-halid molecules and the azines molecules, these two molecular types are without and with only one higher fuzzy rotation axis, respectively.

5.1. The Fuzzy D_{2h} Symmetry Characteristics – Ethylene Tetra-Halid Molecules

The prototypical planer molecules without higher fuzzy rotation axis we have chosen to study are the ethylene tetra-halid molecules. These molecules relate to the fuzzy symmetry in connection with the fuzzy $\underset{\sim}{D}_{2h}$ subgroup or set. According to the condition of fuzzy subgroup [21,34], in relation to D2h group for the tetra-halid molecular skeleton and MOs, the corresponding fuzzy symmetry transformation sets sometimes only the fuzzy sets but not the fuzzy subgroups. As we known, the well-defined D_{2h} point group includes an identity transformation and seven two-fold symmetry transformations but without higher-fold ones. Meanwhile, it is related only to some one-dimensional irreducible representations, but there is not to any multi-dimensional irreducible representation.

However, there are some two-fold symmetry transformations with various membership functions. It is different from the C_i point group, in which included only one two-fold symmetry transformation and the identity one. Especially, for D_{2h} point group, some different irreducible representations are included. When we analyze the fuzzy representation of MO, not only the overlap between the symmetrical and anti-symmetrical components, but also the overlap touching to some more kinds of various irreducible representation components, should be considered. Therefore, for the ethylene tetra-halide, the analysis of molecular fuzzy symmetry ought to be much complex than that of simple linear molecules.

In this section, the MO theoretical computation base often at the AM1 level. Of course, we can compute at other levels. As based on the some other levels, the molecular fuzzy symmetry of some linear molecules had been analyzed, the results are similar. As various MO theorem, the complicacy for the treatment method of the molecular fuzzy symmetry ought to be some differ in degree. However, we may establish the method at another MO theorem, in principle.

According to Noether's theorem[4], there are two important concepts: the symmetry and invariable. They are relative but different concepts. The symmetry is connected with the molecular point group, but the invariable is connected with the representation or generalized parity of MO[5]. The symmetry denoted the invariance of the molecule or MO as follow a certain transformation. But the invariable denotes the invariance of generalized parity or irreducible representation for MO. However, they are confused in W-H theorem [1]. Corresponding to these two concepts, there are two important amounts for the topic of molecular fuzzy symmetry [24,25]: one is the membership function, which measures the perfection degree of the molecular symmetry; the other is the irreducible representation components, which measures the perfection degree of the generalized parity or representation of the MOs. Now we turn to analyze how to calculate the membership functions and the irreducible representation components, for ethylene tetra-halides. The membership function of the symmetry transformations, $\hat{G}_j$, and it can be obtained by eq(5) for molecular skeleton or by eq(13) for MO. It should be noted that there are some various symmetry transformations $\hat{G}_j$ in the D_{2h} point group to which ethylene tetra-halide molecule belongs. And corresponding to various symmetry transformations, the membership functions of them are different.

As for the irreducible representation components, When we analyze the fuzzy point group ($\underset{\sim}{C}_i$) of the linear molecular MO, the concepts, symmetry components and anti-symmetry components are introduced. As the ethylene tetra-halides, the fuzzy symmetries should be connected with the $\underset{\sim}{D}_{2h}$ fuzzy point group. And correspondingly, more various irreducible representations should be taken into consideration, only the symmetrical or asymmetrical representation would be not enough. According to eq(10), for the MO-Ψ_ρ, the corresponding linear combination coefficient of the ith AO of the Jth atom, $a_\rho(J,i)$, will be decomposed to denoted as the summation of some other linear combination coefficients each of which only relates to one single irreducible representation, respectively, i.e.

$$a_\rho(J,i) = a_\rho(J,i;\Gamma_1) + a_\rho(J,i;\Gamma_2) + a_\rho(J,i;\Gamma_3) + \ldots\ldots + a_\rho(J,i;\Gamma_r) \tag{19}$$

Where the $\Gamma_1, \Gamma_2, \ldots\ldots, \Gamma_r$ are the various irreducible representations in regard to $a_\rho(J,i)$.

The ethylene tetra-halide molecules, we may analyze their fuzzy symmetry in related to the $\underset{\sim}{D}_{2h}$. There are two carbon atoms and four halide atoms. These six atoms and corresponding serial number are shown in Figure 11, together with the space axes. If all of the four halide atoms in such a molecule are the same, such molecule ought to be provided with the common D2h point group symmetry, otherwise with certain subgroup symmetry. Certainly, the symmetry of mirror with the molecular plane ought to be always existed. As for other symmetries, some of them may be existed and other may be omitted.

There are four AOs for each atomic outmost shell, and they will combined to form 24 MOs. Taking C2F4 as the example, using methods at AM1 level within Gaussian program [35], we can get the orbital irreducible representations of these 24 MOs as follows:

Occupied (Ag) (B1u) (B2u) (B3g) (Ag) (B1u) (Ag) (B2u) (B3u)

(B3g) (B2g) (B1u) (B1g) (Au) (B3g) (B2u) (Ag)(B3u)

Virtual (B2g) (B1u) (B2u) (Ag) (B1u) (B3g)

The occupied MOs are MO-1 to MO-18 and virtual MOs are MO-19 to MO-24 respectively. Sometimes, the σMO and πMO are analyzed, separately. The relative occupied MOs may be denoted as σOMO-*j* and πOMO-*j*, and virtual MOs as σVMO-*j* and πVMO-*j*. Where, *j* is a number and it indicated the sequence of the MO separated from the nonbonding energy level.

Because the C2F4 molecule possesses the common D_{2h} point group symmetry, the representations of the MO above-mentioned are "pure", i.e. such representation composition ought to be 1 and other will be 0. For six π-MOs composed from six p_x -AOs of six atoms, the representations are 1[B1g], 2[B2g], 1[Au], 2[B3u], and for 18 σ MO from *s*, p_y, p_z -AOs of six atoms, the representations are 5[Ag], 4[B3g], 5[B1u], 4[B2u]. As concerns some other ethylene tetra-halide molecules, there are six π-MOs composed from six p_x-AOs and 18 σ MOs from *s*, p_y, p_z-AOs of six atoms, too. However, these irreducible representations are not "pure", but the overlap of some representations. The [B1g], [B2g], [Au] and [B3u] representations may overlap to compose the six π-MOs by six p_x-AOs, and we can get the LCAO coefficients in relation with such four irreducible representations. Similarly, we may get the LCAO coefficients in relation with these four irreducible representations, [Ag], [B3g], [B1u] and [B2u] representations, in related to the 18 σ -MOs, *s*, p_y, p_z-AOs. Further, we can obtain the corresponding representation-components for various MOs.

Now, we analyze the method to calculate the LC- p_x-AOs coefficients of the π- MOs. For the four irreducible representations in relation to π-MOs, all of the LC- *s*, p_y, p_z-AOs coefficients are null, but for the LC- p_x-AOs coefficients, we can get:

$$a(1x;Au) = a(2x;Au) = 0 \tag{20-1}$$

$$a(3x;Au) = a(5x;Au) = -a(4x;Au) = -a(6x;Au) \tag{20-2}$$

$$a(1x;B1g) = a(2x;B1g) = 0 \tag{20-3}$$

$$a(3x;B1g) = a(6x;B1g) = -a(4x;B1g) = -a(5x;B1g) \tag{20-4}$$

$$a(1x;B2g) = -a(2x;B2g) \tag{20-5}$$

$$a(3x;B2g) = a(4x;B2g) = -a(5x;B2g) = -a(6x;B2g) \tag{20-6}$$

$$a(1x;B3u) = a(2x;B3u) \tag{20-7}$$

$$a(3x;B3u) = a(4x;B3u) = a(5x;B3u) = a(6x;B3u) \tag{20-8}$$

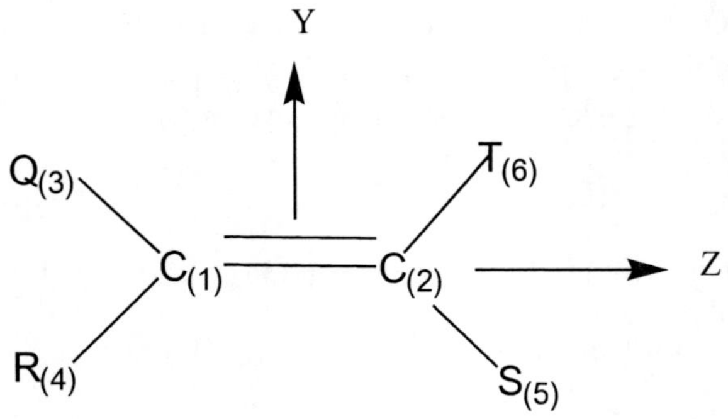

Figure 11. The space axes and atomic serial numbers (in bracket) of ethylene tetra- halide molecule. X-axis is vertical to the molecule plane and cross the CC bond. The Q, R, S and T denote four halogen atoms (where these atoms may be different or the same).

Where a(ix;Γ)s are the LC- p_x-AOs coefficients for the i-atom (the serial number i = 1 - 6; as shown in Figure 11) within the Γ (Γ = Au, B1g, B2g or B3u) irreducible representation. The relationship equations (5) depend on the characters of various representations according to the D_{2h} point group and the phases. For the MOs with "pure" representation, these relationship equations are obviously. As for the MO with non-pure representation Γ, the corresponding equations are also true. On the other hand, the π-MO may be expressed as:

$$\Psi=a(1x)\Phi(1x)+a(2x)\Phi(2x)+a(3x)\Phi(3x)+a(4x)\Phi(4x)+a(5x)\Phi(5x)+a(6x)\Phi(6x) \quad (21)$$

Where a(ix)s represent the LCAO coefficients of the i-atomic(i=1,……,6) p_x-AOs Φ(ix). The π-MOs belong to the fuzzy representation overlapped by the Au, B1g, B2g and B3u. From equation (20), we can get:

$$a(1x) = a(1x;B3u) + a(1x;B2g) \quad (22\text{-}1)$$

$$a(2x) = a(2x;B3u) + a(2x;B2g) \quad (22\text{-}2)$$

$$a(3x) = a(3x;B3u) + a(3x;B2g) + a(3x;Au) + a(3x;B1g) \quad (22\text{-}3)$$

$$a(4x) = a(4x;B3u) + a(4x;B2g) + a(4x;Au) + a(4x;B1g) \quad (22\text{-}4)$$

$$a(5x) = a(5x;B3u) + a(5x;B2g) + a(5x;Au) + a(5x;B1g) \quad (22\text{-}5)$$

$$a(6x) = a(6x;B3u) + a(6x;B2g) + a(6x;Au) + a(6x;B1g) \quad (22\text{-}6)$$

Further, by using the eqs(21) and (22) and primary algebra, we may obtain that:

$$a(1x;B3u) = a(2x;B3u) =[a(1x) + a(2x)]/2 \quad (23\text{-}1)$$

$$a(1x;B2g) = -a(2x;B2g) =[a(1x) - a(2x)]/2 \quad (23\text{-}2)$$

$$a(3x;B3u) = a(4x;B3u) = a(5x;B3u) = a(6x;B3u) \\ = [a(3x) + a(4x) + a(5x) + a(6x)]/4 \quad (23\text{-}3)$$

$$a(3x;B2g) = a(4x;B2g) = -a(5x;B2g) = -a(6x;B2g) \\ = [a(3x) + a(4x) - a(5x) - a(6x)]/4 \quad (23\text{-}4)$$

$$a(3x;Au) = -a(4x;Au) = a(5x;Au) = -a(6x;Au)$$

$$= [a(3x) - a(4x) + a(5x) - a(6x)]/4 \qquad (23\text{-}5)$$

$$a(3x;B1g) = -a(4x;B1g) = -a(5x;B1g) = a(6x;B1g)$$

$$= [a(3x) - a(4x) - a(5x) + a(6x)]/4 \qquad (23\text{-}6)$$

These results we can also get by means of projection operators [36]. Therefore, the irreducible representation compositions of these π-MOs may be calculated as follows:

$$X(B3u) = \Sigma[a^2(B3u)]/\Sigma(a^2) = [a^2(1x;B3u) + a^2(2x;B3u) + a^2(3x;B3u) + a^2(4x;B3u) + a^2(5x;B3u) + a^2(6x;B3u)] / \Sigma(a^2) \qquad (24\text{-}1)$$

$$X(B2g) = \Sigma[a^2(B2g)]/\Sigma(a^2) = [a^2(1x;B2g) + a^2(2x;B2g) + a^2(3x;B2g) + a^2(4x;B2g) + a^2(5x;B2g) + a^2(6x;B2g)] / \Sigma(a^2) \qquad (24\text{-}2)$$

$$X(Au) = \Sigma[a^2(Au)]/\Sigma(a^2) = [a^2(3x;Au) + a^2(4x;Au) + a^2(5x;Au) + a^2(6x;Au)] / \Sigma(a^2) \qquad (24\text{-}3)$$

$$X(B1g) = \Sigma[a^2(B1g)]/\Sigma(a^2) = [a^2(3x;B1g) + a^2(4x;B1g) + a^2(5x;B1g) + a^2(6x;B1g)] / \Sigma(a^2) \qquad (24\text{-}4)$$

where :

$$\Sigma(a^2) \equiv a^2(1x) + a^2(2x) + a^2(3x) + a^2(4x) + a^2(5x) + a^2(6x) \qquad (25)$$

It is no difficult to prove that :

$$X_{(B3u)} + X_{(B2g)} + X_{(Au)} + X_{(B1g)} = 1 \qquad (26\text{-}1)$$

The fuzzy representation in which the π-MO belongs to may be shown as:

$$\underset{\sim}{\Gamma}(\pi) = X_{(B3u)}B3u + X_{(B2g)}B2g + X_{(Au)}Au + X_{(B1g)}B1g \qquad (26\text{-}2)$$

Now we turn to analyze the LCAO coefficients of σ -MO. For the four irreducible representations in relation to σ -MOs, all of the LC- p_x-AOs coefficients vanish. The LC- s, p_y,p_z-AOs coefficients may be obtained by similar way of dealing π-MOs. We may get the a(is;Γ), a(iy;Γ) and a(iz;Γ), where the atomic serial number i = 1 - 6 , and the irreducible representation Γ = Ag, B1u, B2u or B3g. However since all kinds of the s,p_y,p_z-AOs.can affect the relative irreducible representation Γ, the calculation program for σ MOs, should be similar to eq(9), which is in connection with various irreducible representations, Γ. Thus, we must consider the contribution of all the three kinds of s,p_y,p_z-AOs to each irreducible representation. For the σ -MO, it may also prove that:

$$X_{(B3g)} + X_{(B2u)} + X_{(Ag)} + X_{(B1u)} = 1 \qquad (27\text{-}1)$$

The fuzzy representation in which the σ-MO belongs to may be shown as:

$$\underset{\sim}{\Gamma}(\sigma) = X_{(B3g)}B3g + X_{(B2u)}B2u + X_{(Ag)}Ag + X_{(B1u)}B1u \qquad (27\text{-}2)$$

5.1.1. The Membership Functions and Representation Components of Ethylene Tetra-Halide

As mentioned above, we can get the membership functions and representation components of relative molecules and MOs by the diverse and complicated but no objected calculation. The calculation ought to be simplified as we make it program.

For the ethylene tetra-halide molecular skeletons, using the eqs(17) and (5) and setting the atomic number as the criterion, we can get the membership functions in relation to all symmetry transformations in D_{2h}. The results of some ethylene tetra-halide molecular skeletons are shown in Table 4. Utilizing this Table, we may show the relative fuzzy symmetry point group (or set) of these molecular skeletons according to the ideal of Zadeh [34], easily. Such as the fuzzy point groups of the C2F3Cl and CF2CCl2 molecular skeletons can be written as:

$$\underset{\sim}{D}_{2h} = \frac{1}{\hat{E}} + \frac{1}{\hat{M}x} + \frac{0.8571}{\hat{M}y} + \frac{0.8571}{\hat{M}z} + \frac{0.8571}{\hat{P}} + \frac{0.8571}{\hat{C}x} + \frac{0.8571}{\hat{C}y} + \frac{0.8571}{\hat{C}z} \tag{28-1}$$

for C2F3Cl and

$$\underset{\sim}{D}_{2h} = \frac{1}{\hat{E}} + \frac{1}{\hat{M}x} + \frac{1}{\hat{M}y} + \frac{0.7500}{\hat{M}z} + \frac{0.7500}{\hat{P}} + \frac{0.7500}{\hat{C}x} + \frac{0.7500}{\hat{C}y} + \frac{1}{\hat{C}z} \tag{28-2}$$

for CF2CCl2 (*meta*-C2F2Cl2), respectively. Where the Mw and Cw($w = x,y,z$) are the mirror reflection (w-axis as the normal) and the two-fold rotation(w-axis as the rotation axis) transform operations, respectively.

By the way, according to the conditions of fuzzy subgroup[22,34], in relation to D2h group for the these molecular skeletons and MOs, the corresponding fuzzy symmetry transformation sets sometimes only the fuzzy sets but not the fuzzy groups. For example, if based on the atomic number as the atomic criteria (Y) for C2F3Cl molecular skeleton (Table 4), the fuzzy symmetry transformation set in relation to D2h group may be the fuzzy subgroup, however if based on the electronic charge, it ought to be fuzzy set but not fuzzy subgroup. As for C2F3Cl MOs, some of them are not fuzzy subgroup, yet.

For the ethylene tetra-halide MOs, using the eq(13) and setting the electron population on atom J with $a^*_\rho(J,i)a_\rho(J,i)$ of MO-Ψ_ρ as the criterion, we can get the membership functions in relation to all symmetry transformations in D_{2h}. Meanwhile, according to eq(17), the fuzzy point group of the relative MOs may be obtained. For the fuzzy point groups of the MOs in the C2F3X(X =F, Cl, Br, I) may be shown as:

$$\underset{\sim}{D}_{2h} = \frac{1}{\hat{E}} + \frac{1}{\hat{M}x} + \frac{\mu_1}{\hat{M}y} + \frac{\mu_2}{\hat{M}z} + \frac{\mu_3}{\hat{P}} + \frac{\mu_3}{\hat{C}x} + \frac{\mu_2}{\hat{C}y} + \frac{\mu_1}{\hat{C}z} \tag{29-1}$$

Table 4. The membership functions in relation to various symmetry transformations of some tetra-halide molecule skeletons

Ethylene tetra-halide	$\hat{P}$ & $\hat{C}x$	$\hat{M}z$ & $\hat{C}y$	$\hat{M}y$ & $\hat{C}z$	$\hat{M}x$ & $\hat{E}$
C2F4	1	1	1	1
C2F3Cl	0.8571	0.8571	0.8571	1
C2F3Br	0.6487	0.6487	0.6487	1
C2F3I	0.5217	0.5217	0.5217	1
CF2CCl2	0.7500	0.7500	1	1
cis-C2F2Cl2	0.7500	1	0.7500	1
trans-C2F2Cl2(Z)	1	0.7500	0.7500	1

As X = F, then $\mu_1 = \mu_2 = \mu_3$. For the *meta-*, *cis-* and *trans-* C2F2Cl2, the fuzzy point groups of the relative MOs may be respectively shown as:

$$\underset{\sim}{D}_{2h} = \frac{1}{\hat{E}} + \frac{1}{\hat{M}x} + \frac{1}{\hat{M}y} + \frac{\mu_m}{\hat{M}z} + \frac{\mu_m}{\hat{P}} + \frac{\mu_m}{\hat{C}x} + \frac{\mu_m}{\hat{C}y} + \frac{1}{\hat{C}z} \quad (29\text{-}2)$$

$$\underset{\sim}{D}_{2h} = \frac{1}{\hat{E}} + \frac{1}{\hat{M}x} + \frac{\mu_c}{\hat{M}y} + \frac{1}{\hat{M}z} + \frac{\mu_c}{\hat{P}} + \frac{\mu_c}{\hat{C}x} + \frac{1}{\hat{C}y} + \frac{\mu_c}{\hat{C}z} \quad (29\text{-}3)$$

$$\underset{\sim}{D}_{2h} = \frac{1}{\hat{E}} + \frac{1}{\hat{M}x} + \frac{\mu_t}{\hat{M}y} + \frac{\mu_t}{\hat{M}z} + \frac{1}{\hat{P}} + \frac{1}{\hat{C}x} + \frac{\mu_t}{\hat{C}y} + \frac{\mu_t}{\hat{C}z} \quad (29\text{-}4)$$

The membership functions of the skeleton and various MOs of a certain molecule are difference. Using eq(13), we can get all membership functions in above equations.

Table 5. The fuzzy representations in which the MOs of ethylene tetra-halide belongs to.

ethylene tetra-halide	Fuzzy representation	
	σ-MO	**π -MO**
$C_2F_3X(X\neq F)$	$X_{(B3g)}$B3g+ $X_{(B2u)}$B2u + $X_{(Ag)}$Ag + $X_{(B1u)}$B1u	$X_{(B3u)}$B3u+ $X_{(B2g)}$B2g + $X_{(Au)}$Au + $X_{(B1g)}$B1g
meta-$C_2F_2Cl_2$	$X_{(B3g)}$B3g+ $X_{(B2u)}$B2u or $X_{(Ag)}$Ag + $X_{(B1u)}$B1u	$X_{(B3u)}$B3u+ $X_{(B2g)}$B2g or $X_{(Au)}$Au + $X_{(B1g)}$B1g
cis-$C_2F_2Cl_2$	$X_{(B2u)}$B2u + $X_{(Ag)}$Ag or $X_{(B1u)}$B1u + $X_{(B3g)}$B3g	$X_{(B2g)}$B2g + $X_{(Au)}$Au or $X_{(B1g)}$B1g + $X_{(B3u)}$B3u
trans-$C_2F_2Cl_2$	$X_{(B3g)}$B3g + $X_{(Ag)}$Ag or $X_{(B2u)}$B2u + $X_{(B1u)}$B1u	$X_{(B3u)}$B3u + $X_{(Au)}$Au or $X_{(B2g)}$B2g + $X_{(B1g)}$B1g

The fuzzy representations of relative σ-MO and π-MO may be shown as eqs(27-2) and (26-2), respectively. The MO fuzzy representation of C_2F_3X ($X\neq F$) may be composed in connection to four irreducible representations, but for the $C_2F_2Cl_2$ only to two irreducible representations, as shown in Table 5.

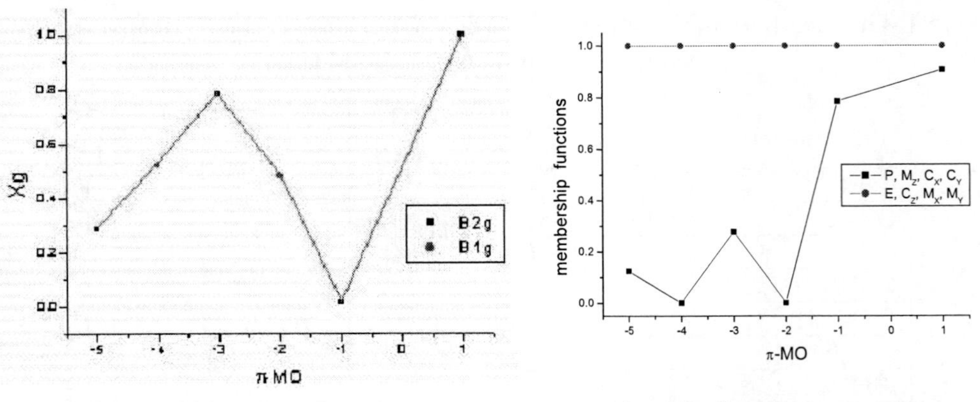

(A) Xg vs π-MO

(B) membership function vs π-MO

Figure 12. The membership functions and the symmetric representation components of the *meta*-C2F2Cl2 MOs.

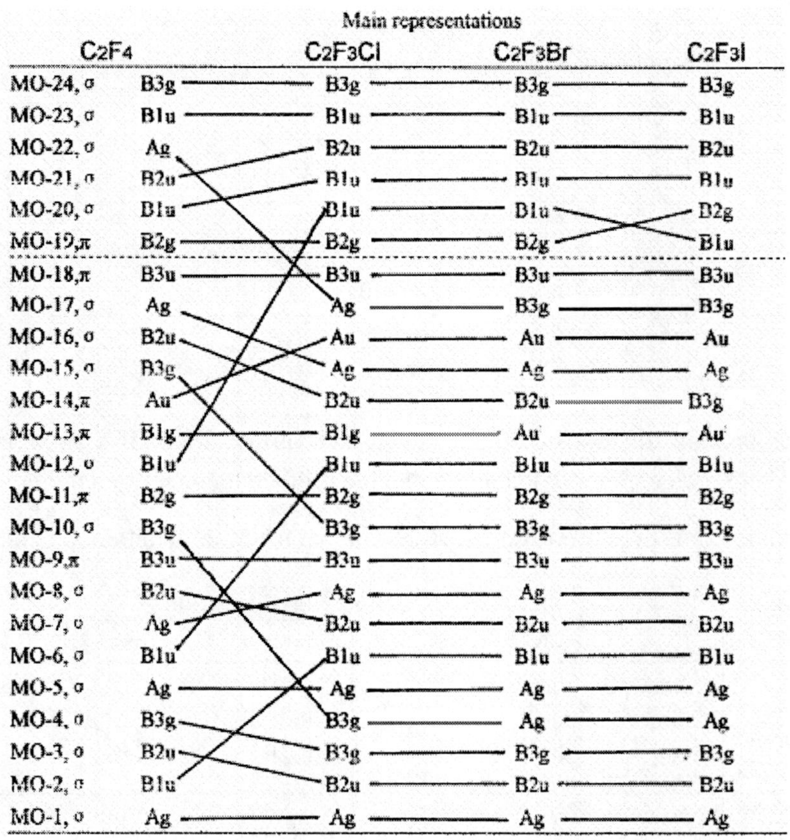

Figure 13. MO fuzzy correlation diagram between C_2F_4 , C_2F_3Cl, C_2F_3Br and C_2F_3I

For example, as shown in Figure 12, there are such membership functions and the symmetric representation components (Xg) of the *meta*-C2F2Cl2, where the $Xg = X_{(B1g)}$ or $X_{(B2g)}$ for π-MOs the corresponding anti-symmetric representation components $Xu = 1\text{-}Xg = X_{(Au)}$ or $X_{(B3u)}$, respectively. Similarly, the $Xg = X_{(Ag)}$ or $X_{(B3g)}$ for σ-MOs the corresponding anti-symmetric representation components Xu = 1-Xg =$X_{(B1u)}$ or $X_{(B2u)}$, respectively. We can also get the similar figure for σ-MOs, but it is omitted here. For *cis*- and *trans*-C2F2Cl2, we can get the similar figures, but corresponding relationship between Xg and Xu will be somewhat difference, it may be found from Table 5, easily. It is notable that there are only two representation components for each MO in such molecule.

5.1.2. The MO Fuzzy Correlation Diagram of Ethylene Tetra-Halide

Though the MOs of C2X4 (here these four X atoms are not all the same) are not provided with the 'pure' irreducible representations in relation to the common D2h point group, but with the mixture of some irreducible representations. Among these irreducible representations, the one with the maximum components may be called main irreducible representation. However, the cases for various MOs are quite different. Sometimes, some irreducible representation components close to one. However, in other cases, may not be near by one. The frontier orbitals, HOMO(π-OMO-1, *i.e.* MO-18) and LUMO(π-VMO-1, *i.e.* MO-10), are more 'pure', their main irreducible representation components may quite close to one. Such phenomena appear not only in the C2F2Cl_2 molecule, but also in some other molecules not a rare. Such as for the HOMOs of C2F4, C2F3Cl, C2F3Br and C2F3I, the main irreducible representations of all the HOMOs are B3u. The corresponding representation components are all more then 0.98, and it is 1 for the main and pure representation of C2F4. For the LUMOs of the C2F4, C2F3Cl and C2F3Br molecules, the main irreducible representations are B2g, and the corresponding representation components are more than 0.99. For this reason, the frontier orbital theorem [37] may be in effect to un-perfect symmetry condition. By means of the main representation for the fuzzy symmetry molecules, we may construct the correlation diagram similarly as the common symmetry molecules. Such correlation diagram, as shown in Figure 13, can be called fuzzy correlation diagram. The MO fuzzy correlation diagram differs somewhat from common MO correlation diagram. In the common MO correlation diagram, the symmetries between MOs are either the same or not the same. All the correlation lines are equi-weight, and the corresponding symmetries between MOs linked by a correlation line are the same. If the symmetries are not the same, the MOs can not be linked by a correlation line. However for the molecular fuzzy symmetry, the symmetries may be partial but not whole the same. It means that the membership functions and the representations components relative to the symmetries may be a certain value between the 0 and 1, while in a common MO correlation diagram, they are either 0 or 1. All of the MO representation components in C2F4 are 1 , but those in C2F3X (X=Cl, Br or I) are between 0 and 1 . Therefore, the correlation degree of various correlation lines may be different. Moreover, one MO may be correlated with more MOs, but usually only the main correlation lines are shown.

5.2. The Fuzzy D_{6h} Symmetry Characteristics --Azines Molecules

The prototypical planer molecules with only one higher multi-fold fuzzy rotation axis we have chosen to study are the benzene and azines molecules. These molecules relate to the fuzzy symmetry in connection with the fuzzy $\underset{\sim}{D}_{6h}$ subgroup or set. Now we will start to study the fuzzy symmetry related to more complex point groups that include and only include one multi-fold symmetry element. The MOs of a molecule with such a symmetry element may also belong to two-dimensional but not higher dimensional irreducible representations, along with the one-dimensional ones. The benzene molecule is a typical model that possesses the D_{6h} point group symmetry. Its MOs belong to certain irreducible representations of this group, which may be analyzed according to the usual quantum chemistry methods. As a MO belongs to a two-dimensional irreducible representation, such a MO cannot be the eigenstates of the all symmetry transformation in D_{6h} point group, simultaneously [5]. However, the two MOs complete set of the corresponding two-dimensional irreducible representation will be the eigenstate of the all transformations in D_{6h} point group, simultaneously. Such a single MO is the eigenstate of some but not all transformations in D_{6h} point group, and then the membership functions of such a MO will be equal to one only for some of but not for all of the symmetry transformations in the D_{6h} point group. The relative membership functions in relation to some transformations may be less one.

The molecules similar to benzene (that is, benzene-like molecules) ought to belong to the fuzzy D_{6h} symmetry, and their symmetry characterization is more complicated. A single MO may be assigned to either the one-dimensional or two-dimensional irreducible representations. There are two-MO fuzzy sets in the benzene-like molecules that corresponding to two-MO complete sets of a certain two-dimensional irreducible representation of D_{6h} point group in benzene. For the fuzzy set, the membership function in general should be less than one, although the functions related to the symmetry transformations of D_{6h} point group may be larger than those of the single MO. This is different from the benzene molecule with the complete D_{6h} point group symmetry. How to analyze the molecular fuzzy symmetry when a multi-fold symmetry element exists? Now, we will discuss this problem in detail. To stress the main point, we may only analyze one fuzzy symmetry element sometimes, i.e., the relative fuzzy symmetry by means of the C_6 point group. Here we will use azine, pyridine, diazines and the pyridine hydride as the examples. To study the symmetry and other structure characterization of these molecules, it may be useful for the research on these derivatives.

The atomic numbering and the coordinate axes of these molecules are shown in Figure 14. Using the quantum chemical calculation at various levels within Gaussian program [35], we can obtain the relevant MOs which are combined by the atomic orbitals (AOs) in the atomic valance shell. The π-MOs are combined by the *p*-AOs which are perpendicular to the molecular plane, whereas the σ -MOs by the sp^2-AOs in the molecular plane.

The membership functions of symmetry transformations $\hat{G}$ in a point group may be obtained by the method as described in the previous text. If any atom J of a molecule M moves to the position of the atom GJ after the operation $\hat{G}$ of the point group G is carried out, then the membership function of this molecule is defined as eq(5). For the whole molecule skeleton, the criteria (Y_J and Y_{GJ}) of atoms may be the atomic number, still. As for the MO, according to the LCAO-MO theory, the ρth MO Ψ_ρ may be denoted as eq(10). Then the membership functions of the MOs may be obtained by eq(13).

As for the benzene-like molecule, although the membership function may also be calculated as mentioned above, the overlap case will be more complex owing to the existence of two-dimensional irreducible representations. For the benzene molecule (as shown in Figure

14, the cycles in the hexatomic ring represent the carbon atoms, and others the hydrogen atoms), the π-MO will belong to one of the A2u, B2g, E1g or E2u irreducible representations in D_{6h} point group, whereas the σ -MO will belong to one of the A1g, A2g, B1u, B2u, E1u or E2g irreducible representations in this group. However, for the benzene-like molecule, the π-MO will belong to the fuzzy representations combined with a certain component ratio by the overlap of some of the A2u, B2g, E1g and E1u irreducible representations, but the σ -MO will belong to the fuzzy representation combined with a certain component ratio by the overlap of some of the A1g, A2g, B1u, B2u, E1u and E2g irreducible representations in D_{6h} point group. After we obtained the MOs using the quantum chemical calculation at various levels within Gaussian program [35], we calculated the irreducible representation components, utilizing the projection operator [36].

The $a_\rho(J,i)$ in MO may be as follows:

$$a_\rho(J,i) = a_\rho(J,i;\Gamma_1) + a_\rho(J,i;\Gamma_2) + a_\rho(J,i;\Gamma_3) +\ldots+ a_\rho(J,i;\Gamma_r) =\sum_R a_\rho(J,i;\Gamma_R) \tag{30}$$

where the $\Gamma_R = \Gamma_1,\Gamma_2 ,\ldots; \Gamma_r$ are the various irreducible representations (they may not all be one-dimensional) in relation to $a_\rho(J,i)$. The $a_\rho(J,i;\Gamma_R)$ should be related to the irreducible representation Γ_R. Therefore, for the ρth MO, Ψ_ρ may be obtained as follows :

$$\Psi_\rho =\sum_J\sum_i[\sum_R a_\rho(J,i;\Gamma_R)]\phi(J,i)=\sum_R[\sum_J\sum_i a_\rho(J,i;\Gamma_R)\phi(J,i)]\equiv\sum_R\Psi_{\rho R} \tag{31}$$

where the $\Psi_{\rho R}=\sum_J\sum_i a_\rho(J,i;\Gamma_R)\phi(J,i)$ is only the part in connection with the irreducible representation Γ_R. As the projection operator $\hat{P}(\Gamma_R)$ method, such operator may operate to a certain function and retain the part-function only related to the irreducible representation Γ_R. As to above certain MO Ψ_ρ, we can get $\Psi_{\rho R}$, i.e.

$$\hat{P}(\Gamma_R)\,\Psi_\rho = \Psi_{\rho R} = [\sum_J\sum_i a_\rho(J,i;\Gamma_R)\,\phi(J,i)\,] \tag{32}$$

The irreducible representation Γ_R component of Ψ_ρ :

$$X(\Psi_\rho;\Gamma_R)= [\sum_J\sum_i a_\rho^2(J,i;\Gamma_R)\,] \,/\, [\sum_J\sum_i a_\rho^2(J,i)\,] \tag{33}$$

It would conform to the normalized condition :

$$\sum_R X(\Psi_\rho;\Gamma_R) = 1 \tag{34}$$

We may use the projection operator method in quantum chemistry to obtain the corresponding irreducible representation components of MOs with the fuzzy symmetry related to various point groups. By means of this method, obviously, we can also obtain the same results with respect to the MOs for ethylene tetra-halide molecules with the D_{2h} fuzzy symmetry. The projection operator method is more general and convenient for programming, and then we analyze some typical molecular systems further.

We consider the π-MOs in the hexatomic ring of some simple benzene-like molecules. To start with, we utilize some suitable sub-group to replace the complete point group (D_{6h}) and the un-normalized projection operator before the further calculate. For the π-MOs in the hexatomic ring of some simple benzene-like molecules, we use the C_6 group to replace the

D_{6h} point group. For the four irreducible representations that the related π-MOs belong, A2u, B2g, E1g and E2u of D_{6h} point group are replaced by the four irreducible representations, A, B, E1 and E2 of C_6 point group, respectively. We may then transform the irreducible representations of C_6 point group to those of D_{6h}.

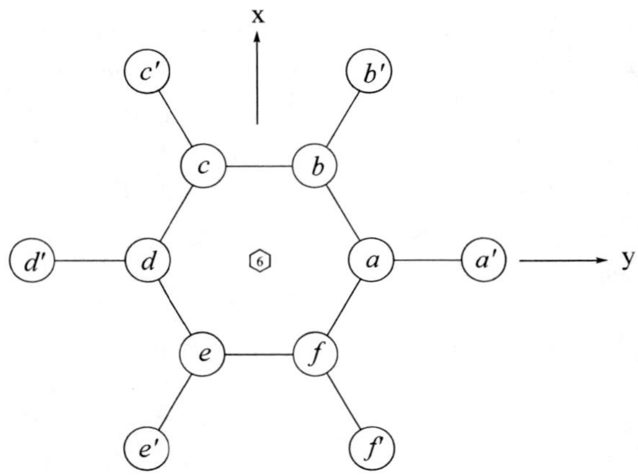

Figure 14. The coordinate axes and the atomic numbering of the benzene-like molecule. Where *a, b, c, d, e* and *f* denote six atoms of the hexatomic ring, they may be the carbon atoms or hetero-atoms. *a', b', c', d', e'* and *f'* denote the atoms that link to the hexatomic ring, and may be a hydrogen atom or an empty position. The molecular plane lies on the xy coordinate plane, and the (fuzzy) six-fold axis is taken as the z axis.

As regards this four irreducible representations Γ_R of C_6 point group, the corresponding the un-normalized projection operators $\hat{P}(\Gamma_R^{\#})$ may be obtained as follows by means of the character values of C_6:

$$\hat{P}(A^{\#}) = \hat{C}_6^0 + \hat{C}_6^1 + \hat{C}_6^2 + \hat{C}_6^3 + \hat{C}_6^4 + \hat{C}_6^5 \quad (35a)$$

$$\hat{P}(B^{\#}) = \hat{C}_6^0 - \hat{C}_6^1 + \hat{C}_6^2 - \hat{C}_6^3 + \hat{C}_6^4 - \hat{C}_6^5 \quad (35b)$$

$$\hat{P}(E1^{\#}) = 2\hat{C}_6^0 + \hat{C}_6^1 - \hat{C}_6^2 - 2\hat{C}_6^3 - \hat{C}_6^4 + \hat{C}_6^5 \quad (35c)$$

$$\hat{P}(E2^{\#}) = 2\hat{C}_6^0 - \hat{C}_6^1 - \hat{C}_6^2 + 2\hat{C}_6^3 - \hat{C}_6^4 - \hat{C}_6^5 \quad (35d)$$

The un-normalized $\Psi_{\rho R}$ can be obtained, as these projection operators $\hat{P}(\Gamma_R^{\#})$ denoted in eq(35) operate on Ψ_ρ. The normalizing factors depend on the characterization and dimension of the irreducible representation for the relevant MO. The πMOs in various hexatomic rings with the same dimension of the MO representation should have the same normalizing factor. Therefore through calculation, we may find the normalizing factor of a certain πMO with the irreducible representation Γ_R in the hexatomic ring of benzene molecule, and then use it as a normalizing factor of the other πMO with the same irreducible representation Γ_R in the hexatomic ring of the benzene-like molecule. In this way, we can obtain the corresponding normalized $\Psi_{\rho R}$. Then we can get the relative irreducible representation (Γ_R) component of Ψ_ρ. Meanwhile, we may assign the A, B, E1 and E2 irreducible representations of C_6 point group to the A2u, B2g, E1g and E2u irreducible representations of D_{6h} point group, respectively and directly.

The σ -MOs for the hexatomic ring of the benzene-like molecule may also be analyzed similarly. However it must be noticed: (1) owing to the combined components of the σ -MOs that includes the sp^2-AOs in the valance shell of the hexatomic ring, it is more complex than that of the π-MO because the calculation will be more prolix; (2) only four irreducible representations of D_{6h} are adopted by the πMOs, but six irreducible representations of D_{6h}, including the A1g, A2g, B1u, B2u, E1u and E2g by the σ -MOs. Of cause, we may use the D_{6h} point group to construct the projection operators, directly. As for the fuzzy symmetry of the σ -MO, the calculation will be lengthier than that for the π-MO, although the basic calculation rules are the same.

5.2.1. The Fuzzy Symmetry of Benzene and Pyridine

As is well known, benzene molecule has the D_{6h} point group symmetry. Its MOs belong to certain irreducible representations of this group. However, any single MO that belongs to a multi-dimensional irreducible representation should be not the common eigenstate of the all symmetry transformation in D_{6h}. Only the MO complete set of this multi-dimensional irreducible representation may be the common eigenstate of all the symmetry transformations in D_{6h} point group. That means for such single MO, the membership functions corresponding to various symmetry transformations in D_{6h} are not all equal to one. Therefore we may analyze their fuzzy symmetry.

Every carbon atom of benzene molecule has four AOs in the valence shell, and 24 AOs for all ring carbons; these AOs with the six 1s AOs of related hydrogen atoms may combine to form 30 MOs. According to calculation at a certain level, we can obtain the relative MOs and corresponding irreducible representations of D_{6h}. At various levels, there are always six π-MOs, and they may be denoted as follows:

$$\Psi_1(\mathrm{A2u}) = \alpha_1[\Phi_a + \Phi_b + \Phi_c + \Phi_d + \Phi_e + \Phi_f] \tag{36-1}$$

$$\Psi_2(\mathrm{E1g}) = \alpha_2[\Phi_b + \Phi_c - \Phi_e - \Phi_f] \tag{36-2}$$

$$\Psi_3(\mathrm{E1g}) = \alpha_3[2\Phi_a + \Phi_b - \Phi_c - 2\Phi_d - \Phi_e + \Phi_f] \tag{36-3}$$

$$\Psi_4(\mathrm{E2u}) = \alpha_4[2\Phi_a - \Phi_b - \Phi_c + 2\Phi_d - \Phi_e - \Phi_f] \tag{36-4}$$

$$\Psi_5(\mathrm{E2u}) = \alpha_5[\Phi_b - \Phi_c + \Phi_e - \Phi_f] \tag{36-5}$$

$$\Psi_6(\mathrm{B2g}) = \alpha_6[\Phi_a - \Phi_b + \Phi_c - \Phi_d + \Phi_e - \Phi_f] \tag{36-6}$$

where Ψ_2 (E1g) and Ψ_3 (E1g) are the degenerate HOMOs, whereas Ψ_4 (E2u) and Ψ_5 (E2u) are the degenerate LUMOs. Here Φ's are corresponding *pz*-AOs. For all these levels of calculations, eq(36) should be ture, but the values of α_i(i=1 to 6) may be different at various levels (*e. g.* Table 6).

The MOs, Ψ_1(A2u) and Ψ_6(B2g), which belong to the one-dimensional irreducible representation of D_{6h} group, and so all of the membership functions of related various symmetry transformations are equal to one. Here the square of the LCAO coefficient is used as the atomic criterion. Consequently, in such MOs, each atomic criterion would be the same, i.e., $\alpha_a^2 = \alpha_b^2 = \alpha_c^2 = \alpha_d^2 = \alpha_e^2 = \alpha_f^2$, to ensure any atomic criterion unchanged after any symmetry transformation in relation to D_{6h}, and then the corresponding membership function must be one.

However, the degenerate MOs Ψ_2(E1g) and Ψ_3(E1g) belong to a two-dimensional irreducible representation E1g, and form a complete basis set. Similarly, MOs Ψ_4 (E2u) and Ψ_5 (E2u) also belong to another two-dimensional irreducible representation, E2u, and form a complete basis set. As we select such, the single MO will not be the eigenstates of all transformations in D_{6h}, but may be so of the all transformations in D_{2h}. The membership functions of all corresponding symmetry transformations in D_{2h} are equal to one, but the membership functions of other symmetry transformations in the D_{6h} point group would be less to one. For the Ψ_2(E1g) MO, where both atomic criteria of atoms *a* and *d* are null, but all other atomic criteria are equal to α_2^2. According to the symmetry transformation of the rotation of 60^0 around the C_6 axis, the membership function equals 1/2. The Ψ_3(E1g) MO may be analyzed in the same fashion, and the membership function for the symmetry transformation of the rotation of 60^0 around the six-fold axis equals 1/2, too. This means that for such degenerate MOs, both their orbital energies and their membership function are equal. However, for the above degenerate MO complete set, all of the membership functions are equal to one, where the atomic criterion is the summation of the square of LCAO coefficients. As for the MO set {Ψ_2(E1g),Ψ_3(E1g)}, the atomic criteria of atom *a,b,c,d,e* and *f* are $4\alpha_3^2$, $\alpha_2^2+\alpha_3^2$, $\alpha_2^2+\alpha_3^2$, $4\alpha_3^2$, $\alpha_2^2+\alpha_3^2$ and $\alpha_2^2+\alpha_3^2$, respectively. Since membership functions of all these symmetry transformations in D_{6h} point group equal to one, the necessary and sufficient condition should be that all of these criteria must be the same, i.e. $\alpha_2^2=3\alpha_3^2$ should be true. As such we known, the coefficients α_2 and α_3 must depend on each other. Such necessary and sufficient condition will be true at various levels, and it may be checked by the data in Table 6. We may analyze the MO set {Ψ_4 (E2u),Ψ_5 (E2u)} in the same way. The results are the same at various MO theoretical levels. It should be noted that when a single MO belongs to a multi-dimensional irreducible representation, its irreducible representation is pure, but some membership functions may be less one. Furthermore, as to other similar π-MOs in the hexatomic ring (including the hetero-atoms and the substituents), we may analyze their fuzzy symmetry based on above results. Their fuzzy representations may overlap by means of some of above four irreducible representations.

Table 6. The LC*pz*-AO coefficients of the π MOs in the valance shell of benzene.

MO-calculated level	α_1	α_2	α_3	α_4	α_5	α_6
HMO	0.4083	0.5000	0.2887	0.2887	0.5000	0.4083
AM1	0.3311	0.4594	0.2652	0.3316	0.5743	0.5300
HF/STO-3G	0.3333	0.4601	0.2656	0.3299	0.5713	0.5244

The σ-MOs may be discussed similarly, even though the AOs that combine to form them include *py*-, *px*- and *s*-AOs. The irreducible representations to which the σ-MOs belong may be the A1g, A2g, B1u, B2u, E1u, and E2g. As for other similar hexatomic ring molecules, the fuzzy symmetry of σ-MOs may be analyzed further, too. The fuzzy representations of their σ-MOs may be the overlap of these six-irreducible representations. According to the chemical view, the π-MOs are more active than σ-MOs, therefore, we will not analyze the σ-MOs here.

Different from benzene molecule, pyridine molecule is composed of the five carbon atoms and one nitrogen atom in its ring, and then the fuzzy D_{6h} symmetry ought to be used. The geometrical structure of this molecule may be shown as in Figure 14, and the *d* atom in the ring is a nitrogen atom, whereas others are the carbon atoms. The atoms that connect the hexatomic ring are the hydrogen atoms except position *d'* where there is no atom. In addition to the mirror of molecular plane, this molecule has also a two-fold axis and another mirror plane that is perpendicular to the molecular plane, through the nitrogen atom and its *para*-carbon atom. Thus pyridine molecule should belong to C_{2v} point group, with the *y*-axis being its main-axis. According to the fuzzy D_{6h} symmetry, using the eq(1), we may calculate the membership function of pyridine molecular skeleton. Using the atomic number (Z) as the criterion of the atom and zero as the criterion of an empty atom (*d'*), we can easily find that the membership functions of the all symmetry transformations in C_{2v} point group as above, are equal to one, whereas those of other symmetry transformations in D_{6h} point group, are equal to 0.952. For this fuzzy D_{6h} point group, according to the idea of Zadeh [34], it may be symbolized as:

$$\underset{\sim}{D_{6h}} = \sum_{G\subset C2v} (1/\hat{G}) + \sum_{G\subsetneq C2v} (0.952/\hat{G}) \tag{37}$$

where $G\subset C_{2v}$ and $G\subsetneq C_{2v}$ represent the summation in relation to the symmetry transformation $\hat{G}$ which belong to or not to the symmetry transformation set in the C_{2v} point group above, respectively. It may show that the molecular symmetry of this molecule would be near that of D_{6h} point group because the membership functions of all the symmetry transformations in D_{6h} point group are more than 0.95. There are four symmetry transformations $\hat{G}$ with $G\subset C2v$, and twenty with $G\subsetneq C2v$. Some symmetry transformations $\hat{G}$ are equivalent to those in benzene (D_{6h} point group), but not equivalent in benzene-like molecule. For example, three two-fold axes through two *para*-atoms of benzene molecule are equivalent elements in D_{6h}. However, in pyridine molecule, the two-fold axis through the *para*-CN-atoms and the two two-fold axes through the *para*-CC-atoms will not be equivalent in fuzzy D_{6h}. We would like to introduce some different symbols to distinguish these symmetry transformations. The reflection transformations about the mirrors perpendicular to molecular plane can be labeled as $\hat{M}$ (P;*qr*) and $\hat{M}$ (N;*qr*), where *qr* may be *ad, be* or *cf* to denote the two *para*-atoms. $\hat{M}$ (P;*qr*) means the cross-line between the mirror and molecular plane through the points *q* and *r*. $\hat{M}$ (N;*qr*) means the normal line of the mirror lying on the molecular plane through the points *q* and *r*. The rotations 180^0 about the two-fold axes lain on the molecular plane, can be symbolized as $\hat{C}$(P;*qr*) and $\hat{C}$(N;*qr*), where $\hat{C}$ (P;*qr*) and $\hat{C}$ (N;*qr*) denoted the relative two-fold axes through and orthogonal *qr*, respectively. $\hat{C}(6,l)$ and $\hat{S}(6,l)$ denote the symmetry transformations of l($l = \pm1,\pm2,\pm3$) times rotation about the six-fold rotation axis and rotation-reflection about the six-fold roto-reflection axis, respectively. $\hat{E}$ and $\hat{M}$(h) are the identity transformation and the reflection about the molecular plane as mirror, respectively. As for the center inversion symmetry transformation, $\hat{i}$, it is $\hat{S}(6,3)$. After defining these symmetry transformations, we may start to analyze the fuzzy symmetry of pyridine MOs.

Using the above methods, we have been calculated at AM1 level the various irreducible representation components included in the six π-MOs of the hexatomic ring in pyridine molecule, and results are shown in Table 7.

For benzene molecule, πOMO-1 and 2 are the degenerate HOMO, πVMO-1 and 2 are the degenerate LVMO. Comparison of the πMOs of pyridine and those of benzene molecule it is shown that the main representations (the irreducible representation with maximal components) of pyridine π-MOs are the same as the 'pure' ones of benzene π-MOs.

Meanwhile, all the corresponding irreducible representation components are much near one. The fuzzy correlation diagram between the πMOs of pyridine and those of benzene may be matched. It is noteworthy that for benzene the π-OMOs belonging to the two-dimensional irreducible representation are degenerate, but they are not for pyridine. That is to say, the pyridine π-MO set that belong to a same two-dimensional irreducible main-representation these two MOs will still not be degenerate. Accordingly, their energy and some membership functions will not be the same. Therefore the fuzzy irreducible representations will be different from those of the "pure" ones. Such fuzzy representations that both the pyridine π-OMO-2 and 1 belong, according to data in Table 7, may expressed as,

$$\underset{\sim}{\Gamma}(\pi\text{OMO-2}) = 0.01972\text{A2u} + 0.00005\text{B1g} + 0.97748\text{E1g} + 0.00275\text{E2u} \tag{38a}$$

$$\underset{\sim}{\Gamma}(\pi\text{OMO-1}) = 0.99967\text{E1g} + 0.00033\text{E2u} \tag{38b}$$

Table 7. The irreducible representation components of the π MOs in the valance shell of C_5H_5N in relation to D_{6h} at AM1 level.

πMO	irreducible representation components				
serial order	A2u	B1g	E1g	E2u	The main representation
πOMO-3	0.95384	0.00047	0.04169	0.00399	A2u
πOMO-2	0.01972	0.00005	0.97748	0.00275	E1g
πOMO-1	0	0	0.99967	0.00033	E1g
πVMO-1	0	0	0.00023	0.99977	E2u
πVMO-2	0.00025	0.01501	0.00232	0.98242	E2u
πVMO-3	0.00001	0.99706	0.00010	0.00283	B2g

Although the fuzzy representations of πOMO-2 and 1 for pyridine are similar, and both are close to the pure E1g irreducible representation, this may be different for some other molecules. The results as above at HF/STO-3G level are similar.

Now we cut the symmetry transformations in the D_{6h} point group into four subsets:

$$G0 = \{\hat{G}_0\} = \{\hat{E},\ \hat{M}(h),\ \hat{M}(P, ad),\ \hat{C}(P, ad)\} \tag{39a}$$

$$G1 = \{\hat{G}_1\} = \{\hat{M}(L, ab), \hat{M}(L, cd), \hat{C}(L, ab), \hat{C}(L, cd), \hat{C}(6, \pm 1), \hat{S}(6, \pm 1)\} \tag{39b}$$

$$G2 = \{\hat{G}_2\} = \{\hat{M}(P, be), \hat{M}(P, cf), \hat{C}(P, be), \hat{C}(P, cf), \hat{C}(6, \pm 2), \hat{S}(6, \pm 2)\} \tag{39c}$$

$$G3 = \{\hat{G}_3\} = \{\hat{M}(L, bc), \hat{C}(6, 3), \hat{C}(L, bc), \hat{i}\} \tag{39d}$$

The membership functions of these symmetry transformations for pyridine and benzene πMOs may be obtained. For eq(39), we may designate a certain transformation as the typical transformation, and for the subset Gm(m=0,1,2,3), the $\hat{C}(6, m)$ may be taken as the representative and called the typical symmetry transformation. In fact, it corresponds to the analysis of the fuzzy symmetry with C_6 point group. Sometimes, this is convenient, but not necessary.

The membership functions in relation to all symmetry transformations in G0 subset are equal to one. As to other subsets, the membership functions in relation to all symmetry transformations of the same subset should have the same values, but those of different subsets may be different.

5.2.2. The Fuzzy Symmetry of Diazine

For above analysis about the fuzzy symmetry characteristics of benzene and pyridine, it may be used to examine the other hexatomic ring molecules. Now, we turn to analyze the fuzzy symmetry for diazine. There are three diazine isomers: *o*-diazine (pyridazine), *m*-diazine (pyrimidine) and *p*-diazine (pyrazine). The molecular geometries are still arranged as in Figure 14, where the four carbon atoms and two nitrogen atoms constitute the hexatomic ring. The nitrogen atoms occupy the (*b,c*), (*b,f*) and (*a,d*) positions of *o*-, *m*- and *p*- isomers, respectively. For the hexatomic ring, each carbon atom will be adjacent to a hydrogen atom, but every nitrogen atom is adjacent to none. We may now analyze the membership functions of these three isomers in relation to various symmetry transformations. By means of the C_6 point group and using the atomic numbers as the criteria, we can calculate the membership functions of relative molecular skeletons via eq(5). Results are shown in Table 8. Observation of the data in this table indicates that the membership functions are still a bit different are even though they may be equal to or close to one.

It is important that the membership function for the *o*-$C_4H_4N_2$ molecule in relation to the $\hat{C}(6, \pm 1)$ is bigger than that to the $\hat{C}(6, \pm 2)$ because two nitrogen atoms in this molecule will move from *bc*-positions to *cd*-positions (Figure 14) under the operation of $\hat{C}(6, \pm 1)$ transformation, where *c*-position is occupied by nitrogen atom both before and after the operation. However, the nitrogen atom will move to *de*-positions after the operation of $\hat{C}(6, \pm 2)$ transformation, whereas both *d*- and *e*-positions are not the positions of nitrogen atoms before the operation. The less membership function means the less similarity of the molecular geometries before and after the operation. On the other hand, the membership function μ of any elements *x* and *y* in a fuzzy group must satisfy eq(9a-c)[21,34]. Usually, eqs (9a) and (9c) may be true. If both *x* and *y* are the $\hat{C}(6, \pm 1)$ transformation, then the *xy* ought to be the $\hat{C}(6, \pm 2)$ transformation. Thus, the relative membership function μ may not satisfy eq(9b). It means that for the *o*-$C_4H_4N_2$, in relation to C_6 or D_{6h} point group transformation, the corresponding fuzzy symmetry set is not a fuzzy group. As two carbon atoms (or their adjoined hydrogen atoms) in *o*-positions of benzene are replaced by some other atoms, the relative fuzzy symmetry set would not form a fuzzy group. By the way, as the tri-atomic ring, this case will not appear owing to $\hat{C}(3, 1) = \hat{C}(3, -1)$.

Table 8. The membership functions of some azines molecular skeletons in relation to the C_6 point group symmetry transformations.

symmetry transformation	membership function				
	C_6H_6	C_5H_5N	*o*-$C_4H_4N_2$	*m*-$C_4H_4N_2$	*p*-$C_4H_4N_2$
$\hat{C}(6, 0) = \hat{E}$	1	1	1	1	1
$\hat{C}(6, \pm 1)$	1	0.9524	0.9524	0.9048	0.9048
$\hat{C}(6, \pm 2)$	1	0.9524	0.9048	0.9524	0.9048
$\hat{C}(6, 3)$	1	0.9524	0.9048	0.9048	1

Similarly, the irreducible representation components of these diazine, pyridine and benzene were obtained and shown in Figure 15. These results are near close to those of benzene. Usually, we may neglect their symmetry defect for these MOs. Thus, their main irreducible representations would be the same as the pure ones of benzene πMOs. Their fuzzy correlation diagram may be matched.

According to the above analysis, we may inquire into the membership functions in relation to various symmetry transformations for the πMOs of these three isomers. Although the representations which these πMOs belong to are more ‘pure’, but the relative membership functions may be somewhat different, each other. According to eq(39), we may classify the symmetry transformations into four subsets, by grouping the equal membership functions into the same subset. This methods may be used to all these three isomers, *m*-, *o*- and *p*-$C_4H_4N_2$, however for the G1 and G2 subsets the contents in relation to the *m*- and *o*-$C_4H_4N_2$ isomers are different. As to the *p*-$C_4H_4N_2$, their membership function values of these two subsets are the same. Obviously, the reason is that both the common point groups for both *m*-$C_4H_4N_2$ and *o*-$C_4H_4N_2$ are the proper subgroups for that of *p*-$C_4H_4N_2$. Although both *o*-$C_4H_4N_2$ and *m*-$C_4H_4N_2$ belong to the C_{2v} point group, their main-axis directions are different, and the symmetry transformations in G1 and G2 are classified differently. On the other hand, the union subsets, $G1 \cup G2$, are the same for all of these isomers.

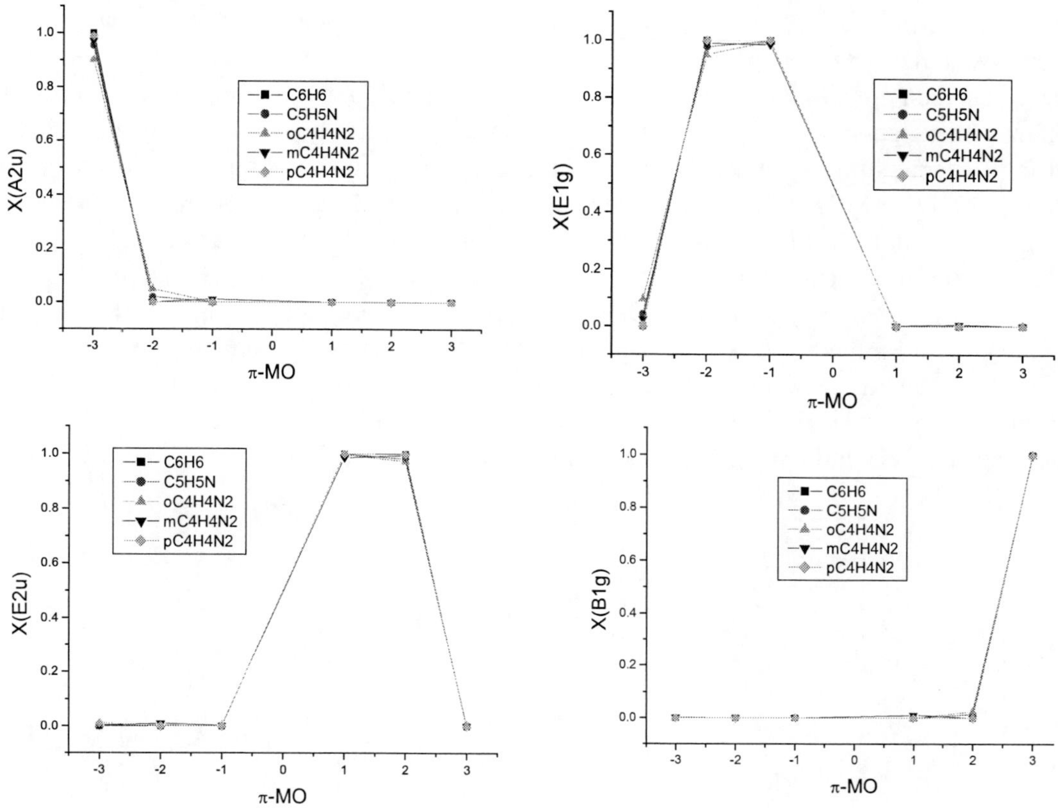

Figure 15. The irreducible representation components for the πMOs of daizine, pyridine and benzene in relation to the D_{6h} point group. Where the vertical axes X(Γ); Γ=, denote the components of the Γ(A2u,B1g,E1g or E2u) representations that the MOs belong to; and the horizontal axis denotes the serial number of the πMOs, at AM1 level.

Besides the typical symmetry transformations, $\hat{C}$(6,m), for the subsets Gm (m=0,1,2,3), others included in these subsets may differ for various isomers, depending on the common molecular symmetry point group of these isomers. The symmetry transformations for pyridine and *m*-diazine are the same, but they are different for *o*-diazine. As for the *p*-diazine, the symmetry transformations may be classified to the subsets, according to both the ways as for *m*- and *o*-diazine. However for these two methods, the union subsets G0UG3 of *p*-diazine are the same, and so are its union subsets G1UG2.

For comparing, the membership functions of the symmetry transformations in the subset Gm (m=0,1,2,3) which include the $\hat{C}$(6,m) are shown in Figure 16. It is evident that the membership functions of single πOMO-1 or πOMO-2 are less than those of the {πMO-E1g} set, similarly the membership functions of single πVMO1 or πVMO2 are less than those of the {πMO-E2u} set. According to the D_{6h} point group, the main-irreducible representations which the πMOs of the azine hexatomic ring belong to are as follows: (A2u), (E1g), (E1g), (E2u), (E2u), (B2g). They are the same as the irreducible representations of the πMOs in benzene hexatomic ring. Therefore, the correlation diagram may be drawn and ought to be mapped by the D_{6h} irreducible representation of one-to-one correspondence. It is noticed that the MO set of benzene, which belongs to D_{6h} two-dimensional irreducible representation in real number field, is combined by two degenerate and orthogonal MOs. However, the corresponding MO set of azine in relation to a fuzzy two-dimensional irreducible representation are two non-degenerate MOs. Now we need to analyze how such two π-MOs of azine hexatomic ring correlate with and match each other. In general, we may analyze this using certain symmetry transformations, i.e., some sub-groups. As the example shown in Figure 17, the π-MOs of benzene which belong to E1g and E2u irreducible representation are degenerate, but the π-MOs of azine are not. By means of the symmetry transformations about the mirrors of xz- and yz- coordinate planes as shown in Figure 14, we may analyze the representations of related π-MOs, these representations are either symmetrical or anti-symmetrical. Furthermore, we may attain the MO fuzzy correlation relationship. In general cases, the energies of the MOs belong to a same two-dimensional fuzzy irreducible representation will only differ a little, whereas their energetic order may differ at various theoretical levels. In Figure 17, the result at the AM1 level is shown.

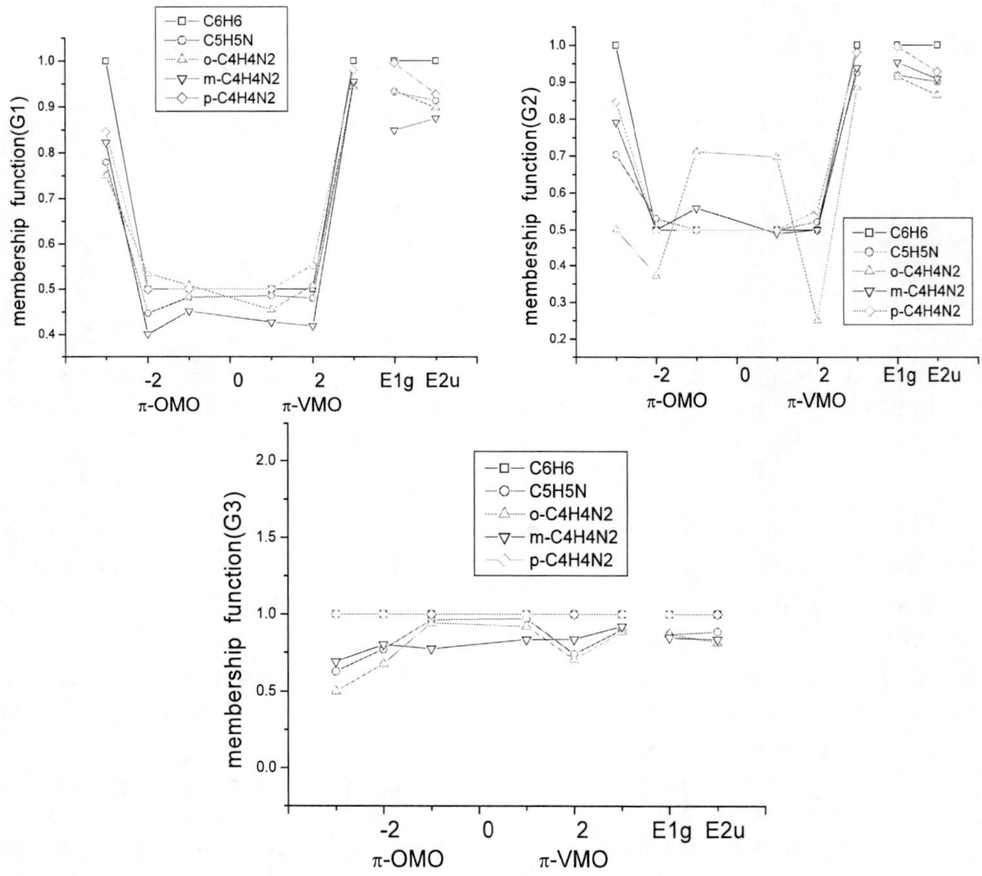

Figure 16. The membership functions of the AM1 πMOs of azines in relation to the symmetry transformations in the D_{6h} point group. Where the vertical axes denote the membership functions in relation to the symmetry transformations, the horizontal axis denots the π-OMO, □-VMO and the relative certain two-dimensional representation MO-sets, at AM1 level. C_6H_6 denotes the result of benzene for reference.

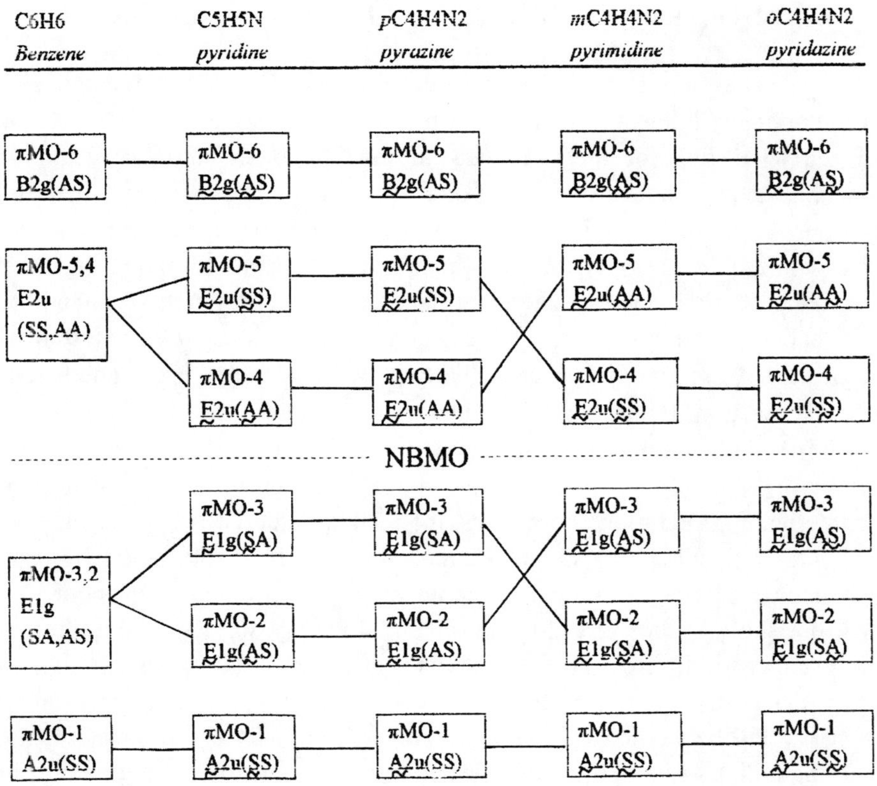

Figure 17. The πMO fuzzy correlation diagram for azines. The first row in every quad denotes the πMO serial number in the same order as the orbital energy, following with the irreducible representation in relation to the D_{6h} point group. The symbols in brackets denote the irreducible representations (S, symmetrical, A, anti-symmetrical) about the mirrors with xz and yz coordinate planes (Figure 14), respectively. The '~' below the irreducible representation symbol denotes related fuzzy irreducible representation.

6. The Fuzzy Symmetry Characteristics of Dynamic Molecular System

As is known to all, according to the W-H rule [1] there are the selection rules of the chemical dynamic processes in connection with the symmetry, successfully. Owing to the imperfection of the symmetry will be completely neglected in these rules, the quantitative relationship between the chemical dynamic property and symmetry can not be established. By means of the fuzzy symmetry method, perhaps such relationship may be obtained. As regards the fuzzy symmetry, there are two important quanta, the membership functions and the irreducible representation components in relative two important concepts, symmetry and invariant in common symmetric selection rules, respectively. Now, we will analyze the fuzzy symmetry characteristic of the molecular dynamic system, especially above important quanta. To start with we consider the simple tri-atomic dynamic system, and then examine some more complex molecular systems. Meanwhile some new symmetry transformations will encounter.

6.1. The Fuzzy Symmetry for Simple Tri-Atomic Dynamic System

Now we will study the fuzzy symmetry related to simple reaction dynamic systems and start with the simplest linear tri-atomic system. By means of the quantum field theory, we have analyzed W-H rule for the symmetry in relation to sigmatropic reaction[5,8]. We pointed out that there is not any perfect point group symmetry, but only the well-defined symmetry transformation in relation to the joint (combined or union) reaction reversal and some point symmetry transformation. The simplest system with such kind of symmetry is the H_3-system. Since the first potential energy surface of linear H_3-system was announced, there are a lot of papers in relation to this system and other linear tri-atomic system B…A…C published [39]. They are mainly concerned with the work of the potential energy surface that relates with the reaction: B+AC→BA+C. Now we will analyze the subject in relation to this symmetry. For the H_3 system or the tri-atomic B…A…C system with the same B and C atom, there are not any point group symmetry in general. But using the atomic distances of AB and AC as the internal configuration coordinates, the related potential energy surface ought to be provided with the reflection symmetry about the dividing surface (dividing line—the diagonal line in above internal configuration coordinate plane). Such symmetry about the dividing surface will not show in the corresponding potential energy surface of the linear tri-atomic B...A...C system with different B and C atoms. For linear tri-atomic B…A…C system, the point group symmetry transformation, the reaction reversal transformation and their combination [5] will be studied. Here reaction reversal transformation means that the relate system with the initial state ought to be reflected to a new state about the dividing surface (or line) [39,40] as a mirror. Generally speaking, such transformation is a kind of symmetry transformation for the whole potential surface, but not for such dynamic system, itself. The reaction reversal transformation may be denoted as $\hat{R}$. In some cases, it may often unite with a certain point group transformation $\hat{G}$ to form a new joint symmetrical transformation $\hat{G}\hat{R}$ for the dynamic system-self. Here, $\hat{G}$ is usually the space inversion transformation $\hat{P}$. Now, in relation to the reaction B+AC→BA+C, three related transformations will mainly be studied in detail. There are three transformations: 1) the space inversion transformation about the mid-atom as the center, 2) the reaction reversal transformation and 3) the joint transformation of the above two.

R_{12} R_{13}

B.........A.........C

2 1 3

Figure 18. The linear tri-atomic B…A…C system and atomic numbering.

The atomic positions in the linear B…A…C system are specified from A, B and C by 1, 2 and 3, respectively, as shown in Figure 18, and the atomic distances by R_{12} and R_{13} , respectively.

Since the atomic intervals in the system may be varied, the atomic criteria (Y) are depended not only on the atomic intrinsic property but also on the atomic related position (internal configuration coordinate). For the linear B...A...C tri-atomic system, the atomic related position may be determined by atomic intervals R_{12} and R_{13}. The atomic criteria may thus be denoted as follows:

$$Y_A = Y_A(R_{12},R_{13}),\ Y_B = Y_B(R_{12},R_{13}) \text{ and } Y_C = Y_C(R_{12},R_{13}). \tag{40}$$

For this system, both atomic related positions (R_{12} and R_{13}) and atomic criteria (Y_A, Y_B, Y_C) ought to change under the operation of transformation $\hat{G}$:

$$\hat{G}(R_{12}, R_{13}) = (R_{G12}, R_{G13}) \tag{41a}$$

$$\hat{G}(Y_A, Y_B, Y_C) = (Y_{GA}, Y_{GB}, Y_{GC}) \tag{41b}$$

As $\hat{G}$ being the space inversion transformation $\hat{P}$ (about the A atom as the centre) and the reaction reversal transformation $\hat{R}$, then we can get:

$$\hat{P}(R_{12}, R_{13}) = (R_{P12}, R_{P13}) = (R_{13}, R_{12}), \tag{42a}$$

$$\hat{P}(Y_A, Y_B, Y_C) = (Y_{PA}, Y_{PB}, Y_{PC}) = (Y_A, Y_C, Y_B). \tag{42b}$$

and

$$\hat{R}(R_{12}, R_{13}) = (R_{R12}, R_{R13}) \tag{43a}$$

$$\hat{R}(Y_A, Y_B, Y_C) = (Y_{RA}, Y_{RB}, Y_{RC}) \tag{43b}$$

In general, the internal configuration coordinates and the atomic criteria of initial and final state can not be interrelated by simple ways due to the operation of $\hat{P}$ or $\hat{R}$, alone. On the other hand, for the union transformation, $\hat{P}\hat{R} = \hat{R}\hat{P}$:

$$\hat{P}\hat{R}(R_{12}, R_{13}) = (R_{PR12}, R_{PR13}) = (R_{R13}, R_{R12}) \tag{44a}$$

$$\hat{P}\hat{R}(Y_A, Y_B, Y_C) = (Y_{PRA}, Y_{PRB}, Y_{PRC}) = (Y_{RA}, Y_{RC}, Y_{RB}) \tag{44b}$$

As we known [25, 26], the membership functions (μ) for a certain molecule or MO system in relation to the operation of transformation $\hat{G}$ may be calculated. If the normalization criterion is introduced, for our system, $\mu(\hat{G})$ may be calculated as:

$$\mu(\hat{G}) = (Y_A \wedge Y_{GA} + Y_B \wedge Y_{GB} + Y_C \wedge Y_{GC}) \tag{45}$$

Therefore, the membership functions of the transformations $\hat{P}$, $\hat{R}$ and $\hat{R}\hat{P}$, would become, respectively:

$$\mu(\hat{P}) = Y_A \wedge Y_{PA} + Y_B \wedge Y_{PB} + Y_C \wedge Y_{PC} = Y_A + 2\, Y_B \wedge Y_C \tag{45a}$$

$$\mu(\hat{R}) = Y_A \wedge Y_{RA} + Y_B \wedge Y_{RB} + Y_C \wedge Y_{RC} \tag{45b}$$

$$\begin{aligned}\mu(\hat{P}\hat{R}) &= Y_A \wedge Y_{PRA} + Y_B \wedge Y_{PRB} + Y_C \wedge Y_{PRC} \\ &= Y_A \wedge Y_{RA} + Y_B \wedge Y_{RC} + Y_C \wedge Y_{RB}\end{aligned} \tag{45c}$$

If atoms B and C are different, all of these membership functions would be less than one. If B and C are the same while $R_{12} \neq R_{13}$ and $Y_B \neq Y_C$ are true, the membership functions would not be the same, but if $Y_B = Y_{RC}$ and $Y_C = Y_{RB}$ are true, the corresponding membership functions in relation to transformation $\hat{P}\hat{R}$ would be equal to one. In other words, there is some symmetry in relation to the joint reaction reversal and parity transformation ($\hat{P}\hat{R}$) for such system [5], but the membership functions for both $\hat{R}$ and $\hat{P}$ are less one. That is, there is no accurate symmetry for $\hat{R}$ and $\hat{P}$, but only fuzzy symmetry for them. If B and C are different, the system has only some related fuzzy symmetry in relation to $\hat{P}\hat{R}$.

Now we turn to determine the various atomic criteria for the dynamic system. In theoretic chemistry region, σ MO occupied by single electron (σ SMO) and its neighboring σ MOs, occupied and virtual (noted as σ OMO-1 and σ VMO-1, respectively) are more interesting. We suggested that the atomic criterion for a certain MO may be assigned to the square or summation of square for one or more LCAO coefficients according to the MO composition of the AOs [25,26]. This scheme may also be applied to our present system. However, it is noteworthy that in the dynamic systems the LCAO coefficients depend on the internal configuration coordinate of the system. Therefore, the atomic criteria for various internal configuration coordinate conditions will be different. Now we consider a certain state J, corresponding to the point (R_{12},R_{13}) of the internal configuration coordinate plane for the linear tri-atomic B…A…C system, and the state J will be changed to state GJ under transformation $\hat{G}$, *i.e,* the state J transform to state GJ. As shown in Figure 19:

As we analyze the space inversion transformation ($\hat{G} = \hat{P}$) about atom A as a symmetrical centre. After the operation of $\hat{P}$, the linear tri-atomic system B…A…C ought to be changed from the initial state J to the final state PJ, and B and C will commute, and so will R_{12} and R_{13} (*i.e.*, $R_{P12}=R_{13}$; $R_{P13}=R_{12}$). If B and C are the same moreover the R_{12} equals R_{13}, the final state and the initial state may be identical; such system will have the usual space inversion symmetry, and the membership function in relation to transformation $\hat{P}$ will equal unity. For the majority cases, however, the membership functions in relation to $\hat{P}$ will be less than one. The reaction reversal transformation ($\hat{G} = \hat{R}$) is also shown in Figure 19. From the initial state J to the final state RJ, no commutation exists between neither B and C nor R_{12} and R_{13}. If the initial state J belongs to intrinsic coordinate (IRC) path, the intrinsic reaction coordinate value (x_J-IRC) may be obtained using quantum chemical programs such as Gaussian [35]. The intrinsic reaction coordinate value (x_{RJ}-IRC) for the final state RJ ought to be the opposite number of x_J-IRC, *i.e.*

$$(x_{RJ}\text{-IRC}) = -(x_J\text{-IRC}). \tag{46}$$

According to state RJ, we may obtain R_{R12} and R_{R13}. In general, R_{R12} and R_{R13} are different from R_{P12} and R_{P13}, respectively. That is:

$$R_{R12} \neq R_{P12} = R_{13}, \tag{46a}$$

$$R_{R13} \neq R_{P13} = R_{12}. \tag{46b}$$

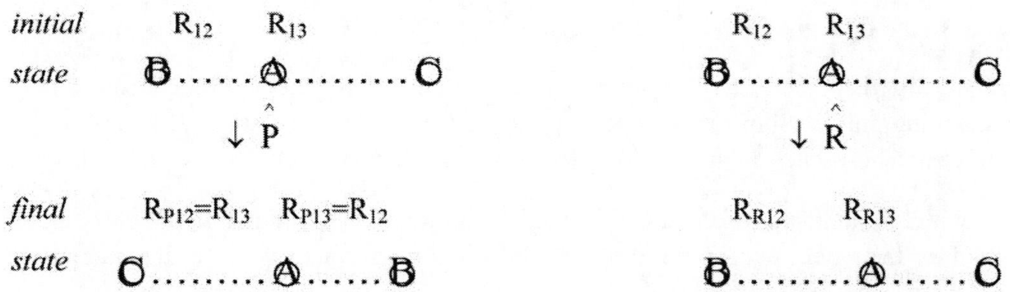

Figure 19. The linear tri-atomic B…A…C system under operations $\hat{P}$ and $\hat{R}$.

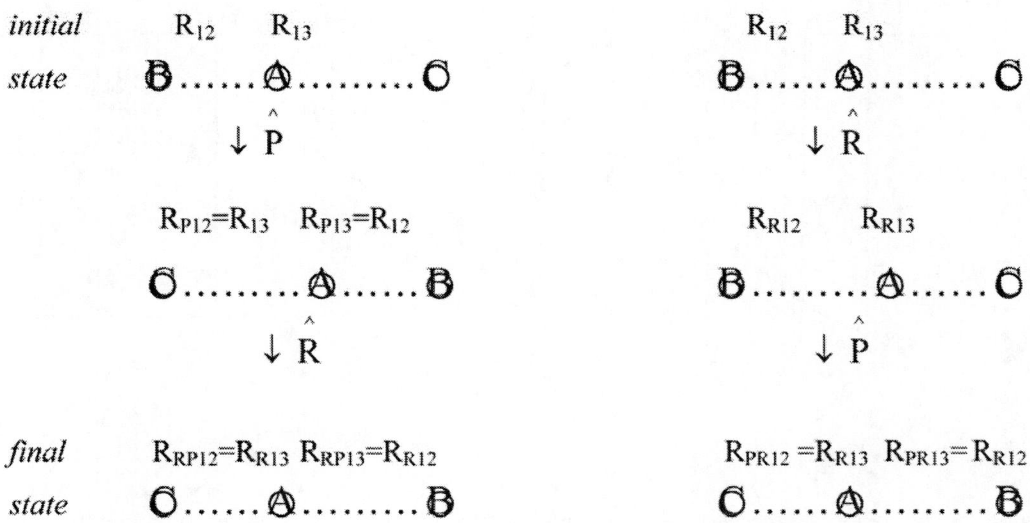

Figure 20. The linear tri-atomic B…A…C system under joint operation $\hat{P}\hat{R}$.

If B and C are same atoms, (R_{R12}, R_{R13}) and (R_{P12}, R_{P13}) may be the equivalence. For such case, the dividing curve (line) would be the diagonal line in the internal configuration coordinate (R_{R12}, R_{R13}) space. For states in the dividing line, membership functions in relation to both transformations $\hat{P}$ and $\hat{R}$ ought to equal one, whereas for states elsewhere, the membership functions in relation to both $\hat{P}$ and $\hat{R}$ ought to be less than one even though they may be equal.

As shown in Figure 20 clearly, the operation results for the joint transformation $\hat{P}\,\hat{R}$ are the same as those for $\hat{R}\,\hat{P}$. For the linear tri-atomic system B…A…C with the same atoms B and C, we have:

$$R_{R12} = R_{P12} = R_{13} \tag{47a}$$

$$R_{R13} = R_{P13} = R_{12}, \tag{47b}$$

And so the final and initial states in Figure 20 are the same.

It means that there is the joint transformation $\hat{P}\,\hat{R}$ for such system, and the corresponding membership function equals one, as B and C are the same. If atoms B and C are different, the related membership function is less than one. On the other hand, for the system with the same atoms B and C the membership function in relation to $\hat{R}$ or $\hat{P}$ alone is usually less than one, except for those in the dividing surface, whereas the membership function in relation to the joint transformation $\hat{P}\,\hat{R}$ will be one no matter what the state belongs to dividing line or not.

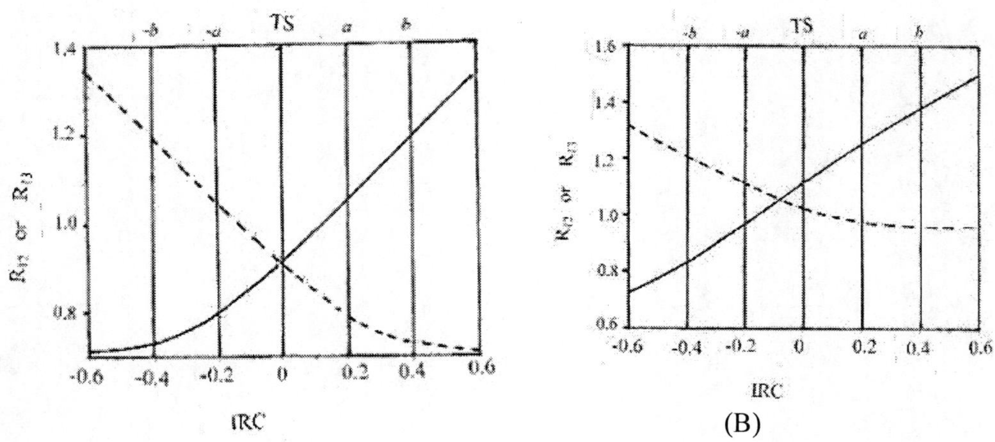

(A) (B)

Figure 21. Variation of the internal configuration coordinates for the linear tri-atomic B…A…C systems: R_{12}(solid curve) and R_{13}(dotted curve) along the IRC at HF/STO-3G. (A) HHH system; (B) HHF system (R_{12}=R_{HH}; R_{13}=R_{HF}).

Owing to the operation of $\hat{R}$, the initial state $\{R_{12},R_{13}\}$ would change into the final state $\{R_{R12},R_{R13}\}$ in general. For the linear tri-atomic system B…A…C with the same atoms B and C, the final state (R_{R12},R_{R13}) can be obtained by eq. (47). Regardless the B and C are the same or not, for the states in the dividing surface, the $\hat{R}$ ought to be equivalent to the identity transformation $\hat{E}$, and we may get the :

$$(R_{R12},R_{R13}) = (R_{12},R_{13}). \tag{48}$$

If the state in the IRC path, the final state (R_{R12},R_{R13}) may be obtained by eq. (46). However, the so-called general cases including states other than in the dividing surface and IRC path must also be examined. Only the IRC path and corresponding IRC value(x-IRC) can be obtained using Gaussian [35], but the dividing surface can not be obtained directly. If B and C are different, the related dividing surface may be not a straight line (*i.e.*, not a one dimensional plan). How can we obtain the state generated by the operation of $\hat{R}$? To start with, we examined Figure 21, where Figure (A) and (B) denote the two typical examples with B and C being and not being the same, respectively.

In Figure 21, the vertical axis denotes the internal configuration coordinate (R_{12} and R_{13}) of the state, and the abscissa axis the IRC value. As shown in the figure, both R_{12} and R_{13} are the monotonic functions of the IRC value. R_{12} and R_{13} are monotonically increasing and decreasing, respectively. We may use the IRC values to denote R_{12} and R_{13}, and call them the IRC-scale of R_{12} and R_{13}. The IRC value is also the IRC-scale of IRC path itself. Corresponding to a certain state in the IRC path, R_{12} and R_{13} ought to be the same IRC-scale value. For the state outside of the IRC path, R_{12} and R_{13} IRC-scale values ought to be different. When both of these two IRC-scale values equal zero, this corresponds to the transition state. When B and C are the same, the curves of R_{12} and R_{13} ought to cross in the transition state. However, when B and C are different, the curves of R_{12} and R_{13} ought to cross not in the transition state.

We may use $R_{12}(a)$ and $R_{13}(a')$ to denote R_{12} and R_{13} both of which have the IRC-scale values a and a', respectively. And use $\{R_{12},R_{13}\}$ to denote any state of the linear tri-atomic B…A…C system. Therefore the $\{R_{12}(a),R_{13}(a)\}$ and $\{R_{12}(a),R_{13}(b)\}$ ($b \neq a$) denote states in the IRC path and the outside of the IRC path, respectively. Under the operation of $\hat{R}$, the initial state $\{R_{12}(a),R_{13}(b)\}$ of the linear tri-atomic B…A…C system will be changed to:

$$\hat{R}\{R_{12}(a),R_{13}(b)\} = \{R_{R12}(a),R_{R13}(b)\} = \{R_{12}(-b),R_{13}(-a)\} \tag{49a}$$

If B and C are the same atoms, from inspection of Figure 21(A) we have :

$$\hat{R}\{R_{12}(a),R_{13}(b)\} = \{R_{12}(-b),R_{13}(-a)\} = \{R_{13}(b),R_{12}(a)\} \tag{49b}$$

That is, $R_{12}(a)$ and $R_{13}(b)$ of the initial state will be commute under operation $\hat{R}$ for systems with the same B and C atoms. For the initial state in the IRC path, as B and C are the same, we have:

$$\hat{R}\{R_{12}(a),R_{13}(a)\} = \{R_{12}(-a),R_{13}(-a)\} = \{R_{13}(a),R_{12}(a)\} \tag{49c}$$

The state in the dividing surface (line) ought to be unchanged due to operation $\hat{R}$. Using eq. (49a), we found that state $\{R_{12}(a),R_{13}(b)\}$ in the dividing line satisfies a condition:

$$a = -b \tag{50a}$$

On the other hand, state $\{R_{12}(a),R_{13}(b)\}$ in the IRC path ought to satisfy a different condition:

$$a = b. \tag{50b}$$

As for the transition state (TS), i.e., the state at the point of intersection between the dividing line and the IRC path, this state must satisfy the following condition:

$$a = b = 0 \tag{50c}$$

Therefore, for TS, the reaction reversal transformation $\hat{R}$ is equivalent to the identity transformation $\hat{E}$. Various states $\{R_{12}(a),R_{13}(a)\}$ in the IRC path may be obtained using program Gaussian[35]. According to eq. (50a), the state in dividing line ought to be $\{R_{12}(a),R_{13}(-a)\}$. This may be a simplification and approximation method for the calculation of dividing surfaces. It is noticed that eq. (50) is true whether B and C are the same or not.

For the initial state $\{R_{12}(a),R_{13}(b)\}$ of the linear tri-atomic B…A…C system, R_{12} and R_{13} will be commute under operation of the space inversion $\hat{P}$ about A, *i.e.*,

$$\hat{P}\{R_{12}(a),R_{13}(b)\} = \{R_{P12}(a),R_{P13}(b)\} = \{R_{13}(b),R_{12}(a)\}. \tag{51}$$

This equation is similar to eq. (49b). However, eq. (49b) may be true only for systems with the same B and C atoms, but eq. (51) is true whether or no. In general, we have,

$$\{R_{P12}(a),R_{P13}(b)\} \neq \{R_{R12}(a),R_{R13}(b)\}. \tag{52a}$$

Only for the cases with the same B and C do we have,

$$\{R_{P12}(a),R_{P13}(b)\} = \{R_{R12}(a),R_{R13}(b)\} \tag{52b}$$

It must be pointed out that $\hat{R}$ and $\hat{P}$ are different transformations even if B and C are the same. Since in such case, the same B and C would commute under transformation $\hat{P}$, but they would not commute under transformation $\hat{R}$. This is important for the analysis of the fuzzy symmetry of MOs. The membership function in relation to a certain transformation ought to be connected with only the square term of_the phase, the membership functions in related to $\hat{R}$ and $\hat{P}$ would be equal for systems with the same B and C. The analysis in connection with the representation component may differ evident. On the other hand, the membership functions of the system with different B and C ought to be unequal under $\hat{R}$ and $\hat{P}$.

As for joint transformation $\hat{P}\hat{R}$, the initial state $\{R_{12}(a),R_{13}(b)\}$ of linear tri-atomic B...A...C system may be transformed as :

$$\hat{P}\hat{R}\{R_{12}(a),R_{13}(b)\} = \hat{P}\{R_{R12}(a),R_{R13}(b)\} = \{R_{R13}(b),R_{R12}(a)\}$$
$$= \hat{P}\{R_{12}(-b),R_{13}(-a)\} = \{R_{13}(-a),R_{12}(-b)\}. \qquad (53a)$$

For the system with the same B and C, using eq. (52b), we have,

$$\hat{P}\hat{R}\{R_{12}(a),R_{13}(b)\} = \{R_{12}(a),R_{13}(b)\}. \qquad (53b)$$

Therefore, for such system, the membership function related to the joint transformation $\hat{P}\hat{R}$ would be unity. In other words, the system stay unchanged under transformation $\hat{P}\hat{R}$, and the related representation would be pure (i.e., pure symmetrical or pure anti-symmetrical). For the system whose B and C are different, the membership function of corresponding joint transformation $\hat{P}\hat{R}$ will be less than one, and such system has only some fuzzy symmetry in relation to the joint transformation $\hat{P}\hat{R}$. The related representations would be not pure.

By the way, as shown in Figure 21, both R_{12} and R_{13} are monotonic functions of the IRC value, so we may introduce the IRC-scale to substitute the common nature scale (!) . Some examples will be given in the section below. It should be noted that using the IRC-scale to denote R_{12} and R_{13} will require them not too small in common nature scale, otherwise R_{12} and R_{13} may not correspond to any IRC-scale values.

In additional, since the B...A...C tri-atomic system ought to be the open shell system, the relative MOs would be various spin states, α- and β-spin, and we may obtain them by means of Gaussian[35]. For various spin-states, the fuzzy symmetry characteristics may be somewhat differ but the relative calculation methods are similar.

6.1.1. The Fuzzy Symmetry of B...A...C (with the same B and C) System

For the linear B...A...C system, we may find the transition state (TS) using the Gaussian[35]. We first examine the fuzzy symmetry of TS. Since for TS, $\hat{R}$ is equivalent as $\hat{E}$, so we have:

$$\hat{R}(R_{12TS}, R_{13TS}) = (R_{R12TS}, R_{R13TS}) = (R_{12TS}, R_{13TS}) \qquad (54a)$$

$$\hat{R}(Y_{ATS}, Y_{BTS}, Y_{CTS}) = (Y_{RATS}, Y_{RBTS}, Y_{RCTS}) = (Y_{ATS}, Y_{BTS}, Y_{CTS}) \qquad (54b)$$

and as $\hat{P}\hat{R}$ is equivalent to $\hat{P}$, we obtain:

$$\hat{P}\hat{R}(R_{12TS}, R_{13TS}) = \hat{P}(R_{12TS}, R_{13TS}) = (R_{13TS}, R_{12TS}) \quad (55a)$$

$$\hat{P}\hat{R}(Y_{ATS}, Y_{BTS}, Y_{CTS}) = \hat{P}(Y_{ATS}, Y_{BTS}, Y_{CTS}) = (Y_{ATS}, Y_{CTS}, Y_{BTS}) \quad (55b)$$

For the same B and C, since, in general, $R_{12} \neq R_{13}$, and thus Y_B and Y_C are also different. However for TS, both $Y_{BTS} = Y_{RCTS} = Y_{CTS} = Y_{RBTS}$ and $R_{12TS} = R_{13TS}$ may be true. Therefore, for the whole skeleton and MOs, the membership functions in relation to $\hat{P}\hat{R} = \hat{P}$ ought to be unity, and the irreducible representations of the MOs would be pure. For example, σ SMO of TS of H...H...H system belongs to the pure anti-symmetrical irreducible representation Au, whereas σ OMO-1 and σ VMO-1 belong to the pure symmetrical one Ag, as shown in Figure 22.

The linear tri-atomic H...H...H system the simplest system with joint point group transformation and reaction reversal symmetry [5]. As mentioned above, its TS has the usual symmetrical center and the membership function in relation to $\hat{P}$ ought to be unity and the corresponding MO irreducible representation will be pure symmetrical or pure anti-symmetrical. The irreducible representation of σ SMO is pure anti-symmetrical, but those of σ OMO-1 and σ VMO-1 are pure symmetrical. Besides the TS, the states in IRC path, their membership functions in relation to $\hat{P}$ will be less than one, and the corresponding irreducible representations will be not pure symmetrical nor pure anti-symmetrical. On the other hand, owing to such system has the perfect symmetry in relation to the joint transformation $\hat{P}\hat{R}$, the related membership function ought to be one for the states in the IRC path, and the corresponding MOs belong to pure symmetrical or pure anti-symmetrical irreducible representations. Different spin states of σ SMO belong to the pure asymmetrical irreducible representations, but those of σ OMO-1 and σ VMO-1 belong to the pure symmetrical one.

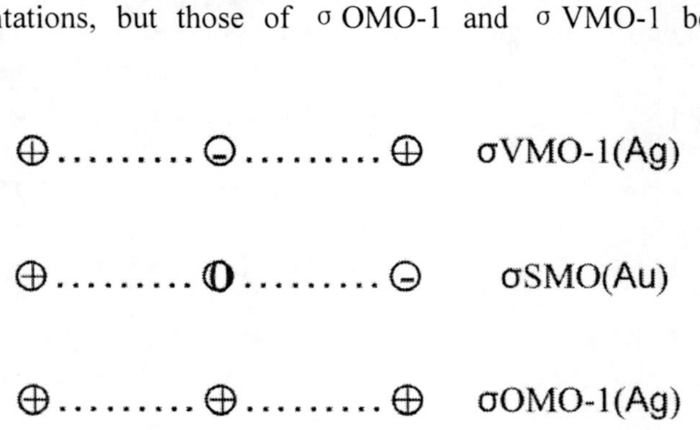

Figure 22. MOs of the transition state of H...H...H system and corresponding irreducible representations.

For the MOs of the state $\{R_{12}(a),R_{13}(a)\}$ in the IRC path calculated above, we may obtain their membership functions in relation to $\hat{R}$ and $\hat{P}$. According to the calculation results for linear tri-atomic H_3 - system using Gaussian [35] at the HF/STO-3G level, we may further calculate these membership functions. As shown in the Figure 23, all the membership functions in relation to $\hat{P}\hat{R}$ are unity, and the membership functions in relation to $\hat{R}$ and $\hat{P}$ are equal for the same MO with the same spin state, although there are somewhat difference for the same MO with different spin states. For TS (IRC=0), all these membership functions are unity. This is always true for the linear tri-atomic B…A…C system with the same B and C.

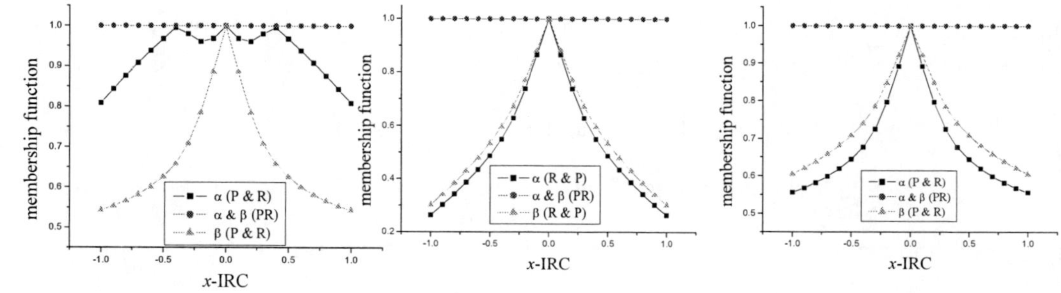

(A) σ OMO-1 (BMO) (B) σ SMO(NBMO) (C) σ VMO-1(ABMO)

Figure 23. Variation of the membership functions of H…H…H system with IRC values along the IRC path forσOMO-1(A), σSMO(B),σVMO-1(C) at HF/STO3G level, with various spin states (□-or □-) under transformations $\hat{P}\hat{R}$, $\hat{R}$ and $\hat{P}$.

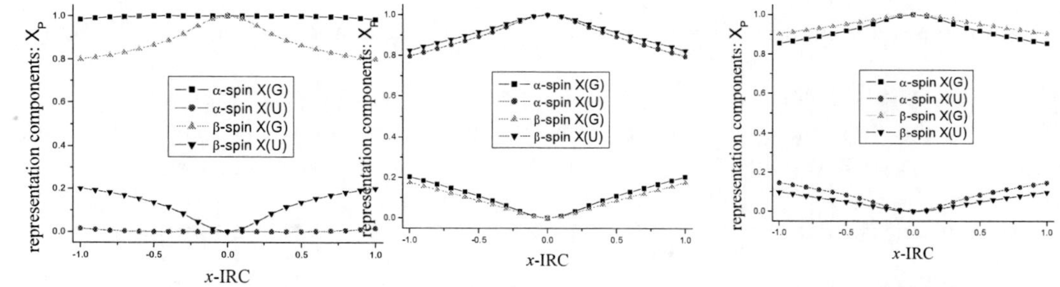

(A) σ OMO-1 (BMO) (B) σ SMO(NBMO) (C) σ VMO-1(ABMO)

Figure 24. Variation of the irreducible representation components of H…H…H system along the IRC path for σ OMO-1(A), σ SMO(B), σ VMO-1(C) at HF/STO3G level, with various spin states (α-or β-) in relation to transformation $\hat{P}$.

For the MOs of the H…H…H system in the states along the IRC path, we may consider their irreducible representation (*i.e.*, symmetrical and anti-symmetrical) components in relation to $\hat{P}$. Figure 24 denotes the variation of the irreducible representation components for the σ OMO-1, σ SMO and σ VMO-1 of this system along the IRC path. For the states in the IRC, the IRC coordinate values (IRC-scale) are denoted as *x*-IRC. For *x*-IRC=0 (transition state), the related irreducible representations ought to be pure. For σ OMO-1 and σ VMO-1, their symmetrical representation component, X(G)=1, and the anti-symmetrical one, X(U)=0, but for the σ SMO, X(G)=0 and X(U)=1. For other states, the irreducible representations will not be pure, yet their main representation components will usually be related to the same as for the transition state. For such system, the irreducible representation components in relation to $\hat{P}\hat{R}$, which the MOs belong, ought to be either pure symmetrical or pure anti-symmetrical, the same as those for corresponding TS. This is, X(G)=1 and X(U)=0 for σ OMO-1 and σ VMO-1, meanwhile, X(G)=0 and X(U)=1 for the σ SOMO. For $\hat{R}$, there is no anti-symmetrical case, and so we will not be analyzed the relative representation components. Interestingly, generally speaking, the H…H…H system (IRC≠0) is provided with the $\hat{P}\hat{R}$ symmetry, but without the alone $\hat{R}$ nor $\hat{P}$ symmetry.

It is notable that transformations $\hat{R}$ and $\hat{P}$ are different. For TS, owing to the operation of $\hat{P}$, both the atoms and phases of MOs will change in relation to space inversion (Figure 22). Under transform $\hat{P}$, atoms 2 and 3 and their orbital phases will be commute, but after operation $\hat{R}$, they will not. For the system with the same atoms 2 and 3, under the two transformations, the MO with symmetrical representation Ag will change in the same way, but the MO with anti-symmetrical representation Au will change differently. For the non-TS system with the same atoms 2 and 3, the changes due to the operation of both transformations include commutation between R_{12} and R_{13}. Under $\hat{P}$, both atoms 2 and 3 and their phases ought to commute, but under $\hat{R}$, atom 1 only moves between atoms 2 and 3, and the LCAO-MO coefficients may change in a way different from that under $\hat{P}$.

As for the states along the dividing line, the linear tri-atomic B…A…C system with the same B and C, such as the H…H…H system, has the perfect symmetry under space inversion transformation $\hat{P}$ and reaction reversal transformation $\hat{R}$, corresponding to the identity transformation $\hat{E}$. Therefore the membership functions in relation to all of $\hat{P}$, $\hat{R}$ and $\hat{P}\hat{R}$, ought to be unity.

Now, we examine the internal configuration states for the linear H_3-system outside of the IRC path and the dividing line. For the linear H_3-system, under space inversion $\hat{P}$ about the middle H atom as the inversion centre, atoms 2 and 3 and their phases will be exchanged. Under operation $\hat{R}$, the initial state along the IRC path will be change about the TS as its centre, and so the *x*-IRC will be multiply with (-1) to form the unique final state. When the state outside of the IRC path, the final state under operation $\hat{R}$ ought to be the reversal about the dividing surface (line) [5] which is a super-surface cross cut to the IRC path orthogonally through the TS. The dividing surface may divide the internal configuration coordinate space into two parts: the reagent region and the product region. For the linear H_3 -system, the dividing surface is a one dimensional line, the diagonal line of R_{12} and R_{13} internal configuration coordinate plane with R_{12}=R_{13}. For such H_3-system under operation $\hat{R}$, R_{12} and R_{13} will commute. For the fuzzy symmetry of this system outside of the IRC path, the MO-LCAO coefficients can be calculated point by point using Gaussian [35], and then the related criteria, membership functions and representation components can be obtained. After that, the corresponding contour map can be obtained using the grid method. It must be noted that for the various grid methods, the degree of approximation ought to be different and the related contour maps will be somewhat different, even if from the same grid method, due to the difference of the number and site (distribution) of the grid points selected in the calculation, the contour maps may be somewhat differ still.

For some MOs (with certain spin state) of the H_3 linear system at HF/STO3G level, the membership functions in relation to $\hat{P}$ may be used to produce the contour maps as shown in Figures 25.

In the membership function contour maps, the fluctuation may appear along the diagonal lines. However, according to the theoretic analysis or calculated for the state in the diagonal line using Gaussian [35], the membership functions always equal unity. It means that such fluctuation may be introduced by the approximation in the grid calculation method. As we increase the grid points in the dividing line, such fluctuation will be weakened and smoothed gradually. As the grid points increase to ten fold for the dividing line, for the fluctuation in the most contour maps will be dampened or even removed completely. Generally speaking, the more grid points are used, the more accurate the grid method and results become, although the more work load is needed. In principle, such contour maps for the linear H_3 system should be symmetric about the diagonal line, but there may be some small deviation due to the approximation of the grid method. Figure 25 shows the membership function contour map in relation to the space inversion transformation $\hat{P}$ for some MOs of the H_3-linear system. The diagonal line in the Figure is the dividing line, and the whole contour map is provided with the symmetry about the dividing line.

This system has the whole perfect symmetry in relation to $\hat{P}\hat{R}$, so the corresponding membership functions would be always unity. The membership functions in relation to $\hat{R}$ and $\hat{P}$ ought to be equality, similar as in the IRC path. For SMOs (with certain spin state) of the H_3 linear system at HF/STO3G level, the irreducible representation components in relation to $\hat{P}$ may be used to produce the contour maps. As shown in Figures 26, there are the contour maps of anti-symmetrical representation component Xu in relation to transformation $\hat{P}$ for the SMO of the H_3-linear system, as for the diagonal line, the corresponding states would always be anti-symmetrical, that is, Xu =1.

For other MOs, *e.g.,* OMO-1 and VMO-1, we may obtain similar main representation component contour map in relation to $\hat{P}$. For OMO-1 and VMO-1, the representations with more components in relation to $\hat{P}$ would be symmetrical. But for SMO, there is anti-symmetrical. Since the H_3-linear system is provided with the perfect symmetry in relation to $\hat{P}\hat{R}$, the corresponding representation ought to be pure symmetrical or pure anti-symmetrical. Moreover, that is true not only for TS but also for the whole internal configuration coordinate space. For $\hat{R}$, there is without the anti-symmetrical representation, we can not analyze the related representation component.

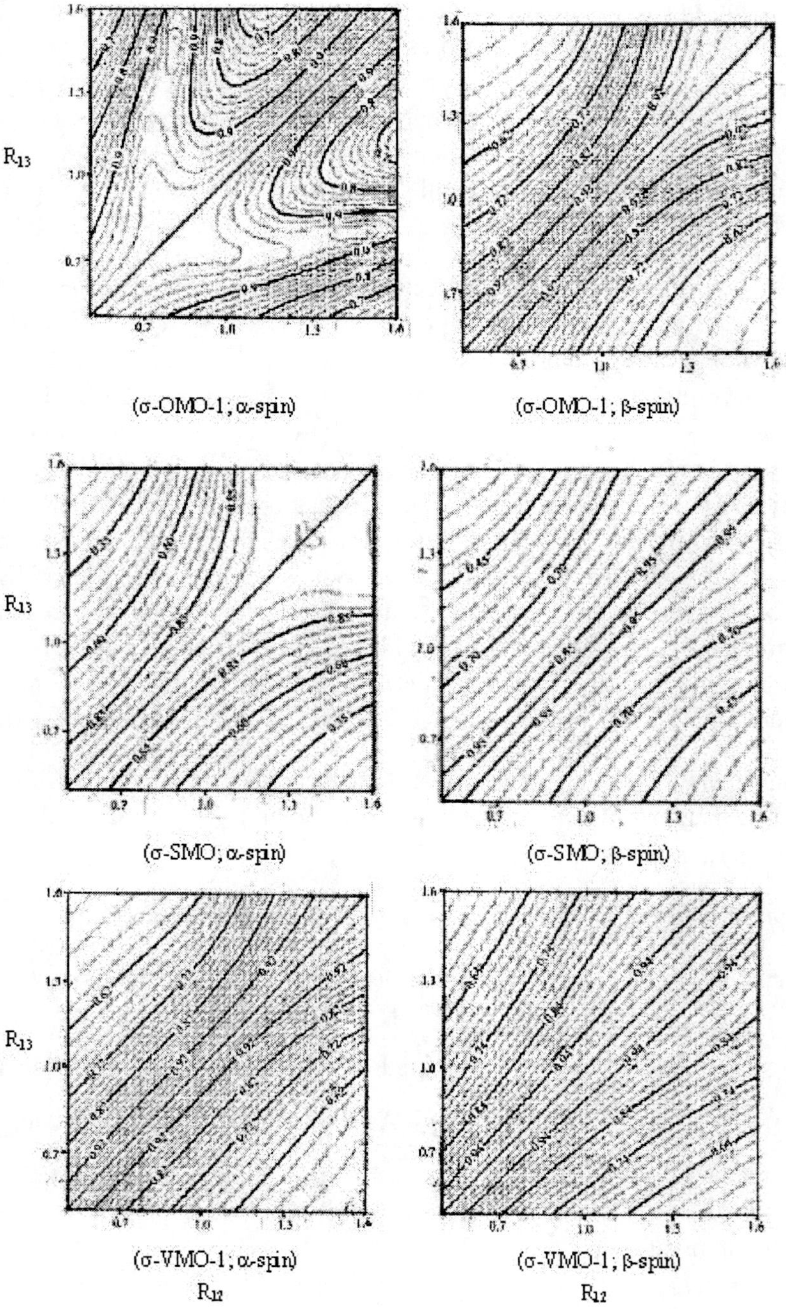

Figure 25. The membership function contour maps of H_3 linear system for certain MOs in relation to $\hat{P}$ (or to $\hat{R}$).

For the other linear B…A…C systems (with the same B and C), we may analyze their fuzzy symmetry in similar way. The similar results ought to be obtained, sometime the somewhat difference from H…H…H may be appear. For example in relation to $\hat{P}$, the main representation of H…H…H SMO is anti-symmetrical, but that of H…F…H SMO is symmetrical. However both of them in TS would be pure.

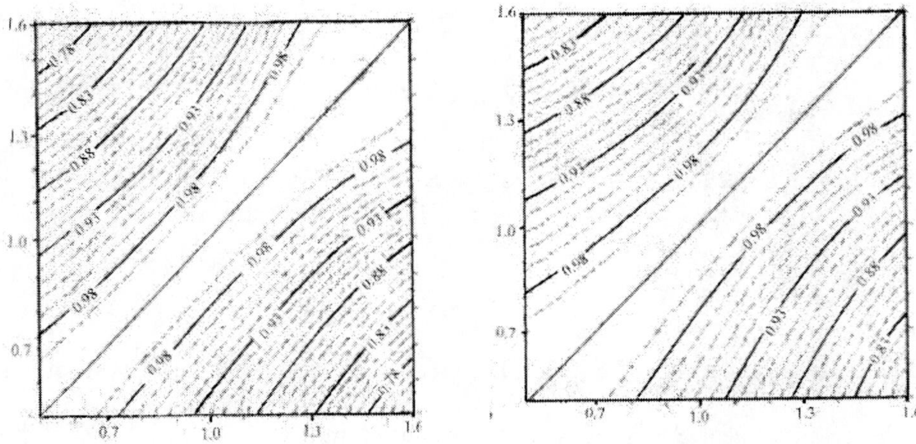

Figure 26. The anti-symmetrical representation component Xu contour maps of H_3 linear system for σ-SMO in relation to $\hat{P}$.

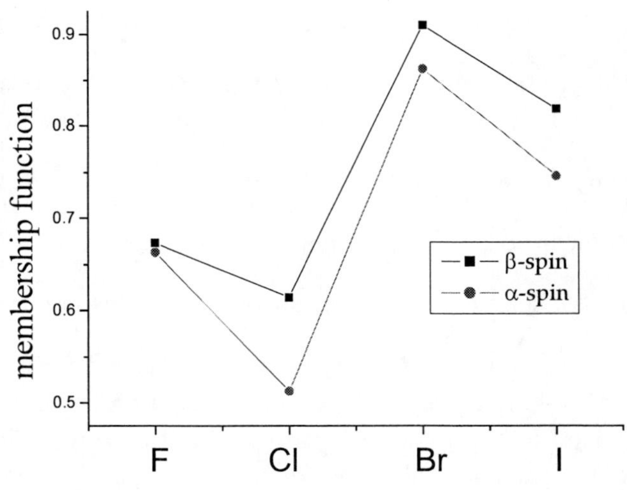

Figure 27. The membership functions of H…H…X (X = halide atom) for σSMO of the transition state at HF/STO-3G in relation to $\hat{P}\,\hat{R}$ (= $\hat{P}$).

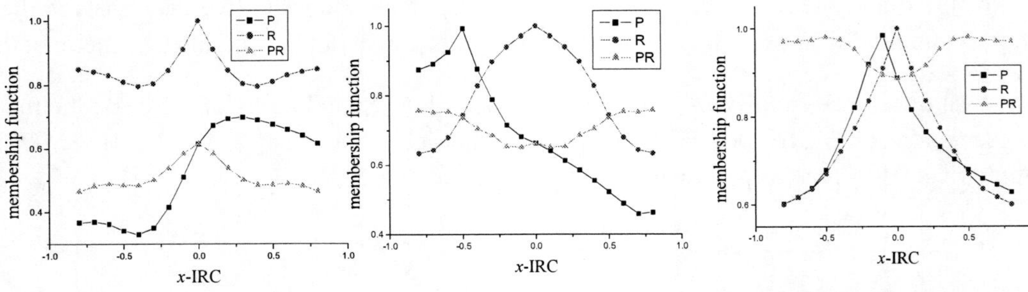

(A) σ OMO-1 (BMO) (B) σ SMO(NBMO) (C) σ VMO-1(ABMO)

Figure 28. Variation of the membership functions of H...H...F system with IRC values along the IRC path forσOMO-1(A), σSMO(B),σVMO-1(C) at HF/STO3G level, with □-spin state under transformations $\hat{P}\hat{R}$, $\hat{R}$ and $\hat{P}$.

6.1.2. The Fuzzy Symmetry of B…A…C (with the various B and C) System

For the TS of a linear B…A…C system with different B and C, *e.g,* H…H…F, it will be more complicated - the representations that σ SMO, σ OMO-1 and σ VMO-1 belong to will be not pure symmetrical nor pure anti-symmetrical. However, $\hat{R}$ and $\hat{E}$ are still equivalent for TS, and the membership functions in relation to $\hat{R}$ ought to be unity. Eqs. (54) and (55) are true, too. The fuzzy symmetry in relation to $\hat{P}$ may be analyzed as the common linear tri-atomic molecule [26]. As shown in Figure 27, there are the membership functions in relation to $\hat{P}\hat{R}=\hat{P}$ for σ SMO (two different spin states) of the H…H…X (X=halide atom) TS.

For the linear tri-atomic B…A…C system with different B and C atoms, there is no perfect $\hat{P}\hat{R}$ symmetry. Therefore, the membership functions for the related MOs in relation to $\hat{P}\hat{R}$, $\hat{R}$ and $\hat{P}$ ought to be less than unity. For TS, the relationships, $\hat{P}\hat{R}=\hat{P}$ and $\hat{R}=\hat{E}$, are still true. For such system, the membership functions in relation to $\hat{R}\hat{P}$ and $\hat{P}$ vs the IRC path curves would intersect at TS, and the membership functions in relation to $\hat{R}$ would be unity at TS. Taking this system, H…H…F, as an example, we show the internal configuration coordinates R_{12} (R_{HH}) and R_{13}(R_{HF}) vs the IRC path in Figure 21(B). For TS, R_{12} and R_{13} are unequal. The potential energy along the IRC path will be maximal in TS, but it is non-symmetrical about TS. For the MOs of such system, there are not perfect symmetry in relation to transformations $\hat{P}\hat{R}$, $\hat{R}$ and $\hat{P}$, and so we may only analyze their fuzzy symmetry.

We first analyze the fuzzy symmetry of the MOs of the H…H…F system along the IRC path in relation to the $\hat{P}\hat{R}$, $\hat{R}$ and $\hat{P}$. For *x*-IRC ≠ 0 (*i.e.* not TS), the corresponding state of such system will change due to operation $\hat{R}$: one state in the IRC path ought to change to another one in the path. The IRC values of the finial state and of initial state would be opposite (different by a merely negative sign). According to Figure 21(B), we may note values of R_{12} and R_{13} corresponding to the finial state from initial state under operation $\hat{R}$, whereas the corresponding system under operation $\hat{P}$ may be treated as the ordinary point symmetrical transformation. We may then calculate the related MOs using Gaussian [35] as well as the atomic criteria. Furthermore, we may obtain the membership functions in relation to $\hat{P}\hat{R}$, $\hat{R}$ and $\hat{P}$. Figure 28 shows the results using HF/STO-3G, and the membership functions for σ SMO, σ OMO-1 and σ VMO-1 (α-spin states) of the linear tri-atomic H…H…F systems along the IRC path in relation to $\hat{P}\hat{R}$, $\hat{R}$ and $\hat{P}$.

It should be noted that there are not perfect symmetry in relation to $\hat{P}\hat{R}$, and so the corresponding membership functions will be less than one. Although the variation curves of various MOs are considerably different, comparing with Figure 28, we note the following common features. First, the membership functions in relation to $\hat{P}\hat{R}$ and $\hat{R}$ vs IRC path are symmetrical about TS (x-IRC=0), but they are not in relation to $\hat{P}$. Moreover, in relation to $\hat{P}\hat{R}$ and $\hat{P}$ the membership functions are equality at TS. Furthermore, the membership function in relation to $\hat{R}$ equals one at TS. For the β-spin state, such common features will exist, too.

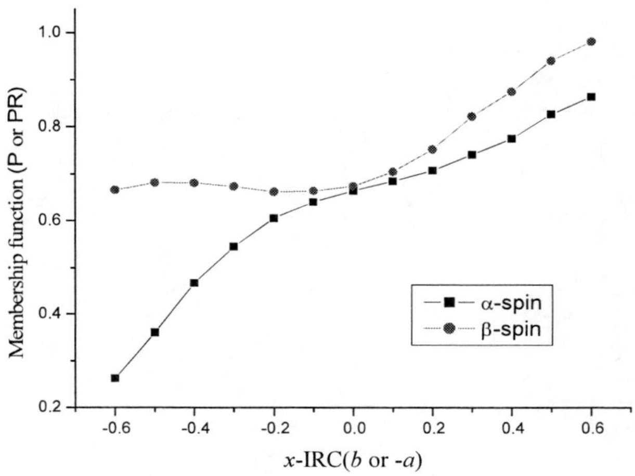

Figure 29. Variation of the membership functions of H…H…F system along the dividing line for σ SMO in α-spin (black) and β-spin (red) states in relation to $\hat{P}$ and $\hat{P}\hat{R}$: the membership function in relation to $\hat{R}$ equal to one along the dividing line.

The membership functions for various MOs along the dividing line in relation to $\hat{R}$ ought to be unity because $\hat{R}$ corresponding to the identity $\hat{E}$, and those in relation to $\hat{P}$ and $\hat{P}\hat{R}$ are equal as shown in Figure 29 for σ SMO.

For the MOs of linear tri-atomic H...H...F system, their irreducible representation components in relation to $\hat{P}$, $\hat{R}$ and $\hat{P}\hat{R}$ may be examined in principle, but there may be some various methods for this purpose, and some of them may obtain some similar but not quite same results.

For the membership functions of any internal configuration state in relation to $\hat{P}$, $\hat{R}$ and $\hat{P}\hat{R}$ for the linear tri-atomic H...H...F system, we may obtain the final internal configuration states about these transformations according to eqs. (49-55), and calculate the MO-LCAO coefficients for the states before and after the operation using Gaussian [35], and then obtain the atomic criteria and the membership functions related to the transformations. Using the calculated results, we can compute point by point the values of these membership functions to form the contour map according to Kriging grid method. The contour maps of the membership function in relation to $\hat{P}$, $\hat{R}$ and $\hat{P}\hat{R}$ for the SMO (in α-spin state) of the system are shown in Figure 30. As for β-spin state and for other MOs, the treatment is similar and omitted here. For convenient to analysis, the IRC-path and dividing line (curve) are also shown in Figure 30.

Comparing with the Figure 25 (the H_3 system as the example for the linear tri-atomic B...A...C system with same kinds atoms B and C), the membership functions in relation to $\hat{R}$ and $\hat{P}$ the contour maps are different for H...H...F in Figure 30, and the membership functions in relation to $\hat{P}\hat{R}$ are not always unity.

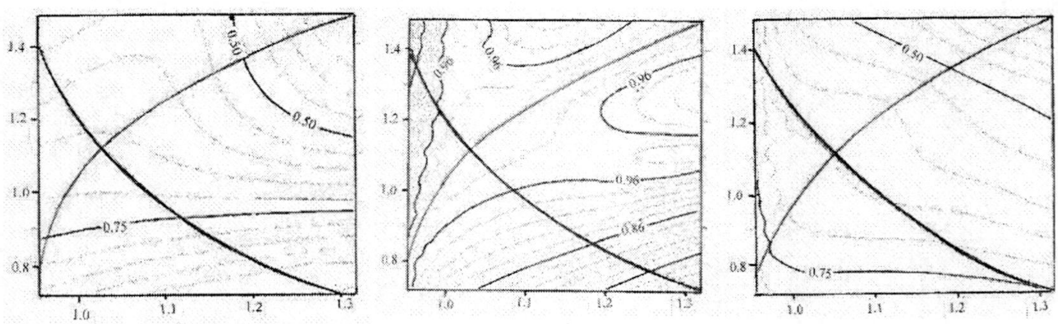

(A) $\hat{P}$ (B) $\hat{R}$ (C) $\hat{P}\hat{R}$

Figure 30. The membership function contour maps in relation to the $\hat{P}$, $\hat{R}$ and $\hat{P}\hat{R}$ for SMO (α-spin states) of linear H...H...F system corresponding to various internal configurations. The common nature scale (the Cartesian coordinate scale !) is used for the internal configuration coordinate R_{12} (ordinate) and R_{13} (abscissa). The blue and red curves denote the IRC-path and dividing line, respectively.

On the other hand, as shown in Figures 28, for the MOs of this system, the membership function curves in relation to $\hat{R}$ and $\hat{P}\hat{R}$ along the IRC path are symmetry about the TS. Why is such symmetry not shown in Figure 30? The cause is that we used the common nature scale (!) in Fig 30 whereas the IRC scales is used in Figure 28. In fact, as shown in Figure 30, although there is not the symmetry obviously, for the membership functions in relation to $\hat{R}$ and $\hat{P}\hat{R}$ the contour maps, the membership function seems somewhat well-distribution about the dividing curve. According to the natural scale, both IRC path and dividing line are not straight lines in Figure 30. If we present the contour maps of the membership functions in relation to $\hat{P}$, $\hat{R}$ and $\hat{P}\hat{R}$ in the IRC scale (!), we can have Figure 31, where the blue diagonal lines are the IRC paths and the red diagonal lines are the dividing lines.

The intersection point of these two diagonal lines and the origin point of the coordinate system ought to be connected with the TS. Since R_{12} and R_{13} are monotonically increasing and decreasing respectively along with the IRC, the initial (reactant) and final (product) states of this reaction system lie in the upper right and lower left parts of the internal configuration coordinate space, respectively. Similar to Figure 28, for the MOs of such system, there is the symmetry about the dividing line for the membership functions in relation to $\hat{R}$ and $\hat{P}\hat{R}$. However, there is not such symmetry for the membership functions in relation to $\hat{P}$.

In addition, the membership functions in relation to $\hat{R}$ in the dividing line would be always unity, but there may be somewhat fluctuation in Figure 31 due to the error introduced in the grid calculation.

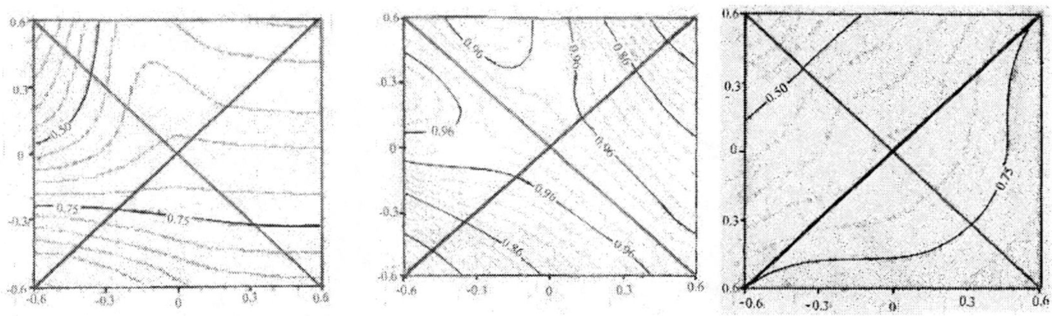

(A) $\hat{P}$ (B) $\hat{R}$ (C) $\hat{P}\hat{R}$

Figure 31. The membership function contour maps of linear H...H...F system in relation to $\hat{P}$, $\hat{R}$ and $\hat{P}\hat{R}$ for SMO in α-spin states corresponding to various internal configurations. The IRC scale is used for the internal configuration coordinates R_{12} (vertical ordinate) and R_{13} (abscissa). The blue and red diagonal lines denoted the IRC-path and dividing line, respectively.

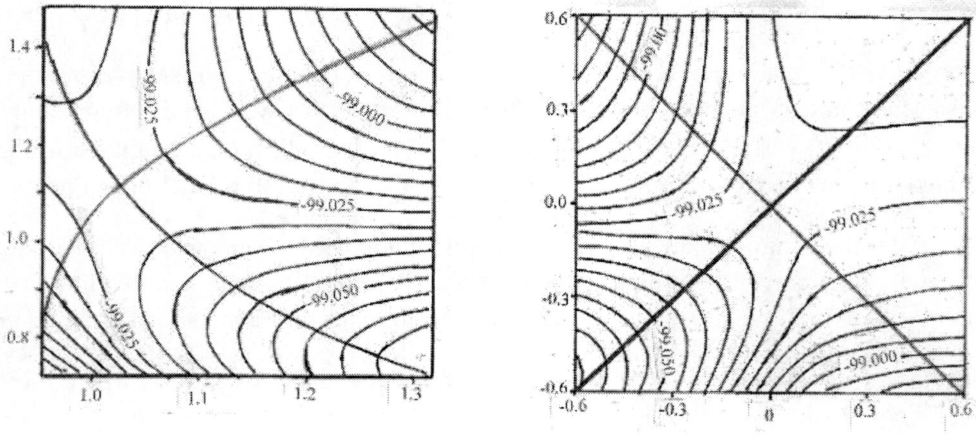

(A) nature scale

(A) IRC scale

Figure 32. The potential energy contour maps of linear H...H...F system corresponding to various internal configurations. The common nature scale (!) and the IRC scale are used in figures (A) and (B), respectively. The intersection point of blue (IRC path) line and red (dividing) line denoted the TS.

By the way, using the nature and IRC scales, we can get the potential energy contour maps of linear HHF system with somewhat different shapes (Figure 32). Although the intersection points of the IRC path and dividing lines are the TS and saddle points for both these two scales, however the IRC-path and dividing lines are the straight lines in IRC-scale case, and the curves (not straight) in nature scale case, respectively.

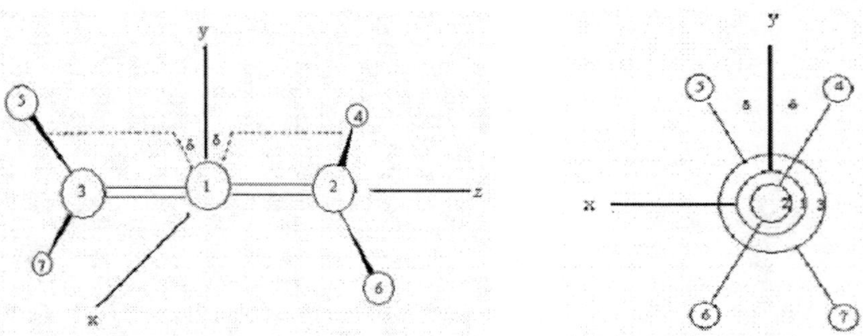

(A)

(B)

Figure 33. Allene molecular geometry and the atomic serial numbers. (A) stereogram, (B) perspective view with z-direction.

6.2. The Fuzzy Symmetry for Internal-Rotation of Allene and Its 1.3-Dihalide

For consider the chemical dynamics real system, the prototypical uni-molecular dynamics systems, we have chosen to study are the internal-rotation systems of allene and its 1,3-dihalide. The internal-rotation of allene will not appear the macroscopic chemical variation. However the internal-rotation of its 1,3-dihalide will accomplish the chiral transition process.

The molecular geometry and atomic serial numbers required to probe the internal-rotation of propadine molecule are shown in Figure 33, where Figure 33(A) and (B) represent the stereogram and the perspective view (with *z*-direction). In Figure 33, the serial numbers 1, 2, and 3 denote carbon atoms and the serial numbers 4, 5, 6, and 7 denote hydrogen atoms. The atoms represented by 1, 2, 4, and 6 are coplanar, the atoms represented by 1, 3, 5, and 7 are also coplanar. We denoted the dihedral angle between these two CH_2 group planes as $\theta = 2\delta$. As $\delta = 45°$, two CH_2 group planes are orthogonal to each other, it is the stable geometry of allene. As $\delta = 0°$, two CH_2 group planes are coplanar, it means the transition state for the internal-rotation process. The Cartesian coordinates system for allene is as follow: origin — the center of atom 1. *z* axis — coincident with the three carbon atom line. *yz* coordinate plane — dividing the dihedral angle between the above two CH_2 group planes.

In the stable geometry of allene, there are two orthogonal CH_2 group planes and dihedral angle $\theta = 2\delta = 90°$, therefore it possess D_{2d} symmetry. When the CH_2 groups rotate around the axis of three C atoms, θ will be changed and the molecule is provided with only D_2 point group, which is the subgroup of D_{2d} symmetry in general. As $\theta = 2\delta = 0°$, *i.e.* the two CH_2 group planes are coplanar, the molecule will correspond to the transition state and provided with D_{2h} symmetry.

As the end hydrogen atoms of allene are substituted by various atoms or groups, the derivative will be chiral. Therefore, the chiral transition of these molecules in the internal rotation can also be important. For allene molecule, the internal rotation does not accompany chiral transition, but for allene-1,3-dihalide the rotation may accompany chiral transition. When allene-1,3-dihalide undergoes chiral transition in internal rotation, the atomic numbering and adjacent relationship may be defined similar as Figure 33. Where 1, 2 and 3 denote three C atoms while 1 numbers the mid-C atom as C(1), but these two CH_2 groups will be replace with CHX and CHY (X and Y may be the same or not same halogen atoms). For allene-1,3-dihalide, three C atoms may be not linear, and there are two kinds of TS, according to the relative position of the halogen atoms they may be denoted as *cis*- and *trans*-TS, respectively. It may be predicted that the symmetry characteristic for internal rotation of allene-1,3-dihalide will be more complex than that of allene.

6.2.1. The Fuzzy Symmetry for Internal-Rotation of Allene

For the stable state, the transition state and the other internal-rotation state of allene, the relative molecules possess D_{2d}, D_{2h} and D_2 point group symmetry. The D_2 point group would be the subgroup of both D_{2d} and D_{2h} point groups. Obviously, D_2 ought to be the intersection set of D_{2d} and D_{2h}, *i.e.*

$$D_2 \subset D_{2d},\ D_2 \subset D_{2h},\ D_2 = D_{2d} \cap D_{2h} \qquad (56-1)$$

On the other hand, both D_{2d} and D_{2h} are the subgroup of D_{4h}. The D_{4h} will be the union set of D_{2d} and D_{2h} , *i.e.*

$$D_{2d} \subset D_{4h},\ D_{2h} \subset D_{4h},\ D_{4h} = D_{2d} \cup D_{2h} \qquad (56-2)$$

Therefore, corresponding to the values of various dihedral angles, we analyze above state molecular fuzzy symmetry using D_{4h} point group. For convenience, the 16 elements included in D_{4h} are classified to four subsets:

$$G0 = \{ \hat{E}, \hat{C}_{2z}, \hat{C}_{2(1)}, \hat{C}_{2(2)} \} \tag{57-1}$$

$$G1 = \{ \hat{P}, \hat{M}_h, \hat{M}_{v(1)}, \hat{M}_{v(2)} \} \tag{57-2}$$

$$G2 = \{ \hat{S}_4^1, \hat{S}_4^3, \hat{M}_{d(1)}, \hat{M}_{d(2)} \} \tag{57-3}$$

$$G3 = \{ \hat{C}_4^1, \hat{C}_4^3, \hat{C}_2', \hat{C}_2'' \} \tag{57-4}$$

As for Ψ_ρ-MO, eq(10) , the corresponding membership function in relation to the symmetry operation $\hat{G}$ can be written as eq (12) or (13). After the symmetry transformation ($\hat{G}$), the J-atom will be changed to the GJ-atom. However, for some J-th atom in the internal-rotation state of allene, the related GJ-th atom may be ambiguous, sometimes, the J-th atom will be changed to the $G^{\#}J$ atom in ideal case through $\hat{G}$, but there is without the true atoms. For the common stable molecule, we may assign the real GJ-atom which near the $G^{\#}J$-atom to replace, the position difference may reflect on the variation of MO LCAO coefficient and then the atomic criteria. However for the dynamic system, especially as to involve the obvious change of angle, the position difference may not reflect on the variation of the atomic criteria, enough. The corresponding membership function in relation to the symmetry operation $\hat{G}$ ought to be modified as:

$$\begin{aligned}\mu_E(\hat{G}/\Psi_\rho) &= [\textstyle\sum_J\sum_i (Y_{Ji} \wedge Y_{GJi})\varphi(GJ,G^{\#}J)] / [\sum_J\sum_i (Y_{Ji})] \\ &= \textstyle\sum_J\sum_i [(a^*_\rho(J,i)\, a_\rho(J,i)) \wedge (a^*_\rho(GJ,i)\, a_\rho(GJ,i))]\, [\varphi(GJ,G^{\#}J)] \\ &\quad / \textstyle\sum_J\sum_i [a^2_\rho(J,i) \wedge a^2_\rho(GJ,i)]\end{aligned} \tag{58}$$

Where $\varphi(GJ,G^{\#}J)$ may stand for the space factor. Following the operation of perfect symmetry transformation($\hat{G}$), there exist some geometric difference between the set of the theoretical imaginary $G^{\#}J$ atom, where J atom should be changed and the calculated set of GJ atom should be not. The $\varphi(GJ,G^{\#}J)$ ought to be the function of such different. As the GJ and $G^{\#}J$ atoms are coincided, the $\varphi(GJ,G^{\#}J)$ would be equal to one. As the GJ and $G^{\#}J$ atoms are near to each other, Eq(84) would be the more perfect approach. For the propadine molecule, as J-th atom is carbon atom, $\varphi(GJ, G^{\#}J)$ equals one. As the *J*-th atom is the hydrogen atom in the CH_2 group, we set $\varphi(GJ,G^{\#}J) = \cos\theta'$, where θ' denotes the difference of the dihedral angles between these two CH_2 group planes for GJ and $G^{\#}J$ atom. As θ' enhances from 0° to 90°, $\varphi(GJ,G^{\#}J)$ conform to the condition of 1 to 0. Corresponding to the different θ values for the internal-rotation process of allene molecule, the membership functions of various MO may be obtained by using Eq(58) in relation to different symmetry transformations.

Now we analyze them according to D_{4h} point group and classify as in Eq. (57), still. It follows that:

For the symmetry elements in G0 subset, *i.e.* the elements included in D_2 point group, as the symmetry in relation to these elements are always maintained through the whole internal-rotation process, all the MOs would have relative symmetry. Therefore, the membership functions in relation to these elements in G0 subset ought to be always one.

For the symmetry elements in G1 subset, *i.e.* the elements, which belong to D_{2h} point group, not to D_{2d} point group, as the dihedral angle(θ) increased from 0° (the transition state with the D_{2h} point group symmetry) to 90° (the stable state with the D_{2d} point group symmetry), the membership functions of various MOs in relation to the elements in G1 subset would decrease monotonously as shown in Figure 34(A).

For the symmetry elements in G2 subset, *i.e.* the elements, which belong to D_{2d} point group, not to D_{2h} point group, the variations in the membership functions of various MOs in relation to the symmetry transformation in G2 subset vs the dihedral angle θ may be shown in Figure 34(B). Two cases appear as follow: (i) The membership functions of MOs are increasing monotonously to one, such MOs in connection to the stable state MOs belong to the one-dimensional irreducible representation of D_{2d} point group. (ii) The membership functions of MOs are increasing roughly, not monotonously, meanwhile two MOs approach to same value. Such two MOs in connection to the stable state MO complete sets belong to the two-dimensional irreducible representations of D_{2d} point group. It is interesting that a certain complete set, which belongs to two-dimensional irreducible representation, may be split to two MOs through various equivalent ways. These two MOs provided the same membership function values, but they may much differ in terms of methods used. For such whole complete set, the membership functions in relation to all the elements in G2 subset ought to be equal to one.

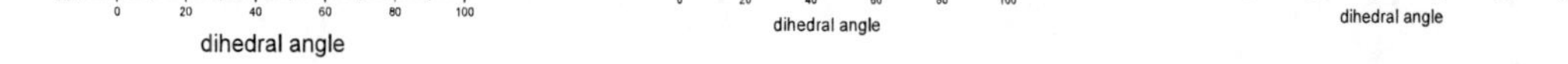

(A) G 1 (B) G2 (C) G3

Figure 34 Relationship between the membership functions of the allene MOs in relation to the symmetry transformations in D_{4h} point group vs the dihedral angle(θ) of two CH_2 group planes through the internal-rotation process calculated at STO-3G level.

For the symmetry elements in G3 subset, *i.e.* the elements, which belong to neither D_{2d} nor D_{2h} point group, the variations in the membership functions of these MOs in relation to the elements in G3 subset vs the dihedral angle θ may be shown in Figure 34(C). There are two types of MOs as follow: (i) Those MOs in connection with the stable state MOs which belong to the one-dimensional irreducible representation of D_{2d} point group. The variation of their membership functions in relation to G3 subset vs the dihedral angle will be somewhat steady. (ii) Those MOs in connection with the stable state MOs, which belong to the two-dimensional irreducible representation of D_{2d} point group. The variation of their membership functions in relation to the elements of G3 subset vs the dihedral angle will be similar as in the case of G2 subset, but the amplitude will be somewhat less.

6.2.2. The Fuzzy Symmetry for Chiral Transition of Allene 1,3-Dihalide

For allene-1,3-dihalide, three C atoms may be not linear, and there are two kinds of TS, according to the relative position of the halogen atoms they may be denoted as *cis*- and *trans*-TS, both of them are planar. As these two halogen atoms are the same, the *cis*-TS and *trans*-TS belong to the well-defined C_{2v} and C_{2h} symmetry, respectively. But other states of such system belong to C_2 symmetry, only. However there are some special symmetry groups in relation to the joint transformations of the reaction reversal and the point symmetry, they may describe the whole chiral transition process of such allene 1,3-dihalide.

As the allene-1,3-dihalide with the same halogen atoms, the well-defined point symmetry groups of the *cis*-IRC and *cis*-TS are C_2 and C_{2V}, respectively, as follows:

$$C_2 : G_{cis\text{-}IRC} = \{\hat{E}, \hat{C}_2\} \tag{59-1}$$

$$\begin{aligned} C_{2V}: G_{cis\text{-}TS} &= G_{cis\text{-}IRC} \cup G'_{cis\text{-}TS} = \{\hat{E}, \hat{C}_2\} \cup \{\hat{M}_{XZ}, \hat{M}_{YZ}\} \\ &= \{\hat{E}, \hat{C}_2, \hat{M}_{XZ}, \hat{M}_{YZ}\} \end{aligned} \tag{59-2}$$

Table 9. Irreducible representations and characters of $G(R_cC_{2v})$ transformation group .

	$\hat{E}$	$\hat{C}_{2Z}$	$\hat{R}_c\hat{M}_{XZ}$	$\hat{R}_c\hat{M}_{YZ}$
rA_1	1	1	1	1
rA_2	1	1	-1	-1
rB_1	1	-1	1	-1
rB_2	1	-1	-1	1

Table 10. Irreducible representations and characters of G(RtC2h) transformation group.

	$\hat{E}$	$\hat{C}_{2Z}$	$\hat{R}_t\hat{M}_{XY}$	$\hat{R}_t\hat{P}$
rA_g	1	1	1	1
rA_u	1	1	-1	-1
rB_g	1	-1	-1	1
rB_u	1	-1	1	-1

Though *cis*-IRC belong to $C_2 = G_{IRC} = \{\hat{E}, \hat{C}_2\}$ group, except for the *cis*-TS, all other states in the *cis*-IRC do not have the well-defined symmetry related to $G'_{cis\text{-}TS}=\{\hat{M}_{XZ}, \hat{M}_{YZ}\}$, but have the symmetry in relation to the joint transformations, $\hat{R}_C\hat{M}_{XZ}$

and $\hat{R}_C\hat{M}_{YZ}$, where $\hat{R}_C$ is the reaction reversal transformation whose operation makes the IRC value of a certain state to its opposite value of another state[5]. Therefore, all *cis*-IRC states belong to well-defined symmetry in relation to group $G(RM)_C$ defined below,

$$G(RM)_C \equiv G(R_cC_{2v}) = \{\hat{E}, \hat{C}_2, \hat{R}_C\hat{M}_{XZ}, \hat{R}_C\hat{M}_{YZ}\} \quad (60)$$

As the *cis*-TS, the $\hat{R}_C = \hat{E}$ and eq(60) ought to reduce to eq(59-2), the relative characters are shown in Table 9.

Similarly, as the allene-1,3-dihalide with the same halogen atoms, the well-defined point symmetry groups of the *trans*-IRC and *tran*-TS are C_2 and C_{2h}, respectively, as follows:

$$C_2 : G_{trans\text{-}IRC} = \{\hat{E}, \hat{C}_2\} \quad (61\text{-}1)$$

$$C_{2h}: G_{trans\text{-}TS} = G_{trans\text{-}IRC} \cup G'_{trans\text{-}TS} = \{\hat{E}, \hat{C}_2\} \cup \{\hat{P}, \hat{M}_h\}$$
$$= \{\hat{E}, \hat{C}_2, \hat{P}, \hat{M}_h\} \quad (61\text{-}2)$$

And all *tran*-IRC states belong to well-defined symmetry in relation to group $G(RM)_C$ defined below,

$$G(RM)_T \equiv G(RC_{2h}) = \{\hat{E}, \hat{C}_2, \hat{R}_t\hat{P}, \hat{R}_t\hat{M}_{XY}\} \quad (62)$$

As the *trans*-TS, the $\hat{R}_t = \hat{E}$ and eq(62) ought to reduce to eq(61-2), the relative characters are shown in Table 10.

Owing to except for the TS all other states in the IRC do not have the well-defined symmetry related to G'_{TS}, then in relation to the element of G'_{TS} set, the membership functions ought to be less 1, on the other hand, there are the well-defined symmetry related to the joint transformation set, $\hat{R}G'_{TS}$, in relation to the element of set $\hat{R}G'_{TS}$, the membership function of the system with a certain IRC scale will be equal that of the system with the opposite IRC scale. As for the $\hat{C}_2$ in G_{IRC} group, the relative membership function MFC will be always one. As shown in Figure 35, there are the diagrams in connection to the OMO membership functions of HFC=C=CHF, the prototypical allene-1,3-dihalide with the same halogen atoms in relation to such symmetry transformations vs the *cis*- and *trans*-IRC. For save the space, the similar characteristics of VMO are omitted. It is the similar case as $H_2C{=}C{=}CH_2$ in relation to the elements in G1 set. It is notable that the dihedral angle of TS equals 0^0 and that of reagent and product will be 90^0, in Figure 34(A).

For the allene-1,3-dihalide with the different halogen atoms, through the internal rotation the molecule will also experience the optical chiral transition, an isomerization between the enantiomers. In this process, the curve of energy vs. IRC would be symmetric about the TS point, too. As with the different halogen atoms, TS of allene-1,3-dihalide should belong to C_S

group, while the other states only to C_1, no symmetry indeed. These two point groups include the following symmetric transformations:

$$C1 : GIRC = \{\hat{E}\} \quad (63\text{-}1)$$

$$CS: G\,TS = GIRC \cup G'TS = \{\hat{E}\}\cup\{\hat{M}\} = \{\hat{E}, \hat{M}\} \quad (63\text{-}2)$$

where $G'_{TS} = \{\hat{M}\}$ is the symmetric transformation set of the TS only, and $\hat{M}$ is the mirror reflection. Certainly, for the *cis*-IRC and *trans*-IRC, $\hat{M}$ should be different. Although both mirrors are the TS molecular plane, the *cis*-TS mirror includes the fuzzy two-fold axis ($\tilde{C}_2$), whereas the *trans*-TS mirror would be orthogonal to $\tilde{C}_2$. For the IRC states other than TS, the space relationship between the fuzzy mirror and $\tilde{C}_2$ is maintained in TS. The IRC states other than TS do not belong to any point group, but they may have the joint symmetry of $\hat{R}\hat{M}$, where $\hat{R}$ and $\hat{M}$ are the reaction reversal and reflection, respectively. The corresponding symmetric transformation group is:

$$G(RM) = \{\hat{E}, \hat{R}\hat{M}\} \quad (64)$$

Of course, for the *cis*-IRC and *trans*-IRC, their own $\hat{R}$ and $\hat{M}$ are both different; they may be denoted as follows:

$$G(RM)_C = \{\hat{E}, \hat{R}_c\hat{M}_{YZ}\} \quad (65\text{-}1)$$

$$G(RM)_T = \{\hat{E}, \hat{R}_t\hat{M}_{XY}\} \quad (65\text{-}2)$$

For the optical chiral transition of the allene-1,3-dihalide with the different halogen atoms, there is not the well-defined symmetry in relation to the $\hat{C}_2$, therefore the relative membership function MFC will be less one. According to the well-defined symmetry transformation group G(RM), the membership function of the system with a certain IRC scale will be equal that of the system with the opposite IRC scale. Here the membership function less one owing to the non-space and intrinsic factor, we may get the relative irreducible representation components, Xg or Xu, the value of the system with a certain IRC scale will be equal that of the system with the opposite IRC scale, too. As shown in Figure 36, there are the diagrams of the OMO membership functions and Xg of HFC=C=CHCl, the prototypical allene-1,3-dihalide with the different halogen atoms in relative the symmetry transformations $\hat{C}_2$ vs the *cis*-IRC.

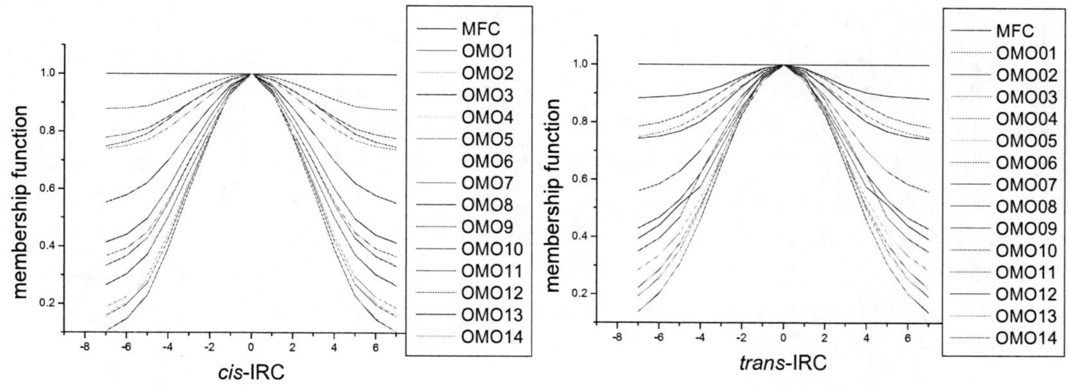

(A) (B)

Figure 35. The membership function of CHF=C=CHF OMOs vs. IRC through the internal rotation IRC path at AM1 level. (A) *cis*-IRC (B) *trans*-IRC.

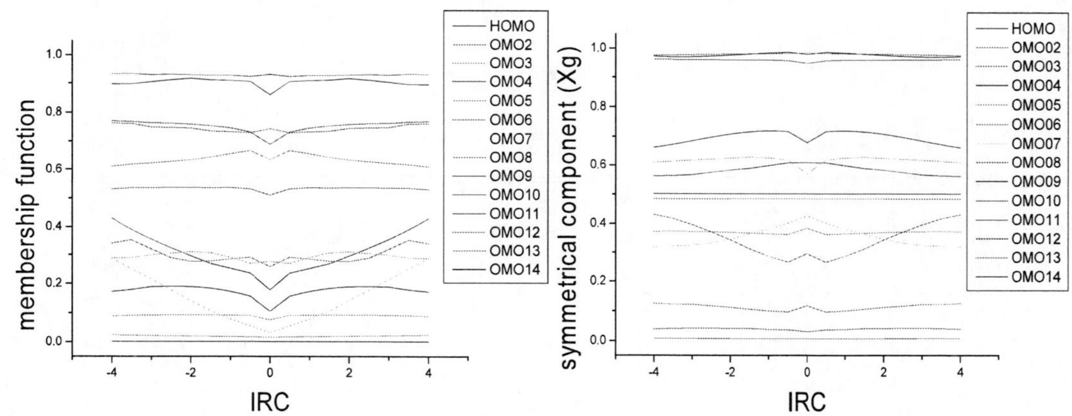

(A) (B)

Figure 36. The membership function(A) and Xg(B) of CHF=C=CHCl OMOs vs. IRC through the internal rotation *cis*-IRC path at AM1 level.

Regardless of the same or different halogen atoms, the reagent and product for chiral transition of allene 1,3-dihalide are the enantiomer. The prototypical allene 1,3-dihalide HXC=C=CHY with the same and different halogen atoms (X and Y) we may chosen to study are the HFC=C=CHF and HFC=C=CHCl, respectively. For the chiral transition of these allene 1,3-dihalide, we may construct the relative MO correlation diagram (for HFC=C=CHF) or the relative fuzzy MO correlation diagram (for HFC=C=CHCl) in relation to the $\hat{C}_2$ transformation, as shown in Figure 37(A) and (B), respectively.

For all MOs in HFC=C=CHF, Figure 37(A), the irreducible representations may be adapted to make the processes symmetry allowed. Usually, W-H rule is used via the MO correlation diagram based on the symmetry through the whole process. Therefore, we also examined the MO correlation diagram as shown in Figure 37(A) in relation to C_2 group. This figure shows the MO irreducible representation, where P and R denote a pair of the stable

CHF=C=CHF enantiomers, which may be the initial or final state of the related internal rotation process, and *cis*-TS and *trans*-TS denote the TS in the process. Although the MO irreducible representations are denoted by the point group (C_{2v} or C_{2h}) *cis*-TS or *trans*-TS belongs to, we link the same first label (A or B) in the irreducible representations in the figure. The bonding OMOs and the anti-bonding VMO are separated with a horizontal double line. It is notable that no correlation lines cross the double line, *i.e.* the NB(non-bonding)MO level, and so all of them are symmetry allowed no matter what the IRC path is. However, MO correlation lines may cross through the *cis*- or *trans*-IRC process. If the line crosses between P and TS, then it must also cross between R and TS, *i.e.*, crossing twice.

(A) HFCCCHF						(B) HFCCCHCl				
cis-TS	*cis*-IRC	PR	*trans*-IRC	*trans*-TS		*cis*-TS	*cis*-IRC	PR	*trans*-IRC	*trans*-TS
B_2	—	B	—	B_u	VMO8	0.022B	—	0.040B	—	0.030B
A_1	—	A	—	A_g	VMO7	0.968A	—	0.674A	—	0.833A
A_1	╳	B	—	B_u	VMO6	0.879A	—	0.878A	╳	0.392B
B_2	╳	A	—	A_g	VMO5	0.119B	—	0.260B	╳	0.575A
B_2	—	B	—	B_u	VMO4	0.654A	—	0.636A	—	0.585A
A_1	—	A	—	A_g	VMO3	0.001B	···	0.561A	—	0.999A
B_1	╳	A	—	A_u	VMO2	0.472B	╳	0.625A	—	0.655A
A_2	╳	B	—	B_g	VMO1	0.998A	╳	0.492B	—	0.002B
A_1	╳	B	—	B_u	OMO1	0.979A	╳	0.451B	—	0.038B
B_1	╳	A	—	A_u	OMO2	0.035B	╳	0.573A	—	0.964A
B_2	—	B	—	B_u	OMO3	0.390B	—	0.370B	—	0.410B
A_1	—	A	—	A_g	OMO4	0.606A	—	0.801A	╳	0.398B
A_2	╳	B	—	B_g	OMO5	0.560A	╳	0.306B	╳	0.602A
B_1	╳	A	—	A_u	OMO6	0.302B	╳	0.451B	···	0.710A
B_2	—	B	╳	A_g	OMO7	0.428B	—	0.433B	╳	0.567A
B_2	╳	A	╳	B_u	OMO8	0.124B	—	0.211B	╳	0.102B
A_1	╳	B	—	B_u	OMO9	0.674A	—	0.537A	╳	0.327B
A_1	—	A	—	A_g	OMO10	0.945A	—	0.906A	—	0.943A
B_2	—	B	—	B_u	OMO11	0.006B	—	0.017B	—	0.008B
A_1	—	A	—	A_g	OMO12	0.982A	—	0.977A	—	0.979A
B_2	╳	A	—	A_g	OMO13	0.486B	—	0.491B	—	0.493B
A_1	╳	B	—	B_u	OMO14	0.502A	—	0.500B	—	0.498B

Figure 37. The MO correlation diagram for chiral transition of HFC=C=CHF and the fuzzy MO correlation diagram for chiral transition of HFC=C=CHCl in relation to $\hat{C}_2$.

For HFC=C=CHCl, as shown in Figure 37(B), there is not the well-defined symmetry related to $\hat{C}_2$ exists, so only fuzzy symmetry presents. For MOs, they are expressed by symmetric representation components and main representation symbols, through the internal rotation process. The red correlation line links different main representations of MOs, at least one of them with the representation component near 0.5, *i.e.* the membership function ought to be small.

In these Figures, the number value before the MO main-representation (A or B) denotes its symmetric representation component Xg. When Xg is larger than 0.5, the main-representation is A; otherwise the main-representation is B. Black correlation lines denote the MOs with the same main-representation are, while red ones denote the minority MOs with different main-representations. On the red correlation line, at least one correlated MO has an Xg close to 0.5 (*e.g.*, 0.4 to 0.6). Since the MO correlation lines of stable molecules P and R are symmetric about the TS, they are analyzed jointly. The correlation lines do not cross the NBMO level, nor do they cross between OMO and VMO, and so these processes should be fuzzy symmetry allowed.

7. The One-Dimensional Space Periodic Fuzzy Symmetry of Some Molecules

In this section, we start to analyze the fuzzy space symmetry group. It is notable that the space group is an infinite group whereas the point group is a finite group. For the fuzzy point group, we focus on considering the fuzzy characterization introduced due to the difference of atomic types in the monomer through point symmetry transformation in the beginning; and then we consider the difference between the infinity of space group and the finite size of real molecules. The difference between the point group and the space group lies in the translation symmetry transformation.

Starting with the simple case, we will only analyze the one-dimensional translation transformation, *i.e.* the case with the one-dimensional periodic fuzzy symmetry, now. The relative space group may be denoted as G_1^n, for the system in the *n*-dimensional space. In the three-dimensional space, the cylindrical group, G_1^3 [41-44] is usually used to study the symmetry of periodic objects, which have periodicity in only one special orientation. Now, we will consider the fuzzy symmetry of $G_1^1$and G_1^2, only. As for more complex space symmetry group, the relative fuzzy symmetry will be studied in our future works.

7.1. The Fuzzy G_1^1 Symmetry of Polyynes and Their Cyano-Compounds

For the prototypical fuzzy G_1^1 symmetry molecules, we consider the polyyne and their cyano-compounds The atomic numbers and Cartesian coordinates for these molecules are shown in Figure 38, where . usually denotes the carbon atom, whereas it may also denotes the nitrogen atom in the terminal (serial number: 0 and n-1). As the terminal atom is a carbon atom, it may bind a hydrogen atom outside; when the terminal atom is a nitrogen atom, it may bind none. In Figure 38, □ may be a single or triple bond. The *z*-axis and the molecular axis are taken to be the same; the *x* and *y* axes are orthogonal to each other but lie in the vertical plane of the *z*-axis.

The MOs, the LCAO of valence shells, were calculated at a certain theoretical level using the Gaussian [35]. The π-MOs are combined using the p_x and p_y AOs. All the π-MOs are two-fold degenerate, *i.e.*, both their energies and membership functions are the same; and thus to analyze one of them should be adequate for our purpose. The σ -MOs are obtained by combination of the *sp*(z)-AOs, and these MOs are further from the frontier orbitals and their chemical activity will be less than that of the π-MOs. In this section we will not analyze the σ -MOs.

The molecular membership function *related to* the symmetry transformation $\hat{G}$ may be shown as eq(5). Where Y_J and Y_{GJ} are the criteria of atoms J and GJ, and GJ is produced from J atom though the symmetry transformation $\hat{G}$. Y may be the atomic number (Z_J) for the molecular skeleton. According to the LCAO-MO scheme, the ρ*th* MO Ψ_ρ may be expressed as eq(10), the relative criterion can be taken as to the electron population on atom J. The MO membership function *related to* the symmetry transformation $\hat{G}$ may be shown as eq(12) or (13). If there is only one AO in each atom to combine the MO, like the π-MO we analyzed in this paper, eq.(12) will be reduced to eq.(5). For the molecule as shown in Figure 38, if J is beyond the region [-1, *n*], the relative criterion Y_J=0. As J=-1 or *n*, the relative atom may be hydrogen or null, and relative criterion may be Y_H or 0. As the π-MOs, Y_H=0, and as the molecular skeleton, Y_H=1.

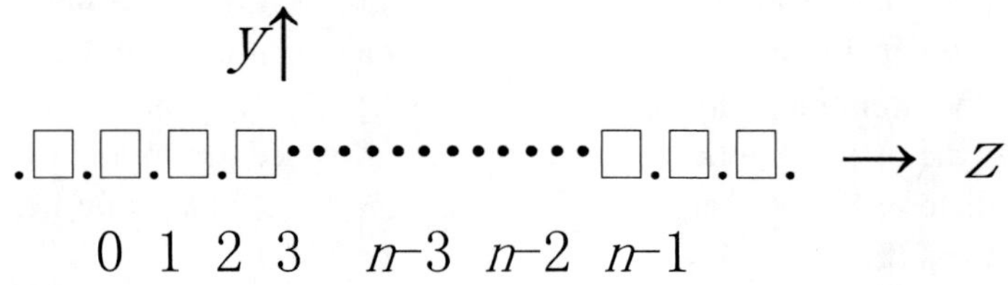

Figure 38. The atomic serial numbers and the Cartesian coordinates for polyyne and cyano-derivative molecules.

7.1.1. The Fuzzy Symmetry of Molecular Skeletons

If $\hat{G}$ is the parallel translation (towards the right or left) of *m* interatomic distance units, $\hat{T}(m)$, the membership function of polyyne C_nH_2 molecular skeleton may be given as :

$$\mu_{YT(m)}=[\sum_J(Y_J \wedge Y_{T(m)J})] /[\sum_J(Y_J)] = [\sum_J(Z_J \wedge Z_{T(m)J})]/[\sum_J(Z_J)]$$
$$=[(n-m)Z_C+2Z_H]/[nZ_C+2Z_H] = (6n-6m+2)/(6n+2) = \mu_{ZT(m)} \qquad (66)$$

where Z_J is the atomic number. As shown in Figure 39, the membership functions of some polyyne C_nH_2 molecular skeletons exist accompanying the transformation of translation of *m* interatomic distance units, $\hat{T}(m)$. Figure 39(A) and (B) show the membership function *vs. n*, and the membership function *vs. m*, respectively.

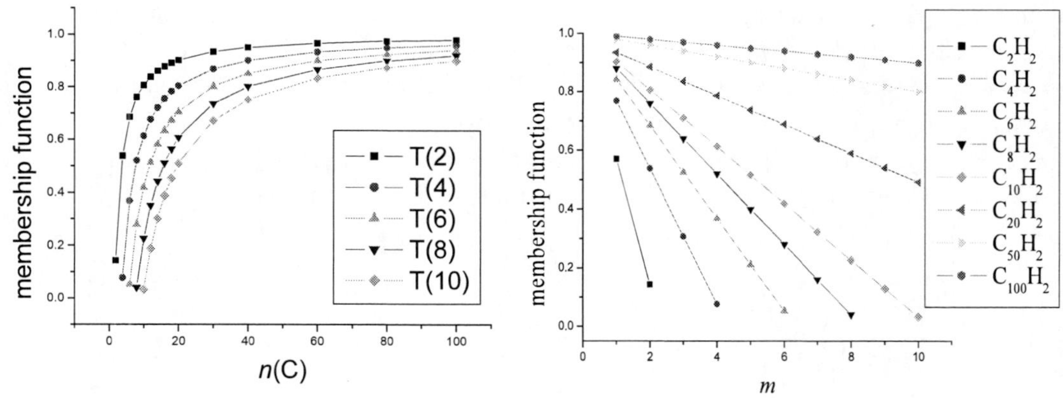

(A) (B)

Figure 39. The membership functions for the some polyyne (C_nH_2) molecular skeletons related to m inter-atomic distance transformation, $\hat{T}$ (m). (A) the membership functions *vs* the numbers of C atoms, n(C); (B) the membership functions *vs* the numbers of translation inter-atomic distances, m.

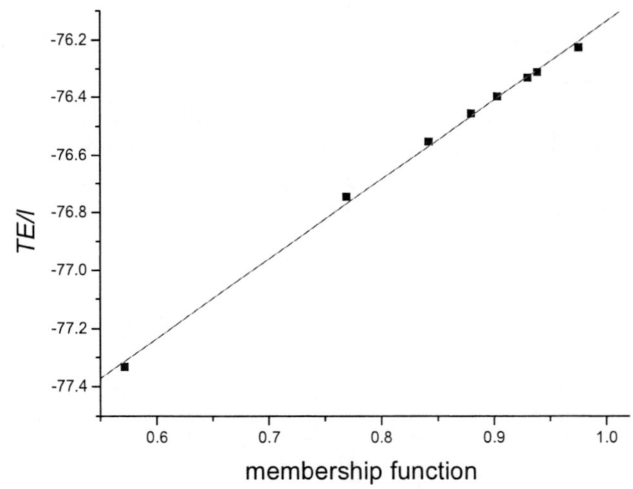

Figure 40. The linear relationship between *TE/l* and membership functions of polyyne molecules for translation operation $\hat{T}$. Where *TE* and *l* are the total energy and the C≡C unit number, respectively.

As shown in Figure39(A), for translation transformation $\hat{T}$-m, when n>10m, the membership function will be larger than 0.8. If we consider the polyyne as a one-dimensional crystal, the molecular crystal cell may have m atoms. As the crystal size is more than tenfold of the cell, the membership functions related to the space parallel translation group will be near or more than 0.9. It ought to be corrected in the ordinary crystal. As shown in Figure 39, the membership function will be cut down as m increases or n decreases, and is the linearly dependent on m. By the way, for the frontier orbilals [37] of $C_{16}H_2$, the dependence between the membership function and m is sawtooth-like, (cf. the following part *7.1.2*). This arises from the alternation of single- and tri- bonds. When we examine the MO fuzzy symmetry, the atomic criteria include such alternation, but the atomic criteria for the molecular skeleton do not. Therefore, we choose m as an even number.

Using the total energies (*TE*) calculated at RB3LYP/cc-pVDZ level [45] and divided by the number ($l = n/2$) of C≡C units in the polyyne, we may obtain the relationship for *TE/l* vs membership function in linearly dependent, as shown in Figure 40. It means that some molecular properties may be in relation to their fuzzy symmetry characteristics.

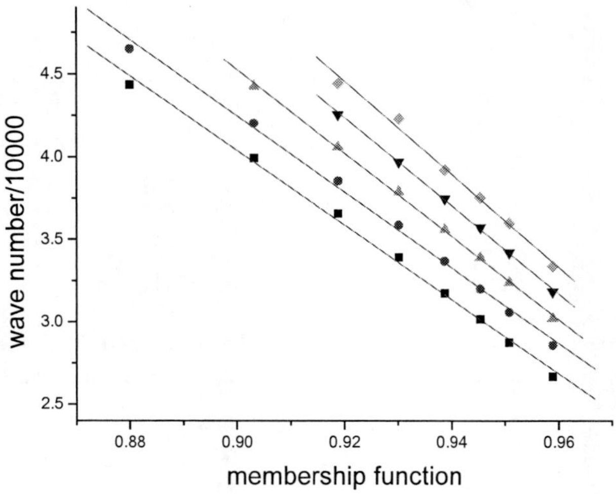

Figure 41. For the polyynes, the dependent relationship between the wave numbers [46] of the molecular electronic spectral bands and the membership functions in relation to the translation fuzzy symmetry for the polyyne molecules. In this figure, the various points relate to various spectral bands, and the corresponding regression lines are denoted.

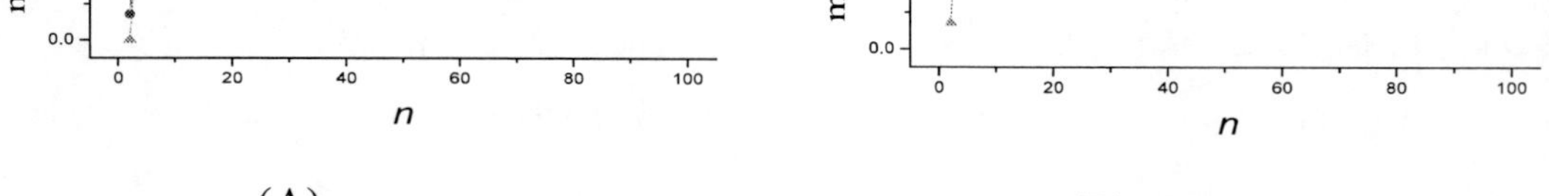

(A) (B)

Figure 42. The membership functions for some polyynes (C_nH_2) and their cyano-derivatives ($C_{n-1}NH$ and $C_{n-2}N_2$) in relation to (A) the translation $\hat{T}(2)$ for C_nH_2, $C_{n-1}NH$ and $C_{n-2}N_2$; (B) the fuzzy space inversion $\hat{P}$ about two various symmetrical centre, P(CC) and P(C), for $C_{n-1}NH$. For reference, the membership functions related to corresponding $\hat{T}(2)$ are also shown.

As shown in Figure 41, such a dependent relationship occurs between the wave numbers of the molecular electronic spectral bands vs the membership functions related to the translation fuzzy symmetry for the polyyne molecules. Jiang has put forward a so-called 'the rule of homologous linearity' [46] to rationalize such relationships. He pointed out that there are homologous factors for homologous compounds, and there is a linear dependency between the properties of these compounds and their homologous factors. In fact, for the polyyne, there is also a linearly dependency between the membership functions of their translation fuzzy symmetry and their homologous factor.

Similar as eq(66), if one or both end carbon atoms of polyyne are substituted by nitrogen atoms, the membership functions for the molecular skeleton related to the transformations $\hat{T}(m)$ are given as follows, respectively:

$$\mu_{YT(m)} = [\sum_J(Z_J \wedge Z_{T(m)J})]/[\sum_J(Z_J)] = [(n-m)Z_C+Z_H]/[(n-1)Z_C+Z_N+Z_H] = (6n-6m+1)/(6n+2) = \mu^1_{ZT(m)} \quad (67a)$$

and

$$\mu_{YT(m)} = [\sum_J(Z_J \wedge Z_{T(m)J})]/[\sum_J(Z_J)] = [(n-m)ZC]/[(n-2)ZC+2Z_N] = (6n-6m)/(6n+2) = \mu^2_{ZT(m)} \quad (67b)$$

It is notable that here $m \neq 0$. Translation transformation towards right, $\hat{T}(m)$, the criterion Y_J will be changed to Y_{J+m}, but that towards left, $\hat{T}(-m)$, Y_J will be changed to Y_{J-m}. Though the $(Y_J \wedge Y_{J+m})$ and $(Y_J \wedge Y_{J-m})$ may be unequal, but $\sum_J(Y_J \wedge Y_{J+m})$ and $\sum_J(Y_J \wedge Y_{J-m})$ are equalization. Figure 42(A) shows the membership functions for the polyyne (C_nH_2) and corresponding cyano-derivatives ($C_{n-1}NH$ and $C_{n-2}N$) related to the translation transformation $\hat{T}(2)$. It may be shown that the membership functions for these compounds are close.

It can be shown that the membership function of the inter-reversible symmetry transformation of *m* (integer) interatomic distance units towards right and left are the same. This satisfies the first condition eq(9a) for membership functions of fuzzy group and the condition eq(9c) of identity transformation (*m*=0). However, another condition eq(9b) could not usually be satisfied, and thus the fuzzy set of the space parallel translation transformation cannot form a fuzzy group.

For the above three molecular skeletons, the corresponding fuzzy parallel translation group may be denoted as follows by means of Zadeh's method [34]:

$$\underset{\sim}{T}(C_nH_2) = \sum_m [(6n-6m+2)/(6n+2)]/[\hat{T}(m)] \quad (68a)$$

$$\underset{\sim}{T}(C_{n-1}NH) = \sum_m [(6n-6m+1)/(6n+2)]/[\hat{T}(m)] \quad (68b)$$

$$\underset{\sim}{T}(C_{n-2}N_2) = \sum_m [(6n-6m)/(6n+2)]/[\hat{T}(m)] \quad (68c)$$

As all of the *m* related to the fuzzy parallel translation group are even, *m* in the summations of eq. (68) will only run over the even number.

Another important fuzzy symmetry transformation for linear molecules is that of space inversion. Polyyne has a symmetric centre. The symmetric element of the well-defined point group is definite, but the fuzzy symmetric element can be selected in various ways (see section 4.2)[26]. Membership functions corresponding to different ways differ, although all of they are between 0 and 1. They may be calculated using eq. (5) with $\hat{G}$ taken as the space inversion transform ($\hat{P}$).

For the polyyne molecule (Figure38), the symmetric centre would be the position between $n/2=q$-th and the $(q+1)$-th C atoms. If q is odd the centre will be on a triple C≡C bond, and if q is even the center is on a single C—C bond. As the polyyne molecule is composed of infinite C≡C units, it can be thought as a one-dimensional infinite crystal formed by such C≡C cells. The symmetric centre may be between any two consecutive C-atoms, and so there are infinite symmetric centre. Since the polyyne molecule is composed of finite C≡C units, the common symmetric centre is unique, but the fuzzy symmetric centre can be selected in various ways. For such fuzzy symmetric centre, the relative membership functions will be between the 0 and 1. Meanwhile, the membership function will become smaller as the distance between fuzzy centre and common symmetric centre increases. For the polyyne molecule composed of n C-atoms, when the fuzzy symmetric centre lies l C-atoms away from the common symmetric centre, the membership function related to the fuzzy symmetric centre can be obtained as follows:

$$\mu_{YP(l)} = [\sum_J(Y_J \wedge Y_{P(l)J})]/[\sum_J(Y_J)] = [\sum_J(Z_J \wedge Z_{P(l)J})]/[\sum_J(Z_J)]$$
$$=[2\{(n/2)-l\}Z_C+2Z_H]/[nZ_C+2Z_H]=[6n-12l+2]/[6n+2] \qquad (69)$$

As $l = 0$, it will be the common symmetric centre, and the membership function is one. On the other hand, the membership function for the identity transformation is also one. If the distance between fuzzy symmetric centre and common symmetric centre is l C-atoms, the membership function of such fuzzy space inversion transformation ought to be equal that related to the translation of a polyyne with m C≡C units (*i.e.* $m = 2l$ atoms). It is clear that this conclusion can only be established when the common symmetric centre exists; otherwise the distance from the common symmetric centre cannot be determined. As an example, for corresponding monocyano-polyyne molecular skeleton, there is no common symmetric centre, we can only analyze its fuzzy symmetric centre. As shown in Figure 38, if the nitrogen atom is assigned serial number J=0, and the carbon atom that links an outside hydrogen atom J=n-1, the fuzzy symmetric centre, P(CC), is between two carbon atoms with serial numbers J=q-1=(n/2)-1 and J=q=(n/2), and its corresponding membership function is $6n/(6n+2)$. However, if the fuzzy symmetric centre, P(C), of the same molecule is set on the carbon atom with J=q=(n/2), the relative membership function is $(6n-4)/(6n+2)$. Both membership functions will be between the 0 and 1, as shown in Figure 42(B) for those related to the fuzzy space inversion transformation $\hat{P}$ about these two fuzzy symmetric centers of monocyano-polyyne molecules. For reference, the membership function corresponding to $\hat{T}(2)$ was shown in this figure, too.

7.1.2. The Membership Function of MO

For the membership function of the MOs may be analyzed similarly as the skeleton, but atomic number Z (as the atomic criteria Y of the molecular skeleton) must be changed to the atomic criteria of the MO. Since the criteria of the same atoms in relation to the MO can be different, therefore the calculation of membership functions for MOs will be more complicated than that for the molecular skeleton. For the polyyne and their cyano-compound MO, the square of the LC-p_x or p_y AO coefficient was used as the atomic criterion of the π-MO, whereas the summation of the square of LC-sAO coefficients and the square of LC-p_z AO coefficients was used as the atomic criterion of the σ -MO.

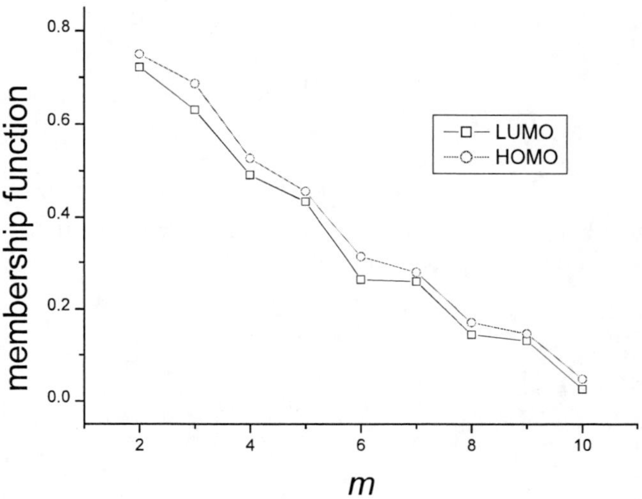

Figure 43. The membership functions for HOMO and LUMO of polynne C16H2, corresponding to the $\hat{T}(m)$ translation operation.

As shown in Figure 43, for the HOMO and LUMO of hexadec-polyyne, $C_{16}H_2$, the membership functions related to transformation $\hat{T}(m)$ decrease saw-toothedly as m increases. This arises from the alternation of single- and triple-bonds. The alternation effect is naturally incorporated in the LCAO coefficients of the MO, and reflected by the membership function although it has not been introduced directly into the function. However, the membership function for the molecular skeleton does not incorporate such effect, and so it depends on m smoothly (*i.e.*, linearly).

Other molecules may also involve in the alteration of bond-lengths, and such effect may be merged into the atomic criteria of MO too. Such phenomenon means that the corresponding one-dimensional translation unit ought to include two inter-atomic distances. In other words, the one-dimensional lattice cell including two atoms, and m will be an even number for transformation $\hat{T}(m)$. Therefore, we often analyze the case of even m.

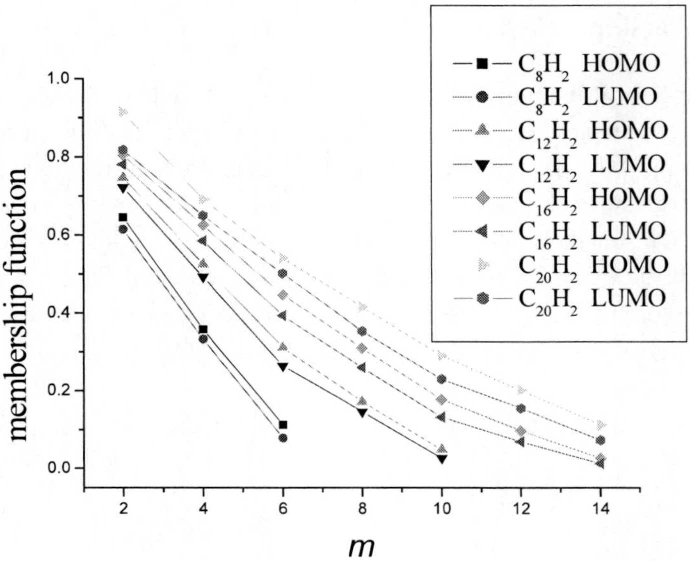

Figure 44. The membership functions in relation to the fuzzy translation $\hat{T}(m)$ for the HOMO and LUMO(at HF/STO-3G level) of various polyyne molecules.

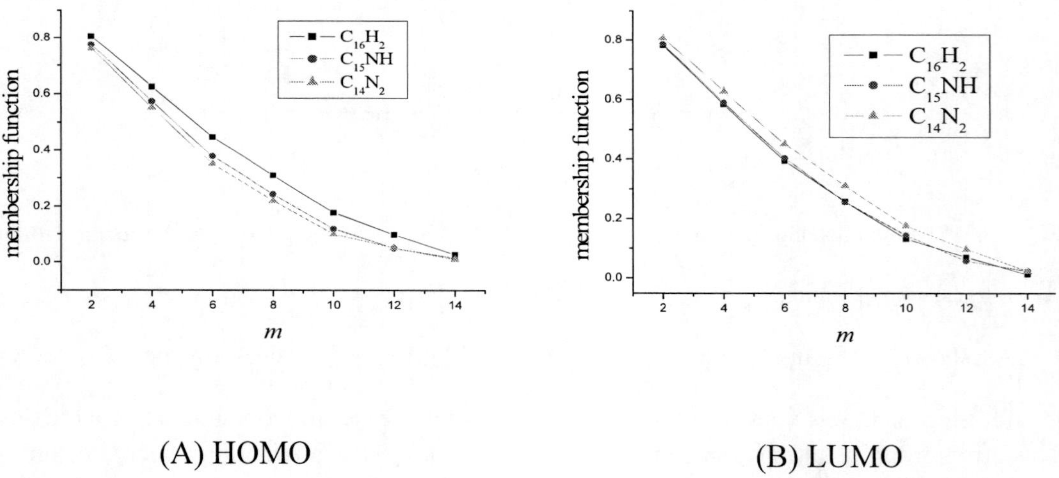

Figure 45. The plots of membership functions related to fuzzy translation transformation *vs.* the translation magnitude *m*, for the frontier MO (at HF/STO-3G level) of $C_{16}H_2$, $C_{15}NH$ and $C_{14}N_2$ for (A) HOMO, and (B) LUMO.

The membership function related to the fuzzy translation transformation for some frontier MOs of the polyyne obtained at HF/STO-3G level is shown in Figure 44. Each π-MO relates only to one AO from every atom. There is no saw-toohed as Figure 43 since all of *m* are even numbers.

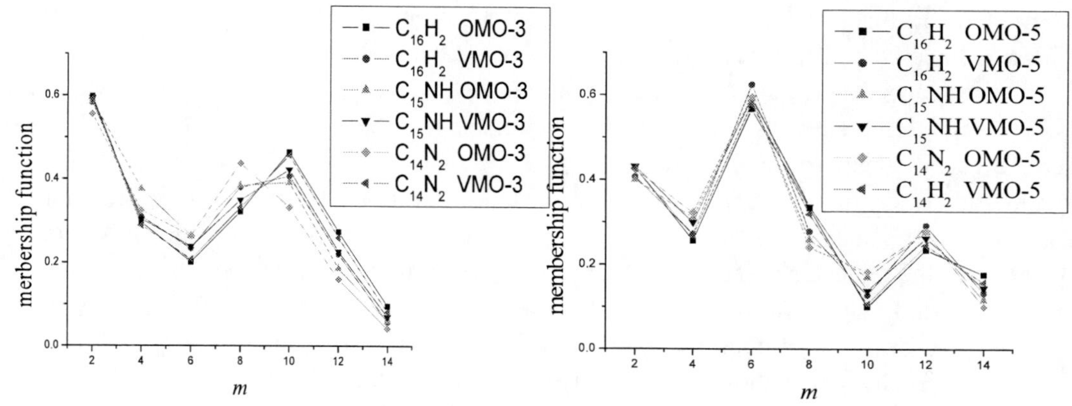

(A) (B)

Figure 46. The plots of the membership functions related to fuzzy translation transformations *vs.* the translation magnitude *m*, for certain frontier MOs of $C_{16}H_2$, $C_{15}NH$ and $C_{14}N_2$ at HF/STO-3G level. (A) OMO3 & VMO3, (B) OMO5 & VMO5,

As shown in Figure 44, for the same polyyne molecule, the membership function of HOMO will be larger than that of LUMO. In addition, the membership function will decrease obviously with the increase of translation magnitude *m*. As the same *m* value, the smaller the polynne molecule (*i.e.* smaller *n*) ought to be the less the membership function.

Here the 16-carbon-polyyne ($C_{16}H_2$) and their cyano-derivatives with one or both ethynyls (-C≡CH) substituted by -C≡N, *i.e.* $H(C{\equiv}C)_7(CN)$ or $(C{\equiv}C)_6(CN)_2$, are taken as the prototypical. We also consider only transformation $\hat{T}(m)$ with even *m*. To start with, as above we calculate and examine the membership functions of certain frontier MOs of $C_{16}H_2$ related to the transformation at HF/STO-3G level. It is notable that these π-MOs are two-fold degenerate (for both their energy and membership function), and so the even suffixes may be omitted. Now we examine polyyne $C_{16}H_2$ and its monocyano-derivative, $C_{15}NH$ and dicyano-derivative, $C_{14}N_2$. Figures 45(A) and (B) show the membership functions related to translation transformation $\hat{T}(m)$ for the HOMO and LUMO of these compounds, respectively.

The correlation curves of membership functions *vs.* *m* are similar to each other for these frontier MOs of these molecules. However, for the same *m*, the membership functions of HOMO decrease from $C_{16}H_2$ to $C_{15}NH$ and $C_{14}N_2$, whereas those of LUMO increase for the same order. Moreover, the membership function for the HOMO of $C_{15}NH$ is close to that of $C_{14}N_2$, but for the LUMO of $C_{15}NH$ is close to that of $C_{16}H_2$. Figure 46 shows the plot of membership functions *vs.* *m* related to translation transformation $\hat{T}(m)$ for certain other MOs in these molecules. It seems that for $C_{16}H_2$ and $C_{14}N_2$ the OMO and VMO with same suffixes are closer than do those for $C_{15}NH$.

As mentioned above, the membership functions related to transformation $\hat{T}(m)$ of the molecular skeleton and the MO is very different, in particular, those of non-frontier MOs.

They often fluctuate *vs. m*. though they decrease monotonically with m for the frontier MO. This non-linearity is different from the dependence for the skeleton.

There is a space inversion symmetric centre in $C_{16}H_2$ and $C_{14}N_2$, and this centre will often be represented in their MOs. Consequently, the related membership function ought to be one. As the fuzzy symmetric centre translates j inter-atomic distance units from the common symmetric centre, the membership function related to the fuzzy symmetry transformation of space inversion is equal to that of $2j$ inter-atomic distance unit translation, which is similar to the case for the molecular skeleton with a symmetric centre. $C_{15}NH$ does not have a common space inversion symmetric centre, and so we can only analyzed the fuzzy space inversion symmetry according to a chosen symmetric centre. The membership functions of a certain MO would depend on the choice of the fuzzy symmetric centre.

7.1.3. The Representation Component of MO

As regards to analyze the irreducible representation component for MOs, the first question is, what representation space will be expanded? In addition to the point group representation, we need to consider the space group representation. It is important that now we must consider the parallel translation group, its corresponding states and representations. In solid physics, there are already some important results, in particular, the Bloch's theorem [10]. According to this theorem, the state of the periodic space symmetry may be described by the Bloch function.

In the one-dimensional space of a periodic unit length a, the eigenstate Φ_k related to the translation of m periodic transformation $\hat{T}(m)$ is given below [9],

$$\hat{T}(m)\,\Phi_k = \exp(kmai)\,\Phi_k \qquad (69)$$

where k is the wave vector, a number or a scalar in one-dimensional case. According to the Bloch theorem, the so-called Born-Karman boundary condition[47] is used. For the one-dimensional lattice composed of n lattice points, the value of (kma) would be in $(-\pi, \pi)$. According to such boundary continuity, k will be $\nu(2\pi/na)$, where ν is a positive integer, taking values of 0,1,2, …, n-1. In eq. (69), term $\exp(kmai)$ becomes $\exp[m\nu(2\pi i/n)] \equiv \varepsilon^{m\nu}$, and is an eigenvalue of the wave-vector state Φ_k related to the $\hat{T}(m)$ transformation, maybe called wave-parity [5,9]. The term Φ_k will then depend on ν uniquely. The wave-vector state Φ_k with discrete ν belongs to an irreducible representation, $\Gamma_{\nu'}$, related to the one-dimensional periodic parallel translation group. This group is an Abelian group, and so all of its irreducible representations are one-dimensional. The corresponding characters are shown in Table 11. It is clear that this group and the C_n point group are isomorphic. This result means that the Born-Karman boundary condition have been introduced implicitly.

Of course, the Born-Karman boundary condition is strictly correct only when n is infinite, and it is only an approximation for the finite molecule, *i.e.*, a finite n. It should also be noteworthy that the value field is a complex one. The state of wave vector corresponding to the one-dimensional irreducible representation ought to be the common eigenstate related to all symmetry transformations included in this parallel translation group. The eigenvalue (character) may be a complex.

Some states belonging to various irreducible representations may be degenerate. For example, two states belonging to Γ_ν and $\Gamma_{n-\nu}$ irreducible representations (Table 11) may be degenerate. Both states are the eigen-states related to all symmetry transformations in the group, and some of their eigenvalues are indeed complex. However, for the real field when it is not the eigenstate for the symmetry transformation in corresponding parallel translation group to which it belongs, the two-dimensional irreducible representation will be introduced necessarily. As the degenerate states belong to Γ_ν and $\Gamma_{n-\nu}$ irreducible representation, the may be composed of states belonging to the two dimensional irreducible representation E_ν as the case for the C_6 point group [28] we examined. Since we are examining the irreducible representation in the complex field, the relative projection operators [36] ought to be a complex formula. We may start with the projection and then do the normalization for further analysis and calculation.

According Table 11, the projection operator related to Γ_ν irreducible representation may be shown as,

$$\hat{P}(\Gamma_\nu) = \sum_{m=0}^{n-1} \varepsilon^{-m\nu}\, \hat{T}\text{-}m, \tag{70}$$

where $\varepsilon^{-m\nu}$ is the complex conjugate of the eigenvalue related to $\hat{T}(m)$, and the normalizing factor is ignored. We consider only one AO for each atom that is combined into the MO, Ψ_ρ as shown in eq. (10). As the projection operator, $\hat{P}(\Gamma_\nu)$ in eq. (70), acts on MO Ψ_ρ, the part of the MO that belongs to Γ_ν irreducible representation, $\Psi_\rho(\Gamma_\nu)$, can be obtained,

$$\begin{aligned}\Psi_\rho(\Gamma_\nu) = \hat{P}(\Gamma_\nu)\,\Psi_\rho &= \sum_{m=0}^{n-1} \varepsilon^{-m\nu}\, \hat{T}\text{-}m\, \Psi_\rho = \sum_{m=0}^{n-1} \varepsilon^{-m\nu}\, \hat{T}\text{-}m \left[\sum_{J=0}^{n-1} a_\rho(J)\, \phi(J)\right] \\ &= \sum_{m=0}^{n-1} \sum_{J=0}^{n-1} a_\rho(J)\, \varepsilon^{-m\nu}\, \hat{T}\text{-}m\, \phi(J)\end{aligned} \tag{71}$$

where,

$$\Psi_\rho \equiv \sum_{J=0}^{n-1} a_\rho(J)\, \phi(J) = \sum_{\nu=0}^{n-1} \Psi_\rho(\Gamma_\nu) \tag{72a}$$

$$\Psi_\rho(\Gamma_\nu) \equiv \sum_{J=0}^{n-1} a_\rho(J;\Gamma_\nu)\phi(J) \tag{72b}$$

Here $a_\rho(J)$ and $a_\rho(J;\Gamma_\nu)$ are the LCAO coefficients of Ψ_ρ and $\Psi_\rho(\Gamma_\nu)$, respectively. The irreducible representation component of Ψ_ρ in connection to $\Psi_\rho(\Gamma_\nu)$ ought to be:

$$X_\rho(\Gamma_\nu) = \frac{\sum_{J=0}^{n-1} a_\rho^*(J;\Gamma_\nu)a_\rho(J;\Gamma_\nu)}{\sum_{\nu=0}^{n-1}\sum_{J=0}^{n-1} a_\rho^*(J;\Gamma_\nu)a_\rho(J;\Gamma_\nu)} \tag{73}$$

where $a_\rho{}^*(J;\Gamma_\nu)$ is the conjugate of $a_\rho(J;\Gamma_\nu)$. For simplifying the formula, we introduce the conjugate amount,

$$A_\rho(\Gamma_\nu) = \sum_{J=0}^{n-1} \varepsilon^{-J\nu} a_\rho(J) \tag{74a}$$

$$A^*{}_\rho(\Gamma_\nu) = \sum_{J=0}^{n-1} \varepsilon^{J\nu} a_\rho(J) \tag{74b}$$

Table 11. The characters of parallel translation transformation group for one-dimensional lattice.

Parallel translation	$\hat{T}(0)$	$\hat{T}(1)$	$\hat{T}(2)$	……	$\hat{T}(m)$	……	$\hat{T}(n\text{-}1)$
Irreducible representation		[where: ε =	exp(2π*i*/*n*)]				
Γ_0	1	1	1	……	1	……	1
Γ_1	1	ε	ε^2	……	ε^m	……	$\varepsilon^{(n-1)}=\varepsilon^*$
Γ_2	1	ε^2	ε^4	……	ε^{2m}	……	$\varepsilon^{2(n-1)}=\varepsilon^{2*}$
……	……	……	……	……	……	……	……
Γ_ν	1	ε^ν	$\varepsilon^{2\nu}$	……	$\varepsilon^{m\nu}$	……	$\varepsilon^{(n-1)\nu}=\varepsilon^{\nu*}$
……	……	……	……	……	……	……	……
Γ_{n-1}	1	$\varepsilon^{(n-1)}=\varepsilon^*$	$\varepsilon^{2(n-1)}=\varepsilon^{2*}$	……	$\varepsilon^{(n-1)m}=\varepsilon^{m*}$	……	$[\varepsilon^{(n-1)}]\times[\varepsilon^{(n-1)}]=\varepsilon$

Eq. (73) may be simplified and denoted as,

$$X_\rho(\Gamma_\nu) = \frac{A_\rho{}^*(\Gamma_\nu)A_\rho(\Gamma_\nu)}{\sum_{\nu=0}^{n-1} A_\rho{}^*(\Gamma_\nu)A_\rho(\Gamma_\nu)} \tag{75}$$

where the MO is not necessarily normalized. To start with the case of the common symmetric centre, we set n (Figure 38) to an even number. As $n=2q$, the centre will be between (q-1)-th and q-th atoms, and the LCAO-MO coefficients for the J-th and (n-J-1)-th atoms ought to be equal in absolute value. Accordingly, another pair of conjugate amounts, $B_\rho(\Gamma_\nu)$ are $B^*{}_\rho(\Gamma_\nu)$, can be introduced,

$$A_\rho(\Gamma_\nu) = \varepsilon^{\nu/2}\sum_{J=0}^{n-1} \varepsilon^{-(J+\frac{1}{2})\nu} a_\rho(J) \equiv \varepsilon^{\nu/2} B_\rho(\Gamma_\nu) \tag{76a}$$

$$A^*{}_\rho(\Gamma_\nu) = \varepsilon^{-\nu/2}\sum_{J=0}^{n-1} \varepsilon^{(J+\frac{1}{2})\nu} a_\rho(J) \equiv \varepsilon^{-\nu/2} B^*{}_\rho(\Gamma_\nu) \tag{76b}$$

In the summation of $B_\rho(\Gamma_\nu)$, the complex coefficient before $a_\rho(n\text{-}J\text{-}1)$ would be $\varepsilon^{-(n-J-\frac{1}{2})\nu} = \varepsilon^{(J+\frac{1}{2})\nu}$; it is conjugated with that before $a_\rho(n\text{-}J\text{-}1)$. We can then readily collate these two terms and combine them into a real term and a pure imaginary term.

$$B_\rho(\Gamma_\nu)=\mathrm{Re}[B_\rho(\Gamma_\nu)] - \mathrm{Im}[B_\rho(\Gamma_\nu)]i \tag{77a}$$

$$B^*_\rho(\Gamma_\nu)=\mathrm{Re}[B_\rho(\Gamma_\nu)] + \mathrm{Im}[B_\rho(\Gamma_\nu)]I \tag{77b}$$

The irreducible representation component $X_\rho(\Gamma_\nu)$ may then become the following,

$$X_\rho(\Gamma_\nu) = \frac{\mathrm{Re}^2[B_\rho(\Gamma_\nu)] + \mathrm{Im}^2[B_\rho(\Gamma_\nu)]}{\sum_{\nu=0}^{n-1} \{\mathrm{Re}^2[B_\rho(\Gamma_\nu)] + \mathrm{Im}^2[B_\rho(\Gamma_\nu)]\}} \tag{78}$$

where the real and imaginary parts of $B_\rho(\Gamma_\nu)$ may be respectively denoted as:

$$\mathrm{Re}[B_\rho(\Gamma_\nu)]=\sum_{L=0}^{q-1} a_\rho(L;\mathrm{Re})\cos[(L+\tfrac{1}{2})\pi\nu/q] \tag{79a}$$

$$\mathrm{Im}[B_\rho(\Gamma_\nu)]=\sum_{L=0}^{q-1} a_\rho(L;\mathrm{Im})\sin[(L+\tfrac{1}{2})\pi\nu/q] \tag{79b}$$

Here, the coefficients before the trigonometric functions reflect the symmetrical and asymmetrical components related to the common space inversion transformation, and may be obtained from the LCAO-MO coefficients a_ρ (L) and a_ρ (GL). When the symmetrical representation is related to the common inversion transformation, $a_\rho(L;\mathrm{Re})$ would be zero; and when the anti-symmetrical one is related to, $a_\rho(L;\mathrm{Im})$ would be zero. According to the symmetrical and anti-symmetrical states, $\Psi_\rho(\Gamma_\nu;g)$ and $\Psi_\rho(\Gamma_\nu;u)$, in connection with representation Γ_ν, corresponding representation components for Ψ_ρ ought to be respectively as follows:

$$X_\rho(\Gamma_\nu;g) = \frac{\mathrm{Im}^2[B_\rho(\Gamma_\nu)]}{\sum_{\nu=0}^{n-1} \{\mathrm{Re}^2[B_\rho(\Gamma_\nu)] + \mathrm{Im}^2[B_\rho(\Gamma_\nu)]\}} \tag{80a}$$

$$X_\rho(\Gamma_\nu;u) = \frac{\mathrm{Re}^2[B_\rho(\Gamma_\nu)]}{\sum_{\nu=0}^{n-1} \{\mathrm{Re}^2[B_\rho(\Gamma_\nu)] + \mathrm{Im}^2[B_\rho(\Gamma_\nu)]\}} \tag{80b}$$

As above, we consider the polyyne of linear polymer molecule as a one- dimensional fuzzy lattice in which the unit cell contains only one atom. If each lattice is constructed by *s* atoms, and the LCAO coefficient for the *i-th* AO (*i*=1 to A*i*) of the A-*th* atom (A=1 to *s*) in the J-*th* lattice (J=0 to *n'*-1) in the ρ-*th* MO (Ψ_ρ) is $a_\rho(J,A,i)$, then replacing eq. (72), this MO may be denoted as:

$$\Psi_\rho \equiv \sum_{J=0}^{n'-1} \sum_{A=1}^{s} \sum_{i=1}^{Ai} a_\rho(J,A,i)\ \phi(J) = \sum_{\nu'=0}^{n'-1} \Psi_\rho(\Gamma_{\nu'}) \tag{81a}$$

$$\Psi_\rho(\Gamma_{\nu'}) \equiv \sum_{J=0}^{n'-1} \sum_{A=1}^{s} \sum_{i=1}^{Ai} a_\rho(J,A,i;\Gamma_{\nu'})\phi 1$$

The irreducible representation components related to the fuzzy parallel translation group above may be used further, but the summation running over J will be replaced by over J, A, and *i*. It is notable that for such case there are *sn'* atoms, whereas the number of both lattice points and irreducible representations for the fuzzy group is only *n'*. Of course, we may still examine the irreducible representation components related to the fuzzy space inversion symmetrical transformation.

Now we will examine some prototypical molecules. The results will be strictly correct for the infinite lattice owing to the Born-Karman boundary condition. Since the molecule is not very small, the desirable results are often approximate.

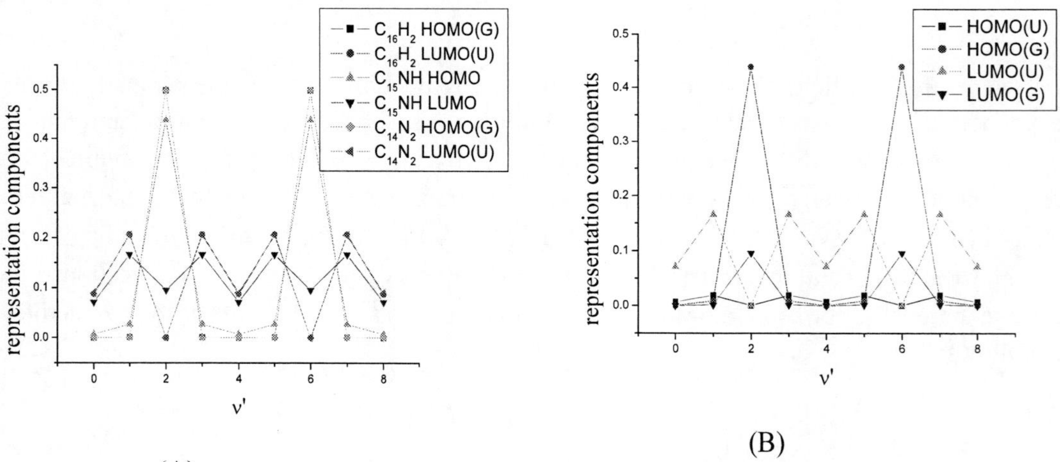

(A)

(B)

Figure 47. The plots of irreducible representation components *vs.* ν': (G) and (U) denote symmetry and anti-symmetry corresponding to space inversion, respectively. (A) related to the fuzzy parallel translation for frontier MOs of $C_{16}H_2$, $C_{15}NH$ and $C_{14}N_2$, (B) related to both fuzzy space inversion and fuzzy parallel translation for frontier MOs of $C_{15}NH$.

We took $C_{16}H_2$, $C_{15}NH$ and $C_{14}N_2$ as the example of the one-dimensional lattice in which each unit lattice contains two atoms (CC or CN), and considered the πMO. This reduces eq. (81) as A_i is 1 and no summation over *i* is needed, and *s* and *n'* are 2 and 8, respectively. In this parallel translation group, the minimum unit length is one single and one triplet bond lengths. There are eight cells including 16 AOs to form a fuzzy one-dimensional crystal. For the corresponding fuzzy parallel translation group, there are eight irreducible representations, $\Gamma_{v'}$ (v'=0 to7 ; $\Gamma_0=\Gamma_8$). Figure 47 shows the irreducible representation components for the frontier πMOs of these molecules.

As shown in Figure 47(A), for the HOMOs of these three molecules, the representation components related to $\Gamma_{v'}$ with v'=2 and 6 are maxima, but for their LUMOs, the representation components with v'=1, 3, 5 and 7 are maxima. As the space inversion transformation for $C_{16}H_2$ and $C_{14}N_2$, the HOMOs belong to the pure symmetrical representation (G), but the LUMOs belong to the pure asymmetrical representation (U). For $C_{15}NH$ without the common space inversion symmetry, the MOs may include both symmetrical and the asymmetrical components, simultaneously, and their main representations ought to be the same as the pure representations of the corresponding MOs of $C_{16}H_2$ and $C_{14}N_2$ related to the space inversion. Figure 47(B) shows the representation components related to the fuzzy parallel translation and the fuzzy space inversion for the frontier MOs of $C_{15}NH$, *i.e.*, the products of the representation components related to these two fuzzy groups (subsets). We may get the irreducible representation components related to the parallel translation group for some other π-MOs of these three molecules, similarly.

7.2. The Fuzzy G_1^2 Symmetry of Cis-Trans-Conjugate Polyenes

The planar molecules with periodicity in a special orientation are probed as to possess fuzzy G_1^2 symmetry. In the G_1^1 symmetrical system, the only space symmetry transformation is the parallel translation. While in the G_1^2 symmetrical system, there may also be other two kinds of space symmetry transformation, the screw rotation and the glide plane. Now, we take conjugate polyene as the example with fuzzy *ta/2* (where the *ta/2* is a sort of G_1^2 group) symmetry to analyze[41-44]. The more complex system with fuzzy G_1^2 and G_1^3 symmetry will be studied in our future works.

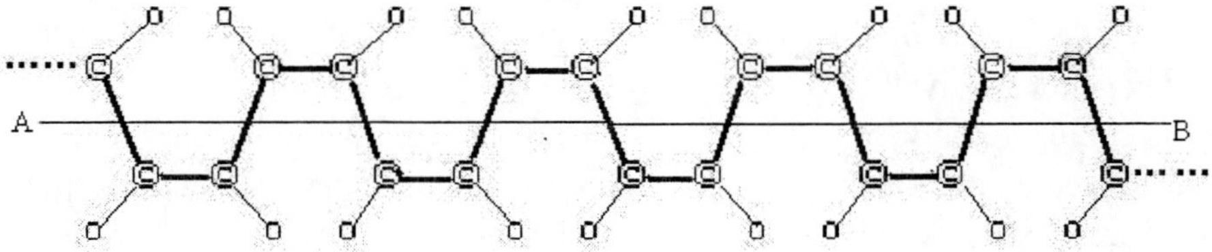

(A)

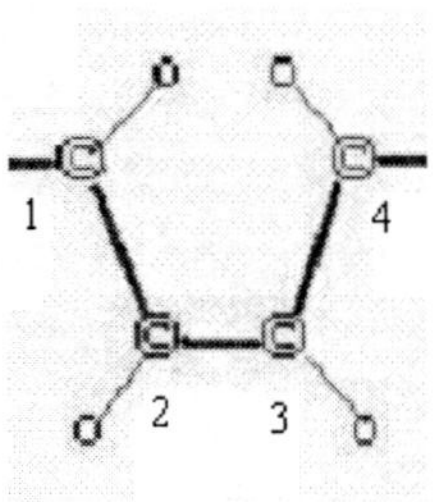

(B)

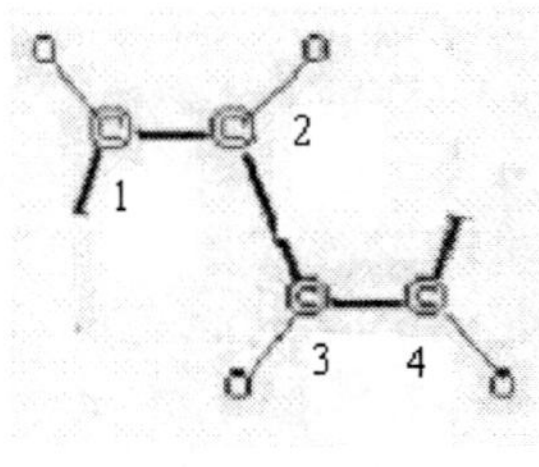

(C)

Figure 48. The fraction and one- dimensional cells of conjugate polyene molecule: (A) a fraction molecule; (B) *cis*-cell ; (C) *trans*-cell. In the figure, where © denotes C-atom, the ringlets denote H-atom,

In the conjugate polyene molecules, there may be two different ways, *cis*- and *trans*-, for the alternation of single- and double-bonds, and the molecules may have various molecular conformation. Convenient for the discussion, the fuzzy one-dimensional space parallel translation symmetry is examined. Thus, here we select the *cis-trans*-polyene molecules with the approximate linear structures. In figure 48(A), a fraction of the long straight chain conjugate polyene molecule is shown. All the atoms are in the same plane, there is provided with the approximate *ta/2* group symmetry.

As shown in the figure, the carbon-hydrogen bond is the C-H single-bond and the carbon-carbon bond is the alternation of C-C single-bond and C=C double-bond. It should be noteworthy that the horizontal carbon-carbon bond here may be total single-bond or total double-bond, and they must be the same type, while the skew carbon-carbon bonds are the other type ones. The binding for all the horizontal carbon-carbon bonds and their neighbor C-atoms adopts *cis*-, while it adopts *trans*- for all the skew carbon-carbon bonds and their neighbor ones. Moreover, there may be different selected ways for its one-dimensional holonomic cells. Two of the different selected cells are shown in figure 48(B) and (C), of course, the cell can be selected in other ways. Here, we mostly take the way of figure 48(B). The same serial number stands for the same atom in each cell. There is the fuzzy space group symmetry of parallel transformation $\hat{T}(m)$. Hereon, m is a positive or negative integer. Figure 48(B) and (C) show that the four atoms in the cell bind in the ways of *cis*- and *trans*-, respectively. The two kinds of cells possess the C_{2v} and C_{2h} point group symmetry, respectively. For the Cartesian coordinates about the description of the molecular geometries, we can choose the molecular plane as the coordinate plane, and take the direction of the line AB to be the orientation of one coordinate. The origin of the coordinate can be the molecular center, and it should be in the line AB. Furthermore, it can be the center of one skew carbon-carbon bond in figure 48(A), and it can be also the midpoint of the line that connects the centers of the two neighbor skew carbon-carbon bonds. Whereas, the two directions of passing the origin and orthogonal to the line AB are determined to be the other two coordinate axes. Meanwhile, the two directions are line in the molecular plane and in the vertical molecular plane, respectively.

For the conjugate polyene, $H(C_2H_2)_nH$, when n is even, the molecular center is a single bond. The molecule possesses the C_{2v} point group symmetry when taking the parallel carbon-carbon in figure 48(A) as the single-bond, while it possesses the C_{2h} point group symmetry when considering the skew carbon-carbon in figure 48(A) as the single-bond. On the other hand, if n is odd, the molecular center is double-bond, we may analyze such system, too. For the conjugate polyene molecules composed of finite atoms, their directional structures can maintain the well-defined symmetry of C_{2v} or C_{2h} point group, but they can't keep the well-defined space group symmetry, only the fuzzy symmetry can be probed. For the conjugate polyene molecules holding the C_{2v} or C_{2h} point group symmetry, the selected way to the symmetrical elements is unique about the corresponding point group transformation. Although we can select the symmetrical elements in other ways, then we can only discuss the corresponding fuzzy symmetry. Membership functions that are corresponding to point group symmetry transformation of the molecular skeleton and MO are one.

The one-dimensional crystals that include infinite cells and arrange according to figure 48(A) have the space parallel translation transformation $\hat{T}(m)$ symmetry, m is integer.

What's more, they have symmetry of other space transformation. For the half-integer m', the system does not have such space parallel translation transformation $\hat{T}(m')$ symmetry, alone. However, it has the combined transformation symmetry in jointed the point one, such combined transformation may be the glide reflection or the screw rotation transformation. When the cell in figure 48(A) is finite crystal, the symmetries are all fuzzy ones. Taking the line AB as the two fold axis, and $\hat{C}_{2\,(AB)}$ is denoted as the twofold rotation transformation of AB. Considering the plane that passes the line AB and is perpendicular to the molecular plane is the mirror plane, and the reflection transformation of the mirror plane is denoted as $\hat{M}_{(AB)}$. The system can not be recovered under the transformation of $\hat{C}_{2\,(AB)}$ and $\hat{M}_{(AB)}$. That is to say, such system would be not provided the symmetry in relation to the $\hat{C}_{2\,(AB)}$, $\hat{M}_{(AB)}$ nor the parallel translate transformation $\hat{T}(m')$, where m' is a half-integer number, alone. For this one-dimensional infinite crystal, it may have the combined transformation symmetry:

$$\hat{T}(m')\,\hat{M}_{(AB)} = \hat{M}_{(AB)}\,\hat{T}(m') \equiv \hat{M}_{G(AB)} \tag{82a}$$

and

$$\hat{T}(m')\,\hat{C}_{2\,(AB)} = \hat{C}_{2\,(AB)}\,\hat{T}(m') \equiv \hat{C}_{S2\,(AB)} \tag{82b}$$

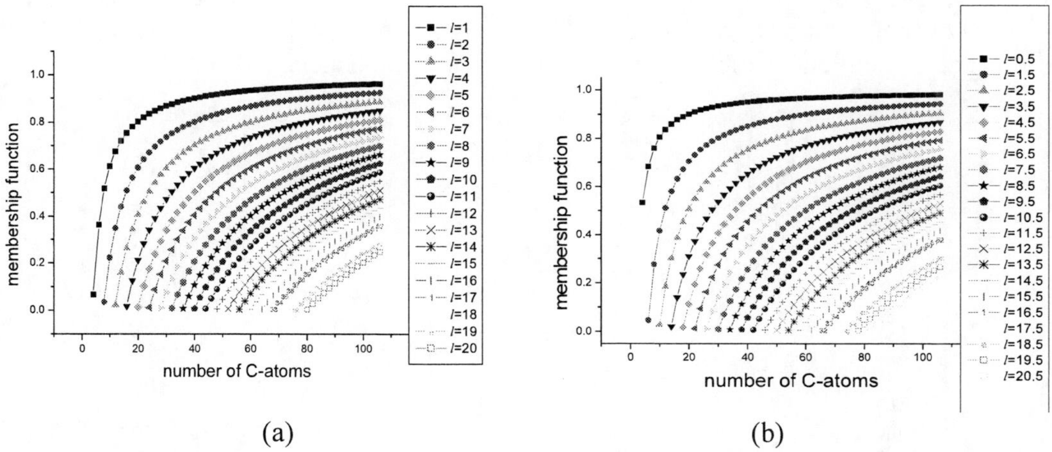

Figure 49. the membership functions of some symmetry transformation for the skeleton of conjugate polyene molecules, $H(C_2H_2)_nH$, (A) the membership functions of the parallel translation symmetry transformation $(\hat{T})^l$ vs. number of C-atoms (B) the membership functions of the twofold screw rotation $(\hat{T}(1/2)\,\hat{C}_{2\,(AB)})^{ls}$ or the glide plane $(\hat{T}(1/2)\,\hat{M}_{(AB)})^{ls}$ transformation vs. number of C-atoms.

The two combined transformations are called glide reflection and screw rotation in turn. This belongs to one-dimensional parallel translation combined point symmetry transformation. It is notable that there are no such symmetry about the pure parallel transformation, $\hat{T}$ (*m'*), or point transformation, $\hat{C}_2$ (AB) and $\hat{M}$ (AB), while only the combined transformation symmetry is existed. Consequently, to the conjugate polyene of finite "one-dimensional periodic planar crystal", there is only the fuzzy symmetry characterization about the two combined-transformations. It possesses G_1^2 symmetry for the one-dimensional periodic planar molecules, and there may be seven symmetrical groups that belong to G_1^2, and in the complex *ta/2* [41-44] group is related. It is notable that as shown in Fig 48(a) there is the polyacetylene (a lower dimensional organic conductor) molecule with *cis-trans* or *trans-cis* conformation.

Based on the above analyze in relation to the *ta/2* symmetry transformation group, we my study the fuzzy symmetry of *trans-cis-* polyenes[33], similar as that for polyynes. Some important results may be obtained and shown as follows.

7.2.1. The Fuzzy Symmetry of Molecular Skeleons

Now we consider the conjugate polyene molecular skeleton, H(C2H2)*n*H. There is a holonomic cell in the center of the H(C2H2)*n*H molecule and there are *m* holonomic cells and a terminal cell on its left and right side, respectively. Thereby, there are 2*m*+1 holonomic cells and 2 terminal cells in total. The total number of the cells is 2*m*+3, and the total C-atoms is 2*n*=4(2*m*+1)+2*i*, therein, *i* is the C-atom number of each terminal cell. As for H-atoms, the number is 2*n*+2. For the molecular skeletons of the conjugate polyene, using the atomic number as the criterion, we can get the relative membership function in relation to a certain symmetry transformation, $\hat{G}$ (*l*), according to eq(5). By the way, there are not only the space symmetry of translating integer (*l*) cell lengths along AB but also the space symmetries relate to some glide plane and twofold screw rotation transformation along AB.

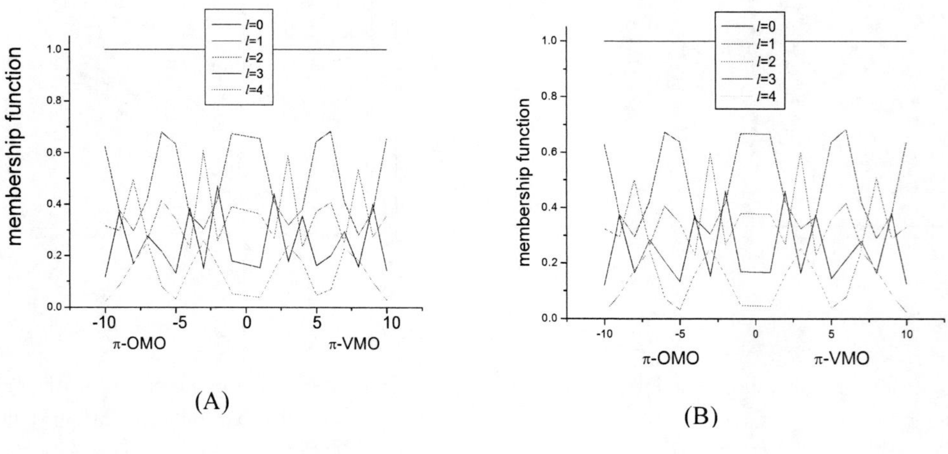

Figure 50. The plots of membership function vs MO serial numbers for $C_{20}H_{22}$, (A) at AM1 level, (B) at STO-3G level.

Figure 49(A) shows the membership function changes for the different conjugate polyene molecular skeletons (the abscissa coordinate shows the C-atom number) relate to the parallel symmetry transformation $\hat{G} = \hat{T}(l) = (\hat{T})^l$, where l represents the integer, and it indicates the transformation that the whole molecule translates l cell length. As shown in figure 49(B), the membership function changes about that the different conjugate polyene molecular skeletons relate to the ls times glide plane $\hat{G} = (\hat{T}(1/2)\,\hat{M}_{(AB)})^{ls}$ or the twofold screw rotation transformation $\hat{G} = (\hat{T}(1/2)\,\hat{C}_{2\,(AB)})^{ls}$. It includes the transformation of translating odd ls structure unit length (that is half-integer $l=ls/2$ cell length).

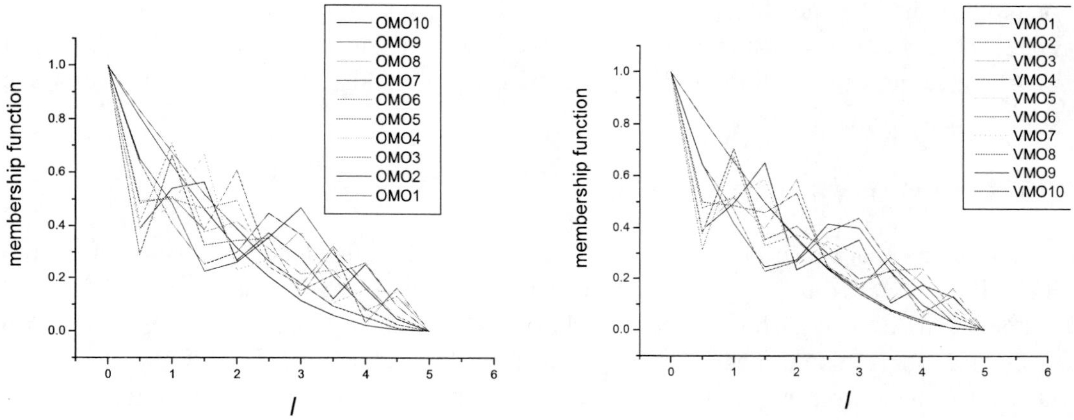

(A) **π-OMOs,** (B) **π-VMOs,**

Figure 51. The plots of membership function vs l for $C_{20}H_{22}$, at AM1 level.

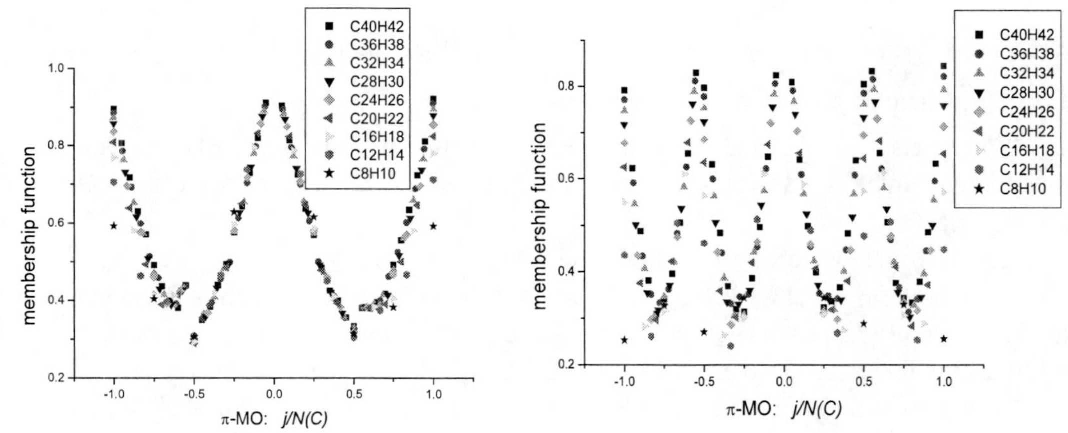

(A) l=0.5 (B) l=1.0

Figure 52. The plots of membership function vs $j/N(C)$ in relation to some space transformations with l = 0.5(A) and 1.0(B), respectively, for C_nH_{n+2}, at AM1 level.

In figure 49(A) and (B), for the invariable l, the curve of the membership function vs C-atom number monotonically increases. With the increasing of the C-atom numbers in the conjugate polyene molecules, it can be anticipated that the space symmetry transformation is more and more close to the well-defined space group transformation and the corresponding membership function ascends monotonically and tends to one. In figure 49(B), when the numbers of C-atoms > 20 (namely, more than 10 structure units are involved in the molecule), the membership function is close to one, for l=0.5. While in figure 49(A), for l=1, the membership function is near to one when numbers of C-atoms > 40 (namely, more than 10 cells are involved in the molecule). These results are in agreement with our studies on the polyyne.

There is the rule of homologous linearity in the organic homologue [46]. We have pointed out that there is the homologue linearity rule between the properties of the polyyne homologue and the membership function of their molecular skeleton parallel translation symmetry transformation. Such rule also exists in the conjugate polyene, and it relates to not only the membership function of the parallel translation but also to that of some other space symmetry transformation [33].

7.2.2. The Fuzzy Symmetry of MOs

The molecular skeleton discussed here has the well-defined C_{2v} symmetry character, although the atomic criteria of MO can't be denoted simply by atomic number, the criteria of its atoms and AO may also have the well-defined C_{2v} symmetry character. The membership functions relates to some space symmetry transformations should be equal to those obtained via displacement from some point symmetry transformation elements. Thus, they can be studied tighter. Some membership functions of $C_{20}H_{22}$ π-MOs in relation to $\hat{T}(l) = (\hat{T})^l$, are shown in figure50 (A) and (B), at the AM1 and STO-3G levels, respectively, there are very close. The abscissa coordinate shows the serial numbers of MO separated from the nonbonding energy level, OMO is negative and VMO is positive. 0 denotes nonbonding without MO. For the membership functions in relation to $(\hat{T}(1/2)\hat{M}_{(AB)})^{l'}$ or $(\hat{T}(1/2)\hat{C}_{2\,(AB)})^{l'}$, where odd $l' = 2l$, we can get the similar results.

We can also examine the appointed MO to discuss the dependence of the membership function of symmetry transformation vs. l. Figure 51 shows the dependence of the conjugate polyene π-MO.

Figure A and B indicate the OMO and VMO, respectively, and they are very similar. These results from the close value of OMO-j and VMO-j in π-MO, which is the membership functions, appear approximate symmetry distributing about nonbonding energy level. The membership functions decrease with the increasing of l, roughly. However, it is neither monotone nor linear decreasing as the fuzzy symmetry of the molecular skeletons. By the way, for OMO-j and OMO-(10-j), they are similar about the membership functions of symmetry transformation relates to the same l. (here 10 is the number of π-OMOs in the conjugate polyene molecule). Thus, some lines almost overlap in figure 51A. So do the VMO-j and VMO-(10-j), in Figure 51B.

In addition, we can see an interesting phenomenon that is for different conjugate polyene, although the referred π-MOs number is increasing it is not more for the change of the membership function distributing with the same l. The phenomena also exist in the membership functions relate to the tiny parallel transformation of glide reflection and twofold screw rotation. There will be interesting results if the space symmetry transformation membership functions of different conjugate ployene with the same l (integer or half-integer) are drawn in the same figure. The space transformation with the confirmed l indicates that the glide reflection or the twofold screw rotation $(\hat{T}(1/2)\hat{M}_{(AB)})^{2l}$ or $(\hat{T}(1/2)\hat{C}_{2(AB)})^{2l}$, which translates $2l$ times and each times translates one structure unit(*i.e.* half cell). The membership functions of the two space transformations are identical. In addition, when l is integer, they are all equal to the space parallel symmetry transformation and the translating length is l cell length. The membership functions of the different conjugate polyene π-MOs vs $j/N(C)$ diagrams are shown in figure 52(A & B), with l is 0.5 and 1.0, respectively. Included in the abscissa coordinate, the serial number j in π-MO-j, and taking π-OMO is negative and π-VMO is positive and the conjugate polyene molecule including n=2$N(C)$ conjugate C-atoms, the value range of abscissa coordinate is from -1 to +1. For different conjugate polyene π-MOs, the majority membership functions of the symmetry transformation with confirmed l fall into the same curve, approximately. Both π-OMOs and π-VMOs have the similar opening upward curves.

For the irreducible representation components of the MO in these conjugate polyene molecules, we may analyze them as for polyynes molecules, similarly. But the relative results are omitted here to save the space.

8. Conclusion

The molecular symmetry is one of the important regions in theoretical chemistry. There are two interesting titles in region: 1. What is the molecular symmetry? 2. How to describe the molecular symmetrical ‘quantity’? Last century, one of the most important achievements in this regions was the conservation principle of orbital symmetry, i. e. the Woodward-Hoffmann (W-H) rules. For above two titles, there are some contents ought to be studied further more in W-H rules.

As regard to the first title, the traditional statement of W-H rules may be expressed: orbital symmetry is conserved in concerted reaction. By means of the field theory, according to Noether’s theorem, the symmetry and invariant are two different concepts, but they are not distinguished in W-H rules. In fact, the symmetry is a property of the molecular system, *i.e.* the environment space in which the MO being (the invariance of the molecular system under the action of relative symmetric transformation group element with membership function = 1) and the invariant is a characteristic of the MO (the eigenvalue of MO under the corresponding symmetry operation group element, for all the group elements, they will be relative the irreducible representation).

As regard to the second title, the imperfection of the symmetry is neglected in W-H rules, and so-called the topological considerations [48], sometimes. The prior way to analyze this title ought to be using the fuzzy subset theory. Some theoretical chemists, in particular, Mezey P.G. and his coworkers get some important results [13-20.]. But we only introduce our works with the relative logic structure, here.

Owing to the symmetry denotes the variation rules: the invariance of the molecular system under the action of relative symmetric transformation group element with membership function = 1, we design the fuzzy symmetry to denote the variation rules: the variance of the molecular system under the action of relative symmetric transformation fuzzy subset element with membership function less 1. As the membership function near 1, the fuzzy symmetry will be the approximate (perfect) symmetry. As the membership function near 0, the fuzzy symmetry will be not the approximate (perfect) symmetry. Although the fuzzy symmetry is a disputable terminology, we still introduce it.

8.1. Summary

In this paper, we sum up our works in relation to the molecular symmetry and fuzzy symmetry. The main points are as follows:

1. According to Noether's theorem, there are two different concepts (symmetry and invariant) they are not distinguished in Woodward-Hoffmann rules. However, they ought to be distinguished. The symmetry ought to be the invariance of the molecular system under the action of relative symmetry transformation and the invariant is the eigenvalue of MO under the corresponding symmetry operation. For the discrete symmetry the relative invariant may be called the generalized parity. The invariance in relation to the point symmetry, space symmetry and the symmetry joint the reaction reversal and point transformation are analyzed.
2. Based on the fuzzy subset theory, the molecular fuzzy symmetry has been supposed. Corresponding to the symmetry and invariant for the well-defined cases, the relative observable quanta are membership function and representation component (of MO), respectively. The calculation method of these quanta has been supposed, too. As the membership function near one, the fuzzy symmetry ought to be the similar as approximate symmetry.
3. The fuzzy symmetry of some prototypical linear and plane static molecules has been analyzed. We may get these molecular membership functions and representation components (of MO). For more complex molecules, the MO representation components may be obtained by means of the projection operator method and the representation components can refer to not only the one- but also the multi-dimensional irreducible representation. For various molecules, we may analyze the relative characteristics of these membership functions and representation components. For the similar molecules, the MO fuzzy correlation diagram may be established.
4. For simple linear tri-atomic dynamic system, the fuzzy symmetry characteristic has been shown. The membership function contour maps and representation component contour maps of such dynamic systems are constructed. In these maps, the IRC-scale may be introduced and some interesting new symmetric graphs will be appeared. The symmetry and fuzzy symmetry of such dynamic system in relation to the joint transformation of reaction reversal and space inversion are analyzed in more detail.
5. The fuzzy symmetry characteristics for the internal-rotation of allene and its 1,3-dihalides are studied as the prototypical dynamic system. Along the IRC-path the all

states, especially, the stable and transition state molecules are examined. For 1,3-dihalides, both the *cis*- and *trans*-IRC paths are considered. According to these two IRC paths, the space factor and intrinsic factor in relation to the fuzzy symmetry are analyzed. Some special symmetry transformation groups are introduced. For these IRC paths, the relative MO correlation diagram and fuzzy MO correlation diagram are suggested.

6. The one-dimensional space periodic fuzzy symmetry of polynnes and *cis-trans*-conjugate polyenes as the prototypical molecules has been analyzed. It is the example of the more simple fuzzy space symmetry. Such fuzzy space symmetry may be owing to the difference between the infinite space group and the finite real molecular system. The rule of homologous linearity between the molecular property and fuzzy symmetry has been examined. Some interesting results in relation to the fuzzy space symmetry of homologous compounds may be appeared.

8.2. Outlook

The molecular symmetry and fuzzy symmetry may be the interesting and important title for theoretical chemist. Although the 'fuzzy symmetry' is a disputed terminology, the works in relation to molecular symmetry based on fuzzy subset theory are not rare. Recently, we have done some thing in the field, too. However, comparing with the contents of the field, there are a lot of works ought to be explored.

1. The fuzzy symmetry of some prototypical non-plane molecules would be analyzed. For these molecules, the fuzzy symmetry would be related the senior point symmetry group included more multi-fold symmetric elements. The irreducible representation of MOs may be the multi-(more than two-) dimensional, the relative representation components will be more complex. Such molecules include the complex compounds and hetero-fullerenes *etc*.
2. The fuzzy G_1^3 symmetry of some prototypical non-plane molecules would be analyzed. The fuzzy multi-fold screw rotation symmetry would be considered. Such molecules include the carbon nano-tubes and DNA *etc*.
3. The multi-dimensional space periodic fuzzy symmetry of some molecules would be analyzed. For these molecules, the fuzzy symmetry would be related the multi-dimensional space symmetry group included more multi-fold symmetric elements. The fuzzy Brillouin zone may be introduced, perhaps. Such molecules include the fused ring compounds and some polymers *etc*.
4. The fuzzy symmetry of some more complex dynamic molecular system would be analyzed. For these dynamic molecular systems the relationship between the fuzzy symmetry observable quanta and the dynamic parameters would be explored. The selection rules in relation to the fuzzy symmetry ought to be not only "allowed or forbidden" but also the "probability".
5. Although the MO calculation may be realized at various levels by means of Gaussian easy. However the calculation of both the membership function and representation component will be mach lengthy and tedious, meanwhile calculation at the higher level the more lengthy and tedious. Such lengthy and tedious calculation may

introduce the more error and even some mistakes, sometimes. The computer current program in relation would be established, impatiently. Until we do finish that, we may study the fuzzy symmetry at higher levels.

References

[1] Woodward, R. B. & Hoffmann, R. (1970). "The conservation of orbital symmetry", Weinheis, see also Woodward R.B. & Hoffmann R. (1969). *Angew. Chem.*, **8**, 781-853.

[2] Hargittai, I. & Hargitta, M. (2000). "In our own image: Personal symmetry in discovery", *Kluwer Academic, Plenum*, **235**.

[3] Rouvray, D. H. (1997). "Are the concepts of chemsitry all fuzzy?" in: *Concepts in Chemistry,* Edited by Rouvray D.H., Research Studies Press, *Taunton, Somerset*, 1-15.

[4] Roman, P. (1964). "Theory of Elementary Particles", North-Holland Publishing Company. 246-248.; see also Noether E. (1918). *Nach d Kel. Ges. D. Wiss..* Gottingen.

[5] Zhao, X. Z, Cai, Z. S., Wang, G. C., Pan, Y. M. & Wu, B. X. (2002). "The conservation of generalized parity: molecular symmetry and conservation rules in chemistry" *J. Mol. Structure (THEOCHEM),* **586**, 209-223. see also Zhao X.Z. (1986) "The Application of Symmetry Principle in Field Theory to Chemistry." Science Press, Beijing, PRC. (in Chinese),,

[6] Tang, A. Q. (1982). "Quantum chemistry", Science Press, Beijing, (in Chinese), 2-10. see also: In: M. Born, & J. P. Oppenheimer, (1917), *Ann. Physik.,* **84**, 457.

[7] Weissbluth, M. (1987). "Atoms and molecules", Academic Press, New York, 551-555.

[8] Zhao, X. Z. (1979). "The symmetry of sigmatropic reaction" *Acta Chimica Scinica,* (in Chinese), **37**, 247-254. *CA,* **92**, 117175b.

[9] Zhao, X. Z, Yi, X. Z., Guan, D. R., Xu, X. F., Wang, G. C., Shang, Z. F., Cai, Z. S., Pan, Y. M. (2005). "The characteristics and the relative conservation rules of the environment time-space with periodic symmetry" *J. Mol. Structure (THEOCHEM),* **713**, 87-91.

[10] Callaway, J. (1991). "Quantum theory of solid state" (2nd Ed), Academic Press, Inc. New York., 2-4; see also F. Bloch,1928, *Z. Physik.,* **52**, 555.

[11] Boer, K. W. (1990). "Survey of semiconductor physics, electrons and other particles in bulk semiconductors," *Van Nostrand Reinhold.*, New York. 697-749.

[12] Xu, G. X., Li, L. M. (1980). "*Quantum chemistry(***1***)*", Science Press, Beijing, (in Chinese), 115-119.

[13] Mezey, P. G. & Maruani, J. (1990). "A continuous extension of the symmetry concept for quasi-symmetric structures using fuzzy-set theory." *Mol. Phys.*, **69**(1), 97-113.

[14] Mezey, P. G. & Maruani, J. (1993). "The fundamental syntopy of quasi-symmetric systems: Geometric criteria and the underlying syntopy of a nuclear configuration space." *Int. J. Quantum Chem.*, **45**(2), 177-187.

[15] Mezey, P. G. (1998). "Generalized chirality and symmetry deficiency." *J. Math. Chem.*, **23**(1), 65-84.

[16] Maruani, J. & Mezey, P. G. (1987). "Le concept de《 syntopie》: une extension continue du concept de symétrie pour des structures quasi−symétriques à l'aide de la méthode des ensembles flous." *Compt. Rend. Acad. Sci. Paris* (Série II), 305,1051- 1054.

[17] Maruani, J. & Toro-Labbé, A. (1996). " Le modèle de la syntopie et l'état de transition de réactions chimiques: fonctions d'appartenance et coefficients de Brönsted pour l'isomérisation cis–trans." *Compt. Rend. Acad. Sci. Paris* (Série IIb), 323, 609-615.

[18] Mezey, P. G. (1997). "Quantum chemistry of macromolecular shape." *Int. Rev. Phys. Chem.,* **16**, 361-388.

[19] Zabrodsky, H., Peleg, S. & Avnir, D. (1993) "Continuous symmetry measures. 2. symmetry groups and the tetrahedron" *J. Am. Chem. Soc.*, **115**, 8278-8289, and references therein.

[20] Avnir, D., Zabrodsky, H., Hel-Or, H. & Mezey, P. G. (1998). "Symmetry and chirality: continuous measures" *ncyclopaedia of Computational Chemistry*, vol, 4, 2890-2901. Ed. Paul von Ragué Schleyer, Wiley: Chichester.

[21] Chauvin, R. (1994). "Chemical algebra. I: Fuzzy subgroups." *J. Math. Chem.*, **16**(1), 245-256.

[22] Chauvin, R. (1994). "Chemical algebra. II: Discriminating pairing products". *J. Math. Chem.*, **16**(1), 257-258.

[23] Zhou, X. Z, Fan, Z. X. & Zhan, J. J. (2002). "Application of fuzzy mathematics in chemistry." Changsha: *National University of Defence Technology Press*,. 325-349.

[24] Zhao, X. Z. & Xu, X. F. (2004). "The molecular fuzzy symmetry." *Acta Phys. Chim. Sci.*, 20(10): 1175-1178 (in Chinese).

[25] Zhao, X. Z., Xu, X. F., Wang, G. C., Pan, Y. M. & Cai, Z. S. (2005). "Fuzzy symmetries of molecule and molecular orbital: characterization and simple application." *Mol. Phys.*, **103**(24), 3233-3241

[26] Xu, X. F., Wang, G. C., Zhao, X. Z., Pan, Y. M., Liang, Y. X. & Shang, Z. F. (2007). "Fuzzy symmetries for linear molecules and their molecular orbitals." *J. Math. Chem.*, **41**(2), 143-160.

[27] Zhao, X. Z., Xu, X. F., Wang, G. C., Pan, Y. M., Shang, Z. F. & Li, R. F. (2007). "The fuzzy D_{2h}-symmetries of ethylene tetra-halide molecules and their molecular orbitals." *J. Math. Chem.*, **42**(2), 265-288.

[28] Zhao, X. Z., Wang, G. C., Xu, X. F., Pan, Y. M., Shang, Z. F., Li, R. F. & Li, Z. C. (2008). "The fuzzy D_{6h}-symmetries of azines molecules and their molecular orbitals." *J. Math. Chem.*, **43**(2), 485-507.

[29] Zhao, X. Z., Shang, Z. F., Wang, G. C., Xu, X. F., Li, R. F., Pan, Y. M. & Li, Z. C. (2008). "Fuzzy space periodic symmetries for polyynes and their cyano-compounds." *J. Math. Chem.*, **43**(3), 1141-1162.

[30] Zhao, X. Z., Shang, Z. F., Sun, H. W., Chen, L., Wang, G. C., Xu, X. F., Li, R. F., Pan, Y. M., Li, Z. C. (2008). "The fuzzy symmetries for linear tri-atomic B···A···C dynamic systems." *J. Math. Chem.*, **44**(1), 46-74.

[31] Zhao, X. Z., Xu, X. F., Shang, Z. F., Wang, G. C. & Li, R. F. (2008). "Fuzzy Symmetry Characteristics of Propadine Molecule." *Acta Phys. Chim. Sci.*, **24**(5), 772-780.

[32] Zhao, X. Z., Shang, Z. F., Li, Z. C., Xu, X. F., Wang, G. C., Li, R. F. & Li, Y. "Approximate symmetry characteristics using fuzzy-subset theory study for chiral transitions of allene-1,3-dihalides" *J. Math. Chemistry,* (submitted, 2009) file No.: JOMC 207R1.

[33] Li, Y., Zhao, X. Z., Xu, X. F., Shang, Z. F., Cai, Z. S., Wang, G. C. & Li, R. F. (2009). "The fuzzy *ta*/2 symmetry of straight chain conjugate polyene molecules" *Science in China* Ser. B (submitted, 2009) file No.: 032009-491.

[34] Xiao, S. W. (1992). "Basic fuzzy mathematics and its application", Aircraft Industry Press, Beijing, 201-204 (in Chinese). see also Zadeh L.A. (1965), *Inform. and Control,* **8B** , 338-353.

[35] Frisch, M. J., Trucks, G. W., Schlegel, H. B., Scuseria, G. E., Robb, M. A., Cheeseman, J. R., Zakrzewski, V. G., Montgomery, J. r. J. A., Stramann, R. E., Burant, J. C., Dapprich, J. M., Daniels, A. D., Kudin, K. N., Strain, M. C., Farkas, O., Tomasi, J., Barone, V., Cossi, M., Cammi, R., Mennuggi, B., Pomelli, C., Adamo, C., Clifford, S., Ochterski, J., Petersson, G. A., Ayala, P. Y., Cui, Q., Morokuma, K., Malick, D. K., Rabuck, A. D., Raghavachari, K., Foresman, J. B., Cioslowski, J., Ortiz, J. V., Baboul, A. G., Stefanov, B. B., Liu, G., Liashenko, A., Piskorz, P., Komaromi, I., Gomperts, R., Martin, L. R., Fox, D. J., Keith, T., Al-Laham, M. A., Peng, C. Y., Nanayakkara, A., Gonzalez, C., Challacombe, M., Gill, P. M. W., Johnson, B., Chen, W., Wong, M. W., Andres, J. L., Gonzalez, C., Head-Gordon, M., Replogle, E. S. & Pople, J. A. (1998). *Gaussian* **98**, Revision A.3, Gaussian, Inc., Pittsburgh PA.

[36] Cotton, F. A . (1999). *"Chemical application of group Theory"* John Wiley, New York, Chapter 6.

[37] Liu, J. J. (2004). "Basic quantum chemistry and its application" Higher Education Press, Beijing. (in Chinese) 121-131. see also: K. Fukui, T. Yonezawa and H. Shingu, (1952). *J. Chem. Phys.*, **20**, 722.

[38] Eyring, H., Lin, S. H. & Lin, S. M. (1980). *"Basic chemical kinetics"*, John Wiley, New York, 75.

[39] Zhao, X. Z. (1990). "Theorem of chemical reaction kinetics (II) ", chapter 7, Higher Education Press, Beijing, (in Chinese). See also: H. Eyring, H. Gershinowitz and C.E. Sun, (1935), *J. Chem. Phys.,* **3**, 786.

[40] Zhao, X. S., (2003). " *Introduction to the theory of chemical reaction,"* Chapter 3, Peking University Press, Beijing, (in Chinese).

[41] Vainshtein, B. K. (1981). *" Modern crystallography I. Symmetry of crystals, methods of structural crystals, methods of structural crystallography"* Springer-Verlag, Berlin, Heidelberg, New York.

[42] Vainshtein, B. K., Fridkin, V. M. & Indenbom, V. I. (1981). *" Modern crystallography II. Structural of crystals"*. Springer-Verlag, Berlin, Heidelberg, New York.

[43] Zhao, C. D. (1997). *"Solid state quantum chemistry"*. Higher Education Press, Beijing. (in Chinese).

[44] Wang, R. H. & Gao, K. X. (1990). *"Symmetry group of crystallography",* Science Press, Beijing. (in Chinese).

[45] Molder, U., Burk, P. & Koppel, I. A. (2001). "Quantum chemical calculations of geometries and gas-phase linear polyyne chains" *International Journal of Quantum Chemistry*, **43**(**2**), 73-85.

[46] Jiang, M. Q. (1980). *"The Rule of Homologous Linearity of Organic Compounds"*, pp 123-210 & 232-250, Science Press, Beijing, (in Chinese).

[47] Li, Z. Z. (1985). *"Solid state theory",* 4-6, Higher Education Press, Beijing. (in Chinese).

[48] Pearson, R. G. (1976). *"Symmetry rules for chemical reactions".* 91-104, John Wiley, New York.

In: Mathematical Chemistry
Editor: W.I. Hong, pp. 247-325
ISBN: 978-1-60876-894-3

Chapter 3

INFORMATION PERSPECTIVE ON MOLECULAR ELECTRONIC STRUCTURE

F. Roman Nalewajski*

Faculty of Chemistry, Jagiellonian University, Cracow, Poland

Abstract

Recent *Information-Theory* (IT) probes of the electronic structure and chemical bonds in molecules are surveyed. The electron redistribution accompanying the bond formation, from the (molecularly placed) free atoms $\{X^0\}$ of the "promolecule", exhibiting the ground-state densities $\{\rho_X^0\}$, to the bonded *Atoms-in-Molecules* (AIM) with densities $\{\rho_X\}$, is marked by the difference $\Delta\rho = \rho - \rho^0$ between the molecular ($\rho = \Sigma_X \rho_X$) and promolecular ($\rho^0 = \Sigma_X \rho_X^0$) electron densities, which mark the process final and initial stages, respectively. The entropy/information content of these distributions provides a basis for a novel IT perspective on the molecular electronic structure. The Schrödinger equations of quantum mechanics, which determine the equilibrium distribution of electrons and its time dependence for the fixed external potential due to the system nuclei, are shown to result from the constrained-extremum principles of the Fisher information, related to the system average kinetic energy. The densities of displacements in the Shannon entropy and missing information (cross entropy) relative to the promolecular reference are used as sensitive diagnostic tools for detecting the chemical bonds, and to monitor the promotion/hybridization changes the bonded atoms undergo in the molecular environment. When applied to a classical problem of density partition into AIM pieces the IT approach gives the familiar "stockholder" division scheme of Hirshfeld. The non-additive Fisher information density in the *Molecular Orbital* (MO) and *Atomic Orbital* (AO) resolutions respectively generate the *Electron Localization Function* (ELF) and *Contra-Gradience* (CG) probes for localizing electrons and for detecting the bonding regions in molecules, respectively. Representative examples of such an exploration of molecular ground-states will be presented. The *Orbital Communication Theory* (OCT) of the chemical bond uses the standard entropy/information descriptors of the Shannon theory of communication to characterize the AO scattering of electron probabilities, throughout the network of chemical bonds generated by the system occupied MO. The molecule is thus treated as an information system which propagates from the channel AO "inputs" to AO "outputs" the "signals" of electron allocations to these basis functions. The underlying conditional probabilities generated from the bond-projected superposition principle of

* E-mail address: nalewajs@chemia.uj.edu.pl (Corresponding author)

quantum mechanics are shown to be proportional to the squares of the corresponding elements of the *first*-order density matrix in the AO representation, thus being related to Wiberg's quadratic indices of the chemical bond multiplicity. Such information propagation in molecules exhibits the communication "noise", due to electron delocalization via the system chemical bonds, which effectively lowers the information content in the output-signal distribution, compared to that contained in probabilities determining the channel input-signal, molecular or promolecular. The orbital information systems are used to generate the entropic measures of the chemical bond multiplicity and their covalent/ionic composition for both the molecule as a whole and its diatomic fragments. The average *conditional-entropy*, which measures the channel communication-noise due to electron delocalization, measures the IT-*covalency* in the molecule, while the complementary descriptor of the network *mutual-information* (information-capacity) reflects the electron localization effects and measures the system IT-*ionic* component. The illustrative examples of applying these IT tools for exploring the electronic and bonding structures of representative molecules are reported and discussed. Other applications of the IT concepts to classical issues in quantum chemistry are also briefly summarized. The illustrative *Restricted-Hartree-Fock* (RHF) OCT results for diatomic fragments of selected molecules, which reproduce the Wiberg bond-order in diatomic molecules, are examined.

Keywords: Bonded atoms; Chemical bonds; Communication theory of chemical bond; Contra-gradience bond localization; Electron localization function; Entropic bond indices; Information theory; Local information probes; Non-additive Fisher information; Schrödinger equation as information rule; Variational principles of quantum mechanics.

1. Introduction

The *Information Theory* (IT) [1-6] is one of the youngest branches of the applied probability theory, in which the probability ideas have been introduced into the field of communication, control, and data processing. It has originated from the needs of practice, to create a theoretical model for the transmission of information, and evolved into an important chapter of the general theory of probability. Its foundations have been laid in 1920s by Sir R. A. Fisher [1] in his classical measurement theory, and in 1940s by C. E. Shannon [3] in his mathematical theory of communication. The electronic quantum mechanical state of a molecule is determined by the system wave function, the amplitude of the particle probability distribution, which carries the information. It is thus intriguing to explore the information content of electronic probability distribution in a molecule, and to extract from it the pattern of chemical bonds, trends in chemical reactivity, and other molecular descriptors, e.g., bond multiplicities ("orders"). In particular, it is of interest to examine the information representation of the subtle electron redistributions in chemical processes, e.g., those accompanying formation or breaking of bonds in chemical reactions. It has been amply demonstrated, e.g., [7-18], that many classical problems in chemistry can be approached anew using this novel IT perspective.

The concepts and techniques of IT have been successfully used to explore the chemical properties of molecules and their fragments and to examine the bonding patterns in both the molecular and reactive systems. For example, the displacements in the information distribution in molecules, relative to the *promolecular* reference consisting of the non-bonded constituent atoms, has been investigated [8-10] and the least-biased partition of the molecular electron distributions into the subsystem contributions, e.g., densities of bonded *Atoms-In-*

Molecules (AIM), have been investigated [8,11-17]. The IT approach has been shown to lead to the *"stockholder"* molecular fragments of Hirshfeld [18], the density pieces of which have been derived from alternative global and local variational principles of IT.

The spatial localization of specific bonds, not to mention some qualitative questions about the very existence of some chemical bonds, e.g., between the bridgehead carbon atoms in propellanes, presents another challenging problem for this novel IT treatment of molecular systems. Another important aspect of the molecular electronic structure deals with the shell-structure and electron localization in atoms and molecules. The non-additive Fisher information in *Atomic Orbital* (AO) resolution has been recently used as the *Contra-Gradience* (CG) criterion for localizing the bonding regions in molecules [19,20], while the related information density in the *Molecular Orbital* (MO) resolution has been shown [8,21] to determine the vital ingredient of the *Electron-Localization Function* (ELF) [22].

The *Communication Theory of the Chemical Bond* (CTCB) has been developed using the basic entropy/information descriptors of the *molecular information* (communication) *channels* in the AIM, orbital and local resolution levels of the electron probability distributions [7,8,23-28]. The same bond descriptors have been used to provide the information scattering perspective on the intermediate stages in the electron redistribution processes [29a], including the atom promotion via the orbital hybridization [29b], and the communication theory for excited electron configurations has been developed [30]. Moreover, the phenomenological description of equilibria in molecular subsystems has been developed [8,31], which formally follows that in ordinary thermodynamics. To summarize, the entropic probes of the molecular electronic structure have provided novel attractive tools for describing the chemical bond phenomenon in information terms.

It is the main purpose of this survey to review some of the recent developments in the orbital formulation of CTCB, called the *Orbital Communication Theory* (OCT), and in alternative local entropy/information probes of the molecular electronic structure. The importance of the non-additive effects in the chemical bond phenomena will be emphasized. We start with a short summary of the key IT concepts and techniques.

2. Entropy, Information and Communication Systems

2.1. Alternative Measures of Information

The *Shannon* (S) entropy $S[p]$ [3] in the (normalized) spatial probability distribution $p(\boldsymbol{r})$,

$$S[p] \equiv I^S[p] = -\int p(\boldsymbol{r}) \log p(\boldsymbol{r})\, d\boldsymbol{r},\ \int p(\boldsymbol{r})\, d\boldsymbol{r} = 1, \tag{1}$$

where the definite integration is over the whole range of the random (position) variable $\boldsymbol{r}$, provides a measure of the average *indeterminacy* in $p(\boldsymbol{r})$ for the locality events $\{\boldsymbol{r}\}$. As also indicated above, it also measures the average amount of information $I^S[p]$ obtained when this spatial uncertainty is removed by an appropriate localization measurement (experiment). Here the logarithm is taken to an arbitrary but fixed base: when taken to base 2, $\log = \log_2$, the information is measured in *bits*, while selecting $\log = \ln$ expresses the information in *nats*: 1 nat = 1.44 bits.

The *Fisher* (F) information [1,2] measure for locality, called the *intrinsic accuracy*, historically predates the Shannon entropy by about 25 years, being proposed in about the same time when the final form of the quantum mechanics was shaped. It emerges as an expected error in a "smart" measurement, in the context of the efficient estimators of a parameter. Fisher was the first to suggest that data samples in an experiment together with a given parametric distribution model contain the statistical information about the parameter(s). Let $p(\boldsymbol{r}|\theta)$ be the probability distribution function depending upon the parameter θ. The Fisher measure of the information contained in this probability density is defined as follows:

$$I(\theta) \equiv \int p(\boldsymbol{r}|\theta)\left(\frac{\partial \ln p(\boldsymbol{r}|\theta)}{\partial\theta}\right)^2 d\boldsymbol{r} = \int \frac{[p'(\boldsymbol{r}|\theta)]^2}{p(\boldsymbol{r}|\theta)} d\boldsymbol{r}, \qquad p'(\boldsymbol{r}|\theta) = \frac{\partial p(\boldsymbol{r}|\theta)}{\partial\theta}. \tag{2}$$

The intrinsic accuracy is a special case of this parametric measure, for the locality parameter θ, when $p(r|\,\theta) = p(r + \theta) = p(r')$. In this case the parametric Fisher information provides the information about the probability distribution itself:

$$\frac{\partial p(\boldsymbol{r}|\boldsymbol{\theta})}{\partial\boldsymbol{\theta}} = \frac{\partial \boldsymbol{r}'}{\partial\boldsymbol{\theta}}\frac{\partial p(\boldsymbol{r}')}{\partial \boldsymbol{r}'} = \frac{\partial p(\boldsymbol{r}')}{\partial \boldsymbol{r}'} = \nabla p(\boldsymbol{r}'). \tag{3}$$

Hence, for the *single*-component probability distribution

$$I(\theta) = I[p] = \int p(r')\,[\nabla \ln p(r')]^2\, dr' = \int [\nabla p(r)]^2/p(r)\, dr \equiv I^F[p]. \tag{4}$$

This Fisher information functional, reminiscent of von Weizsäcker's [32] inhomogeneity correction to the electronic kinetic energy in the Thomas-Fermi theory, characterizes the compactness (order) of the probability density $p(\boldsymbol{r})$. For example, the Fisher information in the *normal distribution* measures the inverse of its *variance*, called the *invariance*, while the complementary Shannon entropy is proportional to the logarithm of variance, thus monotonically increasing with the spread of the Gaussian distribution. The Shannon entropy and the Fisher information for locality thus describe the complementary facets of the probability density: the former reflects distribution's "spread" (a measure of uncertainty, "disorder"), while the latter measures its "narrowness" ("order"). The analytical properties of the Shannon and Fisher information functionals are quite different [2]. When extremized through variation of the probability distribution the Shannon entropy gives an exponential solution, while the Fisher information gives rises to the differential equation, and hence to multiple solutions specified by the appropriate boundary conditions.

The form of the intrinsic-accuracy functional can be simplified by expressing it as a functional of the associated classical (real) amplitude $A(\boldsymbol{r}) = \sqrt{p(\boldsymbol{r})}$ of the probability distribution $p(\boldsymbol{r})$:

$$I[p] = 4\int [\nabla A(\boldsymbol{r})]^2\, d\boldsymbol{r} \equiv I[A]. \tag{5}$$

It is naturally generalized into the realm of the complex probability amplitudes one encounters in quantum mechanics, i.e., the system wave-functions [2,19]. For the simplest case of the spinless *one*-particle system, when $A(r) = \psi(r)$ and $p(r) = \psi^*(r)\psi(r) = |\psi(r)|^2$,

$$I[\psi] = 4\int|\nabla \psi(r)|^2\, dr = 4\int \nabla \psi^*(r)\cdot \nabla \psi(r)\, dr \equiv \int f(r)\, dr \tag{6}$$

Therefore, $I[A]$ and $I[\psi]$ measure the gradient content in the amplitude of the probability-density. Its extension to the *multi*-component (vector) probabilities $p(r) \equiv \{p_n(r)\}$, expressed in terms of the associated probability amplitudes $A(r) \equiv \{A_n(r) \equiv p_n(r)^{1/2}\}$ or $\psi \equiv \{\psi_n(r), p_n(r) = |\psi_n(r)|^2\}$ reads [2]:

$$I[\boldsymbol{p}] = \sum_n I[p_n] = 4\sum_n \int[\nabla A_n(\boldsymbol{r})]^2 d\boldsymbol{r} = \sum_n I[A_n] \equiv I[\boldsymbol{A}] \text{ or}$$

$$I[\boldsymbol{p}] = 4\sum_n \int|\nabla \psi_n(\boldsymbol{r})|^2 d\boldsymbol{r} = \sum_n I[\psi_n] \equiv I[\boldsymbol{\psi}]. \tag{7}$$

An important generalization of Shannon's entropy, called the *relative* (*cross*) *entropy*, also known as the *entropy deficiency*, *missing information* or the *directed divergence*, has been proposed by *Kullback* and *Leibler* (KL) [4,5]. It measures the *information "distance"* between the two (normalized) probability distributions for the same set of events. For example, in the discrete probability scheme A = $[a, P(a)]$, involving events, identified by the admissible values of the discrete random variable $a = \{a_i\}$, and their probabilities $P(a) = \{P(a_i) = p_i\} \equiv p$, this discrimination information in p with respect to the *reference* distribution $p^0 = \{P(a_i^0) = p_i^0\}$ reads:

$$\Delta S(p|p^0) \equiv I^{KL}(p|p^0) = \Sigma_i\, p_i \log(p_i/p_i^0) \geq 0, \tag{8}$$

In the continuous distribution case, e.g., A = $\{r, p(r)\}$, this directed-divergence measure of the entropy deficiency in the probability density $p(r)$, relative to the prior distribution $p^0(r)$, is defined by the related functional:

$$\Delta S[p|p^0] \equiv I^{KL}[p|p^0] = \int p(r) \log[p(r)/p^0(r)]\, dr \geq 0. \tag{9}$$

For individual events the logarithm of probability ratio, $I_i = \log(p_i/p_i^0)$, $I(r) = \log[p(r)/p^0(r)]$ or for the *one*-dimensional continuous distribution $I(x) = \log[p(x)/p^0(x)]$, called the *surprisal*, provides a measure of the information in the current distribition relative to the reference distribution. The equality in the last two equations takes place when the surprisal identically vanishes for all events, i.e., when the compared probability schemes are identical.

The non-negative character of the entropy deficiency in $p(x)$ relative to $p^0(x)$,

$$\Delta S[p\,|\,p^0] = \int p(x) \log[p(x)/p^0(x)]\, dx \equiv \int p(x)\, I(x)\, dx, \tag{10}$$

directly follows from the observation that the line $y = z - 1$ lies above the curve $y = \log z$, with two functions having equal (zero) value only for $z = 1$. Taking $z(x) = p^0(x)/p(x)$ and using the condition of the probability normalization then gives:

$$\Delta S[p\,|\,p^0] = -\int p(x) \log z(x)\, dx \geq \int p(x)\, [z(x) - 1]\, dx = \int[p^0(x) - p(x)]\, dx = 0. \tag{11}$$

Therefore, the entropy deficiency provides a measure of an information resemblance between the two compared probability schemes. The more the two probability distributions differ from one another, the larger the information distance. As we have remarked before, the Shannon entropy measures a "*disorder*" (uncertainty, indeterminacy, "smoothness") of the probability distribution. On a finite interval, the distribution possessing the highest entropy is the uniform distribution, and any deviation from uniformity indicates the "perturbing" presence of *"order"*. The Kullback-Leibler measure, i.e., the *referenced* Shannon's entropy, provides a similar description, but in reference to some prior distribution.

Notice, however, that the surprisal itself becomes negative, when the current probability is lower than the reference value. Therefore, the directed divergence functional is not symmetrical with respect to the two probability distributions involved, and exhibits negative values of the integrand. To avoid this limitation Kullback (K) [5] has proposed an alternative measure called the *divergence*, defined by the symmetrized combination of the two admissible entropy deficiencies, which gives rise to a *non*-negative integrand in the continuous case:

$$\Delta S(p, p^0) \equiv I^K(p, p^0) = \Delta S(p|p^0) + \Delta S(p^0|p) = \Sigma_i\,(p_i - p_i^0)\, I[p_i/p_i^0] \equiv \Sigma_i\, \Delta p_i\, I_i,\ \Delta p_i\, I_i \geq 0,$$

$$\Delta S[p, p^0] \equiv I^K[p, p^0] = \Delta S[p|p^0] + \Delta S[p^0|p] = \int [p(r) - p^0(r)]\, I(r)\, dr \equiv \int \Delta p(r)\, I(r)\, dr,\ \Delta p(r)\, I(r) \geq 0. \tag{12}$$

2.2. Dependent Probability Distributions

For two mutually dependent (discrete) probability distributions,

$$P(a) = \{P(a_i) = p_i\} \equiv p \text{ and } P(b) = \{P(b_j) = q_j\} \equiv q, \tag{13}$$

determining the probability schemes A = $[a, P(a)]$ and B = $[b, P(b)]$, respectively, we decompose the joint-probabilities $P(a \otimes b) = \{P(a_i b_j) = \pi_{i,j}\} \equiv \pi$ of the simultaneous events $a \wedge b = \{a_i \wedge b_j \equiv a_i b_j\} = a \otimes b$, as the products $\{\pi_{i,j} = p_i\, P(j|i)\}$ of the *marginal* probability $p_i = P(a_i)$ of ith event in the a set of events, and the *conditional* probability $P(j|i) = P(a_i b_j)/P(a_i)$ of the event b_j of the other set b of outcomes, given that the event a_i has already occurred. The relevant normalization conditions for the joint probabilities π and the conditional probabilities $P(b|a) = \{P(j|i)\}$ then read:

$$\Sigma_j\, \pi_{i,j} = p_i,\ \Sigma_i\, \pi_{i,j} = q_j,\ \Sigma_i \Sigma_j\, \pi_{i,j} = 1,\ \Sigma_j\, P(j|i) = 1,\ i = 1, 2, ..., n. \tag{14}$$

In this scenario the Shannon entropy of the product distribution $\pi = \{\pi_{i,j}\}$,

$$\begin{aligned} S(\pi) &= -\Sigma_i \Sigma_j\, \pi_{i,j} \log \pi_{i,j} = -\Sigma_i \Sigma_j\, p_i\, P(j|i)\, [\log p_i + \log P(j|i)] \\ &= -[\Sigma_j\, P(j|i)]\, \Sigma_i\, p_i \log p_i - \Sigma_i\, p_i\, [\Sigma_j\, P(j|i) \log P(j|i)] \\ &\equiv S(p) + \Sigma_i\, p_i\, S(q|i) \equiv S(p) + S(q|p), \end{aligned} \tag{15}$$

is seen to be given by the sum of the average entropy in the marginal probability distribution, $S(p)$, and the average *conditional entropy* in q given p:

$$S(q|p) = -\Sigma_i\, p_i\, [\Sigma_j\, P(j|i) \log P(j|i)] \equiv \Sigma_i\, p_i\, S(q|i). \tag{16}$$

The latter represents the extra amount of uncertainty about the occurrence of events b, given that the events a are known to have occurred. In other words: the amount of information obtained as a result of simultaneously observing the events a and b of the two discrete probability distributions $p = P(a)$ and $q = P(b)$, respectively, equals to the amount of information observed in one set, say a, supplemented by the extra information provided by the occurrence of events in the other set b, when a are known to have occurred already. This is qualitatively illustrated in Scheme 1 (see also the upper part of Scheme 4).

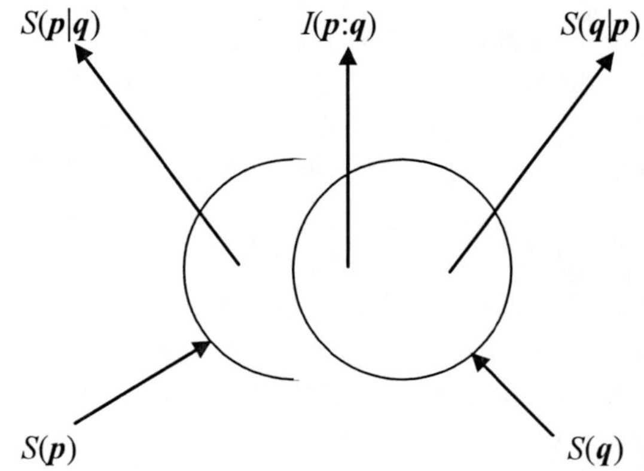

Scheme 1. A qualitative diagram of the conditional entropy and mutual information quantities of two dependent probability distributions p and q. Two circles enclose the areas representing the entropies $S(p)$ and $S(q)$ of the two *separate* distributions. The common (overlap) area of the two circles corresponds to the mutual information $I(p:q)$ in two distributions. The remaining parts of two circles represent the corresponding conditional entropies $S(p|q)$ and $S(q|p)$, measuring the residual uncertainty about events in one set, when one has the full knowledge of the occurrence of the events in the other set of outcomes. The area enclosed by the envelope of the two overlapping circles then represents the entropy of the "product" (joint) distribution: $S(\pi) = S(\mathrm{P}(a \otimes b)) = S(p) + S(q) - I(p:q) = S(p) + S(q|p) = S(q) + S(p|q)$.

Clearly, by using the other probability distribution $q = \{P(b_i)\}$ as the marginal one, one arrives at the alternative expression for $S(\pi)$:

$$S(\pi) = S(q) + \textstyle\sum_j q_j\, S(p|j) \equiv S(q) + S(p|q), \tag{17}$$

where $S(p|q) = \sum_j q_j\, S(p|j)$ and the conditional probabilities $P(a|b) = \{P(i|j)\}$ are normalized to satisfy the conditions: $\sum_i P(i|j) = 1, j = 1, 2, ..., m$. Here, the average conditional entropy $S(p|q)$ represents the residual uncertainty about a, when b are known to have been observed. Properties of the conditional entropy are similar to those of the information entropy itself, since the conditional probability is a probability measure.

The common amount of information in two events a_i and b_j, $I(i:j)$, measuring the information about a_i provided by the occurrence of b_j, or the information about b_j provided by the occurrence of a_i, is called the *mutual information* in two events:

$$I(i:j) = \log[P(a_ib_j)/P(a_i)P(b_j)] = \log[\pi_{i,j}/(p_i\, q_j)] \equiv \log[P(i|j)/p_i] \equiv \log[P(j|i)/q_j] = I(j:i). \tag{18}$$

The mutual information in two events may take on any real value, positive, negative, or zero. It vanishes, when both events are independent, i.e., when the occurrence of one event does not influence (or condition) the probability of the occurrence of the other event, and it is negative when the occurrence of one event makes the nonoccurrence of the other event more likely.

It also follows from the preceding equation that

$$I(i:j) = I(i) - I(i|j) = I(j) - I(j|i) = I(i) + I(j) - I(i\wedge j) \text{ or } I(i\wedge j) = I(i) + I(j) - I(i:j), \quad (19)$$

where the self-information of the joint event $I(i\wedge j) = -\log\pi_{i,j}$. Thus, the information in the joint occurrence of two events a_i and b_j is the information in the occurrence of a_i plus that in the occurrence of b_j minus the mutual information. Clearly, for independent events $I(i, j) = I(i) + I(j)$, so that $I(i:j) = 0$.

The mutual information of an event with itself defines its *self-information*: $I(i:i) \equiv I(i) = \log[P(i|i)/p_i] = -\log p_i$, since $P(i|i) = 1$. It vanishes when $p_i = 1$, i.e., when there is no uncertainty about the occurrence of a_i, so that the occurrence of this event removes no uncertainty, hence conveys no information. This quantity provides a measure of the uncertainty about the occurrence of the event, i.e., the information received when the event occurs. The Shannon entropy of Eq. (1) can be thus interpreted as the mean value of the self-informations in all individual events defining the probability distribution, e.g., $S(p) = -\sum_i p_i \log p_i \equiv \sum_i p_i\, I(i)$.

One defines the *average* mutual information in two probability distributions, $I(p:q)$, as the π-weighted mean value of the mutual information quantities for the individual joint events:

$$I(p:q) = \sum_i \sum_j \pi_{i,j}\, I(i:j) = \sum_i \sum_j \pi_{i,j} \log(\pi_{i,j}/\pi_{i,j}^{0})$$
$$= S(p) + S(q) - S(\pi) = S(p) - S(p|q) = S(q) - S(q|p) \geq 0. \quad (20)$$

where the equality holds only for the independent distributions, when $\pi_{i,j} = \pi_{i,j}^{0} \equiv p_i q_j$. These average entropy/information relations are also illustrated in Scheme 1. The non-negative character of this average information measure can again be demonstrated using the inequality $\log z \leq z - 1$: for $z = \pi_{i,j}^{0}/\pi_{i,j}$ the normalization conditions of π and π^0 then give:

$$I(p:q) = \sum_i \sum_j \pi_{i,j} \log(\pi_{i,j}/\pi_{i,j}^{0}) \geq \sum_i \sum_j (\pi_{i,j} - \pi_{i,j}^{0}) = 0. \quad (21)$$

Indeed, the amount of uncertainty in q can only decrease, when p has been known beforehand, $S(q) \geq S(q|p) = S(q) - I(p:q)$, as indeed seen in Scheme 1, with equality being observed only when the two sets of events are independent (non-overlapping). It should be observed that the average mutual information quantity is an example of the entropy deficiency functional, which measures the missing information between the joint probabilities $P(a\otimes b) = \pi$ of the *dependent* events a and b, and the joint probabilities $P^{ind.}(a\otimes b) = \pi^0 \equiv p\otimes q$ for independent events: $I(p:q) = \Delta S(\pi|\pi^0)$. The average mutual information thus measures a degree of a dependence between events defining the two probability schemes. A similar information-distance interpretation can be given to the conditional entropy: $S(p|q) = S(p) - \Delta S(\pi|\pi^0)$.

2.3. Communication Systems

We continue this short overview with the key entropy/information descriptors of a transmission of signals in the communication systems [3,6]. The basic elements of such a "device" are shown in Scheme 2. The *input signal* emitted from n "inputs" $a = (a_1, a_2, \ldots, a_n)$ of the channel *source* (A) is characterized by the input probability distribution $P(a) = p = (p_1, p_2, \ldots, p_n) \equiv P(\mathrm{A})$. It can be received at m "outputs" $b = (b_1, b_2, \ldots, b_m)$ of the system *receiver* (B). The distribution of the output signal among the detection "events" b gives rise to the *output* probability distribution $P(b) = q = (q_1, q_2, \ldots, q_m) \equiv P(\mathrm{B})$. The transmission of signals is randomly disturbed within the communication system, thus exhibiting a typical communication *noise* since the signal sent at the given input can be in general received with a non-zero probability at several outputs. This feature of communication systems is described by the conditional probabilities of the outputs given inputs, $\mathrm{P(B|A)} = \{P(b_j|a_i) = P(a_i \wedge b_j)/P(a_i) \equiv P(j|i)\}$, or the conditional probabilities of the inputs, given outputs, $\mathrm{P(A|B)} = \{P(a_i|b_j) = P(a_i \wedge b_j)/P(b_j) \equiv P(i|j)\}$, where $P(a_i \wedge b_j) \equiv \pi_{i,j}$ stands for the probability of the joint occurrence of the specified pair of the output and input events.

The Shannon entropy of the input (source) probabilities p, $H(\mathrm{A}) \equiv S(p)$, determines the channel *a priori* entropy. The average *conditional entropy* $H(\mathrm{B|A}) \equiv S(q|p)$, of outputs given inputs, is determined by the scattering probabilities $\mathrm{P(B|A)} = \{P(b_j|a_i) \equiv P(j|i)\} \equiv \mathrm{P}(b|a)$. It measures the average noise in the *"forward"* transmission of signals from a to b. The so called *a posteriori* entropy of the input, given output, $H(\mathrm{A|B}) \equiv S(p|q)$, is similarly defined by the conditional probabilities of the *"reverse"* probability scattering from b to a: $\mathrm{P(A|B)} = \{P(a_i|b_j) = P(i|j)\} \equiv \mathrm{P}(a|b)$. It reflects the residual indeterminacy about input signal, when the output signal has already been received.

The average conditional entropy $H(\mathrm{A|B}) \equiv S(p|q)$ thus measures the indeterminacy of the source with respect to the receiver, while the conditional entropy $H(\mathrm{B|A}) \equiv S(q|p)$ reflects the uncertainty of the receiver relative to the source. Hence, an observation of an output signal provides on average the amount of information given by the difference between the *a priori* and *a posteriori* uncertainties, $S(p) - S(p|q) = I(p{:}q) \equiv I(\mathrm{A{:}B})$, which defines the *mutual information* in the source and receiver. In other words, the mutual information measures the net amount of information transmitted through the communication channel, while the conditional information $S(p|q)$ reflects a portion of $S(p)$ transformed into the communication "noise" as a result of the input signal passing through the information channel. Accordingly, $S(q|p)$ reflects the noise part of $S(q)$: $S(q) = S(q|p) + I(p{:}q)$.

Consider, for example, the *Symmetric Binary Channel* (SBC) of Scheme 3, consisting of two inputs and two outputs with the input probabilities $p = (x, 1 - x)$ and the symmetric conditional probability matrix $\mathrm{P(B|A)} = \begin{bmatrix} 1-\omega & \omega \\ \omega & 1-\omega \end{bmatrix}$. Its input entropy is determined by the *Binary Entropy Function* (BEF) $H(x)$, $H(\mathrm{A}) = -x \log x - (1 - x) \log(1 - x) \equiv H(x)$ (see Figure 1), which defines the channel *a priori* entropy. The system output entropy is also generated by BEF, $H(\mathrm{B}) = H(z(x, \omega))$, where $z(x, \omega) \equiv q_2 = x\omega + (1 - x)(1 - \omega)$. One also finds that the channel conditional entropy $H(\mathrm{B|A}) = H(\omega)$ measures its average communication noise in the forward transmission of signals and hence the mutual information between the system inputs and outputs, $I(\mathrm{A{:}B}) = S(\mathrm{B}) - S(\mathrm{B|A}) = H[z(x, \omega)] - H(\omega)$, reflects the net information flow in SBC. These relations are illustrated in Figure 1. It should be observed that z always lies between ω and $1-\omega$, so that $H(z) = H(1 - z) \geq H(\omega) = H(1 - \omega)$. This demonstrates a *non*-negative character of the mutual information, represented by the overlap areas between the two entropy circles in qualitative diagrams of Schemes 1 and 4.

Input (Source): **A** Communication network: **P(B|A)** *Output* (Receiver): **B**

a_1 b_1

a_2 b_2

....

$p_i \rightarrow a_i$ —— $P(b_j|a_i) \equiv P(j|i)$ ——→ $b_j \rightarrow q_j$

....

a_n b_m

Scheme 2. A schematic diagram of the communication system, characterized by the probability vectors: $P(a) = \{P(a_i)\} = p = (p_1, \ldots, p_n) \equiv P(\mathrm{A})$, of the channel *"input"* events $a = (a_1, \ldots, a_n)$ in the system *source* A, and $P(b) = \{P(b_j)\} = q = (q_1, \ldots, q_m) \equiv P(\mathrm{B})$, of the *"output"* events $b = (b_1, \ldots, b_m)$ in the system *receiver* B. The transmission of signals via a network of communication channels is described by the $(n{\times}m)$-matrix of the conditional probabilities $\mathrm{P(B|A)} = \{P(b_j|a_i) \equiv P(j|i)\} \equiv P(b|a)$, of observing different "outputs" (*columns*, $j = 1, 2, \ldots, m$), given the specified "inputs" (*rows*, $i = 1, 2, \ldots, n$). For clarity, only a single "forward" scattering $a_i{\rightarrow}b_j$ is shown in the diagram.

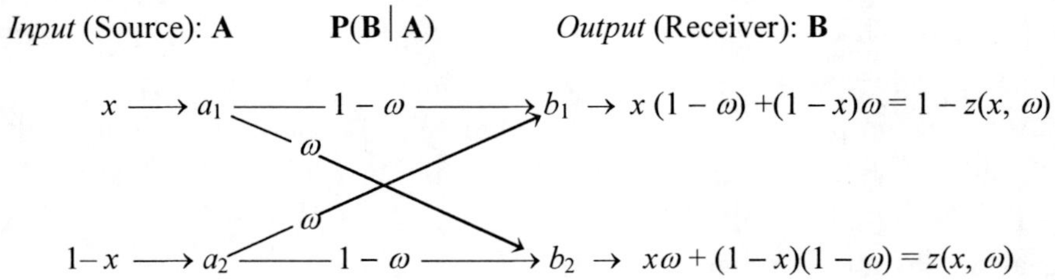

Scheme 3. The symmetric binary channel (SBC).

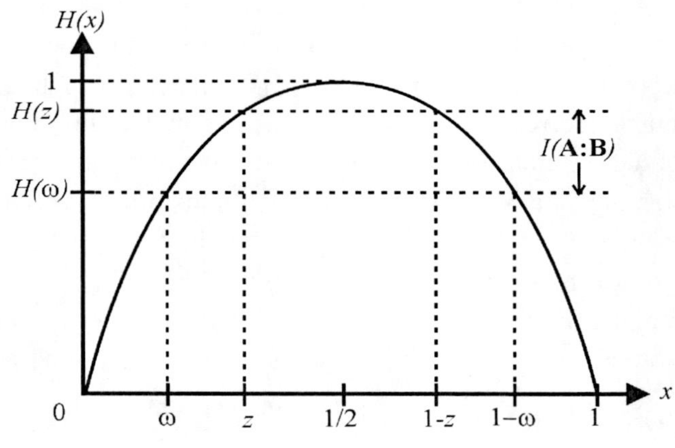

Figure 1. The binary entropy function (in bits), $H(x) = -x\log_2 x - (1-x)\log_2(1-x)$, and the geometric interpretation of the conditional entropy $H(\mathrm{B|A}) = H(\omega)$ and mutual information $I(\mathrm{A{:}B}) = H(z) - H(\omega)$ in the SBC of Scheme 3.

The amount of information $I(\mathrm{A{:}B})$ flowing through SBC thus depends on both the conditional probability parameter ω, characterizing the communication system itself, and on the input probability parameter x, which determines the way the channel is used or probed. For $x = 0$ (or 1) $H(z) = H(\omega)$ and thus $I(\mathrm{A{:}B}) = 0$, i.e., there is no net flow of information from the source to the receiver whatever. For $x = ½$, when the two input signals are equally probable, one finds $H(z) = 1$ bit, thus giving rise to the maximum value of the channel mutual information, determining the system transmission *capacity*: $C(\omega) \equiv \max_A I(A{:}B) = \max_x\{H[z(x, \omega)] - H(\omega)\} = 1 - H(\omega)$ (bits). Hence, for $\omega = ½$, the information capacity of SBC identically vanishes.

2.4. Several Probability Schemes

The lower part of Scheme 4 shows a variety of the entropy/information descriptors of three dependent (overlapping) probability schemes [6]: A = [a, $P(a)$], B = [b, $P(b)$], and C = [c, $P(c)$], while the diagram upper part summarizes quantities characterizing the *two*-distribution case of Scheme 1. In what follows we denote by a, b and c, the representative events in these three probability schemes: $a \in a$, $b \in b$ and $c \in c$. These entropy/information data now involve the conditional (relative) entropies with respect to a single or two probability schemes, the mutual information contained in two or three probability distributions, and the conditional mutual-information characteristics of the three probability schemes. The diagrams in the lower part of Scheme 4 demonstrate the *additivity* of the information quantities in the three-probability scenario, which may now involve two *output*-signals resulting from a single *input*-signal in the communication channel, or the repeated, two *input*-signals transmitted in the noisy information system. For example, the information about the input signal A obtained by observing two output signals B and C is measured by the difference of the Shannon (*a priori*) entropy H(A) and the conditional (double *a posteriori*) entropy in A given B and C, i.e., the indeterminacy of A relative to (B *and* C), which defines the *three*-scheme mutual-information quantity

$$I(\mathrm{A{:}BC}) = \sum_{a\in a}\sum_{b\in b}\sum_{c\in c} P(a\wedge b\wedge c)\log\frac{P(a\wedge b\wedge c)}{p(a)P(b\wedge c)} = \sum_{a\in a}\sum_{b\in b}\sum_{c\in c} P(a\wedge b\wedge c)\log\frac{P(a|b\wedge c)}{p(a)}$$

$$= H(\mathrm{A}) - H(\mathrm{A|BC}) = S(P(a)) - S(P(a)|\mathrm{P}(b\wedge c)) \equiv I(P(a){:}\mathrm{P}(b\wedge c)), \tag{22}$$

where

$$H(\mathrm{A|BC}) = -\sum_{a\in a}\sum_{b\in b}\sum_{c\in c} P(a\wedge b\wedge c)\log P(a|b\wedge c) \equiv S(P(a)|P(b\wedge c)). \tag{23}$$

The mutual information quantity of Eq. (22) can be alternatively expressed in terms of other entropic quantities shown in the qualitative diagrams of Scheme 4. For example, it is seen to measure the sum of elementary amounts of information in the three or two probability schemes or it can be expressed in terms of the relevant entropies:

$$I(\mathrm{A{:}BC}) = I(\mathrm{A{:}B|C}) + I(\mathrm{A{:}C|B}) + I(\mathrm{A{:}B{:}C})$$

$$\equiv I(P(a){:}P(b)|P(c)) + I(P(a){:}P(c)|P(b)) + I(P(a){:}P(b){:}P(c))$$

$$= I(\mathrm{A{:}B}) + I(\mathrm{A{:}C|B}) \equiv I(P(a){:}P(b)) + I(P(a){:}P(c)|P(b))$$

$$= H(\mathrm{AB}) - H(\mathrm{A|BC}) - H(\mathrm{B|A}) \equiv S(\mathrm{P}(a\wedge b)) - S(P(a)|\mathrm{P}(b\wedge c)) - S(P(b)|P(a)). \tag{24}$$

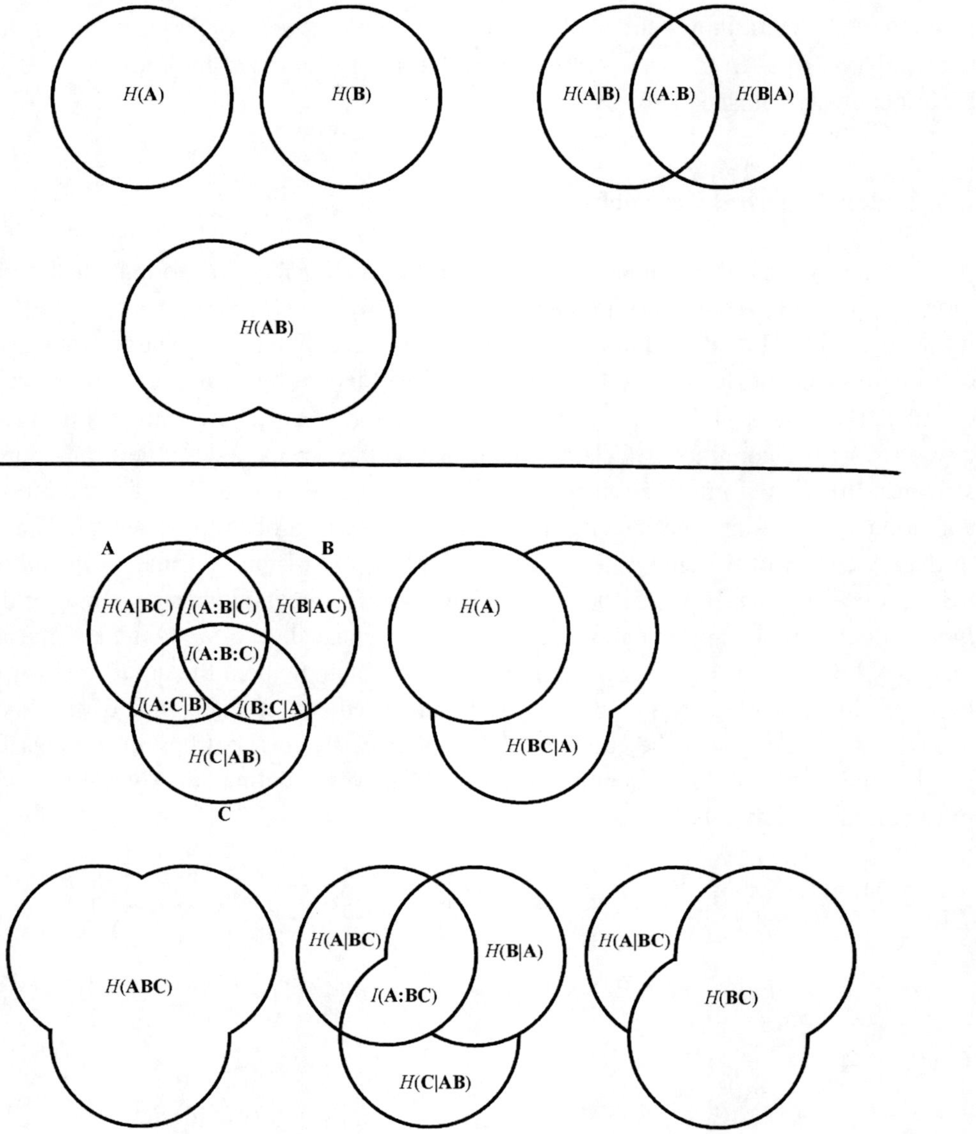

Scheme 4. A variety of the conditional-entropy and mutual-information descriptors of two (upper part) and three (lower part) dependent probability schemes [6]. Here, the individual circles $H(\mathrm{A}) = S(P(a))$, $H(\mathrm{B}) = S(P(b))$ and $H(\mathrm{C}) = S(P(c))$ represent the Shannon entropies of the separate probability distributions, $H(\mathrm{AB}) = S(\mathrm{P}(a\otimes b)) = S(\mathrm{P}(a\wedge b))$, $H(\mathrm{B}|\mathrm{A}) = S(P(b)|P(a))$, $I(\mathrm{A{:}B}) = I(P(a){:}P(b))$, etc.

Here, the mutual information in two probability distributions, given the third distribution, reads:

$$I(\mathrm{A{:}B}|\mathrm{C}) = H(\mathrm{A}|\mathrm{C}) - H(\mathrm{A}|\mathrm{BC}) = \sum_{a\in\boldsymbol{a}}\sum_{b\in\boldsymbol{b}}\sum_{c\in\boldsymbol{c}} P(a\wedge b\wedge c)\log\frac{P(a|b\wedge c)}{P(a|c)}$$

$$= \sum_{a\in\boldsymbol{a}}\sum_{b\in\boldsymbol{b}}\sum_{c\in\boldsymbol{c}} P(a\wedge b\wedge c)\log\frac{P(a\wedge b|c)}{P(a|c)P(b|c)} = \sum_{a\in\boldsymbol{a}}\sum_{b\in\boldsymbol{b}}\sum_{c\in\boldsymbol{c}} P(a\wedge b\wedge c)\log\frac{P(a\wedge b\wedge c)p(c)}{P(a\wedge c)P(b\wedge c)}, \quad (25)$$

The mutual information in three probability schemes,

$$I(\mathbf{A{:}B{:}C}) = I(\mathbf{A{:}B}) - I(\mathbf{A{:}B}|\mathbf{C}) = I(\mathbf{A{:}C}) - I(\mathbf{A{:}C}|\mathbf{B}) = I(\mathbf{B{:}C}) - I(\mathbf{B{:}C}|\mathbf{A})$$
$$= \sum_{a\in\mathbf{a}}\sum_{b\in\mathbf{b}}\sum_{c\in\mathbf{c}} P(a\wedge b\wedge c)\log\frac{P(a\wedge b)P(a\wedge c)P(b\wedge c)}{p(a)p(b)p(c)P(a\wedge b\wedge c)}$$
$$= H(\mathbf{A}) + H(\mathbf{B}) + H(\mathbf{C}) - H(\mathbf{AB}) - H(\mathbf{AC}) - H(\mathbf{BC}) + H(\mathbf{ABC}), \qquad (26)$$

which may assume negative values, is represented in Scheme 4 by the common area of three entropy circles. This expression can be straightforwardly generalized for a larger number of probability schemes. For example, in the *four*-scheme case one finds [6]:

$$I(\mathrm{A{:}B{:}C{:}D}) = I(\mathrm{A{:}B{:}C}) - I(\mathrm{A{:}B{:}C}|\mathrm{D}) = H(\mathrm{A}) + H(\mathrm{B}) + H(\mathrm{C}) + H(\mathrm{D}) - H(\mathrm{AB}) - H(\mathrm{AC}) - H(\mathrm{BC}) - H(\mathrm{AD}) - H(\mathrm{BD}) - H(\mathrm{CD}) + H(\mathrm{ABC}) + H(\mathrm{ABD}) + H(\mathrm{ACD}) + H(\mathrm{BCD}) - H(\mathrm{ABCD}). \qquad (27)$$

The conditional entropy of Eq. (23),

$$H(\mathrm{A}|\mathrm{BC}) \equiv H(\mathrm{A}|\mathrm{D}) = H(\mathrm{AD}) - H(\mathrm{D}) = S(P(a)|\mathrm{P}(b\wedge c)) \equiv S(P(a)|P(d)), \qquad (28)$$

where the "product" probability scheme D = BC = $[d, P(d)] = [b\wedge c, \mathrm{P}(b\wedge c)]$, measures the average noise in the $d\rightarrow a$ propagation of information in the underlying communication system. This entropy can be compared (see Scheme 4) with the two-scheme relative entropy $H(\mathrm{A}|\mathrm{B}) = H(\mathrm{A}) - I(\mathrm{A{:}B})$, which measures the communication noise in the $b\rightarrow a$ probability scattering [see also Eq. (22)]:

$$H(\mathbf{A}|\mathbf{BC}) = -\sum_{a\in\mathbf{a}}\sum_{b\in\mathbf{b}}\sum_{c\in\mathbf{c}} P(a\wedge b\wedge c)\log\frac{P(a\wedge b\wedge c)}{P(b\wedge c)}$$
$$= -\sum_{a\in\mathbf{a}}\sum_{b\in\mathbf{b}}\sum_{c\in\mathbf{c}} P(a\wedge b\wedge c)\log\frac{P(a\wedge b\wedge c)p(a)}{p(a)P(b\wedge c)}$$
$$= H(\mathrm{A}) - I(\mathrm{A{:}BC}) < H(\mathrm{A}|\mathrm{B}), \quad (29)$$

since $I(\mathrm{A{:}D}) > I(\mathrm{A{:}B})$. The information propagation from the product events in the $d\rightarrow a$ channel thus results in less noise (IT-covalency) compared to that characterizing the $b\rightarrow a$ network.

Accordingly, the opposite trend must be detected in the complementary mutual-information quantities of Eq. (22):

$$I(\mathrm{A{:}BC}) = H(\mathrm{A}) - H(\mathrm{A}|\mathrm{D}) > I(\mathrm{A{:}B}) = H(\mathrm{A}) - H(\mathrm{A}|\mathrm{B}). \qquad (30)$$

Therefore, the product-event scattering preserves in D a larger portion of the information content of A, compared to that conserved in the scheme B alone, due to the extra information in C about A given B, as measured by $I(\mathrm{A{:}C}|\mathrm{B})$. In other words, the $d\rightarrow a$ probability propagation results in a larger IT-ionic component, compared to the $b\rightarrow a$ scattering.

To summarize, the $(d = b\wedge c)\rightarrow a$ communications appear less noisy (more deterministic) campared to the $b\rightarrow a$ probability propagation. Therefore, the *a posteriori* indeterminacy of A relative to B, $H(\mathrm{A}|\mathrm{B})$, exceeds the *double a posteriori* entropy $H(\mathrm{A}|\mathrm{BC})$, which measures the indeterminacy of A relative to (B and C), while the information about A provided by B alone,

I(A:B), is lower than that provided by (B and C), I(A:BC), by the extra amount of information about A provided by C, when B is known beforehand, I(A:C|B).

2.5. Variational Principles

Let us summarize the properties of the Shannon entropy, which might be expected of a reasonable measure of uncertainty/information. For the simplest case of a finite, discrete probability distribution p, $S(p) = 0$ if and only if one of the probabilities in p is one and all the others are zero. Indeed, this is the case, where the outcome of the experiment carries no uncertainty, since the result can be predicted beforehand with complete certainty. In all other cases $S(p) > 0$. For the fixed number of n outcomes the probability scheme exhibiting the most uncertainty, i.e., the largest value of the entropy, is the one with equally likely outcomes: $p_i = 1/n$.

This "maximum smoothness" property of the optimum probability distribution is summarized by the constrained variational principle of Shannon and Jaynes, called the *Maximum Entropy* (ME) rule: given data $F^0 = \{F_i^0\}$, the probability distribution p [or $p(r)$], which describes these constraints most objectively, must maximize the entropy $S(p)$ (or $S[p]$) with respect to all p's (p's) satisfying F^0:

$$\delta\{S[p] - \sum_i \lambda_i[F_i(p) - F_i^0]\} = \delta\{S[p] - \sum_i \lambda_i F_i(p)\} = 0 \text{ or } \delta\{S[p] - \sum_i \lambda_i F_i[p] = 0, \quad (31)$$

where $\{\lambda_i\}$ denote the relevant Lagrange multipliers enforcing the auxiliary conditions, to be eventually determined from the constraint values. Thus, the entropy maximization results in the most "evenly spread" of all distributions consistent with the imposed constraints. This principle represents a device allowing one to assimilate in the optimum probability distribution the physical information contained in the constraints in the most unbiased manner possible.

The ME principle represents a powerful method for determining the equilibrium probability distribution of a physical system or process, given some information about it. It has been successfully used to reconstruct the equilibrium thermodynamics, asserting at the same time the principle applicability to far more general problems. The ME rule involves the maximization of Shannon's entropy subject to the imposed constraints and provides a unifying principle in statistical physics, allowing a construction of thermodynamic laws based upon statistical inference and the unbiased assimilation of available data. In other words, the entropy becomes the starting point in the construction of the statistical mechanics, instead of being identified in the end as a byproduct of the energy-centered arguments. The Shannon-Jaynes principle is the uniquely correct method for inductive inference, when a new information is given in the form of the statistical expectation values.

Kullback has proposed the generalization of the ME rule to problems involving the reference probability distribution(s), which can be called the *Maximum-Resemblance* principle or the *Minimum Entropy Deficiency* (MED) rule, as another unbiased procedure for assimilating in the optimum probability distribution the constraints of available experimental data, and thus its highest similarity to the reference distribution afforded by the constraints. Suppose that an experiment has been performed, which yields the expectation values of several functions of the discrete probability distribution p, $\{F_i(p) = F_i^0\}$, or functionals of the probability density $p(r)$ in the continuous case, $\{F_i[p] = F_i^0\}$. The Kullback principle asserts that the task of assimilating this information in the optimum probability distribution, which is to resemble the reference distribution as much as possible, can be accomplished by minimizing the entropy deficiency subject to these constraints:

$$\delta\{\Delta S[p\,|\,p^0] - \Sigma_i \lambda_i F_i(p)\} = 0 \text{ or } \delta\{\Delta S[p\,|\,p^0] - \Sigma_i \lambda_i F_i[p]\} = 0. \tag{32}$$

Alternatively, the Fisher measure of information, which reflects the overall "*order*" ("sharpness") of the probability distribution, can be used in the variational procedure of assimilating the available data in the probability distribution:

$$\delta\{I[p] - \Sigma_i \lambda_i F_i[p]\} = 0. \tag{33}$$

This principle of the *Extreme* (minimum) *Physical Information* (EPI) is closely related to the theory of measurement [2]. The maximum Shannon entropy and the minimum Fisher information principles, for a given set of constraints, may have coincident solutions. The Maxwell-Boltzmann velocity dispersion law of the equilibrium statistical thermodynamics results from both information principles, with the Fisher information principle generating additional, non-equilibrium solutions as subsidiary minima, with the absolute minimum being attained by the Maxwell-Boltzmann solution. The coincidence is observed at the equilibrium level of statistical mechanics, with the Shannon ME rule being unable to cover the non-equilibrium phenomena. We shall briefly argue in the next section that the differential Schrödinger equations of quantum mechanics (stationary and time-dependent) also result from the EPI principle using the Fisher information measure [2,19,33].

3. Schrödinger Equation as Information Principle

Consider the stationary wave-equation of quantum mechanics for the N-electron molecule, the eigenvalue equation

$$\hat{\mathrm{H}}(N)\Psi(N) = E\Psi(N), \tag{34}$$

for the electronic (fixed-nuclei) Hamiltonian in the adiabatic, Born-Oppenheimer approximation,

$$\hat{\mathrm{H}}(N) = -(\hbar^2/2m_e)\sum_{i=1}^{N}\Delta_i + V(N) = \hat{\mathrm{T}}(N) + \hat{\mathrm{V}}(N), \tag{35}$$

where the (multiplicative) potential-energy operator $\hat{\mathrm{V}}(N) = V_{ne}(N) + V_{ee}(N) = V(N)$ includes the nuclear-electron (ne) attraction and electron-electron (ee) repulsions terms, and E stands for the system ground-state electronic energy. It follows from the familiar Schrödinger variational principle,

$$\delta\{\langle E\rangle - E\langle\Psi|\Psi\rangle\} = 0, \tag{36}$$

in which the exact electronic energy E plays the role of the Lagrange multiplier enforcing the constraint of the wave-function normalization.

The energy expectation value contains the electronic kinetic and potential parts,

$$\langle E\rangle = \langle\Psi(N)|\hat{\mathrm{H}}(N)|\Psi(N)\rangle = \langle\Psi(N)|\hat{\mathrm{T}}(N)|\Psi(N)\rangle + \langle\Psi(N)|\hat{\mathrm{V}}(N)|\Psi(N)\rangle = \langle T\rangle + \langle V\rangle, \quad (37)$$

with the kinetic contribution being proportional to the N-electron Fisher information [34]:

$$\begin{aligned}\langle I(N)\rangle &= \int\sum_{i=1}^{N}\frac{|\nabla_i P(N)|^2}{P(N)}d\tau^N \\ &= -4\sum_{i=1}^{N}\langle\Psi(N)|\Delta_i|\Psi(N)\rangle = 4\sum_{i=1}^{N}\langle\nabla_i\Psi^*(N)\cdot\nabla_i\Psi(N)\rangle = \frac{8m_e}{\hbar^2}\langle T(N)\rangle,\end{aligned} \quad (38)$$

where the joint-probability distribution of N electrons,

$$P(N) = \Psi^*(N)\,\Psi(N) = |\Psi(N)|^2, \quad \int P(N)d\tau^N = 1, \quad (39)$$

determines the average value of the potential energy,

$$\langle V(N)\rangle = \int P(N)\,V(N)\,d\tau^N. \quad (40)$$

Therefore, the wave-function principle of Eq. (36) also expresses the stationary character of the constrained Fisher information [8,14,39] in the EPI rule:

$$\delta\left\{\langle I(N)\rangle + \frac{8m_e}{\hbar^2}\left[\int P(N)\,V(N)\,d\tau^N - E\int P(N)\,d\tau^N\right]\right\} = 0. \quad (41)$$

Here, the average Fisher information term $\langle I(N)\rangle$ representing the EPI *intrinsic*-information and the remaining, constraining terms standing for the EPI *bound*-information contribution. The latter consists of two terms: the physical, potential-energy part, with the N-electron Lagrange multiplier function $\lambda(N) = -8m_eV(N)/\hbar^2$, and the "geometric" condition of the normalization of the system probability-amplitude being enforced by the system electronic energy.

Let us briefly examine sufficient physical constraints in the relevant Euler-Lagrange information principle giving rise to the time-dependent Schrödinger equation. For simplicity we assume the simplest case of a single, spin-less particle of mass μ described by the wave function $\psi(\boldsymbol{r})$, for which the time-independent equation results from the Euler-Lagrange EPI principle for the trial functions ψ' including the constraint terms due to the probability/wave-function normalization [19],

$$\|\psi\|^2 = \int\psi^*(\boldsymbol{r})\,\psi(\boldsymbol{r})\,d\boldsymbol{r} = \int p(\boldsymbol{r})\,d\boldsymbol{r} = 1, \quad (42)$$

and the fixed value of V_{ne} due to the external potential $v(\boldsymbol{r})$ due to the nuclei:

$$\langle V \rangle = \int \psi^*(\boldsymbol{r}) v(\boldsymbol{r}) \psi(\boldsymbol{r})\, d\boldsymbol{r} = \int p(\boldsymbol{r}) v(\boldsymbol{r})\, d\boldsymbol{r} = V^0, \tag{43}$$

enforced by the global Lagrange multipliers $\alpha = 8E\mu/\hbar^2$ and $\beta = -8\mu/\hbar^2$, respectively:

$$\delta\{ I[\psi'] - \alpha \int \psi'^*(\boldsymbol{r})\psi'(\boldsymbol{r})\, d\boldsymbol{r} - \beta \int \psi'^*(\boldsymbol{r}) v(\boldsymbol{r}) \psi'(\boldsymbol{r})\, d\boldsymbol{r} \} = 0. \tag{44}$$

The *non*-stationary wave-equation requires the time-dependent constraint of the conservation of the probability normalization in time [19]:

$$\frac{\partial \langle \psi | \psi \rangle}{\partial t} = \frac{\partial}{\partial t} \int p(\boldsymbol{r}) d\boldsymbol{r} = 0 \cdot \tag{45}$$

For the normalized wave-functions, when α = 0 (redundant constraint), the relevant Fisher-information EPI rule reads:

$$\delta \left\{ I[\psi'] - \beta \int \psi'^*(\boldsymbol{r}) v(\boldsymbol{r}) \psi'(\boldsymbol{r})\, d\boldsymbol{r} - \gamma \frac{\partial}{\partial t} \int \psi'^*(\boldsymbol{r}) \psi'(\boldsymbol{r})\, d\boldsymbol{r} \right\} = 0, \tag{46}$$

where β has already been identified before and $\gamma = 8\mu i/\hbar$. These Lagrange multipliers can be also interpreted as derivatives of the optimum Fisher information with respect to the associated constraints [19].

4. Electron Probabilities as Carriers of Information in Molecules

The electron densities $\{\rho_i^0\}$ of the separated atoms define the molecular (isoelectronic) prototype called the atomic "promolecule" [8,18], given by the sum of the free-atom electron densities shifted to the actual locations of atoms in the molecule. The resulting electron density $\rho^0 = \Sigma_i \rho_i^0$ of this collection of the "frozen" atomic electron distributions defines the initial stage in the bond-formation process and determines a natural reference for extracting changes due to the chemical bonds in the *density difference* function $\Delta\rho = \rho - \rho^0$, between the molecular and promolecular electron distributions, which has been widely used to probe the electronic structure of molecular systems. In this section we shall use some local IT probes to explore the molecular ground-state electron distributions $\rho(\boldsymbol{r})$ or their shape (probability)-factors $p(\boldsymbol{r}) = \rho(\boldsymbol{r})/N$ [8-10], generated by the Kohn-Sham (KS) [35] calculations in the *Local Density Approximation* (LDA) of the *Density Functional Theory* (DFT) [36].

Consider first the density $\Delta s(\boldsymbol{r})$ (in nats per unit volume) of the KL entropy-deficiency (directed-divergence) functional, of the information distance between the molecular and promolecular electron distributions,

$$\Delta S[\rho|\rho^0] = \int \rho(\boldsymbol{r}) \ln[\rho(\boldsymbol{r})/\rho^0(\boldsymbol{r})]\, d\boldsymbol{r} = \int \rho(\boldsymbol{r})\, I[w(\boldsymbol{r})]\, d\boldsymbol{r} \equiv \int \Delta s(\boldsymbol{r})\, d\boldsymbol{r}, \tag{47}$$

which also represents the (renormalized) missing information between the shape-factors $p(\boldsymbol{r})$ and $p^0(\boldsymbol{r}) = \rho^0(\boldsymbol{r})/N$ of the two compared electron densities,

$$\Delta s(\boldsymbol{r}) \equiv \Delta s[\rho(\boldsymbol{r})|\rho^0(\boldsymbol{r})] = Np(\boldsymbol{r})\ln[p(\boldsymbol{r})/p^0(\boldsymbol{r})] \equiv N\Delta s[p(\boldsymbol{r})|p^0(\boldsymbol{r})], \tag{48}$$

where $w(\boldsymbol{r})$, and $I[w(\boldsymbol{r})]$ denote the density/probability *enhancement* and *surprisal* functions, respectively. This KL-information density thus measures the electron density/probability weighted surprisal. The related divergence measure of the missing information reads:

$$\Delta S[\rho, \rho^0] = \int[\rho(\boldsymbol{r}) - \rho^0(\boldsymbol{r})] \ln[\rho(\boldsymbol{r})/\rho^0(\boldsymbol{r})]\, d\boldsymbol{r} = \int\Delta\rho(\boldsymbol{r})\, I[w(\boldsymbol{r})]\, d\boldsymbol{r} \equiv \int\Delta D(\boldsymbol{r})\, d\boldsymbol{r}. \tag{49}$$

Its density now represents the $\Delta\rho = \rho - \rho^0$ (or $\Delta p = p - p^0 = \Delta\rho/N$) weighted surprisal:

$$\Delta D(\boldsymbol{r}) \equiv \Delta s[\rho(\boldsymbol{r}), \rho^0(\boldsymbol{r})] = \Delta\rho(\boldsymbol{r})I[w(\boldsymbol{r})] = N\Delta p(\boldsymbol{r})I[w(\boldsymbol{r})] = N\Delta s[p(\boldsymbol{r}), p^0(\boldsymbol{r})]. \tag{50}$$

Therefore, the molecular surprisal $I[w(\boldsymbol{r})]$ measures the density per electron of the local entropy-deficiency relative to the promolecular reference, $I[w(\boldsymbol{r})] = \Delta s(\boldsymbol{r})/\rho(\boldsymbol{r})$, or the density per a single electron displacement of the molecular divergence, $I[w(\boldsymbol{r})] = \Delta D(\boldsymbol{r})/\Delta\rho(\boldsymbol{r})$.

We shall also examine the molecular displacements in the Shannon entropy, relative to the promolecular reference,

$$\mathcal{H}[\rho] \equiv S[\rho] - S[\rho^0] = -\int\rho(\boldsymbol{r}) \ln\rho(\boldsymbol{r})\, d\boldsymbol{r} + \int\rho^0(\boldsymbol{r}) \ln\rho^0(\boldsymbol{r})\, d\boldsymbol{r} \equiv \int h_\rho(\boldsymbol{r})\, d\boldsymbol{r}, \tag{51}$$

and its density $h_\rho(\boldsymbol{r})$ as alternative global and local probes of electron distributions in molecules. Alternatively, the corresponding entropy shifts in terms of the (normalized) probability distributions,

$$\mathcal{H}[p] \equiv S[p] - S[p^0] = -\int p(\boldsymbol{r}) \ln p(\boldsymbol{r})\, d\boldsymbol{r} + \int p^0(\boldsymbol{r}) \ln p^0(\boldsymbol{r})\, d\boldsymbol{r} \equiv \int h_p(\boldsymbol{r})\, d\boldsymbol{r}, \tag{52}$$

can also be used to explore the local relaxation in electron uncertainties, which accompanies the bond formation in molecules. In fact all these densities can be regarded as complementary tools for diagnosing the presence of chemical bonds and monitoring the effective valence states of the bonded atoms in the molecular environment [8-10].

We further observe that the molecular electron density $\rho(\boldsymbol{r})$ is on average only slightly modified relative to the promolecular distribution $\rho^0(\boldsymbol{r})$, $\rho(\boldsymbol{r}) \approx \rho^0(\boldsymbol{r})$ or $w(\boldsymbol{r}) \approx 1$. Indeed, the formation of chemical bonds involves only a minor reconstruction of the electronic structure, mainly in the valence-shells of the constituent atoms, so that $|\Delta\rho(\boldsymbol{r})| \equiv |\rho(\boldsymbol{r}) - \rho^0(\boldsymbol{r})| << \rho(\boldsymbol{r}) \approx \rho^0(\boldsymbol{r})$ and hence the ratio $\Delta\rho(\boldsymbol{r})/\rho(\boldsymbol{r}) \approx \Delta\rho(\boldsymbol{r})/\rho^0(\boldsymbol{r})$ is generally small in the energetically important regions of large density values. As explicitly shown in the first column of Figure 2, the largest values of the density difference $\Delta\rho(\boldsymbol{r})$ are observed mainly in the bond region, between the nuclei of chemically bonded atoms; the reconstruction of atomic lone pairs can also lead to an appreciable displacement in the molecular electron density.

By expanding the logarithm of the molecular surprisal $I[w(\boldsymbol{r})]$ around $w(\boldsymbol{r}) = 1$, to first-order in the relative displacement of the electron density, one obtains the following approximate relations between the local value of the molecular surprisal and that of the density-difference function:

$$I[w(\boldsymbol{r})] = \ln[\rho(\boldsymbol{r})/\rho^0(\boldsymbol{r})] = \ln\{[\rho^0(\boldsymbol{r}) + \Delta\rho(\boldsymbol{r})]/\rho^0(\boldsymbol{r})\} \cong \Delta\rho(\boldsymbol{r})/\rho^0(\boldsymbol{r}) \approx \Delta\rho(\boldsymbol{r})/\rho(\boldsymbol{r}). \tag{53}$$

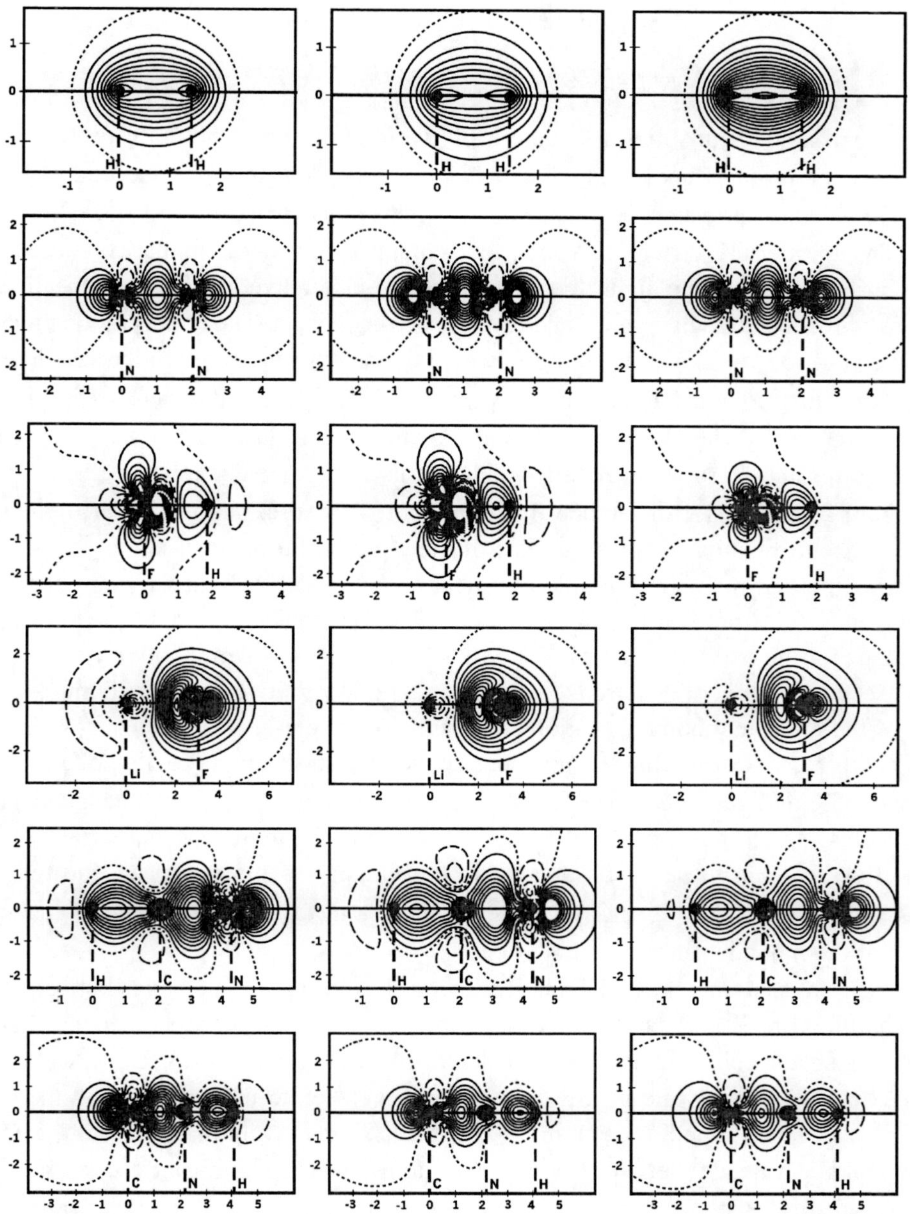

Figure 2. The contour diagrams of the molecular density difference function, $\Delta\rho(r) = \rho(r) - \rho^0(r)$ (first column), the information-distance density, $\Delta s(r) = \rho(r)I[w(r)]$ (second column) and its approximate, first-order expansion, $\Delta s(r) \cong \Delta\rho(r)w(r)$ (third column), for selected diatomic and linear triatomic molecules: H_2, HF, LiF, HCN and HNC. The solid, pointed and broken lines denote the positive, zero and negative values, respectively, of (equally spaced) contours [8,9]; the same convention is applied in Figs. 3, 4, 6, 10, and 11.

This relation provides a semi-quantitative information-theoretic interpretation of the relative density difference diagrams and links the local surprisal of IT to the density difference function of quantum chemistry [8-10]. It also relates the integrands of alternative information-distance functionals with the corresponding displacements in the electron density:

$$\Delta s(\boldsymbol{r}) = \rho(\boldsymbol{r}) I[w(\boldsymbol{r})] \cong \Delta\rho(\boldsymbol{r})\, w(\boldsymbol{r}) \approx \Delta\rho(\boldsymbol{r}), \tag{54}$$

$$\Delta D(\boldsymbol{r}) = \Delta\rho(\boldsymbol{r})\, I[w(\boldsymbol{r})] \cong [\Delta\rho(\boldsymbol{r})]^2/\rho^0(\boldsymbol{r}) \geq 0. \tag{55}$$

These approximate equalities are numerically verified in Figs 2 and 3, respectively. In the former, the contour diagram of the directed divergence density $\Delta s(\boldsymbol{r})$ (second column) is compared with the corresponding map of its first-order approximation, $\Delta\rho(\boldsymbol{r})w(\boldsymbol{r})$ (third column), and the density difference itself (first column). A general similarity between these diagrams in each row of the figure confirms a semi-quantitative character of the first-order expansions of the directed divergence and divergence functionals. The corresponding numerical validation of Eq. (55) is shown in Figure 3, where the contour maps of Kullback's divergence density $\Delta D(\boldsymbol{r})$ (first column) are compared with the corresponding diagrams of its first-order approximation $[\Delta\rho(\boldsymbol{r})]^2/\rho^0(\boldsymbol{r})$ (second column). Again, a semi-quantitative similarity between the two diagrams in each row numerically validates this approximate relation.

In Figure 2 the density difference function $\Delta\rho(\boldsymbol{r})$ for representative linear diatomic and triatomic molecules exhibits typical aspects of the equilibrium reconstructions of free-atoms during formation of single and multiple chemical bonds, which exhibit varying degree of the bond covalency (electron-sharing) and ionicity (electron transfer). Let us examine first the contour maps for the two homonuclear diatomics. The single covalent bond in H_2 gives rise to a relative accumulation of electrons in the bond region, between the two nuclei, at the expense of the outer, nonbonding regions of space. The triple-bond pattern for N_2 is seen to be more complex, reflecting the density accumulations in the bonding region, due to both the σ and π bonds, and the accompanying increase in the density of the lone pairs on both nitrogen atoms, due to their expected *sp*-hybridization in the atomic valence state. One also observes a decrease in the electron density in the vicinity of the nuclei and an outflow of electrons from the $2p_\pi$ AO to their overlapping area, a clear sign of their involvement in a formation of the π bond system.

Both heteronuclear diatomics, HF and LiF, represent partially-ionic bonds between the two atoms exhibiting small and large differences in their electronegativity (chemical hardness) descriptors, respectively. A pattern of density displacement in HF reflects a weakly ionic (strongly covalent) bond, while in LiF the two AIM are seen to be connected by the strongly ionic (weakly covalent) bond. Indeed, in HF one detects a relatively high degree of a common possession of the valence electrons by the two atoms, which significantly contribute to the shared bond-charge located between them, and a comparatively weak H→F polarization. In LiF a substantial Li→F electron transfer can be detected so that an ion-pair picture indeed provides an adequate description of the chemical bond in this diatomic. Finally, in the two triatomic molecules shown in Figure 2 one identifies a strongly covalent pattern of the electron density displacements in the regions of the single N–H and C–H atoms. A typical buildup of the bond charge due to the multiple CN bonds in the two isomers HCN and HNC can be also observed. The increase in the lone-pair electron density on the terminal heavy atom, N in HCN and or C in HNC, can be also detected thus confirming the expected *sp*-hybridization of these bonded atoms in their promoted, valence state in the molecule.

A comparison between the corresponding panels of the first two columns in the Figure 2 shows that the two displacement maps so strongly resemble one another that they are hardly distinguishable. This confirms a close relation between the local density and entropy-deficiency relaxation patterns, thus attributing to the former a complementary information-theoretic interpretation of the latter. A strong resemblance between these two types of molecular diagrams indicates that the local inflow of electrons increases the relative entropy, while the outflow of electrons gives rise to a diminished level of the relative-uncertainty content of the electron distribution in the molecule. The density displacement and the missing-information distribution can be thus viewed as equivalent probes of the system chemical bonds.

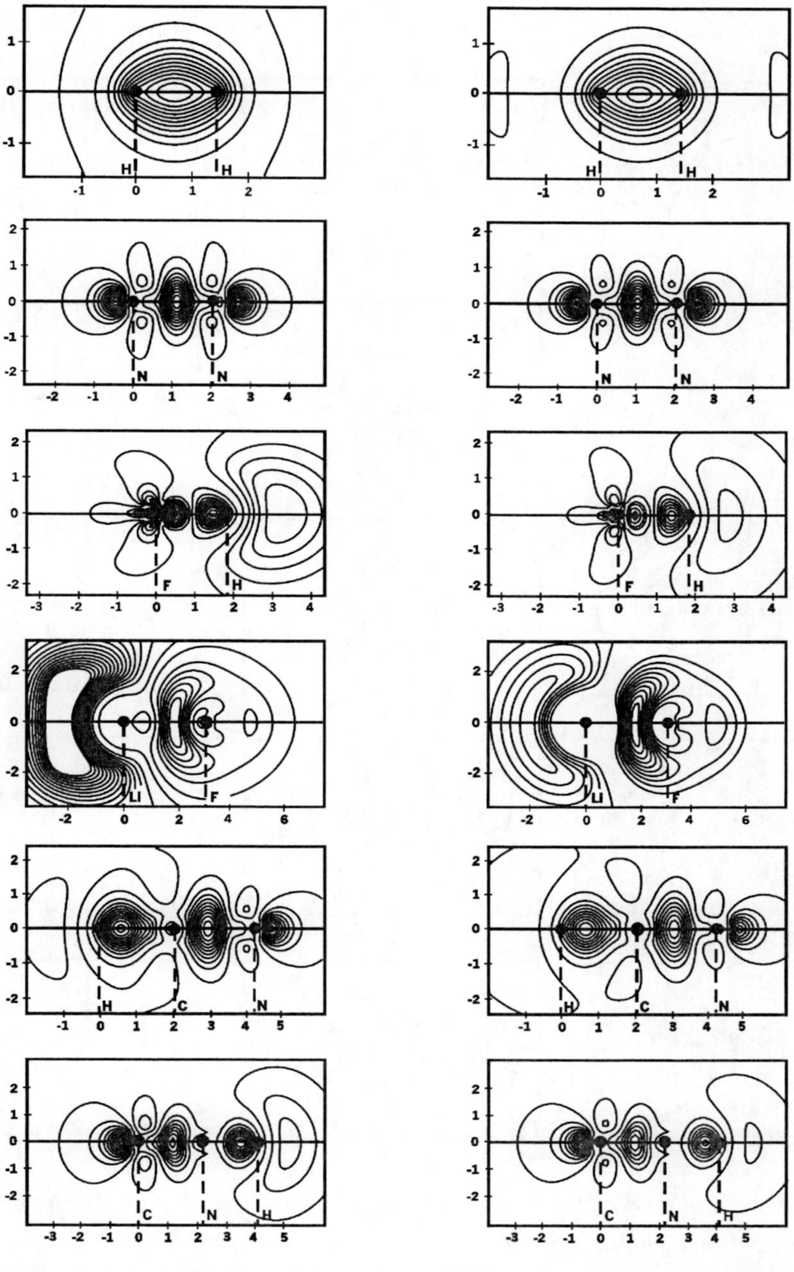

Figure 3. A comparison between the contour maps of the molecular divergence density, $\Delta D(r) = \Delta\rho(r)I[w(r)]$ (first column), $[\Delta\rho(r)]^2/\rho^0(r)$ (second column), for the molecules of Figure 2, which validates the approximate relation of Eq. (55) [8,9].

Similar diagnostic conclusions follow from the divergence density plots of Figure 3, where all crucial bonding and non-bonding regions of space are now identified by the *positive* values of the information density.

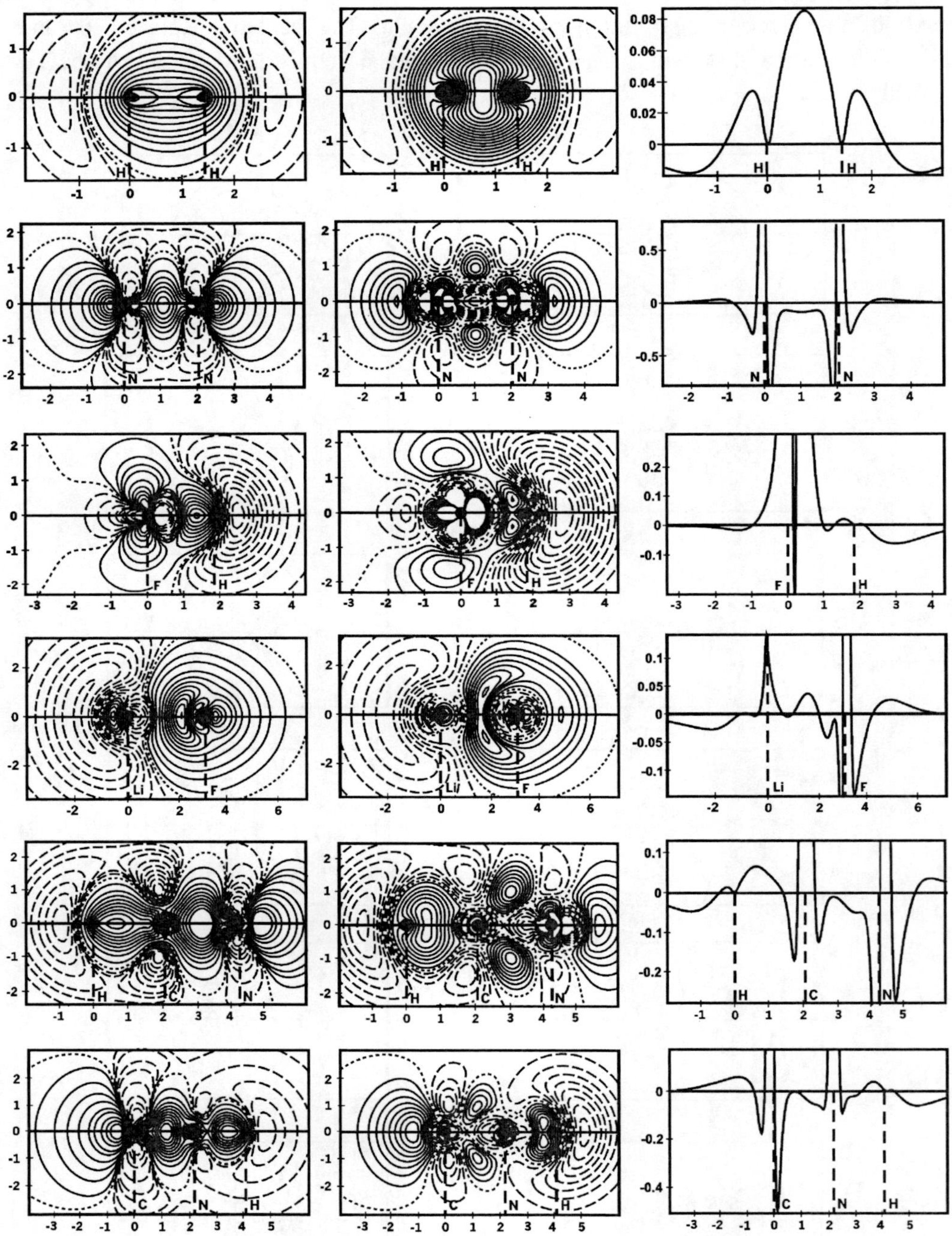

Figure 4. A comparison between the (non-equidistant) contour diagrams of the density-difference $\Delta\rho(r)$ (first column) and entropy-difference $h_\rho(r)$ (second column) functions for the linear molecules of Figure 2. The corresponding profiles of $h_\rho(r)$ for the cuts along the bond axis are shown in the third column of the figure [8,10].

In Figure 4 the contour maps of the entropy-displacement density $h_\rho(r)$ are compared with the corresponding density difference diagrams for representative linear molecules of Figure 2. To better visualize details of the two functions and to facilitate a qualitative comparison between their topographies, *non*-equidistant contour values have been selected. Therefore, only the profile of $h_\rho(r)$, shown in the third column of the figure, reflects the relative importance of each feature. When interpreting these plots one should realize, that a negative (positive) value of $h_\rho(r)$ [or $h_p(r)$] signifies a decrease (increase) in the local electron *uncertainty* in the molecule, relative to the associated promolecular reference value. Again, the $\Delta\rho$ and h_ρ diagrams for H_2 are seen to qualitatively resemble one another and the corresponding map Δs shown in Figure 2. The main feature of the h_ρ diagram, an increase in the electron distribution uncertainty in the bonding region between the two nuclei, is due to the inflow of electrons to this region. This manifests the bond covalency, which can be attributed to the electron-sharing effect and a delocalization of the bonding electrons, now effectively moving in the field of both nuclei. One detects in all these maps a similar nodal structure and finds that the nonbonding regions exhibit a decreased uncertainty, due to a transfer of the electron density from this area to the vicinity of the two nuclei and the region between them.

Therefore, also the molecular entropy difference function displays typical features in the reconstruction of electron distributions in a molecule, relative to those of the corresponding free atoms. Its diagrams thus provide an additional information tool for diagnosing the chemical bonds in terms of displacements in the entropy/information content of the molecular electron densities. In fact, the comparison of Figure 4 demonstrates, that the entropy difference plots provide in many respects a more detailed account of the reorganization of the electronic structure, relative to the free atoms in the promolecule, than do the corresponding density difference diagrams, particularly in the atomic inner shells.

In Table 1 we have listed representative values of the molecular entropy difference of Eq. (51) together with the Shannon entropies for the molecular and promolecular electron densities. These results show that in general the molecular distribution gives rise to a lower level of the information-entropy (less uncertainty) compared to the promolecule.

Table 1. Displacements of the molecular Shannon entropies (in bits) for molecules of Figure 2 [8,10].

Molecule	$\mathcal{H}[\rho] = S[\rho] - S[\rho^0]$	$S[\rho]$	$S[\rho^0]$
H_2	– 0.84	6.61	7.45
N_2	– 0.68	8.95	9.63
HF	– 1.00	3.00	4.00
LiF	– 3.16	5.12	8.28
HCN	– 1.44	12.99	14.45
HNC	– 1.39	13.06	14.45

Thus, the degree of uncertainty contained in the electron distribution on average decreases when the constituent free-atoms form the chemical bonds. Indeed, the dominating overall contraction of atomic electron distributions in the field of all nuclear attractors in the molecule should imply a higher degree of “order” (less uncertainty) in the molecular electron density in comparison to that present in the promolecular distribution. The largest magnitude of this relative decrease in the entropy content of the molecular electron density is observed for LiF, which exhibits the most ionic bond (largest amount of charge transfer) among all molecules in the table.

There is no apparent correlation in Table 1 between the global entropy displacement and the bond multiplicity. For example, a triple covalent bond in N_2 generates less overall entropy loss than does a single bond in H_2. The reason for a low magnitude of the entropy displacement in N_2 is the result of a mutual cancellation of the negative and positive contributions due to valence electrons. Indeed, the orbital hybridization, AO contraction, and the charge transfer should lower the entropy of the atomic electron distribution, since they increase charge inhomogeneity in the molecule, relative to the promolecule. By the same criterion, the effective expansion of atomic densities due to the AIM promotion and the electron delocalization via the system chemical bonds should have the opposite effect, of increasing the uncertainty content of the electron distribution in the molecule. Notice that, should one assume a similar entropy displacement of about –0.7 bits for all triple bonds in a series of isoelectronic molecules N_2, HCN and CNH, one obtains a contribution due to a single C–H or N–H bond of about –0.7, a result close to that found for the H–H bond.

As an additional illustration we now present the combined density-difference, entropy-displacement, and information-distance analyses of the central bond in the propellane systems shown in Figure 5 [8,10]. The main purpose of this study was to examine the effect on the central C'–C' bond, between the (primed) "bridgehead" carbon atoms, of a successive increase in the size of the carbon bridges in the series of the [1.1.1], [2.1.1], [2.2.1], and [2.2.2] propellanes shown in Figure 5.

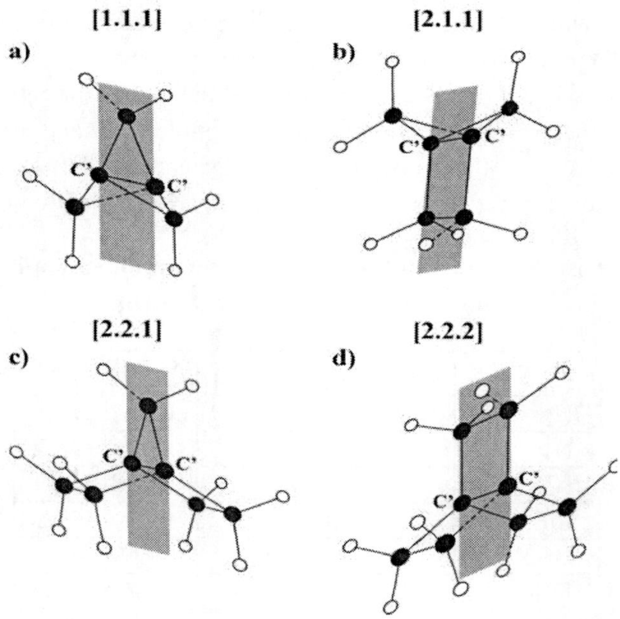

Figure 5. The propellane structures and the planes of sections containing the bridge and bridgehead (C') carbon atoms identified by black circles.

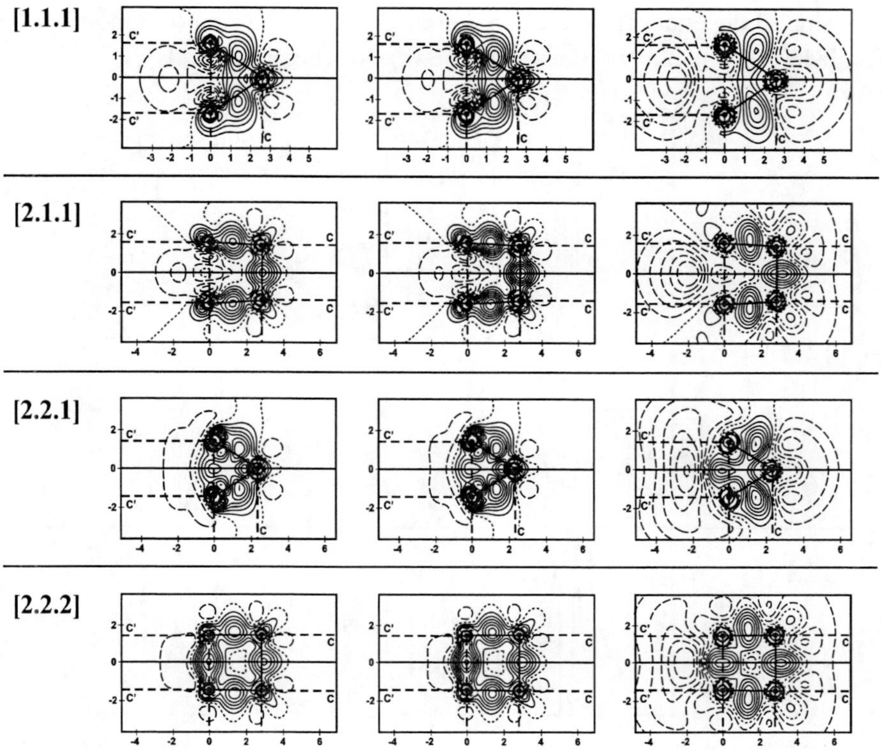

Figure 6. A comparison between the equidistant-contour maps of the density-difference function $\Delta\rho(r)$ (first column), the information-distance density $\Delta s(r)$ (second column), and the entropy-displacement density $h_\rho(r)$ (third column), for the four propellanes of Figure 5 [8,10].

Figure 6 reports the contour maps of the molecular density difference function $\Delta\rho(\boldsymbol{r})$, the Kullback-Leibler integrand $\Delta s(\boldsymbol{r})$, and the entropy displacement function $h_\rho(\boldsymbol{r})$, for the planes of sections shown in Figure 5. The corresponding central-bond profiles of the density- and entropy- difference functions are shown in Figure 7. These contour maps have been generated using the DFT-LDA calculations in the extended (DZVP) basis set. The density difference plots show that there is on average a depletion of the electron density, relative to the promolecule, between the bridgehead carbon atoms in the small [1.1.1] and [2.1.1] propellanes, while larger [2.2.1] and [2.2.2] systems exhibit a net density buildup in this region. A similar conclusion follows from the entropy-displacement and entropy-deficiency plots of these figures. The two entropic diagrams are again seen to be qualitatively similar to the corresponding density-difference plots. This resemblance is again particularly strong between the $\Delta\rho(\boldsymbol{r})$ and $\Delta s(\boldsymbol{r})$ diagrams, of the first two columns in Figure 6.

A more general outlook on the bond-order concept, emerging from both the quadratic-difference approaches [37-41] and the communication theory [8], identifies the chemical bond as a measure of a statistical "*dependence*" (non-additivity) between orbitals on different atomic centers [8,20]. On one hand, this dependence between basis functions of different atoms is partly realized *directly* (through space), by a constructive interference of orbitals (probability amplitudes) on two atoms, which increases the electron density between them. On the other hand, it has a partial *indirect origins*, through the orbitals of the remaining AIM.

The latter is due to dependencies between AO originating from the orthonormality relations of the occupied MO, which determine the framework of chemical bonds.

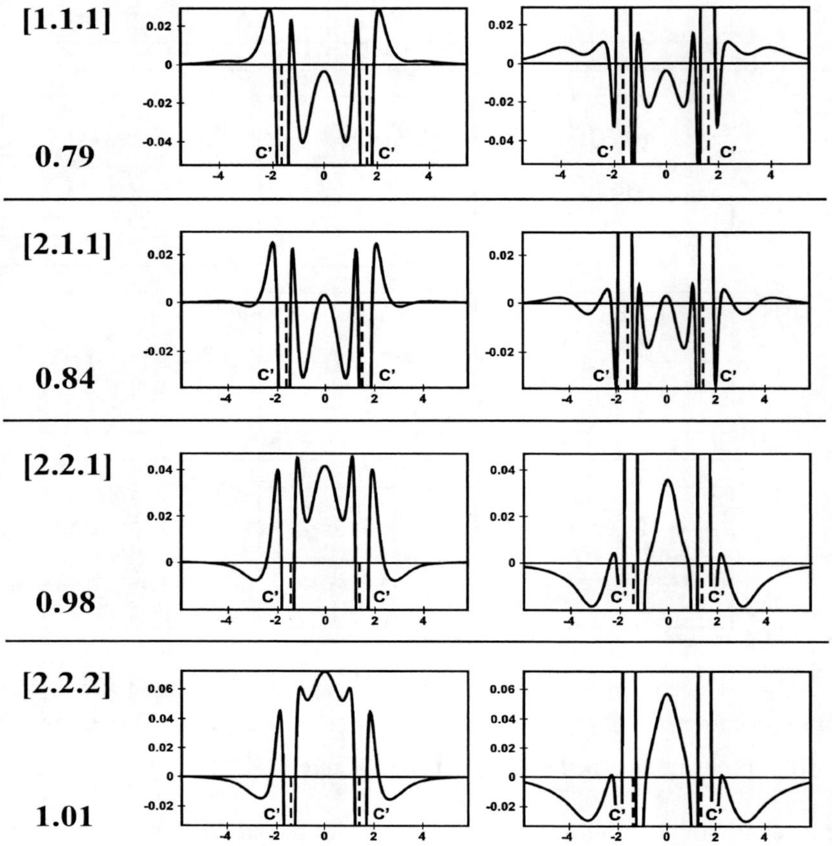

Figure 7. The bridgehead bond profiles of the density difference function (left panel) and molecular entropy displacement (right panel) for the four propellanes shown in Figure 5. For comparison the numerical values of the bond multiplicities from the difference approach [39] are also reported.

The numerical bond-orders reported in Figure 7 originate from the *two*-electron difference approach [37-41], an extension of the original (covalent) bond multiplicity index of Wiberg [42] and its numerous generalizations, e.g., [43]. Together with the corresponding density profiles shown in this figure they reveal a changing nature of the central bond in the four propellanes included in this analysis. The central "bonds" in the smallest propellanes, lacking the accumulation of the electron density or the entropy (entropy-deficiency) density between the bridgehead atoms, are seen to be mostly of the *indirect* character, being realized "through-bridges" rather than "through-space". Clearly, the most important atomic intermediates for this indirect bond mechanism in propellanes are the *bridge* carbons, which strongly overlap with the *bridgehead* carbons. A gradual emergence of the direct, "through-space" component of the central bond, due to accumulation of the electron density and the entropy (entropy-deficiency) density between the bridgehead carbons, is observed when the bridges are enlarged in the two largest propellanes. Using the two-electron difference approach one roughly estimates [39] a full single bond in the [2.2.1] and [2.2.2] propellanes and approximately 0.8 bond-order in the [1.1.1] propellane. Hence, using the latter estimate as a measure of the "through-bridges" component in the largest propellane, one predicts about 0.2 bond-order measure for the "through-space" component of the central bond in the largest [2.2.2] propellane.

5. Bonded Atoms from Information Theory

In chemistry an understanding of the electronic structure of molecules and their preferences in reactions comes from transforming the computed (or experimental) electron distributions into statements in terms of chemical concepts, such as bonded atoms, the building blocks of molecules, their collections defining larger fragments of the whole molecular system, e.g., the fuctional groups, and the chemical bonds, which represent the molecular "connectivities" between atoms and molecular subsystems. Indeed, an understanding of molecules as combinations of atoms is fundamental to chemistry. It is not surprising, then, that the concept of *Atoms-in-Molecules* (AIM) has been much discussed in the scientific literature [44]. In fact, chemistry mainly deals with small changes of bonded atoms and larger molecular fragments, with reasonably well understood and *transferable* molecular invariants, such as AIM, functional groups, molecular subsystems, e.g., reactants and products of an elementary chemical reaction, etc., which tend to maintain their identity in different molecular environments. Most molecular systems may be thought of as consisting of only slightly perturbed atoms (or atomic ions), deformed by the presence of the molecular remainder and exhibiting modified net electric charges. These displacements in the atomic electronic structure are due to the coupled processes of the *intra*-atomic *Polarization* (P), responsible for the *promotion* of bonded atoms to their effective *valence-states* in the molecule, and the *inter*-atomic *Charge Transfer* (CT), which accompany the formation of the system chemical bonds.

An important part of the chemical science is based on such intuitive notions, which escape the rigorous definition in molecular quantum mechanics, thus representing the Kantian *noumenons* of chemistry [15]. The natural question to be addresses next is: can IT help in making these concepts more precisely defined in terms of the molecular and/or promolecular electron probabilities, which together reflect the information redistribution during the bond formation process?

The bonded fragments of a given molecular system represent the mutually-*open* subsystems capable of exchanging electrons with their respective molecular reminders. One would hope to find that a given AIM, like its free (non-bonded) analog, would possess a single cusp at the nucleus in its electron density, linked to the effective atomic number of the nucleus. The bonded atoms are pieces of the molecule so that their electron densities $\{\rho_i\}$ must sum up to the molecular electron density $\rho = \sum_i \rho_i$. However, since each bonded atom preserves to a remarkably high degree the free-atom identity, the AIM distributions should be also closely related to densities $\{\rho_i^0\}$ of the separated atoms, which define the molecular prototype, called the atomic promolecule [18], consisting of the *free*-atom electron densities shifted to the actual AIM locations in the molecule. The resulting electron density, $\rho^0 = \sum_i \rho_i^0$, of this collection of the "frozen" atomic electron distributions, which defines the initial stage in the bond formation process, determines a natural reference for extracting changes due to the chemical bonds. Indeed, the familiar density difference function of Section 4, $\Delta\rho = \rho - \rho^0$, has been widely used to probe chemical bonds in molecules.

The electronic structure of molecular systems is characterized by their *one*-, *two*-and *many*-electron probability distributions in the continuous ("fine-grained") description. To obtain its *chemical* interpretation, e.g., in terms of AIM, functional groups, reactants or other type of chemically significant subsystems, e.g., the σ or π electrons, these overall distributions have to be "discretized" in terms of the relevant pieces of the overall density attributed to the constituent parts of the molecular system under consideration, e.g., the bonded atoms or molecular fragments (clusters of AIM). The densities of molecular subsystems constitute their fine-grained description. By an appropriate integration of the electron/probability densities one can then obtain the corresponding *condensed* descriptors of the electronic structure of molecules and their fragments, providing the discrete (*coarse-grained*) indices of the associated continuous electron distributions. Additional resolution levels are provided by the AO or MO representations of the quantum-mechanical calculations of the molecular electronic structure.

An exhaustive partitioning of a given molecular electron density between constituent (bonded) atoms, which determines the AIM effective net charges (oxidation states) in a given molecular environment, is not unique, since it depends on the adopted criterion for such a division [15]. The imposing variety of published theoretical methods for partitioning the molecular density into *"best"* AIM contributions testifies to the importance of this theme in chemistry. Different methods are based on different principles, some to a degree arbitrary or heuristic, which can produce conflicting trends in the associated AIM charges. Methods differ in the theoretical techniques used, e.g., topological analysis of the density, wave-function description, or the density-functional approach. They also differ in the physical/heuristic principles invoked, e.g., electronegativity equalization, zero flux, the minimum promotion-energy rules, and the minimum entropy-deficiency (information-distance, missing-information) in the IT approach.

The historically first scheme of the Mulliken/Löwdin *population analysis* have used the *functional-space* partitioning, in which one distributes electrons between AO, which form the basis set for expanding MO of the Hartree-Fock (SCF LCAO MO) theory and its *Configuration Interaction* (CI) extension, non-orthogonal in the Mulliken approach and the symmetrically orthogonalized in the Löwdin variant. Another popular approach of Bader (1990), with a solid topological and quantum-mechanical basis, uses the *physical space* partitioning, i.e., a division of space into the exclusive atomic basins, with the boundaries determined by the zero-flux surfaces, on which the flow of electrons between subsystems vanishes. In the latter approach the spherical, spatially non-confined bonded atoms of the population analysis are replaced by the topological, non-spherical pieces of the molecular density, obtained as cuts along the zero-flux surfaces. As a result the topological AIM represent the spatially-confined (non-overlapping) and strongly non-symmetrical atoms. They represent the stationary pieces of molecular v-representable density, which individually are not v-representable [8].

5.1. Stockholder AIM

Yet another *stockholder* division scheme of Hirshfeld [18], which is widely used in crystallography, uses the "common-sense" *local* partitioning principle, which parallels the stock-market rule: in forming a molecule each bonded atom locally partakes of a molecular density ("profit") gain or loss in the molecule in proportion to its share in the promolecular density ("investment"). Thus, by construction, the (overlapping) stockholder AIM are infinitely extending (not confined spatially) and known to be only slightly polarized relative to the free-atom reference. Both the topological and stockholder atoms are derived from the molecular electron density, and as such they preserve a "memory" of the original overall distribution of electrons in a molecule. This *one*-electron stockholder division scheme has recently been shown to have a strong basis in IT [8,11-17]. However, the *two*-electron generalization of the stockholder principle [13] generates in molecules involving light atoms, e.g., hydrogen and lithium, slightly different effective one-electron distributions of the associated bonded hydrogens, which emphasize more strongly the bonding (overlap) regions between the bond partners [17], relative to the corresponding *one*-electron stockholder atoms of Hirshfeld.

It has been shown by Hirshfeld [18] that the electron density $\rho(\boldsymbol{r})$ of the molecular system $\mathrm{M} = (\mathrm{A}^{\mathrm{H}} \,\vdots\, \mathrm{B}^{\mathrm{H}} \,\vdots\, \ldots)$, consisting of the mutually-*open* atoms $\mathrm{X}^{\mathrm{H}} = (\mathrm{A}^{\mathrm{H}}, \mathrm{B}^{\mathrm{H}}, \ldots)$, as marked by the perpendicular *broken* lines separating the AIM symbols in M, can be exhaustively partitioned, into the "stockholder" AIM densities $\{\rho_{\mathrm{X}}^{\mathrm{H}}(\boldsymbol{r})\} \equiv \boldsymbol{\rho}^{\mathrm{H}}(\boldsymbol{r})$:

$$\rho_{\mathrm{X}}^{\mathrm{H}}(\boldsymbol{r}) = \rho_{\mathrm{X}}^{0}(\boldsymbol{r})\,[\rho(\boldsymbol{r})/\rho^{0}(\boldsymbol{r})] \equiv \rho_{\mathrm{X}}^{0}(\boldsymbol{r})\,w(\boldsymbol{r}) = \rho(\boldsymbol{r})\,[\rho_{\mathrm{X}}^{0}(\boldsymbol{r})/\rho^{0}(\boldsymbol{r})] \equiv \rho(\boldsymbol{r})\,d_{\mathrm{X}}^{\mathrm{H}}(\boldsymbol{r}),\ \mathrm{X} = \mathrm{A}, \mathrm{B}, \ldots$$
$$\textstyle\sum_{\mathrm{X}} d_{\mathrm{X}}^{\mathrm{H}}(\boldsymbol{r}) = 1,\ \rho(\boldsymbol{r}) = \sum_{\mathrm{X}} \rho_{\mathrm{X}}^{\mathrm{H}}(\boldsymbol{r}). \qquad (56)$$

Here $\boldsymbol{\rho}^{0}(\boldsymbol{r}) = \{\rho_{\mathrm{X}}^{0}(\boldsymbol{r})\}$ groups the densities of the free atoms, giving rise to the reference electron density $\rho^{0}(\boldsymbol{r}) = \sum_{\mathrm{X}}\rho_{\mathrm{X}}^{0}(\boldsymbol{r})$ of the (isoelectronic) promolecule $\mathrm{M}^{0} = (\mathrm{A}^{0}|\ \mathrm{B}^{0}|\ \ldots)$, consisting of the non-bonded (mutually-*closed*) atoms $\mathrm{X}^{0} = (\mathrm{A}^{0}, \mathrm{B}^{0}, \ldots)$, as marked by the perpendicular *solid* lines separating the free atom symbols in M^{0}:

$$N^{0} = \int \rho^{0}(\boldsymbol{r})\,d\boldsymbol{r} = \textstyle\sum_{\mathrm{X}} \int \rho_{\mathrm{X}}^{0}(\boldsymbol{r})\,d\boldsymbol{r} = \sum_{\mathrm{X}} N_{\mathrm{X}}^{0} = N = \int \rho(\boldsymbol{r})\,d\boldsymbol{r} = \sum_{\mathrm{X}} \int \rho_{\mathrm{X}}^{\mathrm{H}}(\boldsymbol{r})\,d\boldsymbol{r} = \sum_{\mathrm{X}} N_{\mathrm{X}}^{\mathrm{H}}. \quad (57)$$

The free-atom densities $\boldsymbol{\rho}^{0}$ in M^{0} are shifted to the respective atomic positions in the molecule and the vectors $\boldsymbol{N}^{\mathrm{H}} = \{N_{\mathrm{X}}^{\mathrm{H}}\}$ and $\boldsymbol{N}^{0} = \{N_{\mathrm{X}}^{0}\}$ group the atomic average numbers of electrons of the bonded and free atoms, respectively. As we have already observed in Section 4, the same promolecular reference is used to determine the density difference function $\Delta\rho(\boldsymbol{r}) = \rho(\boldsymbol{r}) - \rho^{0}(\boldsymbol{r})$.

A reference to Eq. (56) shows that the Hirshfeld AIM densities satisfy the local principle of the *one*-electron stockholder division, which can be stated as the following equality between the local molecular and promolecular *conditional* probabilities:

$$d_{\mathrm{X}}^{\mathrm{H}}(\boldsymbol{r}) = \rho_{\mathrm{X}}^{\mathrm{H}}(\boldsymbol{r})/\rho(\boldsymbol{r}) \equiv P^{\mathrm{H}}(\mathrm{X}|\boldsymbol{r}) = d_{\mathrm{X}}^{0}(\boldsymbol{r}) = \rho_{\mathrm{X}}^{0}(\boldsymbol{r})/\rho^{0}(\boldsymbol{r}) \equiv P^{0}(\mathrm{X}|\boldsymbol{r}),$$
$$\textstyle\sum_{\mathrm{X}} P^{\mathrm{H}}(\mathrm{X}|\boldsymbol{r}) = \sum_{\mathrm{X}} P^{0}(\mathrm{X}|\boldsymbol{r}) = 1. \qquad (58)$$

As we have already remarked before, it has been interpreted by Hirshfeld (1977) using the stock-market analogy: each atom participates locally in the molecular "profit" $\rho(\boldsymbol{r})$ in proportion to its "share" $d_{\mathrm{X}}^{0}(\boldsymbol{r}) = P^{0}(\mathrm{X}|\boldsymbol{r})$ in the promolecular "investment" $\rho^{0}(\boldsymbol{r})$. In the remaining part of this section we shall demonstrate that this common-sense division rule has a solid basis in IT.

By extracting the overall number of electrons $N = N^0$ from the molecular and subsystem densities one introduces the associated probability distributions:

$$\rho(\boldsymbol{r}) = Np(\boldsymbol{r}) = N\Sigma_X\, p_X{}^H(\boldsymbol{r}) \text{ and } \boldsymbol{\rho}^H(\boldsymbol{r}) = N\boldsymbol{p}^H(\boldsymbol{r}) = N\{p_X{}^H(\boldsymbol{r})\}, \tag{59}$$

where $p(\boldsymbol{r})$ and $\boldsymbol{p}^H(\boldsymbol{r})$ stand for the *molecularly* normalized *shape*-factors of the system as a whole and of its Hirshfeld atoms, respectively,

$$\Sigma_X \int p_X{}^H(\boldsymbol{r})\, d\boldsymbol{r} = \Sigma_X (N_X{}^H/N) \equiv \Sigma_X P_X{}^H = 1, \tag{60}$$

while $\boldsymbol{P}^H = \{P_X{}^H\}$ groups the condensed probabilities of finding an electron of M on the specified stockholder AIM.

This normalization reflects the important fact that bonded atoms are constituent parts of the molecule, so that the full normalization condition has to involve the summation/integration over the complete set of *one*-electron events, consisting of all possible "values" of the discrete argument X (atomic label) and all spatial locations of an electron, identified by continuous coordinates $\boldsymbol{r}$ in the subsystem probability distributions: $\boldsymbol{p}^H(\boldsymbol{r}) = \{p_X{}^H(\boldsymbol{r}) \equiv P^H(X\wedge \boldsymbol{r}) = p(\boldsymbol{r})P^H(X|\boldsymbol{r})\}$.The same, global normalization has to be adopted for the *free*-atom pieces of the *one*-electron probability distribution in the isoelectronic promolecule and for its *free*-atom components, respectively,

$$p^0(\boldsymbol{r}) = \rho^0(\boldsymbol{r})/N = \Sigma_X\, p_X{}^0(\boldsymbol{r}),\ \boldsymbol{p}^0(\boldsymbol{r}) = \boldsymbol{\rho}^0(\boldsymbol{r})/N = \{p_X{}^H(\boldsymbol{r})\};$$
$$\Sigma_X \int p_X{}^0(\boldsymbol{r})\, d\boldsymbol{r} = \Sigma_X (N_X{}^0/N) \equiv \Sigma_X P_X{}^0 = 1, \tag{61}$$

where $\boldsymbol{P}^0 = \{P_X{}^0\}$ collects the condensed probabilities of observing an electron of M^0 on the specified free atom. Again the full normalization of the shape (probability) factors $\boldsymbol{p}^0(\boldsymbol{r}) = \{p_X{}^0(\boldsymbol{r}) \equiv P^0(X\wedge \boldsymbol{r})\}$ of the non-bonded atoms in the promolecular system involves summation over the discrete atomic "variable" X and integration over all positions $\boldsymbol{r}$ of an electron, the latter representing a continuous-event argument of the probability distributions of atomic fragments in M^0.

It follows from Eq. (56) that in the *one*-electron stockholder division each free subsystem density (or its shape factor) is locally modified in accordance with the *molecular* (subsystem-independent) enhancement factor $w(\boldsymbol{r})$:

$$w_X{}^H(\boldsymbol{r}) \equiv \rho_X{}^H(\boldsymbol{r})/\rho_X{}^0(\boldsymbol{r}) = p_X{}^H(\boldsymbol{r})/p_X{}^0(\boldsymbol{r}) = \rho(\boldsymbol{r})/\rho^0(\boldsymbol{r}) = p(\boldsymbol{r})/p^0(\boldsymbol{r}) \equiv w(\boldsymbol{r}). \tag{62}$$

Therefore, this procedure is devoid of any subsystem bias and as such appears to be fully objective.

Representative plots of the overlapping electron densities of bonded hydrogen atoms in H_2 are shown in Figure 8. They are seen to be distributed all over the physical space, decaying exponentially at large distances from the molecule and exhibiting in the bond density profile a single cusp at the atomic nucleus. They also display the expected polarization towards the bonding partner. These subsystem densities are highly transferable [18] and their overlap in the molecule accords with the classical interpretation of the origin of the chemical bond. One also observes a higher AIM density at the atomic nucleus, in comparison to the free-hydrogen density, i.e., a contraction of the AIM distribution at the expense of the nonbonding part of the free-atom density. This is due to the presence of the other atom causing an effective lowering of the molecular external potential relative to the external potential of the separated atom.

Let us briefly examine the asymptotic properties of the stockholder atomic densities. For simplicity we consider a diatomic system $M = (A^H|B^H)$, $\rho = \rho_A^H + \rho_B^H$, consisting of two Hirshfeld atoms A^H and B^H, the free analogs of which exhibit the relative electron *acceptor* (acidic) and *donor* (basic) properties respectively. This implies $\mu_A^0 = -I_A^0 < \mu_B^0 = -I_B^0$, where μ_X^0 and I_X^0 denote the *chemical potential* (negative electronegativity) and the *ionization potential* of X^0. Rewriting Eq. (56) in terms of the local density ratio $x = \rho_B^0/\rho_A^0$ gives:

$$\rho_A^H = (1 + x)^{-1}\rho \text{and } \rho_B^H = (1 + x^{-1})^{-1}\rho. \tag{63}$$

Hence, for $r \to \infty$, when the distances from both nuclei become large compared to inter-atomic distance R_{AB}, $x \to \infty$ since the asymptotic behavior of the free subsystems of the promolecule are determined by the their ionization potentials, i.e., negative energies of the highest occupied Kohn-Sham orbitals (atomic chemical potentials):

$$\rho_X^0 \to \exp[-2(2I_X^0)\, r]\ (r \to \infty),\ X = A, B. \tag{64}$$

Thus, for $r\to\infty$, $\rho_A^H\to 0$ and $\rho_B^H \to \rho$, so that the density of the softer (donor) atom B has a dominant contribution to the molecular density at distances from the molecule large compared to R_{AB}.

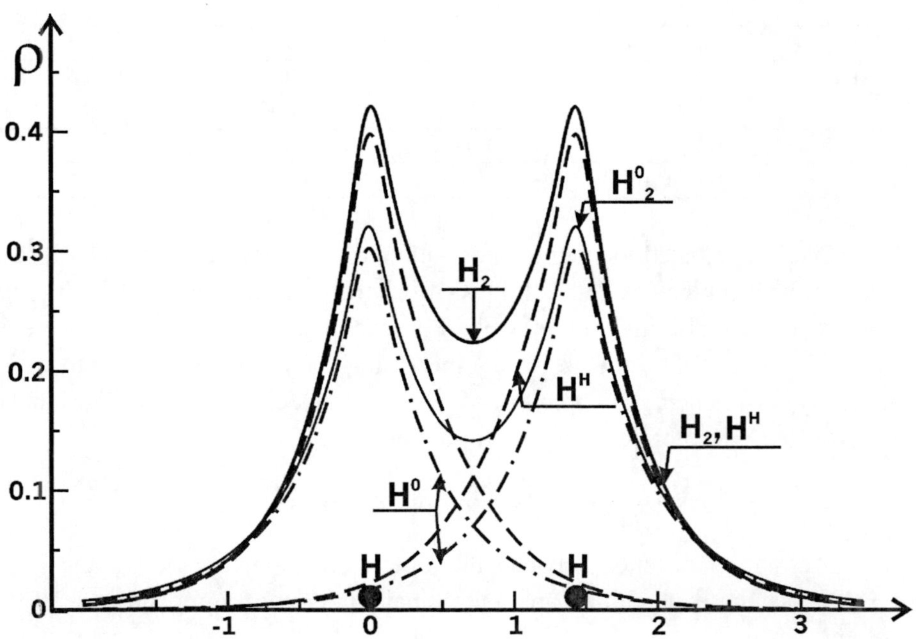

Figure 8. The Hirshfeld electron densities of bonded hydrogen atoms (H^H) obtained from the molecular density (H_2). The free-hydrogen densities (H^0) and the resulting electron density of the promolecule (H_2^0) are also shown for comparison. The density and inter-nuclear distance are in a.u. The zero cusps at nuclear positions are the artifacts of the Gaussian basis set used in DFT calculations.

Table 2. Representative net charges $\{q_X^H = N_X^H - Z_X\}$ (a.u.) of the stockholder AIM, where Z_X denotes the charge of nucleus X, the AIM entropy deficiencies $\Delta S_X[p_X^H|p_X^0]$, and global entropy deficiency $\Delta S[p|p^0]$ (in bits), for the linear molecules of Figure 2 [9].

Molecule	X	q_X^H	$\Delta S_X[p_X^H \vert p_X^0]$	$\Delta S[p \vert p^0]$
H_2	H	0.00	0.056	0.056
N_2	N	0.00	0.006	0.006
HF	H	0.24	0.144	0.020
	F	−0.24	0.005	
LiH	Li	0.35	0.157	0.136
	H	−0.35	0.012	
LiF	Li	0.58	0.244	0.063
	F	−0.58	0.007	
LiCl	Li	0.53	0.212	0.033
	Cl	−0.53	0.003	
HCN	H	0.14	0.104	0.017
	C	0.03	0.015	
	N	−0.17	0.005	
HNC	H	0.20	0.110	0.018
	N	−0.10	0.008	
	C	−0.10	0.011	
HNCS	H	0.19	0.114	0.008
	N	−0.13	0.007	
	C	0.05	0.008	
	S	−0.11	0.002	
HSCN	H	0.22	0.088	0.008
	S	−0.07	0.002	
	C	0.04	0.008	
	N	−0.19	0.004	

In Table 2 we have listed the net charges and entropy-deficiencies of the Hirshfeld AIM for illustrative linear molecules [9]. It follows from these results that the Hirshfeld charges represent the chemical intuition quite well. For example, in the series of heteronuclear diatomics, HF, LiH, LiCl, and LiF, of increasing bond ionicity due to a growing electronegativity difference, the amount of charge transfer monotonically increases, as intuitively expected. The atomic missing information

$$\Delta S_X[p_X^H|p_X^0] = \int p_X^H(\boldsymbol{r}) \log[p_X^H(\boldsymbol{r})/p_X^0(\boldsymbol{r})]\, d\boldsymbol{r}\,, \tag{65}$$

reflects the information distance between the atomic shape-factors $p_X^H(\boldsymbol{r})$ and $p_X^0(\boldsymbol{r})$. The reported values of these quantities are quite small, thus confirming numerically that the bonded atoms strongly resemble their free-atom analogs. The same general conclusion follows from examining the reported global entropy deficiencies:

$$\Delta S[p\,|\,p^0] = \sum_X \Delta S_X[p_X^H|p_X^0] = \int p(\boldsymbol{r}) \log[p(\boldsymbol{r})/p^0(\boldsymbol{r})]\, d\boldsymbol{r} \equiv \int p(\boldsymbol{r})\, I[w(\boldsymbol{r})]\, d\boldsymbol{r}. \tag{66}$$

This strong similarity between the molecular and promolecular electron distributions is also seen in Figure 8: the appreciable changes of the free-atom densities in the molecule are only observed around the nuclei (a contraction of the *free*-atom density) and in the bond region between the two nuclei (a polarization of the free atoms towards the bonding partner). In heavier atoms only a slight distortion of the valence (external) electrons is observed in the stockholder (bonded) atoms, with the inner shell structure left practically intact.

5.2. Information-Theoretic Justification

The optimum *local* partition of the molecular density can be best formulated in terms of the unknown *conditional* probabilities $d(r) = \{d_X(r) \equiv P(X|r)\}$, which uniquely determine the AIM pieces of $p(r)$, $\{p_X(r) = d_X(r)p(r)\}$[8,13,14]. The relevant local KL function of the unknown share factors $d(r)$, which measures the overall local information distance relative to the promolecular reference values $d^0(r) = \{P^0(X|r)\}$, is then given (in nats) by the sum of AIM contributions:

$$\Delta S[d(r)|d^0(r)] = \Sigma_X P(X|r) \ln[P(X|r)/P^0(X|r)]. \quad (67)$$

The best (unbiased) share factors of subsystems, giving rise to the maximum similarity of the bonded fragment densities to their non-bonded (free) analogs, must minimize this missing information function subject to the normalization of the local conditional probabilities,

$$\delta\{\Delta S[d(r)|d^0(r)] - \lambda(r)\, \Sigma_X P(X|r)\} = 0, \quad (68)$$

where $\lambda(r)$ stands for the Lagrange multiplier of probability normalization at r. The resulting Euler equation for the optimum local conditional probability $P(X|r)$,

$$\ln[P^{opt.}(X|r)/P^0(X|r)] + [1 - \lambda(r)] \equiv \ln\{P^{opt.}(X|r)/[CP^0(X|r)]\} = 0, \quad (69)$$

or $P^{opt.}(X|r) = CP^0(X|r)$, when combined with the normalization constraint, $\Sigma_X P^{opt.}(X|r) = C\Sigma_X P^0(X|r) = C = 1$, is then seen to give the Hirshfeld solution of Eqs. (56) and (62): $d^{opt.}(r) = d^0(r) = d^H(r)$. Therefore, the Hirshfeld choice of the local share factors minimizes the local information distance of Eq. (67) to the lowest value possible: $\Delta S[d^H(r)|d^0(r)] = 0$.

The same answer follows from the relevant *global* information principles formulated in terms of either the electron densities or their shape factors [11], in which one seeks the optimum atomic (or fragment) distributions exhibiting the strongest resemblance to the corresponding non-bonded, reference distributions. Let us define the KL information distance functional (see Section 2.1) between the trial *one*-electron densities of atomic fragments $\{\rho_X\} \equiv \rho$ (or the associated probability distributions $\{p_X\} \equiv p = \rho/N$) of the bonded atoms and the corresponding reference densities $\{\rho_X^0\} \equiv \rho^0$ (or $\{p_X^0\} \equiv p^0 = \rho^0/N$) of the free-atoms:

$$\begin{aligned}\Delta S[\rho|\rho^0] &= \Sigma_X \int \rho_X(r) \ln[\rho_X(r)/\rho_X^0(r)]\, dr \equiv \Sigma_X \int \rho_X(r)\, I_X[w_X(r)]\, dr \equiv \Sigma_X \Delta S_X[\rho_X|\rho_X^0] \\ &= N\Delta S[p|p^0] = N\Sigma_X \int p_X(r) \ln[p_X(r)/p_X^0(r)]\, dr \\ &\equiv N\Sigma_X \int p_X(r)\, I_X[w_X(r)]\, dr \equiv N\Sigma_X \Delta S_X[p_X|p_X^0]. \end{aligned} \quad (70)$$

Here, $I_X[w_X(r)] = \ln[p_X(r)/p_X^0(r)] = \ln[\rho_X(r)/\rho_X^0(r)]$, stands for the atomic (variational) surprisal, for the current value of the local enhancement factor relative to the *free*-subsystem reference: $w_X(r) = \rho_X(r)/\rho_X^0(r) = p_X(r)/p_X^0(r)$. We have also indicated in the preceding equation that the directed divergence of ρ relative to ρ^0, $\Delta S[\rho|\rho^0]$ is just N times the entropy deficiency $\Delta S[p|p^0]$. The same relation holds between the subsystem missing-information quantities: $\Delta S_X[\rho_X|\rho_X^0] = N\Delta S_X[p_X|p_X^0]$.

Therefore, for the isoelectronic molecular and promolecular systems, the problem of normalization of the compared electronic densities does not influence the corresponding constrained variational principle for determining the optimum densities:

$$\delta\{\Delta S[\rho|\rho^0] - \int \lambda(r)\,\Sigma_X\,\rho_X(r)\,dr\} = N\delta\{\Delta S[p|p^0] - \int \lambda'(r)\,\Sigma_X\,p_X(r)\,dr\} = 0, \tag{71}$$

where the local Lagrange multipliers $\lambda(r)$ and $\lambda'(r)$ enforce the exhaustive-division conditions at point r: $\Sigma_X\rho_X(r) = \rho(r)$ and $\Sigma_X\,p_X(r) = p(r)$, respectively. The above variational problem in terms of electron densities is then indeed equivalent to the associated principle in terms of the probability distributions (shape-factors):

$$\delta\{\Delta S[p|p^0] - \int \lambda(r)\,\Sigma_X\,p_X(r)\,dr\} = 0, \tag{72}$$

with $\lambda(r)$ now multiplying the local exhaustive-division constraint: $\Sigma_X\,p_X(r) = p(r)$. It can be easily verified by a straightforward functional differentiation that both these variational principles give the same answer of the local stockholder division: $p_X^{opt.}(r) = p_X^H(r) = \rho_X^H(r)/N$. In these global minimum entropy-deficiency principle the missing-information term provides an entropy "penalty" for the AIM densities deviating from the free atom densities.

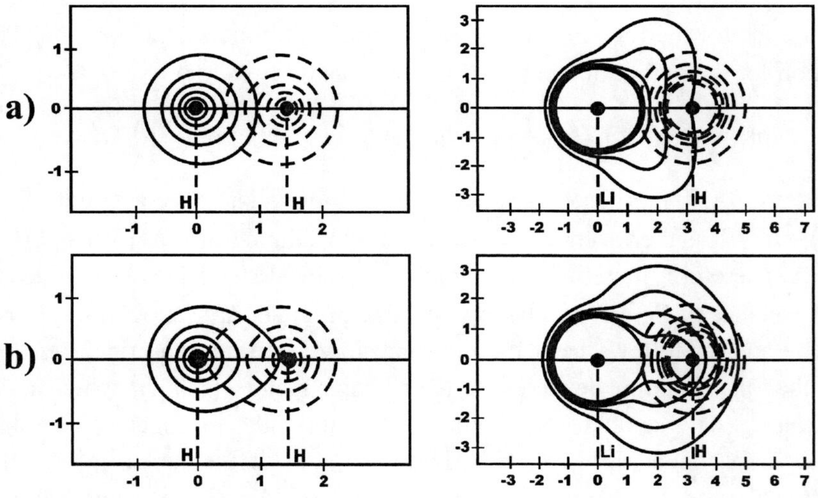

Figure 9. A comparison of the contour diagrams of the $\rho_X^H(r)$ (1-*S*, Panel *a*) and $\rho_X^{eff}(r)$ (2-*S*, Panel *b*) distributions of bonded atoms in H_2 (left column) and LiH (right column) [8,17].

This Hirshfeld solution of the IT density-partitioning problem is independent of the adopted entropy/information measure. For example, the symmetrized divergence of Kullback (1959),

$$\Delta S[\rho^0, \rho] = \Delta S[\rho|\rho^0] + \Delta S[\rho^0|\rho] = \Sigma_X \int \Delta\rho_X(r)\ln[\rho_X(r)/\rho_X^0(r)]\,dr = N\Delta S[p^0, p], \tag{73}$$

where $\Delta\rho_X(r) = \rho_X(r) - \rho_X^0(r) = N\Delta p_X(r)$, gives rise to the same partition. The Fisher intrinsic accuracy generates to the differential Euler equation for the optimum subsystem distributions. One of its particular solutions is the Hirshfeld prescription for the bonded atoms.

The entropy-deficiency principle can be similarly applied to a division of the molecular joint distribution of k electrons into the corresponding pieces corresponding to the AIM k-clusters, giving rise to the associated *many*-electron stockholder principle [8,13,16,17]. By an appropriate summation and integration these cluster distriubutions can be used to determine the associated effective densities $\{\rho_X^{eff.}(\boldsymbol{r};\, k)\}$ of the k-electron "stockholder" (k-S) atoms, which in general slightly differ from their *one*-electron (1-S) analogs of Hirshfeld. The largest differences (Figure 9) between densities of the 2-S and 1-S bonded atoms have been observed for the lightest hydrogen and lithium atoms [8,17]. As seen in this figure the 2-S H and Li atoms exhibit a more bonding (overlapping) character, compared to their 1-S analogs.

5.3. Information Densities of Bonded Atoms

The same entropic descriptors, which have been used in Section 4 to probe the molecular electron densities/probabilities, can be applied to diagnose changes the bonded atoms undergo in molecules, relative to the free atoms [8,10]. The entropy deficiency of the stockholder AIM,

$$\begin{aligned}\Delta S[\rho_X^H|\rho_X^0] &= \int \rho_X^H(\boldsymbol{r}) \log[\rho_X^H(\boldsymbol{r})/\rho_X^0(\boldsymbol{r})]\, d\boldsymbol{r} \equiv \int \rho_X^H(\boldsymbol{r})\, I_X^H(\boldsymbol{r}) \\ &= \int P^H(X|\boldsymbol{r})\rho(\boldsymbol{r})\, I(\boldsymbol{r})\, d\boldsymbol{r} \equiv \int P^H(X|\boldsymbol{r})\, \Delta s(\boldsymbol{r})\, d\boldsymbol{r} \equiv \int \Delta s_X^H(\boldsymbol{r})\, d\boldsymbol{r},\ \textstyle\sum_X P^H(X|\boldsymbol{r}) = 1, \qquad (74)\end{aligned}$$

defines the density of the atomic cross-entropy (shown in Figs. 10 and 11), $\Delta s_X^H(\boldsymbol{r}) = P^H(X|\boldsymbol{r})\Delta s(\boldsymbol{r})$, where the conditional probability $P^H(X|\boldsymbol{r})$ is defined by the stockholder share-factor $d_X^H(\boldsymbol{r})$ of Eq. (58), and the atomic surprisals $\{I_X^H(\boldsymbol{r})\}$ are all equalized at the global (molecular) value: $I(\boldsymbol{r}) = I(w(\boldsymbol{r}))$ [see Eq. (62)].

The global values of the entropy displacement of the Hirshfeld AIM,

$$\mathcal{H}_X^H = S[\rho_X^H] - S[\rho_X^0] = -\int \rho_X^H(\boldsymbol{r}) \log\rho_X^H(\boldsymbol{r})\, d\boldsymbol{r} + \int \rho_X^0(\boldsymbol{r}) \log\rho_X^0(\boldsymbol{r})\, d\boldsymbol{r} \equiv \int h_X^H(\boldsymbol{r})\, d\boldsymbol{r}, \quad (75)$$

for selected linear molecules of Table 1 are listed in Table 3 together with the corresponding stockholder- and free-atom values of the Shannon entropy. The preceding equation also defines the atomic entropy-displacement density $h_X^H(\boldsymbol{r})$, shown in Figs. 10 and 11. A reference to H_2 and N_2 entries of Tables 1 and 3 shows that the atomic entropy displacements for bonded hydrogen atoms are approximately additive: $2\mathcal{H}_X^H \cong \mathcal{H}[\rho(H_2)]$. A similar near-additivity is observed for most of the remaining molecules, with the largest deviation from such an uncoupled (independent) behavior of changes in the AIM entropy being observed for the most ionic Li–F bond.

In Figure 10 we have compared the contour maps of the Hirshfeld AIM density-difference function,

In general, the bonded atom exhibits a lower degree of uncertainty compared to the free-atom value, a clear sign of the dominating effect of a relatively more compact electron distribution. In strong electron acceptors, e.g., for F in LiF and N in HCN, one detects positive displacements due to the dominating CT contribution, which should result in a softer atomic distribution of electrons, thus exhibiting more "disorder" (uncertainty).

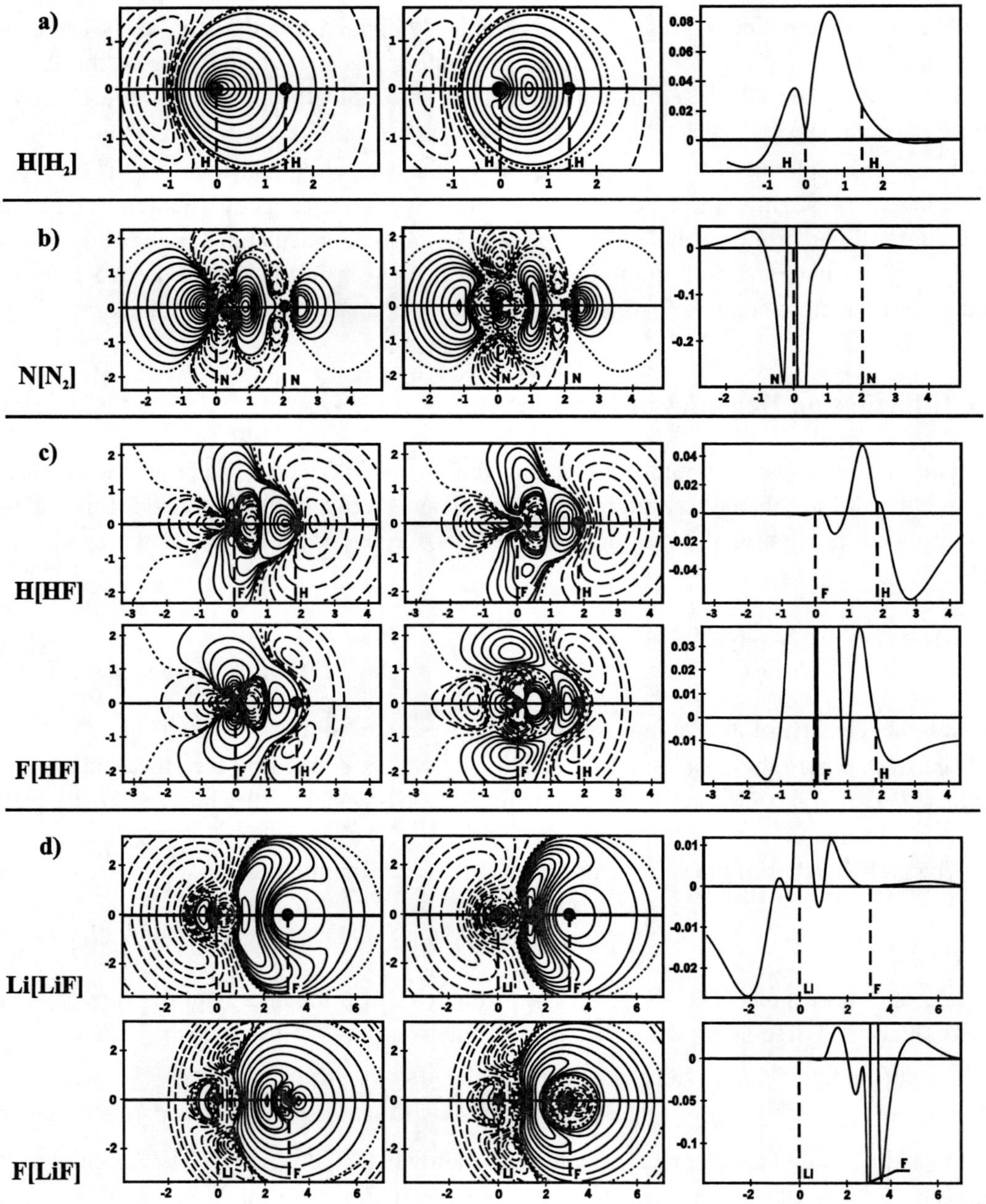

Figure 10. Representative contour maps of the atomic electron density-difference function $\Delta\rho_X^H(r)$ (first column) and the entropy-displacement density $h_X^H(r)$ (second column), for the constituent atoms of diatomic molecules in Figure 2: H_2 (*a*), N_2 (*b*), HF (*c*), and LiF (*d*). The contour values are not equidistant, having been selected only for the purpose of revealing the topographic features of the compared quantities. In the third column the bond axis profiles of $h_X^H(r)$ (a.u.) are reported [8,10].

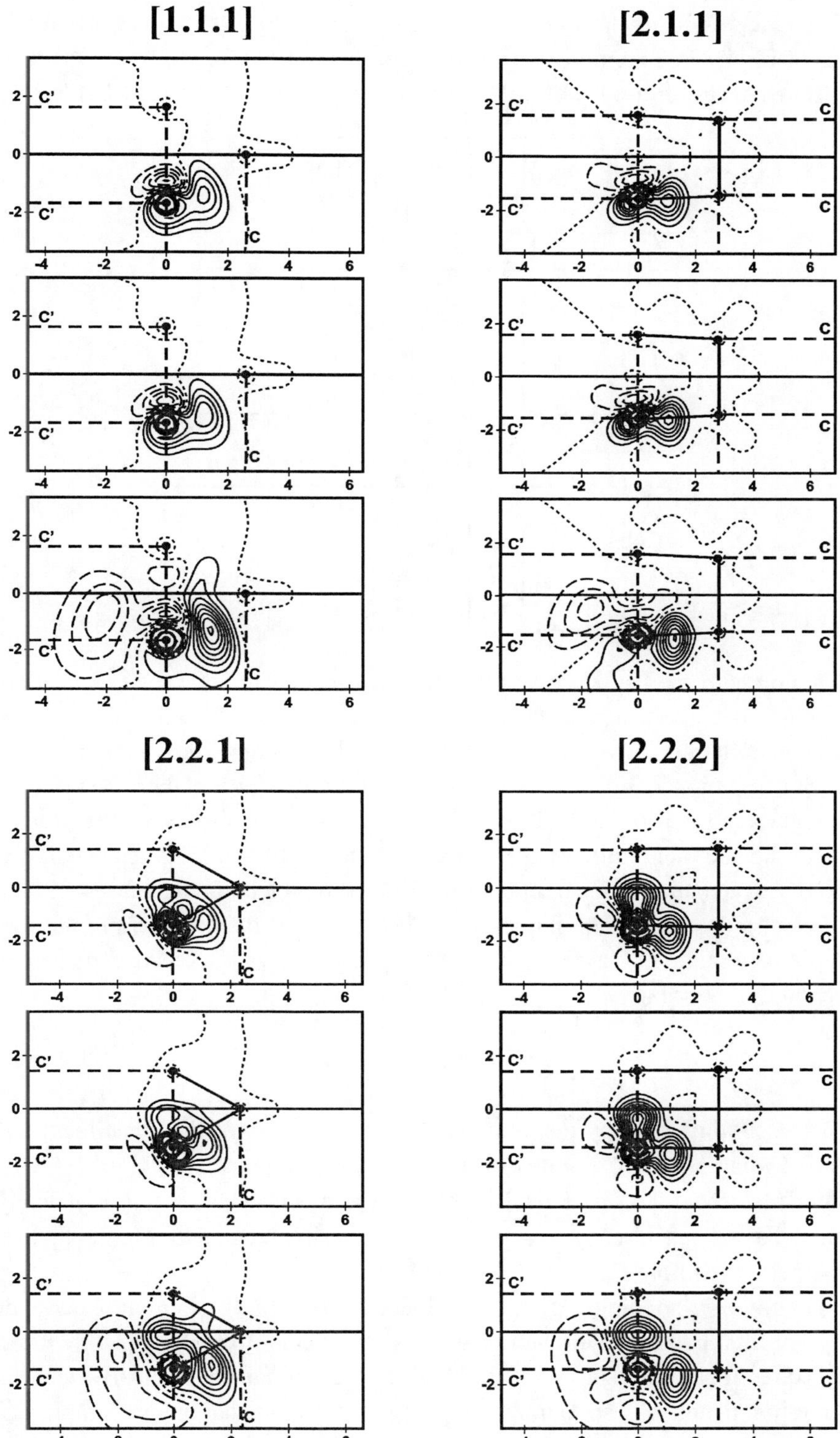

Figure 11. Contour diagrams of displacements in the electron density (upper panel), information-distance density (medium panel) and of the atomic Shannon entropy (lowest panel), for the stockholder bridgehead carbon atoms in the four propellanes of Figure 5.

Table 3. Displacements of the Hirshfeld AIM entropies (in bits) for representative linear molecules of Figure 2. The molecular and promolecular entropy data are also reported. The corresponding molecular entropy shifts are reported in Table 1.

Molecule	X	$\mathcal{H}_X^H$	$S[\rho_X^H]$	$S[\rho_X^0]$
H_2	H	− 0.41	3.77	4.18
N_2	N	− 0.34	5.86	6.20
HF	H	− 1.09	3.09	4.18
	F	0.03	1.22	1.19
LiF	Li	− 4.02	3.87	7.89
	F	0.97	2.14	1.17
HCN	H	− 0.87	3.31	4.18
	C	− 0.73	7.29	8.03
	N	0.15	6.35	6.20
CNH	C	0.01	8.04	8.03
	N	− 0.44	5.76	6.20
	H	− 0.98	3.20	4.18

The atomic entropy displacements for HF and LiF indicate that the *donor* atom exhibits the dominating (negative) displacement, while the *acceptor* AIM only slightly increases its entropy (see also the net AIM charges in Table 2). The triatomic data in Table 3 provide an additional confirmation of this rule, with the exception of N in HNC. A reference to the atomic charges reported in Table 2 again shows a strong sensitivity of the atomic entropy displacements to the magnitude of the atomic net charge transfer. As expected, a degree of this sensitivity to a change in the atomic overall electron population decreases with the overall number of electrons on the atom in question. Indeed, a given displacement in the AIM charge is seen to produce relatively large reconstructions of the free-atom electron distributions in H or Li, compared to N or F.

$$\Delta\rho_X^H(\boldsymbol{r}) = \rho_X^H(\boldsymbol{r}) - \rho_X^0(\boldsymbol{r}), \qquad (76)$$

with the corresponding entropy displacement density $h_X^H(r)$, for the constituent AIM of the representative diatomics of Figure 2. The density-difference plot (left panel) for the "stockholder" hydrogen in H_2, H[H_2], exhibits typical changes for the covalently-bonded atom, which have been already observed in the density profile of Figure 8: the electron density buildup around the nucleus and in the bond region between nuclei, at the expense of the outer, mainly *non*-bonding regions of the atomic density distribution. This is due to the contraction of the bonded hydrogen and its polarization towards the other atom. This cylindrical, directional polarization of the atomic electron density is clearly seen in the left, electron-density panel. A similar pattern, with a somewhat more emphasized bond polarization, is seen in the entropy-difference plot of the right contour map. The bonding part of the AIM entropy displacement is positive, thus marking an increase in the local uncertainty due to the electron delocalization towards the bonding partner. A reference to part *b* of the figure, devoted to N in N_2, again shows a general similarity between the two compared contour diagrams. The displacements observed in the valence shell of both panels accord with the known density-deformation changes in the triply-bonded nitrogen: the $(2s,2p_\sigma)$-hybridization along the molecular axis, a transfer of the $2p_\pi$ electrons to the bond-charge region between two nuclei, with the buildup (lowering) of the electron density of the left panel giving rise to the associated increase (decrease) in the local electron uncertainty of the right panel. Both these panels confirm the molecular character of the Hirshfels atoms, with the atom displacement "tails" extending all over the molecule.

In both density and entropy panels for H[HF], again strongly resembling one another, the hydrogen is seen to be strongly polarized towards the fluorine atom, with the transferred electron density being channeled to both the σ bond region and to the $2p_\pi$ lone-pair regions around fluorine. The valence-shell part of the F[HF] panels reveals a similar polarization of the fluorine towards the hydrogen, with the accompanying increase in the lone pair $(2p_\pi)$ density, in the direction perpendicular to the bond axis. These observations confirm a relatively strong covalency of H—F bond. The localization of the fluorine bond-charge close to the proton position provides an additional support to this observation. A different CT pattern is found in the strongly ionic Li—F bond, for which the difference diagrams of the AIM density and entropy consistently show a transfer of electrons from the peripheral part of the Li[LiF] electron distribution toward F as a whole, giving rise to a much lower degree of localization of the density displacement. This reflects a partially covalent bond character, with much lower covalent (electron-sharing) contribution though, compared to that observed in HF. This observation accords with the relative hardness of the two atoms in these molecules: the two hard atoms in HF give rise to strongly covalent bond, while the soft (Li) and hard (F) atoms in LiF generate a relatively more ionic (CT) bond.

The observed features of the local entropy displacement thus indicate that it can indeed serve as an alternative, sensitive diagnostic tool for monitoring the bonding condition of AIM.

As a final illustration of this section we examine the bridgehead carbon atoms of the four propellanes of Figure 5. The contour maps of Figure 11 report the density difference function, $\Delta\rho_X^H$, the KL missing-information density, Δs_X^H, and the entropy-displacement density, h_X^H. They fully support the conclusions already drawn from the molecular plots of Figures 6 and 7.

The three contour diagrams for the given bridgehead carbon atom in Figure 11 are seen to be qualitatively similar, thus further validating the usefulness of all these quantities as sensitive probes into changes the free atoms undergo in the molecule. These alternative diagnostic tools are seen to be to a large extent equivalent. For the small [1.1.1] and [2.1.1] propellanes, which are lacking the "through-space" component of the central bond between the bridgehead carbons, the atomic density/information buildups observed in the three panels indeed reflect the "through-bridge" chemical bonds only, with a distinct lowering of the AIM densities in the direction of the other bridgehead atom. In the two larger systems of [2.2.1] and [2.2.2] propellanes, in which the presence of the "through-space" component of the central bond has been inferred from both the molecular density difference and the direct bond-order measures, the displacement in AIM electron density and the corresponding information-distance/entropy densities exhibit increases in directions of both the neighboring bridgehead and bridge carbon atoms.

6. Importance of Non-Additive Information Measures

Each resolution of the molecular electron density/probability distribution, e.g., $\rho = \sum_\alpha \rho_\alpha$, implies the associated division of the molecular (*total*) physical quantity $A[\rho]$ into its additive, $A^{add.}[\rho]$, and *non-additive*, $A^{nadd.}[\rho]$, contributions:

$$A[\rho] \equiv A^{total}[\rho] = A^{add.}[\rho] + A^{nadd.}[\rho], A^{add.}[\rho] = \sum_\alpha A[\rho_\alpha]. \tag{77}$$

We have indicated above that in the underlying "multi-component" system $A[\rho]$ becomes the functional of the whole vector of the subsystem densities: $A[\rho] = A^{total}[\rho]$ [45]. For example, this Gordon-Kim-type division [46] of the kinetic energy functional defines the non-additive contribution, which constitutes the basis of the DFT-embedding concept of Cortona and Wesołowski [47,48]. Such division can be also used to partition the information quantities themselves [8,20,21]. In particular, the inverse of the non-additive Fisher information in the MO resolution has been shown to define the IT-ELF concept [21], in the spirit of the original Becke and Edgecombe formulation [22], while the related quantity in the AO resolution of the SCF MO theory offers the key CG criterion for localization of chemical bonds in molecular systems [19,20].

As an illustration consider again the *local* variational problem of Eq. (68), in partitioning the molecular electron density into the AIM components, in which the molecular probability density at point r is divided into atomic contributions in accordance with the conditional probabilities (*share* factors) $d(r) = \{P(X|r) = p_X(r)/p(r) = \rho_X(r)/\rho(r)\}$. The promolecular reference similarly determines the initial conditional probabilities of free atoms: $d^0(r) = \{P(X^0|r) = p_X{}^0(r)/p^0(r) = \rho_X{}^0(r)/\rho^0(r)\}$. In this local scheme the overall distributions $p(r)$ and $p^0(r)$ as well as the promolecular conditional probabilities $d^0(r)$ are fixed. Therefore, only the molecular conditional probabilities of AIM, $d(r)$, are to be determined from the corresponding (entropic) Euler equations, which define the optimum shapes of bonded atoms. The optimized information density $\Delta S[d(r)|d^0(r)]$ in this local partition is additive in AIM contributions,

$$\begin{gathered}\Delta S[d(r)|d^0(r)] = \Delta S^{add.}[d(r)|d^0(r)] = \sum_X \Delta S_X[P(X|r)|P(X^0|r)],\\ \Delta S_X[P(X|r)|P(X^0|r)] = P(X|r)\log[P(X|r)/P(X^0|r)],\end{gathered} \tag{78}$$

and hence $\Delta S^{nadd.}[d^H(r)|d^0(r)] = 0$. The non-additive entropy deficiency reaches the (absolute) minimum value $\Delta S^{add.}[d^H(r)|d^0(r)] = 0$, for the optimum, Hirshfeld solutions, when $d^H(r) = d^0(r)$.

The same feature is observed in the KL information densities of the (molecularly-normalized) probability distributions $p^H(r) = \{p_X{}^H(r)\}$, relative to those characterizing the free atoms: $p^0(r) = \{p_X{}^0(r)\}$. One of the remarkable features of this division is that the density of the overall entropy deficiency between the molecular and promolecular electron probability densities, which determines the *total* missing-information density in the stockholder-AIM resolution,

$$\begin{gathered}\Delta S[p(r)|p^0(r)] = p(r)\log[p(r)/p^0(r)] = p(r)\, I(r) = p(r)\, I_X{}^H(r)\\ \equiv \Delta S^{total}[p^H(r)|p^0(r)] = \Delta S^{add.}[p^H(r)|p^0(r)] + \Delta S^{nadd.}[p^H(r)|p^0(r)],\end{gathered} \tag{79}$$

is equal to its *additive* part:

$$\Delta S[p(r)|p^0(r)] = p(r)I(r) = \sum_X p_X{}^H(r)\, I_X{}^H(r) = \sum_X \Delta S_X[p_X(r)|p_X{}^0(r)] \equiv \Delta S^{add.}[p^H(r)|p^0(r)], \tag{80}$$

due to the common, molecular surprisal for all stockholder AIM. In other words, the equality of the preceding equation implies that for this particular partitioning the *non-additive* information distance between $p^{H}(r)$ and $p^{0}(r)$ also vanishes identically, $\Delta S^{nadd.}[p^{H}(r)|p^{0}(r)] = 0$, and hence also the overall functional $\Delta S^{nadd.}[p^{H}|p^{0}] = 0$. The normalization-constrained minimum principle of $\Delta S^{add.}[p|p^{0}]$ thus implies the associated minimum of the (positive) $\Delta S[p|p^{0}] = \Delta S^{total}[p|p^{0}]$ at the optimum, stockholder solutions.

6.1. Electron Localization Function

Consider next the *non*-additivities of the Fisher measure of the information [Eqs. (4) and (6)] in the MO resolution, say, defined by the occupied *Kohn-Sham* (KS) orbitals of DFT calculations [35]. In the spin-resolved approach one examines the energy functionals of spin densities $\{\rho_\sigma(\boldsymbol{r})\}$, for the spin-like electrons occupying the $\boldsymbol{\psi} = \left\{ \boldsymbol{\psi}_\sigma = \{\psi_i^\sigma\} \right\}$ MO,

$$\rho_\sigma(\boldsymbol{r}) = \sum_{i\in\sigma}^{occd.} \left|\psi_i^\sigma(\boldsymbol{r})\right|^2 . \tag{81}$$

The kinetic energy density,

$$\tau_\sigma(\boldsymbol{r}) = \sum_{i\in\sigma}^{occd.} \left|\nabla\psi_i^\sigma(\boldsymbol{r})\right|^2 , \tag{82}$$

is then proportional to the sum of *additive* Fisher information densities in the amplitude representation [see Eqs. (6) and (7)]:

$$\tau_\sigma(\boldsymbol{r}) = \frac{1}{4} f_\sigma^{add.}(\boldsymbol{r}) . \tag{83}$$

The leading term of the Taylor expansion of the spherically averaged (Hartree-Fock) conditional pair-probability of finding in distance s from the reference electron of spin σ at position $\boldsymbol{r}$ the other spin-like electron then reads [22],

$$P_c^{\sigma\sigma}(s|\boldsymbol{r}) = \frac{1}{3} D_\sigma(\boldsymbol{r}) s^2 + ..., \quad D_\sigma(\boldsymbol{r}) =$$
$$\tau_\sigma(\boldsymbol{r}) - \frac{\left|\nabla\rho_\sigma(\boldsymbol{r})\right|^2}{4\rho_\sigma(\boldsymbol{r})} = -\frac{1}{4}[f_\sigma^{total}(\boldsymbol{r}) - f_\sigma^{add.}(\boldsymbol{r})] = -\frac{1}{4} f_\sigma^{nadd.}(\boldsymbol{r}) \geq 0 . \tag{84}$$

We have explicitly indicated above that the key function $D_\sigma(\boldsymbol{r})$ can be identified as being proportional to the negative non-additive component of the Fisher information in this spin-resolved MO resolution [8,21]. The appropriately calibrated square of its inverse, which determines ELF [22], has been successfully used as the probe of the electron localization patterns in molecules [49]. A somewhat improved behavior is observed in the (DFT-tailored) simple inverse of this function, known as IT-ELF [8,21]. In Figure 12 representative graphs of both functions are presented for Ne, Ar, Kr and Xe. A qualitative behavior of the two curves is seen to be very similar. In general, IT-ELF exhibits smaller outer amplitudes and thus a larger spatial extension than the original ELF.

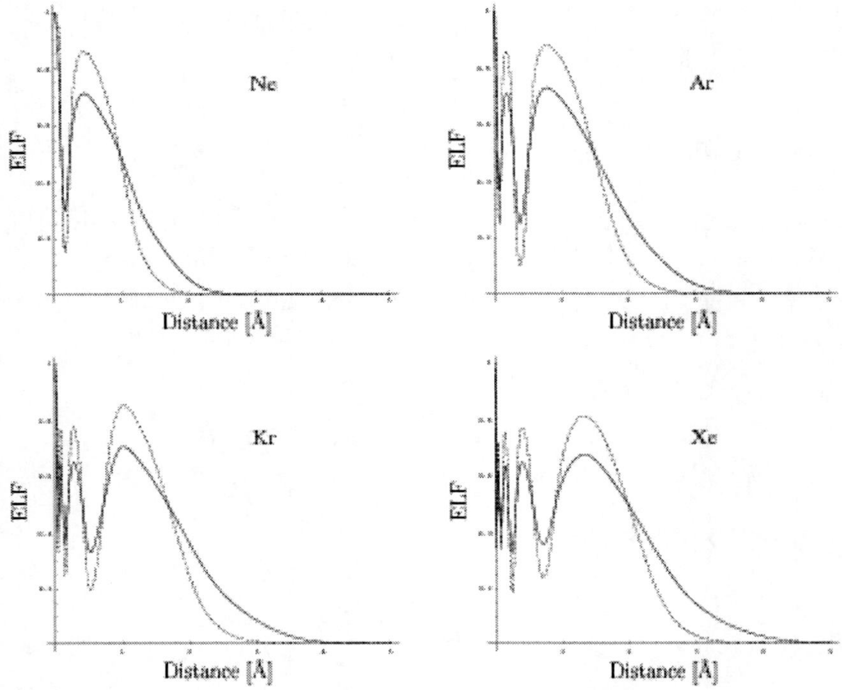

Figure 12. Plots of ELF (dashed line) and IT-ELF (solid line) for Ne, Ar, Kr and Xe.

In Figure 13 we have compared the perspective views of these functions for N_2, NH_3, PH_3 and B_2H_6. These ELF functions convincingly validate the use of this local probe as an indicator of the localization of the valence electrons in the bonding and non-bonding (lone-pair) regions of these illustrative molecular systems. Indeed in the homonuclear diatomic N_2 one detects in both plots the typical accumulation of electrons between the nuclei, due to the formation of the triple covalent bond, and the accompanying increase in the localization of the lone pair electrons in the non-bonding regions of both atoms, a clear manifestation of the accompanying *sp*-hybridization. The three localized N—H bonds are also clearly visualized in both NH_3 panels of the figure and a similar ELF features are detected in its PH_3 part. The final B_2H_6 plots are also seen to successfully locate the bonding electrons of the four terminal B—H bonds.

The final application, shown in Figure 14, again investigates the central and bridge bonds between the carbon atoms in the smallest, [1.1.1] and largest [2.2.2] propellanes of Figure 5. In these contour maps the absence (presence) of the through-space component of the central bond between the bridgehead carbon atoms is clearly seen in the upper (lower) panel for both versions of ELF, while the structure of the C—C bonds in the bridges is also transparently revealed. An interesting feature of the bridge bonds in the upper diagrams, also seen in Figure 6, is a slight displacement of the bonding electrons away from the line connecting the bridge and bridgehead carbons, due to the near sp^3-hybridization of the bridge atom which is required to accommodate the two hydrogen atoms.

N_2

ELF

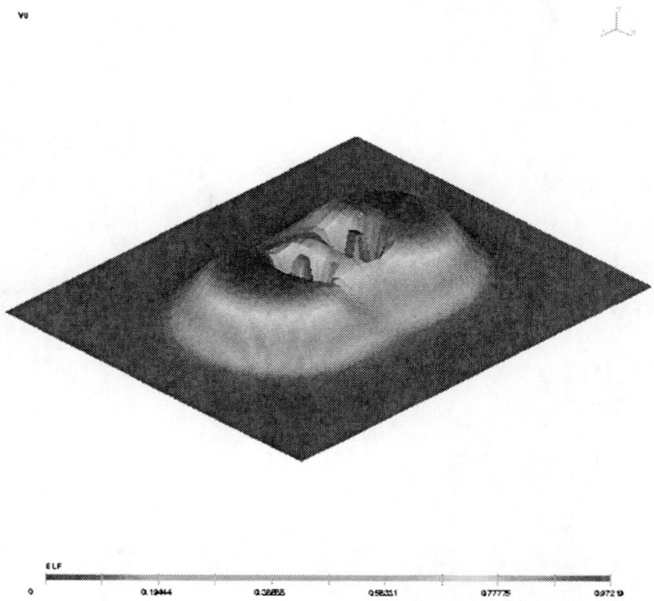

IT-ELF

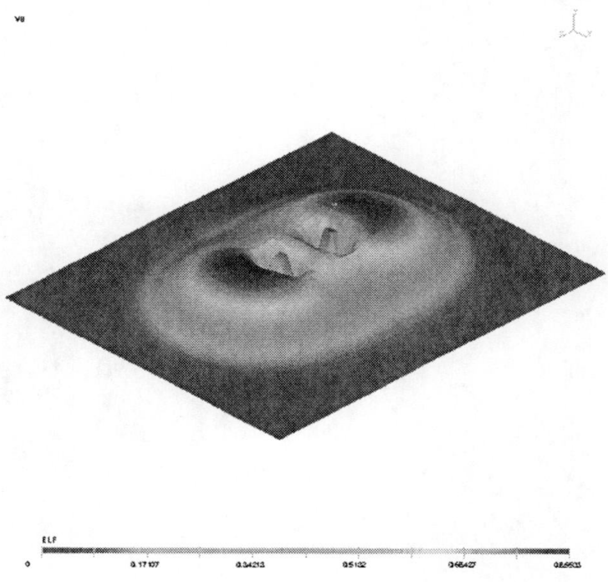

Figure 13. Continued on next page.

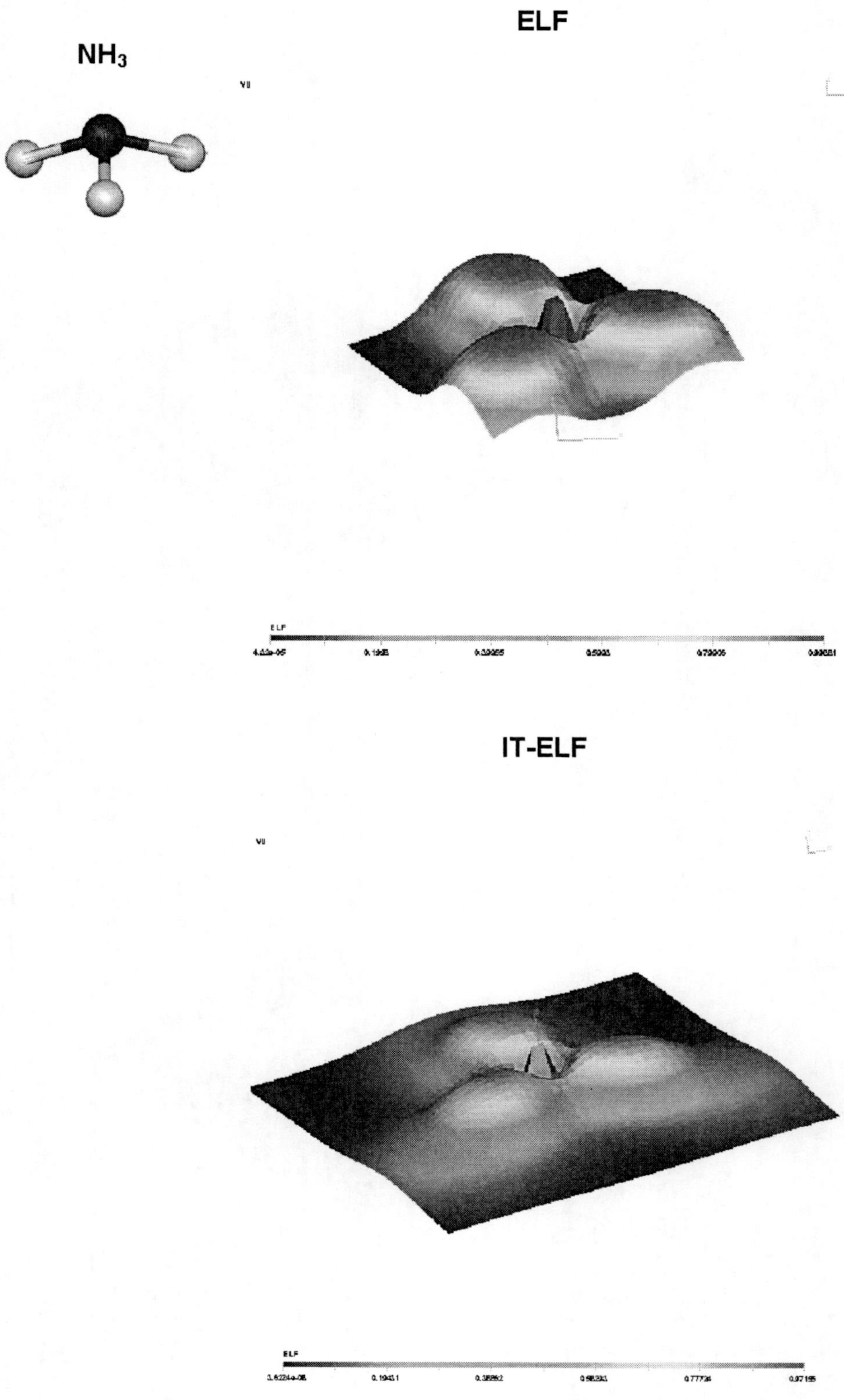

Figure 13. Continued on next page.

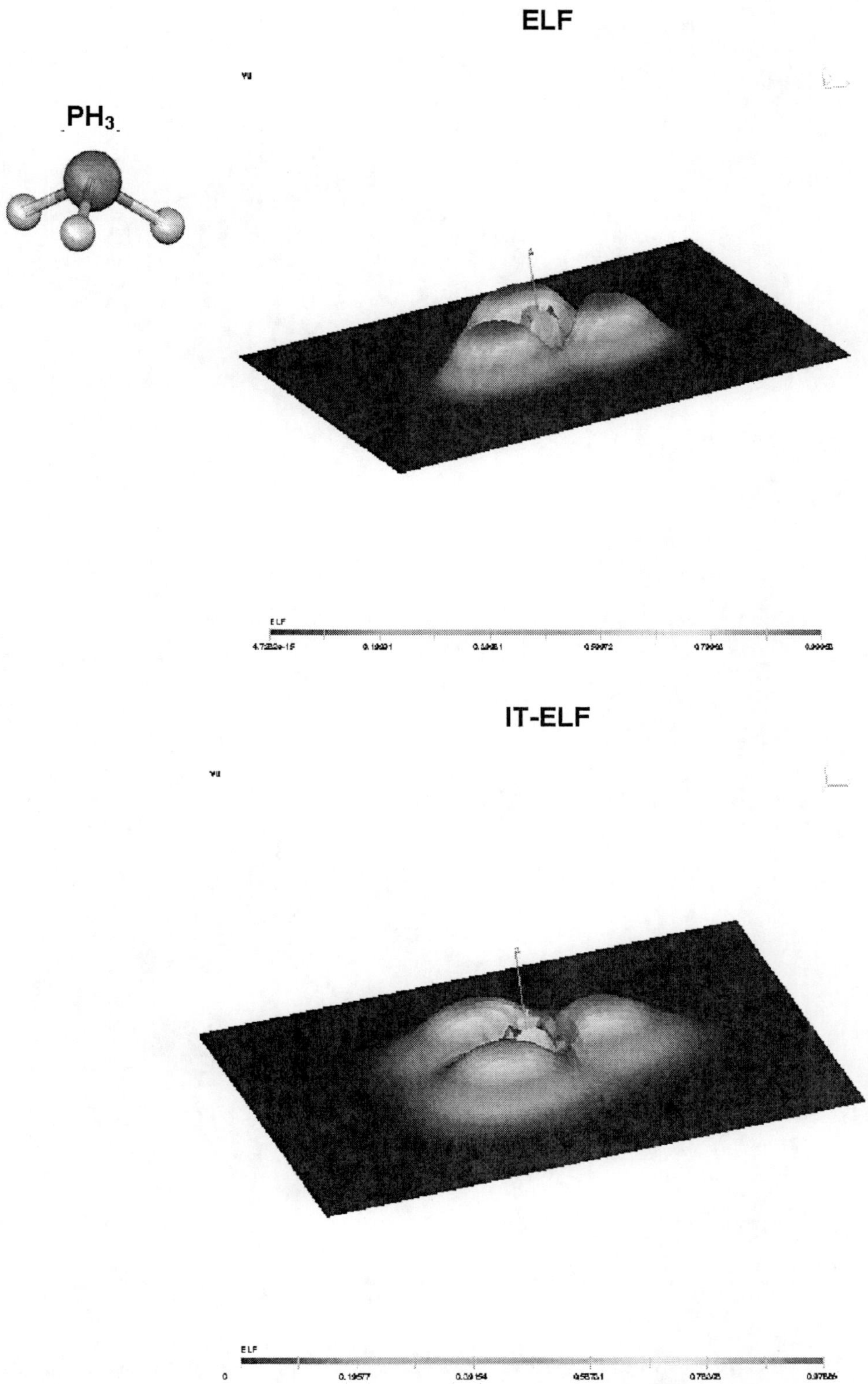

Figure 13. Continued on next page.

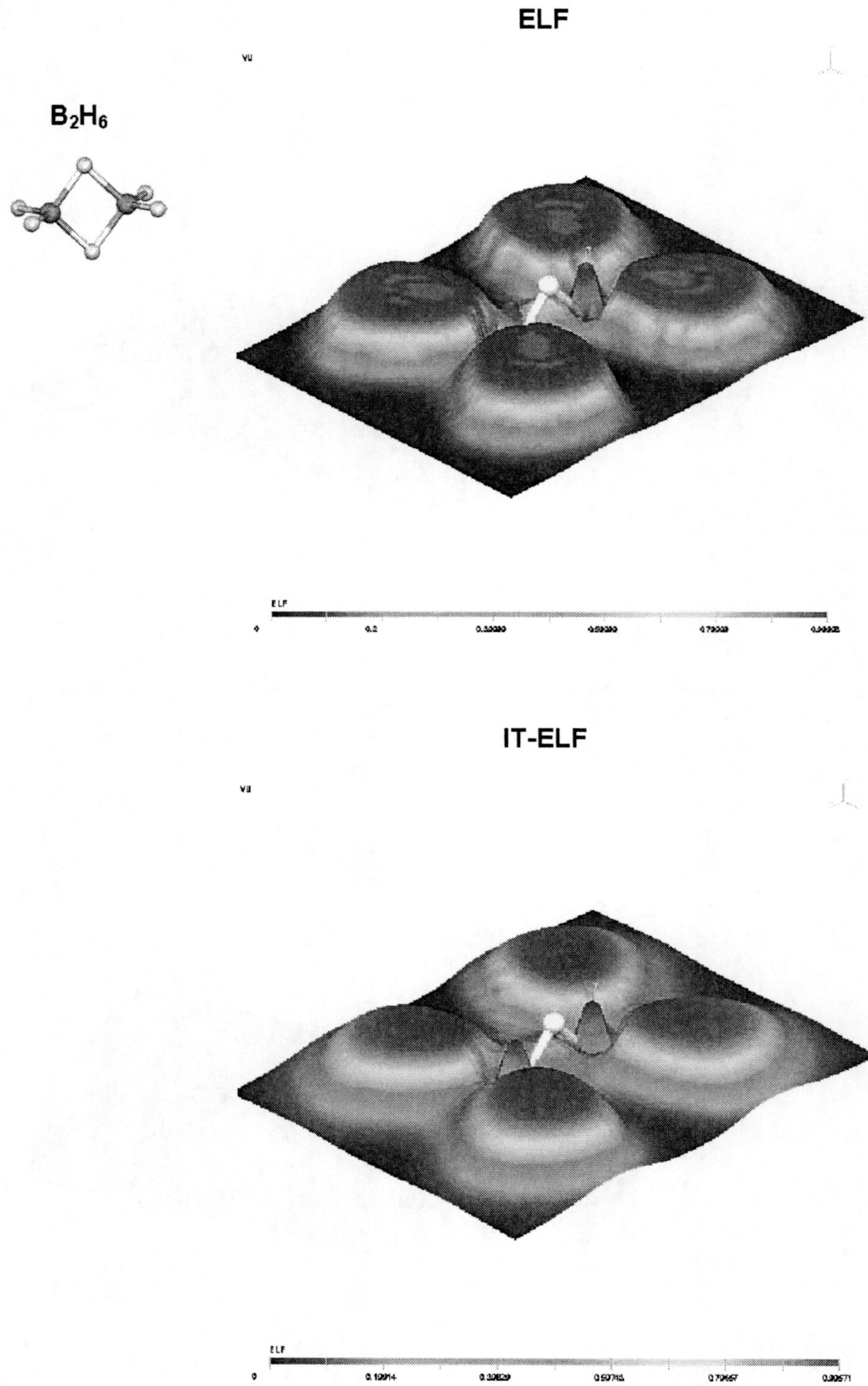

Figure. 13. Comparison between perspective views of the ELF and IT-ELF functiona for N_2, NH_3, PH_3, and B_2H_6, on the following planes of section: along the bond axis (N_2), determined by three hydrogen atoms (NH_3 and PH_3), and passing through both terminal BH_2 groups (B_2H_6).

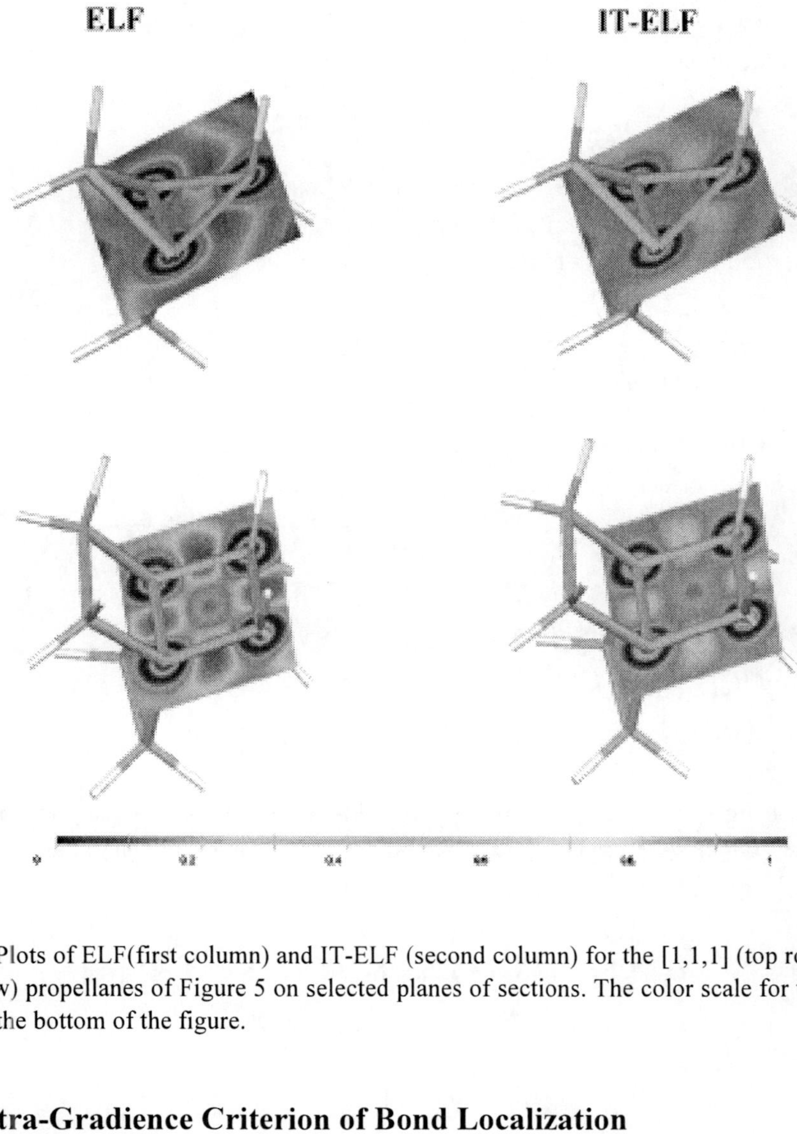

Figure 14. Plots of ELF(first column) and IT-ELF (second column) for the [1,1,1] (top row) and (2,2,2) (bottom row) propellanes of Figure 5 on selected planes of sections. The color scale for the ELF values is given in the bottom of the figure.

6.2. Contra-Gradience Criterion of Bond Localization

Consider next the familiar problem of combining the two Löwdin-orthogonalized AO (OAO), $A(\boldsymbol{r})$ and $B(\boldsymbol{r})$, say two 1*s* orbitals centered on atoms A and B, respectively, which contribute a single electron each to form the chemical bond A—B. The two basis functions $\boldsymbol{\chi} = (A, B)$ then form the bonding (φ_b) and anti-bonding (φ_a) MO combinations, $\boldsymbol{\varphi} = (\varphi_b, \varphi_a) = \boldsymbol{\chi}\mathbf{C}$:

$$\varphi_b = \sqrt{P}A + \sqrt{Q}B \equiv \Sigma_{k=A,B}\, \chi_k C_{k,b},\ \varphi_a = -\sqrt{Q}A + \sqrt{P}B \equiv \Sigma_{k=A,B}\, \chi_k C_{k,a},\ P + Q = 1, \quad (85)$$

where the square matrix $\mathbf{C}$ groups the LCAO MO expansion coefficients expressed in terms of the complementary probabilities P and $Q = 1 - P$, with P marking the conditional probabilities $P(A|\varphi_b) = P(B|\varphi_a) = P$, and Q measuring the remaining matrix elements $P(B|\varphi_b) = P(A|\varphi_a) = Q$. We examine the (*bonding*) ground-state, $\Psi_0 = [\varphi_b^{\,2}]$, the (*non-bonding*) first-excited configuration $\Psi_1 = [\varphi_b^{\,1}\varphi_a^{\,1}]$, and the (*anti-bonding*) doubly-excited state $\Psi_2 = [\varphi_a^{\,2}]$.

The corresponding *charge–and–bond-order* (CBO) density matrix of the SCF LCAO MO theory for each model configuration,

$$\gamma_i = \mathbf{C}\,\mathbf{n}(\Psi_i)\,\mathbf{C}^{\mathrm{T}},\ i = 0, 1, 2, \tag{86}$$

with the diagonal matrix of the MO electron occupations $\mathbf{n}(\Psi_i) = \{n_s(\Psi_i)\delta_{s,s'}\}$, which constitutes the OAO representation of the projection operator on the occupied MO subspace of the configuration in question, then read:

$$\gamma_0 = 2\begin{bmatrix} P & \sqrt{PQ} \\ \sqrt{PQ} & Q \end{bmatrix},\ \gamma_1 = \begin{bmatrix} 1 & 0 \\ 0 & 1 \end{bmatrix}, \gamma_2 = 2\begin{bmatrix} Q & -\sqrt{PQ} \\ -\sqrt{PQ} & P \end{bmatrix}. \tag{87}$$

They are seen to reflect the configuration chemical character by the positive, zero, and negative values, respectively, of the off-diagonal *bond-order* $\gamma_{1,2}$.

In the 2-OAO (real) model the partition of the Fisher information densities of the bonding and anti-bonding MO give:

$$\begin{aligned} f_b &= 4(\nabla\varphi_b)^2 \equiv f_\chi^{total}[\varphi_b] = 4[P(\nabla A)^2 + Q(\nabla B)^2] + 8\sqrt{PQ}\,\nabla A\bullet\nabla B \\ &\equiv f_\chi^{add.}[\varphi_b] + f_\chi^{nadd.}[\varphi_b], f_a = 4(\nabla\varphi_a)^2 \equiv f_\chi^{total}[\varphi_a] \\ &= 4[Q(\nabla A)^2 + P(\nabla B)^2] - 8\sqrt{PQ}\,\nabla A\bullet\nabla B \equiv f_\chi^{add.}[\varphi_a] + f_\chi^{nadd.}[\varphi_a]. \end{aligned} \tag{88}$$

It should be further realized that along the bond-axis in the AO-overlap region between the two atoms, which is decisive for the bonding or anti-bonding character of MO (see Appendix A), $\nabla A\bullet\nabla B < 0$, so that for these locations

$$\begin{aligned} f_\chi^{nadd.}[\varphi_b] &= 8\sqrt{PQ}\,\nabla A\bullet\nabla B \equiv 8\sqrt{PQ}i^{c-g} < 0,\ f_\chi^{nadd.}[\varphi_a] = \\ &-8\sqrt{PQ}\,\nabla A\bullet\nabla B = -8\sqrt{PQ}i^{c-g} > 0. \end{aligned} \tag{89}$$

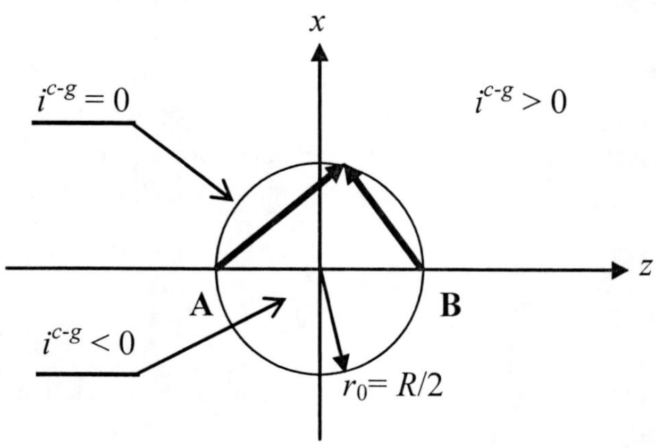

Figure 15. The circular contour of the vanishing contra-gradience integrand of two *s* orbitals on atoms A and B, $i^{cg}(r) = 0$, passing through both nuclei, which separates the bonding region (inside the circle), where $i^{c\text{-}g}(r) < 0$, from the region of positive contributions $i^{c\text{-}g}(r) > 0$ (outside the circle).

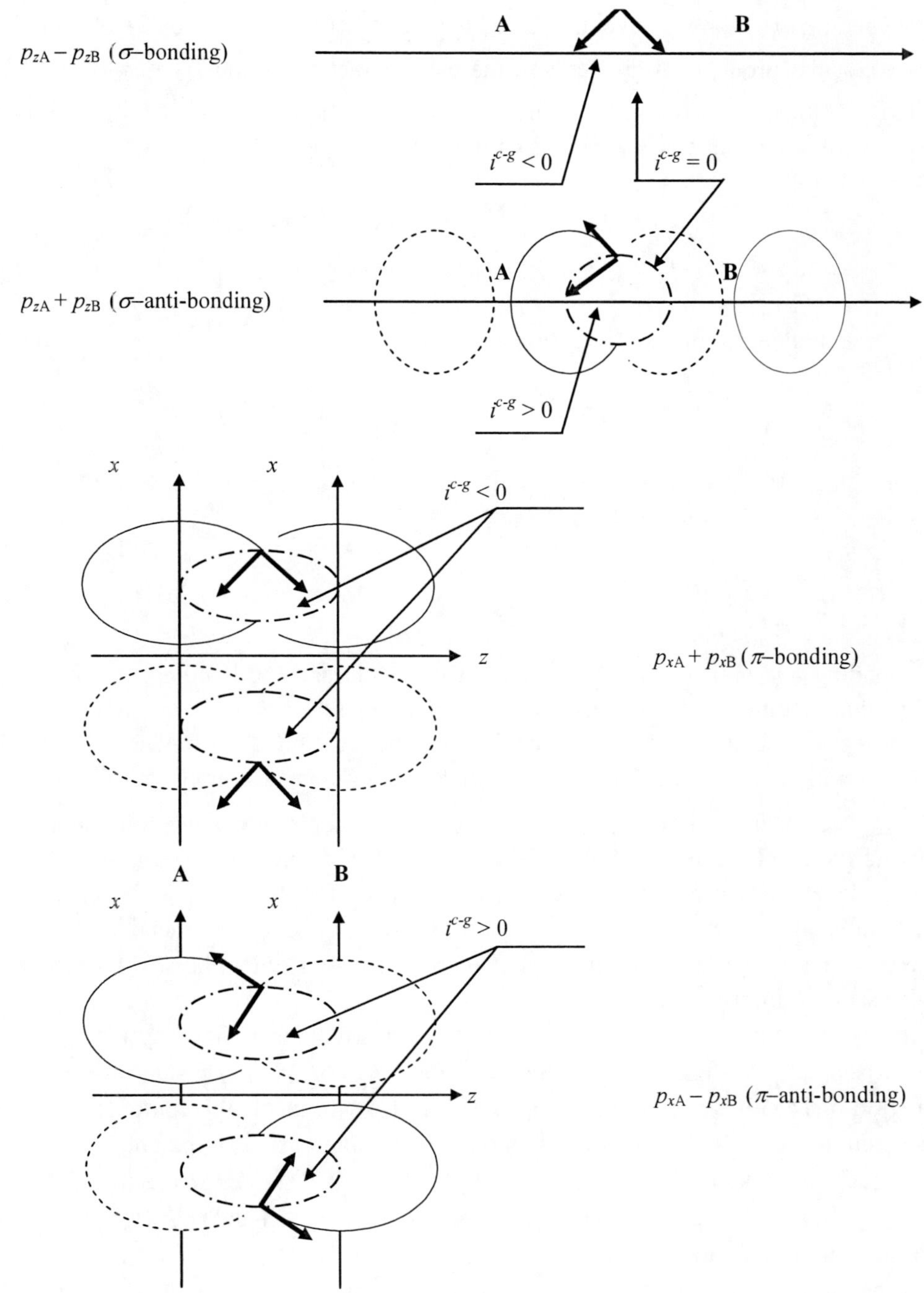

Figure 16. A schematic representation of the bonding [$i^{c\text{-}g}(r) < 0$] and anti-bonding [$i^{c\text{-}g}(r) > 0$] regions of the strong orbital overlap in the chemical interactions between the valence orbitals p_σ and p_π of atoms A and B, enclosed by the vanishing contra-gradience contours $i^{c\text{-}g}(r) = 0$, represented by the pointed broken line. The AO gradients, each in the direction perpendicular to the orbital contour, are mutually perpendicular on the $i^{c\text{-}g}(r) = 0$ surface (see also Figure 15). It again follows from these qualitative diagrams that the negative density of the non-additive Fisher information accompanies the electron delocalization in the bonding MO-combination of AO, while the positive density of this information contribution is associated with an effective electron localization, which characterizes the anti-bonding MO-combination of AO.

In the preceding equation we have introduced the CG density, which in the 2-OAO model is given by the scalar products of gradients of the two interacting orbitals, $i^{c-g} = \nabla A \bullet \nabla B$. One similarly defines the average densities of the *non*-additive and additive Fisher information contributions per electron for the configuration in question:

$$f_{\chi}^{nadd.}[\Psi_i] = \Sigma_s [n_s(\Psi_i)/N]\, f_{\chi}^{nadd.}[\varphi_s] \equiv \Sigma_s p_s(\Psi_i)\, f_{\chi}^{nadd.}[\varphi_s],$$
$$f_{\chi}^{add.}[\Psi_i] = \Sigma_s p_s(\Psi_i)\, f_{\chi}^{add.}[\varphi_s]. \qquad (90)$$

For the three model electron configurations one thus obtains:

$$f_{\chi}^{nadd.}[\Psi_0] = 8\sqrt{PQ}\, i^{c-g},\ f_{\chi}^{nadd.}[\Psi_1] = 0,\ f_{\chi}^{nadd.}[\Psi_2] = -8\sqrt{PQ}\, i^{c-g}. \qquad (91)$$

Therefore, the AO-phase dependent, non-additive contribution to the MO Fisher information density, proportional to the CG integral

$$I^{c\text{-}g} = \int i^{c-g}(\boldsymbol{r})\, d\boldsymbol{r} = \int \nabla A(\boldsymbol{r}) \bullet \nabla B(\boldsymbol{r})\, d\boldsymbol{r} = -\int A(\boldsymbol{r}) \Delta B(\boldsymbol{r})\, d\boldsymbol{r} = \frac{2\mu}{\hbar^2}\langle A|\hat{\mathrm{T}}|B\rangle \equiv \frac{2\mu}{\hbar^2} T_{A,B}, \qquad (92)$$

reflect the bonding and anti-bonding characters of both MO and the bonding, non-bonding and anti-bonding nature of the three model configurations. As such they may provide attractive concepts for both locating and indexing the chemical bonds in molecules [19,20]. As also indicated in the preceding equation, this integral measures the coupling (off-diagonal) element $T_{A,B}$ of the electronic kinetic energy operator $\hat{\mathrm{T}}$. The contra-gradience integral is thus proportional to the kinetic-energy coupling between AO (basis functions). Such integrals are routinely calculated in typical quantum-chemical packages for determining the electronic structure of molecular systems. This observation also emphasizes the crucial role of the kinetic energy terms in the IT interpretations of the origins of chemical bonding, which use the Fisher measure of information.

A reference to Figure 15 indicates that the $i^{c\text{-}g}(r) = 0$ contour is defined by the equation $r_A \bullet r_B = 0$. It separates the region of positive contributions $i^{c\text{-}g}(r) > 0$, outside this contour, from the region of negative contra-gradience density $i^{c\text{-}g}(r) < 0$, inside the contour. Consider its *xz*-plane section, for $y = 0$. For the Cartesian reference frame located at the bond mid-point and the two nuclei in positions $R_A = (0, 0, -\frac{1}{2}R)$ and $R_B = (0, 0, \frac{1}{2}R)$, the electron-position vectors $\{r_X = r - R_X\}$ in this plane of section are: $r_A = (x, 0, z + \frac{1}{2}R)$ and $r_B = (x, 0, z - \frac{1}{2}R)$. The equation determining the $i^{c\text{-}g}(r) = 0$ contour then reads: $x^2 + z^2 = (\frac{1}{2}R)^2 \equiv r_0^2$. It determines the circle centered at the bond mid-point, passing through both nuclei (see Figure 15).

As argued in the qualitative diagram of Figure 16, the negative contra-gradience is also essential for the bonding interaction between the *p*-orbitals. Therefore, it is essential for the *bonding* chemical interaction between a given pair of basis functions that the gradient of one orbital exhibits a non-vanishing *opposite* component along on the direction of the gradient of the other orbital, which justifies the name of the "contra-gradience" criterion itself.

In a general case of the N-electron system and the OAO basis set $\chi = (\chi_1, \chi_2, ..., \chi_m)$ the non-additive Fisher information density in the AO resolution for the electron configuration defined by the *singly*-occupied molecular *spin*-orbitals $\psi = \{\psi_k\}$, with the spatial MO parts $\varphi = \boldsymbol{\chi}\mathbf{C} = \{\varphi_k, k = 1,2,..., N\}$, of the real (interacting) molecular system, e.g., those from the SCF MO or KS calculations, reads:

$$I^{nadd.}[\chi] = 4\sum_{k=1}^{m}\sum_{l=1}^{m} \int \gamma_{k,l}(1-\delta_{k,l})\nabla\chi_l^*(\boldsymbol{r})\cdot\nabla\chi_k(\boldsymbol{r})\,d\boldsymbol{r} \equiv 2\int f^{nadd.}(\boldsymbol{r})\,d\boldsymbol{r} = 8T^{nadd.}[\chi], \quad (93)$$

where the CBO matrix:

$$\gamma = \langle\chi|\left(\sum_{k=1}^{N}|\psi_k\rangle\langle\psi_k|\right)|\chi\rangle = \langle\chi|\psi\rangle\langle\psi|\chi\rangle = \mathbf{CC}^{\dagger} = \{\gamma_{u,w}\}. \quad (94)$$

It is proportional to the non-additive component $T^{nadd.}[\chi]$ of the system average kinetic energy: $T[\chi]$ = tr(γ**T**), where the kinetic energy matrix in AO representation $\mathbf{T} = \{T_{k,l} = \langle\chi_k|\hat{\mathrm{T}}|\chi_l\rangle\}$. In this general molecular scenario of the orbital approximation one uses the most extended (valence) basins of the negative CG density, $f^{nadd.}(\boldsymbol{r}) < 0$ [Eq. (93)], enclosed by the corresponding $f^{nadd.}(\boldsymbol{r}) = 0$ surface, as location of chemical bonds. This proposition has been recently validated numerically [20]. In the remaining part of this section we presents representative results of this extensive study. These results have been obtained using standard SCF MO calculations (GAMESS software) in the minimum Gaussian (STO-3G) basis set. The contour maps, for the optimized geometries of all molecules, will be reported in atomic units. The negative contra-gradience basins, also shown in the perspective views, are identified by the broken-line contours. For the visualization purposes we have used the Matpack and DISLIN graphic libraries.

The contour map of Figure 17 confirms the qualitative predictions of Figure 15. In this axial cut of $f^{nadd.}(\boldsymbol{r})$ for H_2 the *non*-additive Fisher information is seen to be lowered in the spherical *bonding* region between the two nuclei. At the same time the accompanying increases in this quantity are observed in the *non-bonding* regions of each hydrogen atom, signifying the increased localization/structure in this homonuclear diatomic due to the axial polarization of the initially spherical atomic densities. The integral of the contra-gradience density over the whole physical space must be positive, giving rise to the overall *"production"* of the non-additive Fisher information in molecular hydrogen [19,20]. This familiar result from the virial-theorem reflects a net increase in the average kinetic energy relative to the separated-atom value at the equilibrium bond length.

A similar analysis for HF is presented in Figure 18. The perspective view of the $f^{nadd.}(\boldsymbol{r}) < 0$ volumes (left panel) and the contour map of the axial cut of $f^{nadd.}(\boldsymbol{r})$ (right panel) indicate existence of three basins of a decreased non-additive Fisher information: a large, dominating valence (bonding) region Ω_1 between the two atoms, and two small volumes in the inner-shell of F. The shape of the bonding volume exhibits its polarization towards the more electronegative fluorine atom.

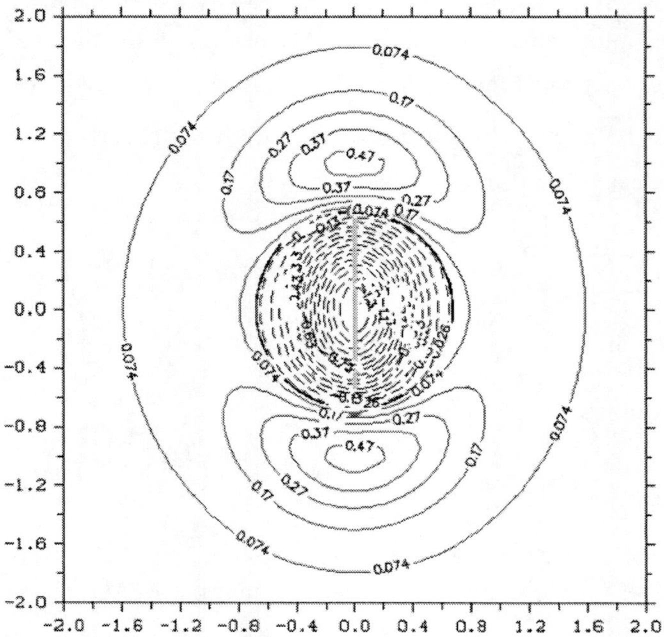

Figure 17. The contour map of the contra-gradience density $f^{nadd.}(\boldsymbol{r})$ for H_2 (see also Figure 15).

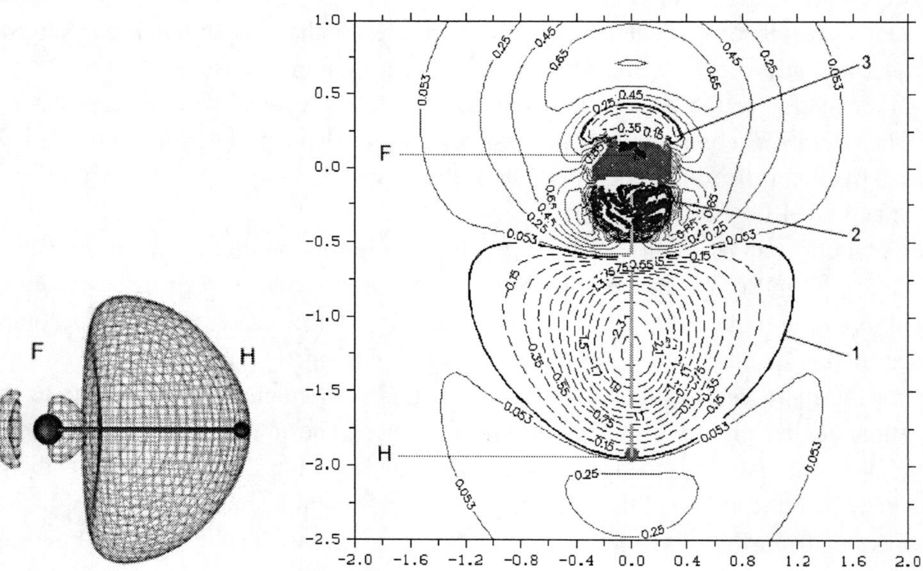

Figure 18. The perspective view of the negative basins of CG and the contour map of the CG density for HF.

In HCl (Figure 19) one again observes two smaller (inner shell) and a large (valence shell) basins of the negative CG density. The softer heavy atom is now seen to undergo a more substantial inner-shell reconstruction of the non-additive Fisher information. Again, there is an axial build-up of $f^{nadd.}(\boldsymbol{r})$ in the non-bonding regions of two atoms, particularly on the hydrogen atom. It should be realized that, compared to HF, where both atoms exhibit the "hard" (difficult to polarize) electron distributions in their valence shells, the (soft) chlorine atom combined with the (hard) hydrogen generates a stronger ionic (electron-transfer) component H→Cl, ultimately giving rise to the ionic pair H^+Cl^-, and hence a smaller covalent (electron-sharring) component H—Cl in the resultant chemical bond.

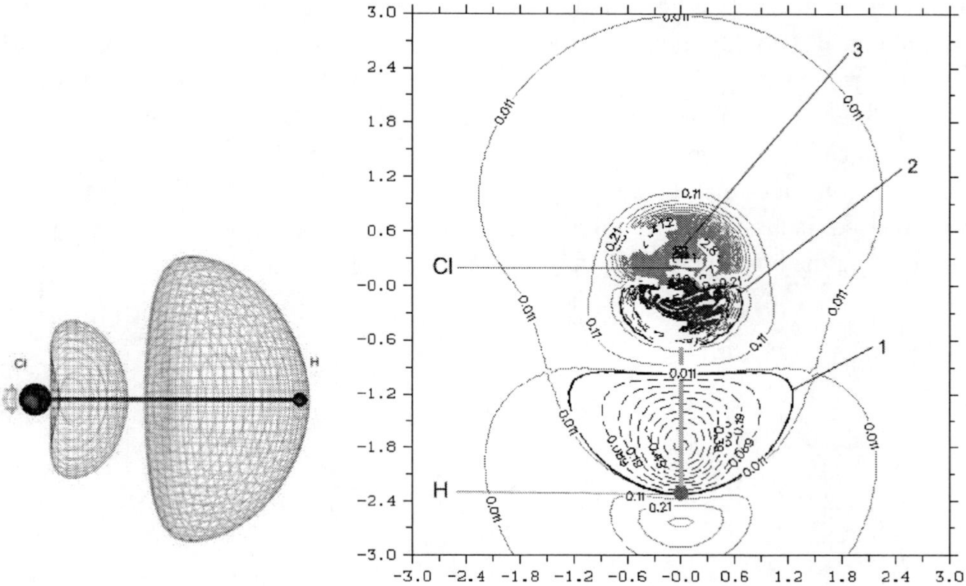

Figure 19. The same as in Figure 18 for HCl.

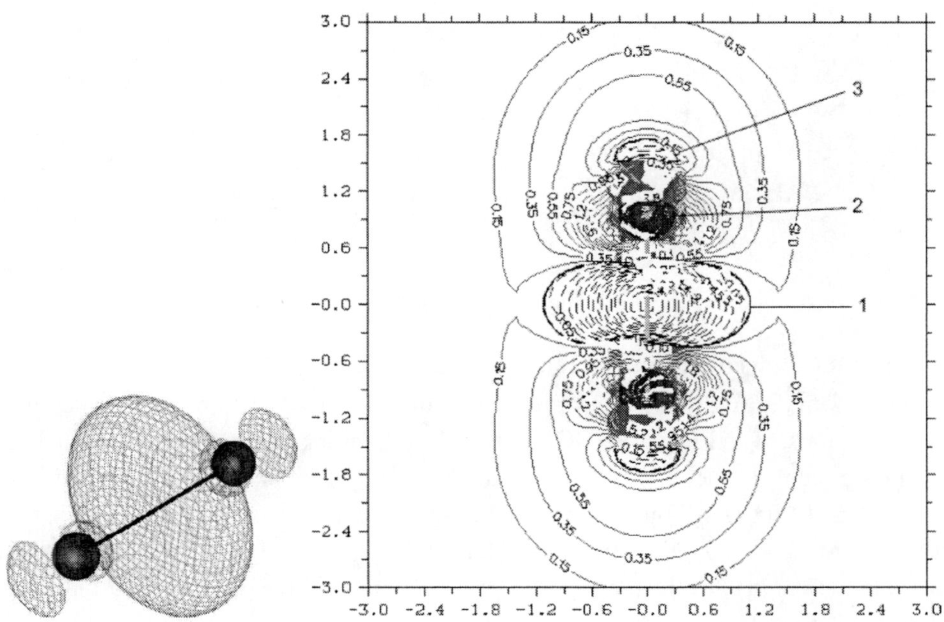

Figure 20. The same as in Figure 18 for N_2.

Consider next the triple chemical bond in N_2 (Figure 20), where the bonding (valence) basin is now distinctly extended away from the bond axis, due to the presence of two π bonds accompanying the central σ bond. Small core-polarization basins, now a symmetrically distributed near each constituent atom along the bond axis, are again observed in the perspective view, while the *sp*-hybridization reconstruction of the non-bonding regions on both atoms is again much in evidence in the accompanying contour map. The dominating (bonding) region around the bond middle-point is now "squeezed" between the two cores of nitrogen atoms. The small, axially placed core regions of the depleted contra-gradience are seen to be surrounded by the volumes of the positive values of this information density in transverse directions. They reflect the charge displacements accompanying the π-bond formation, which is also seen in the density difference diagram of Figure 2.

One thus concludes that the contra-gradience criterion for detecting the valence basins of a diminished non-additive Fisher information in AO-resolution indeed provides an efficient tool for locating the bonding regions in typical diatomics.

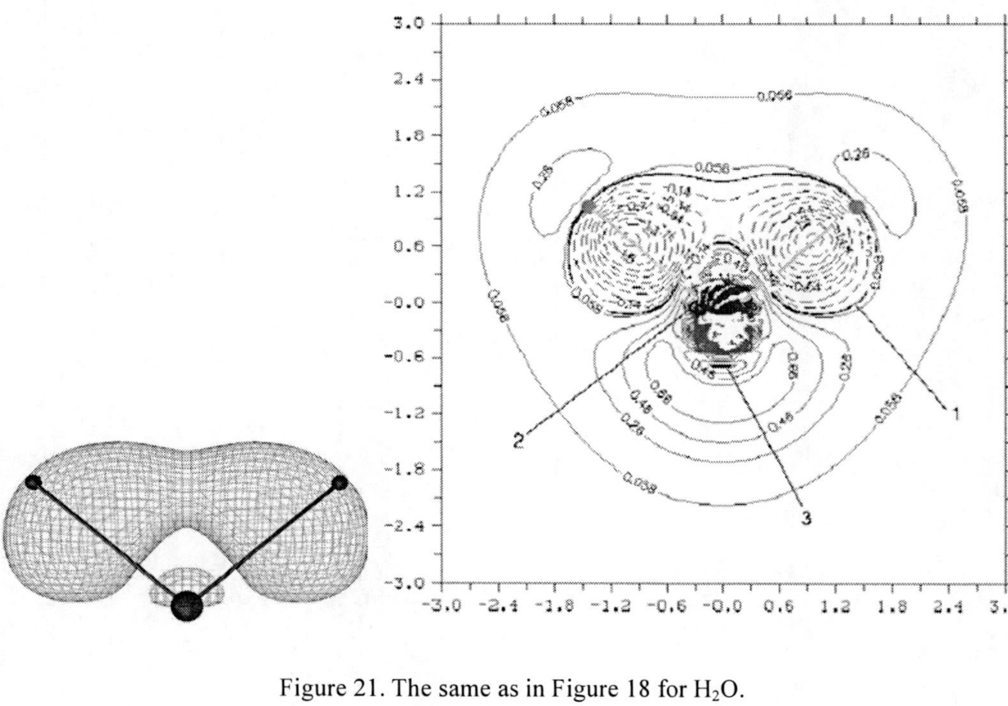

Figure 21. The same as in Figure 18 for H_2O.

In the water molecule (Figure 21) one detects two, slightly overlapping outer-basins of the negative non-additive Fisher information in the O—H bonding regions, and two small inner shell-regions of the negative CG density. The bonding basins are located between the corresponding pairs of nuclei, which define the two localized single bonds, and the lowering of the CG density in each bond is seen to be the strongest in the direction linking the two nuclear attractors. The overlapping character of these two regions of the negative non-additive kinetic-energy, reflected by the present non-additive Fisher-information probe, indicates delocalization of the bonding electrons of one O—H bond into the bonding region of the other chemical bond, as indeed implied by the delocalized character of the occupied canonical MO. The contour map for the cut in the molecular plane also reveals a strong buildup of this information/kinetic-energy quantity in the lone-pair region of oxygen, and – to a lesser degree – in the non-bonding regions of two hydrogens. This effect on the heavy atom should indeed be expected for its promotion due to the nearly tetrahedral sp^3-hybridization.

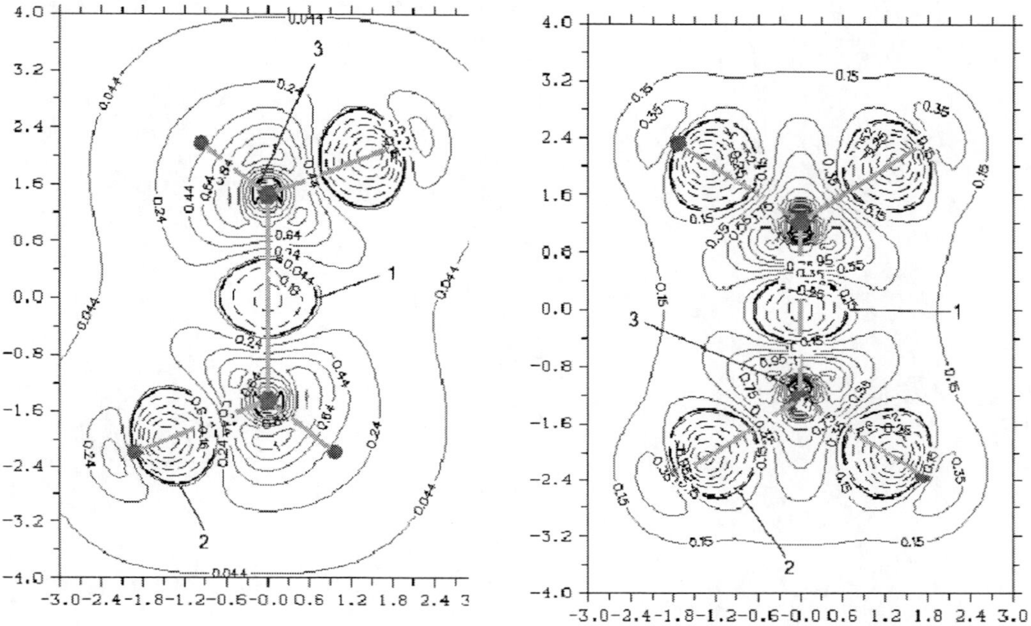

Figure 22. The same as in Figure 17 for ethane (left panel) and ethylene (right panel).

The chemical bonds in small hydrocarbons are investigated in contour maps of Figs. (22)-(24). These diagrams testify to the efficiency of the CG criterion in localizing all C—C and C—H bonding regions in ethane, ethylene, acetylene, butadiene, and benzene. The CG pattern of the triple bond between the carbon atoms in acetylene strongly resembles that observed in N_2 (Figure 20). It also follows from the right panel of Figure 23 that the π bond between the neighboring peripheral carbons in butadiene is much stronger than its central counterpart.

In acetylene the two cylindrical bonding regions of the C—H bonds, axially extended due to the linear (*sp*-hybridization) promotion of both carbon atoms, and the central bonding basin due to the triple C—C bond, now transversely extended in the directions perpendicular to the bond axis, can be clearly seen in the contour map. The $f^{nadd.}(\boldsymbol{r}) > 0$ regions on each carbon atom, very much resembling the atomic $2p_\pi$ distributions, reflect the presence of the bond π component. The depletion of the $2p_\pi$ electron density near the carbon nuclei generates more structure in the electron π-donating (non-bonding) regions of both carbon atoms, and hence less structure (more delocalization) in the π-accepting (bonding) volume between the two nuclei.

Finally, the bonding patterns in a series of four small propellanes of Figure 5 are examined in CG contour maps of Figure 25. Each row of the figure is devoted to a different propellane, arranged from the smallest [1.1.1] molecule, exhibiting three single-carbon bridges, to the largest [2.2.2] system, consisting of three double-carbon bridges; the left panels of each row correspond to the plane of section perpendicular to the central bond between the bridgehead carbons, at the bond midpoint, while the axial cuts of the right panels involve one of the system carbon bridges.

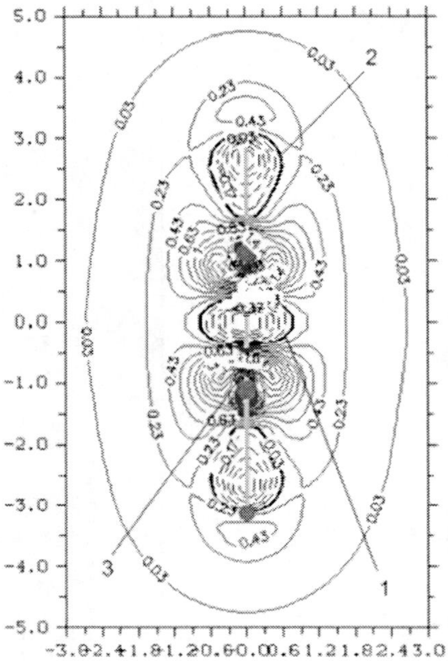

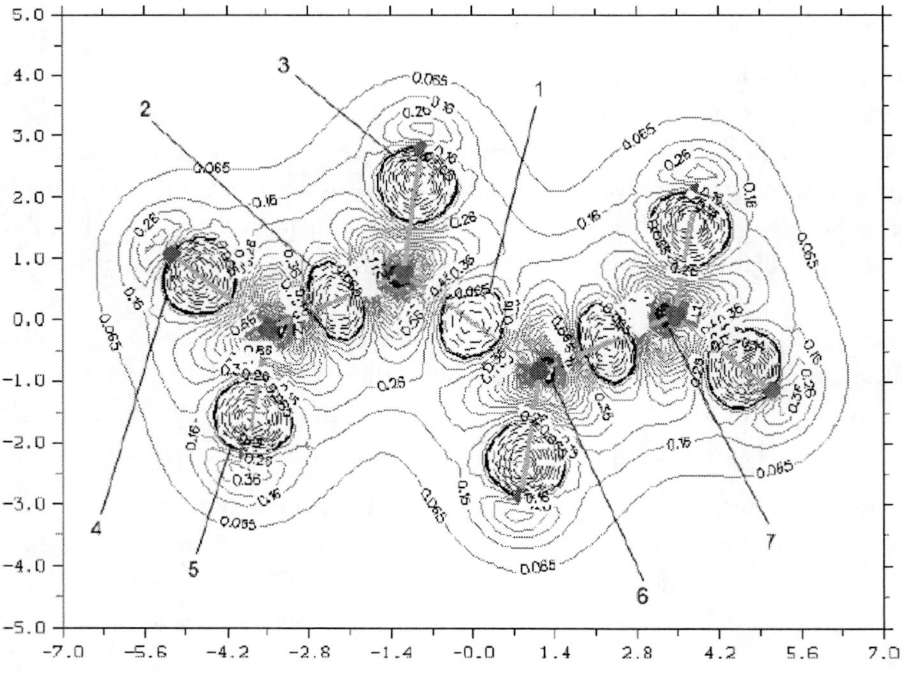

Figure 23. The same as in Figure 17 for acetylene (left panel) and butadiene (right panel).

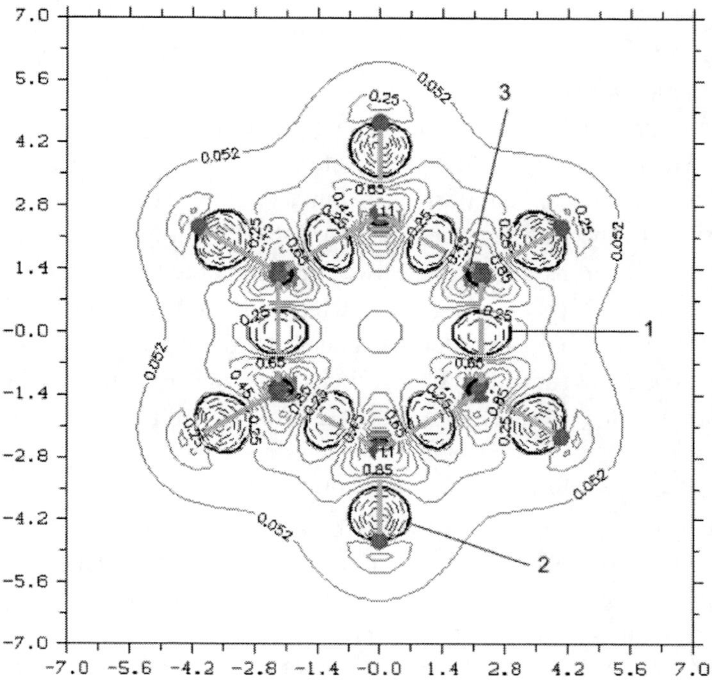

Figure 24. The same as in Figure 2 for benzene.

The main result of the previous studies, the apparent lack of the direct (through-space) bond between the carbon bridgeheads in the [1.1.1] and [2.1.1] systems, and the presence of practically a single bond in the [2.2.1] and [2.2.2] propellanes, remain generally confirmed by the present CG probe, but this transition is now seen to be less sharp, with very small bonding basins between bridgeheads being also observed in the two smallest molecules. The transition from the missing direct bonding in [1.1.1] system to the full central bond in [2.2.2] propellane is now less abrupt: a very small bonding basin identified in the former case is steadily evolving into that attributed to the full bond in both [2.2.1] and [2.2.2] propellanes. The bridge C—C and C—H chemical bonds are again seen to be perfectly localized by the valence surfaces of the vanishing CG density.

The pattern of the non-additive Fisher information density is always very much polarizational in character, with the closed bonding-regions of the negative CG being separated by the molecular environment of the positive values of this quantity, marking the system non-bonding regions. Thus, the depletion of CG in the bonding region is being accompanied by the increase in this information quantity in the nearby regions, so there is but little net increase in the overall Fisher information of the molecular system compared to the atomic promolecule. The overall Fisher information content in the molecule is not conserved [24]. Since the molecule exhibits a higher degree of "structure", due to the overall contraction of atomic electron distributions, it exhibits a net increase in the Fisher information content, and hence the average kinetic energy of electrons, as indeed implied by the molecular virial theorem. More specifically, for the equilibrium geometry of the molecule, i.e., the vanishing forces acting on the system nuclei, the simple virial relation requires $\langle T \rangle = -\langle E \rangle$. Therefore, the negative displacement in the average electronic energy of the given molecule M relative to its atomic promolecule M^0, due to formation of chemical bonds, $\Delta\langle E \rangle = \langle E \rangle - \langle E \rangle_0 < 0$, indeed implies the associated positive displacement in the system average kinetic energy, $\Delta\langle T\rangle = \langle T\rangle - \langle T\rangle_0 > 0$, which is proportional to the overall Fisher information: $\Delta\langle T\rangle \propto \Delta\langle I\rangle = \langle I\rangle - \langle I\rangle_0$; in the minimum basis-set description of H_2 [19], $\Delta\langle T\rangle \propto I^{c\text{-}g}$.

One of the primary goals of quantum chemistry is to identify the physical sources of the chemical bond. Most of existing theoretical interpretations of the origins of the covalent chemical bond at molecular equilibrium geometry emphasize, almost exclusively, the potential (interaction) aspect of this phenomenon, focusing on the mutual attraction between the accumulation of electrons between the two atoms (the negative "bond-charge") and the partially screened (positively charged) nuclei. Indeed, the familiar virial-theorem decomposition of the diatomic Born-Oppenheimer potential indicates that for the equilibrium bond length it is the overall (negative) change in its potential component, due to a *contraction* of constituent atoms in the presence of each other, which is ultimately responsible for the net stabilizing (bonding) effect. The associated change in the overall kinetic energy is negative only at an earlier stage of the mutual approach by both atoms, when it is dominated by the *longitudinal* contribution associated with the gradient component along the bond axis; it ultimately assumes the destabilizing (antibonding) character at the equilibrium inter-nuclear separation, mainly due to its *transverse* contribution associated with the gradient components in the directions perpendicular to the bond axis.

Nonetheless, this overall virial theorem perspective also indicates that the kinetic energy constitutes the driving force of the bond-formation process. It follows from the classical analysis by Ruedenberg and co-workers [50] that the contraction of the atomic electron distributions is possible in molecule due to the relative lowering of the kinetic energy in the bonding region between the two nuclei, as reflected by both the longitudinal and transverse gradient components of the average electronic kinetic energy at large internuclear separations. Therefore, the whole process of redistributing electrons during formation of the chemical bond can be also regarded as being "catalyzed" by the gradient effect of the kinetic-energy. A similar conclusion follows from the theoretical analysis by Goddard and Wilson [51].

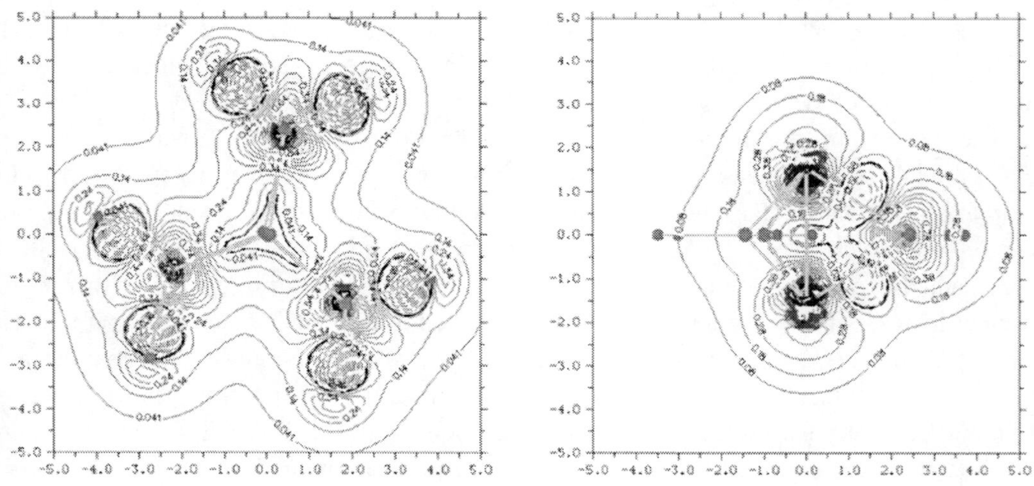

[1.1.1]

Figure 25. Continued on next page.

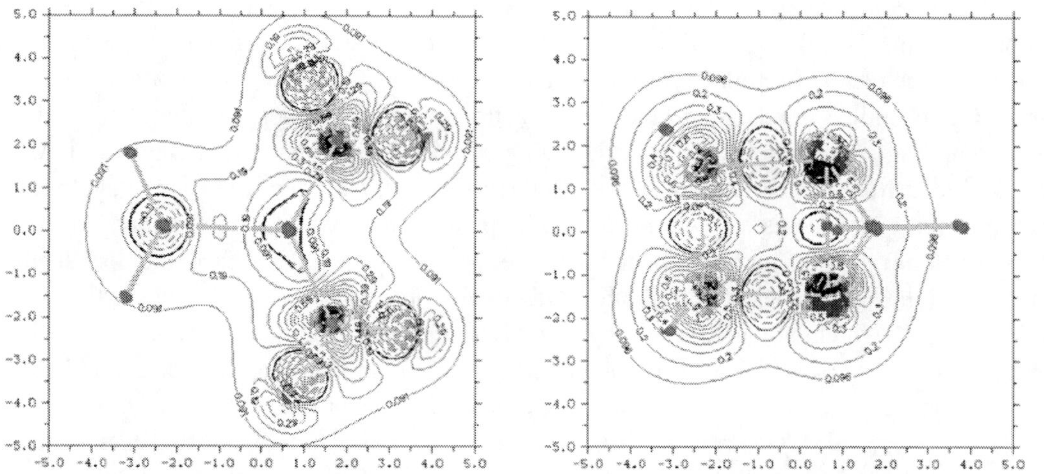

[2.1.1]

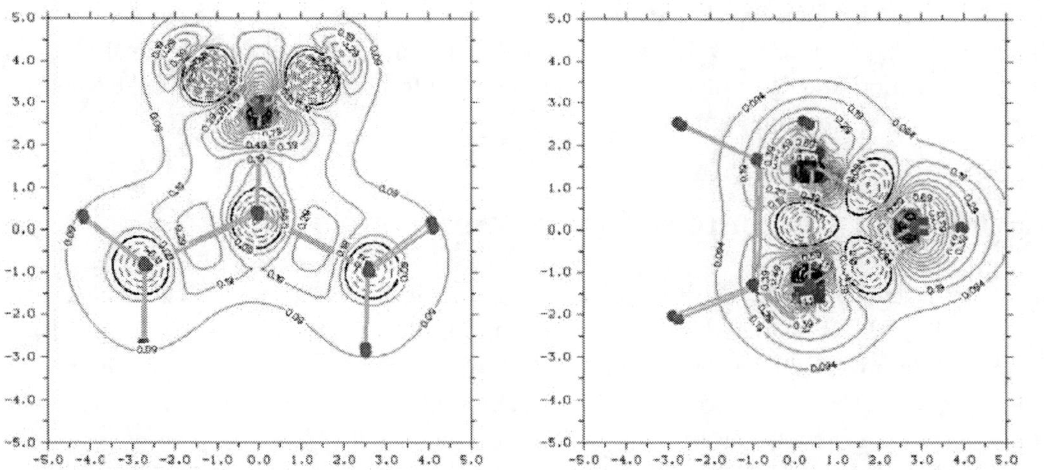

[2.2.1]

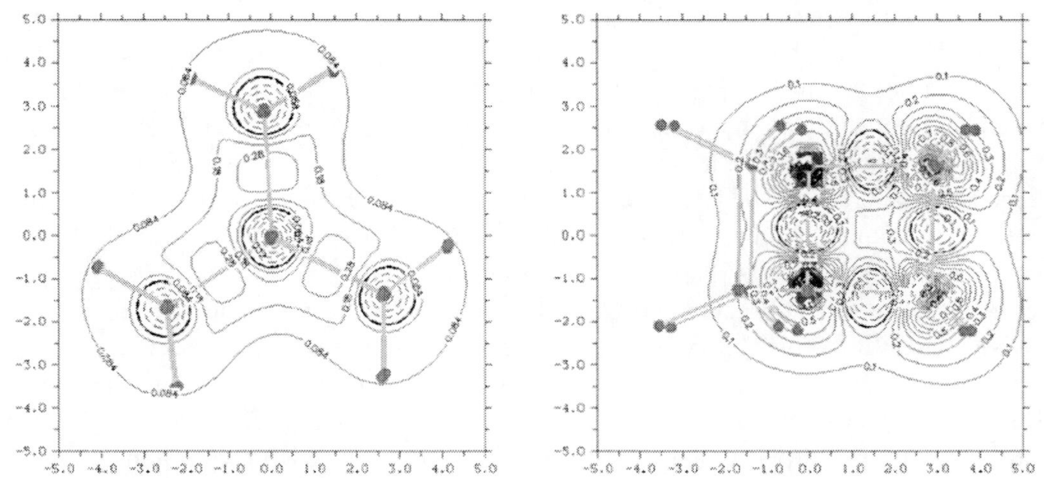

[2.2.2]

Figure 25. (beginning) The same as in Figure 17 for the four propellanes of Figure 5.

The overall kinetic/Fisher-information change due to chemical bond formation emphasizes the overall contraction of the electronic density in the presence of the remaining nuclear attractors in the molecule and it only blurs the picture of the subtle information origins of chemical bonds, since the *total* kinetic energy component combines the delicate (truly bonding) *inter*-atomic effects originating from the stabilizing combinations of AO in the occupied MO, which determine the effective bonding patterns in the molecule, and the accompanying processes of the *intra*-atomic polarization, which involve both the non-bonding (lone) pairs of both the inner- and outer-shell electrons. Therefore, the displacement in the overall kinetic energy contribution effectively hides the minute changes in the system valence-shell, which are associated by chemists with the chemical bond concept. Some partitioning of this overall energy component is called for to separate the subtle *bonding* phenomena from the associated *non-bonding* promotion mechanism.

The ELF and CG criteria, which reflect the non-additive kinetic-energy/Fisher-information terms, adopt the complementary view by stressing the importance of the kinetic-energy bond component and allow for an identification of the bonding regions in the molecule. Only by focusing on the *non-additive* part of the electronic kinetic energy can one uncover the real information origins of chemical bonds, and define the useful local probes for their localization and eventually - the quantitative information descriptors based upon the kinetic energy/Fisher-information.

7. Molecular Communication Systems

The key concept of CTCB is the molecular information system, which can be constructed at alternative levels of resolving the electron probabilities into the underlying elementary "events" determining the channel inputs $a = \{a_i\}$ and outputs $b = \{b_j\}$. For example, they may involve a finding of an electron on the basis-set orbital, AIM, molecular fragment, *etc*. Such communication channels can be generated within both the *local* and *condensed* descriptions of electronic probabilities. These networks describe the probability/information propagation in the molecule and can be characterized by the standard quantities developed in IT for real communication devices (see Section 2.3). Due to the electron delocalization throughout the network of chemical bonds in a molecule the transmission of "signals" about the electron-assignment to the underlying events of the resolution in question becomes randomly disturbed, thus exhibiting the communication "noise". Indeed, an electron initially attributed to the given atom/orbital in the channel "input" a (molecular, promolecular, or the "ensemble" tailored) can be later found with a non-zero probability at several locations in the molecular "output" b. This feature of the electron delocalization is embodied in the conditional probabilities of the outputs-given-inputs, $\mathrm{P}(b\,|\,a) = \{P(b_j|a_i) \equiv P(\,j|i)\}$, which define the molecular information network. Both *one*- and *two*-electron approaches have been devised to construct this matrix. The latter, see e.g., [8], have used the joint probabilities of simultaneous events involving two electrons in a molecule, in the AIM input and output, respectively, to determine the network conditional probabilities, while the former [28-30,52] constructs the orbital-pair probabilities using the bond-projected superposition-principle of quantum mechanics.

The *two*-electron treatment has been found to give rise to rather poor representation of the bond-differentiation patterns in molecules, which is decisively improved in the OCT approach using the AO resolution. The latter scheme complements its earlier orbital implementation using the effective AO-promotion channel generated from the sequential cascade of the elementary orbital-transformation stages [29]. Such consecutive cascades of elementary information systems have been used to represent the underlying orbital-transformations and electron-excitations in the resultant propagations of the electron probabilities in molecules. The information-cascade approach also provides the probability-scattering perspective on atomic promotion due to the orbital hybridization. In OCT the conditional probabilities determining the molecular communication channel in the basis-function resolution follow from the quantum-mechanical superposition principle [53] supplemented by the "physical" projection onto the subspace of the system occupied MO, which determine the molecular network of chemical bonds [28]. Both the molecule as a whole and its constituent subsystems can be adequately described using the OCT bond indices. The internal and external indices of molecular fragments (groups of AO) can be efficiently generated using the appropriate *reduction* of the molecular channel, by combining selected outputs into larger fragment(s).

In OCT formulation of CTCB the off-diagonal orbital communications have been shown to be proportional to the corresponding the Wiberg [42] or related quadratic indices of the chemical bond [37-41,43]. The Wiberg-callibrated IT indices of diatomic interactions in molecules, generated using the *input*-weighted approach, which adopts the flexible ("ensemble") input probabilities to probe the localized bond in the molecule, have been successfully implemented in the RHF MO theory [30c]. The resulting IT descriptors been shown to account for the chemical intuition quite well, at the same time providing the resolution of diatomic bond-multiplicities into the complementary IT-covalent IT-ionic components. In the same study the need for recognizing the signs of the off-diagonal matrix elements of the CBO matrix has been stressed, in order to properly account for the so called "occupation" decoupling, when the *anti*-bonding MO become populated, e.g., in the excited electron configurations. In this section we shall briefly summarize the theoretical basis and representative applications of this OCT for the localized chemical bonds to representative molecules.

7.1. Information Channels in Atomic Orbital Resolution

In MO theory the network of chemical bonds is determined by the occupied MO in the system ground-state. Let us assume, for simplicity, the closed-shell (*cs*) electron configuration of $N = 2n$ electronic system, in the standard RHF description, which involves the n lowest (doubly occupied, orthonormal) MO. In the familiar LCAO MO approach they are generated as linear combinations (LC) of the appropriate (orthogonalized) AO (basis functions) $\chi = (\chi_1, \chi_2, \ldots, \chi_m) = \{\chi_i\}$, $\langle \chi | \chi \rangle = \{\delta_{i,j}\} \equiv \mathrm{I}$, e.g., Löwdin's OAO, $\varphi = (\varphi_1, \varphi_2, \ldots, \varphi_n) = \{\varphi_s\} = \chi\mathrm{C}$, where the rectangular matrix $\mathrm{C} = \{C_{i,s}\} = \langle \chi | \varphi \rangle$ groups the relevant expansion coefficients, is to be determined using the iterative SCF procedure.

The system electron density $\rho(r)$, and hence also the *one*-electron probability distribution $p(r) = \rho(r)/N$, the *shape*-factor of ρ, are determined by the density (CBO) matrix γ of Eq. (94),

$$\rho(r) = 2\varphi(r)\varphi^{\dagger}(r) = \chi(r)\,[2\mathrm{CC}^{\dagger}]\,\chi^{\dagger}(r) \equiv \chi(r)\,\gamma\,\chi^{\dagger}(r) = Np(r), \tag{95}$$

i.e., the AO representation of the projection operator $\hat{\mathrm{P}}_\varphi = |\varphi\rangle\langle\varphi| = \Sigma_s |\varphi_s\rangle\langle\varphi_s| \equiv \Sigma_s \hat{\mathrm{P}}_s$ onto the subspace of all doubly-occupied MO:

$$\gamma = 2\langle\chi|\varphi\rangle\langle\varphi|\chi\rangle = 2\mathbf{CC}^\dagger \equiv 2\langle\chi|\hat{\mathrm{P}}_\varphi|\chi\rangle = \{\gamma_{i,j} = 2\langle\chi_i|\hat{\mathrm{P}}_\varphi|\chi_j\rangle \equiv 2\langle i|\hat{\mathrm{P}}_\varphi|j\rangle\}, \quad (96)$$

thus satisfying the appropriate idempotency relation:

$$(\gamma)^2 = 4\langle\chi|\hat{\mathrm{P}}_\varphi|\chi\rangle\langle\chi|\hat{\mathrm{P}}_\varphi|\chi\rangle = 4\langle\chi|\hat{\mathrm{P}}_\varphi^2|\chi\rangle = 4\langle\chi|\hat{\mathrm{P}}_\varphi|\chi\rangle = 2\gamma. \quad (97)$$

The CBO matrix reflects the promoted, *valence*-state of AO in the molecule, with the diagonal elements measuring the effective electron occupations of the basis functions, $\{N_i = \gamma_{i,i} = Np_i\}$, with the probabilities $p = \{p_i = \gamma_{i,i}/N\}$ of the AO being occupied in the molecule, $\Sigma_i p_i = 1$.

The molecular information system in the (condensed) orbital resolution involves the AO events χ in its input $a = \{\chi_i\}$ and output $b = \{\chi_j\}$. It represents the effective communication promotion of these basis functions in the molecule *via* the probability/information scattering described by the conditional probabilities of the AO-outputs given the AO-inputs, identified by the *row* (input) and *column* (output) indices, respectively. In this *one*-electron description the AO→AO communication network is determined by the conditional probabilities of the output AO-events, given the input AO-events,

$$\mathrm{P}(b|a) = \{P(\chi_j|\chi_i) \equiv P(j|i) = P(i\wedge j)/p_i\},\ \Sigma_j P(j|i) = 1, \quad (98)$$

where the associated joint probabilities of simultaneously observing two AO in the system chemical bonds $\mathrm{P}(a\wedge b) = \{P(i\wedge j)\}$ satisfy the usual partial and total normalization relations:

$$\Sigma_i P(i\wedge j) = p_j,\ \Sigma_j P(i\wedge j) = p_i,\ \Sigma_i\Sigma_j P(i\wedge j) = 1. \quad (99)$$

These conditional probabilities define the probability scattering in the AO-promotion channel of the molecule, in which the "signals" of the molecular electron allocations to basis functions are transmitted between the AO inputs and outputs. Such information system constitutes the basis of OCT of the chemical bond.

As argued elsewhere [28,30c], by using the generalized superposition principle of quantum mechanics [53], this matrix of the (*physical*) conditional probabilities involves the squares of corresponding elements of the CBO matrix:

$$\mathrm{P}(b|a) = \{P(j|i) = \mathcal{N}_i \left|\langle i|\hat{\mathrm{P}}_\varphi|j\rangle\right|^2 = (2\gamma_{i,i})^{-1}\gamma_{i,j}\gamma_{j,i}\}, \quad (100)$$

where the closed-shell normalization constant $\mathcal{N}_i = (2\gamma_{i,i})^{-1}$ follows directly from Eq. (97) (for the open-shell generalization see [30c]). These probabilities explore the dependencies between AO resulting from their participation in the framework of the occupied MO, i.e. their involvement in the entire network of chemical bonds in the molecule. This molecular channel can be probed using both the promolecular ($p^0 = \{p_i^0\}$), molecular (p), or arbitrary input probabilities, in order to extract the desired IT indices of the fragment or global bond multiplicities and their *ionic* and *covalent* components [8].

The off-diagonal conditional probability of jth AO-output given ith AO-input is thus proportional to the squared element of the CBO matrix linking the two AO, $\gamma_{j,i} = \gamma_{i,j}$, thus being also proportional to the corresponding AO contribution $\mathcal{M}_{i,j} = \gamma_{i,j}^2$ to the Wiberg [42] index of the overall chemical bond order between two atoms A and B in the molecule,

$$\mathcal{M}(\mathrm{A, B}) = \sum_{i\in \mathrm{A}} \sum_{j\in \mathrm{B}} \mathcal{M}_{i,j}, \tag{101}$$

or to related, generalized quadratic descriptors of molecular bond-multiplicities [37-41,43].

It can be straightforwardly verified using the idempotency relation of Eq. (97) that the associated joint probability matrix,

$$\mathrm{P}(a\wedge b) = \{P(i\wedge j) = p_i\, P(j|i) = (2N)^{-1}\gamma_{i,j}\gamma_{j,i} = (2N)^{-1}\langle i|\hat{\mathrm{P}}_{\varphi}|j\rangle\langle j|\hat{\mathrm{P}}_{\varphi}|i\rangle\}, \tag{102}$$

indeed satisfies the normalization conditions of Eq. (99), e.g.,

$$\sum_i P(i\wedge j) = (2N)^{-1}\sum_i \gamma_{j,i}\gamma_{i,j} = (2N)^{-1}\, 2\gamma_{j,j} = p_j. \tag{103}$$

In OCT the entropy/information indices of the covalent/ionic components of all chemical bonds in the given molecular system represent the complementary descriptors of the average communication *noise* and the amount of information *flow* in the molecular information channel. One also observes that the molecular input $P(a) \equiv p$ generates the same distribution in the output of the molecular channel,

$$p\,\mathrm{P}(b|a) = \{\sum_i p_i\, P(j|i) \equiv \sum_i P(i\wedge j) = p_j\} = p, \tag{104}$$

thus identifying p as the *stationary* probability vector for the molecular ground state, while the promolecular input $P(a^0) \equiv p^0$ in general produces different output probability. The purely molecular communication channel, with p defining its input signal, is devoid of any reference (history) of the chemical bond formation and generates the average-noise index of the molecular IT bond-covalency, measured by the *conditional-entropy* of the system outputs given inputs (see Section 2.3):

$$\begin{aligned} S(P(b)|P(a)) &\equiv H(\mathrm{B}|\mathrm{A}) = -\sum_i \sum_j P(i\wedge j)\log[P(i\wedge j)/p_i] = \\ &\sum_i p_i[-\sum_j P(j|i)\log P(j|i)] \equiv \sum_i p_i\, S_i \equiv S(p|p) \equiv S. \end{aligned} \tag{105}$$

This average-noise descriptor thus expresses the difference between the Shannon entropies of the molecular *one*- and *two*-orbital probabilities [see Eq. (17) and Schemes 1, 4],

$$\begin{aligned} &S = H(\mathrm{AB}) - H(\mathrm{A}); \\ &H(\mathrm{A}) = -\sum_i p_i \log p_i \equiv S(P(a)) = H(\mathrm{B}) \equiv S(P(b)), \\ &H(\mathrm{AB}) \equiv S(\mathrm{P}(a\wedge b)) = -\sum_i \sum_j P(i\wedge j)\log P(i\wedge j). \end{aligned} \tag{106}$$

Hence, for the *independent* input and output events, when $\mathrm{P}^{ind.}(a\wedge b) = \{p_i p_j\}$, $S(\mathrm{P}^{ind.}(a\wedge b)) = 2S(p)$, and hence $S^{ind.} = S(p)$.

The AO channel with $P(a^0) = p^0$ determining its input "signal" probability refers to the initial-state in the bond-formation process, i.e., the system *promolecule*. In other words, it corresponds to the ground-state (fractional) occupations of the AO contributed by the system constituent free atoms, before their mixing into MO. This input signal gives rise to the average information-flow descriptor of the system IT bond-ionicity, given by the *mutual-information* in the channel inputs and outputs:

$$I(P(a^0){:}P(b)) \equiv I(A^0{:}B) = \sum_i\sum_j P(i\wedge j)\log[P(i\wedge j)/(p_j p_i^0)] = \sum_i p_i\{\sum_j P(j|i)\log[P(i|j)/p_i^0]\} \equiv \sum_i p_i I_i = S(P(b)) + S(P(a^0)) - S(P(a\wedge b)) = S(p^0) - S \equiv I(p^0{:}p) \equiv I. \quad (107)$$

This *amount of information* reflects the fraction of the initial (promolecular) information content $S(p^0)$, which has not been dissipated as noise in the molecular communication system. In particlular, for the molecular input, when $p^0 = p$,

$$I(P(a) : P(b)) = \sum_i\sum_j P(i, j)\log[P(i, j)/(p_j p_i)] = S(p) - S \equiv I(p : p). \quad (108)$$

Hence, for the independent input and output events $I^{ind.}(P(a) : P(b)) = 0$.
Finally, the sum of these two bond components,

$$\mathcal{N}(P(a^0);\, P(b)) = S + I \equiv \mathcal{N}(p^0;\, p) \equiv \mathcal{N} = S(p^0) = \sum_i p_i(S_i + I_i) \equiv \sum_i p_i \mathcal{N}_i, \quad (109)$$

where $\mathcal{N}_i = -\log p_i^0$ stands for the *self-information* in the promolecular AO-intput event χ_i, measures the overall IT bond-multiplicity of all bonds in the molecular system under consideration. It is seen to be conserved at the promolecular-entropy level, which marks the initial information content of AO probabilities. Alternatively, for the molecular input, when $P(a) = p$, this quantity preserves the Shannon entropy of the molecular input probabilities:

$$\mathcal{N}(P(a);P(b)) = S(P(b)|\, P(a)) + I(P(a){:}\, P(b)) = S(P(a)) = S(p)\,. \quad (110)$$

We recall (see Section 2.2) that in the qualitative diagrams of Scheme 1 (see also the upper part of Scheme 4) for two dependent probability schemes, the common (overlap) area of the associated entropy circles corresponds to the mutual information in both distributions, $I(P(a){:}P(b))$, while the remaining parts of individual circles represent the corresponding conditional entropies $S(P(b)|P(a))$ or $S(P(a)|P(b))$. The latter measure the residual uncertainty about events in one set, when one has the full knowledge of the occurrence of the events in the other set of events. Accordingly, the area enclosed by the envelope of these two overlapping circles represents the entropy in the joint distribution of these two sets of outcomes:

$$S(P(a)\wedge P(b)) = S(P(a)) + S(P(b)) - I(P(a){:}P(b)) = S(P(a)) + S(P(b)|P(a)) = S(P(b)) + S(P(a)|P(b)). \quad (111)$$

7.2. Many-Orbital Extension

To determine the entropy/information quantities of three dependent probability distributions, which are schematically shown in the diagrams of the lower part of Scheme 4, the *joint* probabilities

$$P(a\wedge b\wedge c) = \{P(i\wedge j\wedge k)\},\ \sum_i P(i\wedge j\wedge k) = P(j\wedge k),\ \sum_j P(i\wedge j\wedge k) = P(i\wedge k),\ \sum_k P(i\wedge j\wedge k) = P(i\wedge j), \quad (112)$$

of the occurrence of the simultaneous *three*-orbital events in the molecular chemical bond system are required [28b]. They define the associated *conditional* probabilities through the equalities:

$$\mathrm{P}(a \wedge b \wedge c) = \{P(i \wedge j \wedge k) = p_k\, P(i \wedge j|k) = p_j\, P(i \wedge k|j) = p_i\, P(j \wedge k|i)\}.$$
$$\Sigma_i \Sigma_j P(i \wedge j|k) = \Sigma_i \Sigma_k P(i \wedge k|j) = \Sigma_j \Sigma_k P(j \wedge k|i) = 1, \tag{113}$$

A straightforward extension of Eq. (102) gives the following expression for the probability of the simultaneous three-AO event in the molecular framework of chemical bonds [28b]:

$$\mathrm{P}(a \wedge b \wedge c) = \{P(i \wedge j \wedge k) = (4N)^{-1} \gamma_{i,j}\gamma_{j,k}\gamma_{k,i} = (4N)^{-1} \langle i|\hat{\mathrm{P}}_\varphi|j\rangle\langle j|\hat{\mathrm{P}}_\varphi|k\rangle\langle k|\hat{\mathrm{P}}_\varphi|i\rangle\}, \tag{114}$$

and hence the associated conditional probabilities [compare Eq. (100)]:

$$\mathrm{P}(a \wedge b|c) = \{P(i \wedge j|k) = P(i \wedge j \wedge k)/p_k = (4\gamma_{k,k})^{-1} \gamma_{k,i}\gamma_{i,j}\gamma_{j,k}\},$$
$$\mathrm{P}(b \wedge c|a) = \{P(j \wedge k|i) = P(i \wedge j \wedge k)/p_i = (4\gamma_{i,i})^{-1} \gamma_{i,j}\gamma_{j,k}\gamma_{k,i}\},$$
$$\mathrm{P}(a \wedge c|b) = \{P(i \wedge k|j) = P(i \wedge j \wedge k)/p_j = (4\gamma_{j,j})^{-1} \gamma_{j,k}\gamma_{k,i}\gamma_{i,j}\}. \tag{115}$$

It should be realized, that the electron occupying a given MO $\varphi_s = \chi C_s$, where $C_s = \{C_{i,s}\}$ stands for the sth column in C = $\{C_s\}$, is simultaneously observed in all AO which participate in this linear combination, i.e., on all such basis functions. The resultant effect from all occupied MO, which determine the network of chemical bonds in the molecule, such simultaneous probabilities of several AO are obtained by applying the projector $\hat{\mathrm{P}}_\varphi$ onto the occupied subspace of MO, which then links such (physical) probabilities to the relevant product of the CBO matrix elements.

In a similar manner one generates the joint probabilities of the simultaneous *four*-AO events or the associated AO conditional probabilities,

$$\mathrm{P}(a \wedge b \wedge c \wedge d) = \{P(i \wedge j \wedge k \wedge l) = p_l\, P(i \wedge j \wedge k|l)\},\ \Sigma_i \Sigma_j \Sigma_k P(i \wedge j \wedge k|l) = 1,$$
$$\Sigma_l P(i \wedge j \wedge k \wedge l) = P(i \wedge j \wedge k), \text{ etc.,} \tag{116}$$

$$\mathrm{P}(a \wedge b \wedge c \wedge d) = \{P(i \wedge j \wedge k \wedge l) = (8N)^{-1} \gamma_{i,j}\gamma_{j,k}\gamma_{k,l}\gamma_{l,i} =$$
$$(8N)^{-1} \langle i|\hat{\mathrm{P}}_\varphi|j\rangle\langle j|\hat{\mathrm{P}}_\varphi|k\rangle\langle k|\hat{\mathrm{P}}_\varphi|l\rangle\langle l|\hat{\mathrm{P}}_\varphi|i\rangle\}, \tag{117}$$

$$\mathrm{P}(a \wedge b \wedge c|d) = \{P(i \wedge j \wedge k|l) = P(i \wedge j \wedge k \wedge l)/p_l = (8\gamma_{l,l})^{-1} \gamma_{i,j}\gamma_{j,k}\gamma_{k,l}\gamma_{l,i}\}. \tag{118}$$

This development can be straightforwardly extended to cover a general case of the n-orbital distributions $\mathrm{P}(a_1 \wedge a_2 \wedge \ldots \wedge a_n)$ and $\mathrm{P}(a_1 \wedge a_2 \wedge \ldots \wedge a_{n-1}|a_n)$:

$$\mathrm{P}(a_1 \wedge a_2 \wedge \ldots \wedge a_n) = \{P(i_1 \wedge i_2 \wedge \ldots \wedge i_n) = (2^{n-1}N)^{-1} \gamma_{i_1,i_2}\gamma_{i_2,i_3} \cdots \gamma_{i_{n-1},i_n}\gamma_{i_n,i_1}$$
$$= (2^{n-1}N)^{-1} \langle i_1|\hat{\mathrm{P}}_\varphi|i_2\rangle\langle i_2|\hat{\mathrm{P}}_\varphi|i_3\rangle \ldots \langle i_{n-1}|\hat{\mathrm{P}}_\varphi|i_n\rangle\langle i_n|\hat{\mathrm{P}}_\varphi|i_1\rangle\}, \tag{119}$$

$$\mathrm{P}(a_1 \wedge a_2 \wedge \ldots \wedge a_{n-1}|a_n) = \{P(i_1 \wedge i_2 \wedge \ldots \wedge i_{n-1}|i_n) = (2^{n-1}\gamma_{i_n,i_n})^{-1} \gamma_{i_1,i_2}\gamma_{i_2,i_3} \cdots \gamma_{i_{n-1},i_n}\gamma_{i_n,i_1}\}. \tag{120}$$

The *many*-scheme IT quantities have recently been used to characterize the molecular similarity trends in the sequential-cascade approach of CTCB [27c]. The ensemble IT indices generated by several probability distributions have also emerged within the *flexible-input* generalization of OCT [30c]. In the remaining part of this section we shall briefly examine other potential applications of these generalized bond descriptors in describing the bonding patterns of molecular/reactive systems and their fragments. For simplicity, we shall limit out discussion to IT quantities involving three probability schemes in the AO resolution (see the lower part of Scheme 4), for which the joint probabilities of Eq. (114) are required.

The molecular scenario invoking such three-scheme quantities may involve three separate species A, B and C, e.g., two reactants A and B and a catalyst/surface C, with the corresponding sets of the AO-events (a, b, c) of the associated probability schemes (A, B, C) then referring to the basis functions provided by the constituent atoms of each subsystem; alternatively, three molecular fragments can be involved. The three-orbital development then enables one to discuss the influence of one reactant/fragment, say C, on the bond structure (or reactivity) of two remaining fragments A and B. For example, one could then address a natural question: how the electronic/bonding structure of the catalyst affects the structure/reactivity patterns of two adsorbed species, and ultimately assess the cooperation effects between the catalyst-adsorbate bonds (A—C, B—C) and the A—B bond linking the two adsorbates.

More specifically, the conditional three-scheme entropy [Eq. (23), see also Scheme 4], $S(P(b)|\mathrm{P}(a\wedge c)) = H(\mathrm{B}|\mathrm{AC}) = H(\mathrm{B}|\mathrm{A}) - I(\mathrm{B}:\mathrm{C}|\mathrm{A})$, i.e., the molecular indeterminacy of the AO-output events b with respect to the double AO-input (product) events $a\wedge c$, now reflects only a *part* of the overall IT-covalency (noise) in the $a\to b$ communications, measured by the indeterminacy of b with respect to a alone, $H(\mathrm{B}|\mathrm{A}) = S(P(b)|P(a))$, which reflects the *overall* information loss in the $a\to b$ probability scattering. The mutual-information contribution $I(\mathrm{B}:\mathrm{C}|\mathrm{A})$, by which $H(\mathrm{B}|\mathrm{AC})$ is decreased relative to $H(\mathrm{B}|\mathrm{A})$ and by which $I(\mathrm{B}:\mathrm{AC})$ is increased relative to $I(\mathrm{A}:\mathrm{B})$, should now be attributed to the influence of the catalyst C. In fact, the mutual information in A and C, given B, describes the effect of an increased ionic (deterministic) character of the chemical interactions between the chemisorbed reactants A and B, compared to the AO communications in the separate species.

This IT-ionic "activation" of adsorbates, as a result of forming the partial A—C and B—C bonds on active sites of the catalytic surface, also manifests the *competition effect* between these surface bonds and the *inter*-adsorbate bond A—B in the catalytic system: the more heavily are the valence electrons of A and B involved in chemical bonds with C, the less noisy (more deterministic) are their mutual communications, thus giving rise to less IT-covalent (more IT-ionic) interactions between the *chemisorbed* species. The *physical* adsorption of these reactants should be marked by a relatively small value of $I(\mathrm{B}:\mathrm{C}|\mathrm{A})$, since then the dependencies ("overlaps") between (B, A) and C, should be small, thus grossly diminishing the IT-ionic activation effect generated by the presence of the catalyst. Indeed, the surface *chemical* bond between a given adsorbate and the catalyst should be strongly felt at the position of the other adsorbate, and hence the information "coupling" between probability distributions of the chemisorbed species and that of the catalyst should be relatively strong.

Another molecular scenario, in which the *three*-scheme entropy/information descriptors are expected to be useful, is the influence of one reactive site in a molecule upon another, e.g., in the contexts of subtle reactivity preferences in the *donor-acceptor* (DA) reactive systems implied by the *Hard and Soft Acids and Bases* (HSAB) rule [54-56], and particularly - its regional formulation for predicting the regioselectivity trends in cyclization reactions [57], the *Maximum Complementarity* principle [58], and the bond-length variation rules of Gutmann [59]. Such cooperative interaction between different sites in a molecule is also responsible for the directing *trans/cis influence* of ligands in transition metal complexes and the familiar *substituent effect* in aromatic systems. The adequate IT description of the AIM cooperation in *many*-centre bonds, e.g., in boron hydrides or propellanes, may also require the entropy/information indices involving several probability schemes.

7.3. Localized Bonds in Diatomic Fragments

In typical SCF LCAO MO calculations the lone pairs of the valence and/or inner shell electrons can strongly affect the IT descriptors of the chemical bond. Therefore, the contributions due to each AO input should be appropriately weighted using the joint (*two*-orbital) probabilities, which reflect the actual simultaneous participation of the given pair of basis functions centered on different atoms in the system chemical bonds. This probability-weighting procedure, known as the *flexible input approach*, has been shown to be capable of reproducing the Wiberg bond order in diatomics [30c], at the same time providing the IT-covalent/ionic resolution of this index. In this section we shall briefly summarize the theoretical basis of this approach to diatomic interactions in molecules and present representative numerical results from the standard RHF calculations for a selection of representative closed-shell systems.

The molecular probability scattering in the specified diatomic fragment (A, B), involving the basis functions $\chi_{AB} = (\chi_A, \chi_B)$ contributed by these two bonded atoms to the overall set of AO, $\chi = \{\chi_X\}$, is fully characterized by the corresponding $P(\chi_{AB}|\chi_{AB})$ block of the molecular conditional probability matrix of Eq. (100), which determines the molecular communication system in OCT,

$$P(\chi_{AB}|\chi_{AB}) \equiv [P(\chi_Y|\chi_X);\ (X, Y) \in (A, B)] \equiv \{P(j|\chi_{AB});\ \chi_j \in \chi_{AB}\} \equiv \{P(j|i);\ (\chi_i, \chi_j) \in \chi_{AB})\}. \quad (121)$$

Indeed, the square matrix $P(\chi_{AB}|\chi_{AB})$ contains only the *intra*-diatomic communications, missing the probability propagations originating from AO of the remaining constituent atoms $\chi_Z \notin \chi_{AB}$, thus being perfectly capable of describing the localized chemical interactions between A and B.

The atomic *output*-reduction [8] of $P(\chi_{AB}|\chi_{AB})$, by combining the AO events χ_X into a single atomic event X in the output of the molecular channel, gives the associated *condensed* conditional probabilities of the partially reduced information system,

$$P(X_{AB}|\chi_{AB}) = [P(A|\chi_{AB}), P(B|\chi_{AB})] = \{P(X|\chi_{AB}) \equiv \{P(X|i)\} = \sum_{j\in X} \boldsymbol{P}(j|\chi_{AB});\ \chi_i \in \chi_{AB},\ X = A, B\}. \quad (122)$$

Here, $P(Y|i)$ measures the conditional probability that an electron on χ_i will be found on atom Y in the molecule. The sum of these conditional probabilities over all AO contributed by the two atoms then determines the communication connections $\{P(\mathrm{AB}|i)\}$ linking the condensed diatomic output AB and the given AO input χ_i in the communication system of the diatomic fragment under consideration:

$$P(\mathrm{A}|\chi_{\mathrm{AB}}) + P(\mathrm{B}|\chi_{\mathrm{AB}}) = P(\mathrm{AB}|\chi_{\mathrm{AB}}) = \{P(\mathrm{AB}|i) = P(\mathrm{A}|i) + P(\mathrm{B}|i) = \sum_{j\in(\mathrm{A,B})} P(j|i) \le 1\}. \quad (123)$$

In other words, $P(\mathrm{AB}|i)$ measures the probability that an electron occupying χ_i will be detected in the diatomic fragment AB of the molecule. The inequality in the preceding equation reflects the fact that the atomic basis functions participate in chemical bonds with all constituent atoms, with the equality sign corresponding only to the diatomic molecule, when $\chi_{\mathrm{AB}} = \chi$.

The associated *fragment*-normalized AO probabilities,

$$\tilde{\boldsymbol{p}}(\mathrm{AB}) = \{\tilde{p}_i(\mathrm{AB}) = \gamma_{i,i} / N_{\mathrm{AB}},\quad \chi_i \in \chi_{\mathrm{AB}}\},\ \sum_{i\in(\mathrm{A,B})} \tilde{p}_i(\mathrm{AB}) = 1, \quad (124)$$

where $N_{\mathrm{AB}} = \sum_{i\in(\mathrm{A,B})} \gamma_{i,i}$ stands for the number of electrons found in the molecule on the specified diatomic fragment and $\tilde{p}_i(\mathrm{AB})$ denotes the probability that one of them occupies $\chi_{i\in(\mathrm{A,B})}$, then determine the simultaneous probabilities of the joint two-orbital events:

$$\mathrm{P}_{\mathrm{AB}}(\chi_{\mathrm{AB}}\wedge\chi_{\mathrm{AB}}) = \{P_{\mathrm{AB}}(i\wedge j) = \tilde{p}_i(\mathrm{AB})\, P(j|i) = \gamma_{i,j}\gamma_{j,i}/(2N_{\mathrm{AB}})\}. \quad (125)$$

They generate, *via* the relevant partial summations, the joint atom-orbital probabilities in AB, $\{P_{\mathrm{AB}}(\mathrm{X}, i)\}$:

$$\begin{aligned}\mathrm{P}_{\mathrm{AB}}(X_{\mathrm{AB}}\wedge\chi_{\mathrm{AB}}) &= [P_{\mathrm{AB}}(\mathrm{A}\wedge\chi_{\mathrm{AB}}), P_{\mathrm{AB}}(\mathrm{B}\wedge\chi_{\mathrm{AB}})] \\ &= \{P_{\mathrm{AB}}(\mathrm{X}\wedge i) = \sum_{j\in\mathrm{X}} P_{\mathrm{AB}}(i\wedge j) \equiv \tilde{p}_i(\mathrm{AB})\, P(\mathrm{X}|i),\ \mathrm{X} = \mathrm{A, B}\}. \end{aligned} \quad (126)$$

For the closed-shell molecular systems one thus finds:

$$\begin{aligned} P_{\mathrm{AB}}(\mathrm{X}\wedge\chi_{\mathrm{AB}}) &= \left\{P_{\mathrm{AB}}(\mathrm{X}\wedge i) = \tilde{p}_i(\mathrm{AB})\sum_{j\in\mathrm{X}} P(j|i) = \sum_{j\in\mathrm{X}} \frac{\gamma_{i,j}\gamma_{j,i}}{2N_{\mathrm{AB}}}\right\} \equiv \\ & P_{\mathrm{AB}}(\chi_{\mathrm{AB}}\wedge\mathrm{X})^{\mathrm{T}},\ \mathrm{X} = \mathrm{A, B}. \end{aligned} \quad (127)$$

These vectors of AO probabilities in the diatomic fragment AB subsequently define the condensed probabilities $\{P_{\mathrm{X}}(\mathrm{AB})\}$ of both bonded atoms in this subsystem:

$$P_{\mathrm{X}}(\mathrm{AB}) = \frac{N_{\mathrm{X}}(\mathrm{AB})}{N_{\mathrm{AB}}} = \sum_{i\in(\mathrm{A,B})} P_{\mathrm{AB}}(\mathrm{X}\wedge i) = \sum_{i\in(\mathrm{A,B})}\sum_{j\in\mathrm{X}} \frac{\gamma_{i,j}\gamma_{j,i}}{2N_{\mathrm{AB}}},\ \mathrm{X} = \mathrm{A, B}, \quad (128)$$

where the effective number of electrons $N_{\mathrm{X}}(\mathrm{AB})$ on atom X = (A, B) reads:

$$N_X(AB) = \sum_{i\in(A,B)} \sum_{j\in X} \frac{\gamma_{i,j}\gamma_{j,i}}{2}. \tag{129}$$

Therefore, in diatomic molecules, for which $\chi_{AB} = \chi$, one finds using the idempotency relations of Eq. (97),

$$P_X(AB) = \sum_{j\in X}\left(\sum_{i} \frac{\gamma_{j,i}\gamma_{i,j}}{2N_{AB}}\right) = \sum_{j\in X} \frac{\gamma_{j,j}}{N_{AB}} = \sum_{j\in X} \tilde{p}_j(AB), \; X = A, B, \tag{130}$$

and hence $P_A(AB) + P_B(AB) = 1$. Clearly, the last relation does not hold for diatomic fragments in larger molecular systems, when $\chi_{AB} \neq \chi$: $P_X(AB) \neq \sum_{j\in X} \tilde{p}_j$ and $P_A(AB) + P_B(AB) \neq 1$.

We finally observe that the effective orbital probabilities of Eqs (125-127) and the associated condensed probabilities of bonded atoms [Eq. (130)] do not reflect the actual AO participation in all chemical bond(s) in AB, giving rise to comparable values for the *bonding* and *non-bonding* (lone-pair) AO in the valence and inner shells. The relative importance of basis functions of one atom in forming the chemical bonds with the other atom of the specified diatomic fragment is reflected by the joint *bond* (*b*) *probabilities* of the two atoms, defined only by the *diatomic* components of the *simultaneous* probabilities of Eqs (125)-(127):

$$P_b(A\wedge B) \equiv \sum_{j\in B} P_{AB}(A\wedge j) = \sum_{i\in A} P_{AB}(i\wedge B) = P_b(B\wedge A) = \sum_{i\in A}\sum_{j\in B} \frac{\gamma_{i,j}\gamma_{j,i}}{2N_{AB}}. \tag{131}$$

The joint atom-orbital bond probabilities, $\{P_{AB}(A\wedge j), j\in B\}$ and $\{P_{AB}(i\wedge B,), i\in A\}$, to be used as weighting factors in the average conditional-entropy (covalency) and mutual-information (ionicity) descriptors of the chemical bond(s) between A and B, indeed assume appreciable magnitudes only when the electron occupying the atomic orbital χ_i of one atom is simultaneously found with a significant probability on the other atom, thus effectively excluding the contributions to the entropy/information bond descriptors due to the lone-pair electrons. Such joint bond probabilities thus emphasize that AO of both atoms are simultaneously involved in the system of occupied MO in the specified diatomic fragment.

The *reference* bond probabilities of AO, to be used to calculate the mutual-information (IT-ionicity) bond index of the diatomic channel, have to be normalized to the corresponding sums $\boldsymbol{P}(AB|\chi_{AB}) = \{P(AB|i)\}$ of Eq. (123). Since the bond probability concept of Eq. (131) symmetrically involves the two bonded atoms, we apply the same symmetry requirement in determining the associated reference bond probabilities of AO:

$$\{p_b(i) = P(AB|i)/2;\; i\in(A, B)\}, \tag{132}$$

where $P(AB|i)$ denotes the probability that an electron originating from orbital χ_i will be found on atom A or B in the molecule.

As we have already remarked in Section 7.1, in OCT the complementary quantities characterizing the average *noise* (conditional entropy of the channel "output" given "input") and the *information flow* (mutual information in the channel "output" and "input") in the diatomic communication system defined by the conditional AO probabilities of Eq. (100) provide the overall descriptors of the fragment bond covalency and ionicity, respectively. Both molecular and promolecular reference (input) probability distributions have been used in the past to determine the information index characterizing the displacement (ionicity) aspect of the system chemical bonds. In the present development the equal bond probabilities of Eq.(132) will be used as the input reference values.

In the A–B fragment development we similarly define the following ("ensemble") average contributions of both constituent atoms to the diatomic covalency (delocalization) entropy:

$$H_{AB}(B|\chi_A) = \sum_{i\in A} P_{AB}(i \wedge B)\; H(\chi_{AB}|i),\; H_{AB}(A|\chi_B) = \sum_{j\in B} P_{AB}(A \wedge j)\; H(\chi_{AB}|j), \tag{133}$$

where the Shannon entropy (in bits) of the conditional probabilities for the given AO input $\chi_i \in \chi_{AB} = (\chi_A, \chi_B)$ in the diatomic channel:

$$H(\chi_{AB}|i) = -\sum_{j\in(A,B)} P(j|i)\log_2 P(j|i). \tag{134}$$

In Eq. (133), the conditional entropy $H_{AB}(Y|\chi_X)$ quantifies (in bits) the X→Y delocalization per electron, so that the total covalency in the diatomic fragment A–B reads:

$$\mathcal{S}_{AB} = N_{AB}\,[H_{AB}(B|\chi_A) + H_{AB}(A|\chi_B)]. \tag{135}$$

It should be stressed again that the simultaneous (diatomic) probabilities $\{P_{AB}(X \wedge i \in Y), Y \neq X\}$, used in Eq. (133) as weighting factors of contributions due to the specified input AO, effectively eliminate contributions due to the inner- and valence-shell lone pairs, since these weighting factors reflect the actual orbital participation in the fragment chemical bonds. Accordingly, the bond-probability weighted contributions to the average mutual information quantities (in bits) of the two bonded atoms are defined in reference to the unbiased bond probabilities of AO [Eq. (132)]:

$$I_{AB}(\chi_A : B) = \sum_{i\in A} P_{AB}(i \wedge B)\; I(i : \chi_{AB}),\; I_{AB}(A : \chi_B) = \sum_{j\in B} P_{AB}(A \wedge j)\; I(\chi_{AB} : j),$$

$$I(\chi_{AB} : i) = \sum_{j\in(A,B)} P(j|i)\log_2\left(\frac{P(j|i)}{p_b(j)}\right), \text{ etc.} \tag{136}$$

They generate the total information ionicity of all chemical bonds in the diatomic fragment:

$$\mathcal{I}_{AB} = N_{AB}[I_{AB}(\chi_A : B) + I_{AB}(A : \chi_B)]. \tag{137}$$

Finally, the sum of the above total (diatomic) entropy-covalency and information-ionicity indices determines the overall information-theoretic bond multiplicity in the molecular fragment in question:

$$\mathcal{N}_{AB} = \mathcal{S}_{AB} + \mathcal{I}_{AB}. \tag{138}$$

This IT bond descriptor can be compared with the diatomic bond-order $\mathcal{M}(A, B)$ of Wiberg [32], formulated in the standard SCF LCAO MO theory [Eq. (101)], which has been previously shown to adequately reflect the chemical intuition in the ground state of typical molecular systems. This total bond multiplicity is customarily regarded as being of the "covalent" origin, but the LCAO MO coefficients carry the information about both the electron-sharing (covalent) and electron-separation/transfer (ionic) phenomena in the chemical bond [37-41]. Therefore, this overall index in fact combines the covalent and ionic contributions, which remain to be separated. The present IT approach provides such a component resolution of these resultant bond-orders. Such a comparison is performed in Tables 4 and 5, reporting the numerical RHF data of bond-orders in diatomic fragments of representative molecules for their equilibrium geometries in the minimum (STO-3G) and extended (6-31G*) basis set descriptions, respectively. It follows from both these tables that the applied weighting procedure gives rise to an excellent agreement with both the Wiberg bond-orders and chemical intuition. A comparison between corresponding entries in Table 4 and the upper part of Table 5 also reveals generally weak dependence on the adopted AO representation, with the minimum basis set being slightly closer to the familiar chemical estimates of the localized bond multiplicities in these typical molecules.

Table 4. Comparison of the diatomic Wiberg and entropy/information bond-multiplicity descriptors in selected molecules; these RHF results have been obtained in the minimum (STO-3G) basis set.

Molecule	A—B	$\mathcal{M}(A, B)$	$\mathcal{N}_{AB}$	$\mathcal{S}_{AB}$	$\mathcal{I}_{AB}$
H_2	H—H	1.000	1.000	1.000	0.000
F_2	F—F	1.000	1.000	0.947	0.053
HF	H—F	0.980	0.980	0.887	0.093
LiH	Li—H	1.000	1.000	0.997	0.003
LiF	Li—F	1.592	1.592	0.973	0.619
CO	C—O	2.605	2.605	2.094	0.511
H_2O	O—H	0.986	1.009	0.859	0.151
AlF_3	Al—F	1.071	1.093	0.781	0.311
CH_4	C—H	0.998	1.025	0.934	0.091
C_2H_6	C—C	1.023	1.069	0.998	0.071
	C—H	0.991	1.018	0.939	0.079
C_2H_4	C—C	2.028	2.086	1.999	0.087
	C—H	0.984	1.013	0.947	0.066
C_2H_2	C—C	3.003	3.063	2.980	0.062
	C—H	0.991	1.021	0.976	0.045
C_6H_6*	C_1—C_2	1.444	1.526	1.412	0.144
	C_1—C_3	0.000	0.000	0.000	0.000
	C_1—C_4	0.116	0.119	0.084	0.035

* For the sequential numbering of carbon atoms in the benzene ring.

In a series of related compounds, e.g., in hydrides or halides, the trends exhibited by the entropic covalent and ionic components of a roughly conserved overall bond order also agree with intuitive expectations. For example, the single chemical bond between two "hard" atoms in HF appears predominantly covalent, while a substantial ionicity is detected for LiF, for which both Wiberg and information-theoretic results predicts roughly 3/2 bond in the minimum basis set, consisting of approximately 1 covalent and ½ ionic contributions; in the extended basis set both approaches give approximately a single bond estimate, with the information theory again predicting the ionic dominance of this overall bond multiplicity.

The significant information-ionicity contribution is also detected for all halides in the lower part of Table 5. One also finds that all carbon-carbon interactions in the benzene ring are properly differentiated. The chemical orders of the multiple bonds in ethane, ethylene and acetylene are also properly reproduced and the triple bond in CO is accounted for correctly. Even more subtle bond differentiation effects are adequately reflected by the present information-theoretic results. The differentiation of the "equatorial" and "axial" S–F bonds in the irregular tetrahedron of SF_4 is correctly reproduced, and an increase in the strength of the central bond in propellanes with increase of sizes of the bridges is correctly predicted.

Moreover, as intuitively expected, the C–H bonds are seen to slightly increase their information-ionicity, when the number of these terminal bonds increases in a series: acetylene, ethylene, ethane. In B_2H_6 the correct, ≈ ½ bond-order of the bridging B–H bond is predicted and approximately single terminal bond multiplicity is detected. For the alkali metal fluorides the increase in the bond entropy covalency (decrease in information ionicity) with increasing size (softness) of the metal is also observed. For the fixed alkali metal in halides, e.g., in a series consisting LiF, LiCl and LiBr (Table 5), the overall bond order is increased for larger (softer) halogen atoms, mainly due to a higher entropy-covalency (delocalization), noise-component of the molecular communication channel in AO resolution.

The identity $\mathcal{N}_{AB} = \mathcal{M}(A, B)$ for diatomic molecules, for which $\chi_{AB} = \chi$ and the reference probabilities $\{p_b(k) = P(AB|k)/2 = ½\}$, can be readily demonstrated:

$$\mathcal{N}_{AB} = \mathcal{S}_{AB} + \mathcal{I}_{AB} = N_{AB}\{\sum_{i\in A} P_{AB}(i \wedge B)\,[H(\chi|i) + I(i:\chi)] + \sum_{j\in B} P_{AB}(A \wedge j)\,[H(\chi|j) + I(\chi:j)]\}$$

$$\equiv N_{AB}\{\sum_{i\in A} P_{AB}(i \wedge B)\mathcal{N}(\chi;i) + \sum_{j\in B} P_{AB}(A \wedge j)N(\chi;j)\}$$

$$= N_{AB}\{\sum_{i\in A} P_{AB}(i \wedge B) + \sum_{j\in B} P_{AB}(A \wedge j)\} = 2\, N_{AB}\, P_b(A \wedge B) = \mathcal{M}(A, B), \quad (139)$$

where we have observed that the conditional IT bond multiplicity due to input χ_k per single electron

$$N(\chi;k) = \sum_{l\in\chi}\{-P(l|k)\log_2 P(l|k) + P(l|k)\log_2[P(l|k)/p_b(l)]\}$$

$$= [\sum_{l\in\chi} P(l|k)]\log_2 2 = 1. \quad (140)$$

8. Additive and Non-Additive Components of Information Channels

Let us again combine the molecular basis functions of typical LCAO MO calculations into the atomic subsets of the preceding section: $\chi = \{\chi_X\} = (\chi_A, \chi_B, \chi_C, \ldots) \equiv \chi^{AIM}$. This arrangement determines the associated block structure of the AO conditional probability matrix $P(\chi|\chi) = \{P(\chi_X|\chi_Y)\}$, (X, Y) $\in$ A, B, C, ... The diagonal block $P(\chi_X|\chi_X)$ then determines the *internal* (*one*-center) communications X→X in atom X, which are responsible for the AIM promotion to its bonding (*valence*) state in the molecule. Accordingly, as we have already argued in the preceding section, the off-diagonal blocks $P(\chi_X|\chi_Y)$ and $P(\chi_Y|\chi_X)$, X ≠ Y, generate the external (*two*-center) communications, Y→X and X→Y, respectively, between the contributed AO of both atoms, which are ultimately responsible for the truly bonding effects determining the IT multiplicities of localized chemical bonds between the given pair of atoms in the molecule.

The *inter*-atomic communications in the molecular channel reflect the chemical interactions between atoms [8], so that the collection of *non*-bonded (separated) atoms of the promolecule exhibits only the intra-atomic probability propagations. The same principle can be used to naturally partition the molecular AO communication system of the AIM-arranged basis set χ^{AIM} into its *additive* and *non-additive* sub-channels [compare Eq. (77)]:

$$\mathbf{P}^{total}(\chi^{AIM}|\chi^{AIM}) = \mathbf{P}^{add.}(\chi^{AIM}|\chi^{AIM}) + \mathbf{P}^{nadd.}(\chi^{AIM}|\chi^{AIM}). \tag{141}$$

The former combines the internal (*intra*-atomic) communications of each (separable) AIM, thus being solely determined by the diagonal blocks of $P(\chi|\chi)$,

$$P^{int.}(\chi|\chi) = \{P(\chi_X|\chi_X)\delta_{X,Y}\} \equiv \mathbf{P}^{add.}(\chi^{AIM}|\chi^{AIM}), \tag{142}$$

while the latter groups all external (*inter*-atomic) probability propagations in the (non-separable, chemically bonded) diatomic fragments of the molecular system under consideration:

$$P^{ext.}(\chi|\chi) = \{P(\chi_X|\chi_Y)(1 - \delta_{X,Y})\} \equiv \mathbf{P}^{nadd.}(\chi^{AIM}|\chi^{AIM}), \tag{143}$$

Table 5. The same as in Table 4 for the extended 6-31G* basis set.

Molecule	A—B	M(A, B)	N_{AB}	S_{AB}	I_{AB}.
F_2	F—F	1.228	1.228	1.014	0.273
HF	H—F	0.816	0.816	0.598	0.218
LiH	Li—H	1.005	1.005	1.002	0.004
LiF	Li—F	1.121	1.121	0.494	0.627
CO	C—O	2.904	2.904	2.371	0.533
H_2O	O—H	0.878	0.896	0.662	0.234
AlF_3	Al—F	1.147	1.154	0.748	0.406
CH_4	C—H	0.976	1.002	0.921	0.081
C_2H_6	C—C	1.129	1.184	1.078	0.106

Table 5. (Continued)

Molecule	A—B	M(A, B)	N_{AB}	S_{AB}	I_{AB}.
	C—H	0.955	0.985	0.979	0.106
C_2H_4	C—C	2.162	2.226	2.118	0.108
	C—H	0.935	0.967	0.878	0.089
C_2H_2	C—C	3.128	3.192	3.095	0.097
	C—H	0.908	0.943	0.878	0.065
C_6H_6*	C_1—C_2	1.507	1.592	1.473	0.119
	C_1—C_3	0.061	0.059	0.035	0.024
	C_1—C_4	0.114	0.117	0.081	0.035
LiCl	Li—Cl	1.391	1.391	0.729	0.662
LiBr	Li—Br	1.394	1.394	0.732	0.662
NaF	Na—F	0.906	0.906	0.429	0.476
KF	K—F	0.834	0.834	0.371	0.463
SF_2	S—F	1.060	1.085	0.681	0.404
SF_4	S —F_a	1.055	1.064	0.670	0.394
	S—F_b	0.912	0.926	0.603	0.323
SF_6	S—F	0.978	0.979	0.726	0.254
B_2H_6†	B—B	0.823	0.851	0.787	0.063
	B—H_t	0.967	0.995	0.938	0.057
	B—H_b	0.476	0.490	0.462	0.028
Propellanes‡: [1.1.1]	C_b—C_b	0.797	0.829	0.757	0.072
[2.1.1]	C_b—C_b	0.827	0.860	0.794	0.066
[2.2.1]	C_b—C_b	0.946	0.986	0.874	0.112
[2.2.2]	C_b—C_b	1.009	1.049	0.986	0.063

* For the sequential numbering of carbon atoms in the benzene ring. ‡ Central bonds between the *bridgehead* carbon atoms C_b. † H_t and H_b denote the *terminal* and *bridge* hydrogen atoms, respectively.

As we have argued in the preceding section (see also [30c]), the polarized state of each bonded atom, as determined by the associated diagonal block of conditional probabilities, important though it is for the full characterization of the AIM valence preparation in the molecule, has no direct relevance for the pattern of chemical "connectivities" between bonded atoms, which are reflected by the resultant IT bond-multiplicities and their covalent/ionic components. Therefore, the partition of Eqs (141)-(143) again emphasizes the importance of separating the additive and non-additive sub-channels, to distinguish the one-center atom promotion and the two-center chemical bond phenomena in OCT.

Clearly, there are also some internal-covalency ("noise") and internal-ionicity (information flow) contributions involved in the atomic promotion processes [29b]. However, as in the CG probe of Section 6.2, the entropic bond-orders of the inter-atomic chemical bonds are again seen to be solely determined by the non-additive sub-channel, which combines the communications between AO originating from different atoms. This separation removes the internal effects in AIM and it has been shown to fully account for the bond-differentiation patterns in diatomic fragments of typical molecules. It also exactly reproduces the Wiberg bond-orders in diatomic molecules.

The partitioning of the conditional probabilities in Eqs. (142) and (143) is in the spirit of a related division of the AO matrix representations of quantum-mechanical operators in a similar context of their *internal* and *external eigenvalue problems*, respectively [56,60]. The latter approach has been successfully applied in identifying the partially-decoupled channels of the collective electron displacements in reactants [56], and it has been used to determine the *inter*-atomic flows of electrons in molecules [61].

9. Conclusion

In this survey we have introduced the key concepts and techniques of IT, which have been used to explore the electronic structure of molecules. The recent progress and representative applications of such IT of molecular systems has been summarized. Among other developments the use of information densities as local probes of the equilibrium electronic distributions and of the structure of bonded atoms has been emphasized. IT provides a novel perspective on both the location and the information origins of chemical bonds. The Fisher information principle has been used to generate the Schödinger equations of quantum mechanics and the variational principle of the entropy-deficiency, relative to the atomic promolecule, has been shown to give rise to the IT-unbiased, Hirshfeld scheme of partitioning the molecular electron density into the infinitely-extending and the subsystem *v*-representable "stockholder" atoms. The importance of the *non*-additive entropy/information measures in extracting subtle changes due to the chemical bond formation has been stressed. In particular the densities of the *non*-additive Fisher information in the MO and AO resolutions, which emphasize the importance of the electronic kinetic energy in the bond-formation phenomena, have been advocated as useful electron (ELF) and bond (CG) localization probes, respectively. Illustrative numerical results have been presented to validate this claim.

It should be emphasized, however, that the "*vertical*", *sub*-molecular reality of the subsystem resolution [8], which is so important for the *"language"* and understanding in chemistry, cannot be established by a direct experiment, since AIM cannot be formulated as the unique quantum-mechanical "observables". Therefore, the bonded atoms, functional groups and chemical bonds have to be ultimately classified as *Kantian noumenons* [15]. Nonetheless, their partial understanding and indirect probes are available from several different perspectives. The close analogy between the IT phenomenological description of molecules and their fragments and the ordinary thermodynamics further validates these chemical concepts, since it introduces the thermodynamic-like causality into relations between perturbations and responses of molecular subsystems and hence brings more consistency into the theory of chemistry [8].

The novel IT visualization tools should prove useful in exploring bonding patterns of controversial molecular systems, the bonding structure of which still remains a matter for scientific debate, with alternative bond criteria giving conflicting answers to the very qualitative question of the existence or non-existence of the disputed chemical bonds between the specified atoms in the molecular environment under consideration. We have illustrated such a use of these local probes in the representative problem of the central bond in propellanes, between the bridgehead carbon atoms.

The IT approach is very much in the spirit of the Eugene Wigner's observation, often quoted by Walter Kohn, that the understanding in science requires insights from several different points of view. The kinetic-energy probe of the bonding regions in molecules provides such an alternative, Fisher information perspective on the genesis of the chemical bond, which complements the popular *potential*-component interpretation. Only together these complementary tools constitute what one would call the "complete" theory of the this complex phenomenon, which – to paraphrase yet another famous citation from Samuel Beckett – is one of old, good problems that never die out.

Of similar character are recent developments in CTCB, and particularly in OCT, which introduce a novel communication (entropy/information) perspective on several classical issues in the electronic structure theory. Until recently, a wider use of CTCB in probing the molecular electronic structure has been hindered by the originally *two*-electron conditional probabilities, which blur a diversity of chemical bonds. We have demonstrated in this and previous [30c] reviews that the AO-resolved OCT using the flexible-input probabilities and recognizing the bonding/antibonding character of the orbital interactions in a molecule, reflected by the signs of the underlying CBO matrix elements, to a large extent remedies this problem. The off-diagonal conditional probabilities it generates are proportional to the quadratic bond indices of the MO theory, and hence the strong inter-orbital communications correspond to strong Wiberg bond-multiplicities. It also properly accounts for the increasing populational decoupling of AO, when the anti-bonding MO become occupied. It should be emphasized, that the extra computation effort of this IT analysis of the molecular bonding patterns is negligible, compared to the standard computations of the molecular electronic structure, since all quantum-mechanical computations in the orbital approximation already determine the CBO data required by this generalized formulation of OCT.

We have also demonstrated that a dramatic improvement in a differentiation of diatomic bond multiplicities is obtained when one recognizes such diatomic groups of orbitals as the *separate*, mutually-decoupled information systems. Such decoupling process can be satisfactorily described only within the flexible-input approach, which links the specified AO-input distribution to its involvement in communicating (bonding) with the remaining orbitals. The other improvement of the IT description of the bond diversity in molecules and their weakening with the non-zero occupation of anti-bonding MO has been gained by applying the MO-resolved channels supplemented with the extra *sign* convention of their entropy/information bond contributions, linked to those of the associated MO bond-orders. Indeed, the bonding/antibonding classification, although lost in the conditional probabilities, is directly available from the corresponding CBO matrix elements, routinely generated in all LCAO MO calculations and required to generate the information channels themselves.

The diatomic development in OCT extends our understanding of the chemical bond from the complementary viewpoint of the Communication Theory. The purely probabilistic models have previously been shown to be unable to fully reproduce the bond differentiation patterns observed in alternative bond–order measures formulated in the standard MO theory. However, as convincingly demonstrated in Section 7, the bond-probability weighting of contributions due to separate AO inputs gives excellent results, which fully reproduce the bond differentiation in diatomic fragments of the molecule, as implied by the quadratic bond index of Wiberg, and gives rise to its resolution into the associated IT-covalent and IT-ionic components. We have also outlined the many-orbital generalization of OCT, which allows one to describe the inter-bond coupling phenomena and opens a novel, IT-perspective on the origins of catalytical activity.

Acknowledgements

I wish to express my thanks to Dr. E. Broniatowska, for preparing Figs. 1-11 and Tables 1-3, and to Dr. S. Escalante – for drawing Figs. 12-14. Thanks are also due to Mr. P. de Silva, who has prepared Figs. 17-25, and Mr. D. Szczepanik, who has generated the RHF OCT results of Tables 4 and 5.

References

[1] Fisher, R. A. (1925). Proc. *Cambridge Phil. Soc.*, **22**, 700.

[2] Frieden, B. R. (2004). *Physics from the Fisher Information – A Unification*, 2nd Ed. (Cambridge University Press, Cambridge,).

[3] Shannon, C. E. (1948). *Bell System Tech. J.* **27**, 379, 623. In: C. E. Shannon, & W. (1949). Weaver, *The Mathematical Theory of Communication*, (University of Illinois, Urbana,).

[4] Kullback, S. & Leibler, R. A. (1951). *Ann. Math. Stat.*, **22**, 79.

[5] Kullback, S. (1959). *Information Theory and Statistics* (Wiley, New York,).

[6] Abramson, N. (1963). *Information Theory and Coding*, (McGraw-Hill, New York,). see also: P. E. Pfeifer, *Concepts of Probability Theory*, 2nd Ed. (Dover, New York, 1978).

[7] Nalewajski, R. F. (2009). (a) *Adv. Quant. Chem.*, **56**, 217. (b) Nalewajski, R. F. & Szczepanik, D. J. Mrozek, Adv. Quant. Chem., submitted.

[8] Nalewajski, R. F. (2006). *Information Theory of Molecular Systems* (Elsevier, Amsterdam,), and refs. therein.

[9] Nalewajski, R. F., Świtka, E. & Michalak, A. (2002). *Int. J. Quantum. Chem.*, *87*, 198. In: R. F. & Nalewajski, E. Świtka, Phys. *Chem. Chem. Phys.*, **4**, 4952.

[10] Nalewajski, R. F. & Broniatowska, E. (2003). *J. Phys. Chem. A.*, **107**, 6270.

[11] Nalewajski, R. F. & Parr, R. G. (2001). *Proc. Natl. Acad. Sci. USA* **97**, 8879 (2000); *J. Phys. Chem. A*, *105*, 7391.

[12] Nalewajski, R. F. & Loska, R. (2001). *Theoret. Chem. Acc.*, **105**, 374.

[13] Nalewajski, R. F. (2002). *Phys. Chem. Chem. Phys.*, **4,** 1710.

[14] Nalewajski, R. F. (2003). *Chem. Phys. Lett.* **372**, 28.

[15] Parr, R. G., Ayers, P. W. & Nalewajski, R. F. (2005). *J. Phys. Chem.*, A **109**, 3957.

[16] Nalewajski, R. F. (2003). *Adv. Quant. Chem.*, **43**, 119.

[17] Nalewajski, R. F. & Broniatowska, E. (2007). *Theoret. Chem. Acc.*, **117**, 7.

[18] Hirshfeld, F. L. (1977). *Theoret. Chim.* Acta (Berl.), **44**, 129.

[19] Nalewajski, R. F. (2008). Int. *J. Quantum Chem.*, **108**, 2230.

[20] Nalewajski, R. F. & de Silva, P. (2009). *J. Mrozek*, in: *Kinetic Energy Functional*, edited by In: A. Wang, & Wesołowski, T. (World Scientific, Singapore, 2009), in press; (b) Nalewajski, R. F. *J. Math. Chem., submitted.*

[21] Nalewajski, R. F., Köster, A. M. & Escalante, S. (2005). *J. Phys. Chem., A* **109**, 10038.

[22] Becke, A. D. & Edgecombe, K. E. (1990). *J. Chem. Phys.*, *92*, 5397.

[23] Nalewajski, R. F. (2000). *J. Phys. Chem., A* **104**, 11940.

[24] Nalewajski, R. F. & Jug, K. (2004). in: *Reviews of Modern Quantum Chemistry: A Celebration of the Contributions of Robert G. Parr*, edited by K. D. Sen, (World Scientific, Singapore, 2002), Vol. I, p. 148, R. F. *Nalewajski, Struct. Chem.*, **15**, 391.

[25] Nalewajski, R. F. (2004). *Mol. Phys.*, **102**, 531, 547. Ibid., *Mol. Phys.*, **103**, 451 (2005). Ibid., **104**, 365, 493, 1977, 2533 (2006).
[26] Nalewajski, R. F. (2005). *Theoret. Chem. Acc.* **114**, 4 (2005); Ibid., *J. Math. Chem.*, **38**, 43.
[27] Nalewajski, R. F. (a) Math, J. (2008). *Chem.*, **43**, 265. (b) Ibid., 44, 414 (2008), (c) Ibid., (2009). 45, 607. (d) Ibid., 45, 709 (2009). (e) Ibid., *45*, 776 (2009); (f) Ibid., 45, 1041, (2009).
[28] Nalewajski, R. F. (2009). (a) Int. *J. Quantum Chem.*, **109**, 425. (b) *J. Math. Chem.*, submitted.
[29] Nalewajski, R. F. (2008). (a) *J. Math. Chem.*, **43**, 780. (b) Phys. J. (2007). *Chem. A **111***, 4855.
[30] Nalewajski, R. F. (2006). (a) *Mol. Phys.*, **104**, 3339. (b) *Adv. Quant. Chem.*, **56**, 217 (2009). (c) Nalewajski, R. F. & Szczepanik, D. J. Mrozek, *Adv. Quant. Chem.*, submitted.
[31] Nalewajski, R. F. & *Phys. J.* (2003). *Chem*, **A107**, 3792. Ibid., *Mol. Phys.* **104**, 255 (2006); Ibid., *Ann. Phys. (Leipzig)* **13**, 201 (2004).
[32] von Weizsäcker, C. F. (1935). *Z. Phys.*, **96**, 431.
[33] Frieden, B. R. (1989). *Am. J. Phys.*, **57**, 1004. M. Reginatto, *Phys. Rev.* **A58**, 1775 (1998); *Erratum: Phys. Rev.* **A60**, 1730 (1999).
[34] Sears, S. B. (1980). Ph. D. Thesis, The University of North Carolina at Chapel Hill (1980); S. B. Sears, R. G. Parr, U. Dinur, *Israel J. Chem.*, **19**, 165.
[35] Kohn, W. & Sham, L. J. (1965). *Phys. Rev.*, **140A**, 1133.
[36] Hohenberg, P. & Kohn, W. (1964). *Phys. Rev.* **136B**, 864.
[37] Nalewajski, R. F., Köster, A. M. & Jug, K. (1993). *Theoret. Chim. Acta (Berl.)*, **85**, 463.
[38] Nalewajski, R. F. & Mrozek, J. (1994). Int. *J. Quantum Chem.*, **51**, 187. R. F. Nalewajski, S. J. Formosinho, A. J. C. Varandas, J. Mrozek, *Int. J. Quantum Chem.* **52**, 1153.
[39] Nalewajski, R. F., Mrozek, J. & Mazur, G. (1996). *Can. J. Chem.*, **100**, 1121.
[40] Nalewajski, R. F., Mrozek, J. & Michalak, A. (1997). Int. *J. Quantum Chem.*, **61**, 589. In: J., Mrozek, R. F. & Nalewajski, A. (1998). *Michalak, Polish J. Chem.* **72**, 1779.
[41] Nalewajski, R. F. (2004). *Chem. Phys. Lett.*, **386**, 265.
[42] Wiberg, K. A. (1968). *Tetrahedron*, **24**, 1083.
[43] Gopinathan, M. S. & Jug, K. (1983). *Theor. Chim. Acta (Berl.)*, **63**, 497, 511. see also: In: K., Jug, M. S. Gopinathan, (1983). in *Theoretical Models of Chemical Bonding*, edited by Z. B. Maksić, (Springer, Heidelberg, 1990), Vol. II, p. 77; I. Mayer, *Chem. Phys. Lett.*, **97**, 270.
[44] Bader, R. F. W. (1990). *Atoms in Molecules* (Oxford University Press, Oxford,).
[45] Capitani, J. F., Nalewajski, R. F. & Parr, R. G. (1982). *J. Chem. Phys.*, **76**, 568.
[46] Gordon, R. G. & Kim, Y. S. (1972). *J. Chem. Phys.*, **56**, 3122.
[47] Cortona, P. (1991). *Phys. Rev.*, **B 44**, 8454.
[48] Wesołowski, T. & Warshel, A. (1993). *J. Phys. Chem.*, **97**, 8050. T. Wesołowski, *J. Am. Chem. Soc.*, **126**, 11444. Ibid., *Chimia* **58**, 311 (2004), and refs. therein.
[49] Silvi, B. & Savin, A.(1994). *Nature*, **371**, 683, Savin, A., Nesper, R., Wengert, S. & Fässler, T. F. (1997). *Angew. Chem. Int. Ed. Engl.* **36**, 1808.
[50] Feinberg, M. J., Ruedenberg, K. & Mehler, E. L. (1970). Adv. Quant. *Chem.*, **5,** 28. Feinberg, M. J. & Ruedenberg, K. (1971). *J. Chem. Phys.*, **54**, 1495. Ibid. 55, 5804.

[51] Goddard, W. A. & Wilson, C. W. (1972). *Theoret. Chim. Acta, (Berl.)* **26**, 195, 211.
[52] Nalewajski, R. F. (2009). Int. *J. Quantum Chem., (I. Mayer issue),* **109**, 245.
[53] Dirac, P. A. M. (1958). *The Principles of Quantum Mechanics*, 4th ed. (Clarendon, Oxford,).
[54] Pearson, R. G. (1963). *J. Am. Chem.*, Soc. **85**, 3533; *Science* **151**, 172 (1966); *Hard and Soft Acids and Bases* (Dowden, Hatchinson and Ross, Stroudsburg, 1973); C. K. *Jørgensen, Inorg. Chem.*, **3**, 1201 (1964).
[55] Parr, R. G. & Pearson, R. G. (1983). *J. Am. Chem. Soc.*, **105**, 7512 see also: Nalewajski, R. F. (1984). *J. Am. Chem. Soc.*, **106**, 944.
[56] Nalewajski, R. F., Korchowiec, J. & Michalak, A. (2006). (a) *Topics in Current Chemistry*, **183**, 25 (1996). (b) Nalewajski, R. F. & Korchowiec, J. (1997). *Charge Sensitivity Approach to Electronic Structure and Chemical Reactivity* (World-Scientific, Singapore, 1997); (c) R. F. Nalewajski, Adv. *Quant. Chem.*, **51**, 235.
[57] Chandra, A. K., Michalak, A., Nguyen, M. T. & Nalewajski, R. F. (1998). *J. Phys. Chem., A* **102**, 10182.
[58] Nalewajski, R. F. (2000). *Topics in Catalysis*, **11/12**, 469.
[59] Gutmann, V. (1978). *The Donor Acceptor Approach to Molecular Interactions* (Plenum, New York,).
[60] Nalewajski, R. F. (2008). *J. Math. Chem.*, **44**, 802.
[61] Mitoraj, M. & Michalak, A., *J. Mol. Model.*, **11**, 341. Ibid., 13, 347 (2007); M. Mitoraj, Ph. D. Thesis, Jagiellonian University, 2007; M. Mitoraj, H. Zhu, A. Michalak, and T. Ziegler, *J. Org. Chem.* **71**, 9208 (2006); Ibid., *Organometallics* **26**, 1627 (2007).

In: Mathematical Chemistry
Editor: W.I. Hong, pp. 327-366

ISBN: 978-1-60876-894-3

Chapter 4

THE CHEMISTRY AND MATHEMATICS OF DNA POLYHEDRA

Wen-Yuan Qiu[1,*], Ze Wang and Guang Hu
Department of Chemistry, State Key Laboratory of Applied Organic Chemistry, Lanzhou University, Lanzhou, People's Republic of China

Abstract

Polyhedra have attracted scientists' attentions due to their high-symmetric architectures since ancient times, and even served as common modes in natural world, for example, the polyhedral skeletons of molecular structures. Chemists, in the past few centuries, have strived to synthesize these polyhedral targets. Despite achievements that have been arrived, there is still a gap between the great varieties of polyhedral shapes observed in nature (such as viral protein capsids) and the relatively limited molecular polyhedra constructed by small organic molecules. Fortunately, DNA was shown to be an excellent material in molecular construction. The construction of polyhedral structures with DNA improves the development of synthetic chemistry, and meanwhile, produces some complicated molecules with novel topological and geometrical structures, such as the DNA tetrahedron, DNA cube, DNA octahedron, DNA dodecahedron, and DNA icosahedron, as well as the tetrahedral DNA cage and the hexahedral DNA box, and so on. Interests on these novel architectures have inspired mathematical considerations and modeling of a new theory of polyhedra, including structure construction, symmetry analysis, chirality detection and invariant computaion. Herein, we have summarized advances in the chemical syntheses of DNA polyhedra, which also have been envisaged and classified by underlying principles of geometry and symmetry. Moreover, this paper reviews our recent progresses in the theoretical investigation of a new mathematical theory of DNA polyhedra.

1. Introduction

Once the great geometer Coxeter wrote: '...the chief reason for studying regular polyhedra is still the same as in the times of the Pythagoreans, namely, that their symmetrical shapes appeal to one's artistic sense'.[1] These beautiful objects attract continuously

[*]E-mail address: wyqiu@lzu.edu.cn, Tel.: +86-931-891-1356, Fax: +86-931-891-2582 (Corresponding author)

consideration and discussion of mathematicians from ancient times, writing them as one of the oldest chapters in mathematics; and arise the interests of artists, architects and philosophers, making them as an icon of human civilization.[2] Meanwhile, these symmetric structures are also favored in the laboratory and nature.[2, 3] Chemists, molecular architects, fascinated by these symmetrical molecular structures, have been making great efforts in synthesizing these beautiful molecules, by chemically constructing the Platonic solids such as tetrahedron, cube, dodecahedron with small organic molecules and other polyhedra from supermolecular self-assembly.[3-9] The preference for highly symmetric objects with a spherical shape is also found in the case of life, such as the icosahedral shape of a viral capsid protein, the polyhedral structure of Clathrin protein and the cuboctahedral structure of the COPII protein, for examples.[10-16] These endeavors would enlarge and improve our knowledge and skills on synthetic chemistry, and moreover, bring new insights into ubiquitous polyhedral structures of molecules in nature. On the other hand, chemical syntheses of these elegant molecules and their underlying biological complexity have been turned out to be of great challenges.

Fortunately, the life molecule DNA, with amazing abilities, could be utilized as a fantastic material in constructing polyhedral structures.[17, 18] Such a groundbreaking idea, synthesizing polyhedra with genetic material DNA, was first proposed by Ned Seeman and even hoped to fabricate regular 3D architectures extending into 3-space at the very beginning.[19, 20] In practice, this magic molecule really performs amazing self-assembly and programmable abilities in laboratories.[17, 18] During the past two decades, chemists have created a series of polyhedral molecules, far more complicated and delicate ever before.[21-25] These fulfillments open the door to structural DNA nanotechnology and take chemists a further step towards mimicking nature.[26-33] And now, constructing and synthesizing various polyhedron-shaped molecules is a fast growing area in current chemistry. The DNA nanopolyhedra or DNA nanocages, which use DNA as a building material to synthesize polyhedra with nanometer precision, have strongly evoked the research interests of the prodigious and distinguished group of chemists. Although there is much to probe, the stakes are high and the odds for success are excellent.

A number of interesting problems, on the other hand, on characterization and analysis of polyhedral molecules in the syntheses have been proposed, which are now challenging mathematical theory of the nanopolyhedra. Novel topological structures, interlocked and interlinked due to DNA double helices, emerge from DNA nanopolyhedra, differing them from polyhedra. Detailed background information on several newer results can be found in Refs. [21-25]. It is now clear that this is an area ripe for development and deserves special encouragement. However, a great deal of essential feature of the nanopolyhedra is either unknown or, at best, poorly understood. Therefore, systematic understandings of these curious objects in details need new mathematics. Scientists are now vigorously studying DNA nanopolyhedra architectures and, little by little, piecing together a new branch of mathematics, although the full story has yet to be written. Based on graph theory and knot theory, polyhedral links, a novel mathematical model, have been proposed to phrase and answer their structural characteristics.[34] It is believed that polyhedral links are new forms of polyhedra, which open enormous challenges and opportunities for scientists in at least two disciplines. One is that they are potential targets in synthetic chemistry, and the other is that they might lay chances for development of new theories in mathematics. As we will, now see, nanopolyhedra not only provide a unifying framework for chemistry, but it may well forge an equally deep union with mathematics as well.

2. Building Polyhedra with DNA

2.1. Why Polyhedra?

Polyhedron is a model of the eyes and the hands as well as of the mind.[35] In fact, connections between polyhedra and chemistry have been established since ancient times.[36] Plato made systematic considerations on relationships between the real material world and the idea of geometry world. He described firstly five regular polyhedra, known as Platonic polyhedra, and imagined that the element fire could be represented by tetrahedron, earth by cube, air by octahedron, water by icosahedron, and the universe by dodecahedron. Later, the idea of representing elements with five Platonic polyhedra was shown to be impractical. However, it did not disturb chemists' constantly exploring polyhedral models for chemical structures of material world. Without respect to the important tetrahedron and octahedron model in stereochemistry, several unexpected complicated polyhedra are proved to be basic molecular structural model. In 1937, mathematician M. Goldberg, when solving isoperimetric problems of polyhedra, discovered a family of complex polyhedra, namely Goldberg polyhedra, which consists of twelve pentagons and certain numbers of hexagons only.[37] Later in the laboratory, one of the exotic complex polyhedra were found to exist as the third allotrope of carbon, buckminsterfullerenes.[38, 39] More surprisingly, Goldberg polyhedra could be even modeled as frameworks of spherical virus capsid, affording a structural classification for viruses.[11, 40]

Interested in molecules of complex polyhedral structures, biological or nonbiological, chemists are making efforts on synthesizing the beautiful structures in laboratory. As progresses it would bring to synthetic chemistry, some new insights into biomacromolecules, such as viral capsids, might also break out as benefits. Undoubtedly, the polyhedral world is nice and alive in all ages, included ours!

2.2. When DNA Meets Polyhedra

As a fundamental material of life, DNA is functioned as the basis of replication and transcription. In 1953, Watson and Crick proposed the famous double helix structure as the molecular model of genetics, which describes DNA as two polymers of nucleotides run in opposite directions with backbones held together by the hydrogen bond between purine and pyrimidine, with the specific interactions as Adenine (A) bonded to Thymine (T) and Cytosine (C) to Guanine (G).[41, 42] This perfect complementarity provides a clear molecular picture of genetic replication, and uncovers the powerful molecular self-recognization and self-assembly ability of DNA. It is the amazing property that makes DNA a wonderful material for nanoconstruction. Moreover, other features such as a persistence length of 50nm under normal conditions,[43] geometric regularity regardless of particular sequences of nucleotides[44] and proper length scale of 2nm in width (B-DNA)[45] are also critical for constructing three-dimensional nano-materials with DNA.[17, 18]

With the help of branched DNA, Seeman [20] proposed that DNA could be reconstructed by modern biotechnology, sticky-ended cohesion technique, into three-dimensional structures. Branched DNA, also DNA junction, differs from DNA of natural configuration, extending not in one but higher dimensions.[46, 47] During the process of genetic recombination, information exchanges between two DNA helices by a branched DNA known as Holliday junction.[48] However, the Holliday junction is mobile that four strands will eventually separate into two new duplexes by sliding, and therefore not suitable for DNA nanoconstruction. In order to stabilize the structure, Seeman proposed that one can break the sequence symmetry of DNA junctions by designing artificial DNA sequences using a computer program.[20, 46, 47] Indeed, he and co-workers obtained a series of immobile DNA junctions with different arms in laboratory, which made the synthesis of DNA polyhedra richer.[49, 50]

With various stable junctions in hand, it is natural to construct three-dimensional structures by synthesizing vertices of certain valence with DNA junctions of appropriate arm numbers and joining these junctions by sticky-ended cohesion further (Figure 1).[20] In early 1990s, Seeman's group made the ground-breaking progress in DNA three-dimensional construction by subtle synthesis of DNA cube and truncated octahedron.[51, 52]

The synthesis of DNA cube and truncated octahedron is step-wise and provides great control at each step during the construction. However, the protocol is time-consuming, of quite low yield and poor rigidity, difficult for asymmetry sequence design, and not replicable. Therefore, chemists later were dedicated to develop strategies of high efficiency, excellent stereoselectivity, good rigidity, and acceptable sequence design task. Over the past 20 years, these strategies have been successfully performed in laboratory by constructing series of DNA polyhedra, providing an effective approach towards mimicking nature and refreshing knowledge and technology of synthetic chemistry![21-25, 33] These represent a tremendous challenge and opportunity for modern chemistry, and for the chemists of the future.

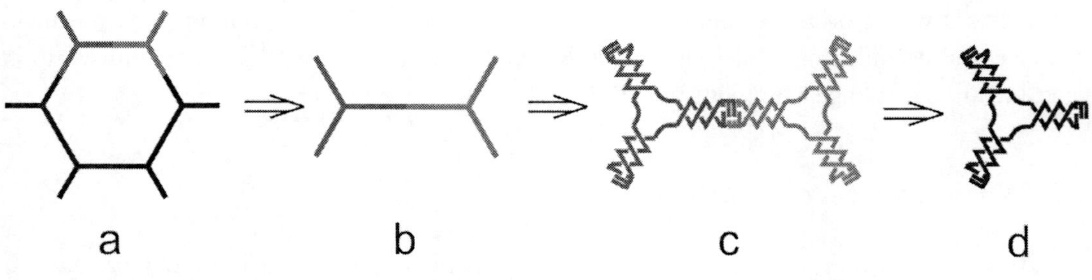

Figure 1. Schematic illustration of the principle of DNA nanoconstruction. Any geometric motifs (a) can be decomposed into the joint of vertices (b), which can be reconstructed with DNA junctions through sticky-ended cohesion (c).

3. DNA Polyhedra

3.1. Platonic Solids

It has been mathematically proved that there are only five Platonic polyhedra, i.e. tetrahedron, cube, octahedron, dodecahedron, and icosahedron. Each of these solids consists of identical regular faces, and possesses the highest symmetry of all polyhedra (Figure 2a).[2,35]

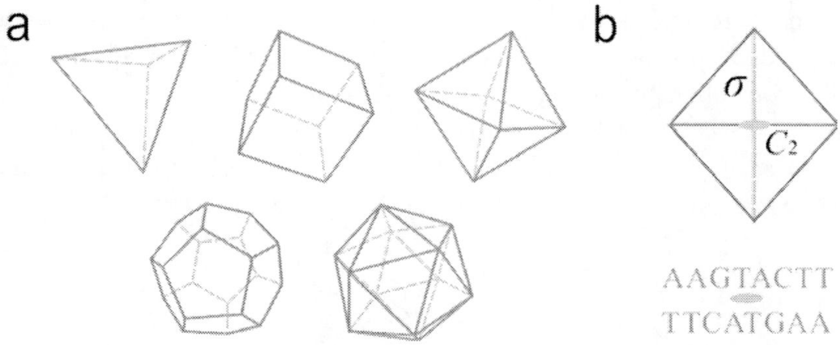

Figure 2. Five Platonic polyhedra. Each is composed of same regular faces, and thus is highly symmetrical.

Point symmetry group analysis shows that the tetrahedron is T_d symmetry, the cube and octahedron are O_h, and the dodecahedron and icosahedron are I_h, respectively. Specifically, without respect to the geometrical and topological structure of DNA helix, the symmetry property of DNA polyhedra is decided exclusively by the symmetric characteristics of DNA sequences on each edge. Therefore, it is convenient to characterize point symmetry group of DNA polyhedra through the analysis of possible symmetry elements of DNA sequences on edges, i.e., 2-fold rotation axis C_2 (palindrome) and reflection mirror σ (Figure 2b). If, by some reasons, the mirror symmetry element σ of DNA sequences was broken, the point group T_d, O_h, I_h of polyhedra will be transformed to T, O, I, respectively. These lower point group (T, O, I) will be degenerated into C_1 (no symmetry) in the case of the absence of rotation axis C_2, which means the DNA sequence is not a palindrome.

The beautiful simplicity of the symmetry principles pursues an inner harmony, where elegance, uniqueness and beauty define molecular structures.

3.1.1. DNA Tetrahedron

The chemical synthesis of DNA tetrahedron was initiated by Turberfield's group in 2004.[53] With great interests on this molecular tetrahedron, several groups redesigned the synthetic strategy with a number of DNA tetrahedra and their derivatives (Turberfield *et al.*, 2005 [54]; Mao *et al.*, 2008 [55]; Hao *et al.*, 2009 [56;57]; Matroianni *et al.*, 2009 [58]; Armitage *et al.*, 2009 [59]).

In 2004, Turberfield and colleagues in Oxford reported the synthesis of a DNA tetrahedron, by assembling four appropriately designed oligonucleotides in an annealing process.[53] Each oligonucleotide contains 55 nucleotides (nt), and is consisted of three subsequences of 17 basepairs (bp) hinged by two nucleotides, which are designed to complement uniquely with subsequences of other strands forming DNA duplexes, as edges of a tetrahedron.[53] Four strands were combined with equimolar quantities, and after annealed the system from 95 to 20 they obtained the DNA tetrahedra within only about four minutes.[53] The DNA tetrahedron contains four nicks at each vertex where the 3' and 5' ends of an ssDNA meet (Figure 3a). And a couple of stereoisomers may exist due to the different subsequences on each edge.[53] The high-resolution structure of DNA tetrahedron was recently confirmed by Turberfield *et al.* with the help of cryoEM technique. [60]

In the successive experiment, with a few improvements, the group showed the synthesis has very good stereoselectivity.[54] They reduced two hinge nucleotides to one and moved four nicks to edges of tetrahedron. Therefore, the improved DNA tetrahedron differs from the former one, containing two edges of intact DNA double helix, which will form easily in intramolecular reaction. The synthesis produces only one stereoisomer with certain configuration at a vertex with high yield (>95%) in only several seconds, indicating both laudable stereochemical control and efficiency of the strategy.[54]

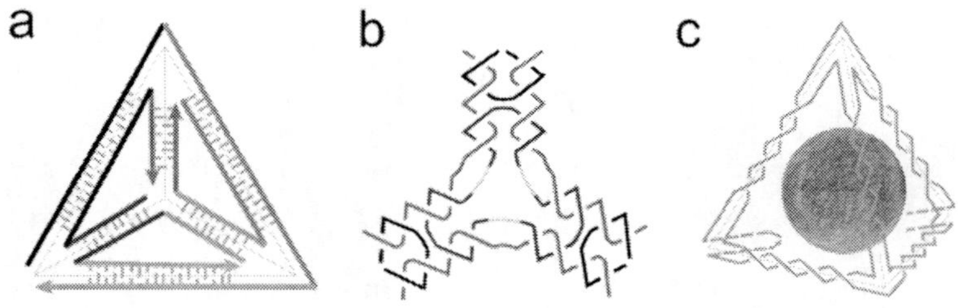

Figure 3. (a) Schematic representations of DNA tetrahedra. DNA subsequences in the same colors are complementary to form edges of tetrahedra.[54] (b) '3-point star' motif developed by Mao *et al.* [55] to synthesize DNA tetrahedron. Red, green, black colors indicate long, medium and short DNA strands, respectively. Yellow colors in the medial represent three unpaired loops. (c) DNA tetrahedron (orange) that encapsulates a protein (yellow) inside.[29].

Based on this strategy, the group synthesized a series of DNA tetrahedra, ones of different edge length,[54] one with a reconfigurable edge,[31] and one with protein molecule encaged (Figure 3c).[29] The geometric structures and parameters of these DNA tetrahedra are summarized in Table 1.

The synthetic strategy developed by Turberfield's group combines several strands by self-assembly in a single step process, proceeds with high yield and good stereochemical selectivity, providing a useful method in the synthesis of DNA polyhedra. This is a tremendous achievement, but it is only part of the reason nanopolyhedra have generated such excitement.

Table 1. Edge parameters of some DNA tetrahedra.

DNA tetrahedra	**Edge parameters**
A series of tetrahedra of different edge lengths	3×20bp/3×30bp; 5×20bp/1×(10,15,20,25,30)bp; 4×20bp/1×10bp/1×(10,15,20,25,30)bp*
Tetrahedron with a reconfigurable edge	5×20bp/1×10bp; 5×20bp/1×30bp
Tetrahedron encapsulating a protein inside	6×20bp

*The notation, for example '4×20bp/1×10bp/1×(10,15,20,25,30)bp' means four edges of tetrahedron contain 20bp, one 10bp, and the left either 10, 15, 20, 25, or 30bp.[29, 53, 54].

Recently, Mao and his fellow researchers reconstructed DNA tetrahedra by an extraordinary effective strategy, which assembles many copies of identical building blocks in a hierarchical fashion.[55] The building block is designed as intermediate tile that self-assembles into final structures in the next step in the hierarchy process.[61-65] The tile '*n*-

point star' is a n-arm branched structure, which is actually a combination of DNA duplexes connected with several unpaired loops, providing both a better stiffness of the branched arms and a controllable flexibility over the whole motif that could be used in a variety of DNA nanostructures (Figure 3b).[66, 67]

In the synthesis of DNA tetrahedra, '3-point star' was used according to 3-valent vertex of tetrahedra. It consists of a long strand (Figure 3b, red), three identical medium strands (Figure 3b, green), and three identical short strands (Figure 3b, black), forming a 3-fold symmetrical structure.[55] Each of its arms is 21bp in length and 4nm in width, as a pair of DNA double helix joined side by side.[55] Meanwhile, the tile contains three unpaired loops with length variable, making different flexibility possible (Figure 3b, yellow).[55]

In the hierarchical self-assembly process, three types of DNA strands combine first into '3-point star' tiles, and then assemble into DNA tetrahedra. By adjusting the concentration of the tiles (75 nanomolar (nM)) and the length of loops (5nt each), DNA tetrahedron were obtained in a one-pot reaction with a yield of 90%.[55] Each tetrahedron consists of four '3-point star' at vertices, contains 42bp in length per edge.[55]

The hierarchical self-assembly strategy provides an effective way to synthesize DNA polyhedra with high yield. Furthermore, the process involves only identical building blocks, avoiding difficulty of sequences design task.[25, 55]

Most recently, Hao and co-workers [56] constructed a DNA tetrahedral container by self-assembling sheets of scaffolded DNA, which is enclosed by four triangular faces (Figure 4a). Matroianni's group [58] have successfully assembled gold colloid to four vertices of a DNA tetrahedron (Figure 4b), and Armitage and his research group [59] have prepared fluorescent DNA tetrahedron with tunable wavelength. It is believed that the explorations of constructing DNA tetrahedron will find their powerful applications in the near future. More curiously, Hao and his team [57] have reported the design and construction of a nanometer-sized tetrahedron from a single strand of DNA that is 286 nucleotides long. The potential gains from these exciting results are considerable.

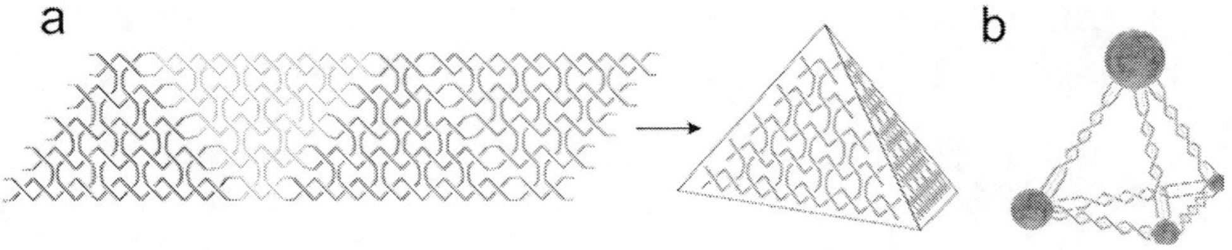

Figure 4. (a) The tetrahedral DNA cage.[56] Self-assembly of 4 DNA triangular sheets (left) into a DNA tetrahedral container (right). Different colors of DNA sheets indicate different triangular faces of DNA tetrahedron. (b) Gold colloid (yellow balls) assembled at each corner of a DNA tetrahedron.[58]

Symmetry in nanopolyhedra has a very concrete and precise meaning, and it is a key part of aesthetics. The symmetry of DNA polyhedra differs from geometric polyhedra due to primary structures of DNA sequences on edges, which is referred as molecular symmetry breaking.[68] The molecular symmetry breaking is a mechanism by which a system of itself goes from a high symmetric conformation to a non-symmetric one. We say that some molecules are more symmetric than others are, or that some molecules have high symmetry while others have low symmetry or no symmetry. Most artificial DNAs may have high symmetry or no symmetry, whereas the natural DNAs have no symmetry.

Turberfield synthesized the DNA tetrahedron [53] which contains asymmetric DNA sequences on edges. For example, one of these sequences can be represented as:

$$\frac{\text{ACATTCCTAAGTCTGAA}}{\text{TGTAAGGATTCAGACTT}},$$

where a central line means two single strands (upper and below) are bonded complementarily to form a DNA helix. It can be intuitively seen that the sequence (the helix) is lack of either a reflection mirror σ or a rotation axis C_2, i.e., not a palindrome. As a result, according to the analysis in Section 3.1, the DNA tetrahedron possesses only C_1 symmetry, the least symmetry or in other words, no symmetry. And, it is impossible to recover the symmetry state of DNA tetrahedron by redesigning DNA palindrome sequences because there are odd numbers of bases (17bp) in the sequence. For example, the following redesigned sequence cannot be a palindrome to reserve the C_2 symmetry due to the central paired bases A and T. Obviously, a rotation axis does not exist between these two bases.

$$\frac{\text{ACATTCCTAAGGAATGT}}{\text{TGTAAGGATTCCTTACA}}$$

Thus, the DNA tetrahedron synthesized by Turberfield *et al.* [17] can not possess higher point symmetry group through DNA sequences redesign without changing the number of bases on edges.

On the other hand, the DNA tetrahedron constructed by Mao *et al.* [55] contains parallel DNA helices on each edge, which can be written as follows:

$$\frac{\text{CCTACGATGGACACGGTAACGCCTAGCAACCTGCCTGGCAAG}}{\text{GGATGCTAC C TGTGCCATTGCGGATCGTTGGACGGACCGTTC}}$$

$$\frac{\text{CTTGCCAGGCAGGTTGCTAGGCGTTACCGTGTCCATCGTAGG}}{\text{GAACGGTCCGTCCAACGATCCGCAATGGCACAGGTAGCATCC}}$$

According to this representation, it can be easily discovered a rotation axis C_2 exist at the center of these sequences, which ensures that the DNA tetrahedron constructed by Mao *et al.* [55] belongs to the point group *T*. Symmetry analysis tell us, intuitively at least, that two states of the system must differ in their DNA sequences.

3.1.2. DNA Cube

The possibility of synthesizing a cube with DNA was first investigated by Seeman's group in 1991.[51] Later, researchers reconstructed DNA cube with improved techniques (Mao *et al.*, 2008 [69]; Andersen *et al.*, 2009 [70]; Kuzuya *et al.*, 2009 [71]). Until now, there exist three independent approaches to construct DNA cube. [51, 69, 70]

DNA cube was first synthesized by Seeman's group in 1991 with 3-arm DNA junctions.[51] They first constructed two squares representing a pair of opposite faces of a cube, and then connected them by joining the sticky ends of corresponding vertices.[51] Belt-like intermediate in the reaction might enclose itself in two opposite ways, resulting in a pair of stereoisomers (Figure 5a).[51] The final cube was compose of six DNA strands, intertwining together to form an interesting molecular catenane with novel topology.[72] Each edge of the cube contains 20bp, 6.8nm, i.e. two helical turns (Figure 5b).[51]

As a milestone, the synthesis of DNA cube opens the door of DNA nanotechnology and provides the fundamental strategy in DNA nanoconstruction by fabricating 3D objects by joining each DNA junction together. However, the protocol is step-wise, time-consuming, and proceeds with quite low yield. Therefore, chemists later were dedicated to develop strategies of high efficiency.

DNA cube was rebuilt recently by Mao *et al.* by self-assembly of '3-point star' motifs.[69] Different from the synthesis of DNA tetrahedra, two types of motifs are used as building units in the manner that different motifs A and B are linked whereas same motifs are not.[69] Consequently, the geometrical objects constructed must contain even number of vertices on each face in order to form a closed structure.[69] Obviously, cube is the smallest and simplest one in this class. So, it can be formed first at low concentration.[69]

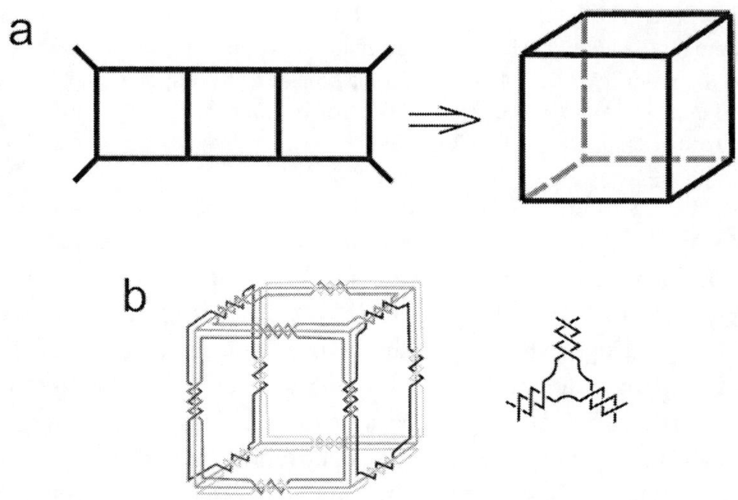

Figure 5. Synthesis of DNA cube. (a) DNA cube (right) can be reconstructed from belt-like intermediate (left). Enclosure of the intermediate, insides and outsides, will produce two different isomers. (b) The catenaned structure of DNA cube (right) and 3-armed DNA junction used in the synthesis (left). DNA cube is consisted of six different strands, each running around a face of the cube.[51]

Mao's group [69] designed two complementary ways to assemble two types of motifs into a cube. One way is to design two completely different types of '3-point star' motifs consisted of same backbones but different DNA sequences, so that DNA sequences at the sticky-ends of motifs are complementary only among the different types.[69] As a result, one type of motifs can bind only to another different type rather than the same.[69] In other words, association between different types of motifs, say type A and B, is possible, whereas interactions between same types, A-A or B-B, is impossible. Using the idea, they constructed DNA cube with two types of motifs, each containing 5nt unpaired loop and of concentration 200nM.[69] The DNA cube consists of two completely different types of motifs arranged alternatively at vertices, two DNA double helices on each edge, each of 42bp, 4 turns (Figure 6 above).[69]

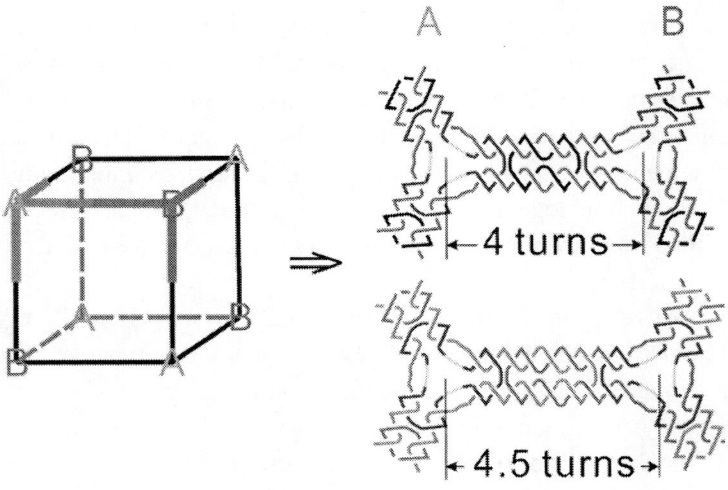

Figure 6. Schematic illustration of the synthesis of DNA cube with '3-point star' motif. DNA cube (left) can be built with two different types of motifs in the manner that different motifs A and B are linked whereas same motifs are not. Mao *et al.* designed the two motifs A and B by two complementary approaches, either by using two completely different types of '3-point star' motifs consisted of same backbones but different DNA sequences (left, above) or by connecting two motifs of same sequences but with different face orientations (left, below).[69]

Except for the straightforward method, another strategy was designed as a complementary way. It is more subtle in that two types of motifs are actually composed of identical DNA sequences. However, two different interactions come out on the consideration of two opposite faces of '3-point star' motifs.[69] The faces of two motifs linked through DNA helices will be on opposite sides if the helix is odd numbers of half-turns, and vice versa.[62] And thus, two different sides of '3-point star' are connected alternatively as two different types of motifs. Therefore, they designed the helices of each edge to be 4.5 turns rather than 4 turns usually involved.[62, 69] By controlling the concentration to be 50nM and unpaired loop of 5nt, DNA cubes were obtained in the hierarchical self-assembly process of a single type of '3-point star'.[69] Different from the former one, the cube contains 47bp on each DNA helical edge, that is, 4.5 helical turns (Figure 6, below).[62, 69]

In general, the method developed in the synthesis of DNA cube with two types of tiles makes good control at the face geometry of DNA polyhedra in that only the one with even number of vertices on each face can be possibly produced.[69] Consequently, it complements the strategy of hierarchical self-assembly, and thus benefits DNA polyhedra construction.[69] But even this is not quite the end of the story.

Recently, based on the idea of single stranded DNA origami, Andersen and co-workers have constructed a DNA cube enclosed by six DNA sheets (Figure 7).[70] This DNA box, with 42×36×36 nm3 in size, can be utilized as a DNA nanocargo that delivers small molecules, as well as a logic sensor for specific signals.[70] Also, Kuzuya and his team have reported an analogous idea but somewhat different effort to a box-shaped 3D origami motif. [71] This approach shows a sophisticated skill, that is, from the initial formation of an open motif, through a stepwise closing process can result in the final box structure.

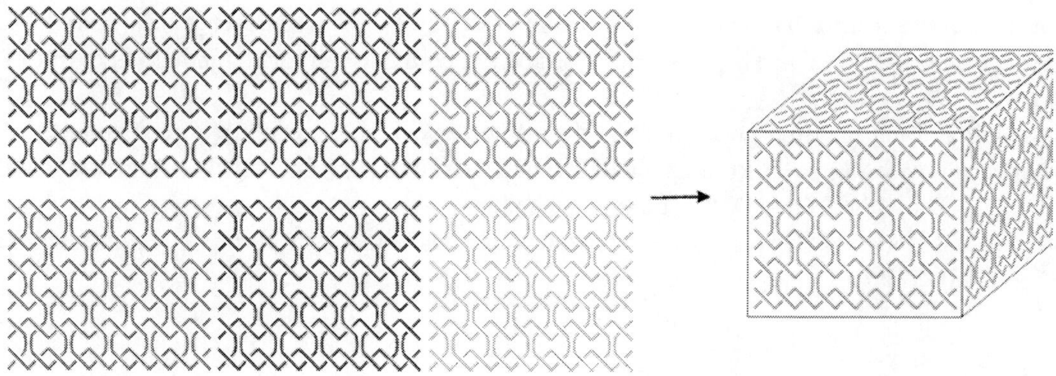

Figure 7. The hexahedral DNA box. Six DNA sheets (left) self-assembled into a DNA box.[70]

The symmetry properties of DNA cube synthesized by Seeman *et al.* [51] can be conveniently analyzed if the DNA sequences were properly considered. In order to avoid unwanted pairing between DNA strands in the synthesis, the DNA sequences on edges of the DNA cube were designed to have the least symmetry. For example, one of these sequences can be represented as:

$$\frac{\text{CTGCATTCGGCCAGCCTGAC}}{\text{GACGTAAGCCGGTCGGACGT}}.$$

Obviously, the DNA sequence does not contain any mirror σ and rotation axis C_2. Thus, the DNA cube possesses only C_1 symmetry, i.e., no symmetry. However, the DNA sequence, due to even number of bases involved (20bp), can be redesigned to hold rotational symmetry C_2 by the following way. Reserve the first half sequence and then rotate the helix 180 degrees through the center will generate a DNA palindrome:

$$\frac{\text{CTGCATTCGGCCGAATGCAG}}{\text{GACGTAAGCCGGCTTACGTC}}.$$

One can easily find that DNA cube consisted of this DNA sequence will conserve O symmetry. Symmetry considerations alone can give us a completely and rigorous answer to the question 'what is possible and what is completely impossible?'. Above results lead us to infer that the DNA octahedra have some features which are regular, and some which are irregular. The symmetry O is very regular whereas the symmetry C_1 is completely irregular. The artificial DNAs are either regular or irregular, whereas the natural DNAs must be irregular. Symmetry means balanced proportions and the high symmetry of molecular structure arising from such balanced proportions. Therefore, artificial DNA with symmetry is easily formed in thermodynamic equilibrium. However, a life system will be open and far from equilibrium, ensuring that molecular symmetry breaking can take place. This explains why the natural DNAs must be irregular and asymmetric.

3.1.3. DNA Octahedron

Octahedron is a good model for DNA nanoconstruction because it provides both regularity and 4-valent connectivity. By now, three complementary methods have been developed to construct DNA octahedron (Shih *et al.*, 2004 [73]; Knudsen *et al.*, 2008 [74]; Mao *et al.*, 2010 [75]). They can be described roughly as 'single-stranded' and 'multi-stranded'.

Shih and his research group have performed an excellent experiment to construct DNA octahedron by folding a 1.7kb long single strand DNA in the presence of five short strands of 40nt.[73] The single long DNA strand first collapse into a branched-tree-like intermediate with the assistance of five short strands, and then fold its branches to form a DNA octahedron (Figure 8a).[73]

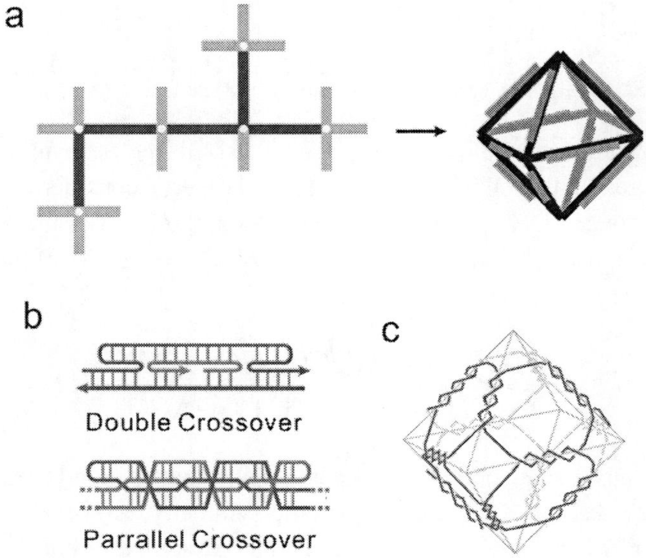

Figure 8. The synthetic strategy and structure of DNA octahedron. (a) DNA octahedron can be formed by folding a branched-tree-like intermediate. Red edges of the intermediate represent DXs and pink edges stand for half PXs. The octahedron contains two layers. The core layer (black) is connected by the 4-armed junctions at vertices. (b) DNA DX (below) and PX (above) can be roughly seen as is composed of parallel DNA double helices. Note that the blue strand in the DX represents short assistant strand. (c) DNA octahedron synthesized by Knudsen *et al.*[74] At each vertex, a hole exists due to the longer unpaired nucleotides (6nt).[73, 74]

The intermediate consists of 5 backbones and 14 branches, each of which is composed of a DNA double-crossover motif (DX)[66] or a half parallel-crossover motif (PX),[75] respectively (Figure 8b). During the annealing process of intermediate, 14 half PXs are paired complementarily to construct seven PXs, as 7 edges of an octahedron, and, associated with 5 DXs, as another 5 edges, to produce the final octahedron.[73] Due to the complex structure of DX and PX motifs, the DNA octahedron can be considered as consisting of two layers.[73] The core layer inside is composed of DNA double helix connected through 4-arm junctions at each vertex, covered with the peripheral layer outside the separated DNA helix on each edge.[73] Thus, each edge of the octahedron consists of a pair of DNA helices, with the one inside containing 40bp and outside 30~35bp, which makes the octahedron much stiffer.[73]

The DNA octahedron contains no catenanes or knots, and can be easily unfolded to a single long DNA strand.[73] Therefore, it can be reproduced by the polymerase chain reaction (PCR), which is crucial for replication of DNA nanostructures.[18]

The construction of DNA octahedron from single-strand-folding strategy wonderfully demonstrates the amazing self-assembly property of DNA in the three-dimensional construction. The main idea of the strategy was improved and demonstrated by Rothemund as a general way to construct DNA polyhedra by folding a single long DNA strand.[77] It is hoped that the strategy will work in laboratory and bring new breakthroughs in DNA nanotechnology.

Recently, Knudsen and colleagues in Denmark constructed DNA octahedron by another approach.[74] They designed 8 DNA strands, each containing three subsequences of 18nt linked by 7 unpaired nucleotides.[74] The DNA subsequences are complementary to each other to form DNA helices as edges of the octahedron in the self-assembly process. Because of a larger number of nucleotide linkers involved, the octahedron contains a hole at each vertex (Figure 8c).[74] The method is reminiscent of the synthesis of DNA tetrahedron, and shown to be effective in constructing larger objects, which may serve as DNA cages that suit for drug delivery. These studies are elegant and exciting!

Very recently, Mao and coworkers have synthesized DNA octahedra by assembling six well-designed '4-point star' motifs.[75] DNA octahedron with 42bp on each edge was constructed through this method. In addition, length difference of central loops was introduced as a 'experimental reporter' to investigate chiral preference of the self-assembly process.[75] This delicate experiment proposes a new way to surve DNA nanopolyhedra as a model of studying stereochemical process of supermoleular assembly.

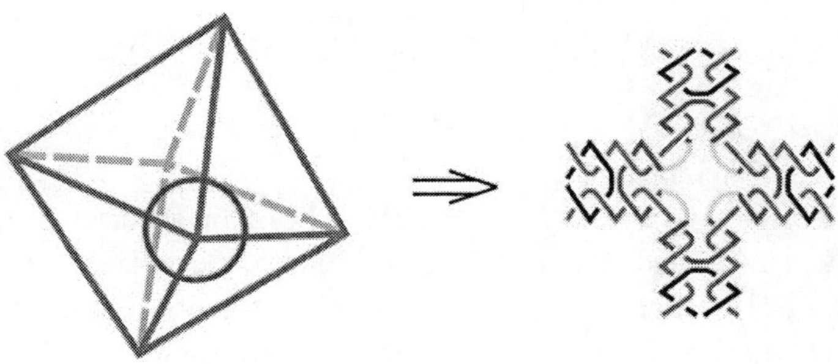

Figure 9. Construction principle of DNA octahedron with '4-point star' motif. [75]

The DNA octahedron constructed by Shih *et al.* [73] contains two layers as a core and a peripheral layer. So the DNA sequence on each edge can be represented as:

$$\frac{\text{GCACTTATCCGGACTAGATGCGCTGATCTGGGACGTCGAT}}{\text{CGTGAATAGGCCTGATCTACGCGACTAGACCCTGCAGCTA}}$$

$$\frac{\text{CAGGAGCAGGTGCCTCTGATAGCAACCAGGTGAGGA}}{\text{GTCCTCGTCCACGGAGACTATCGTTGGTCCACTCCT}},$$

where the core layer (upper) contains 40bp and the peripheral layer 36bp.[73]

Obviously, the DNA sequence does not contain reflection mirror and rotation axis, and can not be redesigned to be a palindrome due to different lengths of the core and peripheral layer. As a result, the DNA octahedron constructed by Shih et al. can only possess the least symmetry C_1.

Another DNA octahedron synthesized by Knudsen's group [74] consists of asymmetric sequence such as:

$$\frac{\text{CGATGTCTAAGCTGACCG}}{\text{GCTACAGATTCGACTGGC}}.$$

As a consequence, the DNA octahedron of this sequence is O symmetry. However, due to an even number of bases, it is possible to preserve octahedral symmetry, at least in some extent, by redesigning DNA sequence as follows:

$$\frac{\text{CGATGTCTATAGACATCG}}{\text{GCTACAGATATCTGTAGC}}.$$

It can be proved that DNA octahedron constructed from the above sequence will possess O symmetry.

The edge sequence of DNA octahedron synthesized by Mao *et al.* [75] is represented as:

$$\frac{\text{CCTACGATGGACACGGTAACGCCTAGCAACCTGCCTGGCAAG}}{\text{GGATGCTAC C TGTGCCATTGCGGATCGTTGGACGGACCGTTC}}$$

$$\frac{\text{CTTGCCAGGCAGGTTGCTAGGCGTTACCGTGTCCATCGTAGG}}{\text{GAACGGTCCGTCCAACGATCCGCAATGGCACAGGTAGCATCC}}.$$

Obviously, each edge contains a two-fold rotational axis, which means the DNA octahedron possesses O symmetry.

Symmetry analysis shows that the symmetry of DNA octahedron is transferred from a higher state (O_h) to a lower one (O or C_1). The formation of these lower symmetrical products largely depends on DNA sequences on edges of DNA octahedra. Thus, DNA sequence with C_2 symmetry produces regular DNA octahedron (O), whereas C_1 only gives irregular one (C_1). This is in part because the regular DNA octahedron is formed in quasi-equilibrium process, when the irregular octahedron is produced in thermodynamic process that is far from equilibrium.

3.1.4. DNA Dodecahedron

The synthetic strategy of dodecahedron was proposed by Seeman 10 years ago,[78] however, until recently, DNA dodecahedron was constructed through two different approaches (von Kiedrowski *et al.*, 2008 [79]; Mao *et al.*, 2008 [55]). These syntheses of DNA dodecahedron successfully demonstrated basic strategies of DNA nanoconstruction.

Von Kiedrowski and his fellow researchers have been studying on properties of nonnatural DNA for long time.[80-82] They developed a new branched DNA structure, trisoligonucleotide, which is composed of three DNA single strands covalently linked by an organic molecule, trislinker (Figure 10a).[80, 81] Each trisoligonucleotide, as a novel type of 3-armed DNA junction, contains three unpaired DNA arms that can be joined by complementary DNA arms of others to form larger structures, and thus can be used to construct 3-valent polyhedra (Figure 10).[80, 81] Surprisingly, von Kiedrowski *et al.* have shown that the trisoligonucleotide can replicate itself, as natural DNA, with the existence of complementary DNA strands and organic molecular linkers. [82]

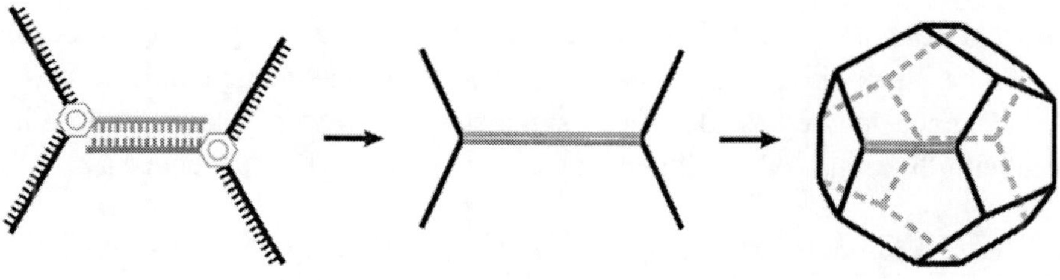

Figure 10. (a) Trisoligonucleotide is composed of three DNA single strands linked to an organic molecule (red). By complementary interactions between ssDNA (a, b), DNA dodecahedron can be formed (c).[79]

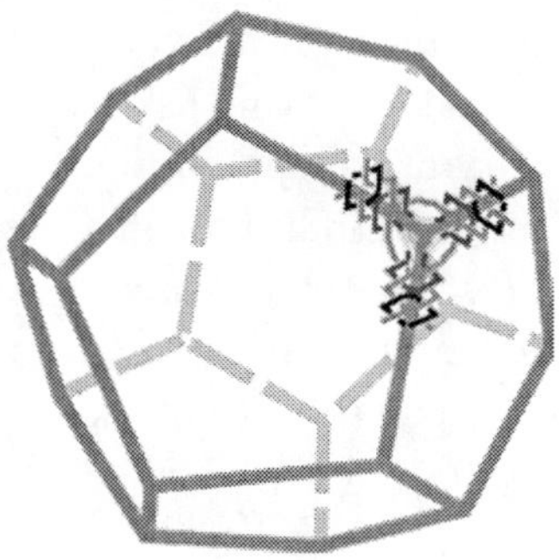

Figure 11. Schematic representation of DNA dodecahedron constructed by the hierarchical self-assembly process of '3-point star' motifs.[55]

In order to construct DNA dodecahedron, 20 trisoligonucleotides were synthesized. Each DNA strand of a trisoligonucleotide contains 15nt and is complementary properly to strands of other trisoligonucleotides to form DNA double helical edges of the dodecahedron.[79] After the annealing process, DNA dodecahedron is produced by the self-assembly of the trisoligonucleotides. Because of the replicable ability of the trisoligonucleotide,[82] it is proposed that the DNA dodecahedron could be reproduced in large amount conveniently, suggesting a possible approach towards the replication of DNA nanostructure.[18, 33]

Another approach towards the synthesis of DNA dodecahedron, as a 3-regular polyhedron, is to assemble twenty '3-point star' motifs, the same tiles involved in DNA tetrahedron, by hierarchical self-assembly process (Figure 11).[55] By adjusting the unpaired loop to be 3nt and the concentration to be 50nM, Mao and colleagues successfully synthesized DNA dodecahedra in the annealing process with 76% yield.[55] The DNA dodecahedron here contains parallel DNA double helices of 42bp, 4nm on each edge, which is consistent with the observation in cryo-EM experiment.[55] This is still an exciting challenging, and there is a window for new ideas and new theories interested in this important field.

The DNA sequence of DNA dodecahedron synthesized by von Kiedrowski *et al.*[79] can be written as:

$$\frac{\text{CCTGTCTTGAACGAG}}{\text{GGACAGAACTTGCTC}}.$$

It does not contain any reflection mirror and rotation axis, and can not be redesigned because of 15 base pairs involved. As a result, DNA dodecahedron synthesized with this kind of DNA sequence possesses only C_1 symmetry. However, the DNA dodecahedron constructed with '3-point star' motif contains each edge with DNA sequence as:

$$\frac{\text{CCTACGATGGACACGGTAACGCCTAGCAACCTGCCTGGCAAG}}{\text{GGATGCTACC TGTGCCATTGCGGATCGTTGGACGGACCGTTC}}$$

$$\frac{\text{CTTGCCAGGCAGGTTGCTAGGCGTTACCGTGTCCATCGTAGG}}{\text{GAACGGTCCGTCCAACGATCCGCAATGGCACAGGTAGCATCC}}.$$

Apparently, a two-fold rotational axis exists at the center of the sequence. And thus, we can conclude that DNA dodecahedron made by Mao *et al.*[55] has I symmetry.

Consideration of the point symmetry groups has led to the discovery of a new way to understand symmetry and asymmetry of nanopolyhedra. The point symmetry group of DNA dodecahedra belongs to I or C_1 group, indicating the loss of either reflection mirror or rotational axis. The process $I \rightarrow C_1$ is referred to as molecular symmetry breaking,[68] which means a system goes from a symmetric state to an asymmetric one. Molecular symmetry breaking may be a basis of characterizing the complexity and variety of the medically and biologically most interesting DNAs.

3.1.5. DNA Icosahedron

Icosahedron is the most complex structure in Platonic polyhedra, and, due to its 5-valent vertex, beyond the binding abilities of common chemical bond, is a formidable challenge for chemical synthesis. However, with the help of DNA, three groups have constructed icosahedron at molecular scale (Mao *et al.*, 2009 [83]; Shih *et al.*, 2009 [84]; Krishnan *et al.*, 2009 [85]), suggesting a laudable approach towards mimicking nature.

It is believed before that, in order to construct complex polyhedron, each motif must have enough rigidity so that the geometric object could be built like tinker-toys.[83] However, Mao's group designed a more flexible tile, '5-point star' with unpaired loop of 5nt, and utilized the tiles to perform the amazing synthesis (Figure 12).[83] Surprisingly, DNA icosahedron was successfully integrated, and after AFM test, the icosahedron was proved to possess good rigidity, which demonstrates appropriate flexibility of DNA motifs is helpful in the DNA nanoconstruction of complex structures.[83]

More recently, Shih's group have extended two-dimensional scaffold DNA origami to build complex three-dimensional structures.[84] They have constructed a DNA icosahedron with three 'double-triangule' intermediates in a two-step hierarchical self-assembly process (Figure 13).[84] Besides, a series of three-dimensional nanostructures have been also created by the same method, which indicates an effective approach towards systematic three-dimensional DNA nanoconstrucion. These important successes of DNA polyhedra lead to new goals for future research in chemical synthesis, but we have only explored a minuscule fraction of this new domain!

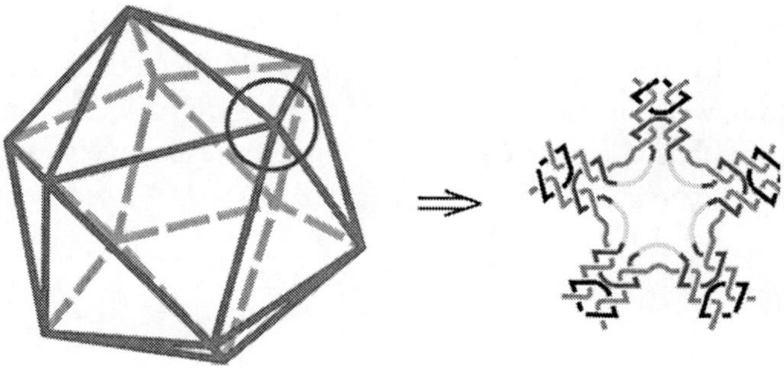

Figure 12. Synthetic principle of DNA icosahedron.[83]

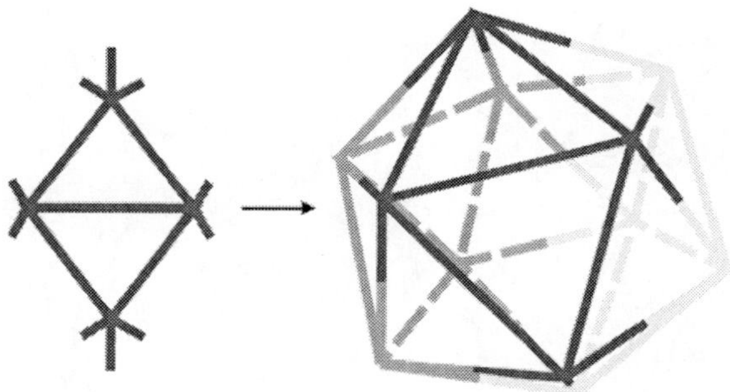

Figure 13. DNA icosahedron constructed from three 'double-triangle' structures (left).[84]

The DNA icosahedron synthesized by Mao *et al.* [83] contains 12 identical '5-point star' motifs and identical DNA sequence on each edge as:

CCTACGATGGACACGGTAACGCCTAGCAACCTGCCTGGCAAG
GGATGCTAC C TGTGCCATTGCGGATCGTTGGACGGACCGTTC

CTTGCCAGGCAGGTTGCTAGGCGTTACCGTGTCCATCGTAGG
GAACGGTCCGTCCAACGATCCGCAATGGCACAGGTAGCATCC

Because the sequence contains a C_2 axis, it can be proved that the DNA icosahedron has the point group *I*. The most highly symmetrical presentation has led to the discovery of a new way to consider molecular design of the novel nanopolyhedra.

In general case, mathematical polyhedra are highly symmetric, and this is why they stand for pure beauties. DNA polyhedra, due to the mechanism of molecular symmetry breaking, are usually in a lower symmetric state, or even completely asymmetric. DNA icosahedron analyzed here belongs to *I* group, because it contains no reflection mirrors. This *I* symmetry group tells us that molecular model of DNA nanopolyhedra always lacks a plane of symmetry, and therefore is said to possess chirality or handedness in a chemical word.

3.2. Archimedean Solids

Archimedean polyhedron, different from Platonic polyhedron, is semi-regular in that each face of the polyhedron is regular but several kinds of faces exist (Figure 14). There are 13 Archimedean solids. However, until now, only two of them, truncated octahedron and truncated icosahedron, were successfully synthesized with DNA.[2] The truncated octahedron and truncated icosahedron possesses O_h and I_h symmetry respectively, and the symmetry of corresponding DNA polyhedra are, of course, decided by the DNA sequences involved.

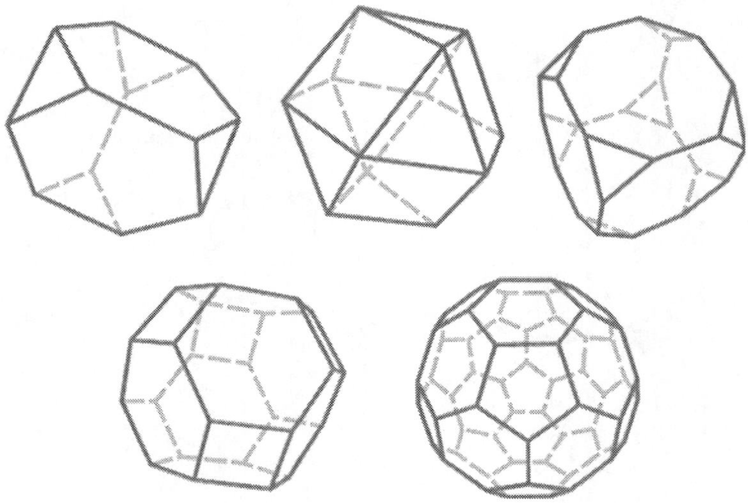

Figure 14. Some of the Archimedean solids. They are, in successive, truncated tetrahedron, truncated octahedron, truncated cube, truncated dodecahedron, and truncated icosahedron.

3.2.1. DNA Truncated Octahedron

The truncated octahedron is derived from octahedron by truncating each vertex, which is composed of eight triangles and six hexagons (Figure 15a).

Although the object contains only 3-valent vertices, however, in order to obtain larger structures in 3-space, Seeman's group [52] utilized 4-armed DNA junctions in the construction, leaving an extra arm at each vertex. Six squares of the truncated octahedron were synthesized first to obtain an intermediate, which can be used to construct the final structure (Figure 15b).[52] If the extra arms were properly ligated, it may produce a three-dimensional lattice structure with the truncated octahedron as repeating units.[52] And with the help of the extra arms, it is believed that the DNA truncated octahedron might prefer a configuration with the extra arms face outside, indicating good stereochemical control (Figure 15c).[52] The final DNA truncated octahedron contains 20bp on each edge, i.e. 6.8nm, and consists of 14 rings, each of ~1000nt.[52] This is a monumental achievement in the history of the DNA nanopolyhedra world!

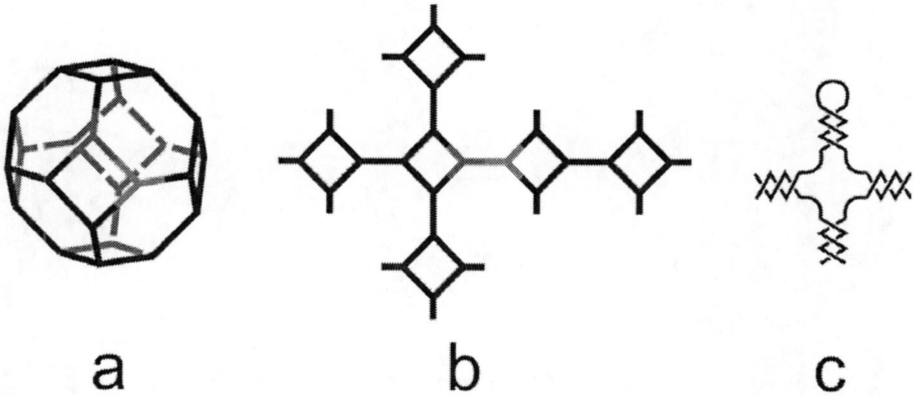

Figure 15. Schematic illustrations of DNA truncated octahedron (a), intermediate with six squares (b) and DNA junctions used in the synthesis (c). [52]

The DNA sequence of DNA truncated octahedron can be written as:

$$\frac{\text{GCAGAGTGGTTCTCACTAGT}}{\text{CGTCTCACCAAGAGTGATCA}}.$$

Because of lacking reflection mirror and rotation axis, the point symmetry group of DNA truncated octahedron degenerates from O_h to C_1. However, if reconstructed the DNA truncated octahedron with the sequence:

$$\frac{\text{GCAGAGTGGTACCACTCTGC}}{\text{CGTCTCACCATGGTGAGACG}},$$

an O symmetry group will be reserved at least. Such symmetry highlights an order and coherence in the working of nanopolyhedra.

3.2.2. DNA Truncated Icosahedron

The truncated icosahedron, which contains 32 faces and 60 vertices, is a fascinating structure that has arisen many chemical interests. Such a structure is well-known by soccer ball and molecular buckminsterfullerene (buckyball, or fullerene C_{60}).[38] Synthesizing DNA buckyball is a particular challenge that could have very important results for chemists.

Mao and colleagues have constructed the DNA buckyball,[55] or DNA truncated icosahedron, with the help of the hierarchical self-assembly strategy. They have employed sixty '3-point star' motifs in the synthesis, as vertices of the truncated icosahedron (Figure 16). By controlling the concentration of motifs at 500nM and the unpaired loop of 3nt, the DNA truncated icosahedra were obtained with a yield of 69%.[55] The structure of the products was examined with the AFM analysis and cryo-EM technique. The possibility of molecular realization of larger geometric polyhedra may provide basis for drug delivery and nanoreactors. This is an especially exciting development for chemistry, and this discovery would mark a beginning, not an end. The most creative act in chemistry is the design and creation of new nanopolyhedra. The challenge is picking the right targets, as well as figuring out how to make them.

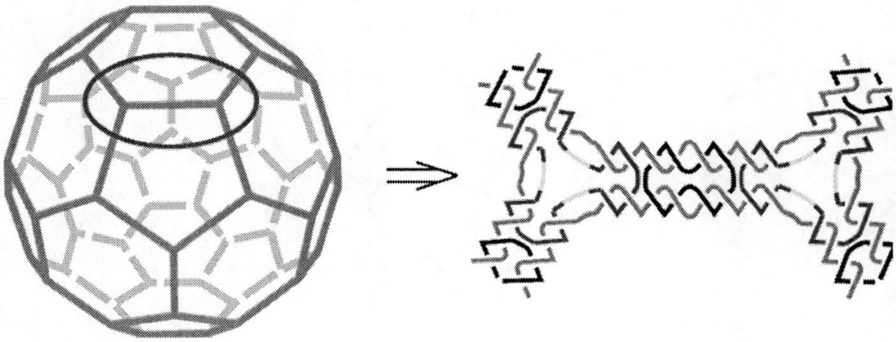

Figure 16. DNA truncated icosahedron (left) is built by the linkages of '3-point star' motifs (right). [55]

The DNA buckyball possesses *I* symmetry. This can be seen by analyzing its sequence structure, which can be represented as:

$$\frac{\text{CCTACGATGGACACGGTAACGCCTAGCAACCTGCCTGGCAAG}}{\text{GGATGCTAC C TGTGCCATTGCGGATCGTTGGACGGACCGTTC}}$$

$$\frac{\text{CTTGCCAGGCAGGTTGCTAGGCGTTACCGTGTCCATCGTAGG}}{\text{GAACGGTCCGTCCAACGATCCGCAATGGCACAGGTAGCATCC}}.$$

Obviously, the sequence contains a C_2 axis at its center. So, the symmetry analysis of DNA truncated icosahedron is valuable according to the discussion in Section 3.1.

3.3. More Complicated Polyhedra

3.3.1. DNA Prisms and Bipyramid

The chemical synthesis of DNA prisms and bipyramid was one of the biggest challenges in DNA nanotechnology due to lower point symmetry group, i.e. D_{nd} group (Figure 17).[2] Sleiman's group and Turberfield's group have been working on this topic, and, both using their strategies, synthesized a series of DNA prisms and bipyramid for a demonstration (Sleiman *et al.*, 2007 [86]; Turberfield *et al.*, 2007 [87]).

Sleiman and colleagues have developed a systematic strategy to construct a library of prisms.[86] They have constructed a few DNA polygons that contain several DNA strands as edges linked by organic molecules at vertices, as molecular modules for further construction.[86] With help of several DNA single-strands as linkers, these polygonal modules can be combined to produce a series of prisms (Figure 18). Furthermore, the DNA polygons with organic molecules involved were shown to be more rigid than branched DNA junctions, which made DNA prisms more stable.[88, 89]

They have utilized DNA strands of 10nt and certain number of organic molecules to synthesize four structure modules, triangles, squares, pentagons and hexagons.[86] And with the assistant of DNA linker strands of 30nt and 10nt, a library of DNA prisms have been constructed, including DNA triangular prism, cube, pentagonal and hexagonal prism, biprisms and heteroprisms, and structurally swichable prism.[86] The geometry parameters of these objects are summarized in Table 2. It is notable that, in principle, any structures that can be retrosynthetically broke down into these DNA modules can be reconstructed, which provides a systematic synthetic approach to DNA nanoconstruction.[86]

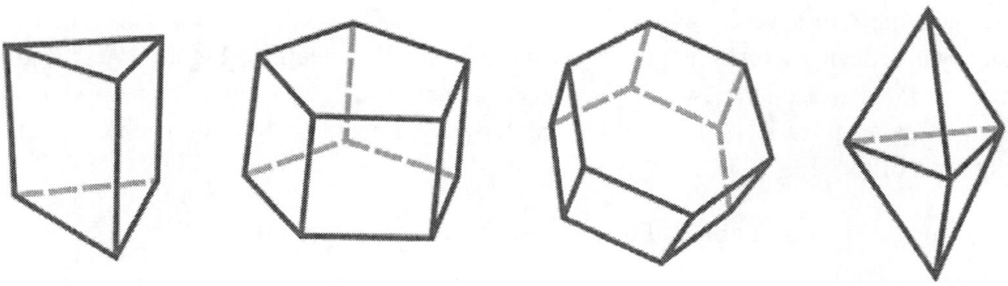

Figure 17. Several prisms and bipyramid. From left to right, they are triangular prism, pentagonal prism, hexagonal prism and bipyramid.

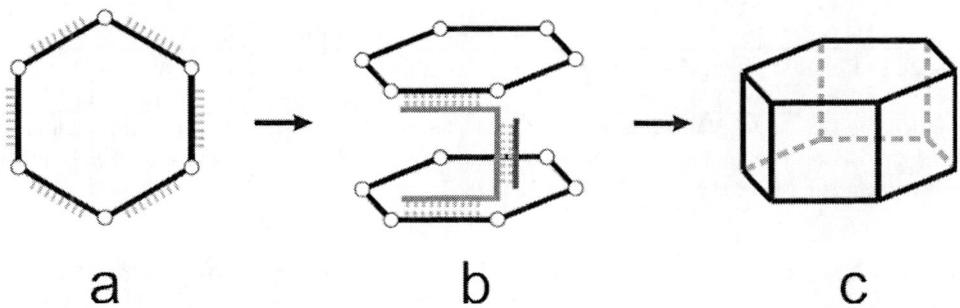

Figure 18. DNA polygon, hexagons for example here, is consisted of several ssDNAs connected through organic molecules (small ball) (a). DNA prisms (c) can be generated by joining polygonal modules with long strand DNA as linker (b, green) and short DNA as stiffer (b, red).[86]

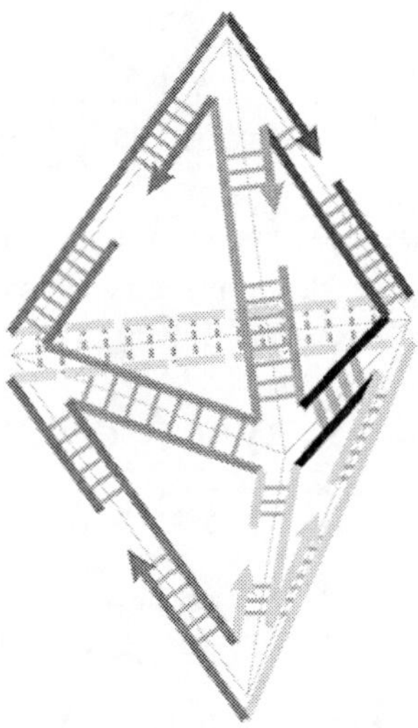

Figure 19. DNA bipyramid is consisted of six ssDNAs (different colors) with each running around a triangular face. Six of the edges contain a nick.[87]

Another polyhedron with rotation symmetry, bipyramid, is a common model in chemistry, serving as a geometric model for molecular orbital structure. Although the bipyramid contains only nine edges and five vertices, it is still difficult to construct such a structure because of its irregular vertex valences, two of which are 3-valent and the other three are 4. However, Turberfield's work [87] have recently constructed DNA bipyramid by the self-assembly of several DNA strands. They designed six single strands, each consisting of three subsequences of 20nt joined by one unpaired nucleotide (Figure 19).[87] In the annealing process, six strands combined each other through the complementarity of DNA subsequences, to form DNA bipyramid. The bipyramid contains 20bp on each edge and six nicks on six of its edges.[87]

Table 2. Edge parameters of DNA prisms.

DNA prisms	Geometer parameters
Triangular, quadrangular, pentagonal, hexagonal prism	face edge: 10bp side edge: 10bp
Heteroprism and biprism*	face edge: 10bp side edge: 20bp
Edge switchable prism	face edge: 10bp side edge: (10,14,20)bp**

*Heteroprism here is composed of a triangular face and a hexagonal face, while biprism is the combination of two heteroprisms, by gluing the hexagonal faces. ** (10, 14, 20)bp means the edge length of the switchable prism is varied within 10, 14, or 20bp.[86]

The DNA sequences of DNA prisms and DNA bipyramid can be represented as:

$$\frac{\text{CCGATTTGTG}}{\text{GGCTAAACAC}}$$

and

$$\frac{\text{CGAACATTCCTAAGTCTGAA}}{\text{GCTTGTAAGGATTCAGACTT}},$$

respectively. Because they both have the least symmetry, the DNA prisms and bipyramid will thus possess only C_1 symmetry or no symmetry. However, the replacement of these sequences with

$$\frac{\text{CCGATATCGG}}{\text{GGCTATAGCC}}$$

and

$$\frac{\text{CGAACATTCCGGAATGTTCG}}{\text{GCTTGTAAGGCCTTACAAGC}}$$

will reserve D_{nd} and D_3 symmetry for prisms and bipyramid, respectively. Since the symmetry has provided a powerful and insightful guidance, chemists usually rely on such an approach to help themselves understand molecular structures.

3.3.2. Goldberg Polyhedra

Goldberg polyhedra are a class of complex polyhedra, which can be generated and constructed in an infinite way (Figure 20). They are composed of pentagons and hexagons only, and the number of its faces is described by following formulas:[37]

$$F_5 = 12\,;$$

$$F_6 = 10 \times (h^2 + hk + k^2 - 1)\,,$$

where F_5 and F_6 represent the number of pentagons and hexagons, and h, k are integers.

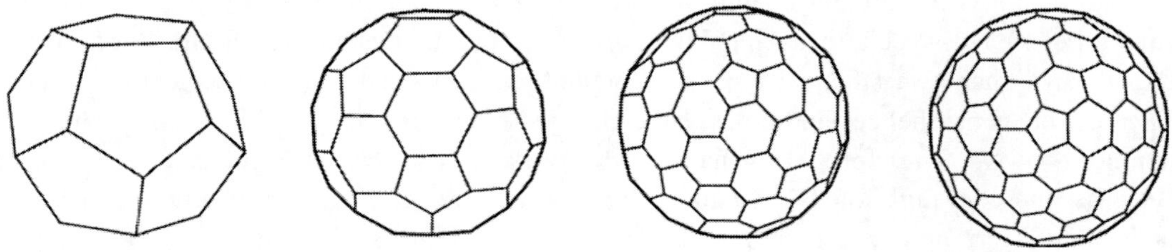

Figure 20. Some examples of Goldberg polyhedra. They are composed of 12 pentagons and certain numbers of hexagons. From left to right are 12-hedron, 32-hedron, 72-hedron, and 92-hedron separately.

The construction of a series of Goldberg polyhedra in a systematic and efficient way is a charming but formidable challenge in chemistry. Fortunately, the two smallest Goldberg polyhedra, i.e. dodecahedron (h=1, k=0; Figure 11) and truncated icosahedron (h=1, k=1; Figure 16) have been constructed (von Kiedrowski *et al.*, 2008 [79]; Mao *et al.*, 2008 [55]). It is an interesting phenomenon that the geometry of Goldberg polyhedra can be used to describe and classify the most number of spherical viral capsid structure.[11, 40] As a example, the virus HK97 of particular scientific interests is a 72-hedron (h=2, k=1).[15, 16] Therefore, chemical synthesis with Goldberg polyhedra as a structural framework is considered to be a natural and infinite way towards mimicking nature, and may lead to our further understanding of viral capsids. Progress is real, but the goal line is still some distance down the field.

4. Polyhedral Links – Novel Structures from DNA Polyhedra

DNA nanopolyhedra have extended our idea of what is possible in the chemical and biological world, since no such structures exist in nature. The construction of polyhedral structures with DNA improves the development of synthetic chemistry, and meanwhile, produces some complicated molecules with novel topological and geometrical structures, such as the DNA tetrahedron [53-55], DNA cube [51, 69], DNA octahedron [73-75], DNA dodecahedron [55, 79], DNA icosahedron [81, 83, 84], and the DNA buckyball [55], as well as the tetrahedron DNA cage [56] and the cube DNA box [70,71], etc.... These are some monumental feats and achievements! A challenge that is just now being addressed concerns how to ascertain and comprehend some of the mysterious characteristics of the DNA polyhedral folding.

Studies with DNA nanopolyhedra indicate that ordinary geometry is replaced by something known as nanopolyhedral geometry and topology. In this geometrical framework, the conventional notions of space and of distance between points melt away, leaving us in a vastly different mathematical apparatus and conceptual innovations. It is always better to know exactly that a polyhedron is a solid of the three-dimensional space limited by a finite number of the faces, the edges and the vertices. More curiously, a DNA nanopolyhedron is an interlocked and interlinked structure with the polyhedral skeleton: the edges are assembled with 2-helix and 4-helix, and so on; the vertices and the faces are filled with 'holes' that are small, and the others large. It is surprising to discover that the small 'holes' are the molecular skeletons, whereas the large 'holes' are a network of tunnels as if some 'creatures' travel through them.[34, 90, 91, 95] Within this new framework, the marriage of the large and the small 'holes' will result in a stunning revision of our understanding of the DNA three-dimensional shapes, aesthetically pleasing sculptures, and functional devices and materials. The new elements, helices and holes, have changed the properties of original polyhedra in a number of ways. Therefore, DNA nanopolyhedra have no faces, no edges, and no vertices. Because these feature of DNA nanopolyhedra require that we drastically change our understanding of new architectures. These interesting objects challenge the mathematical theory of polyhedra and indicate an undiscovered geometry and topology. There are many important questions waiting to be answered, and further explorations are needed! The progress will require the creation of better tools and better theories.

Figure 21. Schematic illustrations of [6]-catenanes (DNA cube), [8]-catenanes (DNA octahedron) and [14]-catenanes (DNA truncated octahedron).

The DNA nanopolyhedra world is so vast and complex that it is now necessary to clarify the definitions to describe topologically interlinked and interlocked cases (Figure 21). The best way to gain a good understanding of the different types of DNA nanopolyhedral architectures is to use a set of models. Since 2005, Qiu's group has proposed and constructed a novel topological and geometrical structure, i.e. polyhedral links, to model the molecular structure of DNA polyhedra, and, with the help of graph theory and knot theory, characterized and analyzed the mathematical properties of these polyhedral links.[34] Some interesting properties, such as chirality and duality, have been uncovered. As a result, these beautiful architectures are now updating our knowledge of the geometry of polyhedra, and improving pure mathematical theory, such as graph theory and knot theory.[92-94] Meanwhile, the systematic knowledge of polyhedral links may also provide new insights into biomolecular structures, such as viral capsids and DNA polyhedra.[15, 16, 21-25] The architecture of the polyhedra links builds an inexhaustible spring of inspiration for geometry thinking and to illustrate the research in topology. Although there is much to learn, the idea is simple but brilliant, as many excellent academic ideas are.

4.1. Platonic Polyhedral Links

Platonic solids are the simplest ones in the polyhedral world. However, they present some of the basic concepts of polyhedra, such as vertex degree and duality. Qiu's group has developed two basic transformations, '*n*-branched curves and *k*-twisted line covering' and '*n*-crossover and *k*-twisted line covering', to systematically generate and construct a series of polyhedral links based on the geometry of Platonic polyhedra.[34, 90, 95, 96] Furthermore, interesting properties such as knot invariant and dual transformation are investigated and explored.[95-97]

It is now clear that polyhedral link is an interlocked cage with holes or tunnels, whose boundaries, of course, are the double- or multi-helix that has the even or odd number of half-twists. More curiously, the vertex area with holes of the polyhedral link is a novel position of topology which controls molecular shape or size.[34, 90, 95]

4.1.1. Type I Platonic Polyhedral Links and Knot Invariants

The first type of Platonic polyhedral links was generated on the Platonic polyhedra by the method of '*n*-branched curves and *k*-twisted line covering'. The basic geometric elements that composed of Platonic polyhedra, vertices, edges and faces, were replaced by novel complex motifs, '*n*-branched curves', '*k*-twisted line' and holes that connect the inside and outside of the polyhedra, respectively.[91] After the transformation, the Type I Platonic polyhedral links were obtained (Figure 22).[95]

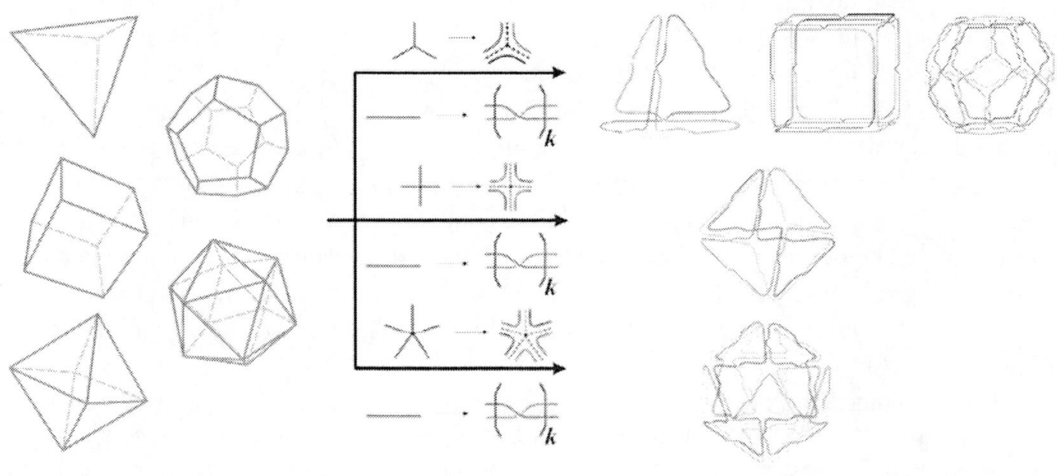

Figure 22. The first type of Platonic polyhedral links are obtained by using '3-branched curves and *k*-twisted line covering', '4-branched curves and *k*-twisted line covering', and '5-branched curves and *k*-twisted line covering' according to the corresponding vertex degree of Platonic polyhedra.[34, 90, 95] Here, polyhedral links are presented with k=1.

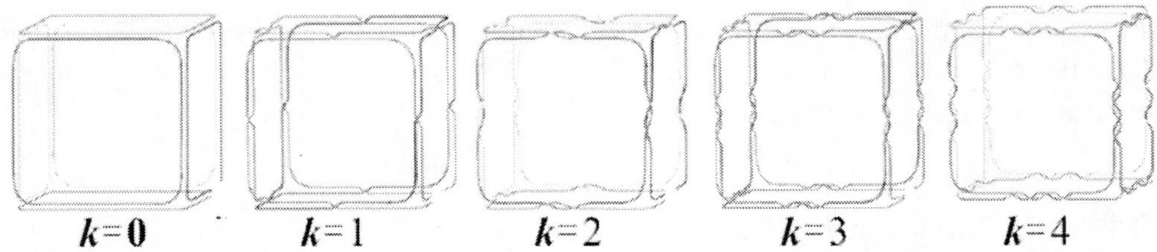

Figure 23. Representations of hexahedral links in the case of k=0, 1, 2, 3, and 4. Platonic polyhedral links with different twist number k result in a series of different 'topological isomers'.[90, 95]

The vertex degree of the first type of Platonic polyhedral links is varied only among 3-, 4- and 5-valent. However, the twist number on edge motif 'k-twisted line' can be changed infinitely in both directions, which will produce a series of 'topological isomers' (Figure 23). To classify all these possible isomers need the structural characterization by knot invariants. Knot invariant is a mathematical quantity that can depict the characters of topological isomers, such as crossing number, writhe number, linking number, and HOMFLY polynomial.[92-94, 98-99] It can be used to, at least in principle, identify tiny differences between similar structures, and therefore, distinguish and classify the 'topological isomers'. Qiu and co-workers have computed these knot invariants of the type I Platonic polyhedral links, which may be helpful for further classification and characterization.[95]

4.1.2. Type II Platonic Polyhedral Links and Dual Transformation

In the second type of Platonic polyhedral links, edges and vertices of polyhedra are replaced by 'n-crossover' and 'k-twisted line', respectively (Figure 24b). Because of different motifs used in generation, the topological structure of the second type of polyhedral links is completely different from the first type. And interestingly, five type II Platonic polyhedral links can be converted from each other by dual transformations (Figure 24a, b).[96]

In geometry, duality means the vertices and faces can be exchanged between a pair of polyhedra. So, in the Platonic polyhedra, the octahedron and the cube, the dodecahedron and the icosahedron, the tetrahedron and itself, are considered to be dual polyhedra respectively (Figure 24a).[2] Surprisingly, dual relationships also exist in the Platonic polyhedral links, and can be classified as two types, i.e. trivial and nontrivial.[96] Dual transformation is said to be trivial if the geometric motifs of dual polyhedral links remain unchanged, otherwise, nontrivial. Therefore, the dual transformation between the tetrahedral links and itself is trivial. And the others belong to the realm of nontrivial, for the different construction methods, i.e. '4-crossover and 2-twisted line covering', '5-crossover and 2-twisted line covering' and '3-crossover and 2-twisted line covering', are connected through the dual transformations (Figure 24b).[96]

Based on these interesting results, Qiu's group has generalized the dual transformations of polyhedral links.[97] In graph theory, a three-dimensional polyhedron can be represented by a two-dimensional graph, Schlegel diagram, and furthermore, a pair of dual polyhedra can be connected through medial graphs on Schlegel diagrams.[100] Medial graphs, according to mathematical graph theory, have such properties that each vertex in the graph is 4-valent and the faces of the graph can be painted with two different colors. This is somewhat like a chessboard that contains two colors of squares and has four directions on each vertex without respect to the outmost lines. In the case of polyhedra, the medial graphs can be obtained by connecting medial points on edges if two edges are neighboured through a vertex. And it can be proved by mathematical techniques that the medial graphs of a pair of dual polyhedra are the same, and vice versa.[101]

Covering each vertex of the medial graph of polyhedra with '*k*-twisted line' motifs, and connecting the ends of adjoined motifs will produce catenaned structures, medial graph links (Figure 24c). It is notable that there are two choices of the arrangement of '*k*-twisted line', either crossing through the white or black faces, as well as two types of medial graph links. And these two types of links can be topologically transformed into a pair of dual polyhedral links (Figure 24c). The dual polyhedral links constructed by the method can be classified into two categories, which are *O*-dual if the twist number *k* is odd and *E*-dual otherwise.[97]

Although the knowledge of dual polyhedral links is still limited now, it is believed that duality, one of the most profound concepts in nature, will bring some new ideas in the study of polyhedral links, such as classification and topological transformation.

4.2. Truncated Platonic Polyhedral Links, Archimedean Polyhedral Links and Chirality

Archimedean polyhedra and truncated Platonic polyhedra are two classes of 3-regular polyhedra, which can be generated by truncating vertices of Platonic polyhedra once or several times respectively (Figure 25a).[1, 2] Qiu's group has used the method of '3-crossover and *k*-twisted line covering' to construct a series of polyhedral links (Figure 25b), which can be characterized by some mathematical recursive formulas.[90]

All these polyhedral links are showed to be chiral by symmetry analysis. A polyhedron is said to be chiral if it cannot be superimposed onto its mirror image. Because of the rigidity of polyhedra, it is easy to decide whether a polyhedron is chiral or not, by the analysis of its improper rotation symmetry. For a catenaned structure with sufficient flexibility like rubber, however, the criterion can no longer insist due to the definition of rotation axis in topological sense. Some of the catenanes seemed lacking of improper rotations could be deformed into its 'mirror image' by stretching and shrinking, just like a transformation between left and right hand as a topological rubber glove (Figure 26a).[68] Therefore, deciding whether topologically catenaned structures, say polyhedral links, possessing chirality is a more difficult task.[68, 102, 103]

In fact, mathematicians have been working for years to develop techniques to probe the topological chirality of molecular catenanes and graphs. A number of techniques, such as knot polynomial calculation, point symmetry group analysis of Seifert construction, 2-fold branched covering and Möbius ladder analysis, are useful in simple knots with a few crossovers.[68, 102-113] However, it is not convenient to apply these techniques in determining whether a polyhedral links is chiral.

Qiu and his team have proposed a new technique in chriality determination by analyzing point symmetry group of polyhedral links. It is believed that point group of polyhedral links and corresponding polyhedra are the same except for the mirror reflection elements. So, the symmetry group of a polyhedral link is usually *T*, *O* or *I*. Thus, it can be concluded empirically that polyhedral links are chiral in general cases (Figure 26b).[34, 90] Recent studies on polynomials of some of the polyhedral links give the same results.[114] This empirical method, if proved mathematically, will be convenient and helpful for the characterization of molecular symmetry breaking of DNA polyhedra.[68]

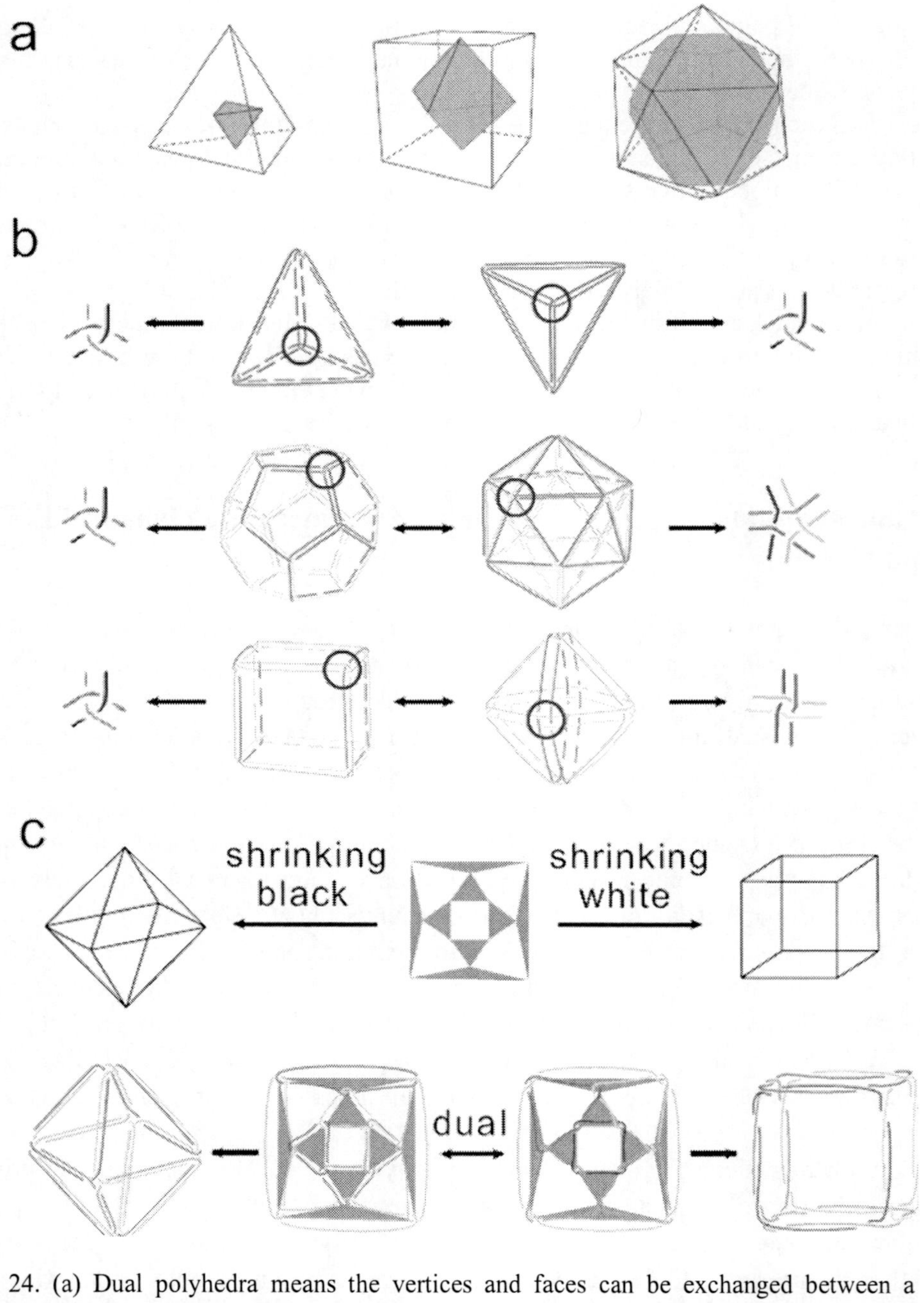

Figure 24. (a) Dual polyhedra means the vertices and faces can be exchanged between a pair of polyhedra. (b) Dual Platonic polyhedral links can be classified as trivial (above) and nontrivial (medium and below), with the difference that the former does not change the geometric motifs that composed of polyhedral links while nontrivial not. (c) Construction principle of dual polyhedral links. A pair of dual polyhedra, for example cube and octahedron, can be connected by the medial graph of their Schlegel diagram (above). Fill the opposite faces of the medial graph with two colors, say black and white. And covering the 4-valent vertex of the graph with '2-twisted line' in the manner either across the white (below, left) or black area (below, right) will produce two different kinds of medial graph links. Shrinking the white or black area then gives hexahedral or octahedral links respectively.[96]

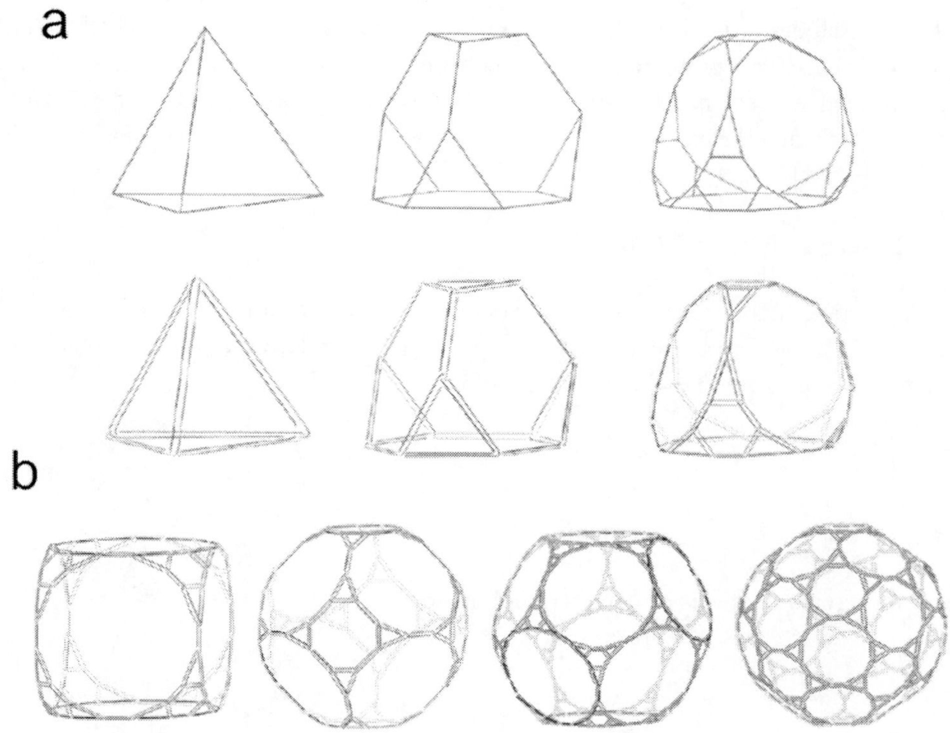

Figure 25. (a) Truncated tetrahedra generated by a series of truncating operations on tetrahedron (above), and their corresponding polyhedral links. (b) Polyhedral links obtained by truncating cube, octahedron, dodecahedron and icosahedron twice, and then with '3-crossover and double twisted line covering'.[90]

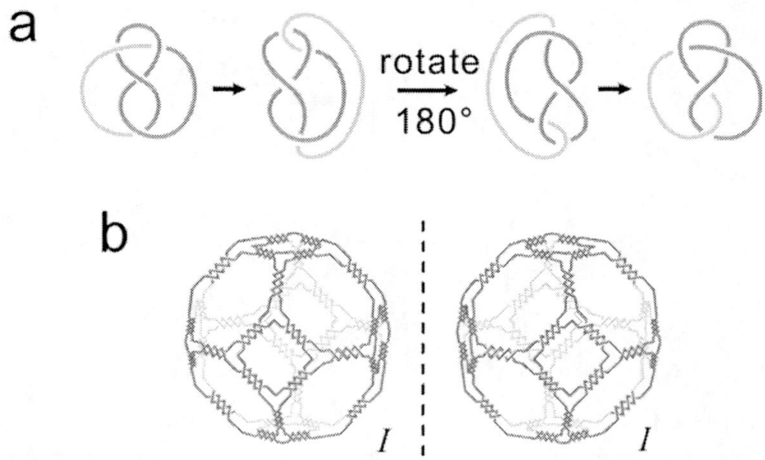

Figure 26. (a) Under the topological transformations, the figure-eight knot can be transformed into its 'mirror image'. (b) truncated octahedral links exist as two mirror images.

4.3. Goldberg Polyhedral Links

Novel geometric objects such as Platonic polyhedral links and Archimedean polyhedral links provide a foundation for the characterization and analysis of DNA polyhedra synthesized in laboratory. Meanwhile, it opens a door for the realm of polyhedral links. In order to explore general principles and mechanisms of these fantastic structures, Qiu *et al.* have constructed a great number of polyhedral links based on the geometry of Goldberg polyhedra, which may be useful to our understanding of biomacromolecules, such as viral capsids.[15, 16, 34]

4.3.1. Goldberg Polyhedral Links

Goldberg polyhedra are complex polyhedra and thus have infinite numbers of different structures.[37] However, there exist two types that are relatively simple, which can be described by the following recursive formulas:[34]

$$\text{Type I:} \quad F_{n+1} = F_n + 10(2n+1);$$

$$\text{Type II:} \quad F_{n+1} = F_n + 20n,$$

where F_n represents the faces of Goldberg polyhedra.

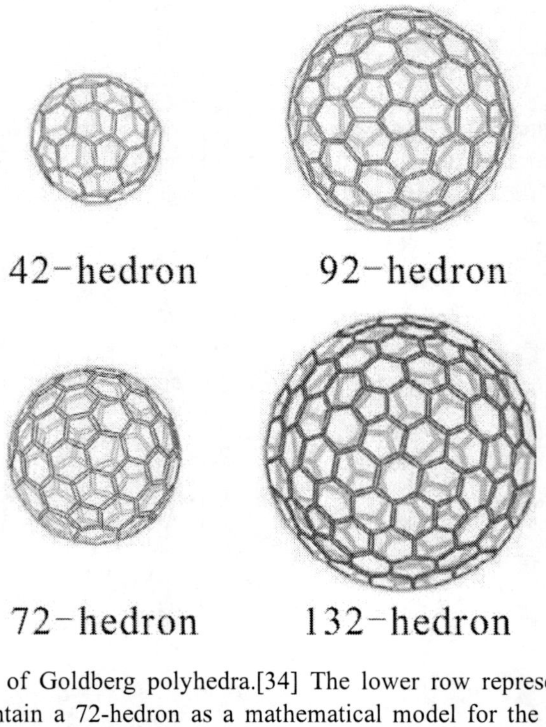

Figure 27. Some examples of Goldberg polyhedra.[34] The lower row represents type II Goldberg polyhedral links, which contain a 72-hedron as a mathematical model for the molecular topological structure of the capsid protein of virus HK97.

Based on the geometry of the two types of Goldberg polyhedra and the method of '3-crossover and *k*-twisted line covering', Qiu's group has constructed two types of Goldberg polyhedral links. Structural analysis shows that they are interlocked catenanes of certain numbers of pentagonal and hexagonal rings. The numbers of component rings satisfy the following relations (Figure 27):[34]

$$\text{Type I:} \quad R_{n+1} = R_n + 10(2n+1);$$

$$\text{Type II:} \quad R_{n+1} = R_n + 20n,$$

where R_n is the number of interlocked rings.

Symmetry analysis indicates that the two types of Goldberg polyhedral links are all chiral, which is the property different from Goldberg polyhedra.

Amazingly, the structure of Goldberg polyhedral links with n=2 in the second type is coincident with the topological architecture of capsid protein of virus HK97, consisting of 12 pentagonal and 60 hexagonal rings (Figure 27).[15, 16] This interesting discovery indicates some common mechanisms of molecular structures. Further investigations will bring additional insights into the structures of biomacromolecules.

4.3.2. Extended Goldberg Polyhedra and Polyhedral Links

Base on the first and second type of Goldberg polyhedra, Qiu *et al.* have used two operations, namely 'spherical rotating' and 'spherical stretching', to construct Rotate-Extended Goldberg Polyhedra (REGP) and Stretch-Extended Goldberg Polyhedra (SEGP), respectively (Figure 28).[115] Except for the pentagons and hexagons contained in the original Goldberg polyhedra, the REGPs are also composed of certain numbers of triangles, whereas the SEGPs consist of triangles and squares. These two types of extended Goldberg polyhedra can be characterized by the following recursive relations:[115]

Type I:

$$\text{REGP:} \quad F_n = 12 + 10(n^2 + 2n) + 20(n+1)^2;$$

$$\text{SEGP:} \quad F_n = 12 + 10(n^2 + 2n) + 50(n+1)^2;$$

Type II:

$$\text{REGP:} \quad F_n = 12 + 10(n^2 + n) + 20(n^2 + n + 1);$$

$$\text{SEGP:} \quad F_n = 12 + 10(n^2 + n) + 50(n^2 + n + 1),$$

where F_n represents the faces of extended Goldberg polyhedra.

The vertex valences of all the two types of extended Goldberg polyhedra are 4, which can be covered with 'k-twisted line' motifs to produce two classes of extended Goldberg polyhedral links, i.e. Rotate-Extended Goldberg Polyhedral Links (REGPL) and Stretch-Extended Goldberg Polyhedral links (SEGPL) (Figure 28).[116, 117] Different from Goldberg polyhedral links, the number of the interlocked rings depends on the twist number k of the motifs.

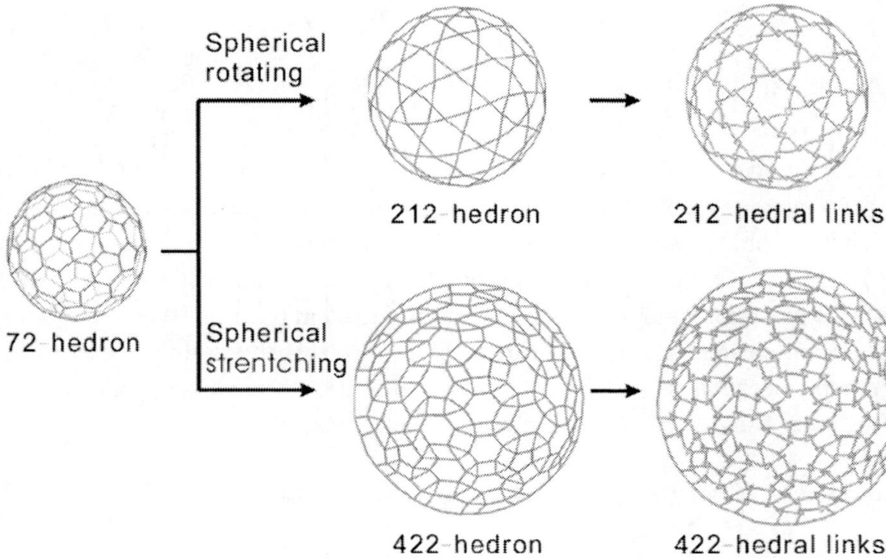

Figure 28. By applying the 'spherical rotating' and 'spherical stretching' operations on Goldberg polyhedra, Rotate-Extended Goldberg Polyhedra (REGP) and Stretch-Extended Goldberg Polyhedra (SEGP) are obtained.[115] Then, by the following '*k*-twisted line covering', Rotate-Extended Goldberg Polyhedral Links (REGPL) and Stretch-Extended Goldberg Polyhedral links will be generated respectively.[116] Here, *k*=2, for examples.

In the case that *k* is even, the each component ring runs along a face in the extended Goldberg polyhedron. And thus, the number of rings can be calculated by the following deductions:[116, 117]

Type I:

$$\text{REGPL: } R_n = 12 + 10(n^2 + 2n) + 20(n+1)^2;$$

$$\text{SEGPL: } R_n = 12 + 10(n^2 + 2n) + 50(n+1)^2;$$

Type II:

$$\text{REGPL: } R_n = 12 + 10(n^2 + n) + 20(n^2 + n + 1);$$

$$\text{SEGPL: } R_n = 12 + 10(n^2 + n) + 50(n^2 + n + 1),$$

where R_n represents the component rings of extended Goldberg polyhedral links.

In the case of an odd *k*, it can be only ascertained that the component rings are less than that when *k* is even. However, to calculate the number of rings in analytic manner is quite difficult for the reason that each ring will encompass through certain number of faces.[118-121] Qiu *et al.* have proposed an algorithm to compute the component rings in some special cases.[117] Meanwhile, Twarock *et al.* are seeking a different mathematical tool, topological graph theory, to solve such a problem. [122] With the help of a computer, they have established relationships between component numbers and some other topological invariants and then give an interesting result. For a dodecahedral and an icosidodecahedral link (32-hedral link), the least numbers of rings are both two.[122-124]

The computation and exploration of the number of component rings is not merely a mathematical issue, but can also work as theoretical blueprints in the design of experiments. DNA polyhedra with minimal number of strands can be easily replicated by PCR, which is meaningful in DNA nanotechnology and even crucial to some RNA virus genome packing. [18, 73, 75, 125, 126]

5. Conclusion

The study of polyhedra is a very old subject. However, chemists have refreshed this issue in laboratory while employing smart DNA as a nanoconstruction material.[1, 2, 21-25] DNA cube synthesized by Seeman et al. [51] has greatly inspired chemists' imagination. In the past two decades, they have utilized DNA to construct a great number of polyhedra from a simple cube to a more complex buckyball. These novel structures of DNA polyhedra have improved our knowledge on the basic forms of molecular structures, bringing insightful ideas to understand mechanisms of biomacromolecules, such as viral capsids. The future interest in these species needs urgent investigation, because of not only their application in nanotechnology, but also their intriguing topologies, which remind us of the reflections on the new theories of polyhedra.

Polyhedral links, the interlinked and interlocked architectures, have extended our ideas of what the potential shape of polyhedral molecules is. The review that presents the theoretical study of polyhedral links involves the following five major endeavors: (1) the exploration of a systematic mathematical transformation to construct polyhedral links on any given polyhedra; (2) the elucidation of duality in the new form of polyhedra, i.e. polyhedral links; (3) the detection of the chirality of polyhedral links and their applications in the synthesis and characterization of new forms of DNA and protein molecules; (4) the usage of topology and graph theory to understand the symmetry and asymmetry of polyhedral links; and (5) the elucidation and computation of the number of component rings with the help of topological graph theory.

However, the story has just begun. There is still a gap between relatively simple theoretical systems and the complicated varieties of DNA supramolecular polyhedra, such as DNA buckyball. Some problems of general interests have not been solved and even proposed yet. For instances, how to generate more complex polyhedra on the simple polyhedra (e.g. Platonic solids) and what the general principle of the geometry and topology of polyhedral links is.[34, 90, 95-97, 114-117, 127] Such problems, though they are strange enough, still lead us to a higher level of the understanding to the characteristics of polyhedral molecules (e.g. DNA polyhedra and viral capsids), such as intramolecular symmetry breaking, providing available opportunities for the development of a mathematical theory of new geometry.

In conclusion, the interactions between chemical and mathematical concerns on DNA polyhedra benefit and promote each other well.[128] It is our belief that further study on these beautiful molecules will serve an arena for the interdisciplinary improvement between chemistry and mathematics. In fact, the scope for new research in mathematical chemistry is so broad that full and cooperative endeavors from mathematics [129-135] and chemistry are warranted and required. Mathematical chemistry, or biomathematics, must provide the basic knowledge for the next cycle of new discovery and development.

Acknowledgements

This work was supported by Grants from the National Natural Science Foundation of China (No. 20973085 and No.10831001) and Specialized Research Fund for the Doctoral Program of Higher Education of China (No. 20090211110006).

References

[1] Coxeter, H. M. S. (1973). *Regular Polytopes*. Dover: Dover Publications.
[2] Cromwell, P. R. (1999). *Polyhedra: "One of the Most Charming Chapters of Geometry"*. Cambridge: Cambridge University Press.
[3] Hargittai, I. & Hargittai, M. (1995). *Symmetry Through the Eyes of a Chemist*. New York: Plenum Press.
[4] Maier, G., Pfriem, S., Schäfer, U. & Matusch, R. (1978). Tetra-tert-butyltetrahedrane. *Angew. Chem., Int. Ed.*, **17**, 520-521.
[5] Eaton, P. E. & Cole, T. W. (1964). The cubane system. *J. Am. Chem. Soc.*, *86*, 962-964.
[6] Eaton, P. E. & Cole, T. W. (1964). Cubane. *J. Am. Chem. Soc.*, **86**, 3157-3158.
[7] Paquette, L. A., Balogh, D. W., Usha, R., Kountz, D. & Christoph, G. G. (1981). Crystal and molecular structure of a pentagonal dodecahedrane. *Science*, **211**, 575-576.
[8] MacGillivray, L. R. & Atwood, J. L. (1999). Structural classification and general principles for the design of spherical molecular hosts. *Angew. Chem., Int. Ed.*, **38**, 1018-1033.
[9] Leininger, S., Olenyuk, B. & Stang, P. J. (2000). Self-assembly of discrete cyclic nanostructures mediated by transition metals. *Chem. Rev.*, **100**, 853-908.
[10] D'Arcy W. Thompson (1992). *On Growth and Form*. Cambridge: Cambridge University Press.
[11] Caspar, D. L. D. & Klug, A. (1962). Physical principles in the construction of regular viruses. *Cold Spring Harb. Symp. Quant. Biol.*, **27**, 1-24.
[12] Stagg, S. M., Gürkan, C., Fowler, D. M., LaPointe, P., Foss, T. R., Potter, C. S., Carragher, B. & Balch, W. E. (2006). Structure of the Sec13/31 COPII coat cage. *Nature*, **439**, 234-238.
[13] Coulibaly, F., Chiu, E., Ikeda, K., Gutmann, S., Haebel, P. W., Schulze-Briese, C., Mori, H. & Metcalf, P. (2007). The molecular organization of cypovirus polyhedra. *Nature*, **446**, 97-101.
[14] Fotin, A., Cheng, Y., Sliz, P., Grigorieff, N., Harrison, S. C., Kirchhausen, T., Walz, T. (2004). Molecular model for a complete clathrin lattice from electron cryomicroscopy. *Nature*, **432**, 573-579.
[15] Duda, R. L. (1998). Protein chainmail: catenated protein in viral capsids. *Cell*, **94**, 55-60.

[16] Wikoff, W. R., Liljas, L., Duda, R. L., Tsuruta, H., Hendrix, R. W. & Johnson, J. E. (2000). Topologically linked protein rings in the bacteriophage HK97 capsid. *Science*, **289**, 2129-2133.

[17] Seeman, N. C. (2003). DNA in a material world. *Nature*, **421**, 427-431.

[18] Seeman, N. C. (2005). Nucleic acid nanostructures: bottom-up control of geometry on the nanoscale. *Rep. Prog. Phys.*, **68**, 237-270.

[19] Seeman, N. C. (2004). Nanotechnology and the double helix. *Sci. Am.*, **290**, 64-75.

[20] Seeman, N. C. (1982). Nucleic acid junctions and lattices. *J. Theor. Biol.*, **99**, 237-247.

[21] Aldaye, F. A., Palmer, A. L. & Sleiman, H. F. (2008). Assembling materials with DNA as the guide. *Science*, **321**, 1795-1799.

[22] Heckel, A. & Famulok, M. (2008). Building objects from nucleic acids for a nanometer world. *Biochimie*, **90**, 1096-1107.

[23] Simmel, F. C. (2008). Three-dimensional nanoconstruction with DNA. *Angew. Chem., Int. Ed.*, **47**, 5884-5887.

[24] Lin, C., Liu, Y. & Yan, H. (2009). Designer DNA nanoarchitectures. *Biochemistry*, **48**, 1663-1674.

[25] Zhang, C., He, Y., Su, M., Ko, S. H., Ye, T., Leng, Y., Sun, X., Ribbe, A. E., Jiang, W. & Mao, C. (2009). DNA self-assembly: from 2D to 3D. *Faraday Discuss.*, **143**, 1-13.

[26] Adleman, L. M. (1994). Molecular computation of solutions to combinatorial problems. *Science*, **266**, 1021-1024.

[27] Jonoska, N., Karl, S. A. & Saito, M. (1999). Three dimensional DNA structures in computing. *BioSystems*, **52**, 143-153.

[28] LaVan, D. A., McGuire, T. & Langer, R. (2003). Small-scale systems for in vivo drug delivery. *Nat. Biotechnol.*, **21**, 1184-1191.

[29] Erben, C. M., Goodman, R. P. & Turberfield, A. J. (2006). Single-molecule protein encapsulation in a rigid DNA cage. *Angew. Chem., Int. Ed.*, **45**, 7414-7417.

[30] Bath, J. & Turberfield, A. J. (2007). DNA nanomachines. *Nat. Nanotechnol.*, **2**, 275-284.

[31] Goodman, R. P., Heilemann, M., Doose, S., Erben, C. M., Kapanidis, A. N. & Turberfield, A. J. (2008). Reconfigurable, braced, three-dimensional DNA nanostructures. *Nat. Nanotechnol.*, **3,** 93-96.

[32] Mitchell, N., Schlapak, R., Kastner, M., Armitage, D., Chrzanowski, W., Riener, J., Hinterdorfer, P., Ebner, A. & Howorka, S. (2009). A DNA nanostructure for the functional assembly of chemical groups with tunable stoichiometry and defined nanoscale geometry. *Angew. Chem., Int. Ed.*, **48**, 525-527.

[33] Zhang, C. & Mao, C. (2008). DNA Nanotechnology: Bacteria as factories. *Nat. Nanotechnol.*, *3*, 707-708.

[34] Qiu, W.-Y. & Zhai, X.-D. (2005). Molecular design of Goldberg polyhedral links. *J. Mol. Struct. THEOCHEM*, **756**, 163-166.

[35] Starck, M. (2007). *A Ride Through the Polyhedra World*. http://www.ac-noumea.nc/maths/polyhedr/index_.htm.

[36] Alvarez, S. (2005). Polyhedra in (inorganic) chemistry. *Dalton Trans.*, 2005, 2209-2233.

[37] Goldberg, M. (1937). A class of multi-symmetryic polyhedra. *Tohoku Math. J.*, **43**, 104-108.

[38] Kroto, H. W., Health, J. R., O'Brien, S. C., Curl, R. F. & Smalley, R. E. (1985). C_{60}: buckminsterfullerene. *Nature*, **318**, 162-163.
[39] Fowler, P. W. & Manolopoulos, D. E. (2007). *An Atlas of Fullerenes*. Dover: Dover Publications.
[40] Johnson, J. E. & Speir, J. A. (1997). Quasi-equivalent viruses: a paradigm for protein assemblies. *J. Mol. Biol.*, **269**, 665-675.
[41] Watson, J. D. & Crick, F. H. C. (1953). A structure for deoxyribose nucleic acid. *Nature*, **171**, 737-738.
[42] Watson, J. D. & Crick, F. H. C. (1953). Genetical implications of the structure of deoxyribonucleic acid. *Nature*, **171**, 964-967.
[43] Hagerman, P. J. (1988). Flexibility of DNA. *Annu. Rev. Biophys. Biomol. Struct.*, **17**, 265-286.
[44] Qiu, H., Dewan, J. C. & Seeman N. C. (1997). A DNA decamer with a sticky end: the crystal structure of d-CGACGATCGT. *J. Mol. Biol.*, **267**, 881-898.
[45] Dickerson, R. E. (1982). The anatomy of A-, B-, and Z-DNA. *Science*, **216**, 475-485.
[46] Seeman, N. C. & Kallenbach, N. R. (1994). DNA Branched Junctions. *Annu. Rev. Biophys. Biomol. Struct.*, **23**, 53-86.
[47] Seeman, N. C. & Kallenbach, N. R. (1983). Design of immobile nucleic acid junctions. *Biophys. J.*, **44**, 201-209.
[48] Broker, T. R. & Lehman, I. R. (1971). Branched DNA molecules: intermediates in T4 recombination. *J. Mol. Biol.*, **60**, 131-149.
[49] Wang, Y., Mueller, J. E., Kemper, B. & Seeman, N. C. (1991). Assembly and characterization of five-arm and six-arm DNA branched junctions. *Biochemistry*, **30**, 5667-5674.
[50] Wang, X. & Seeman, N. C. (2007). Assembly and Characterization of 8-Arm and 12-Arm DNA Branched Junctions. *J. Am. Chem. Soc.*, **129**, 8169-8176.
[51] Chen, J. & Seeman, N. C. (1991). Synthesis from DNA of a molecule with the connectivity of a cube. *Nature*, **350**, 631-633.
[52] Zhang, Y. & Seeman, N. C. (1994). Construction of a DNA-truncated octahedron. *J. Am. Chem. Soc.*, **116**, 1661-1669.
[53] Goodman, R. P., Berry, R. M. & Turberfield, A. J. (2004). The single-step synthesis of a DNA tetrahedron. *Chem. Commun.*, 1372-1373.
[54] Goodman, R. P., Schaap, I. A. T., Tardin, C. F., Erben, C. M., Berry, R. M., Schmidt, C. F., Turberfield, A. J. (2005). Rapid Chiral Assembly of Rigid DNA Building Blocks for Molecular Nanofabrication. *Science*, **310**, 1661-1665.
[55] He, Y., Ye, T., Su, M., Zhang, C., Ribbe, A. E., Jiang, W. & Mao, C. (2008). Hierarchical self-assembly of DNA into symmetric supramolecular polyhedra. *Nature*, **452**, 198-201.
[56] Ke, Y., Sharma, J., Liu, M., Jahn, K., Liu, Y. & Yan, H. (2009). Scaffolded DNA origami of a DNA tetrahedron molecular container. *Nano Lett.*, **9,** 2445-2447.
[57] Li, Z., Wei, B., Nangreave, J., Lin,C., Liu,Y., Mi,Y., & Yan, H. (2009). A replicable tetrahedral nanostructure self-assembled from a single DNA strand, *J. Am. Chem. Soc.*, **131**, 13093–13098.
[58] Mastroianni, A. J., Claridge, S. A. & Alivisatos, A. P. (2009). Pyramidal and chiral groupings of gold nanocrystals assembled using DNA scaffolds. *J. Am. Chem. Soc.*, **131**, 8455-8459.

[59] Özhalıcı-Ünal, H. & Armitage, B. A. (2009). Fluorescent DNA nanotags based on a self-assembled DNA tetrahedron. *ACS Nano*, **3**, 425-433.

[60] Kato, T., Goodman, R. P., Erben, C. M., Turberfield, A. J. & Namba, K. (2009). High-resolution structural analysis of a DNA nanostructure by cryoEM. *Nano Lett.*, **9**, 2747-2750.

[61] He, Y. & Mao, C. (2006). Balancing flexibility and stress in DNA nanostructures. *Chem. Commun.*, 968-969.

[62] He, Y., Chen, Y., Liu, H. P., Ribbe, A. E. & Mao, C. (2005). Self-Assembly of Hexagonal DNA Two-Dimensional (2D) Arrays. *J. Am. Chem. Soc.*, **127**, 12202-12203.

[63] Yan, H., Park, S. H., Finkelstein, G., Reif, J. H. & LaBean, T. H. (2003). DNA-Templated Self-Assembly of Protein Arrays and Highly Conductive Nanowires. *Science*, **301**, 1882-1884.

[64] He, Y., Tian, Y., Chen, Y., Deng, Z., Ribbe, A. E. & Mao, C. (2005). Sequence Symmetry as a Tool for Designing DNA Nanostructures. *Angew. Chem., Int. Ed.*, **44**, 6694-6696.

[65] He, Y., Tian, Y., Ribbe, A. E. & Mao, C. (2006). Highly Connected Two-Dimensional Crystals of DNA Six-Point-Stars. *J. Am. Chem. Soc.*, **128**, 15978-15979.

[66] Li, X., Yang, X., Qi, J. & Seeman, N. C. (1996). Antiparallel DNA Double Crossover Molecules As Components for Nanoconstruction. *J. Am. Chem. Soc.*, **118**, 6131-6140.

[67] Sa-Ardyen, P., Vologodskii, A. V. & Seeman, N. C. (2003). The Flexibility of DNA Double Crossover Molecules. *Biophys. J.*, **84**, 3829-3837.

[68] Qiu, W.-Y. (2000). *Knot theory, DNA topology, and molecular symmetry breaking*. In: D. Bonchev & D. H. Rouvray (Eds.), Chemical topology—applications and techniques, Mathematical Chemistry Series (Vol. 6, pp. 175-237). Amsterdam: Gordon and Breach Science Publishers.

[69] Zhang, C., Ko, S. H., Su, M., Leng, Y., Ribbe, A. E., Jiang, W. & Mao, C. (2009). Symmetry Controls the Face Geometry of DNA Polyhedra. *J. Am. Chem. Soc.*, **131**, 1413-1315.

[70] Andersen, E. S., Dong, M., Nielsen, M. M., Jahn, K., Subramani, R., Mamdouh, W., Golas, M. M., Sander, B., Stark, H., Oliveira, C. L. P., Pedersen, J. S., Birkedal, V., Besenbacher, F., Gothelf, K. V. & Kjems, J. (2009). Self-assembly of a nanoscale DNA box with a controllable lid. *Nature*, **459**, 73-76.

[71] Kuzuya, A. & Komiyama, M. (2009). Design and construction of a box-shaped 3D-DNA origami. *Chem. Commun.*, 4182-4184.

[72] Frisch, H. L. & Wasserman, E. (1961). Chemical Topology. *J. Am. Chem. Soc.*, **83**, 3789-3795.

[73] Shih, W. M., Quispe, J. D. & Joyce, G. F. (2004). A 1.7-kilobase single-stranded DNA that folds into a nanoscale octahedron. *Nature*, **427**, 618-621.

[74] Andersen, F. F., Knudsen, B., Oliveira, C. L. P., Frøhlich, R. F., Krüger, D., Bungert, J., Agbandje-McKenna, M., McKenna, R., Juul, S., Veigaard, C., Koch, J., Rubinstein, J. L., Guldbrandtsen, B., Hede, M. S., Karlsson, G., Andersen, A. H., Pedersen, J. S. & Knudsen, B. R. (2008). Assembly and structural analysis of a covalently closed nano-scale DNA cage. *Nucleic Acids Res.*, **36**, 1113-1119.

[75] He, Y., Su, M., Fang, P., Zhang, C., Ribbe, A. E., Jiang, W., & Mao, C. (2010). On the chirality of self-assembled DNA octahedral. *Angew. Chem., Int. Ed.*, **49**, 748-751.

[76] Zhang, X., Yan, H., Shen, Z. & Seeman, N. C. (2002). Paranemic cohesion of topologically-closed DNA molecules. *J. Am. Chem. Soc.*, **124**, 12940-12941.

[77] Rothemund, P. W. K. (2006). Scaffolded DNA origami: from generalized multicrossovers to polygonal networks. In J. H. Chen, N. Jonoska & G. Rozenberg (Eds.), Nanotechnology: science and computation (pp. 3-21). Heidelberg: Springer.

[78] Seeman, N. C. (1998). Nucleic acid nanostructures and topology. *Angew. Chem., Int. Ed.*, **37**, 3220-3238.

[79] Zimmermann, J., Cebulla, M. P. J., Mönninghoff, S. & von Kiedrowski, G. (2008). Self-assembly of a DNA dodecahedron from 20 trisoligonucleotides with C_{3h} linkers. *Angew. Chem., Int. Ed.*, **47**, 3626-3630.

[80] Chandra, M., Keller, S., Gloeckner, C., Bornemann, B. & Marx, A. (2007). New branched DNA constructs. *Chem. Eur. J.*, **13**, 3558-3564.

[81] Scheffler, M., Dorenbeck, A., Jordan, S., Wüstefeld, M. & von Kiedrowski, G. (1999). Self-assembly of trisoligonucleotidyls: the case for nano-acetylene and nano-cyclobutadiene. *Angew. Chem., Int. Ed.*, **38**, 3311-3315.

[82] Eckardt, L. H., Naumann, K., Pankau, W. M., Rein, M., Schweitzer, M., Windhab, N. & von Kiedrowski, G. (2002). DNA nanotechnology: chemical copying of connectivity. *Nature*, **420**, 286-286.

[83] Zhang, C., Su, M., He, Y., Zhao, X., Fang, P., Ribbe, A. E., Jiang, W. & Mao, C. (2008). Conformational flexibility facilitates self-assembly of complex DNA nanostructures. *Proc. Natl. Acad. Sci. U. S. A.*, **105**, 10665-10669.

[84] Douglas, S. M., Dietz, H., Liedl, T., Högberg, B., Graf, F. & Shih, W. M. (2009). Self-assembly of DNA into nanoscale three-dimensional shapes. *Nature*, **459**, 414-418.

[85] Bhatia, D., Mehtab, S., Krishnan, R., Indi, S. S., Basu, A. & Krishnan, Y. (2009). Icosahedral DNA nanocapsules by modular assembly. *Angew. Chem., Int. Ed.*, **48**, 4134-4137.

[86] Aldaye, F. A. & Sleiman, H. F. (2007). Modular access to structurally switchable 3D discrete DNA assemblies. *J. Am. Chem. Soc.*, **129**, 13376-13377.

[87] Erben, C. M., Goodman, R. P. & Turberfield, A. J. (2007). A self-assembled DNA bipyramid. *J. Am. Chem. Soc.*, **129**, 6992-6993.

[88] Rakotondradany, F., Sleiman, H. F. & Whitehead, M. A. (2007). Theoretical study of self-assembled hydrogen-bonded azodibenzoic acid tapes and rosettes. *J. Mol. Struct. THEOCHEM*, **806**, 39-50.

[89] Aldaye, F. A. & Sleiman, H. F. (2007). Dynamic DNA templates for discrete gold nanoparticle assemblies: control of geometry, modularity, write/erase and structural switching. *J. Am. Chem. Soc.*, **129**, 4130-4131.

[90] Qiu, W.-Y., Zhai, X.-D. & Qiu, Y.-Y. (2008). Architecture of Platonic and Archimedean polyhedral links. *Sci. China Ser. B-Chem.*, **51**, 13-18.

[91] Seife, C. (2003). Polyhedral model gives the universe an unexpected twist. *Science*, **302**, 209-210.

[92] Adams, C. C. (1994). *The Knot Book-an Elementary Introduction to the Mathematical Theory of Knots*. New York: W. H. Freeman and Company.

[93] Cromwell, P. R. (2004). *Knots and Links*. Cambridge: Cambridge University Press.

[94] Jablan, S. & Sazdanovic, R. (2008). *Linknot: Knot Theory by Computer*. Singapore: World Scientific.

[95] Hu, G., Zhai, X.-D., Lu, D. & Qiu, W.-Y., (2008). The architecture of Platonic polyhedral links. *J. Math. Chem.*, **46**, 592-602.
[96] Lu, D., Hu, G., Qiu, Y.-Y. & Qiu, W.-Y. (2010). Topological transformation of dual polyhedral links. *MATCH Commun. Math. Comput. Chem.*, **63**, 67-78.
[97] Lu, D. & Qiu, W.-Y. (2010). The keeping and reversal of chirality for dual links. *MATCH Commun. Math. Comput. Chem.*, **63**, 79-90.
[98] Walba, D. M. (1985). Topological stereochemistry. *Tetrahedron*, **41**, 3161-3212.
[99] Sumners, D. W. (1987). The knot theory of molecules. *J. Math. Chem.*, **1**, 1-14.
[100] In geometry, Schlegel diagram remains all topological properties, such as vertex degree and connections, while represents three dimensional graph in two dimensional planes. More about Schlegel diagram see ref. [1].
[101] Archdeacon, D., Siran, J. & Skoviera, M. (1992). Self-dual regular maps from medial graphs. *Acta Math. Univ. Comenianae*, **61**, 57-64.
[102] Flapan, E. (1998). Topological rubber gloves. *J. Math. Chem.*, **23**, 31-49.
[103] Flapan, E. (2000). *When Topology Meets Chemistry: a Topological Look at Molecular Chirality.* Cambridge: Cambridge University Press.
[104] Qiu, W.-Y. & Xin, H.-W. (1997). Molecular design and topological chirality of the Tq-Möbius ladders. *J. Mol. Struct. THEOCHEM*, **401**, 151-156.
[105] Qiu, W.-Y. & Xin, H.-W. (1997). Molecular design and tailor of the doubled knots. *J. Mol. Struct. THEOCHEM*, **397**, 33-37.
[106] Qiu, W.-Y. & Xin, H.-W. (1998). Topological structure of closed circular DNA. *J. Mol. Struct. THEOCHEM*, **428**, 35-39.
[107] Qiu, W.-Y. & Xin, H.-W. (1998). Topological chirality and achirality of DNA knots. *J. Mol. Struct. THEOCHEM*, **429**, 81-86.
[108] Mislow, K. (1996). A commetary on the topological chirality and achirality of molecules. *Croat. Chem. Acta.*, **69**, 485-511.
[109] Simon, J. (1986). Topological chirality of certain molecules. *Topology*, **25,** 229-235.
[110] Liang, C. Z. & Mislow, K. (1994). On amphicheiral knots. *J. Math. Chem.*, **15**, 1-34.
[111] Liang, C. Z. & Mislow, K. (1994). A left-right classification of topologically chiral knots. *J. Math. Chem.*, **15**, 35-62.
[112] Liang, C. Z. & Mislow, K. (1994). Classification of topologically chiral molecules. *J. Math. Chem.*, **15**, 245-260.
[113] Liang, C. Z. & Mislow, K. (1995). Topological chirality and achirality of links. *J. Math. Chem.*, **18**, 1-24.
[114] Cheng, X.-S., Qiu, W.-Y. & Zhang, H.-P. (2009). A novel molecular design of polyhedral links and their chiral analysis. *MATCH Commun. Math. Comput. Chem.*, **62**, 115-130.
[115] Hu, G. & Qiu, W.-Y. (2008). Extended Goldberg polyhedra. *MATCH Commun. Math. Comput. Chem.*, **59**, 585-594.
[116] Hu, G., Qiu, W.-Y. (2009). Extended Goldberg polyhedral links with even tangles. *MATCH Commun. Math. Comput. Chem.*, **61**, 737-752.
[117] Hu, G. & Qiu, W.-Y. (2009). Extended Goldberg polyhedral links with odd tangles. *MATCH Commun. Math. Comput. Chem.*, **61**, 753-766.
[118] In fact, the calculation of the number of component rings of Extended Goldberg polyhedra when k is odd is mathematically equivalent to the problem of central circuits

counting of 4-valent graphs. However, until now, it still remains unsolved. More about central circuits counting problem on 4-valent graphs see ref. [119-121]

[119] Deza, M., Huang, T. & Lih, K.-O. (2002). Central circuit coverings of octahedrites and medial polyhedra. *J. Math. Res. Exposition*, **22**, 49-65.

[120] Dutour, M. & Deza, M. (2004). Goldberg-Coxeter construction for 3- and 4-valent plane graphs. *Electron. J. Combinatorics*, **11**, R20.

[121] Deza, M. & Dutour, M. (2005). Zigzag structure of simple bifaced polyhedra. Combinatorics, *Probability Comput.*, **14**, 31-57.

[122] Jonoska, N. & Twarock, R. (2008). Blueprints for dodecahedral DNA cages. *J. Phys. A: Math. Theor.*, **41**, 304043-304057.

[123] Grayson, N. E., Taorminab, A. & Twarock, R., (2009). DNA duplex cage structures with icosahedral symmetry. *Theor. Comput. Sci.*, **410,** 1440-1447.

[124] Jonoska, N. & Saito, M. (2002). Boundary Components of Thickened Graphs. In: N. Jonoska & N. C. Seeman (Eds.), Lecture notes in computer science: DNA7 (Vol. 2340, pp. 70-81). London: Springer-Verlag.

[125] Tang, L., Johnson, K. N., Ball, L. A., Lin, T., Yeager, M. & Johnson, J. E. (2001). The structure of Pariacoto virus reveals a dodecahedral cage of duplex RNA. *Nat. Struct. Mol. Biol.*, **8**, 77-83.

[126] Rudnick, J. & Bruinsma, R. (2005). Icosahedral packing of RNA viral genomes. *Phys. Rev. Lett.*, **94**, 038101-038105.

[127] Yang, Y.-M. & Qiu, W.-Y. (2007). Molecular design and mathematical analysis of carbon nanotube links. *MATCH Commun. Math. Comput. Chem.*, **58**, 635-646.

[128] National Research Council. (1995). *Commission on Physical Sciences, Mathematics, and Applications.* Washington, D. C.: National Academy Press.

[129] Torquato, S., Jiao, Y. (2009). Dense packings of the Platonic and Archimedean solids. *Nature*, **460**, 876-879.

[130] Hu, G. & Qiu, W.-Y. (2010). Two Combinatorial Operations and a Knot Theoretical Approach for Fullerene Polyhedra. *MATCH Commun. Math. Comput. Chem.*, **63**, 347-362.

[131] Andersson, S. (2008). The structure of virus capsids. *Z. Anorg. Allg. Chem.*, **634**, 2161-2170.

[132] Andersson, S. (2008). Description of virus capsid structures with methods from inorganic solid. *Z. Anorg. Allg. Chem.*, **634**, 2504-2510.

[133] Andersson, S. (2009). Virus evolution and the beginning. *Z. Anorg. Allg. Chem.*, **635**, 717-724.

[134] Andersson, S. (2009). Virus structures, stellations, spikes, and rods. *Z. Anorg. Allg. Chem.*, **635**, 725-731.

[135] Andersson, S. (2008). *Intrinsic Structure of Virus Capsids*. Sweden: Sandforsk.

In: Mathematical Chemistry
Editor: W.I. Hong, pp. 367-416

ISBN: 978-1-60876-894-3

Chapter 5

HIGH-GAIN OBSERVERS FOR ESTIMATION OF KINETICS IN BATCH AND CONTINUOUS BIOREACTORS

Dan Selişteanu[a], Emil Petre[b], Dorin Şendrescu[c] and Monica Roman[d]

Department of Automatic Control, University of Craiova, Craiova, Romania

Abstract

This chapter deals with the design of high-gain observers for the unknown kinetics in two types of biochemical reactors: Batch (and Fed-Batch) Bioreactors, and Continuous Stirred Tank Bioreactors. The general dynamical state-space model of bioprocesses that take place inside these types of bioreactors is presented, and some particular bioprocesses are widely analyzed, such as wastewater biomethanation process, activated sludge bioprocesses, lipase production process. The models of the bioprocesses are obtained by using the classical modeling approach, which utilizes the mass balance of the components inside the bioreactor and a set of modeling rules. Also, a novel modeling procedure for bioprocesses is introduced, based on the pseudo Bond Graph approach, which uses the reaction schemes and the analysis of biochemical phenomena. The nonlinear bioprocesses are very complex and are characterized by the lack of on-line measurements. Therefore, in order to estimate the imprecisely known kinetic rates of these bioprocesses, some high-gain observers are designed. The estimation scheme does not require any model for the kinetic rates. The tuning of the proposed observers is reduced to the calibration of a single parameter. If the high-gain observers require on-line state estimates, these will be provided by asymptotic observers. In order to study the behavior and the performance of the proposed observers, extensive numerical simulations are performed.

---

[a] E-mail address: dansel@automation.ucv.ro
[b] E-mail address: epetre@automation.ucv.ro
[c] E-mail address: dorins@automation.ucv.ro
[d] E-mail address: monica@automation.ucv.ro

1. Introduction

The development of advanced control strategies for bioprocesses, which are highly complex processes that take place inside biochemical reactors (bioreactors), is hampered by major difficulties [1], [2]: nonlinearity, uncertainty of the process parameters, and absence of cheap and reliable instrumentation. These difficulties arise from the presence of living organisms, the high complexity of the interactions between the microorganisms, as well as the high complexity of the metabolic reactions.

The bioreactors can operate in three modes: the continuous mode, the fed-batch mode and the batch mode [1], [3]. A Fed-Batch Bioreactor (FBB) initially contains a small amount of substrates (the nutrients) and micro-organisms and is progressively filled with the influent substrates. When the FBB is full the content is harvested. A Batch Bioreactor is filled with the reactant mixture (substrates and micro-organisms) and allows for a particular time period that the reactions take place inside the reactor; after some time the products are removed from the tank. In a Continuous Stirred Tank Bioreactor (CSTB), the substrates are fed to the bioreactor continuously and an effluent stream is continuously withdrawn such that the culture volume is constant. In practice, the bioprocess control is often limited to regulation of the temperature and pH at constant values favorable to the microbial growth. There is however no doubt that the control of the biological state variables (concentrations of biomass, substrates, products, etc.) can help to increase the performance. To develop and apply advanced control strategies for these biological variables, it is necessary to obtain useful dynamical models and to design estimation and identification strategies.

The bioprocess modeling is a difficult task; however, using the mass balance of the components inside the process and obeying some modeling rules, a dynamical state-space model can be obtained - see [1], [2], [4].

A viable alternative to the classical modeling is the Bond Graph method. This method was first introduced by Paynter in 1961, and further developed by Karnopp and Rosenberg [5], and by Thoma [6]. Over the last four decades there have been a lot of publications regarding the theory and application of Bond Graphs in different engineering domains [7], [8], [9]. This method provides a uniform manner to describe the dynamical behavior for all types of physical systems and illustrates the exchange power in a system, which is normally the product between the effort and flow variables in the true Bond Graph. Besides this representation there is another one, in which the product effort-flow does not have the physical dimension of power, called pseudo Bond Graph [8], [10]. Pseudo Bond Graphs are more suitable for chemical systems due to the physical meaning of the effort and flow variables. The advantages of Bond Graph modeling are the following: offers a unified approach for all types of systems; allows to display the exchange of power in a system by its graphical representation; due to causality assignment it gives the possibility of localization of the state variables and achieving the mathematical model in terms of state space equations in an easier way than using classical methods; provides information regarding the structural properties of the system, in terms of controllability and observability [11], [12]. The Bond Graph modeling of some biological systems was reported in some works, such as [13], [14]. However, the Bond Graph modeling of biotechnological processes is not well exploited yet.

The estimation issues of the bioprocesses can be solved using "software sensors". A software sensor is a combination between a hardware sensor and a software estimator. These software sensors are used not only for the estimation of the concentrations - the state variables - but also for the estimation of the kinetic parameters. Very important is the estimation of kinetic rates inside a bioreactor (the so-called kinetics of the bioprocess) - the estimates of these rates are used for advanced control strategies. The interest for development of software sensors for bioreactors is proved by the big number of publications and applications in this area [4], [15], [16], [17], [18], [19], etc.

Generally speaking, the design of stable and convergent state and parameters estimators for bioprocesses is a complex task and good solutions are given only by studying each particular bioprocess.

Several estimation strategies have been developed to provide accurate on-line estimations of biological state variables. Presently, two classes of state observers for bioprocesses can be found [1], [4]. The first class of observers (including classical observers like Luenberger and Kalman observers, nonlinear observers) is based on a perfect knowledge of the model structure. A second class, called asymptotic observers, is based on the idea that the process uncertainty lies in the kinetics models. The design is based on mass and energy balances without the knowledge of kinetics being necessary.

Another key issue for bioreactors is the estimation of parameters, especially of the kinetic rates. Concerning this issue, an early approach was based on Kalman filter which leads to complex nonlinear algorithms. A well-known technique is the approach based on adaptive systems theory, which consists in the estimation of unmeasured states with asymptotic observers, and after that, the measurements and the state estimates are used for on-line estimation of kinetics. This useful method was applied for bioreactors [1], [4], but in some cases, when many reactions are involved, it requires the calibration of too many parameters. For example, if we have n components' concentrations used for the estimation of m kinetic rates, it is necessary to calibrate $2n$ tuning parameters. In order to overcome this problem, a possibility is to design an estimator using a high-gain approach [15], [20], [21]. The gain expression of high-gain observers involves a single tuning parameter whatever the number of components and reactions.

High-gain observers have evolved over the past two decades as an important tool for the design of output feedback control of nonlinear systems [21]. The early work on high-gain observers appeared in the late 1980's [21], and afterwards the technique was developed independently by two schools of researchers: a French school lead by Gauthier, Hammouri, Farza and others, [15], [20], and a U.S. school lead by Khalil [21].

This chapter is an extended work of the research achieved in some works of the authors: [16], [22], [23], [24], [25], [26], [27], [28], [29], and deals with the design of high-gain observers for the unknown kinetics in two types of bioreactors: Batch Bioreactors and CSTBs. First, the general dynamical state-space model of bioprocesses that take place inside these types of bioreactors is presented, by using the classical modeling approach. Then some particular bioprocesses are widely analyzed: wastewater biomethanation process, two activated sludge bioprocesses, and the lipase production process. These nonlinear bioprocesses are often used in bioindustry; their models are very complex and uncertain. The models of these processes are derived from the reaction schemes and the general dynamical model. Also, by using the pseudo Bond Graph technique, the models of the wastewater biomethanation process (inside a CSTB) and of an activated sludge bioprocess (inside a Batch Bioreactor) are obtained. These models are similar with those obtained via classical approach. The Bond Graphs are depicted using the modeling and simulation environment 20sim (registered trademark of Controllab Products B.V. Enschede, Netherlands). In the next paragraph, high-gain observers for the on-line estimation of imprecisely known kinetic rates for these bioprocesses are designed. The estimation scheme does not require any model for the kinetic rates. The tuning of the proposed observers is reduced to the calibration of a single parameter. The observer design is based on the work of Gauthier *et al.* [20], [30], Farza *et al.* [15], focused on deriving global results under global Lipschitz conditions. If the high-gain observers require on-line state estimates, these will be provided by asymptotic observers. In order to study the behavior and performance of the proposed observers, extensive numerical simulations are performed.

All simulations are achieved by using the development, programming and simulation environment MATLAB (registered trademark of The MathWorks, Inc., USA). Finally, concluding remarks and further research directions are presented.

2. Nonlinear Dynamical Models of Bioprocesses

2.1. The General Dynamical State-Space Model of Bioprocesses

A process carried out in a bioreactor can be defined as a set of *m* biochemical reactions involving *n* components (with $n \geq m$). The reaction scheme of a bioprocess (the reaction network) contains *n* components and *m* reactions. The concentrations of the physical components will be denoted with the notations ξ_i, $i = \overline{1,n}$. The reaction rates will be denoted as φ_j, $j = \overline{1,m}$. The evolution of ith component is described by the differential equation [4]:

$$\frac{d\xi_i}{dt} = \sum_{j \sim i} (\pm) k_{ij} \varphi_j - D\xi_i + F_i - Q_i \,, \tag{1}$$

where ξ_i (*g/l*) is the concentration and the notation $j \sim i$ indicates that the sum is done in accordance with the reactions *j* that involve the component *i*. The positive and dimensionless constants k_{ij} are yield coefficients. The sign of the first term of (1) is given by the type of the component ξ_i : plus (+) when the component is a reaction product and minus (-) otherwise. *D* is the specific volumetric rate (h^{-1}), usually called dilution rate. F_i represents the rate of supply of the component ξ_i (external substrate) to the bioreactor per unit of volume (*g/lh*). When this component is not an external substrate, then $F_i \equiv 0$. Q_i represents the rate of removal of the component ξ_i from the bioreactor in gaseous form per unit of volume (*g/lh*).

In order to obtain a dynamical state-space model of the entire bioprocess, we denote $\xi = [\xi_1 \; \xi_2 \; \cdots \; \xi_n]^T$, where ξ is the *n*-dimensional vector of the instantaneous concentrations, also is the state of the model. The vector of the reaction rates (the reaction kinetics) is denoted $\varphi = [\varphi_1 \; \varphi_2 \; \cdots \; \varphi_m]^T$. The reaction rate vector is *m*-dimensional. Usually, a reaction rate is represented by a non-negative rational function of the state ξ. The yield coefficients can be written as the $(n \times m)$ – dimensional yield matrix $K = [K_{ij}]$, $i = \overline{1,n}; j = \overline{1,m}$, where

$K_{ij} = (\pm)k_{ij}$ if $j \sim i$ and 0 otherwise. Next, we introduce the notations $F = [F_1 \ F_2 \ \cdots \ F_n]^T$, $Q = [Q_1 \ Q_2 \ \cdots \ Q_n]^T$, where F is the vector of rates of supply and Q is the vector of rates of removal of the components in gaseous form.

From (1), with the above notations, the global dynamics can be represented by following the dynamical state-space model [4]:

$$\frac{d\xi}{dt} = K \cdot \varphi(\xi) - D\xi + F - Q. \qquad (2)$$

This model describes in fact the behavior of an entire class of biotechnological processes and is referred to as the general dynamical state-space model of this class of bioprocesses [1], [4]. In (2), the term $K \cdot \varphi(\xi)$ is in fact the rate of consumption and/or production of the components in the reactor, i.e. the reaction kinetics. The term $-D\xi + F - Q$ represents the exchange with the environment, i.e. the dynamics of the component transportation through the bioreactor. The strongly nonlinear character of the model (2) is given by the reaction kinetics. In many situations, the yield coefficients, the structure and the parameters of the reaction rates are partially known or unknown.

Remark 1. In a FBB, the term $D\xi_i$ in (1) represents the dilution of a component due to the increase in volume. In this case D is the specific rate of volume increase ($\dot{V} = D \cdot V$, with V the liquid volume in the FBB and $\dot{V}$ its time derivative). In a CSTB, $D\xi_i$ represents the rate of removal of a component in liquid form (in a CSTB, $\dot{V} = 0$).

2.2. Pseudo Bond Graph Modeling Procedure for Simple Bioprocesses

Bond Graph method uses the effort-flow analogy to describe physical processes. A Bond Graph consists of subsystems linked together by lines representing power bonds. Each process is described by a pair of variables, effort *e* and flow *f*, and their product is the power. The direction of power is depicted by a half arrow. In a dynamic system the effort and the flow variables, and hence the power fluctuate in time.

One of the advantages of Bond Graph method is that models of various systems belonging to different engineering domains can be expressed using a set of only nine elements. A classification of Bond Graph elements can be made up by the number of ports; ports are places where interactions with other processes take place. There are one port elements represented by inertial elements (I), capacitive elements (C), resistive elements (R), effort sources (Se) and flow sources (Sf), two ports elements represented by transformer elements (TF) and gyrator elements (GY), and multi ports elements - effort junctions (J0) and flow junctions (J1). I, C, and R elements are passive elements because they convert the supplied energy into stored or dissipated energy. Se and Sf elements are active elements because they supply power to the system and TF, GY, 0 and 1-junctions are junction elements that serve to connect I, C, R, Se and Sf, and constitute the junction structure of the Bond Graph model. Besides the power variables, two other types of variables are very important in describing dynamic systems and these variables, sometimes called energy variables, are the generalized momentum p as time integral of effort and the generalized displacement q as time integral of flow [5], [6].

One of the simplest biological reactions is the micro-organisms growth process, with the reaction scheme [1], [4] given by:

$$S \overset{\varphi}{\rightarrow} X, \tag{3}$$

where S is the substrate, X is the biomass and φ is the reaction rate.

This simple growth reaction represents in fact a prototype reaction, which can be found in almost every bioprocess. The dynamic of the concentrations of the components from reaction scheme (3) can be modeled considering the mass balance of the components; accordingly, a dynamical model of the form (2) can be obtained. The dynamical model of the bioprocess (3) is quite simple, but if the reaction scheme is more complicated, the achievement of the dynamical model is difficult.

In such cases, the Bond Graph method can be a suitable approach. In order to model bioprocesses, pseudo Bond Graph method is more appropriate because of the meaning of variables involved – effort (concentration) and flow (mass flow). Next, pseudo Bond Graph models for two prototype bioprocesses are obtained: the first one for a bioprocess taking place into a Batch Bioreactor and the second one for a CSTB case.

Batch Bioreactor Case

In the case of the batch bioreactor, there is no influent into or effluent stream from the bioreactor and the biomass X is periodically collected. From the reaction scheme (3) and taking into account the mass transfer through the batch reactor, the pseudo Bond Graph model of the bioprocess is achieved and is depicted in Figure 1 [29]. The directions of half arrows correspond to the run of the reaction, going out from the substrate S towards biomass X.

In the Bond Graph model, the mass balances of the species involved in the bioreactor are represented by two 0-junctions: $0_{1,2}$ (mass balance for substrate), $0_{6,7}$ (mass balance for biomass). The constitutive relations of these junctions are characterized by the equality to zero of the sum of flow variables corresponding to the junction bonds. In order to model the rate of reaction φ, because of the dependency of substrate and biomass concentrations, we have used a modulated R element, denoted MR_4. Thus, the accumulations of species S and X in the bioreactor are represented by bonds 1 and 7 and are modeled using capacitive elements C. The constitutive equations of C-elements are as follows:

$$e_1 = \frac{1}{C_1} q_1 = \frac{1}{C_1} \int_t (-f_2) dt ,\; e_7 = \frac{1}{C_7} q_7 = \frac{1}{C_7} \int_t (f_6) d . \tag{4}$$

From the constitutive relations of 1-junction, we obtain:

$$f_3 = f_4 = f_5 . \tag{5}$$

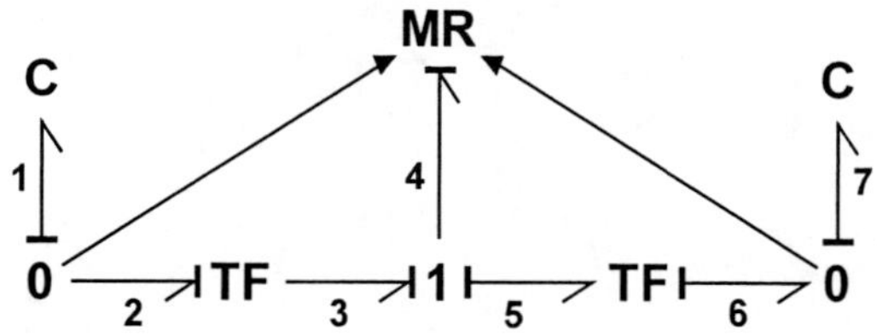

Figure 1. Bond Graph model – the batch bioprocess.

Using the constitutive relations of transformer elements $TF_{2,3}$ and $TF_{5,6}$ we obtain the relations for the flows f_2 and f_6:

$$f_2 = k_{2,3} f_4\,,\ f_6 = (1/k_{5,6})\cdot f_4\,, \tag{6}$$

with $k_{2,3}$ and $k_{5,6}$ the transformers modulus, which are in fact yield coefficients of the bioprocess (their values equal one for this batch bioprocess).

In fact, e_1 is the substrate concentration, which will be denoted with S (*g/l*), e_7 is the biomass concentration X (*g/l*), f_4 is $\varphi\, V$, and $C_1 = C_7 = V$ (*l*) is the volume of the bioreactor. Therefore, from (4)-(6) we will obtain the dynamical model of the batch process:

$$\begin{aligned} V\frac{dS}{dt} &= V\cdot \dot{S}(t) = -\varphi V, \\ V\frac{dX}{dt} &= V\cdot \dot{X}(t) = \varphi V. \end{aligned} \tag{7}$$

The model (7) expresses the equations of mass balance for the reaction scheme (3). The dynamical behavior of the concentrations can be easily obtained from (7):

$$\frac{dS}{dt} = -\varphi,\ \frac{dX}{dt} = \varphi. \tag{8}$$

Continuous Bioreactor Case

In the second case - the continuous bioprocess, the substrate (the nutrient) is fed to the bioreactor continuously and an effluent stream is continuously withdrawn such that the culture volume is constant.

From the reaction scheme (3) and taking into account the mass transfer through the CSTB, using the Bond Graph modeling procedure, the pseudo Bond Graph model of the continuous bioprocess is achieved and is given in Figure 2. The directions of half arrows correspond to the run of the reaction, going out from S to X. For the model presented in Figure 2, the mass balances of the species involved in the bioreactor are represented by two 0-junctions: $0_{1,2,3,4}$ (mass balance for S), $0_{8,9,10}$ (mass balance for biomass X). A modulated resistive element MR_6 was used to model the reaction rate φ. Mass flow of the component entering the reaction is modeled using a source flow element Sf and quantities exiting from the reaction are modeled using resistive elements R represented by bonds 3 and 10.

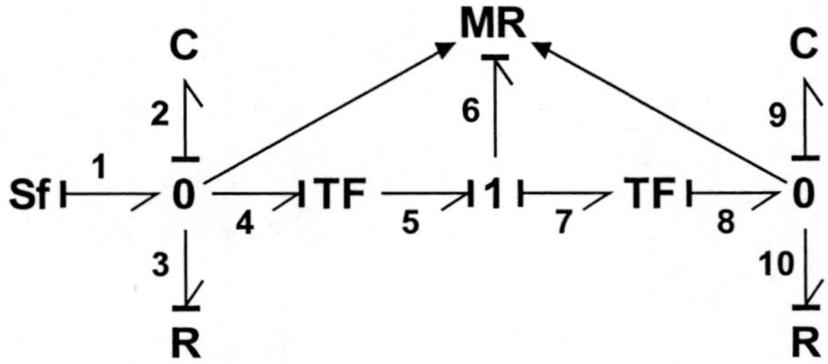

Figure 2. Bond Graph model – the continuous bioprocess.

From the constitutive equations of R-elements we obtain:

$$f_3 = e_3 / R_3,\ f_{10} = e_{10} / R_{10}. \tag{9}$$

The accumulations of substrate and biomass in the CSTB are represented by bonds 2 and 9, and are modeled using capacitive elements C, with the constitutive equations:

$$e_2 = \frac{1}{C_2} q_2 = \frac{1}{C_2} \int_t (f_1 - f_3 - f_4) dt, \tag{10}$$

$$e_9 = \frac{1}{C_9} q_9 = \frac{1}{C_9} \int_t (f_8 - f_{10}) dt. \tag{11}$$

The signification of Bond Graph elements is as follows: e_2 is the substrate concentration S (*g/l*), e_9 is the biomass concentration X (*g/l*), f_6 is proportional to φ and V, $C_2 = C_9 = V$ (*l*) is the volume, and $R_3 = R_{10} = 1/F_0$, where F_0 is the output flow (*l/h*).

Therefore, from (9)-(11) and taking into account the constitutive relations of junction elements, we will obtain the dynamical model of the continuous bioprocess:

$$\begin{aligned} V \cdot \frac{dS}{dt}(t) &= F_{in} S_{in} - F_0 S - \varphi V, \\ V \cdot \frac{dX}{dt}(t) &= -F_0 X + \varphi V, \end{aligned} \tag{12}$$

where F_{in} is the influent substrate flow (*l/h*), S_{in} is the influent substrate concentration (*g/l*).

From the equation of continuity $F_{in} = F_0$ and using the so-called dilution rate $D = F_{in} / V = 1/t_r$, with t_r - medium residence time, the equations (12) become:

$$\begin{aligned} \frac{dS}{dt} &= DS_{in} - DS - \varphi, \\ \frac{dX}{dt} &= -DX + \varphi. \end{aligned} \tag{13}$$

2.3. The Model of a Wastewater Biodegradation Process inside a CSTB

A highly important biotechnological process is the wastewater biodegradation with production of methane gas [4], [22], [23]. This anaerobic degradation process is a frequently method of wastewater treatment, which consists in four metabolic phases: two phases for acid production and two phases for methanation [4] - see Figure 3.

In the first phase, the glucose from the wastewater is decomposed in fat volatile acids (acetates, propionic acid), hydrogen and inorganic carbon under action of the acidogenic bacteria. In the second phase, the ionized hydrogen decomposes the propionic acid CH_3CH_2COOH in acetates, H_2 and carbon dioxide CO_2. In the first methanogenic phase, the acetate is transformed into methane and CO_2, and finally, in the second methanogenic phase, the methane gas CH_4 is obtained from H_2 and CO_2.

The reaction scheme of this complex bioprocess involves 4 reactions and 10 components [4]. From the reaction scheme we can obtain the dynamical model of this bioprocess that takes place inside a CSTB, considering the mass balance of the components (see [4], [23]).

The obtained dynamic model is very complex and the design of useful control strategies is hampered because of this large dimension (ten) of the model. So, it is necessary to reduce the order of this model, taking into consideration some particular aspects of the process and using the singular perturbation theory [31]. A simplified reaction scheme and the corresponding model can be derived for this bioprocess (see [22], [23]):

$$\begin{cases} S_1 \overset{\varphi_1}{\rightarrow} X_1 + S_2 \\ S_2 \overset{\varphi_2}{\rightarrow} X_2 + P \end{cases} \tag{14}$$

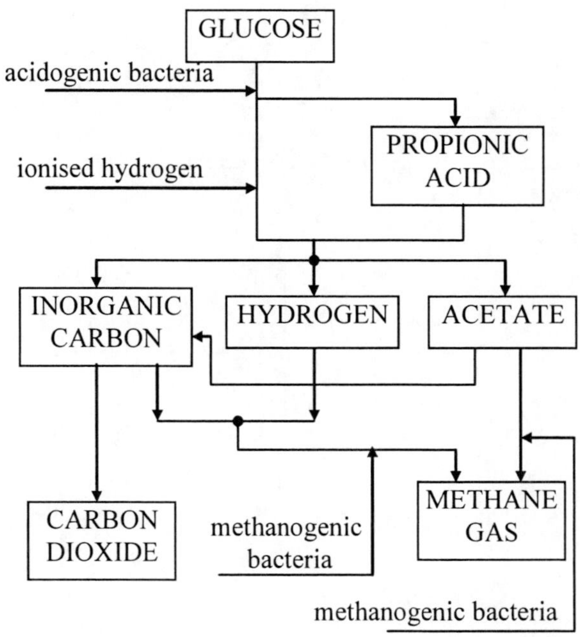

Figure 3. The structure of the biodegradation process.

In the reaction scheme (14) (which is a simply qualitative relation and does not include stoichiometric considerations), S_i, $i = 1, 2$ are substrates: S_1 represents the glucose, S_2 the acetate; X_1 is the acidogenic bacteria, X_2 the acetoclastic methanogenic bacteria and P represents the product, i.e. the methane gas. We remark that X_1, X_2 are auto-catalysts. The reaction rates are $\varphi_i, i = 1, 2$.

Next, using the Bond Graph method, we will derive a model of the bioprocess (14). From the reaction scheme (14), and considering the mass transfer through the CSTB, using the modeling procedure previously described, the pseudo Bond Graph model of the bioprocess is achieved [26]. This model is presented in Figure 4. The directions of the half arrows in the Bond Graph correspond to the progress of the reactions, going out from the component S_1 towards X_1 and S_2, for the first reaction, and from S_2 towards X_2 and P, for the second reaction. In Bond Graph terms, the mass balances of the species involved in the bioreactor are represented by five 0-junctions: $0_{1,2,3,4}$ (mass balance for S_1), $0_{8,9,10}$ (mass balance for X_1), $0_{12,13,14,15}$ (mass balance for S_2), $0_{19,20,21}$ (mass balance for X_2), $0_{23,24,25,26}$ (mass balance for P).

The constitutive relations of these junctions are characterized by the equality to zero of the sum of flow variables corresponding to the junction bonds; therefore, the next relations are obtained:

$$f_2 = f_1 - f_3 - f_4, \; f_9 = f_8 - f_{10}, \; f_{13} = f_{12} - f_{14} - f_{15},$$

$$f_{21} = f_{19} - f_{20}, \; f_{24} = f_{23} - f_{25} - f_{26}.$$

Thus, the accumulations of species S_1, X_1, S_2, X_2, and P in the bioreactor are represented by bonds 2, 9, 13, 21 and respectively 24, and are modeled using capacitive elements C.

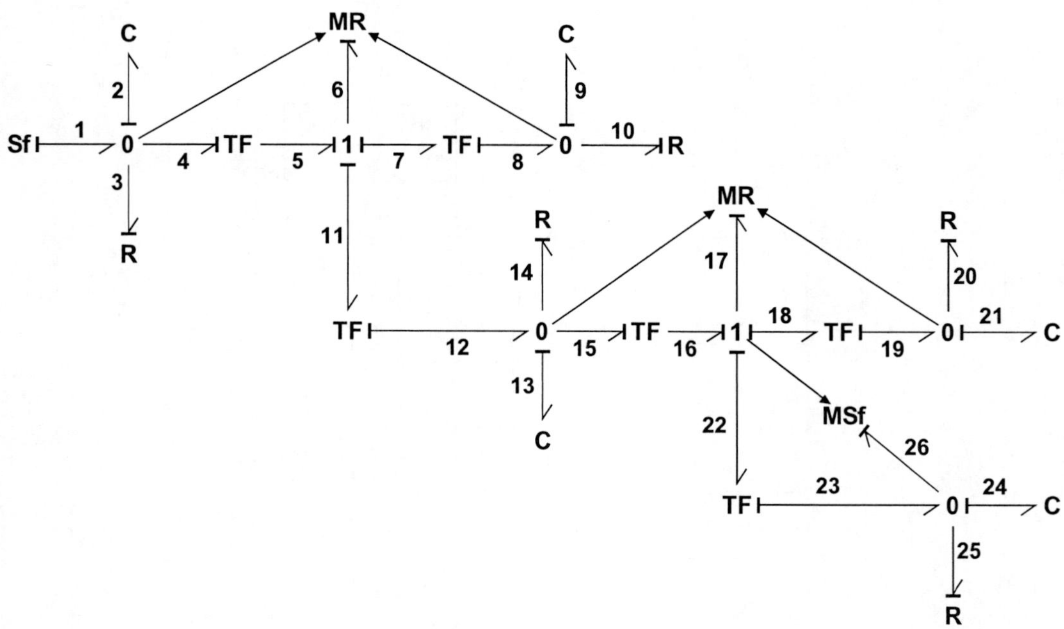

Figure 4. The pseudo Bond Graph model of the wastewater biodegradation process.

The constitutive equations of C-elements are as follows:

$$e_2 = \frac{1}{C_2} q_2 = \frac{1}{C_2} \int_t (f_1 - f_3 - f_4) dt \,, \; e_9 = \frac{1}{C_9} q_9 = \frac{1}{C_9} \int_t (f_8 - f_{10}) dt \,,$$

$$e_{13} = \frac{1}{C_{13}} q_{13} = \frac{1}{C_{13}} \int_t (f_{12} - f_{14} - f_{15}) dt \,, \; e_{21} = \frac{1}{C_{21}} q_{21} = \frac{1}{C_{21}} \int_t (f_{19} - f_{20}) dt \,, \qquad (15)$$

$$e_{24} = \frac{1}{C_{24}} q_{24} = \frac{1}{C_{24}} \int_t (f_{23} - f_{25} - f_{26}) dt \,.$$

where $C_2, C_9, C_{13}, C_{21}, C_{24}$ are the parameters of C-elements:

$$C_2 = C_9 = C_{13} = C_{21} = C_{24} = V \,.$$

The quantities exiting from the reaction are modeled using resistive elements R represented by bonds 3, 10, 14, 20 and 25; the constitutive equations of these elements are:

$$f_3 = e_3 / R_3 ,\; f_{10} = e_{10} / R_{10} ,\; f_{14} = e_{14} / R_{14} ,\; f_{20} = e_{20} / R_{20} ,\; f_{25} = e_{25} / R_{25} ,$$

where $R_3, R_{10}, R_{14}, R_{20}, R_{25}$ are parameters of R-elements:

$$R_3 = R_{10} = R_{14} = R_{20} = R_{25} = 1 / F_0 \,,$$

and F_0 is the output flow (*l/h*).

Mass flow of the component entering the reaction is modeled using a source flow element Sf and the masses that actually react are represented by bonds 4, 8, 12, 19 and 23. The transformer elements $TF_{4,5}$, $TF_{7,8}$, $TF_{11,12}$, $TF_{15,16}$, $TF_{18,19}$, $TF_{22,23}$ were introduced to model the yield coefficients. We used a modulated Sf element in order to model the methane gas outflow rate f_{26}, which will be denoted with Q (*g/h*).

For the modeling of the reaction rates we used two modulated R-elements: MR_6 and MR_{17}. From the constitutive relations of the two 1-junction elements, $1_{5,6,7,11}$ and $1_{16,17,18,22}$, we obtain: $f_5 = f_6 = f_7 = f_{11}$, $f_{16} = f_{17} = f_{18} = f_{22}$, where f_6 is $\varphi_1 V$ and f_{17} is $\varphi_2 V$. From the constitutive relations of transformer elements we obtain the relations for the flows $f_4, f_8, f_{12}, f_{19}, f_{23}$:

$$f_4 = k_{4,5} f_6 ,\; f_8 = f_6 / k_{7,8} ,\; f_{12} = f_6 / k_{12,13} ,\; f_{15} = k_{15,16} f_{17} ,$$
$$f_{19} = f_{17} / k_{18,19} ,\; f_{23} = f_{17} / k_{22,23} ,$$

with $k_{4,5} = k_1$, $k_{7,8} = 1 / k_2$, $k_{12,13} = 1 / k_3$, $k_{15,16} = k_4$, $k_{18,19} = k_5$ and $k_{22,23} = 1 / k_6$ the transformers modulus in Bond Graph terms, which are in fact yield coefficients.

The signification of Bond Graph elements is as follows: e_2 is the glucose concentration S_1 (*g/l*), e_9 - the acidogenic bacteria concentration X_1 (*g/l*), e_{13} - the acetate concentration

S_2 (*g/l*), e_{21} is the acetoclastic methanogenic bacteria concentration X_2 (*g/l*), e_{24} is the product concentration P (*g/l*), $f_1 = F_{in}S_{in}$, where F_{in} is the influent glucose flow (*l/h*) and S_{in} is the influent glucose concentration (*g/l*).

Using these notations, from (15) we will obtain the following dynamical model of the wastewater treatment bioprocess:

$$\begin{aligned}
V\cdot\frac{dS_1}{dt} &= F_{in}S_{in} - F_0S_1 - k_{4,5}\varphi_1 V \\
V\cdot\frac{dX_1}{dt} &= -F_0X_1 + (1/k_{7,8})\varphi_1 V \\
V\cdot\frac{dS_2}{dt} &= -F_0S_2 + (1/k_{12,13})\varphi_1 V - k_{15,16}\varphi_2 V \\
V\cdot\frac{dX_2}{dt} &= -F_0X_2 + k_{18,19}\varphi_2 V \\
V\cdot\frac{dP}{dt} &= -F_0P - Q + (1/k_{22,23})\varphi_2 V
\end{aligned} \tag{16}$$

From the equation of continuity $F_{in} = F_0$, using the dilution rate $D = F_{in}/V = 1/t_r$ (t_r is the medium residence time), and considering that $k_{7,8} = k_{18,19} = 1$, the equations (16) become:

$$\begin{aligned}
\frac{dS_1}{dt} &= DS_{in} - DS_1 - k_1\varphi_1 \\
\frac{dX_1}{dt} &= -DX_1 + \varphi_1 \\
\frac{dS_2}{dt} &= -DS_2 + k_3\varphi_1 - k_4\varphi_2 \\
\frac{dX_2}{dt} &= -DX_2 + \varphi_2 \\
\frac{dP}{dt} &= -DP - Q + k_6\varphi_2
\end{aligned} \tag{17}$$

The dynamical model (16) (or (17)), obtained using the Bond Graph approach, is equivalent with the dynamical state-space model obtained using classical methods in [4], [17], [22], [23]. The model (17) can be written in the following matrix form:

$$\frac{d}{dt}\begin{bmatrix} X_1 \\ S_1 \\ X_2 \\ S_2 \\ P \end{bmatrix} = \underbrace{\begin{bmatrix} 1 & 0 \\ -k_1 & 0 \\ 0 & 1 \\ k_2 & -k_3 \\ 0 & k_4 \end{bmatrix}}_{K}\begin{bmatrix} \varphi_1 \\ \varphi_2 \end{bmatrix} - D\begin{bmatrix} X_1 \\ S_1 \\ X_2 \\ S_2 \\ P \end{bmatrix} + \begin{bmatrix} 0 \\ DS_{in} \\ 0 \\ 0 \\ -Q \end{bmatrix}. \tag{18}$$

The state vector of (18) is $\xi = \begin{bmatrix} X_1 & S_1 & X_2 & S_2 & P \end{bmatrix}^T = \begin{bmatrix} \xi_1 & \xi_2 & \xi_3 & \xi_4 & \xi_5 \end{bmatrix}^T$. The vector of feed-rates and of rates of removal of components (in gaseous form) can be written as $F = \begin{bmatrix} 0 & DS_{in} & 0 & 0 & -Q \end{bmatrix}^T$. The dynamical model (18) can be compactly written as

$$\frac{d\xi}{dt} = K \cdot \varphi(\xi) - D \cdot \xi + F . \tag{19}$$

This model belongs to the class (2); K is the normalized matrix of the yield coefficients, and the vector of reaction rates (the reaction kinetics) is $\varphi(\xi) = \begin{bmatrix} \varphi_1(\xi) & \varphi_2(\xi) \end{bmatrix}^T$.

The most difficult task for the construction of the dynamical model (19) is the modeling of the reaction kinetics. The form of kinetics is complex, nonlinear and in many cases unknown. A general assumption is that a reaction can take place only if all reactants are presented in the bioreactor. Therefore, the reaction rates are necessarily to be zero whenever the concentration of one of the reactants is zero. Thus, the reaction rates can be expressed as follows:

$$\varphi(\xi) = H(\xi) \cdot \alpha(\xi, t) ,$$

where $\alpha(\xi, t) = \begin{bmatrix} \alpha_1(\xi, t) & \alpha_2(\xi, t) \end{bmatrix}^T$ is a vector of time varying parameters. Each $\alpha_i(\xi, t),\ i = 1, 2$ is called the specific reaction rate. $H(\xi)$ is a 2×2 state dependent diagonal matrix, whose elements correspond to the reactions' reactants:

$$H(\xi) = \begin{bmatrix} S_1 X_1 & 0 \\ 0 & S_2 X_2 \end{bmatrix} . \tag{20}$$

X_1 and X_2 appear because they are auto-catalysts. Then the model (19) becomes:

$$\frac{d\xi}{dt} = K \cdot H(\xi) \cdot \alpha(\xi, t) - D \cdot \xi + F . \tag{21}$$

Another formulation for the reaction rates implies the following structure:

$$\varphi(\xi) = \begin{bmatrix} \mu_1(\xi) \cdot X_1 & \mu_2(\xi) \cdot X_2 \end{bmatrix}^T , \tag{22}$$

with $\mu_1(\xi)$, $\mu_2(\xi)$ the specific growth rates.

Now we can reconsider the form of the matrix $H(\xi)$ and we have:

$$H(\xi) = \begin{bmatrix} X_1 & 0 \\ 0 & X_2 \end{bmatrix} . \tag{23}$$

The model of the bioprocess can be written as:

$$\frac{d\xi}{dt} = K \cdot H(\xi) \cdot \mu(\xi, t) - D \cdot \xi + F , \tag{24}$$

with $\mu(\xi,t)=[\mu_1(\xi,t)\ \mu_2(\xi,t)]^T$. For the specific growth rates $\mu(\xi,t)$ it exist many possible models, like Monod's law or Haldane kinetic model [22]:

$$\mu_1(S_1)=\mu_1^*\frac{S_1}{K_{M_1}+S_1},\ \mu_2(S_2)=\mu_2^0\frac{S_2}{K_{M_2}+S_2+S_2^2/K_i} \tag{25}$$

with $\mu_2^0=\mu_2^*(1+\sqrt{K_{M_2}/K_i}$. The first specific growth rate in (25) is modeled by the Monod law and the second specific growth rate by the Haldane kinetic model, which takes into account the substrate inhibition on the growth. K_{M_1}, K_{M_2} are Michaelis-Menten constants; μ_1^*, μ_2^* represent the maxim specific growth rates and K_i is the inhibition constant.

However, in practice, the analytical models of the specific reaction rates $\alpha(t)$ and/or of the specific growth rates $\mu(t)$ are difficult to obtain. Therefore, the uncertain specific reaction rates and/or specific growth rates need to be on-line estimated.

If we consider the vector of unknown parameters $\rho(t)=[\rho_1(t)\ \rho_2(t)]^T$ (with $\rho(t)=[\alpha_1(t)\ \alpha_2(t)]^T$ or $\rho(t)=[\mu_1(t)\ \mu_2(t)]^T$), then the model of the bioprocess can be written as:

$$\frac{d\xi}{dt}=K\cdot H(\xi)\cdot\rho(t)-D\cdot\xi+F \tag{26}$$

2.4. Models of Activated Sludge Bioprocesses

The activated sludge bioprocess is an aerobic process of biological wastewater treatment [4], [17], [32]. In practice, this bioprocess takes place inside CSTBs or in the so-called Sequencing Batch Reactors (SBRs). In spite of the fact that the CSTBs are often used, the characteristics of many industrial discharges can cause operational problems in continuous flow systems [18], [19].

Therefore, discontinuous processes such as SBRs can be considered because they could be better than the continuous processes, in terms of operation costs, process stability, and operation reliability [19], [33]. Next, both types of bioreactors will be considered, and the models of some prototype activated sludge bioprocesses will be derived.

The Model of an Activated Sludge Bioprocess inside a CSTB

In CSTBs, the activated sludge bioprocess usually operates in at least two interconnected tanks, like in Figure 5: an aerator in which the biological degradation of the pollutants takes place and a sedimentation tank (settler) in which the liquid is clarified, that is the biomass is separated from the treated wastewater. Part of the settled biomass is fed back to the bioreactor, while the surplus biomass is removed from the process.

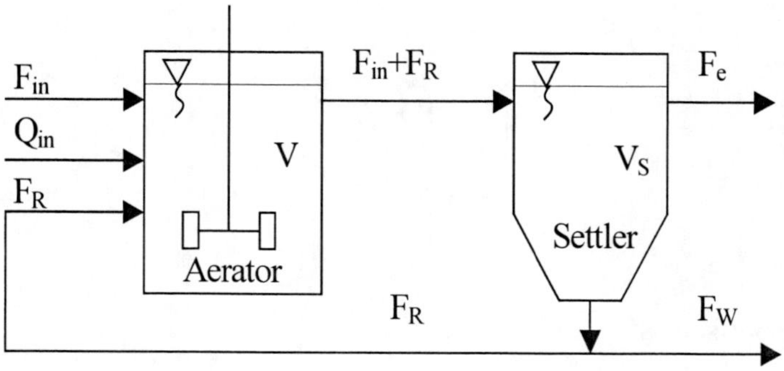

Figure 5. Schema of the activated sludge bioprocess.

The reaction in the aerator may be described by a simple autocatalytic aerobic microbial growth that can be represented by the following reaction scheme:

$$k_1 S + k_2 C \overset{\varphi \leftarrow}{\rightarrow} X \tag{27}$$

where S, X and C are respectively the pollutant, the biomass and the dissolved oxygen, φ is the reaction rate and k_1 and k_2 are the yield coefficients.

It is often assumed that the settler work perfectly, i.e. there is no biomass in the overflow of the settler. Then, the dynamics of the plant (aerator + settler) is described by the following mass balance equations [4], [32], [34], [35]:

$$\begin{aligned}
\frac{dS}{dt} &= -k_1 \mu X - \frac{F_{in}+F_R}{V} S + \frac{F_{in}}{V} S_{in} \\
\frac{dC}{dt} &= -k_2 \mu X - \frac{F_{in}+F_R}{V} C + Q_{in} \\
\frac{dX}{dt} &= \mu X - \frac{F_{in}+F_R}{V} X + \frac{F_R}{V} X_R \\
\frac{dX_R}{dt} &= \frac{F_{in}+F_R}{V_S} X - \frac{F_R+F_W}{V_S} X
\end{aligned} \tag{28}$$

where S_{in} is the concentration of influent substrate (*g/l*), Q_{in} is the oxygen feed rate (*g/lh*), X_R is the concentration of the recycled biomass (*g/l*), F_{in}, F_R and F_W are the influent, recycle and waste flow rates (*l/h*), respectively, V and V_S the aerator and settler volumes (*l*), respectively, and $\mu(\cdot)$ is the specific growth rate (h^{-1}) of reaction φ. If we define by $\xi = [S\ C\ X\ X_R]^T$ the state vector of (28), $\varphi = \mu(\cdot)X$ the reaction rate, $F = [D_{in}S_{in}\ Q_{in}\ 0\ 0]^T$ the feed rate vector, $Q = [0\ 0\ 0\ 0]^T$ the gaseous outflow rate vector and $K = [-k_1\ -k_2\ 1\ 0]^T$ the yield coefficient matrix, then the dynamical model (28) can be compactly written as:

$$\frac{d\xi}{dt} = K\varphi(\xi) - D\xi + F - Q \tag{29}$$

which is in fact like the general dynamical state-space model (2).

However, in (29), D stands for the dilution rate matrix and is given by

$$D = \begin{bmatrix} D_1 & 0 & 0 & 0 \\ 0 & D_1 & 0 & 0 \\ 0 & 0 & D_1 & -D_R \\ 0 & 0 & -D_2 & -D_3 \end{bmatrix} \tag{30}$$

whose entries are defined as:

$$D_{in} = \frac{F_{in}}{V},\ D_R = \frac{F_R}{V},\ D_1 = D_{in} + D_R,\ D_2 = \frac{F_{in} + F_R}{V_S},\ D_3 = \frac{F_R + F_W}{V_S} \tag{31}$$

The oxygen feed rate Q_{in} is usually set equal to the liquid-gas oxygen transfer rate:

$$Q_{in} = k_L a \cdot (C_s - C) \tag{32}$$

where $k_L a$ is the oxygen mass transfer coefficient and C_s the saturation constant.

Next we shall consider that $k_L a$ is a linear function of the air flow rate W [32]:

$$k_L a = a_0 W, \quad a_0 > 0 \tag{33}$$

The most difficult task for the construction of the dynamical model (28) is the modeling of the nonlinear and in many cases partial or completely unknown reaction rate φ. Again, a realistic assumption is that a reaction can take place only if all reactants are presented in the bioreactor.

As a result, the reaction rates are necessarily to be zero whenever the concentration of one of reactants is zero. A common form for the reaction rate is $\varphi = \mu(\cdot)X$. A possible simple structure of the specific growth rate $\mu(\cdot)$ is a Monod-type model, i.e. [34]:

$$\mu(S, C) = \mu_{\max} \frac{S}{K_S + S} \cdot \frac{C}{K_C + C} \tag{34}$$

where $\mu_{\max}$ is the maximum specific growth rate and K_S, K_C are Michaelis-Menten coefficients.

The model order can be reduced using singular perturbation techniques [31]. A systematic approach to model order reduction via singular perturbation theory for bioprocesses is fully described in [4]. Let's consider that there are low-solubility products and/or substrates that occur only in fast reactions. In the general dynamical model (29) each of them is denoted ξ_i and its corresponding dynamical equation is:

$$\frac{d\xi_i}{dt} = -D\xi_i + K_i\varphi + F_i - Q_i \tag{35}$$

where K_i is the line of K corresponding to ξ_i. Then, application of singular perturbation results in setting ξ_i and $d\xi_i / dt$ to zero and replacing the dynamical equation (35) by the algebraic equation:

$$K_i \varphi = Q_i - F_i \tag{36}$$

One more typical situation of application of singular perturbation theory to biosystems control is the multi-reactor case when the volumes of tanks are quite different from one another. For instance, in the activated sludge process, the dynamics of X_R can be written as:

$$\frac{V_S}{V}\frac{dX_R}{dt} = (D_{in} + D_R)X - (D_R + D_{VW})X_R \tag{37}$$

with $D_{VW} = F_W / V$. If V_S is small with respect to V, V_S / V may be considered as being negligible. Application of singular perturbation theory consists of setting V_S / V to zero and replacing (37) by the following algebraic equation:

$$(D_{in} + D_R)X = (D_R + D_{VW})X_R \tag{38}$$

Consequently, the order of the model is reduced from 4 to 3 and the reduced-order model can be used in estimation or control strategies. This reduced-order model consists of the first three equations of the model (28), where the concentration of the recycled biomass X_R is replaced using the algebraic equation (38). Then, the state vector of the reduced-order model becomes $\xi_r = [S\ C\ X\,]^T$, and the reduced-order model of the activated sludge bioprocess will be written as:

$$\frac{d\xi_r}{dt} = K_r \cdot H(\xi_r) \cdot \rho(t) - D_r \cdot \xi_r + F_r \tag{39}$$

with

$$K_r = \begin{bmatrix} -k_1 \\ -k_2 \\ 1 \end{bmatrix};\ D_r = \begin{bmatrix} D_1 & 0 & 0 \\ 0 & D_1 & 0 \\ 0 & 0 & D_1 - D_0 \end{bmatrix};\ F_r = \begin{bmatrix} D_{in} S_{in} \\ a_0 (C_S - C) W \\ 0 \end{bmatrix};\ D_0 = D_R \frac{D_R + D_{in}}{D_R + D_{VW}}.$$

In (39), $\rho(t)$ represents the unknown kinetics of the process. If we suppose that the reaction rate is totally unknown, then $\rho(t) = \varphi(t)$ and $H(\xi_r) = 1$. If the structure of the reaction rate is known: $\varphi = \mu(\cdot)X$, but the specific growth rate is unknown, then $\rho(t) = \mu(t)$ and $H(\xi_r) = X$.

The Model of an Activated Sludge Bioprocess inside a SBR

Next, a bioprocess for the treatment of typical municipal wastewater, polluted with organic carbon and nitrogen will be considered. This bioprocess was intensively studied by

Fibrianto *et al.* [18], Betancour *et al.* [33], Buitron *et al.* [19], etc., in the frame of a European project – EOLI. The model called EM2b, derived in the frame of this project, will be used. The bioprocess is carried out inside a SBR, located at LBE (Laboratoire de Biotechnologie de l'Environnement), Narbonne, France [18], [19], [33].

We will consider the aerobic phase, with one denitrification and one nitrification step, and the reaction scheme as follows [18]:

$$\begin{cases} S_1 + C \overset{\varphi_1}{\rightarrow} X_1 \\ C + S_3 \overset{\varphi_2}{\rightarrow} X_2 + S_2 \end{cases} \tag{40}$$

where X_1, X_2, S_1, S_2, S_3, C hold for the heterotroph and autotroph bacteria, the organic compound, nitrate + nitrite, nitrogen (TKN – Total Kjehdal Nitrogen), and dissolved oxygen, respectively. In (40), φ_1 and φ_2 are the reaction rates.

Next, using the Bond Graph method, we will derive a model of this bioprocess. From the reaction scheme (40), using the modeling procedure previously described, the pseudo Bond Graph model of the bioprocess is achieved. This model is presented in Figure 6. The directions of the half arrows in the Bond Graph correspond to the progress of the reactions, going out from the component S_1 and C towards X_1 for the first reaction, and from C and S_3 towards X_2 and S_2, for the second reaction.

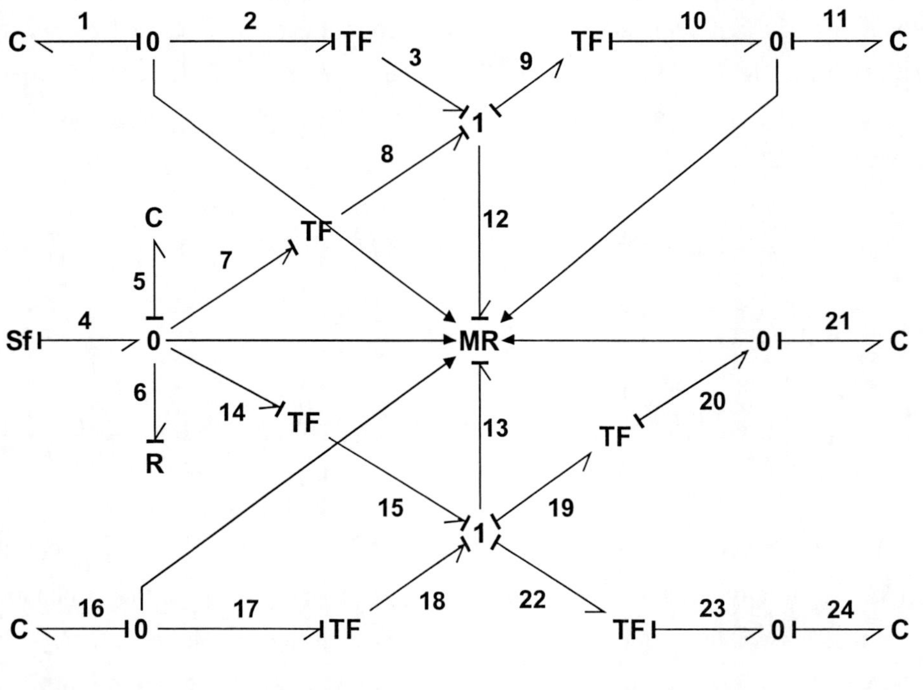

Figure 6. The pseudo Bond Graph model of the activated sludge bioprocess inside SBR.

In Bond Graph terms, the mass balances of the species involved in the bioreactor are represented by six 0-junctions: $0_{1,2}$ (mass balance for S_1), $0_{4,5,6,7,14}$ (mass balance for C), $0_{16,17}$ (mass balance for S_3), $0_{10,11}$ (mass balance for X_1), $0_{20,21}$ (mass balance for X_2), $0_{23,24}$ (mass balance for S_2).

The constitutive relations of these junctions are characterized by the equality to zero of the sum of flow variables corresponding to the junction bonds; therefore, the next relations are obtained:

$$f_1 = -f_2,\ f_5 = f_4 - f_6 - f_7 - f_{14},\ f_{11} = f_{10},\ f_{16} = -f_{17},\ f_{21} = f_{20},\ f_{24} = f_{23}.$$

The accumulations of species S_1, C, X_1, S_3, X_2, and S_2 in the bioreactor are represented by bonds 1, 5, 11, 16, 21, and 24, and they are modeled using capacitive elements C. The constitutive equations of C-elements are as follows:

$$e_1 = \frac{1}{C_1} q_1 = \frac{1}{C_1} \int_t (-f_2) dt,\ e_5 = \frac{1}{C_5} q_5 = \frac{1}{C_5} \int_t (f_4 - f_6 - f_7 - f_{14}) dt,$$

$$e_{11} = \frac{1}{C_{11}} q_{11} = \frac{1}{C_{11}} \int_t (f_{10}) dt,\ e_{16} = \frac{1}{C_{16}} q_{16} = \frac{1}{C_{16}} \int_t (-f_{17}) dt,$$

$$e_{21} = \frac{1}{C_{21}} q_{21} = \frac{1}{C_{21}} \int_t (f_{20}) dt,\ e_{24} = \frac{1}{C_{24}} q_{24} = \frac{1}{C_{24}} \int_t (f_{23}) dt.$$

where C_1, C_5, C_{11}, C_{16}, C_{21}, C_{24} are the parameters of C-elements $C_1 = C_5 = C_{11} = C_{16} = C_{21} = C_{24} = V$, where V is the bioreactor volume (l).

The quantity exiting from the reaction is modeled using resistive elements R represented by bond 6; the constitutive equations of this element is: $f_6 = e_6 / R_6$, where R_6, is the parameter of R-element, $R_6 = 1/(V \cdot K_L a)$, and $K_L a$ is the gas-liquid transfer coefficient.

Mass flow of the component entering the reaction is modeled using a source flow element Sf_4. The transformer elements $TF_{2,3}$, $TF_{7,8}$, $TF_{9,10}$, $TF_{14,15}$, $TF_{17,18}$, $TF_{19,20}$, $TF_{22,23}$ were introduced to model the yield coefficients. For the modeling of the reaction rates we used one di-port modulated R element: $MR_{12,13}$. From the constitutive relations of the two 1-junction elements, $1_{3,8,9,12}$ and $1_{13,15,18,19,22}$, we obtain: $f_3 = f_8 = f_9 = f_{12}$, $f_{13} = f_{15} = f_{18} = f_{19} = f_{22}$, with f_{12} the reaction rate φ_1 and f_{13} the reaction rate φ_2. From the constitutive relations of transformer elements we obtain the relations for the flows $f_2, f_7, f_{10}, f_{14}, f_{17}, f_{20}$, and f_{23}:

$$f_2 = k_{2,3} f_{12},\ f_7 = k_{7,8} f_{12},\ f_{10} = (1/k_{9,10}) f_{12},$$

$$f_{14} = (1/k_{14,15}) f_{13},\ f_{17} = k_{17,18} f_{13},\ f_{20} = f_{13}/k_{19,20},\ f_{23} = f_{13}/k_{22,23}$$

with $k_{2,3}$, $k_{7,8}$, $k_{9,10}$, $k_{14,15}$, $k_{17,18}$ $k_{19,20}$ and $k_{22,23}$ the transformers modulus, which are in fact yield coefficients of the bioprocess. The signification of Bond Graph elements is as follows: e_1 is the concentration of organic compound, S_1 (g/l), e_5 - concentration of the dissolved oxygen C (g/l), e_{11} - the heterotroph bacteria concentration, X_1 (g/l), e_{16} is the nitrogen concentration, S_3 (g/l), e_{21} is the autotroph bacteria concentration, X_2 (g/l) and e_{24} nitrate + nitrite concentration S_2 (g/l), $f_4 = F_{in} C_M$, where $F_{in} = K_L a V$ is the influent flow (l/h), and C_M is the maximum dissolved oxygen concentration, respectively. Using these notations, we will obtain the following dynamical model of the wastewater treatment process:

$$
\begin{aligned}
& V \cdot \frac{dX_1}{dt} = (1/k_{9,10})\varphi_1 V \\
& V \cdot \frac{dS_1}{dt} = -k_{2,3}\varphi_1 V \\
& V \cdot \frac{dX_2}{dt} = (1/k_{19,20})\varphi_2 V \\
& V \cdot \frac{dS_2}{dt} = (1/k_{22,23})\varphi_2 V \\
& V \cdot \frac{dS_3}{dt} = -k_{17,18}\varphi_2 V \\
& V \cdot \frac{dC}{dt} = F_{in}C_M - VK_L aC - k_{7,8}\varphi_1 V - (1/k_{14,15})\varphi_2 V
\end{aligned}
\tag{41}
$$

Taking into account that $k_{2,3} = k_1$, $k_{22,23} = 1/k_2$, $k_{17,18} = k_3$, $k_{7,8} = k_4$, $k_{14,15} = k_5$ $k_{9,10} = k_{19,20} = 1$, and that $\varphi_1 = \mu_1 X_1, \varphi_2 = \mu_2 X_2$, where μ_1, μ_2 are the specific growth rates of heterotroph and of autotroph bacteria, the equations (41) become:

$$
\begin{aligned}
& \frac{dX_1}{dt} = \mu_1 X_1, && \frac{dX_2}{dt} = \mu_2 X_2, \\
& \frac{dS_1}{dt} = -k_1\mu_1 X_1, && \frac{dS_2}{dt} = k_2\mu_2 X_2, \\
& \frac{dS_3}{dt} = -k_3\mu_2 X_2, && \frac{dC}{dt} = -k_4\mu_1 X_1 - k_5\mu_2 X_2 + K_L a(C_M - C),
\end{aligned}
\tag{42}
$$

This comprehensive model of this process, referred to as EM2b, was obtained also in [18], [19], [33], but using classical modeling. By using $\varphi = [\varphi_1 \ \ \varphi_2]^T = [\mu_1 X_1 \ \ \mu_2 X_2]^T$ and:

$$
\xi = \begin{bmatrix} X_1 \\ X_2 \\ S_1 \\ S_2 \\ S_3 \\ C \end{bmatrix}, \ K = \begin{bmatrix} 1 & 0 \\ 0 & 1 \\ -k_1 & 0 \\ 0 & k_2 \\ 0 & -k_3 \\ -k_4 & -k_5 \end{bmatrix}, \ \widetilde{Q} = \begin{bmatrix} 0 \\ 0 \\ 0 \\ 0 \\ 0 \\ -K_L a(C_M - C) \end{bmatrix},
$$

where ξ is the vector of concentrations, φ is the vector of reaction rates, K is the yield matrix and $\widetilde{Q}$ is the vector of removal rates, a batch model is obtained:

$$
\frac{d\xi}{dt} = K \cdot \varphi(\xi) - \widetilde{Q}. \tag{43}
$$

The kinetics for (40), (42) can take the form [18], [33]:

$$\mu_1(S_1, C) = \frac{\mu_1^*(S_1 - S_{1\min})C}{K_{S1} + S_1 - S_{1\min}},$$
$$\mu_2(S_3, C) = \frac{\mu_2^* S_3 C}{K_{S3} + S_3} \tag{44}$$

where $\mu_1^*, \mu_2^*, K_{S1}, K_{S3}$ are the maximum specific growth rates, the Michaelis-Menten constants, respectively, and the term $S_{1\min}$ accounts for the non-biodegradable part of S_1.

The specific rates (44) are directly proportional to the dissolved oxygen concentration, instead of the classical Monod law used in activated sludge models, because the oxygen concentration is not limiting [18], [33]. However, there exists a degree of uncertainty in the expressions of μ_1, μ_2, or equivalently in the knowledge of reaction rates. Therefore, estimation algorithms are needed in order to estimate the imprecisely known bioprocess kinetics. Furthermore, because the design of such estimators may require on-line availability of some states, it is necessary to consider the available measurements and the state estimates provided by state observers. The estimators design will take into account that the available on-line measurements are those of the dissolved oxygen and ammonia nitrogen concentrations [18].

Next, we will suppose that the on-line available states are the dissolved oxygen and ammonia nitrogen concentrations [18]. The model (43) can be rewritten as:

$$\frac{d\xi}{dt} = K \cdot H(\xi) \cdot \rho(t) - \widetilde{Q} = \Psi(\xi) \cdot \rho(t) - \widetilde{Q}, \tag{45}$$

where $\rho(t)$ is the unknown kinetics that will be estimated and $\Psi(\xi)$ is a matrix whose elements are real and scalar functions of C^q class ($q \geq 1$) with respect to their arguments.

2.5. The Model of a Lipase Production Bioprocess Taking Place inside a FBB

The lipase is an enzyme able to split fats and to synthesize glycerides, so the large scale production of lipase is very useful in many applications. This process usually takes place inside FBB, where *Candida Rugosa* is used in order to obtain the lipase. The growth of *Candida Rugosa* is achieved in a special culture medium, by using the oleic acid as substrate [36].

The lipase production bioprocess is very complex, with some simultaneous phases of the culture medium, and the reaction kinetics is uncertain and strongly nonlinear. This bioprocess is widely discussed in [36] and a nonlinear model is designed:

$$\begin{aligned}
\frac{dS_1}{dt} &= -\eta(S_1)X + F \\
\frac{dS_2}{dt} &= \eta(S_1) - \mu(S_2)\cdot(Y + S_2) \\
\frac{dX}{dt} &= \mu(S_2)X \\
\frac{dL_{in}}{dt} &= \nu_p(S_1, X, \mu) - \nu_{ex}(L_{in}) - \mu(S_2)L_{in} \\
\frac{dL_{ex}}{dt} &= \nu_{ex}(L_{in})X \\
Cer &= (a\mu(S_2) + b)X
\end{aligned} \tag{46}$$

S_1 (*g/l*), S_2 (*g/g*), X (*g/l*), L_{in} (*u/mg*), L_{ex} (*u/ml*) are the concentrations of external substrate, internal substrate, biomass, internal and external lipase, respectively, and *Cer* is the carbon dioxide excretion rate (*a*, *b* are excretion parameters) [36]. *F* is the feeding rate of substrate (if $F = 0$ the operating mode is batch) and *Y* is a yield coefficient. η is the absorption rate of external substrate S_1, μ the specific growth rate of biomass *X*, ν_p is the production rate of internal lipase L_{in}, and ν_{ex} the excretion rate of external lipase L_{ex}. These rates are imprecisely known; however, their form can be of Monod type for η, μ, ν_{ex}, but complex for ν_p: Haldane law plus influence of the specific growth rate of biomass [25], [36]:

$$\begin{gathered}
\eta(S_1) = \frac{\eta^* S_1}{K_{M1} + S_1}, \ \mu(S_2) = \frac{\mu^* S_2}{K_{M2} + S_2}, \ \nu_p(S_1, X, \mu) = \tilde{\nu}(S_1/X)\cdot\mu(S_2), \\
\tilde{\nu}(S_1/X) = \frac{\nu_p^* \cdot (S_1/X)}{K_p + (S_1/X) + K_i (S_1/X)^2}, \ \nu_{ex}(L_{in}) = \frac{\nu_{ex}^* L_{in}}{K_{ex} + L_{in}}.
\end{gathered} \tag{47}$$

All coefficients in (46), (47) are strictly positive. $K_{M1}, K_{M2}, K_p, K_{ex}$ are Michaelis-Menten constants, η^*, μ^*, ν_p^*, ν_{ex}^* are maximum specific rates, and K_i is an inhibition constant. It must be noted that the model (46), (47) is valid for biomass and substrate concentrations between 0 *g/l* and 8 *g/l* [25], [36]. As in previous cases, the nonlinear character of the model (46), (47) is given by the kinetics, which are partially known or unknown.

The main purpose of FBB control is to maximize the final lipase product quantity. This objective can be achieved through an optimal control of the bioreactor, i.e. the calculation of the optimal profile for the feeding rate, which is a common solution for FBB control [1], [4]. This optimal control is unsatisfactory when the kinetics is imprecisely known. Feasible suboptimal alternatives are the sliding mode control [3] and the adaptive control [25], [36]. These control strategies require the design of some estimators for the kinetic rates. The kinetic parameters which will be estimated are the absorption rate η, the specific growth rate μ (which enters also in the structure of ν_p), and the excretion rate ν_{ex}. These parameters are modeled by imprecisely known nonlinear functions.

In the lipase production process, often the state variables that are on-line available are the biomass, the external substrate and the external lipase concentrations (X is obtained via the measurements of Cer and L_{ex} via the lipase concentration in aqueous phase). Thus, only this information can be used in the observer design.

The first five differential equations of system (46) can be compactly written in the form:

$$\frac{d\xi}{dt} = \Psi(\xi) \cdot \varphi(\xi) + \widetilde{F}, \tag{48}$$

where $\xi = [S_1 \ S_2 \ X \ L_{in} \ L_{ex}]^T$ is the state vector, $\varphi(\xi) = [\eta(\xi) \ \mu(\xi) \ \nu_{ex}(\xi)]^T$ is the vector of the kinetic rates, and $\widetilde{F} = [F \ 0 \ 0 \ 0 \ 0]^T$ is the vector of feeding rates. $\Psi(\xi)$ is a 5×3 dimensional matrix

$$\Psi(\xi) = \begin{bmatrix} -X & 0 & 0 \\ 1 & -(Y+S_2) & 0 \\ 0 & X & 0 \\ 0 & \widetilde{\nu} - L_{in} & -1 \\ 0 & 0 & X \end{bmatrix}, \tag{49}$$

whose elements are real and scalar functions of C^q class ($q \geq 1$) with respect to their arguments [28]. The model of the lipase production bioprocess (48) can be written as:

$$\frac{d\xi}{dt} = \Psi(\xi) \cdot \rho(t) + \widetilde{F}, \tag{50}$$

In (50), $\rho(t) = [\rho_1(t) \ \rho_2(t) \ \rho_3(t)]^T = [\eta(t) \ \mu(t) \ \nu_{ex}(t)]^T$ represents the unknown kinetics of the process.

3. High-Gain Observers for on-Line Estimation of Kinetics

3.1. The Design of High-Gain Observers

In the process of designing high-gain observers for bioprocess, two types of models derived from the general class (2) will be used. The first type, which applies for the CSTBs, i.e. for the biomethanation process and for the continuous activated sludge bioprocess, is a generalization of the models (26) or (39):

$$\frac{d\xi}{dt} = K \cdot H(\xi) \cdot \rho(t) - D \cdot \xi + F - Q \tag{51}$$

The second type, valid for FBBs, i.e. for the activated sludge bioprocess inside SBR and for the lipase production bioprocess, is a generalization of the models (45) or (50):

$$\frac{d\xi}{dt} = \Psi(\xi)\cdot\rho(t) + \widetilde{F} - \widetilde{Q} \tag{52}$$

In order to design high-gain observers, a factorization of the model (51) or (52) is required [15], [20], [24]. We will consider that the yield matrix K, and the matrix $\Psi(\xi)$ respectively are of full rank. This assumption is true for our particular models, and for general case of the class (2) is a generic property. We shall suppose that all state variables are measured (contrarily, a state estimator can be used).

Since the yield matrix K, and the matrix $\Psi(\xi)$ are of full rank, then the partitions (K_a, K_b), and (Ψ_a, Ψ_b) respectively can be considered, such that the submatrices K_a, and Ψ_a respectively have full rank. Therefore, a partition (ξ_a, ξ_b) of the state vector is obtained, and consequently partitions for $F, Q, \widetilde{F}$, and $\widetilde{Q}$ are achieved: (F_a, F_b), (Q_a, Q_b), $(\widetilde{F}_a, \widetilde{F}_b)$, and $(\widetilde{Q}_a, \widetilde{Q}_b)$. Then, the system (51) can be written as follows:

$$\begin{aligned}\frac{d\xi_a}{dt} &= K_a\cdot H(\xi_a,\xi_b)\cdot\rho(t) - D_a\cdot\xi_a + F_a - Q_a\\ \frac{d\xi_b}{dt} &= K_b\cdot H(\xi_a,\xi_b)\cdot\rho(t) - D_b\cdot\xi_b + F_b - Q_b\end{aligned} \tag{53}$$

Similarly, the system (52) is factorized into:

$$\begin{aligned}\frac{d\xi_a}{dt} &= \Psi_a(\xi_a,\xi_b)\cdot\rho(t) + \widetilde{F}_a - \widetilde{Q}_a\\ \frac{d\xi_b}{dt} &= \Psi_b(\xi_a,\xi_b)\cdot\rho(t) + \widetilde{F}_b - \widetilde{Q}_b\end{aligned} \tag{54}$$

We suppose $\xi_b(t)$ a known (measured) signal, denoted $\sigma(t) = \xi_b(t)$. Then consider the systems [15]:

$$\begin{aligned}\frac{d\xi_a}{dt} &= K_a\cdot H(\xi_a,\sigma)\cdot\rho(t) - D_a\cdot\xi_a + F_a - Q_a\\ \frac{d\rho}{dt} &= g(t)\\ y &= \xi_a\end{aligned} \tag{55}$$

respectively:

$$\begin{aligned}\frac{d\xi_a}{dt} &= \psi_a(\xi_a,\sigma)\cdot\rho(t) + \widetilde{F}_a - \widetilde{Q}_a\\ \frac{d\rho}{dt} &= g(t)\\ y &= \xi_a\end{aligned} \tag{56}$$

where $g(t)$ is a bounded unknown function, which may depend on ξ_a, σ, inputs, noise. The hypothesis of boundedness of the kinetics is in accordance with industrial practice.

The design of nonlinear high-gain observers is done in [15], [20]. The systems (55), (56) belong to the general class of nonlinear systems [15]:

$$\begin{aligned} \frac{dz_1(t)}{dt} &= \Phi(z_1(t),\sigma(t))\cdot z_2(t) + b_1(u(t),z_1(t),\sigma(t)) \\ \frac{dz_2(t)}{dt} &= b_2(u(t),z_1(t),\sigma(t)) + g(t) \\ y(t) &= z_1(t) \end{aligned} \tag{57}$$

where z is a $2n$ – dimensional state vector, $z = [z_1 \;\; z_2]^T \in \Re^{2n}$; $z_1, z_2 \in \Re^n$. In (57), $u \in \Re^p$ is the input, y is the output and Φ is an $n \times n$ dimensional matrix whose elements are real and scalar functions of C^q class ($q \geq 1$) with respect to their arguments.

Remark 2. Usually, the gain of the observers is calculated by solving some Ricatti or Lyapunov differential equations [15], [20]. The high-gain observer design consists in a simple calculation of the gain by using algebraic Lyapunov equation.

The system (57) can be compactly written as

$$\begin{aligned} \frac{dz(t)}{dt} &= \widetilde{\Phi}(z_1(t),\sigma(t))\cdot z(t) + B(u(t),z_1(t),\sigma(t)) + \widetilde{g}(t) \\ y(t) &= Cz(t) \end{aligned} \tag{58}$$

where

$$\widetilde{\Phi}(z_1,\sigma) = \begin{bmatrix} 0 & \Phi(z_1,\sigma) \\ 0 & 0 \end{bmatrix},\; B(u,z_1,\sigma) = \begin{bmatrix} b_1(u,z_1,\sigma) \\ b_2(u,z_1,\sigma) \end{bmatrix},\; \widetilde{g}(t) = \begin{bmatrix} 0 \\ g(t) \end{bmatrix},\; C = [I_n \;\; 0],$$

with I_n the identity matrix.

Next, the following hypotheses will be considered [15], [24], [28]:

H1. There exist $\beta_1, \beta_2 \in \Re$, $0 < \beta_1 \leq \beta_2$ such that $\forall \zeta \in \Re^n, \forall t \geq 0$, next inequality holds:

$$\beta_1^2 I_n \leq \Phi^T(\zeta,\sigma)\Phi(\zeta,\sigma) \leq \beta_2^2 I_n\,;$$

H2. The function $g(t)$ is bounded;

H3. The function $\sigma(t)$ and its derivative with respect to time are bounded;

H4. The function B from (58) is globally Lipschitz with respect to z_1, locally uniformly with respect to u and σ;

H5. The function Φ is globally Lipschitz with respect to z_1, locally uniformly with respect to σ.

Consider now the system:

$$\frac{d\hat{z}}{dt} = \widetilde{\Phi}(\hat{z}_1,\sigma)\hat{z} + B(u,\hat{z}_1,\sigma) - \Omega^{-1}(\hat{z}_1,\sigma)S_\theta^{-1}C^T(C\hat{z}-y) \tag{59}$$

with: $\hat{z} = \left[\hat{z}_1 \ \hat{z}_2\right]^T \in \Re^{2n}$; u and y are the input and respectively the output of the system (58); $\Omega(\hat{z}_1,\sigma) = \begin{bmatrix} I_n & 0 \\ 0 & \Phi(\hat{z}_1,\sigma) \end{bmatrix}$; S_θ is a unique positive defined symmetric matrix, which satisfies the algebraic Lyapunov equation:

$$\theta S_\theta + P^T S_\theta + S_\theta P - C^T C = 0 \tag{60}$$

where $P = \begin{bmatrix} 0 & I_n \\ 0 & 0 \end{bmatrix}$ and $\theta > 0$ is a scalar parameter.

Then, the next theorem can be formulated:

Theorem 1. [15]. Let be the system (58) that satisfies the hypotheses H1–H5. Then, $\forall\delta > 0,\ \exists\theta_0 > 0,\ \forall\theta > \theta_0,\ \exists\mu_\theta,\nu_\theta > 0,\ \exists M_\theta > 0,\ \forall u$ with $\|u\|_\infty \le \delta,\ \forall\hat{z}(0) \in \Re^{2n}$, we have:

$$\|\hat{z}(t) - z(t)\| \le \nu_\theta e^{-\mu_\theta t}\|\hat{z}(0) - z(0)\| + M_\theta\gamma \tag{61}$$

where u is an admissible command, $\|u\|_\infty$ is the upper bound of the command, $z(t)$ is the trajectory of (58) associated to the input u, $\hat{z}(t)$ is any trajectory of (59) and γ is the superior bound of $\|g\|$. Furthermore, we have $\lim\limits_{\theta\to\infty}\mu_\theta = \infty$ and $\lim\limits_{\theta\to\infty} M_\theta = 0$.

Remark 3. When $g(t) = 0$, the nonlinear observer (59) has an exponential convergence of the estimation error $\varepsilon(t) = \hat{z}(t) - z(t)$. Contrariwise, Theorem 1 demonstrates that the estimation error can become arbitrarily small choosing a sufficiently big value for the gain θ.

The gain of the observer (59) can be obtained by solving the Lyapunov algebraic equation (60). The solution of this equation is [15], [24]:

$$S_\theta = \begin{bmatrix} \frac{1}{\theta}I_n & -\frac{1}{\theta^2}I_n \\ -\frac{1}{\theta^2}I_n & \frac{2}{\theta^3}I_n \end{bmatrix} \tag{62}$$

and consequently from (60) the expression of the gain is obtained as:

$$\Omega^{-1}(\hat{z}_1,\sigma)S_\theta^{-1}C^T = \begin{bmatrix} 2\cdot\theta\cdot I_n \\ \theta^2\cdot\Phi^{-1}(\hat{z}_1,\sigma) \end{bmatrix} \tag{63}$$

Then the general equations of the high-gain nonlinear observer [15] can be written using (60) and (63):

$$\begin{aligned} \frac{d\hat{z}_1}{dt} &= \Phi(\hat{z}_1,\sigma)\hat{z}_2 + b_1(u,\hat{z}_1,\sigma) - 2\theta(\hat{z}_1 - z_1) \\ \frac{d\hat{z}_2}{dt} &= -\theta^2\Phi^{-1}(\hat{z}_1,\sigma)(\hat{z}_1 - z_1) \end{aligned} \tag{64}$$

It can be observed that the nonlinear observer is quite simple and moreover the tuning of the gain can be done by modifying only one design parameter.

The high-gain observer equations for the general class of nonlinear systems (57) are applicable also for the bioprocesses described by (51) or (52) (partitioned in the form (53) or (54)). The hypotheses H2 – H5 are fulfilled for this class of bioprocesses. However, the assumption H1 can generate some problems, but usually the modeling rules for the bioprocesses impose that the diagonal elements of the matrix Φ are equal to zero if and only if into the bioreactor do not takes place any reaction [15], [24].

The equations of the high gain observer for the bioprocesses taking place inside CSTBs are obtained using (53), (55) and (64):

$$\begin{aligned} \frac{d\hat{\xi}_a}{dt} &= K_a H(\hat{\xi}_a,\sigma)\hat{\rho}(t) - D_a\hat{\xi}_a + F_a - Q_a - 2\theta(\hat{\xi}_a - \xi_a) \\ \frac{d\hat{\rho}}{dt} &= -\theta^2 \cdot \left[K_a \cdot H(\hat{\xi}_a,\sigma)\right]^{-1} \cdot (\hat{\xi}_a - \xi_a) \end{aligned} \tag{65}$$

In a similar way, the high-gain observer for the bioprocesses inside FBBs is achieved directly from (54), (56) and (64):

$$\begin{aligned} \frac{d\hat{\xi}_a}{dt} &= \Psi_a(\hat{\xi}_a,\xi_b)\cdot\hat{\rho}(t) + \widetilde{F}_a - \widetilde{Q}_a - 2\theta(\hat{\xi}_a - \xi_a) \\ \frac{d\hat{\rho}}{dt} &= -\theta^2\left[\Psi_a(\hat{\xi}_a,\xi_b)\right]^{-1}(\hat{\xi}_a - \xi_a) \end{aligned} \tag{66}$$

The observers (65) and (66) provide on-line estimates $\hat{\rho}$ for the unknown kinetics; these observers are in fact copies of the bioprocesses models, but with the state ξ_a replaced by its estimate $\hat{\xi}_a$, and with a corrective term. The tuning of this observer is very simple because a single parameter is involved: θ.

Remark 4. Note that in (65) and (66), $\hat{\xi}_a$ is an "estimate" of ξ_a, provided by the algorithms in order to be compared with the real state (ξ_a is measured or provided by a state observer), and the resulting error to be used in the estimators (65), and (66) respectively.

3.2. High-Gain Observers for the Wastewater Biodegradation Process inside CSTB

The high-gain observers previously designed will be applied for the biomethanation process, in order to reconstitute the imprecisely known specific reaction rates $\alpha(t)$ or of the specific growth rates $\mu(t)$.

However, there are some alternatives for on-line estimation of these unknown rates, such as an algorithm based on a state observer technique [35], or a linear regressive observer [4]. These algorithms provide good estimates, but the problem is the number of tuning parameters, for example 10 in the algorithm presented in [35].

Simple yet efficient nonlinear high-gain observers can be obtained using the presented design procedure, considering the bioprocess model (26).

For the model (26), the yield matrix K, with $\dim(K) = 5 \times 2$, is of full rank, i.e. rank(K) = 2. This assumption is true for our particular model, and for general case is a generic property.

Remark 5. In some cases, the matrix K is not left invertible. These situations appear for example when the number of unknown parameters is greater than the number of state variables. In these particular cases, the proposed observers cannot be implemented.

We shall suppose that all state variables are measured (contrarily, a state estimator can be used). Since K is of full rank, i.e. is left invertible, a full rank arbitrary submatrix K_a (2×2) of K can be considered. Let K_b be the remaining submatrix of K. The system (26) can be written in the form (53), as follows:

$$\begin{aligned} \frac{d\xi_a}{dt} &= K_a \cdot H(\xi_a, \xi_b) \cdot \rho(t) - D \cdot \xi_a + F_a \\ \frac{d\xi_b}{dt} &= K_b \cdot H(\xi_a, \xi_b) \cdot \rho(t) - D \cdot \xi_b + F_b \end{aligned} \tag{67}$$

where (ξ_a, ξ_b) and (F_a, F_b) are partitions induced by the factorization of K.

For the biomethanation process there are several possibilities of the factorization of yield matrix K. We choose:

$$K_a = \begin{bmatrix} 1 & 0 \\ 0 & 1 \end{bmatrix} \tag{68}$$

for simplicity and consequently we obtain:

$$\xi_a = \begin{bmatrix} X_1 \\ X_2 \end{bmatrix};\ \xi_b = \sigma = \begin{bmatrix} S_1 \\ S_2 \\ P \end{bmatrix};\ K_b = \begin{bmatrix} -k_1 & 0 \\ k_2 & -k_3 \\ 0 & -k_4 \end{bmatrix};\ F_a = \begin{bmatrix} 0 \\ 0 \end{bmatrix};\ F_b = \begin{bmatrix} D \cdot S_{in} \\ 0 \\ -Q \end{bmatrix}. \tag{69}$$

It is clear that K_a is full rank matrix. For this factorization of yield matrix K is possible to design and implement two observers:

(i) An observer for on-line estimation of the specific growth rates $\rho(t) = \mu(t)$

(ii) An observer for on-line estimation of the specific reaction rates $\rho(t) = \alpha(t)$

From equations (65), (67) with $H(\xi)$ of the form (23) and with the factorization (68), (69), the equations of the observer (i) are:

$$\frac{d}{dt}\begin{bmatrix}\hat{X}_1\\ \hat{X}_2\end{bmatrix}=\begin{bmatrix}1 & 0\\ 0 & 1\end{bmatrix}\cdot\begin{bmatrix}\hat{X}_1 & 0\\ 0 & \hat{X}_2\end{bmatrix}\cdot\begin{bmatrix}\hat{\mu}_1\\ \hat{\mu}_2\end{bmatrix}-D\cdot\begin{bmatrix}\hat{X}_1\\ \hat{X}_2\end{bmatrix}-2\cdot\theta\cdot\begin{bmatrix}\hat{X}_1-X_1\\ \hat{X}_2-X_2\end{bmatrix} \tag{70}$$

$$\frac{d}{dt}\begin{bmatrix}\hat{\mu}_1\\ \hat{\mu}_2\end{bmatrix}=-\theta^2\cdot\left[\begin{bmatrix}1 & 0\\ 0 & 1\end{bmatrix}\cdot\begin{bmatrix}\hat{X}_1 & 0\\ 0 & \hat{X}_2\end{bmatrix}\right]^{-1}\cdot\begin{bmatrix}\hat{X}_1-X_1\\ \hat{X}_2-X_2\end{bmatrix} \tag{71}$$

It can be seen that the estimator (70), (71) needs only the measurements of X_1 and X_2.

Similarly, the equations of the second observer (ii) are derived from (65), (67), (68), (69), but with $H(\xi)$ of the form (20):

$$\frac{d}{dt}\begin{bmatrix}\hat{X}_1\\ \hat{X}_2\end{bmatrix}=\begin{bmatrix}1 & 0\\ 0 & 1\end{bmatrix}\cdot\begin{bmatrix}S_1\hat{X}_1 & 0\\ 0 & S_2\hat{X}_2\end{bmatrix}\cdot\begin{bmatrix}\hat{\alpha}_1\\ \hat{\alpha}_2\end{bmatrix}-D\cdot\begin{bmatrix}\hat{X}_1\\ \hat{X}_2\end{bmatrix}-2\cdot\theta\cdot\begin{bmatrix}\hat{X}_1-X_1\\ \hat{X}_2-X_2\end{bmatrix} \tag{72}$$

$$\frac{d}{dt}\begin{bmatrix}\hat{\alpha}_1\\ \hat{\alpha}_2\end{bmatrix}=-\theta^2\cdot\left[\begin{bmatrix}1 & 0\\ 0 & 1\end{bmatrix}\cdot\begin{bmatrix}S_1\hat{X}_1 & 0\\ 0 & S_2\hat{X}_2\end{bmatrix}\right]^{-1}\cdot\begin{bmatrix}\hat{X}_1-X_1\\ \hat{X}_2-X_2\end{bmatrix} \tag{73}$$

For the implementation of the observer (ii) are necessary the measurements of X_1, X_2, S_1 and S_2.

Simulation results

The high-gain observers (i) and (ii) were implemented for the biomethanation process (18). In order to study the performances of this observer, the results are compared with data generated from simulation of the process model (18). The specific growth rates for this simulation are of form (25) – these kinetic expressions are introduced only for simulation; therefore these models are not used in the process of the observer design.

In the case of first estimator (70), (71), the main goal is to reconstitute the time evolution of μ_1 and μ_2 from the measurements of X_1 and X_2 obtained from the simulation. For the second observer (72), (73) the objective is to reconstitute the time evolution of α_1 and α_2 from the measurements of X_1, X_2, S_1 and S_2 obtained from the simulation.

Simulation was performed for a sinusoidal form of the dilution rate and the following bioreactor parameters:

$$S_{in}=30g/l; k_1=5.4; k_2=1; k_3=14.7;\ k_4=10; Q=0; \mu_1^*=0.2h^{-1}; \mu_2^0=0.6h^{-1};$$

$$K_{M_1}=0.75g/l; K_{M_2}=1g/l; K_i=21g/l.$$

Figure 7 shows the time evolution for S_1 and S_2 - the glucose concentration and the acetate concentration. The time profile of X_1 - population of acidogenic bacteria, X_2 - population of methanogenic bacteria and P - the methane gas, are depicted in Figure 8. These "measurements" obtained from simulation are free-noise data.

The performances of proposed estimators are also analyzed for noisy data of X_1 and X_2. The measurements of X_1 and X_2 are vitiated by an additive Gaussian noise. This noise is with zero mean and amplitude equal to 3% of the free-noise values. In Figure 9 the noisy evolution of X_1 and X_2 is presented, and Figure 10 shows the sinusoidal form of the dilution rate.

The estimation and the simulation of the specific growth rates obtained with the observer (i) are depicted in Figure 11 and Figure 12. These pictures correspond to free-noise measurements. Figure 13 and Figure 14 show the results obtained when the noisy data are used. In all figures, the values obtained from simulation are depicted with solid curves and the estimates with dashed curves. The value of the tuning parameter was set to $\theta = 3$.

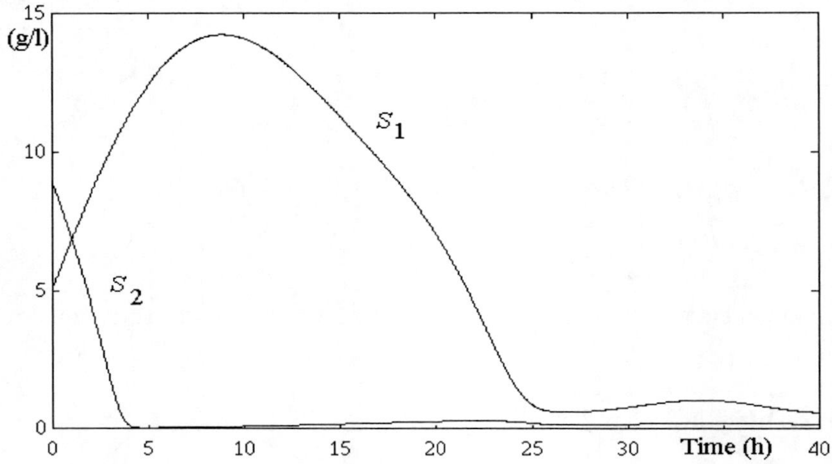

Figure 7. Time profiles of concentrations S_1 and S_2.

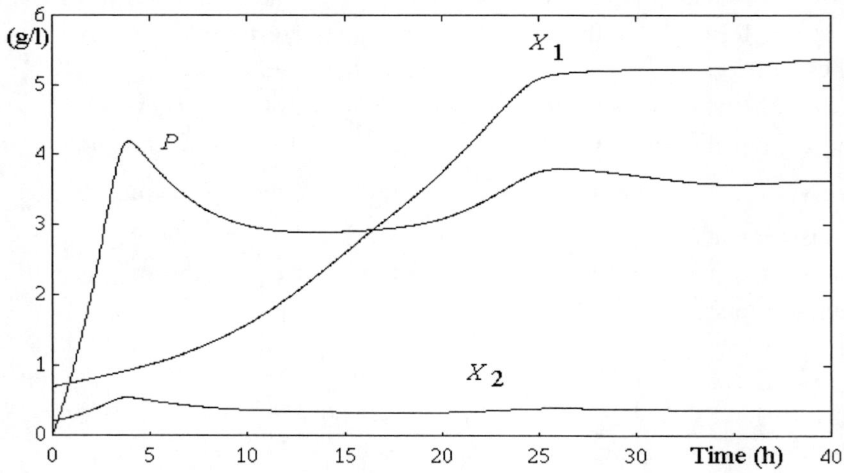

Figure 8. Time evolution of X_1, X_2 and P.

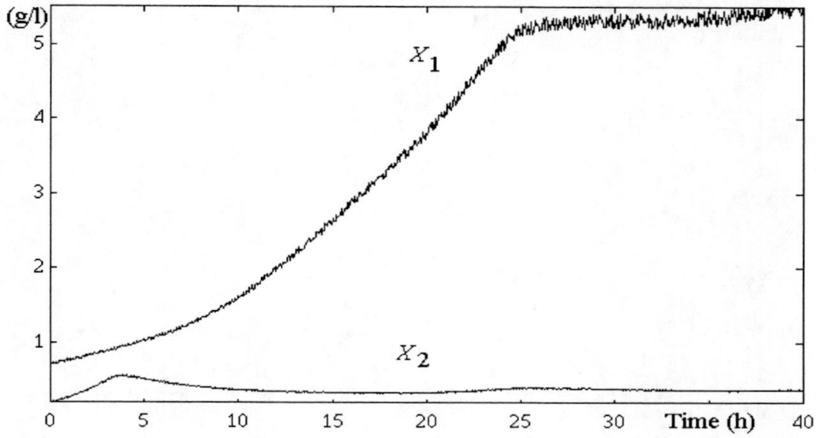

Figure 9. Noisy data of X_1 and X_2.

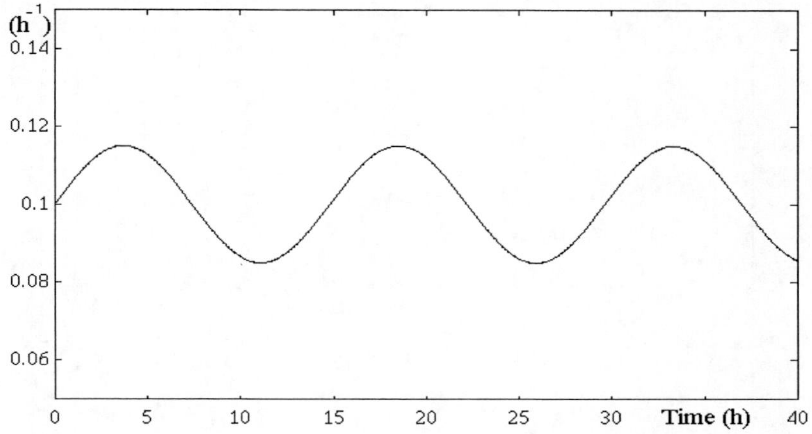

Figure 10. The profile of the dilution rate.

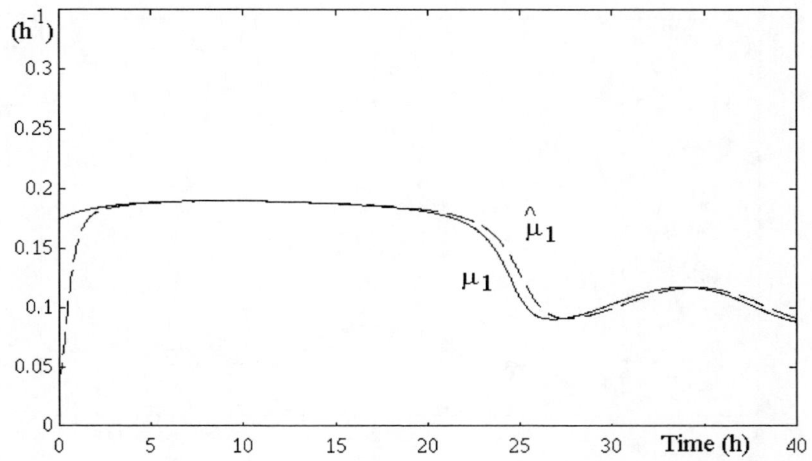

Figure 11. Evolution of μ_1 and $\hat{\mu}_1$ - free noise data.

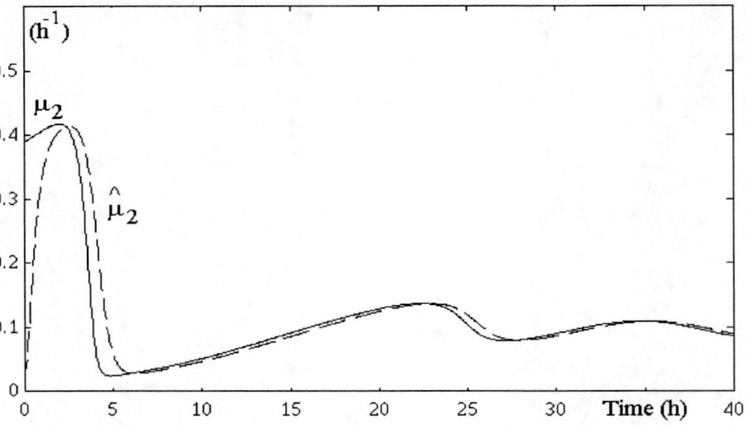

Figure 12. Evolution of μ_2 and $\hat{\mu}_2$ - free noise data.

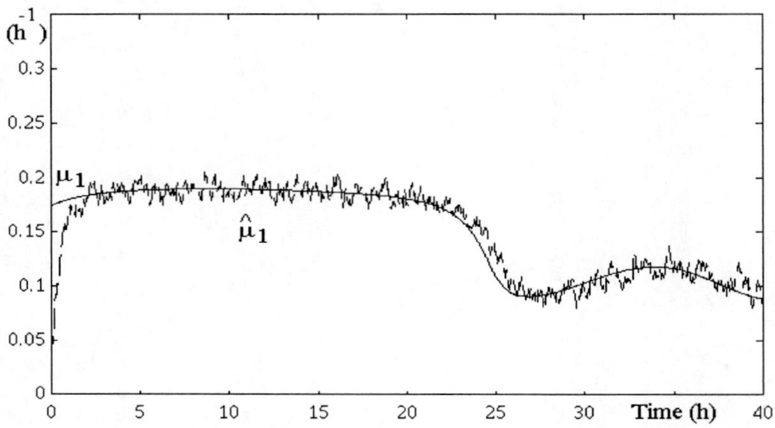

Figure 13. Evolution of μ_1 and $\hat{\mu}_1$ - noisy data.

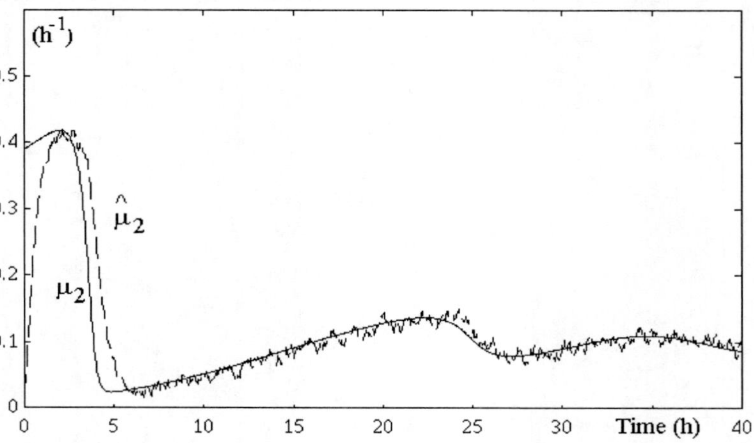

Figure 14. Evolution of μ_2 and $\hat{\mu}_2$ - noisy data.

Notice that the estimation error can be made as small as wished if we choose greater values of θ. The problem for a large value of θ is that the observer becomes noise sensitive. The value of the tuning parameter is therefore a compromise between a good estimation and the noise rejection.

The performances of the second observer (ii) are comparable with the performances obtained for the estimator (i). In the case (ii) we estimate the specific reaction rates α_1 and α_2. The value of the tuning parameter is the same: $\theta = 3$. Figure 15 and Figure 16 show the estimates and the measurements of the specific rates when free-noise data are utilized. Figure 17 and Figure 18 show the results when noisy data are used. We can observe that the effect of noise is important, but only if the measurements have significant variations.

The results obtained for both estimators can be substantially improved if the tuning parameter is higher in value. This relative big value of θ can be used only if the measurements are free-noise. Contrarily it is possible that the estimates of kinetics cannot be utilized. In Figure 19 and Figure 20 the simulation results for the observer (ii) are depicted, for free-noise data and for $\theta = 10$.

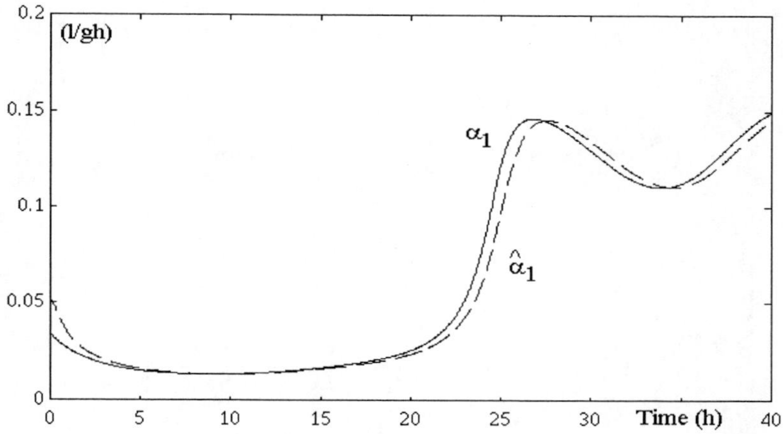

Figure 15. Time profile of α_1 and its estimation.

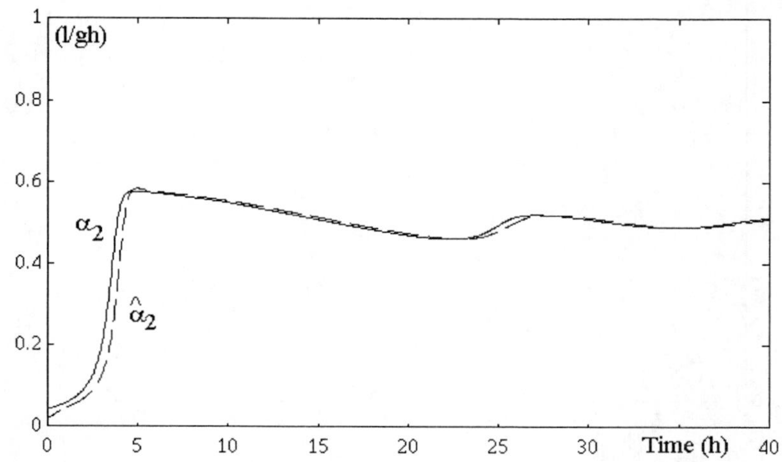

Figure 16. Time profile of α_2 and its estimation.

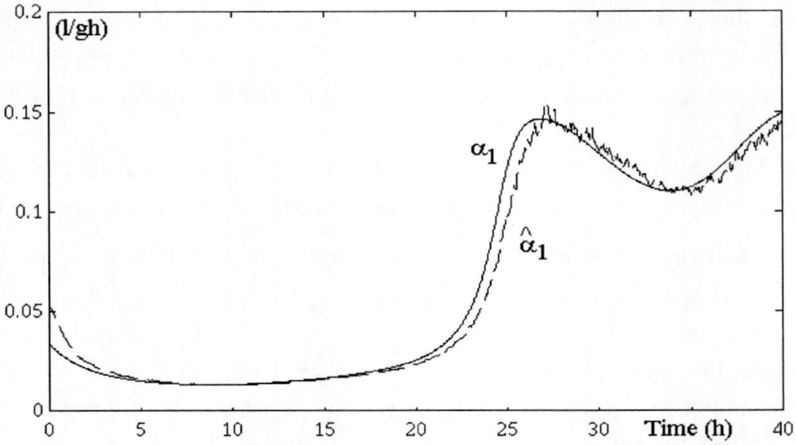

Figure 17. Evolution of α_1 and $\hat{\alpha}_1$ - noisy data.

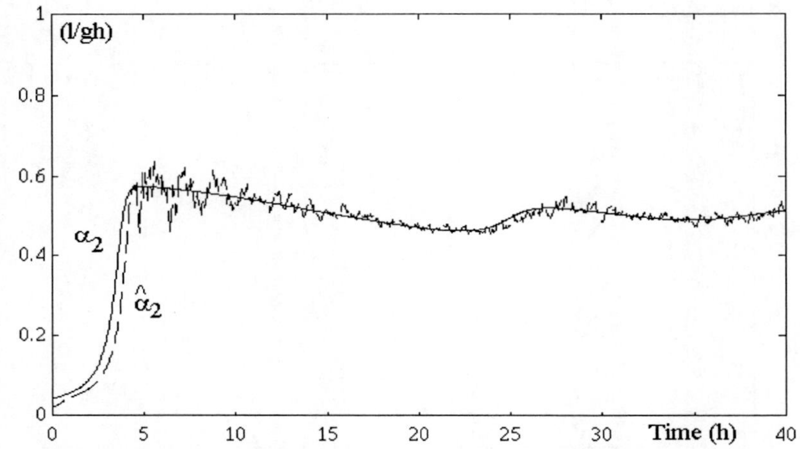

Figure 18. Evolution of α_2 and $\hat{\alpha}_2$ - noisy data.

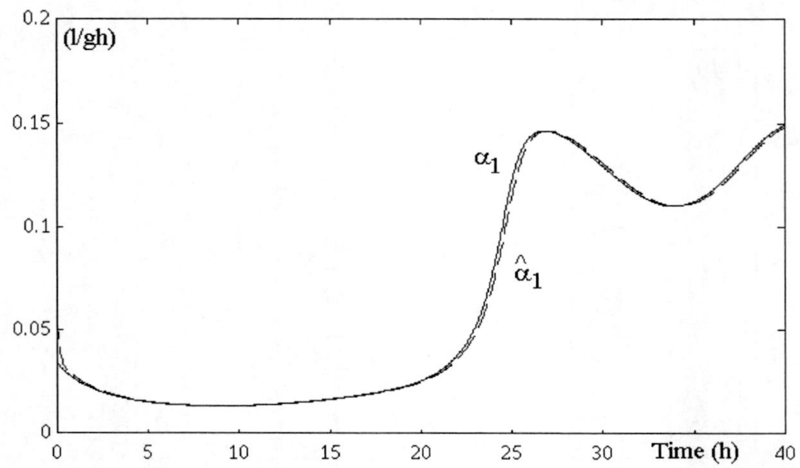

Figure 19. Evolution of α_1 and $\hat{\alpha}_1$ for $\theta = 10$.

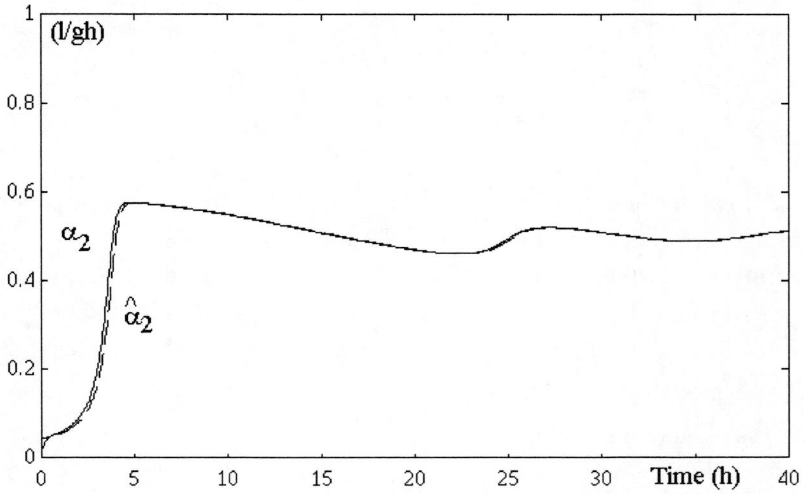

Figure 20. Evolution of α_2 and $\hat{\alpha}_2$ for $\theta = 10$.

3.3. High-Gain Observers for Activated Sludge Processes

The Activated Sludge Bioprocess inside a CSTB

The reduced-order model of the activated sludge bioprocess (39) will be considered. By using the next factorization of yield matrix K_r:

$$K_a = -k_1 \tag{74}$$

we obtain:

$$\xi_a = S;\ \xi_b = \begin{bmatrix} C \\ X \end{bmatrix};\ K_b = \begin{bmatrix} -k_2 \\ 1 \end{bmatrix};\ F_a = D_{in}S_{in};\ F_b = \begin{bmatrix} a_o(C_S - C)W \\ 0 \end{bmatrix};\ D_a = D_1;\ D_b = \begin{bmatrix} D_1 & 0 \\ 0 & D_1 - D_0 \end{bmatrix}$$

From equations (39), (65), and with the above factorization, we can obtain the equations of two high gain observers:

(i) An observer for on-line estimation of the reaction rate $\rho(t) = \varphi(t)$ (with $H(\xi_r) = 1$):

$$\begin{aligned} \frac{d\hat{S}}{dt} &= -k_1 \cdot \hat{\varphi} - D_1\hat{S} + D_{in}S_{in} - 2\cdot\theta\cdot(\hat{S} - S) \\ \frac{d\hat{\varphi}}{dt} &= -\theta^2 \cdot [-k_1]^{-1} \cdot (\hat{S} - S) \end{aligned} \tag{75}$$

(ii) An observer for on-line estimation of the specific growth rate $\rho(t) = \mu(t)$ (with $H(\xi_r) = X$):

$$\begin{aligned}\frac{d\hat{S}}{dt} &= -k_1 \cdot X \cdot \hat{\mu} - D_1\hat{S} + D_{in}S_{in} - 2\cdot\theta\cdot(\hat{S}-S)\\ \frac{d\hat{\mu}}{dt} &= -\theta^2 \cdot \left[-k_1 \cdot X\right]^{-1} \cdot (\hat{S}-S)\end{aligned} \tag{76}$$

Remark 6. It can be seen that the estimator (75) needs only the measurements of *S*, and the estimator (76) needs both measurements of *S* and *X*.

The performances of the nonlinear observers have been tested by performing extensive simulation experiments. The activated sludge process has been simulated by numerical integration of the basic dynamical model equations (39), considering the specific growth rate as a Monod-type model (34) with: $\mu_{max} = 0.2h^{-1}$, $K_S = 75\,mg/l$, $K_C = 2mg/l$, and the following process parameters [32]:

$$k_1 = 1.2, k_2 = 0.565, a_0 = 0.018m^{-3}, C_S = 10\,mg/l, V = 100m^3, V_S = 50\,m^3, F_{in} = 15m^3/h,$$
$$F_W = 0.5m^3/h, F_R = 10m^3/h, W = 100m^3/h.$$

The nonlinear high gain observers (75), (76) were implemented considering that the reaction rate φ for the first observer and the specific growth rate μ for the second observer are unknown. We consider that the influent substrate concentration S_{in} is a disturbance applied to the bioreactor (a combined rectangular and sinusoidal profile).

In the case of first estimator (75), the main goal is to reconstitute the time evolution of φ from the measurements of *S*, and for the second observer (76) the objective is to reconstitute the time evolution of μ using the measurements of *S*, *X* (in fact obtained from simulation). The "true" values of the specific growth rate (34) are used only for the simulation of measured data from the bioprocess. Figure 21 shows the evolutions of state variables, respectively, *S* and *C*; in Figure 22 the time profile of the biomass concentration *X* is depicted. In Figure 23 the estimated parameter $\hat{\rho} = \hat{\varphi}$ and the real reaction rate φ are presented (the observer (i)).

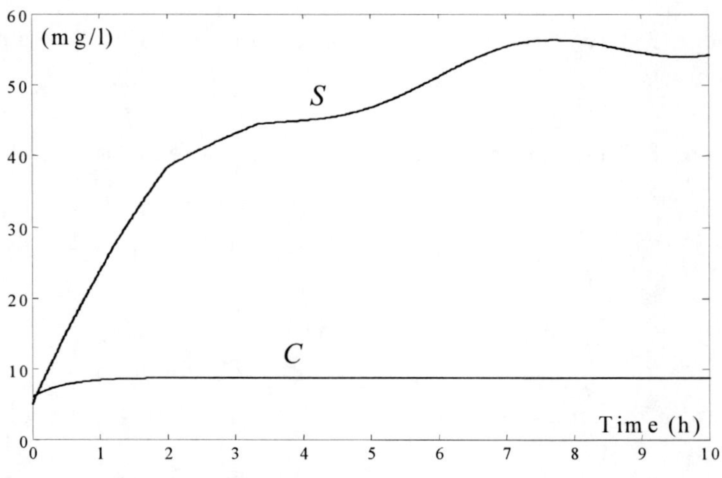

Figure 21. Evolution of substrate and dissolved oxygen concentrations.

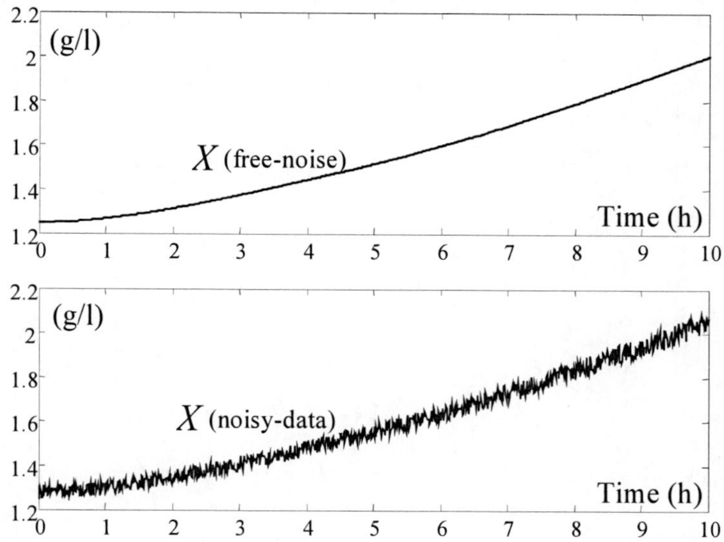

Figure 22. Evolution of the biomass concentration (free-noise and noisy data).

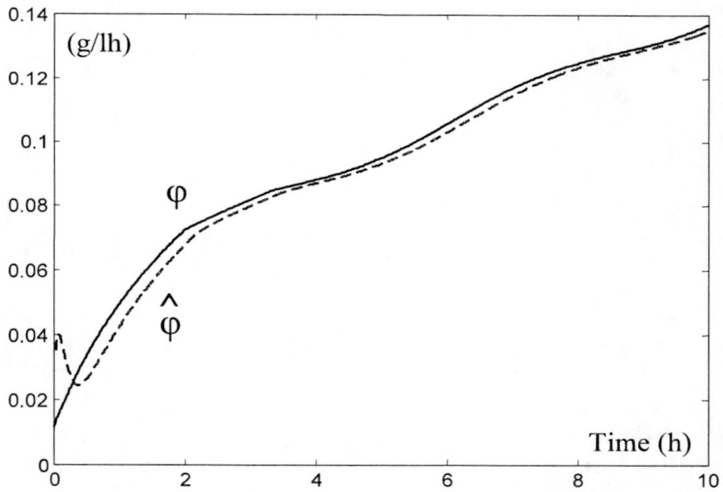

Figure 23. Profiles of reaction rate and its estimate.

Figure 24 depicts the time evolution of the estimated rate $\hat{\rho} = \hat{\mu}$ and the real specific growth rate μ (the observer (ii)). The tuning parameter was set to the value $\theta = 10$.

Finally, in order to test the robustness of the nonlinear observers to noisy measurements, the measurements of X are vitiated by an additive Gaussian noise (see Figure 22). This noise is with zero mean and amplitude equal to 5% of the free noise values. In Figure 25 the specific growth rate and its estimate obtained using noisy data of X are depicted. In all figures, the values obtained from simulation are depicted with solid curves and the estimates with dashed curves. The results obtained for both estimators can be substantially improved if the tuning parameter is chosen higher in value. This relative big value of θ can be used only if the measurements are free-noise. Contrarily is possible that the estimates of kinetics cannot be utilized.

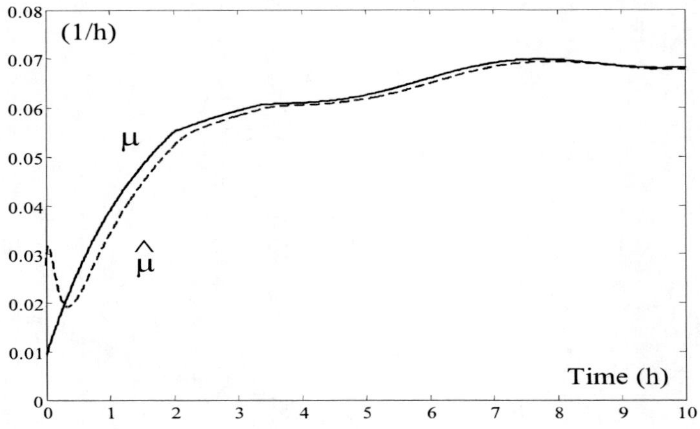

Figure 24. The real and estimated specific growth rate.

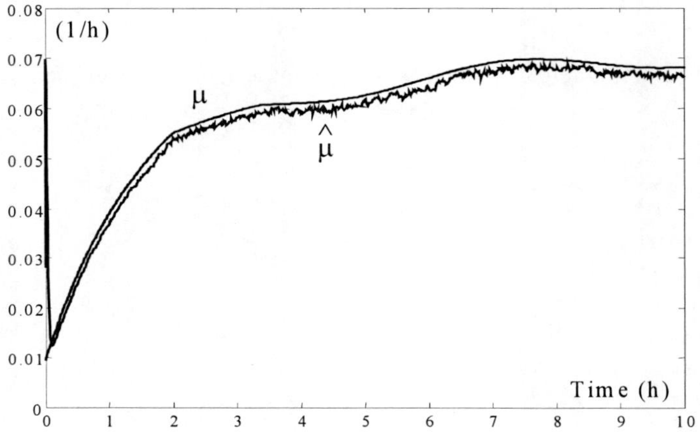

Figure 25. Time evolution of μ , $\hat{\mu}$ – noisy data of *X*.

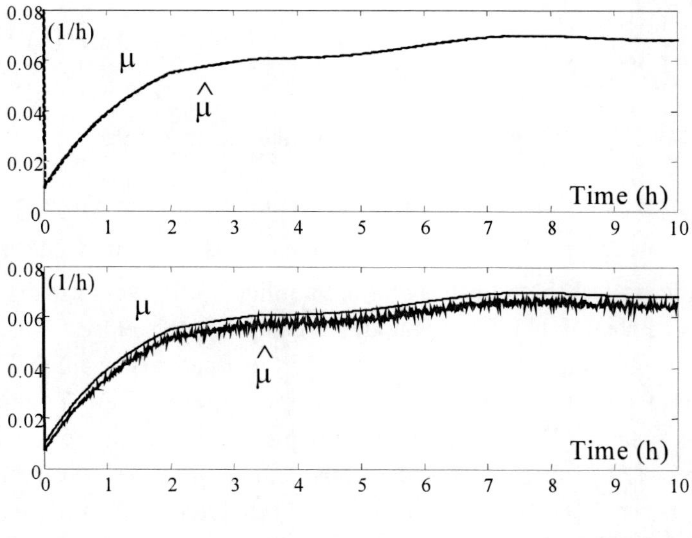

Figure 26. Time evolution of μ , $\hat{\mu}$ for the parameter $\theta = 100$ (free-noise and noisy data of *X*).

The value of the tuning parameter is therefore a compromise between a good estimation and the noise rejection. For instance, if the design parameter was set to $\theta = 100$, the simulation results are plotted in Figure 26 for free-noise measurements (the upper picture) and for noisy measurements of X (the lower picture). Note that the estimation of the specific growth rate for free-noise data is very good, but the presence of the noise corrupts the estimation results (if the noise is more important, it is possible that the estimation algorithm to become divergent).

The Activated Sludge Bioprocess inside a SBR

Concerning the aerobic wastewater treatment process (42), we will suppose that the on-line available states are the dissolved oxygen and ammonia nitrogen concentrations [18]. The model (42) can be rewritten in the form (45), with $\rho(t)$ the unknown kinetics that will be estimated.

In order to design high-gain observers, we will choose the submatrix

$$K_a = \begin{bmatrix} 0 & -k_3 \\ -k_4 & -k_5 \end{bmatrix} \tag{77}$$

of K, and thus we have the state partition $\xi_a = [S_3 \;\; C]^T$, $\xi_b = [X_1 \;\; X_2 \;\; S_1 \;\; S_2]^T$ and:

$$K_b = \begin{bmatrix} 1 & 0 \\ 0 & 1 \\ -k_1 & 0 \\ 0 & k_2 \end{bmatrix}, \; \widetilde{Q}_a = \begin{bmatrix} 0 \\ -K_L a(C_M - C) \end{bmatrix}, \; \widetilde{Q}_b = \begin{bmatrix} 0 \\ 0 \\ 0 \\ 0 \end{bmatrix}.$$

By using this factorization, it is possible to design two high-gain observers for the bioprocess (45):

(i) A high-gain observer for on-line estimation of the reaction rates $\rho(t) = [\rho_1(t) \;\; \rho_2(t)]^T = [\varphi_1(t) \;\; \varphi_2(t)]^T$, with $H(\xi) = \begin{bmatrix} 1 & 0 \\ 0 & 1 \end{bmatrix}$. The high-gain observer is derived from (45) and (66), and has the form:

$$\begin{aligned} \frac{d\hat{S}_3}{dt} &= -k_3\hat{\varphi}_2 + 2\theta(\hat{S}_3 - S_3) \\ \frac{d\hat{C}}{dt} &= -k_4\hat{\varphi}_1 - k_5\hat{\varphi}_2 + K_L a(C_M - \hat{C}) - 2\theta(\hat{C} - C) \\ \frac{d\hat{\varphi}_1}{dt} &= -\theta^2[(k_5 / k_3 k_4)(\hat{S}_3 - S_3) - (1/k_4)(\hat{C} - C)] \\ \frac{d\hat{\varphi}_2}{dt} &= -\theta^2[(-1/k_3)(\hat{S}_3 - S_3)] \end{aligned} \tag{78}$$

(ii) A high-gain observer for on-line estimation of the specific growth rates $\rho(t)=[\rho_1(t)\ \rho_2(t)]^T\ =[\mu_1(t)\ \mu_2(t)]^T$, with $H(\xi)=\begin{bmatrix} X_1 & 0 \\ 0 & X_2 \end{bmatrix}$. Then, the high-gain observer is obtained directly from (45) and (66) as:

$$\begin{aligned}
\frac{d\hat{S}_3}{dt} &= -k_3 X_2 \hat{\mu}_2 + 2\theta(\hat{S}_3 - S_3) \\
\frac{d\hat{C}}{dt} &= -k_4 X_1 \hat{\mu}_1 - k_5 X_2 \hat{\mu}_2 + K_L a(C_M - \hat{C}) - 2\theta(\hat{C} - C) \\
\frac{d\hat{\mu}_1}{dt} &= -\theta^2[(k_5 / k_3 k_4 X_1)(\hat{S}_3 - S_3) - (1/k_4 X_1)(\hat{C} - C)] \\
\frac{d\hat{\mu}_2}{dt} &= -\theta^2[(-1/k_3 X_2)(\hat{S}_3 - S_3)]
\end{aligned} \tag{79}$$

Remark 7. It can be seen that the estimator (i) needs only the measurements of S_3 and C, but the estimator (ii) needs the measurements of S_3, C, X_1 and X_2. If, for example, the concentration of autotroph bacteria X_2 is not on-line measurable, for the implementation of the observer (79), an asymptotic observer can be used. We define an auxiliary variable: $z = S_3 + k_3 X_2$, and we obtain directly from (42) that $\dot{z} = 0$. The asymptotic observer provides the estimate of X_2, i.e. $\hat{X}_2 = (z - S_3)/k_3$, which can be used in (79).

The performances of the nonlinear high-gain observers (78), (79) have been tested by performing extensive simulation experiments. The dynamical model equations (42) were used, considering the specific growth rates (44), and the following process parameters [18]:

$$\begin{aligned}
&\mu_1^* = 1.38h^{-1}, \mu_2^* = 1.02h^{-1}, K_{S1} = 137.6mgCOD/l,\ K_{S3} = 95.6mgN/l, k_1 = 3.38mgCOD/mgVSS, \\
&k_2 = 0.28mgN/mgVSS, k_3 = 0.67mgN/mgVSS, k_4 = 1.74mgO_2/mgN, k_5 = 7.85mgO_2/mgN, \\
&C_M = 7.86mg/l,\ K_L a = 18.75h^{-1}, S_{1\min} = 30.42mgCOD/l.
\end{aligned}$$

where COD stands for Chemical Oxygen Demand.

The nonlinear high gain observers (i) and (ii) were implemented for the bioprocess (42). The simulations are performed considering that the reaction rates φ_1, φ_2 for the first observer and the specific growth rates μ_1, μ_2 for the second observer are unknown.

In the case of first estimator (78), the main goal is to reconstitute the time evolution of φ_1, φ_2 only from the measurements of S_3 and C, and for the second observer (79) the objective is to reconstitute the time evolution of μ_1, μ_2 using the measurements of S_3, C, X_1 and the estimates of X_2 provided by the asymptotic observer. The "true" values of the specific rates (44) are used only for the simulation of measured data from the bioprocess.

Figure 27 illustrates the evolutions of a part of state variables: X_1, S_1, S_3 and C. Figure 28 presents the time evolution of the estimated rates $\hat{\rho}_1 = \hat{\varphi}_1$ and $\hat{\rho}_2 = \hat{\varphi}_2$ compared with the "real" reaction rates φ_1, φ_2. The tuning parameter was set to the value $\theta = 15$.

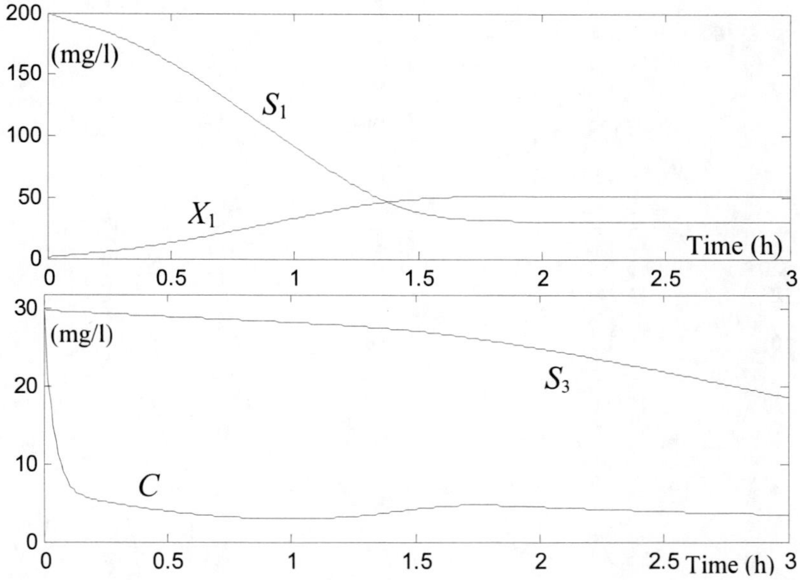

Figure 27. Time evolution of the concentrations X_1 , S_1, S_3 and C.

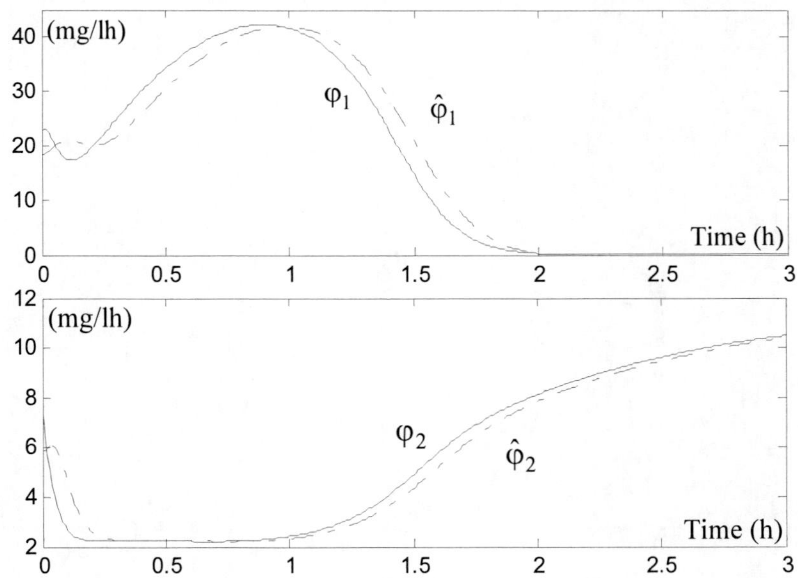

Figure 28. Profiles of the reaction rates and their estimates (free-noise data).

In order to test the robustness of the observer (i) to noisy measurements, the measurements of S_3 are vitiated by an additive Gaussian noise (3%). In Figure 29 the reaction rates and their estimates obtained using noisy data of S_3 are shown.

In Figure 30 the estimated parameters $\hat{\rho}_1 = \hat{\mu}_1$, $\hat{\rho}_2 = \hat{\mu}_2$ and the real specific rates μ_1, μ_2 are presented (obtained with the observer (ii)). The measurements of S_3 used by the observer are vitiated by an additive Gaussian noise (5%).

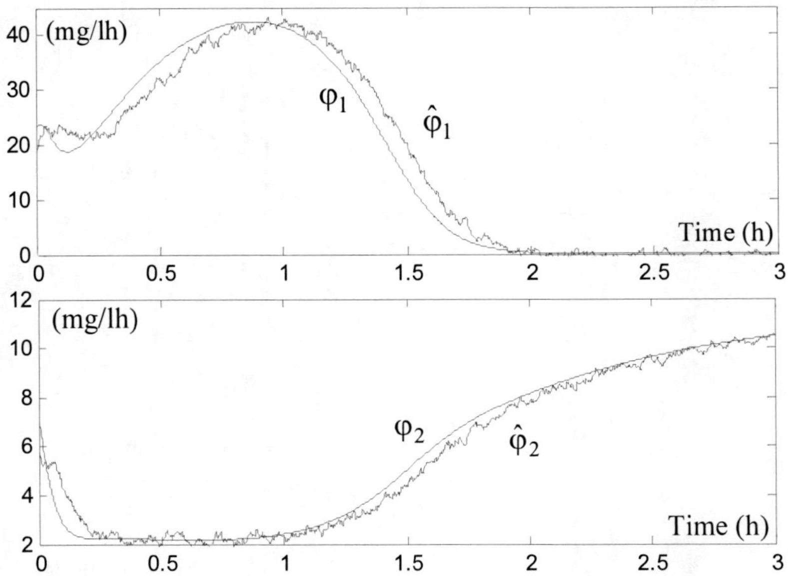

Figure 29. Evolution of the reaction rates and their estimates (noisy data of S_3).

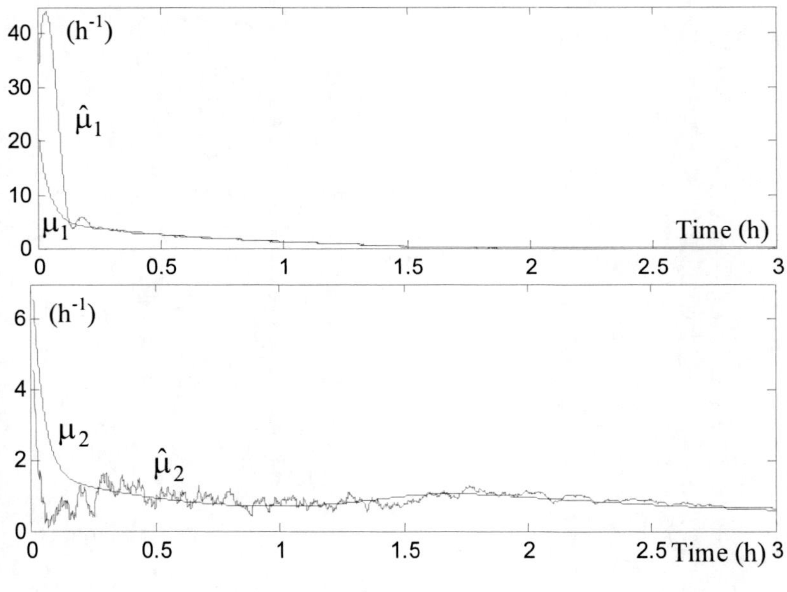

Figure 30. Profiles of specific growth rates and their estimates (noisy data of S_3).

3.4. High-Gain Observers for the Lipase Production Bioprocess

The kinetic parameters which will be estimated are the absorption rate η, the specific growth rate μ (which enters also in the structure of ν_p), and the excretion rate ν_{ex} (see (47)). The state variables that are on-line available are the biomass, the external substrate and external lipase concentrations. Thus, only this information can be used in the observer design.

In order to design the high-gain observer, a factorization of the model (50) is required [15], [24]. Therefore, a partition (ξ_a, ξ_b) of the state vector is considered, with

$\xi_a = [S_1 \; X \; L_{ex}]^T$, $\xi_b = [S_2 \; L_{in}]^T$. Thus, the partitions (Ψ_a, Ψ_b) of the matrix Ψ and $(\widetilde{F}_a, \widetilde{F}_b)$ of the vector $\widetilde{F}$ are obtained:

$$\Psi_a(\xi) = \begin{bmatrix} -X & 0 & 0 \\ 0 & X & 0 \\ 0 & 0 & X \end{bmatrix}, \psi_b(\xi) = \begin{bmatrix} 1 & -(Y+S_2) & 0 \\ 0 & \widetilde{\nu} - L_{in} & -1 \end{bmatrix}, \widetilde{F}_a = [F \;\; 0 \;\; 0]^T, \widetilde{F}_b = [0 \;\; 0]^T \tag{80}$$

The submatrix $\Psi_a(\xi)$ has full rank. Moreover, the necessary hypothesis of the boundedness of $\Psi_a(\xi)$ diagonal elements' away from zero is fulfilled [15], [20].

The detailed equations of the observer can be written from (66) taking into consideration the factorization (80):

$$\begin{aligned} \frac{d}{dt}\begin{bmatrix} \hat{S}_1 \\ \hat{X} \\ \hat{L}_{ex} \end{bmatrix} &= \begin{bmatrix} -\hat{X} & 0 & 0 \\ 0 & \hat{X} & 0 \\ 0 & 0 & \hat{X} \end{bmatrix} \cdot \begin{bmatrix} \hat{\eta} \\ \hat{\mu} \\ \hat{\nu}_{ex} \end{bmatrix} + \begin{bmatrix} F \\ 0 \\ 0 \end{bmatrix} - 2\theta \begin{bmatrix} \hat{S}_1 - S_1 \\ \hat{X} - X \\ \hat{L}_{ex} - L_{ex} \end{bmatrix} \\ \frac{d}{dt}\begin{bmatrix} \hat{\eta} \\ \hat{\mu} \\ \hat{\nu}_{ex} \end{bmatrix} &= -\theta^2 \left[\frac{1}{\hat{X}} \begin{bmatrix} -1 & 0 & 0 \\ 0 & 1 & 0 \\ 0 & 0 & 1 \end{bmatrix} \right] \cdot \begin{bmatrix} \hat{S}_1 - S_1 \\ \hat{X} - X \\ \hat{L}_{ex} - L_{ex} \end{bmatrix} \end{aligned} \tag{81}$$

The partition (Ψ_a, Ψ_b) was chosen such that the estimator (81) needs only the measurements of S_1, X and L_{ex}, which are usually available in the lipase production process [25]. The high-gain observer (81) allows the on-line estimation of the so-called specific reaction rates (the absorption rate, the specific growth rate, and the excretion rate). However, using the structure of estimator (66), if the biomass concentration is considered "embedded" in the unknown kinetics of the process, another observer can be obtained for the unknown reaction rates $\vartheta(t) = [\vartheta_1(t) \; \vartheta_2(t) \; \vartheta_2(t)]^T = [\eta X \; \mu X \; \nu_{ex} X]^T$. This high-gain observer is derived from (66) and can be obtained using (81):

$$\begin{aligned} \frac{d}{dt}\begin{bmatrix} \hat{S}_1 \\ \hat{X} \\ \hat{L}_{ex} \end{bmatrix} &= \begin{bmatrix} -1 & 0 & 0 \\ 0 & 1 & 0 \\ 0 & 0 & 1 \end{bmatrix} \cdot \begin{bmatrix} \hat{\vartheta}_1 \\ \hat{\vartheta}_2 \\ \hat{\vartheta}_3 \end{bmatrix} + \begin{bmatrix} F \\ 0 \\ 0 \end{bmatrix} - 2\theta \begin{bmatrix} \hat{S}_1 - S_1 \\ \hat{X} - X \\ \hat{L}_{ex} - L_{ex} \end{bmatrix} \\ \frac{d}{dt}\begin{bmatrix} \hat{\vartheta}_1 \\ \hat{\vartheta}_2 \\ \hat{\vartheta}_3 \end{bmatrix} &= -\theta^2 \begin{bmatrix} -1 & 0 & 0 \\ 0 & 1 & 0 \\ 0 & 0 & 1 \end{bmatrix} \cdot \begin{bmatrix} \hat{S}_1 - S_1 \\ \hat{X} - X \\ \hat{L}_{ex} - L_{ex} \end{bmatrix} \end{aligned} \tag{82}$$

Both nonlinear high-gain observers are simple and the tuning of the gain can be done by modifying only one design parameter: θ.

Simulation Results

The performance and the behavior of the proposed high-gain observers were tested by performing several simulation experiments. The values of the bioprocess parameters are the following [25], [36]:

$$\eta^* = 0.21\,\mathrm{h}^{-1}; K_{M1} = 0.11\,\mathrm{g/l}; \mu^* = 0.25\,\mathrm{h}^{-1}; K_{M2} = 0.25\,\mathrm{g/l}; \nu_p^* = 123\,\mathrm{u/mg}; b = 0.000216\left(\mathrm{mol/g}\right)\mathrm{h}^{-1};$$

$$a = 0.0185\,\mathrm{mol/g}; K_p = 0.26\,\mathrm{g/g}; K_i = 22.2\,\mathrm{g/g}; \nu_{ex}^* = 4.09\,\mathrm{h}^{-1}; K_{ex} = 19.5\,\mathrm{u/mg};\ Y = 1.16\,\mathrm{g/g}.$$

The bioprocess evolution was examined in open loop. This simulation is considered in order to analyze the lipase production and the performance of high-gain observers when the control law is missing. The following simulation cases are considered:

1) The high-gain observer (81) was implemented and simulated. Figure 31 presents the time evolution of the concentrations (the external substrate and the biomass concentrations in the top panel, and the external lipase concentration in the bottom panel). From this figure it can be seen that after the consumption of external substrate, the growth of external lipase is limited. The simulations considered the following two scenarios: without and with noisy measurements of the state variables. The tuning parameter is set to $\theta = 5$ for free-noise data, and to $\theta = 3$ for noisy data. For the second scenario, in order to test the robustness of the nonlinear observers to noisy measurements, the measurements of S_1, X and L_{ex} are vitiated by an additive Gaussian noise. This noise is with zero mean and amplitude equal to 5% of the free noise values. Figure 32 depicts the time evolution of estimated rate $\hat{\rho}_1 = \hat{\eta}$ and real absorption rate η (top panel – free-noise data; bottom panel – noisy measurements). In Figure 33 the specific growth rate μ and the estimate $\hat{\rho}_2 = \hat{\mu}$ are presented. Figure 34 shows the time profiles of excretion rate ν_{ex} and of estimate $\hat{\rho}_3 = \hat{\nu}_{ex}$.

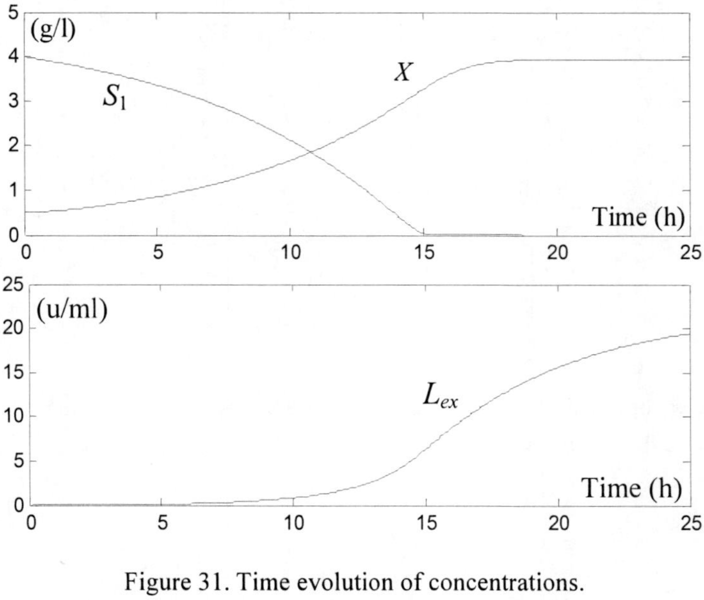

Figure 31. Time evolution of concentrations.

2) The high-gain observer (82) was implemented and simulated, in order to obtain on-line estimations of the unknown reaction rates $\vartheta(t)$. The same scenarios and tuning parameter values as in previous simulation case were used. Figure 35 depicts the time evolution of the estimated rate $\hat{\vartheta}_1$ versus the real profile (again, for all cases: top panel – free-noise data; bottom panel – noisy measurements). Similar results were obtained for the rate $\hat{\vartheta}_2$. Figure 36 shows the estimated rate $\hat{\vartheta}_3$ (free-noise and noisy data).

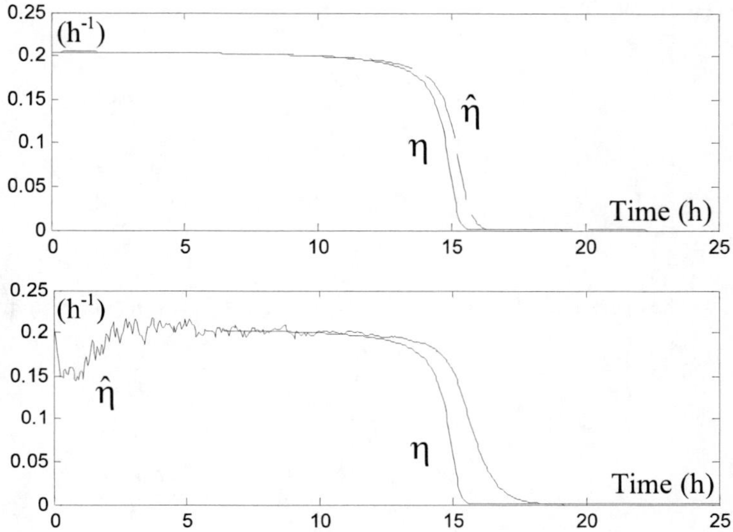

Figure 32. Time profiles of the absorption rate and its estimate (free-noise and noisy measurements).

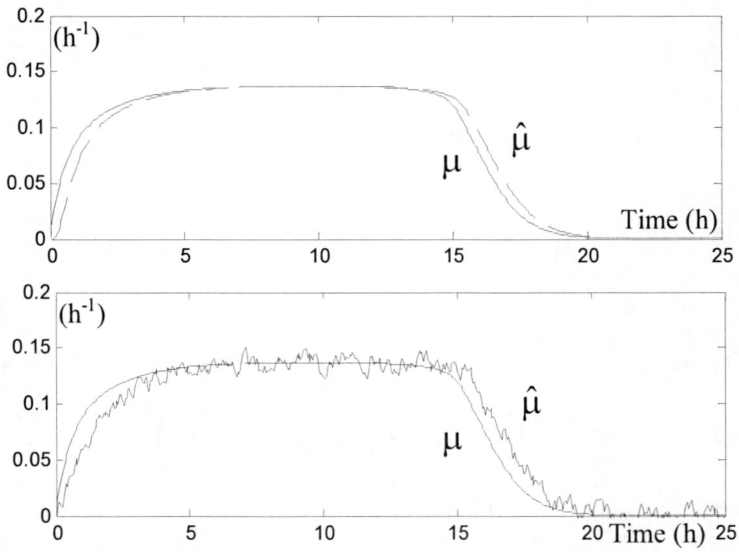

Figure 33. Real and estimated specific growth rates (free-noise and noisy measurements).

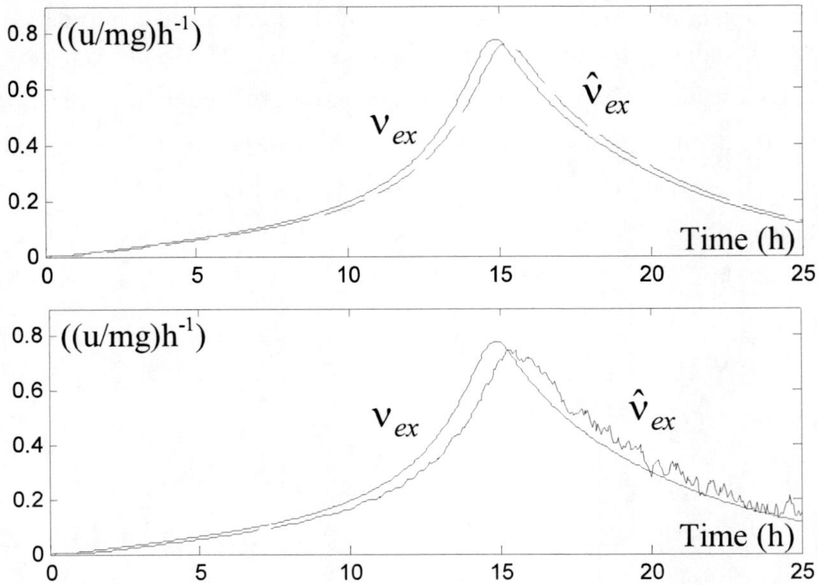

Figure 34. The excretion rate and the estimate (free-noise and noisy measurements).

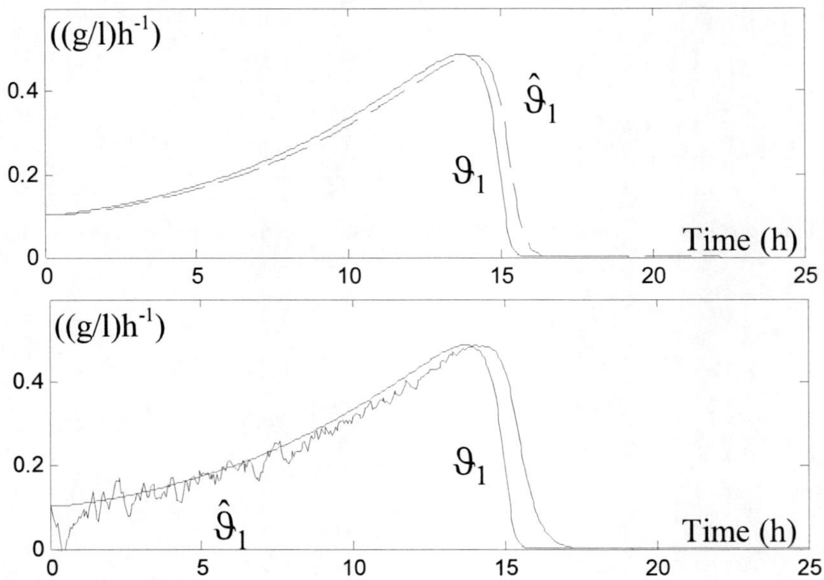

Figure 35. Time profiles of the reaction rates ϑ_1 and $\hat{\vartheta}_1$ (free-noise and noisy measurements).

The obtained results illustrate that both high-gain observers provide accurate estimates of the kinetic rates.

It can be seen that the measurement noise induces some noisy estimates of the kinetics, but the noise effect is limited (this effect can be reduced for lower values of the tuning parameter).

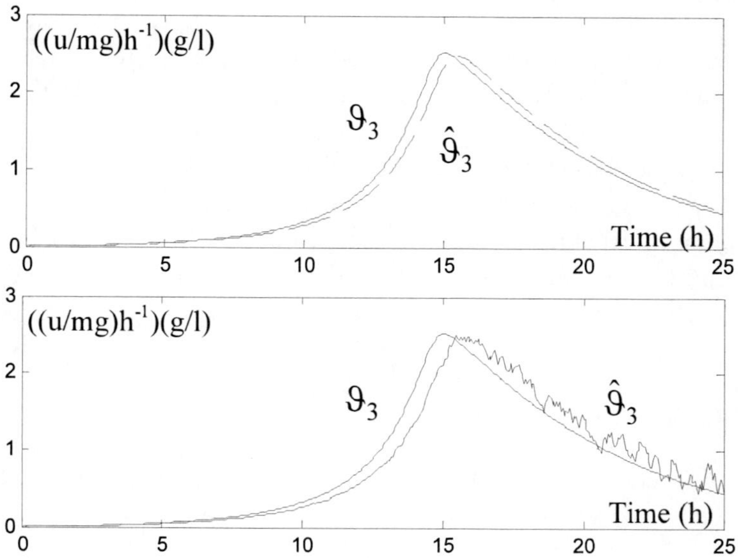

Figure 36. Evolution of reaction rates ϑ_3 and $\hat{\vartheta}_3$ (free-noise and noisy data).

Conclusion

In this chapter, the design of high-gain observers for the unknown kinetics in two types of biochemical reactors: Batch (and Fed-Batch) Bioreactors, and Continuous Stirred Tank Bioreactors was achieved.

The general dynamic nonlinear model of a class of bioprocesses was presented and several particular bioprocesses that take place inside these types of bioreactors were widely analyzed: the wastewater biomethanation process, two activated sludge bioprocesses, and the lipase production process. The models were obtained by using the classical modeling approach, and by using a novel modeling procedure, based on pseudo Bond Graph technique. The models obtained, using the Bond Graph approach, are equivalent to bioprocess' models obtained using classical methods. As a possible further research direction, the use of this technique for bioprocesses will allow the analysis of structural properties, and also the design of observers for control purposes.

To develop and apply advanced control strategies for bioprocesses, it is necessary to obtain useful dynamical models. Therefore, the complexity and uncertainty of the bioprocess models required the use of some reduction techniques and the factorization of models, to emphasize the unknown or imprecisely known kinetics.

In order to overcome problems such as the modeling uncertainties and the lack of on-line measurements, some high-gain observers were designed for the estimation of imprecisely known kinetic rates inside bioreactors. The high-gain observer design does not require any model for the kinetic rates. The calibration of these observers is simple because implies the tuning of a single parameter whatever the number of components and reactions. The proposed observers need measurements of a part of the state variables. Anyhow, if it is no possible to obtain on-line measurements, state estimators are needed, such as asymptotic observers.

The behavior and the performance of the proposed nonlinear observers were tested for four bioprocesses taking place inside Batch and Continuous Stirred Tank Bioreactors, by using realistic numeric simulations, considering some parametric disturbances and noisy measurements. The obtained results can be improved if the tuning parameter is chosen higher in value, but only if the measurements are free-noise. Contrarily, the observers become noise sensitive and it is possible that the estimates of kinetics cannot be utilized.

The results, for all the studied bioprocesses, are quite encouraging from simulation point of view. The proposed high-gain observers can be utilized in further research for the design of some advanced control strategies for bioprocesses.

Acknowledgment

This work was created in the frame of the research project ID 548/2009, which is financed by the National University Research Council, Romania.

References

[1] Dochain, D. (Ed.) (2008). *Automatic control of bioprocesses*. UK: ISTE and John Wiley & Sons.

[2] Schugerl, K. (2001). Progress in monitoring, modelling and control of bioprocesses during the last 20 years. *J. Biotechnology,* vol. 85, no. 2, 149-173.

[3] Selişteanu, D., Petre, E. & Răsvan, V. (2007). Sliding mode and adaptive sliding-mode control of a class of nonlinear bioprocesses. *Int. J. Adaptive Control and Signal Proc.,* vol. 21, no. 8-9, 795-822.

[4] Bastin, G. & Dochain, D. (1990). *On-line estimation and adaptive control of bioreactors.* New York, NY: Elsevier.

[5] Karnopp, D. & Rosenberg, R. (1974). *System dynamics: A unified approach*. New York, NY: John Wiley.

[6] Thoma, J. (1975). *Introduction to Bond Graphs and their applications.* Oxford: Pergamon Press.

[7] Gawthrop, P. & Smith, L. (1996). *Metamodelling: Bond Graphs and dynamic systems.* Hemel Hempstead, UK: Prentice Hall.

[8] Thoma, J. & Ould Bouamama, B. (2000). *Modelling and simulation in thermal and chemical engineering. A Bond Graph Approach*. Berlin, DE: Springer-Verlag.

[9] Păstrăvanu, O. & Ibănescu, R. (2001). *Bond-graph language in modeling and simulation of physical-technical systems* (in Romanian). Iasi, RO: Gh. Asachi.

[10] Heny, C., Simanca, D. & Delgado, M. (2000). Pseudo-bond graph model and simulation of a continuous stirred tank reactor. *J. Franklin Institute*, vol. 337, 21-42.

[11] Dauphin-Tanguy, G. (Ed.) (2000). *Les Bond Graphs*. Paris: Hermes Sci.

[12] Roman, M. & Bobaşu, E. (2006). Structural properties of Bond Graph models. *Trans. of the VŠB – Technical University of Ostrava, Mechanical Series,* vol. 2, 137-143.

[13] Schnakenberg, J. (1981). *Thermodynamics network analysis of biological systems.* 2nd Edition. Berlin: Universitext.

[14] Linkens, D. A. (1990). Bond Graphs for an improved modelling environment in the lifesciences. In: *Proc. of IEE Colloquium on Bond Graphs in Control*, 3/1-3/4. London, UK.

[15] Farza, M., Busawon, K. & Hammouri, H. (1998). Simple nonlinear observers for on-line estimation of kinetic rates in bioreactors. *Automatica,* vol. 34, no. 3, 301-318.

[16] Selişteanu, D., Petre, E. & Ionete, C. (1997). High gain observers for estimation of kinetics in a bioreactor. *Annals Univ. Craiova, Electr. Eng. Series,* no. 21, 378 - 388.

[17] Dochain, D. & Vanrolleghem, P. (2001). *Dynamical modelling and estimation in wastewater treatment processes*. London, UK: IWA Publishing.

[18] Fibrianto, H., Mazouni, D., Ignatova, M., Herveau, M., Harmand, J. & Dochain, D. (2008). Dynamical modelling, identification and software sensors for SBRs. *Math. & Computer Modelling of Dynamical Systems,* vol. 14, no. 1, 17-26.

[19] Buitrón, G., *et al.* (2008). Experiments for modelling the biodegradation of wastewater in sequencing batch reactors. *Math. & Computer Modelling of Dynamical Systems, vol. 14, no. 1,* 3-15.

[20] Gauthier, J. P., Hammouri, H. & Othman, S. (1992). A simple observer for nonlinear systems. Applications to bioreactors. *IEEE Trans. in Aut. Control,* vol. 37, no. 6, 875-880.

[21] Khalil, H. K. (2008). High-gain observers in nonlinear feedback control. In: *Proc. of Int. Conf. on Control, Autom. & Systems ICCAS 2008*, xlvii – lvii, Seoul, Korea.

[22] Selişteanu, D. & Petre, E. (2004). On adaptive control of a wastewater biodegradation bioprocess. *Control Engineering and Applied Informatics,* vol. 6, no. 3, 48-56.

[23] Selişteanu, D., Petre, E. & Şendrescu, D. (2004). Modelling and identification of wastewater biodegradation process. In: *Proc. of the 12th Int. Symposium SIMSIS 12*, 149-154, Galaţi, Romania.

[24] Selişteanu, D., Petre, E., Bobaşu, E. & Popescu, D. (2005). Modelling and on-line estimation of unknown kinetics for an activated sludge bioprocess. *WSEAS Trans. on Systems,* issue 10, vol. 4, 1650-1657.

[25] Selişteanu, D., Petre, E., Marin, C. & Şendrescu, D. (2008). Estimation and adaptive control of a fed-batch bioprocess. In: *Proc. of Int. Conf. on Control, Autom. & Systems ICCAS 2008*, 1349-1354, Seoul, Korea.

[26] Roman, M., Selişteanu, D., Bobaşu, E., Petre, E. & Şendrescu, D. (2009). Bond Graph modelling of a wastewater biodegradation bioprocess. In: *Proc. of IEEE Int. Conf. on Automation and Logistics ICAL 2009*, 1501-1506, Shenyang, China.

[27] Selişteanu, D., Petre, E., Şendrescu, D., Roman, M. & Popescu, D. (2009). High-gain observers for estimation of kinetics in biological sequencing batch reactors. In: *Proc. of the IEEE Int. Conf. on Automation and Logistics ICAL 2009*, 1623-1626, Shenyang, China.

[28] Selişteanu, D., Petre, E., Marin, C. & Şendrescu, D. (2009). High-gain observers for estimation of kinetics in a nonlinear bioprocess. In: *Proc. of the Int. Joint Conference ICCAS-SICE 2009*, 5236-5241, Fukuoka, Japan.

[29] Roman, M., Bobaşu, E., Petre, E., Selişteanu, D. & Şendrescu, D. (2009). Pseudo Bond Graph modelling of some prototype bioprocesses. *Control Engineering and Applied Informatics,* vol. 11, no. 2, 22-29.

[30] Gauthier, J. P., Kupka, I. A. K. (1994). Observability and observers for nonlinear systems, *SIAM J. Contr. and Optim.,* vol. 32, no. 4, 975-994.

[31] Kokotovic, P. V., Khalil, H. K. & O'Reilly, J. (1986). *Singular perturbation methods in control: Analysis and design*. London, UK: Academic Press, Inc.

[32] Dochain, D. & Perrier, M. (1992). Adaptive linearizing control of activated sludge processes. In: *Proc. Control Systems'92*, 211-215, Whistler, Canada.

[33] Betancur, M. J., Mazouni, D., Moreno, J. A. & Harmand, J. (2008). Optimal control of biological SBRs for the treatment of dairy and toxic wastewaters. *Math. & Computer Modelling of Dynamical Systems,* vol. 14, no. 1, 39-52.

[34] Petre, E. (2004). Adaptive control of a recycled depollution bioprocess. In: *Proc. of the 12th Int. Symposium SIMSIS 12*, 127-132, Galați, Romania.

[35] Petre, E. & Selişteanu, D. (2005). *Modelling and identification of depollution bioprocesses* (in Romanian). Craiova, RO: Universitaria.

[36] Charbonnier, S. & Cheruy, A. (1996). Estimation and control strategies for a lipase production process. *Control Eng. Practice,* vol. 4, no. 11, 1521-1534.

In: Mathematical Chemistry
Editor: W.I. Hong, pp. 417-450

ISBN: 978-1-60876-894-3

Chapter 6

REMARKS ON NOVEL EXACT SOLUTION METHODOLOGIES IN WAVEFUNCTION ANALYSIS

Valentino A. Simpao*

Mathematical Consultant Services, Greenville, KY, US

A viable methodology for the exact analytical solution of the multiparticle Schrodinger and Dirac equations has long been considered a holy grail of theoretical chemistry. Since a benchmark work by Torres-Vega and Frederick in the1990's[1], the Quantum Phase Space Representation (QPSR) has been explored as an alternate method for solving various physical systems, including the harmonic oscillator[2], Morse oscillator[3], one-dimensional hydrogen atom[4], and classical Liouville dynamics under the Wigner function[5]. QPSR approaches are particularly challenging because of the complexity of phase space wave functions and the fact that the number of coordinates doubles in the phase space representation.These challenges have heretofore prevented the exact solution of the multiparticle equation in phase space. Recently, Simpao[6] has developed an exact analytical symbolic solution scheme for broad classes of differential equations utilizing the Heaviside Operational Ansatz (HOA). It is proposed to apply this novel methodology to QPSR problems to obtain exact solutions for real chemical systems and their dynamics. In his preliminary work, Simpao[6] has already applied this method to a number of simple systems, including the harmonic oscillator, with solutions in agreement to those obtained by Li [refs.2,3,4,5]. He has also demonstrated the exact solution to the radial Schrodinger Equation for an N-particle system with pairwise Coulomb interaction [7]. In addition to the Schrodinger Equation, the HOA method is capable of treating the Dirac equation[8] as well as differential systems governing both relativistic and non-relativistic particle dynamics. Applying these methods would allow us to pursue further exploration of this methodology, starting with the exact solution of multielectron atoms and moving toward complex molecules and reaction dynamics. It is believed that the coupling of HOA with QPSR represents not only a fundamental breakthrough in theoretical physical chemistry, but it is promising as a basis for exact solution algorithms that would have

* E-mail address: mcs007@muhlon.com, Tel: 1-270-977-0575. (Corresponding author)

tremendous impact on the capabilities of computational chemistry/physics. As the theoretical foundation for spectroscopy is the Schrodinger equation, the significance of this discovery to the enhanced analysis of spectroscopic data is obvious. For example, the analysis of the Compton line in momentum spectroscopy necessitates the consideration of the momentum wavefunction for the molecular system under study. The novel methods refs. [6,7,8] allow the exact determination of the momentum[and configuration] space wavefunction from the QPSR wavefunction by way of a Fourier Transform. For example, the primary focus of [7] is the pairwise 1/rij interaction in context of the radial equation in the nonrelativistic Schrodinger case. This application of the exact solution ansatz developed above corresponds to the problem of n-particles with pairwise Coulomb interaction;scaling the parameters and variables of the problem yields the exact solution of the QPSR Schrodinger equation for the first-principles general polyatomic molecular Hamiltonian. Upon a straightforward slight adaptation of this non-relativistic Schrodinger result, the QPSR Dirac equation addressed in [8] immediately yields the relativistic counterpart for the first –principles general polyatomic molecular Hamiltonian solution. These results form the cornerstone of the exact solution to the quantum dynamics of particular chemical systems, which shall appear elsewhere. Although all of the examples of dynamical systems mentioned above are at once solved as a special case of the general integral method already published in HOA, it is illuminating for applications to write out the solutions as the integrals are evaluated explicitly for the same. The HOA result is currently being used as the primary algorithm in the development of computer programs known as 'solver engines' for quantum chemistry/physics and plethora applications: to be reported elsewhere.

In this note some remarks, examples and further directions, concerning HOA as a tool to solve and provide analytical insight into solutions of dynamical systems occurring in, but not limited to Mathematical Chemistry, are posited. Among these more general considerations are the developement of Novel Exact Analytical solutions of Generalized Hamiltonian/Lagrangian Dynamical systems in the Quantum Phase Space Representation and Classical Connections; also the exact analytical solutions of attendant differential/difference equations. Though detailed treatment of these and many others shall appear elsewhere, the foundations of the HOA scheme will be sketched and some examples given in these Remarks, that will illustrate some of the numerous and sundry analyses that can be enhanced by HOA methods. For convenience, we begin with a recap of the HOA construction following from [ref. 6].

Recap of HOA

Here are the basic relations; x, p, t are respectively the configuration space position, momentum and time variables. The ^ denotes the operators, with H and Ψ denoting the Hamiltonian and wavefunction of the phase space representation[1], respectively. Also, the α, γ, are otherwise free parameters as specified in reference [1].

$$
\begin{aligned}
&H(x,p,t)\rightarrow \hat{H}(\hat{x},\hat{p},t)=\hat{H}\left(i\hbar\partial_p+\alpha x,-i\hbar\partial_x+\gamma p,t\right)\ni\alpha+\gamma=1\\
&x\rightarrow\hat{x}\equiv i\hbar\partial_p+\alpha x,p\rightarrow\hat{p}\equiv -i\hbar\partial_x+\gamma p,t\rightarrow t=t\\
&(x_1,....,x_n)\rightarrow(\hat{x}_1,....,\hat{x}_n)=\left(i\hbar\partial_{p_1}+\alpha_1x_1,....,i\hbar\partial_{p_n}+\alpha_nx_n\right)\ni\alpha_j+\gamma_j=1,j=1,....,n\\
&(p_1,....,p_n)\rightarrow(\hat{p}_1,....,\hat{p}_n)=\left(-i\hbar\partial_{x_1}+\gamma_1p_1,....,-i\hbar\partial_{x_n}+\gamma_np_n\right)\\
&H(x_1,....,x_n;p_1,....,p_n;t)\rightarrow\hat{H}(\hat{x}_1,....,\hat{x}_n;\hat{p}_1,....,\hat{p}_n;t)\\
&\equiv\hat{H}\left(i\hbar\partial_{p_1}+\alpha_1x_1,....,i\hbar\partial_{p_n}+\alpha_nx_n;-i\hbar\partial_{x_1}+\gamma_1p_1,....,-i\hbar\partial_{x_n}+\gamma_np_n;t\right)
\end{aligned}
\tag{1}
$$

Now with the following properties of Heaviside operational methods via Laplace transforms refs. [6]...

$$
\begin{aligned}
&L_{y\rightarrow z}[f(y)]=\int_{y_0}^{\infty}f(y)e^{-yz}dy=\breve{f}(z)\\
&L_{z\rightarrow y}^{-1}\left[\breve{f}(z)\right]=\frac{1}{2\pi i}\oint_{\partial}\breve{f}(z)e^{yz}dz=f(y)\\
&L_{z\rightarrow y}^{-1}\left[\breve{f}(z)\right]=\frac{1}{2\pi i}\oint_{\partial}\breve{f}(z)e^{yz}dz=f(y)=\breve{f}(D_y)U(y)\\
&L_{z\rightarrow y}^{-1}\left[\breve{f}_1(z)\breve{f}_2(z)\right]=f_1(y)*f_2(y)=\int_{y_0}^{y}f_1(y-u)f_2(u)du=\breve{f}_1(D_y)\breve{f}_2(D_y)U(y)=\breve{f}_1(D_y)f_2(y)
\end{aligned}
$$

where $U(y)$ is the Heaviside Unit Step function (2)

$$
\begin{aligned}
&L_{(y_1,....,y_n)\rightarrow(z_1,....,z_n)}[f(y_1,....,y_n)]=\underbrace{\int_{y_{0n}}^{\infty}...\int_{y_{01}}^{\infty}}_{n}f(y_1,....,y_n)e^{-\sum_{j=1}^{n}y_jz_j}dy_1....dy_n=\breve{f}(z_1,....,z_n)\\
&L_{(z_1,....,z_n)\rightarrow(y_1,....,y_n)}^{-1}\left[\breve{f}(z_1,....,z_n)\right]=\left(\frac{1}{2\pi i}\right)^n\oint_{\partial^n}\breve{f}(z_1,....,z_n)e^{\sum_{j=1}^{n}y_jz_j}dz_1....dz_n=f(y_1,....,y_n)\\
&L_{(z_1,....,z_n)\rightarrow(y_1,....,y_n)}^{-1}\left[\breve{f}_1(z_1,....,z_n)\breve{f}_2(z_1,....,z_n)\right]=f_1(y_1,....,y_n)\underbrace{*}_{(y_1,....,y_n)}f_2(y_1,....,y_n)\\
&=\underbrace{\int_{y_{0n}}^{y_n}...\int_{y_{01}}^{y_1}}_{n}f_1(y_1-y_1',....,y_n-y_n')f_2(y_1',....,y_n')dy_1'....dy_n'\\
&=\breve{f}_1(\partial_{y_1},....,\partial_{y_n})\breve{f}_2(\partial_{y_1},....,\partial_{y_n})U(y_1,....,y_n)=\breve{f}_1(\partial_{y_1},....,\partial_{y_n})f_2(y_1,....,y_n)
\end{aligned}
$$

where the zero-subscripted variables (e.g., y_0) are the arbitrarily specified lower limits of integration.

With the phase-space convolution

$$
\begin{aligned}
&f_1(x_1,....,x_n;p_1,....,p_n)\underbrace{*}_{(x_1,....,x_n;p_1,....,p_n)}f_2(x_1,....,x_n;p_1,....,p_n)\\
&=\underbrace{\int_{x_{0n}}^{x_n}...\int_{x_{01}}^{x_1}}_{n}\underbrace{\int_{p_{0n}}^{p_n}...\int_{p_{01}}^{p_1}}_{n}f_1(x_1-x_1',..,x_n-x_n';p_1-p_1',..,p_n-p_n')f_2(x_1',..,x_n';p_1',..,p_n')dx_1'..dx_n'dp_1'..dp_n'
\end{aligned}
\tag{3}
$$

Lower bounds of respective phase space co-ordinates: $(x_{10},....,x_{n0};p_{10},....,p_{n0})$

Also the transform relation

$$L_{z\to y}\left[\breve{f}(az-b)\right]=\frac{1}{a}e^{\frac{by}{a}}f\left(\frac{y}{a}\right)$$

$$L^{-1}_{(z_1,\ldots,z_n)\to(y_1,\ldots,y_n)}\left[\breve{f}(a_1z_1-b_1,\ldots,a_nz_n-b_n)\right]=\prod_{j=1}^{n}\frac{1}{a_j}e^{\frac{b_j}{a_j}y_j}f\left(\frac{y_1}{a_1},\ldots,\frac{y_n}{a_n}\right) \tag{4}$$

From (1) the wave equation becomes

$$\hat{H}\begin{pmatrix} i\hbar\partial_{p_1}+\alpha_1x_1,\ldots,i\hbar\partial_{p_n}+\alpha_nx_n; \\ -i\hbar\partial_{x_1}+\gamma_1p_1,\ldots,-i\hbar\partial_{x_n}+\gamma_np_n;t\end{pmatrix}\Psi(x_1,\ldots,x_n;p_1,\ldots,p_1;t) \\ =i\hbar\partial_t\Psi(x_1,\ldots,x_n;p_1,\ldots,p_1;t) \tag{5}$$

Applying the convolution identity and multivariable inverse transform of (2), the phase space convolution of (3) and relation (4) yields

$$\left[L^{-1}_{\left(\substack{(\partial_{x_1},\ldots,\partial_{x_n})\\ \to(x_1,\ldots,x_n)}\right)}\left[L^{-1}_{\left(\substack{(\partial_{p_1},\ldots,\partial_{p_n})\\ \to(p_1,\ldots,p_n)}\right)}\left[\hat{H}\begin{pmatrix} i\hbar\partial_{p_1}+\alpha_1x_1,\ldots,i\hbar\partial_{p_n}+\alpha_nx_n; \\ -i\hbar\partial_{x_1}+\gamma_1p_1,\ldots,-i\hbar\partial_{x_n}+\gamma_np_n;t\end{pmatrix}\right]\right] \underbrace{*}_{(x_1,\ldots,x_n;p_1,\ldots,p_n)} \Psi(x_1,\ldots,x_n;p_1,\ldots,p_n;t)\right]$$
$$\equiv \tag{6}$$
$$\hat{H}\begin{pmatrix} i\hbar\partial_{p_1}+\alpha_1x_1,\ldots,i\hbar\partial_{p_n}+\alpha_nx_n; \\ -i\hbar\partial_{x_1}+\gamma_1p_1,\ldots,-i\hbar\partial_{x_n}+\gamma_np_n;t\end{pmatrix}\Psi(x_1,\ldots,x_n;p_1,\ldots,p_n;t)$$

Applying (6) to (5) with the convolution identities in (2) and transforming

$$L_{\left(\substack{(p_1,\ldots,p_n)\\ \to(\bar{p}_1,\ldots,\bar{p}_n)}\right)}\left[L_{\left(\substack{(x_1,\ldots,x_n)\\ \to(\bar{x}_1,\ldots,\bar{x}_n)}\right)}\left[\begin{matrix}\hat{H}\begin{pmatrix} i\hbar\partial_{p_1}+\alpha_1x_1,\ldots,i\hbar\partial_{p_n}+\alpha_nx_n; \\ -i\hbar\partial_{x_1}+\gamma_1p_1,\ldots,-i\hbar\partial_{x_n}+\gamma_np_n;t\end{pmatrix}\Psi(x_1,\ldots,x_n;p_1,\ldots,p_n;t)\\ =i\hbar\partial_t\Psi(x_1,\ldots,x_n;p_1,\ldots,p_1;t)\end{matrix}\right]\right]$$
$$\equiv$$
$$L_{\left(\substack{(p_1,\ldots,p_n)\\ \to(\bar{p}_1,\ldots,\bar{p}_n)}\right)}\left[L_{\left(\substack{(x_1,\ldots,x_n)\\ \to(\bar{x}_1,\ldots,\bar{x}_n)}\right)}\left[\begin{matrix}L^{-1}_{\left(\substack{(\partial_{x_1},\ldots,\partial_{x_n})\\ \to(x_1,\ldots,x_n)}\right)}\left[L^{-1}_{\left(\substack{(\partial_{p_1},\ldots,\partial_{p_n})\\ \to(p_1,\ldots,p_n)}\right)}\left[\hat{H}\begin{pmatrix} i\hbar\partial_{p_1}+\alpha_1x_1,\ldots,i\hbar\partial_{p_n}+\alpha_nx_n; \\ -i\hbar\partial_{x_1}+\gamma_1p_1,\ldots,-i\hbar\partial_{x_n}+\gamma_np_n;t\end{pmatrix}\right]\right]\\ \underbrace{*}_{(x_1,\ldots,x_n;p_1,\ldots,p_n)}\Psi(x_1,\ldots,x_n;p_1,\ldots,p_n;t)\\ =i\hbar\partial_t\Psi(x_1,\ldots,x_n;p_1,\ldots,p_n;t)\end{matrix}\right]\right] \tag{7}$$

$$\equiv$$

$$\hat{H}\begin{pmatrix} i\hbar\partial_{p_1}+\alpha_1 x_1,....,i\hbar\partial_{p_n}+\alpha_n x_n; \\ -i\hbar\partial_{x_1}+\gamma_1 p_1,....,-i\hbar\partial_{x_n}+\gamma_n p_n;t \end{pmatrix}_{\begin{pmatrix} \partial_{p_1}\mapsto\overline{p}_1, \\ ....,\partial_{p_n}\mapsto\overline{p}_n; \\ \partial_{x_1}\mapsto\overline{x}_1, \\ ....,\partial_{x_n}\mapsto\overline{x}_n \end{pmatrix}} \breve{\Psi}\left(\overline{x}_1,....,\overline{x}_n;\overline{p}_1,....,\overline{p}_n;t\right)=i\hbar\partial_t\breve{\Psi}\left(\overline{x}_1,....,\overline{x}_n;\overline{p}_1,....,\overline{p}_n;t\right)$$

$$\equiv$$

$$\hat{H}\begin{pmatrix} i\hbar\overline{p}_1+\alpha_1 x_1,....,i\hbar\overline{p}_n+\alpha_n x_n; \\ -i\hbar\overline{x}_1+\gamma_1 p_1,....,-i\hbar\overline{x}_n+\gamma_n p_n;t \end{pmatrix}\breve{\Psi}\left(\overline{x}_1,....,\overline{x}_n;\overline{p}_1,....,\overline{p}_n;t\right)=i\hbar\partial_t\breve{\Psi}\left(\overline{x}_1,....,\overline{x}_n;\overline{p}_1,....,\overline{p}_n;t\right)$$

Hence the wavefunction in phase space may be analytically expressed in exact quadratures, by inverse transforming the above solution $\breve{\Psi}(\overline{x}_1,....,\overline{x}_n;\overline{p}_1,....,\overline{p}_n;t)$ of (7) as

$$\hat{H}\begin{pmatrix} i\hbar\overline{p}_1+\alpha_1 x_1,....,i\hbar\overline{p}_n+\alpha_n x_n; \\ -i\hbar\overline{x}_1+\gamma_1 p_1,....,-i\hbar\overline{x}_n+\gamma_n p_n;t \end{pmatrix}\breve{\Psi}(\overline{x}_1,....,\overline{x}_n;\overline{p}_1,....,\overline{p}_n;t)=i\hbar\partial_t\breve{\Psi}(\overline{x}_1,....,\overline{x}_n;\overline{p}_1,....,\overline{p}_n;t)$$

$$\Psi(x_1,....,x_n;p_1,....,p_n;t)=L^{-1}_{\begin{pmatrix}(\overline{x}_1,....,\overline{x}_n)\\ \to(x_1,....,x_n)\end{pmatrix}}\left[L^{-1}_{\begin{pmatrix}(\overline{p}_1,....,\overline{p}_n)\\ \to(p_1,....,p_n)\end{pmatrix}}\left[\begin{matrix} e^{\frac{-i}{\hbar}\int_0^t \hat{H}\begin{pmatrix} i\hbar\overline{p}_1+\alpha_1 x_1,....,i\hbar\overline{p}_n+\alpha_n x_n; \\ -i\hbar\overline{x}_1+\gamma_1 p_1,....,-i\hbar\overline{x}_n+\gamma_n p_n;u \end{pmatrix}du} \\ \times \\ \breve{\Psi}_0(\overline{x}_1,....,\overline{x}_n;\overline{p}_1,....,\overline{p}_n;t=0) \end{matrix}\right]\right] \tag{8}$$

By definition, solution in one representation implies simultaneous solution in all representations. Along these lines, an interesting consequence arises: By equations (4.18) and (4.19) of reference [1], it is shown that the Fourier projection onto configuration x space of the phase space wavefunction, yields the configuration space wavefunction in the standard Schroedinger configuration space representation. Symbolically

$$\hat{H}_{\text{configuration space}}(x,-i\hbar\partial_x,t)\Psi_{\text{configuration space}}(x,t)=i\hbar\partial_t\Psi_{\text{configuration space}}(x,t)$$

$$\Psi_{\text{configuration space}}(x,t)=\int_{-\infty}^{\infty}\frac{e^{\frac{ixp}{2\hbar}}}{\sqrt{4\pi\hbar}}\Psi(x,p,t)dp$$

$$\hat{H}_{\text{configuration space}}\left(x_1,....,x_n,-i\hbar\partial_{x_1},....,-i\hbar\partial_{x_n},t\right)\Psi_{\text{configuration space}}\left(x_1,....,x_n,t\right) \tag{9}$$

$$=i\hbar\partial_t\Psi_{\text{configuration space}}\left(x_1,....,x_n,t\right)$$

$$\Psi_{\text{configuration space}}\left(x_1,....,x_n,t\right)=\int_{-\infty}^{\infty}\frac{e^{\frac{ix_1p_1}{2\hbar}}}{\sqrt{4\pi\hbar}}....\int_{-\infty}^{\infty}\frac{e^{\frac{ix_np_n}{2\hbar}}}{\sqrt{4\pi\hbar}}\Psi(x_1,....,x_n;p_1,....,p_n;t)dp_1..dp_n$$

In other words, the configuration space wavefunction may be expressed in terms of exact quadratures containing the phase space wavefunction determined herein, Equation (8). Hence, Quantum Dynamics is now reduced to exact quadratures, as are all the associated purely mathematical problems that are abstracted from the physical formalism.

To wit, via HOA the configuration space solution becomes

$$\Psi_{\text{configuration space}}(x_1,....,x_n,t)$$
$$= \int_{-\infty}^{\infty} \frac{e^{\frac{ix_1p_1}{2\hbar}}}{\sqrt{4\pi\hbar}} .... \int_{-\infty}^{\infty} \frac{e^{\frac{ix_np_n}{2\hbar}}}{\sqrt{4\pi\hbar}} L^{-1}_{\left((\bar{x}_1,....,\bar{x}_n;\bar{p}_1,....,\bar{p}_n)\to(x_1,....,x_n;p_1,....,p_n)\right)} \begin{bmatrix} e^{\frac{-i}{\hbar}\int_0^t \hat{H}_{\text{configuration space}}\left(\substack{x_1,....,x_n,\\ -i\hbar\partial_{x_1},....,-i\hbar\partial_{x_n},u}\right)_{\substack{(x_1,....,x_n)\to(i\hbar\bar{p}_1+\alpha_1x_1,....,i\hbar\bar{p}_n+\alpha_nx_n)\\ (-i\hbar\partial_{x_1},....,-i\hbar\partial_{x_n})\to(-i\hbar\bar{x}_1+\gamma_1p_1,....,-i\hbar\bar{x}_n+\gamma_np_n)}} du} \\ \times \\ \breve{\Psi}_{0\,\text{configuration space}}(\bar{x}_1,....,\bar{x}_n;\bar{p}_1,....,\bar{p}_n;t=0) \end{bmatrix} dp_1..dp_n \tag{10}$$

With that said, a relatively simplistic prescription results for actually using the Ansatz to solve the problem,

Given the function $\hat{H}(\hat{x}_1,....,\hat{x}_n;\hat{p}_1,....,\hat{p}_n,t)$

[respectively $\hat{H}(x_1,....,x_n;-i\hbar\partial_{x_1},....,-i\hbar\partial_{x_n},t)$] replace $(\hat{x}_1,....,\hat{x}_n;\hat{p}_1,....,\hat{p}_n,t)$[respectively $(x_1,....,x_n;-i\hbar\partial_{x_1},....,-i\hbar\partial_{x_n},t)$] with $(i\hbar\bar{p}_1+\alpha_1x_1,....,i\hbar\bar{p}_n+\alpha_nx_n;-i\hbar\bar{x}_1+\gamma_1p_1,....,-i\hbar\bar{x}_n+\gamma_1p_n,t)$ in equation (10)

The result of course is the quantum phase space[respectively configuration space] wavefunction for the quantum dynamics wave equation. Just a comment on the α and γ parameters in the above formulae. From the HOA [ref.1], they are otherwise arbitrary except for the condition $\alpha+\gamma=1$. This is explained therein as a consequence of the arbitrary phase shift associated with the quantum phase space wavefunction. Further, any choice of the parameters thus constrained yields a Hamiltonian, which is dynamically equivalent[describes the same physics] as any other choice. However, it is shown therein that the Hamiltonian operator $\hat{H}(i\hbar\partial_p+\alpha x,-i\hbar\partial_x+\gamma p,t)\ni\alpha+\gamma=1$ takes on the symmetric canonical form when $\alpha=\gamma=\frac{1}{2}$ thusly $\hat{H}\left(i\hbar\partial_p+\frac{x}{2},-i\hbar\partial_x+\frac{p}{2},t\right)\ni\alpha+\gamma=1$. Notwithstanding this and with an eye towards computational simplifications for particular classes of applications, it has been found that other choices than $\alpha=\gamma=\frac{1}{2}$ sometime facilitates evaluation of the integral transforms. Unless otherwise directed, the convention for α and γ shall be specified for particular cases, presently and elsewhere. For the problems herein the $\alpha=\gamma=\frac{1}{2}$ sufficeth.

To recall the full details of HOA results, see the original work[EJTP 1 (2004), 10-16]. As pointed out therein,

> 'Notwithstanding its quantum mechanical origins, the HOA scheme takes on a life of its own and transcends the limits of quantum applications to address a wide variety of purely formal mathematical problems as well. Among other things, the result provides a formula for obtaining an exact solution to a wide variety of variable-coefficient integro-differential equations. Since the functional dependence of the Hamiltonian operator as considered is in general arbitrary upon its arguments(i.e., independent variables, derivative operator symbols[including negative powers thereof, thus the possible integral character]), then its multivariable extension can be interpreted as the most general variable coefficient partial differential operator. Moreover, it is not confined to being a scalar or even vector operator, but may be generally construed an arbitrary rank matrix operator. In all cases of course, its rank dictates the matrix rank of the wavefunction solution.'

In the present case of the Schrodinger equation, we shall be dealing with a scalar Hamiltonian structure and the solution wavefunction will be of a scalar character. (elsewhere, e.g. ref[9], the relativistic treatment demands the Dirac equation with such a 4x4 matrix Hamiltonian structure and the solution wavefunction will then be of a 4-dimensional column vector character).

Some Selected HOA Example Calculations

Example 1. 1 dim Simple Harmonic Oscillator

Having completed this brief recap of the HOA from ref.[6], now consider the first example case of the 1 dim. Simple Harmonic Oscillator dynamics in quantum phase space. Here the Hamiltonian has the form

$$\hat{H}(\hat{x}_1,....,\hat{x}_n;\hat{p}_1,....,\hat{p}_n,t)= \frac{\hat{p}^2}{2m}+a\hat{x}^2 \tag{11a}$$

Hence the solution of the phase space quantum dynamics is given by equation (9) as

$$\left(\frac{\left(-i\hbar\partial_x+\frac{p}{2}\right)^2}{2m}+a\left(i\hbar\partial_p+\frac{x}{2}\right)^2\right)\Psi(x;p;t)=i\hbar\partial_t\Psi(x;p;t)$$

$$\left(\frac{\left(-i\hbar\overline{x}+\frac{p}{2}\right)^2}{2m}+a\left(i\hbar\overline{p}+\frac{x}{2}\right)^2\right)\breve{\Psi}(\overline{x};\overline{p};t)=i\hbar\partial_t\breve{\Psi}(\overline{x};\overline{p};t) \tag{11b}$$

$$\Psi(x;p;t)=L^{-1}_{((\overline{x})\to(x))}\left[L^{-1}_{((\overline{p})\to(p))}\left[\begin{array}{l} e^{\frac{-it\left(\frac{\left(-i\hbar\overline{x}+\frac{p}{2}\right)^2}{2m}+a\left(i\hbar\overline{p}+\frac{x}{2}\right)^2\right)}{\hbar}} \\ \times \\ \breve{\Psi}_0(\overline{x};\overline{p};t=0)\end{array}\right]\right]$$

This is the computational construction of the Simple Harmonic Oscillator quantum phase space Schroedinger wavefunction. For the case of a Dirac delta function initial condition in quantum phase space of form $\delta(x,p)$ since the Laplace Transform of this initial condition is $\breve{\Psi}(\bar{x},\bar{p},t=0)=1,\Psi(x,p,t=0)=\delta(x,p)$.

Upon evaluation of (11b) with the above $\delta(x,p)$ initial condition, the quantum phase space wavefunction is

$$\Psi(x,p,t)=\sqrt{\frac{-m}{2a}}\frac{e^{i\left(\frac{p^2+2amx^2}{4a\hbar t}\right)}}{2\pi\hbar t} \tag{11c}$$

For sake of completeness, the case of a generalized to $a(t)$, an arbitrary function, yields

$$\left(\frac{\left(-i\hbar\partial_x+\frac{p}{2}\right)^2}{2m}+a(t)\left(i\hbar\partial_p+\frac{x}{2}\right)^2\right)\Psi(x;p;t)=i\hbar\partial_t\Psi(x;p;t)$$

$$\left(\frac{\left(-i\hbar\bar{x}+\frac{p}{2}\right)^2}{2m}+a(t)\left(i\hbar\bar{p}+\frac{x}{2}\right)^2\right)\breve{\Psi}(\bar{x};\bar{p};t)=i\hbar\partial_t\breve{\Psi}(\bar{x};\bar{p};t) \tag{11d}$$

$$\Psi(x;p;t)=L^{-1}_{((\bar{x})\to(x))}\left[L^{-1}_{((\bar{p})\to(p))}\left[e^{\frac{-i\left(t\frac{\left(-i\hbar\bar{x}+\frac{p}{2}\right)^2}{2m}+\int_0^t a(u)du\left(i\hbar\bar{p}+\frac{x}{2}\right)^2\right)}{\hbar}}\times\breve{\Psi}_0(\bar{x};\bar{p};t=0)\right]\right]$$

$$\Psi(x,p,t)=\Psi_0(x,p,t=0)\underbrace{*}_{x,p}\sqrt{\frac{-m}{2\int_0^t a(u)du}}\frac{e^{i\left(\frac{p^2+2mx^2\int_0^t a(u)du}{4\hbar t\int_0^t a(u)du}\right)}}{2\pi\hbar t} \tag{11e}$$

where $\underbrace{*}_{x,p}$ is the x,p convolution as described in the Recap earlier.

Example 2. Schrödinger Hamiltonian with SHOs having N-Arbitrary Masses in Pairwise Anisotropic Interaction

Now consider the example case of the nonrelativistic Schrödinger Hamiltonian with multiple pairwise interacting anisotropic SHOs of arbitrary masses in quantum phase space. Like the 1-dim example above, the oscillator strengths below may be generalized to arbitrary functions of time $a_{x_{ij}}(t), a_{y_{ij}}(t), a_{z_{ij}}(t)$ in a straightforward fashion, but the calculation is here omitted. Here the Hamiltonian has the form

$$\hat{H}_{\substack{\text{Schrodinger}\\ \text{Anisotropic}\\ \text{SHOs}}} = \sum_i \frac{1}{2m_i}\left(\left(-i\hbar\partial_{x_i} + \frac{p_{x_i}}{2}\right)^2 + \left(-i\hbar\partial_{y_i} + \frac{p_{y_i}}{2}\right)^2 + \left(-i\hbar\partial_{z_i} + \frac{p_{z_i}}{2}\right)^2\right)$$
$$+\sum_{i<j}\left(\begin{array}{l} a_{x_{ij}}\left(\left(i\hbar\partial_{p_{x_i}} + \frac{x_i}{2}\right) - \left(i\hbar\partial_{p_{x_j}} + \frac{x_j}{2}\right)\right)^2 \\ +a_{y_{ij}}\left(\left(i\hbar\partial_{p_{y_i}} + \frac{y_i}{2}\right) - \left(i\hbar\partial_{p_{y_j}} + \frac{y_j}{2}\right)\right)^2 \\ +a_{z_{ij}}\left(\left(i\hbar\partial_{p_{z_i}} + \frac{z_i}{2}\right) - \left(i\hbar\partial_{p_{z_j}} + \frac{z_j}{2}\right)\right)^2 \end{array}\right) \tag{12a}$$

$$\hat{H}_{\substack{\text{Schrodinger}\\ \text{Anisotropic}\\ \text{SHOs}}}\left(\begin{array}{l} -i\hbar\partial_{x_i} + \frac{p_{x_i}}{2}, -i\hbar\partial_{y_i} + \frac{p_{y_i}}{2}, -i\hbar\partial_{z_i} + \frac{p_{z_i}}{2}; \\ i\hbar\partial_{p_{x_i}} + \frac{x_i}{2}, i\hbar\partial_{p_{y_i}} + \frac{y_i}{2}, i\hbar\partial_{p_{z_i}} + \frac{z_i}{2} \end{array}\right)\Psi\left(x_i, y_i, z_i; p_{x_i}, p_{y_i}, p_{z_i}; t\right)$$
$$= i\hbar\partial_t\Psi\left(x_i, y_i, z_i; p_{x_i}, p_{y_i}, p_{z_i}; t\right)$$

$$\hat{H}_{\substack{\text{Schrodinger}\\ \text{Anisotropic}\\ \text{SHOs}}} = \sum_i \frac{1}{2m_i}\left(\left(-i\hbar\overline{x}_i + \frac{p_{x_i}}{2}\right)^2 + \left(-i\hbar\overline{y}_i + \frac{p_{y_i}}{2}\right)^2 + \left(-i\hbar\overline{z}_i + \frac{p_{z_i}}{2}\right)^2\right)$$
$$+\sum_{i<j}\left(\begin{array}{l} a_{x_{ij}}\left(\left(i\hbar\overline{p}_{x_i} + \frac{x_i}{2}\right) - \left(i\hbar\overline{p}_{x_j} + \frac{x_j}{2}\right)\right)^2 \\ +a_{y_{ij}}\left(\left(i\hbar\overline{p}_{y_i} + \frac{y_i}{2}\right) - \left(i\hbar\overline{p}_{y_j} + \frac{y_j}{2}\right)\right)^2 \\ +a_{z_{ij}}\left(\left(i\hbar\overline{p}_{z_i} + \frac{z_i}{2}\right) - \left(i\hbar\overline{p}_{z_j} + \frac{z_j}{2}\right)\right)^2 \end{array}\right) \tag{12b}$$

$$\hat{H}_{\substack{\text{Schrodinger}\\ \text{Anisotropic}\\ \text{SHOs}}}\begin{pmatrix} -i\hbar\overline{x}_i+\frac{p_{x_i}}{2},-i\hbar\overline{y}_i+\frac{p_{y_i}}{2},-i\hbar\overline{z}_i+\frac{p_{z_i}}{2}; \\ i\hbar\overline{p}_{x_i}+\frac{x_i}{2},i\hbar\overline{p}_{y_i}+\frac{y_i}{2},i\hbar\overline{p}_{z_i}+\frac{z_i}{2} \end{pmatrix}\breve{\Psi}\left(\overline{x}_i,\overline{y}_i,\overline{z}_i;\overline{p}_{x_i},\overline{p}_{y_i},\overline{p}_{z_i};t\right)$$

$$=i\hbar\partial_t\breve{\Psi}\left(\overline{x}_i,\overline{y}_i,\overline{z}_i;\overline{p}_{x_i},\overline{p}_{y_i},\overline{p}_{z_i};t\right)$$

$$\Psi\left(x_i,y_i,z_i;p_{x_i},p_{y_i},p_{z_i};t\right)=$$

$$L^{-1}_{\substack{\left(\overline{x}_i,\overline{y}_i,\overline{z}_i;\overline{p}_{x_i},\overline{p}_{y_i},\overline{p}_{z_i}\right)\\ \to\left(x_i,y_i,z_i;p_{x_i},p_{y_i},p_{z_i}\right)}}\left[\begin{matrix} e^{\frac{-i\int_0^t \hat{H}_{\substack{\text{Schrodinger}\\ \text{Anisotropic}\\ \text{SHOs}}}\begin{pmatrix} -i\hbar\overline{x}_i+\frac{p_{x_i}}{2},-i\hbar\overline{y}_i+\frac{p_{y_i}}{2},-i\hbar\overline{z}_i+\frac{p_{z_i}}{2}; \\ i\hbar\overline{p}_{x_i}+\frac{x_i}{2},i\hbar\overline{p}_{y_i}+\frac{y_i}{2},i\hbar\overline{p}_{z_i}+\frac{z_i}{2}\end{pmatrix}du}{\hbar}} \\ \times \\ \breve{\Psi}_0\left(\overline{x}_i,\overline{y}_i,\overline{z}_i;\overline{p}_{x_i},\overline{p}_{y_i},\overline{p}_{z_i};t=0\right)\end{matrix}\right] \tag{12c}$$

Example 3. Schrödinger Molecular Hamiltonian with Pairwise Coulomb Interaction

Now consider the example case of the nonrelativistic Schrödinger Molecular Hamiltonian with Pairwise Coulomb interaction dynamics in quantum phase space [ref.7]. Here the Hamiltonian has the form

(13a)

$$\hat{H}=\sum_A\frac{1}{2M_A}\hat{\mathbf{P}}_A^2+\sum_i\frac{1}{2m_e}\hat{\mathbf{P}}_i^2+\sum_{A<B}\frac{e^2}{8\pi\varepsilon_0}\left(\frac{Z_AZ_B}{\left|\hat{r}_A-\hat{r}_B\right|}\right)+\sum_{i<j}\frac{e^2}{8\pi\varepsilon_0}\left(\frac{1}{\left|\hat{r}_i-\hat{r}_j\right|}\right)+\sum_{A,i}\left(\frac{-Z_Ae^2}{\left|\hat{r}_A-\hat{r}_i\right|}\right)$$

$$\hat{H}_{\substack{\text{Schrodinger}\\ \text{Molecular}}}\begin{pmatrix} -i\hbar\partial_{x_A}+\frac{p_{x_A}}{2},-i\hbar\partial_{y_A}+\frac{p_{y_A}}{2},-i\hbar\partial_{z_A}+\frac{p_{z_A}}{2},-i\hbar\partial_{x_B}+\frac{p_{x_B}}{2},-i\hbar\partial_{y_B}+\frac{p_{y_B}}{2},-i\hbar\partial_{z_B}+\frac{p_{z_B}}{2}, \\ -i\hbar\partial_{x_i}+\frac{p_{x_i}}{2},-i\hbar\partial_{y_i}+\frac{p_{y_i}}{2},-i\hbar\partial_{z_i}+\frac{p_{z_i}}{2},-i\hbar\partial_{x_j}+\frac{p_{x_j}}{2},-i\hbar\partial_{y_j}+\frac{p_{y_j}}{2},-i\hbar\partial_{z_j}+\frac{p_{z_j}}{2}; \\ i\hbar\partial_{p_{xA}}+\frac{x_A}{2},i\hbar\partial_{p_{yA}}+\frac{y_A}{2},i\hbar\partial_{p_{zA}}+\frac{z_A}{2},i\hbar\partial_{p_{xB}}+\frac{x_B}{2},i\hbar\partial_{p_{yB}}+\frac{y_B}{2},i\hbar\partial_{p_{zB}}+\frac{z_B}{2}, \\ i\hbar\partial_{p_{xi}}+\frac{x_i}{2},i\hbar\partial_{p_{yi}}+\frac{y_i}{2},i\hbar\partial_{p_{zi}}+\frac{z_i}{2},i\hbar\partial_{p_{xj}}+\frac{x_j}{2},i\hbar\partial_{p_{yj}}+\frac{y_j}{2},i\hbar\partial_{p_{zj}}+\frac{z_j}{2} \end{pmatrix}$$

$$\hat{H}_{\substack{\text{Schrodinger}\\\text{Molecular}}}=\sum_A\frac{1}{2M_A}\left(\left(-i\hbar\overline{x}_A+\frac{p_{x_A}}{2}\right)^2+\left(-i\hbar\overline{y}_A+\frac{p_{y_A}}{2}\right)^2+\left(-i\hbar\overline{z}_A+\frac{p_{z_A}}{2}\right)^2\right)$$
$$+\sum_i\frac{1}{2m_e}\left(\left(-i\hbar\overline{x}_i+\frac{p_{x_i}}{2}\right)^2+\left(-i\hbar\overline{y}_i+\frac{p_{y_i}}{2}\right)^2+\left(-i\hbar\overline{z}_i+\frac{p_{z_i}}{2}\right)^2\right)$$
$$+\sum_{A<B}\frac{e^2}{8\pi\varepsilon_0}\frac{Z_AZ_B}{\left|\begin{array}{l}\left(\left(i\hbar\overline{p}_{x_A}+\frac{x_A}{2}\right)-\left(i\hbar\overline{p}_{x_B}+\frac{x_B}{2}\right)\right)^2\\+\left(\left(i\hbar\overline{p}_{y_A}+\frac{y_A}{2}\right)-\left(i\hbar\overline{p}_{y_B}+\frac{y_B}{2}\right)\right)^2\\+\left(\left(i\hbar\overline{p}_{z_A}+\frac{z_A}{2}\right)-\left(i\hbar\overline{p}_{z_B}+\frac{z_B}{2}\right)\right)^2\end{array}\right|}+\sum_{i<j}\frac{e^2}{8\pi\varepsilon_0}\frac{1}{\left|\begin{array}{l}\left(\left(i\hbar\overline{p}_{x_i}+\frac{x_i}{2}\right)-\left(i\hbar\overline{p}_{x_j}+\frac{x_j}{2}\right)\right)^2\\+\left(\left(i\hbar\overline{p}_{y_i}+\frac{y_i}{2}\right)-\left(i\hbar\overline{p}_{y_j}+\frac{y_j}{2}\right)\right)^2\\+\left(\left(i\hbar\overline{p}_{z_i}+\frac{z_i}{2}\right)-\left(i\hbar\overline{p}_{z_j}+\frac{z_j}{2}\right)\right)^2\end{array}\right|} \quad (13b)$$
$$+\sum_{A,i}\frac{-Z_Ae^2}{\left|\begin{array}{l}\left(\left(i\hbar\overline{p}_{x_A}+\frac{x_A}{2}\right)-\left(i\hbar\overline{p}_{x_i}+\frac{x_i}{2}\right)\right)^2\\+\left(\left(i\hbar\overline{p}_{y_A}+\frac{y_A}{2}\right)-\left(i\hbar\overline{p}_{y_i}+\frac{y_i}{2}\right)\right)^2\\+\left(\left(i\hbar\overline{p}_{z_A}+\frac{z_A}{2}\right)-\left(i\hbar\overline{p}_{z_i}+\frac{z_i}{2}\right)\right)^2\end{array}\right|}$$

Hence the quantum phase space Hamiltonian is specified by the Ansatz and results in the quantum phase space wave function for a given initial condition via (10) as

$$\Psi\left(\begin{array}{l}x_A,y_A,z_A,x_B,y_B,z_B,\\x_i,y_i,z_i,x_j,y_j,z_j;\\p_{x_A},p_{y_A},p_{z_A},p_{x_B},p_{y_B},p_{z_B},\\p_{x_i},p_{y_i},p_{z_i},p_{x_j},p_{y_j},p_{z_j};t\end{array}\right)=$$
$$L^{-1}_{\left(\left(\begin{array}{l}\overline{x}_A,\overline{y}_A,\overline{z}_A,\overline{x}_B,\overline{y}_B,\overline{z}_B,\\x_i,y_i,\overline{z}_i,x_j,y_j,\overline{z}_j;\\p_{x_A},p_{y_A},p_{z_A},p_{x_B},p_{y_B},p_{z_B},\\p_{x_i},p_{y_i},p_{z_i},p_{x_j},p_{y_j},p_{z_j}\end{array}\right)\to\left(\begin{array}{l}x_A,y_A,z_A,x_B,y_B,z_B,\\x_i,y_i,z_i,x_j,y_j,z_j;\\p_{x_A},p_{y_A},p_{z_A},p_{x_B},p_{y_B},p_{z_B},\\p_{x_i},p_{y_i},p_{z_i},p_{x_j},p_{y_j},p_{z_j}\end{array}\right)\right)}\left[\begin{array}{l}e^{\frac{-i}{\hbar}\int_0^t\hat{H}_{\substack{\text{Schrodinger}\\\text{Molecular}}}\left(\begin{array}{l}-i\hbar\overline{x}_A+\frac{p_{x_A}}{2},-i\hbar\overline{y}_A+\frac{p_{y_A}}{2},-i\hbar\overline{z}_A+\frac{p_{z_A}}{2},-i\hbar\overline{x}_i+\frac{p_{x_i}}{2},-i\hbar\overline{y}_i+\frac{p_{y_i}}{2},-i\hbar\overline{z}_i+\frac{p_{z_i}}{2};\\i\hbar\overline{p}_{x_A}+\frac{x_A}{2},i\hbar\overline{p}_{y_A}+\frac{y_A}{2},i\hbar\overline{p}_{z_A}+\frac{z_A}{2},i\hbar\overline{p}_{x_B}+\frac{x_B}{2},i\hbar\overline{p}_{y_B}+\frac{y_B}{2},i\hbar\overline{p}_{z_B}+\frac{z_B}{2},\\i\hbar\overline{p}_{x_i}+\frac{x_i}{2},i\hbar\overline{p}_{y_i}+\frac{y_i}{2},i\hbar\overline{p}_{z_i}+\frac{z_i}{2},i\hbar\overline{p}_{x_j}+\frac{x_j}{2},i\hbar\overline{p}_{y_j}+\frac{y_j}{2},i\hbar\overline{p}_{z_j}+\frac{z_j}{2}\end{array}\right)du}\\\times\\\breve{\Psi}_0\left(\overline{x}_1,\ldots,\overline{x}_n;\overline{p}_1,\ldots,\overline{p}_n;t=0\right)\end{array}\right] \quad (13c)$$

Where the Schrödinger Molecular Hamiltonian is expressed as

$$\hat{H}_{\substack{\text{Schrodinger}\\ \text{Molecular}}}\begin{pmatrix} -i\hbar\overline{x}_A+\frac{p_{x_A}}{2},-i\hbar\overline{y}_A+\frac{p_{y_A}}{2},-i\hbar\overline{z}_A+\frac{p_{z_A}}{2},-i\hbar\overline{x}_B+\frac{p_{x_B}}{2},-i\hbar\overline{y}_B+\frac{p_{y_B}}{2},-i\hbar\overline{z}_B+\frac{p_{z_B}}{2}, \\ -i\hbar\overline{x}_i+\frac{p_{x_i}}{2},-i\hbar\overline{y}_i+\frac{p_{y_i}}{2},-i\hbar\overline{z}_i+\frac{p_{z_i}}{2},-i\hbar\overline{x}_j+\frac{p_{x_j}}{2},-i\hbar\overline{y}_j+\frac{p_{y_j}}{2},-i\hbar\overline{z}_j+\frac{p_{z_j}}{2}; \\ i\hbar\overline{p}_{x_A}+\frac{x_A}{2},i\hbar\overline{p}_{y_A}+\frac{y_A}{2},i\hbar\overline{p}_{z_A}+\frac{z_A}{2},i\hbar\overline{p}_{x_B}+\frac{x_B}{2},i\hbar\overline{p}_{y_B}+\frac{y_B}{2},,i\hbar\overline{p}_{z_B}+\frac{z_B}{2}, \\ i\hbar\overline{p}_{x_i}+\frac{x_i}{2},i\hbar\overline{p}_{y_i}+\frac{y_i}{2},i\hbar\overline{p}_{z_i}+\frac{z_i}{2},i\hbar\overline{p}_{x_j}+\frac{x_j}{2},i\hbar\overline{p}_{y_j}+\frac{y_j}{2},i\hbar\overline{p}_{z_j}+\frac{z_j}{2} \end{pmatrix}$$

Following from [ref.7] for the radial component of the Molecular Schrodinger equation in Phase Space, the HOA Solution of the Radial Schrödinger Equation for Nonrelativististiic N-particle system with Pairwise $\frac{1}{r_{ij}}$ Radial Potential Interaction where $\frac{1}{r_{ij}}\equiv\frac{1}{|r_i-r_j|}$ in the QPSR. Moreover, the analysis for this multiparticle system is in the Laboratory Reference Frame for the Coordinates[ref.7]. Hence no particle is fixed as a center of motion. As stated earlier the angular components separate in the usual way by [refs.7]. Here a_i,b_i,c_i are scaling constants, $L_i(L_i+1)$ are the angular coupling terms and $\mathbf{r}=(r_1,\dots,r_N)=(r_{i=1,\dots,N})$, $\mathbf{p_r}=(p_{r_1},\dots,p_{r_N})=(p_{r_{i=1,\dots,N}})$. The resulting radial Hamiltonian yields a radial Schrodinger equation of form

$$\begin{pmatrix} \hat{H}=\sum_i\left(\frac{\left(-i\hbar\partial_{r_i}+\frac{p_{r_i}}{2}\right)^2}{2m_i}+\frac{2b_i}{\left(i\hbar\partial_{p_{r_i}}+\frac{r_i}{2}\right)}\left(-i\hbar\partial_{r_i}+\frac{p_{r_i}}{2}\right)+\frac{c_iL_i(L_i+1)}{\left(i\hbar\partial_{p_{r_i}}+\frac{r_i}{2}\right)^2}\right) \\ +\sum_{i<j}a_{ij}\left(\frac{1}{\left|\left(i\hbar\partial_{p_{r_i}}+\frac{r_i}{2}\right)-\left(i\hbar\partial_{p_{rj}}+\frac{r_j}{2}\right)\right|}\right) \end{pmatrix}\Psi_{\mathbf{r}\,\text{phasespace}}(\mathbf{r};\mathbf{p_r};t) \quad (13d)$$

$$=i\hbar\partial_t\Psi_{\mathbf{r}\,\text{phasespace}}(\mathbf{r};\mathbf{p_r};t)$$

Hence via HOA from equation (10),

$$\left(\begin{array}{l}\sum_i\left(\frac{\left(-i\hbar\bar{r}_i+\frac{p_{r_i}}{2}\right)^2}{2m_i}+\frac{2b_i}{\left(i\hbar\bar{p}_{r_i}+\frac{r_i}{2}\right)}\left(-i\hbar\bar{r}_i+\frac{p_{r_i}}{2}\right)+\frac{c_iL_i(L_i+1)}{\left(i\hbar\bar{p}_{r_i}+\frac{r_i}{2}\right)^2}\right)\\+\sum_{i<j}a_{ij}\left(\frac{1}{\left|\left(i\hbar\bar{p}_{r_i}+\frac{r_i}{2}\right)-\left(i\hbar\bar{p}_{r_j}+\frac{r_j}{2}\right)\right|}\right)\end{array}\right)\breve{\Psi}_{\mathbf{r}\text{ phasespace}}(\bar{\mathbf{r}};\bar{\mathbf{p}}_{\mathbf{r}};t) = i\hbar\partial_t\breve{\Psi}_{\mathbf{r}\text{ phasespace}}(\bar{\mathbf{r}};\bar{\mathbf{p}}_{\mathbf{r}};t) \tag{13e}$$

Yielding the exact analytical solution given in quadratures as

$$\Psi_{\mathbf{r}\text{ phasespace}}(\mathbf{r};\mathbf{p}_{\mathbf{r}};t)=L^{-1}_{\left(\substack{\bar{\mathbf{r}}\to\mathbf{r};\\ \bar{\mathbf{p}}_{\mathbf{r}}\to\mathbf{p}_{\mathbf{r}}}\right)}\left(\begin{array}{l}e^{\frac{-it}{\hbar}\left(\begin{array}{l}\sum_i\left(\frac{\left(-i\hbar\bar{r}_i+\frac{p_{r_i}}{2}\right)^2}{2m_i}+\frac{2b_i}{\left(i\hbar\bar{p}_{r_i}+\frac{r_i}{2}\right)}\left(-i\hbar\bar{r}_i+\frac{p_{r_i}}{2}\right)+\frac{c_iL_i(L_i+1)}{\left(i\hbar\bar{p}_{r_i}+\frac{r_i}{2}\right)^2}\right)\\+\sum_{i<j}a_{ij}\left(\frac{1}{\left|\left(i\hbar\bar{p}_{r_i}+\frac{r_i}{2}\right)-\left(i\hbar\bar{p}_{r_j}+\frac{r_j}{2}\right)\right|}\right)\end{array}\right)}\\\times\breve{\Psi}_{0\mathbf{r}\text{ phasespace}}(\bar{\mathbf{r}};\bar{\mathbf{p}}_{\mathbf{r}};t=0)\end{array}\right) \tag{13f}$$

To evaluate (13f), recall that

$$\begin{array}{l}\breve{\Psi}_{\mathbf{r}\text{ phasespace}}(\bar{\mathbf{r}};\bar{\mathbf{p}}_{\mathbf{r}};t)=\\
e^{\frac{-it}{\hbar}\left(\sum_i\left(\frac{\left(-i\hbar\bar{r}_i+\frac{p_{r_i}}{2}\right)^2}{2m_i}\right)\right)}\times e^{\frac{-it}{\hbar}\left(\sum_i\left(\frac{2b_i}{\left(i\hbar\bar{p}_{r_i}+\frac{r_i}{2}\right)}\left(-i\hbar\bar{r}_i+\frac{p_{r_i}}{2}\right)\right)\right)}\\
\times e^{\frac{-i\bar{t}}{\hbar}\left(\sum_i\left(\frac{c_iL_i(L_i+1)}{\left(i\hbar\bar{p}_{r_i}+\frac{r_i}{2}\right)^2}\right)\right)}\times e^{\frac{-it}{\hbar}\left(\sum_{i<j}a_{ij}\left(\frac{1}{\left|\left(i\hbar\bar{p}_{r_i}+\frac{r_i}{2}\right)-\left(i\hbar\bar{p}_{r_j}+\frac{r_j}{2}\right)\right|}\right)\right)}\times\breve{\Psi}_{0\mathbf{r}\text{ phasespace}}(\bar{\mathbf{r}};\bar{\mathbf{p}}_{\mathbf{r}};t=0)\end{array} \tag{13g}$$

Upon evaluating the inverse Laplace transforms and utilizing convolution products herein symbolized as $\underbrace{*}_{\mathbf{r}}$ or $\underbrace{*}_{\mathbf{p}_{\mathbf{r}}}$ or $\underbrace{*}_{\mathbf{r},\mathbf{p}_{\mathbf{r}}}$ [ref 6].

$$L^{-1}_{\left(\substack{\bar{\mathbf{r}}\to\mathbf{r};\\ \bar{\mathbf{p}}_{\mathbf{r}}\to\mathbf{p}_{\mathbf{r}}}\right)}\breve{\Psi}_{\mathbf{r}\,\text{phasespace}}\left(\bar{\mathbf{r}};\bar{\mathbf{p}}_{\mathbf{r}};t\right)=$$

$$L^{-1}_{(\bar{\mathbf{r}}\to\mathbf{r})}\left(e^{\frac{-it}{\hbar}\left(\sum_i\left(\frac{\left(-i\hbar\bar{r}_i+\frac{p_{r_i}}{2}\right)^2}{2m_i}\right)\right)}\right)\underbrace{*}_{\mathbf{r}}L^{-1}_{\left(\substack{\bar{\mathbf{r}}\to\mathbf{r};\\ \bar{\mathbf{p}}_{\mathbf{r}}\to\mathbf{p}_{\mathbf{r}}}\right)}\left(e^{\frac{-it}{\hbar}\left(\sum_i\left(\frac{2b_i}{\left(i\hbar\bar{p}_{r_i}+\frac{r_i}{2}\right)}\left(-i\hbar\bar{r}_i+\frac{p_{r_i}}{2}\right)\right)\right)}\right)$$

$$\underbrace{*}_{\mathbf{p}_{\mathbf{r}}}L^{-1}_{(\bar{\mathbf{p}}_{\mathbf{r}}\to\mathbf{p}_{\mathbf{r}})}\left(e^{\frac{-it}{\hbar}\left(\sum_i\left(\frac{c_iL_i(L_i+1)}{\left(i\hbar\bar{p}_{r_i}+\frac{r_i}{2}\right)^2}\right)\right)}\right)\underbrace{*}_{\mathbf{p}_{\mathbf{r}}}L^{-1}_{(\bar{\mathbf{p}}_{\mathbf{r}}\to\mathbf{p}_{\mathbf{r}})}\left(e^{\frac{-it}{\hbar}\left(\sum_{i<j}a_{ij}\left(\frac{1}{\left|\left(i\hbar\bar{p}_{r_i}+\frac{r_i}{2}\right)-\left(i\hbar\bar{p}_{r_j}+\frac{r_j}{2}\right)\right|}\right)\right)}\right) \quad (13h)$$

$$\underbrace{*}_{\mathbf{r},\mathbf{p}_{\mathbf{r}}}L^{-1}_{\left(\substack{\bar{\mathbf{r}}\to\mathbf{r};\\ \bar{\mathbf{p}}_{\mathbf{r}}\to\mathbf{p}_{\mathbf{r}}}\right)}\left(\breve{\Psi}_{0\mathbf{r}\,\text{phasespace}}\left(\bar{\mathbf{r}};\bar{\mathbf{p}}_{\mathbf{r}};t=0\right)\right)$$

Evaluating (13h) explicitly yields

$$L^{-1}_{(\bar{\mathbf{p}}_{\mathbf{r}}\to\mathbf{p}_{\mathbf{r}})}\left(e^{\frac{-it}{\hbar}\left(\sum_{i<j}a_{ij}\left(\frac{1}{\left|\left(i\hbar\bar{p}_{r_i}+\frac{r_i}{2}\right)-\left(i\hbar\bar{p}_{r_j}+\frac{r_j}{2}\right)\right|}\right)\right)}\right)$$

$$=\underbrace{*}_{i<j}\left(\delta\left[p_{r_i}+p_{r_j}\right]\left(\frac{1}{\pi p_{r_i}\sqrt{\frac{-i\hbar}{a_{ij}t}}}e^{\frac{ip_{r_i}(r_i-r_j)}{2\hbar}}\left(\sqrt{\frac{p_{r_i}}{\hbar}}K_1\left[\frac{2\sqrt{\frac{p_{r_i}}{\hbar}}}{\sqrt{\frac{-i\hbar}{a_{ij}t}}}\right]-\sqrt{\frac{-p_{r_i}}{\hbar}}K_1\left[\frac{2\sqrt{\frac{-p_{r_i}}{\hbar}}}{\sqrt{\frac{-i\hbar}{a_{ij}t}}}\right]\right)\right)\right) \quad (13i)$$

example N = 3, $f_{1,2}\underbrace{*}_{p_{r_1}}f_{1,3}\underbrace{*}_{p_{r_2},p_{r_3}}f_{2,3}$

$$,\ f_{ij}=\delta\left[p_{r_i}+p_{r_j}\right]\left(\frac{1}{\pi p_{r_i}\sqrt{\frac{-i\hbar}{a_{ij}t}}}e^{\frac{ip_{r_i}(r_i-r_j)}{2\hbar}}\left(\sqrt{\frac{p_{r_i}}{\hbar}}K_1\left[\frac{2\sqrt{\frac{p_{r_i}}{\hbar}}}{\sqrt{\frac{-i\hbar}{a_{ij}t}}}\right]-\sqrt{\frac{-p_{r_i}}{\hbar}}K_1\left[\frac{2\sqrt{\frac{-p_{r_i}}{\hbar}}}{\sqrt{\frac{-i\hbar}{a_{ij}t}}}\right]\right)\right)$$

Where $\underbrace{\ast}_{k j}$ represents the convolution between the inverse transforms of the indicated terms as illustrated above in the case of N=3.

$$L^{-1}_{\left(\substack{\bar{\mathbf{r}}\to\mathbf{r};\\ \bar{\mathbf{p}}_{\mathbf{r}}\to\mathbf{p}_{\mathbf{r}}}\right)}\left(e^{\frac{-it}{\hbar}\left(\sum_i\left(\frac{2b_i}{\left(i\hbar\bar{p}_{r_i}+\frac{r_i}{2}\right)}\left(-i\hbar\bar{r}_i+\frac{p_{r_i}}{2}\right)\right)\right)}\right)=\prod_i\left(\delta\left[p_{r_i}\right]\delta\left[r_i\right]+\frac{ib_ie^{\frac{2ib_ip_{r_i}t}{\hbar r_i}}\left(\sqrt{-ip_{r_i}}-\sqrt{ip_{r_i}}\right)t}{\hbar\pi r_i^2\sqrt{\left|p_{r_i}\right|}}\right) \tag{13j}$$

$$L^{-1}_{(\bar{\mathbf{r}}_{\mathbf{r}}\to\mathbf{p}_{\mathbf{r}})}\left(e^{\frac{-it}{\hbar}\left(\sum_i\left(\frac{c_iL_i(L_i+1)}{\left(i\hbar\bar{p}_{r_i}+\frac{r_i}{2}\right)^2}\right)\right)}\right) \tag{13k}$$

$$=\prod_i\left(\frac{-1}{24\pi\hbar^6}\left(c_i^2L_i^2\left(L_i+1\right)^2p_{r_i}^3t^2\left(\left(e^{\frac{ip_{r_i}r_i}{2\hbar}}-1\right)Log\left[\frac{-p_{r_i}}{\hbar}\right]+Log\left[\frac{p_{r_i}}{\hbar}\right]-Log\left[\frac{p_{r_i}}{\hbar}\right]HeavisideStep\left[p_{r_i}\right]\right)\right)\right)$$

$$L^{-1}_{(\bar{\mathbf{r}}\to\mathbf{r})}\left(e^{\frac{-it}{\hbar}\left(\sum_i\left(\frac{\left(-i\hbar\bar{r}_i+\frac{p_{r_i}}{2}\right)^2}{2m_i}\right)\right)}\right)=\prod_i\frac{e^{\frac{i\left(r_i^2m_i-tr_ip_{r_i}\right)}{2\hbar t}}}{\hbar\sqrt{\frac{2\pi it}{\hbar m_i}}} \tag{13l}$$

Since $L^{-1}_{\left(\substack{\bar{\mathbf{r}}\to\mathbf{r};\\ \bar{\mathbf{p}}_{\mathbf{r}}\to\mathbf{p}_{\mathbf{r}}}\right)}\left(\breve{\Psi}_{0\mathbf{r}\text{ phasespace}}\left(\bar{\mathbf{r}};\bar{\mathbf{p}}_{\mathbf{r}};t=0\right)\right)=\Psi_{0\mathbf{r}\text{ phasespace}}\left(\mathbf{r};\mathbf{p}_{\mathbf{r}};t=0\right)$, the evaluated inverse transforms yield the explicit solution in the QPSR

$$\Psi_{\mathbf{r}\ \text{phasespace}}(\mathbf{r};\mathbf{p}_\mathbf{r};t)=$$

$$\Psi_{0\mathbf{r}\ \text{phasespace}}(\mathbf{r};\mathbf{p}_\mathbf{r};t=0)\underbrace{*}_{\mathbf{r},\mathbf{p}_\mathbf{r}}\prod_i \frac{e^{\frac{i\left(r_i^2 m_i - t r_i p_{r_i}\right)}{2\hbar t}}}{\hbar\sqrt{\frac{2\pi i t}{\hbar m_i}}}\underbrace{*}_{\mathbf{r}}\prod_i\left(\delta[p_{r_i}]\delta[r_i]+\frac{i b_i e^{\frac{2 i b_i p_{r_i} t}{\hbar r_i}}\left(\sqrt{-i p_{r_i}}-\sqrt{i p_{r_i}}\right)}{\hbar\pi\ r_i^2\sqrt{\left|p_{r_i}\right|}}\right)$$

$$\underbrace{*}_{\mathbf{p}_\mathbf{r}}\prod_i\left(\frac{-1}{24\pi\hbar^6}\left(c_i^2 L_i^2\left(L_i+1\right)^2 p_{r_i}^3 t^2\left(\begin{array}{l}\left(e^{\frac{i p_{r_i} r_i}{2\hbar}}-1\right)Log\left[\frac{-p_{r_i}}{\hbar}\right]\\ +Log\left[\frac{p_{r_i}}{\hbar}\right]-Log\left[\frac{p_{r_i}}{\hbar}\right]HeavisideStep\left[p_{r_i}\right]\end{array}\right)\right)\right)$$

$$\underbrace{*}_{\mathbf{p}_\mathbf{r}}\left(\underbrace{\ast}_{i<j}\left(\delta\left[p_{r_i}+p_{r_j}\right]\left(\frac{1}{\pi p_{r_i}\sqrt{\frac{-i\hbar}{a_{ij}t}}}e^{\frac{i p_{r_i}(r_i-r_j)}{2\hbar}}\left(\sqrt{\frac{p_{r_i}}{\hbar}}K_1\left[\frac{2\sqrt{\frac{p_{r_i}}{\hbar}}}{\sqrt{\frac{-i\hbar}{a_{ij}t}}}\right]-\sqrt{\frac{-p_{r_i}}{\hbar}}K_1\left[\frac{2\sqrt{\frac{-p_{r_i}}{\hbar}}}{\sqrt{\frac{-i\hbar}{a_{ij}t}}}\right]\right)\right)\right)\right) \tag{13m}$$

Although the above (13m) is explicitly given in the QPSR[down to the level of convolutions of explicitly calculated quantum phase space factors], an illustrative example will now be provided to help illuminate the actual application of the result for the case of three particles: N=3

Now for the solution of the radial Schrodinger equation for the non-relativistiic 3-particle system with pairwise $\frac{1}{r_{ij}}$ radial potential interaction where $\frac{1}{r_{ij}}\equiv\frac{1}{\left|r_i-r_j\right|}$ in the QPSR. Moreover, the analysis for this multiparticle system is in the Laboratory Reference Frame for the Coordinates[ref.7]. Hence no particle is fixed as a center of motion. As stated earlier the angular components separate in the usual way by [refs.7]. Here a_i,b_i,c_i are scaling constants, $L_i(L_i+1)$ are the angular coupling terms and $\mathbf{r}=(r_1,r_2,r_3)=(r_{i=1,2,3})$, $\mathbf{p}_\mathbf{r}=\left(p_{r_1},p_{r_2},p_{r_3}\right)=\left(p_{r_{i=1,2,3}}\right)$. From eq(13d), the resulting radial Hamiltonian yields a radial Schrodinger equation of form

$$\left(\begin{array}{l}\dfrac{\left(-i\hbar\partial_{r_1}+\dfrac{p_{r_1}}{2}\right)^2}{2m_1}+\dfrac{\left(-i\hbar\partial_{r_2}+\dfrac{p_{r_2}}{2}\right)^2}{2m_2}+\dfrac{\left(-i\hbar\partial_{r_3}+\dfrac{p_{r_3}}{2}\right)^2}{2m_3}\\ +\dfrac{2b_1}{\left(i\hbar\partial_{p_{r_1}}+\dfrac{r_1}{2}\right)}\left(-i\hbar\partial_{r_1}+\dfrac{p_{r_1}}{2}\right)+\dfrac{2b_2}{\left(i\hbar\partial_{p_{r_2}}+\dfrac{r_2}{2}\right)}\left(-i\hbar\partial_{r_2}+\dfrac{p_{r_2}}{2}\right)\\ +\dfrac{2b_3}{\left(i\hbar\partial_{p_{r_3}}+\dfrac{r_3}{2}\right)}\left(-i\hbar\partial_{r_3}+\dfrac{p_{r_3}}{2}\right)+\dfrac{c_1L_1(L_1+1)}{\left(i\hbar\partial_{p_{r_1}}+\dfrac{r_1}{2}\right)^2}+\dfrac{c_2L_2(L_2+1)}{\left(i\hbar\partial_{p_{r_2}}+\dfrac{r_2}{2}\right)^2}\\ +\dfrac{c_3L_3(L_3+1)}{\left(i\hbar\partial_{p_{r_3}}+\dfrac{r_3}{2}\right)^2}+a_{12}\left(\dfrac{1}{\left|\left(i\hbar\partial_{p_{r_1}}+\dfrac{r_1}{2}\right)-\left(i\hbar\partial_{p_{r_2}}+\dfrac{r_2}{2}\right)\right|}\right)\\ +a_{13}\left(\dfrac{1}{\left|\left(i\hbar\partial_{p_{r_1}}+\dfrac{r_1}{2}\right)-\left(i\hbar\partial_{p_{r_3}}+\dfrac{r_3}{2}\right)\right|}\right)+a_{23}\left(\dfrac{1}{\left|\left(i\hbar\partial_{p_{r_2}}+\dfrac{r_2}{2}\right)-\left(i\hbar\partial_{p_{r_3}}+\dfrac{r_3}{2}\right)\right|}\right)\end{array}\right)\Psi_{\mathbf{r}\ \text{phasespace}}\left(r_1,r_2,r_3;p_{r_1}.p_{r_2},p_{r_3};t\right)$$
$$=i\hbar\partial_t\Psi_{\mathbf{r}\ \text{phasespace}}\left(r_1,r_2,r_3;p_{r_1}.p_{r_2},p_{r_3};t\right) \tag{13n}$$

Following eq(13e)

$$\left(\begin{array}{l}\dfrac{\left(-i\hbar\overline{r}_1+\dfrac{p_{r_1}}{2}\right)^2}{2m_1}+\dfrac{\left(-i\hbar\overline{r}_2+\dfrac{p_{r_2}}{2}\right)^2}{2m_2}+\dfrac{\left(-i\hbar\overline{r}_3+\dfrac{p_{r_3}}{2}\right)^2}{2m_3}\\ +\dfrac{2b_1}{\left(i\hbar\overline{p}_{r_1}+\dfrac{r_1}{2}\right)}\left(-i\hbar\overline{r}_1+\dfrac{p_{r_1}}{2}\right)+\dfrac{2b_2}{\left(i\hbar\overline{p}_{r_2}+\dfrac{r_2}{2}\right)}\left(-i\hbar\overline{r}_2+\dfrac{p_{r_2}}{2}\right)\\ +\dfrac{2b_3}{\left(i\hbar\overline{p}_{r_3}+\dfrac{r_3}{2}\right)}\left(-i\hbar\overline{r}_3+\dfrac{p_{r_3}}{2}\right)+\dfrac{c_1L_1(L_1+1)}{\left(i\hbar\overline{p}_{r_1}+\dfrac{r_1}{2}\right)^2}+\dfrac{c_2L_2(L_2+1)}{\left(i\hbar\overline{p}_{r_2}+\dfrac{r_2}{2}\right)^2}\\ +\dfrac{c_3L_3(L_3+1)}{\left(i\hbar\overline{p}_{r_3}+\dfrac{r_3}{2}\right)^2}+a_{12}\left(\dfrac{1}{\left|\left(i\hbar\overline{p}_{r_1}+\dfrac{r_1}{2}\right)-\left(i\hbar\overline{p}_{r_2}+\dfrac{r_2}{2}\right)\right|}\right)\\ +a_{13}\left(\dfrac{1}{\left|\left(i\hbar\overline{p}_{r_1}+\dfrac{r_1}{2}\right)-\left(i\hbar\overline{p}_{r_3}+\dfrac{r_3}{2}\right)\right|}\right)+a_{23}\left(\dfrac{1}{\left|\left(i\hbar\overline{p}_{r_2}+\dfrac{r_2}{2}\right)-\left(i\hbar\overline{p}_{r_3}+\dfrac{r_3}{2}\right)\right|}\right)\end{array}\right)\breve{\Psi}_{\mathbf{r}\ \text{phasespace}}\left(\overline{r}_1,\overline{r}_2,\overline{r}_3;\overline{p}_{r_1},\overline{p}_{r_2},\overline{p}_{r_3};t\right)$$
$$=i\hbar\partial_t\breve{\Psi}_{\mathbf{r}\ \text{phasespace}}\left(\overline{r}_1,\overline{r}_2,\overline{r}_3;\overline{p}_{r_1},\overline{p}_{r_2},\overline{p}_{r_3};t\right) \tag{13o}$$

Proceeding as in eq(13f) through eq(13h); for N=3 eqs (13a), (13b), (13c) and (13d) become respectively

$$L^{-1}_{(\bar{\mathbf{p}}_{\mathbf{r}}\to\mathbf{p}_{\mathbf{r}})}\left(e^{\frac{-it}{\hbar}\left(\sum_{i<j}a_{ij}\left(\frac{1}{\left|\left(i\hbar\bar{p}_{r_i}+\frac{r_i}{2}\right)-\left(i\hbar\bar{p}_{r_j}+\frac{r_j}{2}\right)\right|}\right)\right)}\right)$$

$$=\underbrace{\ast}_{i<j}\left(\delta\left[p_{r_i}+p_{r_j}\right]\left(\frac{1}{\pi p_{r_i}\sqrt{\frac{-i\hbar}{a_{ij}t}}}e^{\frac{ip_{r_i}(r_i-r_j)}{2\hbar}}\left(\sqrt{\frac{p_{r_i}}{\hbar}}K_1\left[\frac{2\sqrt{\frac{p_{r_i}}{\hbar}}}{\sqrt{\frac{-i\hbar}{a_{ij}t}}}\right]-\sqrt{\frac{-p_{r_i}}{\hbar}}K_1\left[\frac{2\sqrt{\frac{-p_{r_i}}{\hbar}}}{\sqrt{\frac{-i\hbar}{a_{ij}t}}}\right]\right)\right)\right) \quad (13p)$$

example N = 3, $f_{1,2}\underbrace{\ast}_{p_{r_1}}f_{1,3}\underbrace{\ast}_{p_{r_2},p_{r_3}}f_{2,3}$

$$,\ f_{ij}=\delta\left[p_{r_i}+p_{r_j}\right]\left(\frac{1}{\pi p_{r_i}\sqrt{\frac{-i\hbar}{a_{ij}t}}}e^{\frac{ip_{r_i}(r_i-r_j)}{2\hbar}}\left(\sqrt{\frac{p_{r_i}}{\hbar}}K_1\left[\frac{2\sqrt{\frac{p_{r_i}}{\hbar}}}{\sqrt{\frac{-i\hbar}{a_{ij}t}}}\right]-\sqrt{\frac{-p_{r_i}}{\hbar}}K_1\left[\frac{2\sqrt{\frac{-p_{r_i}}{\hbar}}}{\sqrt{\frac{-i\hbar}{a_{ij}t}}}\right]\right)\right)$$

(13q)

$$\Psi_{\mathbf{r}\,\text{phasespace}}\left(r_1,r_2,r_3;p_{r_1},p_{r_2},p_{r_3};t\right)=$$

$$\Psi_{\mathbf{r}\,\text{phasespace}}\left(r_1,r_2,r_3;p_{r_1},p_{r_2},p_{r_3};t=0\right)\underbrace{\ast}_{\mathbf{r},\mathbf{p}_{\mathbf{r}}}\prod_{i=1}^{3}\left(\frac{e^{\frac{i\left(r_1^2m_1-tr_1p_{r_1}\right)}{2\hbar t}}}{\hbar\sqrt{\frac{2\pi it}{\hbar m_1}}}\right)\underbrace{\ast}_{\mathbf{r}}\prod_{i=1}^{3}\left(\delta\left[p_{r_i}\right]\delta\left[r_i\right]+\frac{ib_ie^{\frac{2ib_ip_{r_i}t}{\hbar r_i}}\left(\sqrt{-ip_{r_i}}-\sqrt{ip_{r_i}}\right)}{\hbar\pi\ r_i^2\sqrt{\left|p_{r_i}\right|}}\right)$$

$$\underbrace{\ast}_{\mathbf{p}_{\mathbf{r}}}\prod_{i=1}^{3}\left(\frac{-1}{24\pi\hbar^6}\left(c_i^2L_i^2\left(L_i+1\right)^2p_{r_i}^3t^2\left(\begin{array}{l}\left(e^{\frac{ip_{r_i}r_i}{2\hbar}}-1\right)Log\left[\frac{-p_{r_i}}{\hbar}\right]\\+Log\left[\frac{p_{r_i}}{\hbar}\right]-Log\left[\frac{p_{r_i}}{\hbar}\right]HeavisideStep\left[p_{r_i}\right]\end{array}\right)\right)\right)$$

$$\underbrace{\ast}_{\mathbf{p}_{\mathbf{r}}}\left(\underbrace{\overset{3}{\ast}}_{i<j}\left(\delta\left[p_{r_i}+p_{r_j}\right]\left(\frac{1}{\pi p_{r_i}\sqrt{\frac{-i\hbar}{a_{ij}t}}}e^{\frac{ip_{r_i}(r_i-r_j)}{2\hbar}}\left(\sqrt{\frac{p_{r_i}}{\hbar}}K_1\left[\frac{2\sqrt{\frac{p_{r_i}}{\hbar}}}{\sqrt{\frac{-i\hbar}{a_{ij}t}}}\right]-\sqrt{\frac{-p_{r_i}}{\hbar}}K_1\left[\frac{2\sqrt{\frac{-p_{r_i}}{\hbar}}}{\sqrt{\frac{-i\hbar}{a_{ij}t}}}\right]\right)\right)\right)\right)$$

Now consider the earlier mentioned particular case of the general nonrelativistic Schrödinger Molecular Hamiltonian with Pairwise Coulomb interaction dynamics

$$\hat{H}=\sum_A\frac{1}{2M_A}\hat{\mathbf{P}}_A^2+\sum_i\frac{1}{2m_e}\hat{\mathbf{P}}_i^2+\sum_{A<B}\frac{e^2}{8\pi\varepsilon_0}\left(\frac{Z_AZ_B}{\left|\hat{r}_A-\hat{r}_B\right|}\right)+\sum_{i<j}\frac{e^2}{8\pi\varepsilon_0}\left(\frac{1}{\left|\hat{r}_i-\hat{r}_j\right|}\right)+\sum_{A,i}\left(\frac{-Z_Ae^2}{\left|\hat{r}_A-\hat{r}_i\right|}\right) \quad (13a)$$

Though we shall not pursue the details in the present note [viz ref. 7], clearly one can see that, after the angular components are separated in the usual way and with a straightforward scaling of the variables, the present N-particle result yields the solution of the radial

Schrodinger equation for the general nonrelativistic molecular Hamiltonian in the QPSR which has the form

$$\hat{H}_{\substack{\text{Schrodinger}\\ \text{Molecular Radial}}} = \sum_A \frac{1}{2M_A}\left(\left(-i\hbar\partial_{r_A} + \frac{p_{r_A}}{2}\right)^2 + \frac{2}{\left(i\hbar\partial_{p_{r_A}} + \frac{r_A}{2}\right)}\left(-i\hbar\partial_{r_A} + \frac{p_{r_A}}{2}\right) + \frac{L_A(L_A+1)}{\left(i\hbar\partial_{p_{r_A}} + \frac{r_A}{2}\right)^2}\right)$$

$$+\sum_i \frac{1}{2m_e}\left(\left(-i\hbar\partial_{r_i} + \frac{p_{r_i}}{2}\right)^2 + \frac{2}{\left(i\hbar\partial_{p_{r_i}} + \frac{r_i}{2}\right)}\left(-i\hbar\partial_{r_i} + \frac{p_{r_i}}{2}\right) + \frac{L_i(L_i+1)}{\left(i\hbar\partial_{p_{r_i}} + \frac{r_i}{2}\right)^2}\right) \tag{13r}$$

$$+\sum_{A<B}\frac{e^2}{8\pi\varepsilon_0}\frac{Z_A Z_B}{\left|\left(i\hbar\partial_{p_{r_A}} + \frac{r_A}{2}\right) - \left(i\hbar\partial_{p_{r_B}} + \frac{r_B}{2}\right)\right|} + \sum_{i<j}\frac{e^2}{8\pi\varepsilon_0}\frac{1}{\left|\left(i\hbar\partial_{p_{r_i}} + \frac{r_i}{2}\right) - \left(i\hbar\partial_{p_{r_j}} + \frac{r_j}{2}\right)\right|}$$

$$+\sum_{A,i}\frac{-Z_A e^2}{\left|\left(i\hbar\partial_{p_{r_A}} + \frac{r_A}{2}\right) - \left(i\hbar\partial_{p_{r_i}} + \frac{r_i}{2}\right)\right|}$$

Hence considering the radial molecular Hamiltonian (13r) as was done in (13a) and applying HOA to (13r) as was done in (13b), the exact analytical solution of the Radial Schrodinger equation for the general nonrelativistic molecular Hamiltonian in QPSR may be expressed in quadratures via the above.

Example 4. Dirac and Majorana Equations with Minimum-Coupled Electromagnetic Gauge Field

Following [ref. 8], first consider the related Dirac equation with minimum-coupled electromagnetic gauge field interaction

$$\mathbf{A}(x_1,x_2,x_3,t) = A_1(x_1,x_2,x_3,t)\mathbf{e}_{\mathbf{x}_1} + A_2(x_1,x_2,x_3,t)\mathbf{e}_{\mathbf{x}_2} + A_3(x_1,x_2,x_3,t)\mathbf{e}_{\mathbf{x}_3},\ A_0(x_1,x_2,x_3,t) \tag{14a}$$

$$\mathbf{H}_{\text{Dirac}_{4\times4}}\Psi_{\text{D}} = \left(m c^2 a_0 + \sum_{j=1}^{3}\left(a_j\left(p_j - eA_j\right)c + eA_0\right)\right)\Psi_{\text{D}} = i\hbar\partial_t\Psi_{\text{D}}$$

$$\mathbf{A}(x_1,x_2,x_3,t) = A_1(x_1,x_2,x_3,t)\mathbf{e}_{\mathbf{x}_1} + A_2(x_1,x_2,x_3,t)\mathbf{e}_{\mathbf{x}_2} + A_3(x_1,x_2,x_3,t)\mathbf{e}_{\mathbf{x}_3},\ A_0(x_1,x_2,x_3,t)$$

$$\Psi_{\text{D}} \equiv \begin{pmatrix}\Psi_{D_1}\\ \Psi_{D_2}\\ \Psi_{D_3}\\ \Psi_{D_4}\end{pmatrix} : \text{4-component Dirac wavefunction}$$

$$\Psi_{D_0} \equiv \begin{pmatrix}\Psi_{D_{10}}\\ \Psi_{D_{20}}\\ \Psi_{D_{30}}\\ \Psi_{D_{40}}\end{pmatrix} : 4-\text{component Dirac Initial-State}$$

$$a_0 = \begin{bmatrix} 1 & 0 & 0 & 0 \\ 0 & 1 & 0 & 0 \\ 0 & 0 & -1 & 0 \\ 0 & 0 & 0 & -1 \end{bmatrix}, a_1 = \begin{bmatrix} 0 & 0 & 0 & 1 \\ 0 & 0 & 1 & 0 \\ 0 & 1 & 0 & 0 \\ 1 & 0 & 0 & 0 \end{bmatrix}$$

$$a_2 = \begin{bmatrix} 0 & 0 & 0 & -i \\ 0 & 0 & i & 0 \\ 0 & -i & 0 & 0 \\ i & 0 & 0 & 0 \end{bmatrix}, a_3 = \begin{bmatrix} 0 & 0 & 1 & 0 \\ 0 & 0 & 0 & -1 \\ 1 & 0 & 0 & 0 \\ 0 & -1 & 0 & 0 \end{bmatrix}$$

Similarly for the Majorana equation minimal-coupled to the EM gauge field

$$\begin{aligned} &-im\, c\sigma^2\rho_M^* + \left(i\hbar\sigma^\mu\partial_\mu - eA_\mu\right)\rho_M = 0 \\ &A_\mu \equiv \begin{pmatrix} \mathbf{A}(x_1,x_2,x_3,t) = A_1(x_1,x_2,x_3,t)\mathbf{e}_{\mathbf{x}_1} + A_2(x_1,x_2,x_3,t)\mathbf{e}_{\mathbf{x}_2} + A_3(x_1,x_2,x_3,t)\mathbf{e}_{\mathbf{x}_3}, \\ A_0(x_1,x_2,x_3,t) \end{pmatrix} \\ &\rho_M = \begin{pmatrix} \rho_{M_1} \\ \rho_{M_2} \end{pmatrix} : 2 - \text{component Majorana wavefunction,} \\ &\rho_{M_0} = \begin{pmatrix} \rho_{M_{10}} \\ \rho_{M_{20}} \end{pmatrix} : 2 - \text{component Majorana Initial-State} \\ &\sigma^\mu : \text{usual } 2\times 2 \text{ Pauli spin matrices } \sigma^{1,2,3}, \sigma^0 = -i\mathbf{I}_{2\times 2} \end{aligned} \tag{14b}$$

Now the connection between the Majorana (14b) ρ_M and Dirac (14a) $\mathbf{\Psi}_\mathrm{D}$ wavefunctions subject to the Majorana self-conjugacy condition $\mathbf{\Psi}_\mathrm{D} = \mathbf{\Psi}_\mathrm{D}^c$ is thoroughly discussed in the bibliography of [ref 8]; only some key relationships between them are reproduced here for convenience

$$\begin{aligned} &\mathbf{\Psi}_\mathrm{D} \equiv \begin{pmatrix} \mathbf{\Psi}_{\mathrm{D}1} \\ \mathbf{\Psi}_{\mathrm{D}2} \\ \mathbf{\Psi}_{\mathrm{D}3} \\ \mathbf{\Psi}_{\mathrm{D}4} \end{pmatrix} = \begin{pmatrix} \chi_D \\ \sigma_2\phi_D^* \end{pmatrix}, \quad \mathbf{\Psi}_\mathrm{D}^c = \begin{pmatrix} \sigma_2 & 0 \\ 0 & \sigma_2 \end{pmatrix} \mathbf{\Psi}_\mathrm{D}^{*Transpose} = \begin{pmatrix} \phi_D \\ \sigma_2\chi_D^* \end{pmatrix}, \\ &\chi_D = \begin{pmatrix} \mathbf{\Psi}_{\mathrm{D}1} \\ \mathbf{\Psi}_{\mathrm{D}2} \end{pmatrix}, \quad \sigma_2\phi_D^* = \begin{pmatrix} \mathbf{\Psi}_{\mathrm{D}3} \\ \mathbf{\Psi}_{\mathrm{D}4} \end{pmatrix}, \phi_\mathrm{D} = \begin{pmatrix} i\mathbf{\Psi}_{\mathrm{D}4}^* \\ -i\mathbf{\Psi}_{\mathrm{D}3}^* \end{pmatrix}, \\ &\text{Majorana Self-Conjugacy condition } \mathbf{\Psi}_\mathrm{D} = \mathbf{\Psi}_\mathrm{D}^c \\ &\chi_D = \frac{1}{\sqrt{2}}\left(\rho_{M_2} + i\rho_{M_1}\right), \rho_{M_2} = \frac{1}{\sqrt{2}}\left(\chi_D + \phi_D\right) = \frac{1}{\sqrt{2}}\begin{pmatrix} \mathbf{\Psi}_{\mathrm{D}1} + i\mathbf{\Psi}_{\mathrm{D}4}^* \\ \mathbf{\Psi}_{\mathrm{D}2} - i\mathbf{\Psi}_{\mathrm{D}3}^* \end{pmatrix} \\ &\phi_D = \frac{1}{\sqrt{2}}\left(\rho_{M_2} - i\rho_{M_1}\right), \rho_{M_1} = \frac{i}{\sqrt{2}}\left(\chi_D - \phi_D\right) = \frac{i}{\sqrt{2}}\begin{pmatrix} \mathbf{\Psi}_{\mathrm{D}1} - i\mathbf{\Psi}_{\mathrm{D}4}^* \\ \mathbf{\Psi}_{\mathrm{D}2} + i\mathbf{\Psi}_{\mathrm{D}3}^* \end{pmatrix} \\ &\rho_M = \begin{pmatrix} \rho_{M_1} \\ \rho_{M_2} \end{pmatrix} \end{aligned} \tag{14c}$$

So by way of (14c), given the related Dirac wavefunction and subject to the Majorana self-conjugacy condition $\mathbf{\Psi}_\mathrm{D} = \mathbf{\Psi}_\mathrm{D}^c$, the Majorana wavefunction ascends naturally. Moreover, by way of the HOA method, substituting the Dirac Hamiltonian of (14a) gives the quantum phase space dynamics of the Dirac system for initial conditions and EM gauge fields of general form.

$$
\begin{aligned}
&\hat{\mathbf{H}}_{\mathrm{Dirac}_{4\times 4}}\begin{pmatrix} i\hbar\partial_{p_1}+\alpha_1 x_1, i\hbar\partial_{p_2}+\alpha_2 x_2, i\hbar\partial_{p_3}+\alpha_3 x_3; \\ -i\hbar\partial_{x_1}+\gamma_1 p_1, -i\hbar\partial_{x_2}+\gamma_2 p_2, -i\hbar\partial_{x_3}+\gamma_3 p_3; t \end{pmatrix}\boldsymbol{\Psi}_{\mathrm{D}}(x_1,x_2,x_3;p_1,p_2,p_3;t) \\
&= i\hbar\partial_t \boldsymbol{\Psi}_{\mathrm{D}}(x_1,x_2,x_3;p_1,p_2,p_3;t) \\
&\left(m\,c^2 a_0 + \sum_{j=1}^{3}\begin{pmatrix} a_j\left(-i\hbar\partial_{x_j}+\gamma_j p_j - eA_j\left(i\hbar\partial_{p_1}+\alpha_1 x_1, i\hbar\partial_{p_2}+\alpha_2 x_2, i\hbar\partial_{p_3}+\alpha_3 x_3, t\right)\right)c \\ +eA_0\left(i\hbar\partial_{p_1}+\alpha_1 x_1, i\hbar\partial_{p_2}+\alpha_2 x_2, i\hbar\partial_{p_3}+\alpha_3 x_3;, t\right) \end{pmatrix}\right)\boldsymbol{\Psi}_{\mathrm{D}} = i\hbar\partial_t\boldsymbol{\Psi}_{\mathrm{D}} \\
&\mathbf{A}\left(i\hbar\partial_{p_1}+\alpha_1 x_1, i\hbar\partial_{p_2}+\alpha_2 x_2, i\hbar\partial_{p_3}+\alpha_3 x_3, t\right)= \\
&A_1\left(i\hbar\partial_{p_1}+\alpha_1 x_1, i\hbar\partial_{p_2}+\alpha_2 x_2, i\hbar\partial_{p_3}+\alpha_3 x_3, t\right)\mathbf{e}_{\mathbf{x}_1} \\
&+A_2\left(i\hbar\partial_{p_1}+\alpha_1 x_1, i\hbar\partial_{p_2}+\alpha_2 x_2, i\hbar\partial_{p_3}+\alpha_3 x_3, t\right)\mathbf{e}_{\mathbf{x}_2} \\
&+A_3\left(i\hbar\partial_{p_1}+\alpha_1 x_1, i\hbar\partial_{p_2}+\alpha_2 x_2, i\hbar\partial_{p_3}+\alpha_3 x_3, t\right)\mathbf{e}_{\mathbf{x}_3} \\
&, A_0\left(i\hbar\partial_{p_1}+\alpha_1 x_1, i\hbar\partial_{p_2}+\alpha_2 x_2, i\hbar\partial_{p_3}+\alpha_3 x_3, t\right) \\
&\boldsymbol{\Psi}_{\mathrm{D}} \equiv \begin{pmatrix} \boldsymbol{\Psi}_{\mathrm{D1}} \\ \boldsymbol{\Psi}_{\mathrm{D2}} \\ \boldsymbol{\Psi}_{\mathrm{D3}} \\ \boldsymbol{\Psi}_{\mathrm{D4}} \end{pmatrix} : \text{4 - component Dirac wavefunction}
\end{aligned}
\tag{14d}
$$

Hence the configuration space dynamics for the related minimal-coupled Dirac system

$$
\begin{aligned}
&\boldsymbol{\Psi}_{\mathrm{D\ configuration\ space}}(x_1,x_2,x_3,t)= \\
&\begin{pmatrix} \boldsymbol{\Psi}_{\mathrm{D1}}(x_1,x_2,x_3,t) \\ \boldsymbol{\Psi}_{\mathrm{D2}}(x_1,x_2,x_3,t) \\ \boldsymbol{\Psi}_{\mathrm{D3}}(x_1,x_2,x_3,t) \\ \boldsymbol{\Psi}_{\mathrm{D4}}(x_1,x_2,x_3,t) \end{pmatrix}_{\text{configuration space}} = \\
&= \int_{-\infty}^{\infty}\frac{e^{\frac{ix_1p_1}{2\hbar}}}{\sqrt{4\pi\hbar}}\int_{-\infty}^{\infty}\frac{e^{\frac{ix_2p_2}{2\hbar}}}{\sqrt{4\pi\hbar}}\int_{-\infty}^{\infty}\frac{e^{\frac{ix_3p_3}{2\hbar}}}{\sqrt{4\pi\hbar}}\, L^{-1}_{\left(\begin{pmatrix}\bar{x}_1,\ldots,\bar{x}_n; \\ \bar{p}_1,\ldots,\bar{p}_n\end{pmatrix} \to \begin{pmatrix} x_1,\ldots,x_n; \\ p_1,\ldots,p_n\end{pmatrix}\right)} e^{\frac{-i}{\hbar}\int_0^t \left(m\,c^2 a_0 + \sum_{j=1}^{3}\left(a_j\left(\begin{matrix} -i\hbar\bar{x}_j+\gamma_j p_j \\ -eA_j\begin{pmatrix} i\hbar\bar{p}_1+\alpha_1x_1, \\ i\hbar\bar{p}_2+\alpha_2x_2, \\ i\hbar\bar{p}_3+\alpha_3x_3, u\end{pmatrix}\end{matrix}\right)c \\ +eA_0\begin{pmatrix} i\hbar\bar{p}_1+\alpha_1x_1, \\ i\hbar\bar{p}_2+\alpha_2x_2, \\ i\hbar\bar{p}_3+\alpha_3x_3, u\end{pmatrix}\right)\right)du} \begin{pmatrix} \breve{\boldsymbol{\Psi}}_{D0_1} \\ \breve{\boldsymbol{\Psi}}_{D0_2} \\ \breve{\boldsymbol{\Psi}}_{D0_3} \\ \breve{\boldsymbol{\Psi}}_{D0_4} \end{pmatrix} dp_1 dp_2 dp_3 \\
&\breve{\boldsymbol{\Psi}}_{D0}\begin{pmatrix} \bar{x}_1,\bar{x}_2,\bar{x}_3; \\ \bar{p}_1,\bar{p}_2,\bar{p}_3; t=0 \end{pmatrix} = \begin{pmatrix} \breve{\boldsymbol{\Psi}}_{D0_1} \\ \breve{\boldsymbol{\Psi}}_{D0_2} \\ \breve{\boldsymbol{\Psi}}_{D0_3} \\ \breve{\boldsymbol{\Psi}}_{D0_4} \end{pmatrix} : \text{Transformed Initial-condition vector}
\end{aligned}
\tag{14e}
$$

where the explicit form of the $\hat{\mathbf{H}}_{\mathrm{Dirac}_{4\times4}}$ in the exponent 4x4 matrix integral is supplied as

$$
\begin{pmatrix}
\begin{pmatrix} mc^2 \\ +ecA_0\begin{pmatrix} i\hbar\bar{p}_1+\alpha_1x_1, \\ i\hbar\bar{p}_2+\alpha_2x_2, \\ i\hbar\bar{p}_3+\alpha_3x_3,t \end{pmatrix} \end{pmatrix} & 0 & \begin{pmatrix} c(-i\hbar\bar{x}_3+\gamma_3p_3) \\ -eA_3\begin{pmatrix} i\hbar\bar{p}_1+\alpha_1x_1, \\ i\hbar\bar{p}_2+\alpha_2x_2, \\ i\hbar\bar{p}_3+\alpha_3x_3,t \end{pmatrix} \end{pmatrix} & \begin{pmatrix} c(-i\hbar\bar{x}_1+\gamma_1p_1) \\ -eA_1\begin{pmatrix} i\hbar\bar{p}_1+\alpha_1x_1, \\ i\hbar\bar{p}_2+\alpha_2x_2, \\ i\hbar\bar{p}_3+\alpha_3x_3,t \end{pmatrix} \\ -i(c(-i\hbar\bar{x}_2+\gamma_2p_2) \\ -eA_2\begin{pmatrix} i\hbar\bar{p}_1+\alpha_1x_1, \\ i\hbar\bar{p}_2+\alpha_2x_2, \\ i\hbar\bar{p}_3+\alpha_3x_3,t \end{pmatrix}) \end{pmatrix} \\
0 & \begin{pmatrix} mc^2 \\ +ecA_0\begin{pmatrix} i\hbar\bar{p}_1+\alpha_1x_1, \\ i\hbar\bar{p}_2+\alpha_2x_2, \\ i\hbar\bar{p}_3+\alpha_3x_3,t \end{pmatrix} \end{pmatrix} & \begin{pmatrix} c(-i\hbar\bar{x}_1+\gamma_1p_1) \\ -eA_1\begin{pmatrix} i\hbar\bar{p}_1+\alpha_1x_1, \\ i\hbar\bar{p}_2+\alpha_2x_2, \\ i\hbar\bar{p}_3+\alpha_3x_3,t \end{pmatrix} \\ +i(c(-i\hbar\bar{x}_2+\gamma_2p_2) \\ -eA_2\begin{pmatrix} i\hbar\bar{p}_1+\alpha_1x_1, \\ i\hbar\bar{p}_2+\alpha_2x_2, \\ i\hbar\bar{p}_3+\alpha_3x_3,t \end{pmatrix}) \end{pmatrix} & -\begin{pmatrix} c(-i\hbar\bar{x}_3+\gamma_3p_3) \\ -eA_3\begin{pmatrix} i\hbar\bar{p}_1+\alpha_1x_1, \\ i\hbar\bar{p}_2+\alpha_2x_2, \\ i\hbar\bar{p}_3+\alpha_3x_3,t \end{pmatrix} \end{pmatrix} \\
\begin{pmatrix} c(-i\hbar\bar{x}_3+\gamma_3p_3) \\ -eA_3\begin{pmatrix} i\hbar\bar{p}_1+\alpha_1x_1, \\ i\hbar\bar{p}_2+\alpha_2x_2, \\ i\hbar\bar{p}_3+\alpha_3x_3,t \end{pmatrix} \end{pmatrix} & \begin{pmatrix} c(-i\hbar\bar{x}_1+\gamma_1p_1) \\ -eA_1\begin{pmatrix} i\hbar\bar{p}_1+\alpha_1x_1, \\ i\hbar\bar{p}_2+\alpha_2x_2, \\ i\hbar\bar{p}_3+\alpha_3x_3,t \end{pmatrix} \\ -i(c(-i\hbar\bar{x}_2+\gamma_2p_2) \\ -eA_2\begin{pmatrix} i\hbar\bar{p}_1+\alpha_1x_1, \\ i\hbar\bar{p}_2+\alpha_2x_2, \\ i\hbar\bar{p}_3+\alpha_3x_3,t \end{pmatrix}) \end{pmatrix} & \begin{pmatrix} -mc^2 \\ +ecA_0\begin{pmatrix} i\hbar\bar{p}_1+\alpha_1x_1, \\ i\hbar\bar{p}_2+\alpha_2x_2, \\ i\hbar\bar{p}_3+\alpha_3x_3,t \end{pmatrix} \end{pmatrix} & 0 \\
\begin{pmatrix} c(-i\hbar\bar{x}_1+\gamma_1p_1) \\ -eA_1\begin{pmatrix} i\hbar\bar{p}_1+\alpha_1x_1, \\ i\hbar\bar{p}_2+\alpha_2x_2, \\ i\hbar\bar{p}_3+\alpha_3x_3,t \end{pmatrix} \\ +i(c(-i\hbar\bar{x}_2+\gamma_2p_2) \\ -eA_2\begin{pmatrix} i\hbar\bar{p}_1+\alpha_1x_1, \\ i\hbar\bar{p}_2+\alpha_2x_2, \\ i\hbar\bar{p}_3+\alpha_3x_3,t \end{pmatrix}) \end{pmatrix} & -\begin{pmatrix} c(-i\hbar\bar{x}_3+\gamma_3p_3) \\ -eA_3\begin{pmatrix} i\hbar\bar{p}_1+\alpha_1x_1, \\ i\hbar\bar{p}_2+\alpha_2x_2, \\ i\hbar\bar{p}_3+\alpha_3x_3,t \end{pmatrix} \end{pmatrix} & 0 & \begin{pmatrix} -mc^2 \\ +ecA_0\begin{pmatrix} i\hbar\bar{p}_1+\alpha_1x_1, \\ i\hbar\bar{p}_2+\alpha_2x_2, \\ i\hbar\bar{p}_3+\alpha_3x_3,t \end{pmatrix} \end{pmatrix}
\end{pmatrix}
\tag{14f}
$$

yields by way of (14c), the associated Majorana wavefunction for the dynamics of the minimal-coupled system with arbitrary profile EM interaction and initial-conditions, in terms of the quadrature solutions for the related Dirac system just calculated above

$$\rho_{M_2} = \frac{1}{\sqrt{2}}\begin{pmatrix} \Psi_{\mathrm{D1}}(x_1,x_2,x_3,t) + i\Psi^*_{\mathrm{D4}}(x_1,x_2,x_3,t) \\ \Psi_{\mathrm{D2}}(x_1,x_2,x_3,t) - i\Psi^*_{\mathrm{D3}}(x_1,x_2,x_3,t) \end{pmatrix}$$

$$\rho_{M_1} = \frac{i}{\sqrt{2}}\begin{pmatrix} \Psi_{\mathrm{D1}}(x_1,x_2,x_3,t) - i\Psi^*_{\mathrm{D4}}(x_1,x_2,x_3,t) \\ \Psi_{\mathrm{D2}}(x_1,x_2,x_3,t) + i\Psi^*_{\mathrm{D3}}(x_1,x_2,x_3,t) \end{pmatrix}$$

$$\rho_{M\,\text{configuration space}} = \begin{pmatrix} \rho_{M_1} \\ \rho_{M_2} \end{pmatrix}_{\text{configuration space}}$$

$$\rho_{M_0\text{configuration space}} = \begin{pmatrix} \rho_{M_{10}} \\ \rho_{M_{20}} \end{pmatrix}_{\text{configuration space}} \quad \text{:Majorana Initial - State}$$

$$\rho_{M_{20}} = \frac{1}{\sqrt{2}}\begin{pmatrix} \Psi_{\mathrm{D1}_0}(x_1,x_2,x_3,t) + i\Psi^*_{\mathrm{D4}_0}(x_1,x_2,x_3,t) \\ \Psi_{\mathrm{D2}_0}(x_1,x_2,x_3,t) - i\Psi^*_{\mathrm{D3}_0}(x_1,x_2,x_3,t) \end{pmatrix}$$

$$\rho_{M_{10}} = \frac{i}{\sqrt{2}}\begin{pmatrix} \Psi_{\mathrm{D1}_0}(x_1,x_2,x_3,t) - i\Psi^*_{\mathrm{D4}_0}(x_1,x_2,x_3,t) \\ \Psi_{\mathrm{D2}_0}(x_1,x_2,x_3,t) + i\Psi^*_{\mathrm{D3}_0}(x_1,x_2,x_3,t) \end{pmatrix} \tag{14g}$$

Example 5. Dirac Molecular Hamiltonian with Pairwise Coulomb-Breit Interaction

Now consider the example case of the relativistic Molecular Hamiltonian with Pairwise Dirac-Coulomb-Briet interaction dynamics in quantum phase space. Here the Hamiltonian has the 4x4 matrix form

$$\hat{\mathbf{H}}_{\substack{\text{DiracCoulombBreit} \\ 4\times4\ \text{Molecular}}} = \sum_A \left(a_{0_A} M_A c^2 + a_{x_A}\left(-i\hbar\partial_{x_A} + \frac{p_{x_A}}{2}\right)c + a_{y_A}\left(-i\hbar\partial_{y_A} + \frac{p_{y_A}}{2}\right)c + a_{z_A}\left(-i\hbar\partial_{z_A} + \frac{p_{z_A}}{2}\right)\right)$$

$$+\sum_i \left(a_{0_i} m_e c^2 + a_{x_i}\left(-i\hbar\partial_{x_i} + \frac{p_{x_i}}{2}\right)c + a_{y_i}\left(-i\hbar\partial_{y_i} + \frac{p_{y_i}}{2}\right)c + a_{z_i}\left(-i\hbar\partial_{z_i} + \frac{p_{zi}}{2}\right)\right)$$

$$+\sum_{A<B} \frac{e^2}{8\pi\varepsilon_0} \frac{Z_A Z_B\left(1 + \frac{1}{2}\left((\mathbf{a}_A\cdot\mathbf{a}_B) + \frac{(\mathbf{a}_A\cdot\mathbf{r}_{AB})(\mathbf{a}_B\cdot\mathbf{r}_{AB})}{r_{AB}^2}\right)\right)}{\left|\begin{array}{l} \left(\left(i\hbar\partial_{p_{xA}} + \frac{x_A}{2}\right) - \left(i\hbar\partial_{p_{xB}} + \frac{x_B}{2}\right)\right)^2 \\ + \left(\left(i\hbar\partial_{p_{yA}} + \frac{y_A}{2}\right) - \left(i\hbar\partial_{p_{yA}} + \frac{y_B}{2}\right)\right)^2 \\ + \left(\left(i\hbar\partial_{p_{zA}} + \frac{z_A}{2}\right) - \left(i\hbar\partial_{p_{zA}} + \frac{z_B}{2}\right)\right)^2 \end{array}\right|} + \sum_{i<j} \frac{e^2}{8\pi\varepsilon_0} \frac{\left(1 + \frac{1}{2}\left((\mathbf{a}_i\cdot\mathbf{a}_j) + \frac{(\mathbf{a}_i\cdot\mathbf{r}_{ij})(\mathbf{a}_j\cdot\mathbf{r}_{ij})}{r_{ij}^2}\right)\right)}{\left|\begin{array}{l} \left(\left(i\hbar\partial_{p_{x_i}} + \frac{x_i}{2}\right) - \left(i\hbar\partial_{p_{x_j}} + \frac{x_j}{2}\right)\right)^2 \\ + \left(\left(i\hbar\partial_{p_{y_i}} + \frac{y_i}{2}\right) - \left(i\hbar\partial_{p_{y_j}} + \frac{y_j}{2}\right)\right)^2 \\ + \left(\left(i\hbar\partial_{p_{z_i}} + \frac{z_i}{2}\right) - \left(i\hbar\partial_{p_{z_j}} + \frac{z_j}{2}\right)\right)^2 \end{array}\right|}$$

$$+\sum_{A,i}\frac{-Z_A e^2\left(1+\frac{1}{2}\left(\left(\mathbf{a}_A\cdot\mathbf{a}_i\right)+\frac{\left(\mathbf{a}_A\cdot\mathbf{r}_{Ai}\right)\left(\mathbf{a}_i\cdot\mathbf{r}_{Ai}\right)}{r_{Ai}^2}\right)\right)}{\left|\begin{array}{l}\left(\left(i\hbar\partial_{p_{x_A}}+\frac{x_A}{2}\right)-\left(i\hbar\partial_{p_{x_i}}+\frac{x_i}{2}\right)\right)^2\\+\left(\left(i\hbar\partial_{p_{y_A}}+\frac{y_A}{2}\right)-\left(i\hbar\partial_{p_{y_i}}+\frac{y_i}{2}\right)\right)^2\\+\left(\left(i\hbar\partial_{p_{z_A}}+\frac{z_A}{2}\right)-\left(i\hbar\partial_{p_{z_i}}+\frac{z_i}{2}\right)\right)^2\end{array}\right|} \tag{15a}$$

$$a_0=\begin{bmatrix}1&0&0&0\\0&1&0&0\\0&0&-1&0\\0&0&0&-1\end{bmatrix} a_x=\begin{bmatrix}0&0&0&1\\0&0&1&0\\0&1&0&0\\1&0&0&0\end{bmatrix}, a_y=\begin{bmatrix}0&0&0&-i\\0&0&i&0\\0&-i&0&0\\i&0&0&0\end{bmatrix}, a_z=\begin{bmatrix}0&0&1&0\\0&0&0&-1\\1&0&0&0\\0&-1&0&0\end{bmatrix}$$

$$\mathbf{a}=a_x\widehat{\mathbf{e}}_\mathbf{x}+a_y\widehat{\mathbf{e}}_\mathbf{y}+a_z\widehat{\mathbf{e}}_\mathbf{z}$$

$$\sum_{i<j}\frac{1}{r_{ij}}\left(1+\frac{1}{2}\left(\left(\mathbf{a}_i\cdot\mathbf{a}_j\right)+\frac{\left(\mathbf{a}_i\cdot\mathbf{r}_{ij}\right)\left(\mathbf{a}_j\cdot\mathbf{r}_{ij}\right)}{r_{ij}^2}\right)\right),$$

$$\mathbf{r}_{ij}=\left(\left(i\hbar\partial_{p_{x_i}}+\frac{x_i}{2}\right)-\left(i\hbar\partial_{p_{x_j}}+\frac{x_j}{2}\right)\right)\widehat{\mathbf{e}}_\mathbf{x}+\left(\left(i\hbar\partial_{p_{y_i}}+\frac{y_i}{2}\right)-\left(i\hbar\partial_{p_{y_j}}+\frac{y_j}{2}\right)\right)\widehat{\mathbf{e}}_\mathbf{y}+\left(\left(i\hbar\partial_{p_{z_i}}+\frac{z_i}{2}\right)-\left(i\hbar\partial_{p_{z_j}}+\frac{z_j}{2}\right)\right)\widehat{\mathbf{e}}_\mathbf{z}$$

$$r_{ij}=\left|\left(\left(i\hbar\partial_{p_{x_i}}+\frac{x_i}{2}\right)-\left(i\hbar\partial_{p_{x_j}}+\frac{x_j}{2}\right)\right)\widehat{\mathbf{e}}_\mathbf{x}+\left(\left(i\hbar\partial_{p_{y_i}}+\frac{y_i}{2}\right)-\left(i\hbar\partial_{p_{y_j}}+\frac{y_j}{2}\right)\right)\widehat{\mathbf{e}}_\mathbf{y}+\left(\left(i\hbar\partial_{p_{z_i}}+\frac{z_i}{2}\right)-\left(i\hbar\partial_{p_{z_j}}+\frac{z_j}{2}\right)\right)\widehat{\mathbf{e}}_\mathbf{z}\right| \tag{15b}$$

Since the Hamiltonian of this system is matrix-valued $\underset{4\times 4}{\hat{\mathbf{H}}}_{\substack{\text{DiracCoulombBreit}\\ \text{Molecular}}}$, the quantum phase space wavefunction.is a column vector $\underset{4\times 1\text{ column vector}}{\mathbf{\Psi}}$

$$\underset{4\times 1\text{ column vector}}{\mathbf{\Psi}}\begin{pmatrix}x_A,y_A,z_A,x_B,y_B,z_B,\\x_i,y_i,z_i,x_j,y_j,z_j;\\p_{x_A},p_{y_A},p_{z_A},p_{x_B},p_{y_B},p_{z_B},\\p_{x_i},p_{y_i},p_{z_i},p_{x_j},p_{y_j},p_{z_j};t\end{pmatrix}=$$

$$L^{-1}_{\left(\begin{pmatrix}\bar{x}_A,\bar{y}_A,\bar{z}_A,\bar{x}_B,\bar{y}_B,\bar{z}_B,\\\bar{x}_i,\bar{y}_i,\bar{z}_i,\bar{x}_j,\bar{y}_j,\bar{z}_j\end{pmatrix}\to\begin{pmatrix}x_A,y_A,z_A,x_B,y_B,z_B,\\x_i,y_i,z_i,x_j,y_j,z_j\end{pmatrix}\right)}\left[L^{-1}_{\left(\begin{matrix}p_{x_A},p_{y_A},p_{z_A},p_{x_B},p_{y_B},p_{z_B},\\p_{x_i},p_{y_i},p_{z_i},p_{x_j},p_{y_j},p_{z_j}\end{matrix}\to\begin{pmatrix}p_{x_A},p_{y_A},p_{z_A},p_{x_B},p_{y_B},p_{z_B},\\p_{x_i},p_{y_i},p_{z_i},p_{x_j},p_{y_j},p_{z_j}\end{pmatrix}\right)}\left[\begin{array}{l}e^{\frac{-i}{\hbar}\int_0^t\underset{4\times 4}{\hat{\mathbf{H}}}_{\substack{\text{DiracCoulombBreit}\\ \text{Molecular}}}\begin{pmatrix}-i\hbar\bar{x}_A+\frac{p_{x_A}}{2},-i\hbar\bar{y}_A+\frac{p_{y_A}}{2},-i\hbar\bar{z}_A+\frac{p_{z_A}}{2},-i\hbar\bar{x}_i+\frac{p_{x_i}}{2},-i\hbar\bar{y}_i+\frac{p_{y_i}}{2},-i\hbar\bar{z}_i+\frac{p_{z_i}}{2};\\i\hbar\bar{p}_{x_A}+\frac{x_A}{2},i\hbar\bar{p}_{y_A}+\frac{y_A}{2},i\hbar\bar{p}_{z_A}+\frac{z_A}{2},i\hbar\bar{p}_{x_B}+\frac{x_B}{2},i\hbar\bar{p}_{y_B}+\frac{y_B}{2},i\hbar\bar{p}_{z_B}+\frac{z_B}{2},\\i\hbar\bar{p}_{x_i}+\frac{x_i}{2},i\hbar\bar{p}_{y_i}+\frac{y_i}{2},i\hbar\bar{p}_{z_i}+\frac{z_i}{2},i\hbar\bar{p}_{x_j}+\frac{x_j}{2},i\hbar\bar{p}_{y_j}+\frac{y_j}{2},i\hbar\bar{p}_{z_j}+\frac{z_j}{2}\end{pmatrix}du}\\\times\\\underset{4\times 1\text{ column vector}}{\breve{\mathbf{\Psi}}_0}\left(\bar{x}_1,\ldots,\bar{x}_n;\bar{p}_1,\ldots,\bar{p}_n;t=0\right)\end{array}\right]\right] \tag{15c}$$

$$\hat{\mathbf{H}}_{\substack{4\times 4\,\text{DiracCoulombBreit}\\ \text{Molecular}}} = \sum_A \left(a_{0_A} M_A c^2 + a_{x_A}\left(-i\hbar\overline{x}_A + \frac{p_{x_A}}{2}\right)c + a_{y_A}\left(-i\hbar\overline{y}_A + \frac{p_{y_A}}{2}\right)c + a_{z_A}\left(-i\hbar\overline{z}_A + \frac{p_{z_A}}{2}\right)\right)$$
$$+\sum_i \left(a_{0_i} m_e c^2 + a_{x_i}\left(-i\hbar\overline{x}_i + \frac{p_{x_i}}{2}\right)c + a_{y_i}\left(-i\hbar\overline{y}_i + \frac{p_{y_i}}{2}\right)c + a_{z_i}\left(-i\hbar\overline{z}_i + \frac{p_{zi}}{2}\right)\right)$$
$$+\sum_{A<B}\frac{e^2}{8\pi\varepsilon_0}\frac{Z_A Z_B\left(1+\frac{1}{2}\left((\mathbf{a}_A\cdot\mathbf{a}_B)+\frac{(\mathbf{a}_A\cdot\mathbf{r}_{AB})(\mathbf{a}_B\cdot\mathbf{r}_{AB})}{r_{AB}^2}\right)\right)}{\left|\begin{array}{l}\left(\left(i\hbar\overline{p}_{x_A}+\frac{x_A}{2}\right)-\left(i\hbar\overline{p}_{x_B}+\frac{x_B}{2}\right)\right)^2\\ +\left(\left(i\hbar\overline{p}_{y_A}+\frac{y_A}{2}\right)-\left(i\hbar\overline{p}_{y_B}+\frac{y_B}{2}\right)\right)^2\\ +\left(\left(i\hbar\overline{p}_{z_A}+\frac{z_A}{2}\right)-\left(i\hbar\overline{p}_{z_B}+\frac{z_B}{2}\right)\right)^2\end{array}\right|} +\sum_{i<j}\frac{e^2}{8\pi\varepsilon_0}\frac{\left(1+\frac{1}{2}\left((\mathbf{a}_i\cdot\mathbf{a}_j)+\frac{(\mathbf{a}_i\cdot\mathbf{r}_{ij})(\mathbf{a}_j\cdot\mathbf{r}_{ij})}{r_{ij}^2}\right)\right)}{\left|\begin{array}{l}\left(\left(i\hbar\overline{p}_{x_i}+\frac{x_i}{2}\right)-\left(i\hbar\overline{p}_{x_j}+\frac{x_j}{2}\right)\right)^2\\ +\left(\left(i\hbar\overline{p}_{y_i}+\frac{y_i}{2}\right)-\left(i\hbar\overline{p}_{y_j}+\frac{y_j}{2}\right)\right)^2\\ +\left(\left(i\hbar\overline{p}_{z_i}+\frac{z_i}{2}\right)-\left(i\hbar\overline{p}_{x_j}+\frac{z_j}{2}\right)\right)^2\end{array}\right|} \tag{15d}$$
$$+\sum_{A,i}\frac{-Z_A e^2\left(1+\frac{1}{2}\left((\mathbf{a}_A\cdot\mathbf{a}_i)+\frac{(\mathbf{a}_A\cdot\mathbf{r}_{Ai})(\mathbf{a}_i\cdot\mathbf{r}_{Ai})}{r_{Ai}^2}\right)\right)}{\left|\begin{array}{l}\left(\left(i\hbar\overline{p}_{x_A}+\frac{x_A}{2}\right)-\left(i\hbar\overline{p}_{x_i}+\frac{x_i}{2}\right)\right)^2\\ +\left(\left(i\hbar\overline{p}_{y_A}+\frac{y_A}{2}\right)-\left(i\hbar\overline{p}_{y_i}+\frac{y_i}{2}\right)\right)^2\\ +\left(\left(i\hbar\overline{p}_{z_A}+\frac{z_A}{2}\right)-\left(i\hbar\overline{p}_{z_i}+\frac{z_i}{2}\right)\right)^2\end{array}\right|}$$

Comment

Though Examples 1-5 are useful applications of HOA outright for various chemical applications, so also are various combinations of these Examples . For instance, mixing the results of Example 2 (Schrödinger Hamiltonian with SHOs having N-Arbitrary Masses in Pairwise Anisotropic Interaction) with Example 3 (Schrödinger Molecular Hamiltonian with Pairwise Coulomb Interaction) utilizing the extension to time-dependent parameters in Example 1 (1dim Simple Harmonic Oscillator) , one arrives at a Quantum Phase Space description of non-relativistic molecular systems interacting with an external time-varying bath environment with anisotropic harmonic constitutive properties. Further, Example 4 (Dirac and Majorana Equations with Minimum-Coupled Electromagnetic Gauge Field) and Example 5 (Dirac Molecular Hamiltonian with Pairwise Coulomb-Breit Interaction) may be mixed to yield a Quantum Phase Space description of relativistic molecular systems in external electromagnetic fields. [ref. 9,10]

Thus far, the Examples of the HOA results with usefulness in Chemical Dynamics have been based on a direct use of particular Hamiltonians to model the particular scenarios in question. We now will consider some additional Examples where the HOA results are used as a formal tool to achieve exact analytical solutions of more general formal mathematical problems of interest in Mathematical Chemistry and Beyond.

Example 6. Exact Quadrature Solution of Linear Eigenvalue Problem for General Class of Variable Coefficient Differential Operators

Recall from the HOA Recap in this note, that

'Notwithstanding its quantum mechanical origins, the HOA scheme takes on a life of its own and transcends the limits of quantum applications to address a wide variety of purely formal mathematical problems as well. Among other things, the result provides a formula for obtaining an exact solution to a wide variety of variable-coefficient integro-differential equations. Since the functional dependence of the Hamiltonian operator as considered is in general arbitrary upon its arguments(i.e., independent variables, derivative operator symbols[including negative powers thereof, thus the possible integral character]), then its multivariable extension can be interpreted as the most general variable coefficient partial differential operator. Moreover, it is not confined to being a scalar or even vector operator, but may be generally construed an arbitrary rank matrix operator. In all cases of course, its rank dictates the matrix rank of the wavefunction solution.'

Recall Eq. (10) of the Recap

$$\hat{H}_{\text{configuration space}}\left(x_1,....,x_n,-i\hbar\partial_{x_1},....,-i\hbar\partial_{x_n},t\right)\Psi_{\text{configuration space}}\left(x_1,....,x_n,t\right)$$
$$=i\hbar\partial_t\Psi_{\text{configuration space}}\left(x_1,....,x_n,t\right)$$

$$\Psi_{\text{configuration space}}\left(x_1,....,x_n,t\right)=\int_{-\infty}^{\infty}\frac{e^{\frac{ix_1p_1}{2\hbar}}}{\sqrt{4\pi\hbar}}....\int_{-\infty}^{\infty}\frac{e^{\frac{ix_np_n}{2\hbar}}}{\sqrt{4\pi\hbar}}\Psi\left(x_1,....,x_n;p_1,....,p_n;t\right)dp_1..dp_n$$

$$\Psi_{\text{configuration space}}\left(x_1,....,x_n,t\right)$$
$$=\int_{-\infty}^{\infty}\frac{e^{\frac{ix_1p_1}{2\hbar}}}{\sqrt{4\pi\hbar}}....\int_{-\infty}^{\infty}\frac{e^{\frac{ix_np_n}{2\hbar}}}{\sqrt{4\pi\hbar}}L^{-1}_{\left(\begin{smallmatrix}(\bar{x}_1,....,\bar{x}_n;\bar{p}_1,....,\bar{p}_n)\\ \to(x_1,....,x_n;p_1,....,p_n)\end{smallmatrix}\right)}\left[\begin{matrix}e^{\frac{-i}{\hbar}\int_0^t\hat{H}_{\text{configuration space}}\left(\begin{smallmatrix}x_1,....,x_n,\\ -i\hbar\partial_{x_1},....,-i\hbar\partial_{x_n},u\end{smallmatrix}\right)_{\begin{smallmatrix}(x_1,....,x_n)\mapsto(i\hbar\bar{p}_1+\alpha_1x_1,....,i\hbar\bar{p}_n+\alpha_nx_n)\\ \left(-i\hbar\partial_{x_1},....,-i\hbar\partial_{x_n}\right)\mapsto\left(-i\hbar\bar{x}_1+\gamma_1p_1,....,-i\hbar\bar{x}_n+\gamma_np_n\right)\end{smallmatrix}}du}\\ \times\\ \breve{\Psi}_{0\,\text{configuration space}}\left(\bar{x}_1,....,\bar{x}_n;\bar{p}_1,....,\bar{p}_n;t=0\right)\end{matrix}\right]dp_1..dp_n$$

Now suppose that

$$\hat{H}_{\text{configuration space}}\left(x_1,....,x_n,-i\hbar\partial_{x_1},....,-i\hbar\partial_{x_n},t\right)=\hat{H}_{\substack{\text{configuration space}\\ \text{scleronomic}}}\left(x_1,....,x_n,-i\hbar\partial_{x_1},....,-i\hbar\partial_{x_n}\right)$$

$$\hat{H}_{\substack{\text{configuration space}\\ \text{scleronomic}}}\left(x_1,....,x_n,-i\hbar\partial_{x_1},....,-i\hbar\partial_{x_n}\right)\Psi_{\text{configuration space}}\left(x_1,....,x_n,t\right) \tag{16a}$$
$$=i\hbar\partial_t\Psi_{\text{configuration space}}\left(x_1,....,x_n,t\right)$$

Hence Fourier transforming (16a) with respect to t

$$F_{t\to\omega}\left[f(t)\right]=\int_{t_0}^{\infty} f(t)e^{-it\omega}dt$$

$$F_{t\to\omega}\begin{bmatrix}\hat{H}_{\substack{\text{configuration space}\\ \text{scleronomic}}}\left(x_1,\ldots,x_n,-i\hbar\partial_{x_1},\ldots,-i\hbar\partial_{x_n}\right)\Psi_{\text{configuration space}}\left(x_1,\ldots,x_n,t\right)\\ =i\hbar\partial_t\Psi_{\text{configuration space}}\left(x_1,\ldots,x_n,t\right)\end{bmatrix}= \tag{16b}$$

$$\hat{H}_{\substack{\text{configuration space}\\ \text{scleronomic}}}\left(x_1,\ldots,x_n,-i\hbar\partial_{x_1},\ldots,-i\hbar\partial_{x_n}\right)\Psi_{\substack{\text{configuration space}\\ \omega\text{ eigenvalue}}}\left(x_1,\ldots,x_n,\omega\right)$$
$$=\hbar\omega\Psi_{\substack{\text{configuration space}\\ \omega\text{ eigenvalue}}}\left(x_1,\ldots,x_n,\omega\right)$$

So

$$\hat{H}_{\substack{\text{configuration space}\\ \text{scleronomic}}}\left(x_1,\ldots,x_n,-i\hbar\partial_{x_1},\ldots,-i\hbar\partial_{x_n}\right)\Psi_{\substack{\text{configuration space}\\ \omega\text{ eigenvalue}}}\left(x_1,\ldots,x_n,\omega\right)$$
$$=\hbar\omega\Psi_{\substack{\text{configuration space}\\ \omega\text{ eigenvalue}}}\left(x_1,\ldots,x_n,\omega\right)$$

∋

$$F_{t\to\omega}\left(\breve{\Psi}_{0\,\text{configuration space}}\begin{pmatrix}\overline{x}_1,\ldots,\overline{x}_n;\\ \overline{p}_1,\ldots,\overline{p}_n;t=0\end{pmatrix}e^{\frac{-i}{\hbar}t\hat{H}_{\text{configuration space}}\begin{pmatrix}x_1,\ldots,x_n,\\ -i\hbar\partial_{x_1},\ldots,-i\hbar\partial_{x_n}\end{pmatrix}\begin{pmatrix}(x_1,\ldots,x_n)\\ \mapsto(i\hbar\overline{p}_1+\alpha_1x_1,\ldots,i\hbar\overline{p}_n+\alpha_nx_n)\end{pmatrix}\begin{pmatrix}(-i\hbar\partial_{x_1},\ldots,-i\hbar\partial_{x_n})\\ \mapsto(-i\hbar\overline{x}_1+\gamma_1p_1,\ldots,-i\hbar\overline{x}_n+\gamma_np_n)\end{pmatrix}}\right)$$

$$=\breve{\Psi}_{0\,\text{configuration space}}\begin{pmatrix}\overline{x}_1,\ldots,\overline{x}_n;\\ \overline{p}_1,\ldots,\overline{p}_n\overline{p}_1,\ldots,\overline{p}_n;t=0\end{pmatrix}\delta\left(\omega-\frac{1}{\hbar}\hat{H}_{\text{configuration space}}\begin{pmatrix}x_1,\ldots,x_n,\\ -i\hbar\partial_{x_1},\ldots,-i\hbar\partial_{x_n}\end{pmatrix}\begin{pmatrix}(x_1,\ldots,x_n)\\ \mapsto(i\hbar\overline{p}_1+\alpha_1x_1,\ldots,i\hbar\overline{p}_n+\alpha_nx_n)\end{pmatrix}\begin{pmatrix}(-i\hbar\partial_{x_1},\ldots,-i\hbar\partial_{x_n})\\ \mapsto(-i\hbar\overline{x}_1+\gamma_1p_1,\ldots,-i\hbar\overline{x}_n+\gamma_np_n)\end{pmatrix}\right) \tag{16c}$$

Hence

$$F_{t\to\omega}\left(\begin{array}{l}\Psi_{\text{configuration space}}\left(x_1,\ldots,x_n,t\right)\\ =\int_{-\infty}^{\infty}\frac{e^{\frac{ix_1p_1}{2\hbar}}}{\sqrt{4\pi\hbar}}\cdots\int_{-\infty}^{\infty}\frac{e^{\frac{ix_np_n}{2\hbar}}}{\sqrt{4\pi\hbar}}L^{-1}_{\begin{pmatrix}(\overline{x}_1,\ldots,\overline{x}_n;\overline{p}_1,\ldots,\overline{p}_n)\\ \to(x_1,\ldots,x_n;p_1,\ldots,p_n)\end{pmatrix}}\begin{bmatrix}e^{\frac{-i}{\hbar}t\hat{H}_{\text{configuration space}}\begin{pmatrix}x_1,\ldots,x_n,\\ -i\hbar\partial_{x_1},\ldots,-i\hbar\partial_{x_n}\end{pmatrix}\begin{pmatrix}(x_1,\ldots,x_n)\\ \mapsto(i\hbar\overline{p}_1+\alpha_1x_1,\ldots,i\hbar\overline{p}_n+\alpha_nx_n)\end{pmatrix}\begin{pmatrix}(-i\hbar\partial_{x_1},\ldots,-i\hbar\partial_{x_n})\\ \mapsto(-i\hbar\overline{x}_1+\gamma_1p_1,\ldots,-i\hbar\overline{x}_n+\gamma_np_n)\end{pmatrix}}\\ \times\breve{\Psi}_{0\,\text{configuration space}}\begin{pmatrix}\overline{x}_1,\ldots,\overline{x}_n;\\ \overline{p}_1,\ldots,\overline{p}_n;t=0\end{pmatrix}\end{bmatrix}dp_1..dp_n\end{array}\right)$$

$$\Psi_{\substack{\text{configuration space}\\ \omega\text{ eigenvalue}}}\left(x_1,\ldots,x_n,\omega\right)$$
$$=\int_{-\infty}^{\infty}\frac{e^{\frac{ix_1p_1}{2\hbar}}}{\sqrt{4\pi\hbar}}\cdots\int_{-\infty}^{\infty}\frac{e^{\frac{ix_np_n}{2\hbar}}}{\sqrt{4\pi\hbar}}L^{-1}_{\begin{pmatrix}\begin{pmatrix}\overline{x}_1,\ldots,\overline{x}_n\\ ;\overline{p}_1,\ldots,\overline{p}_n\end{pmatrix}\\ \to\begin{pmatrix}x_1,\ldots,x_n\\ ;p_1,\ldots,p_n\end{pmatrix}\end{pmatrix}}\begin{bmatrix}\sqrt{2\pi}\delta\left(\omega-\frac{1}{\hbar}\hat{H}_{\text{configuration space}}\begin{pmatrix}x_1,\ldots,x_n,\\ -i\hbar\partial_{x_1},\ldots,-i\hbar\partial_{x_n}\end{pmatrix}\begin{pmatrix}(x_1,\ldots,x_n)\\ \mapsto\begin{pmatrix}i\hbar\overline{p}_1+\alpha_1x_1,\\ \ldots,i\hbar\overline{p}_n+\alpha_nx_n\end{pmatrix}\end{pmatrix}\begin{pmatrix}(-i\hbar\partial_{x_1},\ldots,-i\hbar\partial_{x_n})\\ \mapsto\begin{pmatrix}-i\hbar\overline{x}_1+\gamma_1p_1,\\ \ldots,-i\hbar\overline{x}_n+\gamma_np_n\end{pmatrix}\end{pmatrix}\right)\\ \times\breve{\Psi}_{0\,\text{configuration space}}\begin{pmatrix}\overline{x}_1,\ldots,\overline{x}_n;\\ \overline{p}_1,\ldots,\overline{p}_n;t=0\end{pmatrix}\end{bmatrix}dp_1..dp_n \tag{16d}$$

To facilitate the evaluation of (16e) the Identity for Dirac Delta Functions is useful

$$\delta(g(u))=\sum_{j=1}^{n}\frac{\delta(u-u_j)}{\left|\left(D_u g(u)\right)\right|_{u=u_j}\right|} \quad ,\ni g(u_j)=0\ ,\ \left(D_u g(u)\right)\Big|_{u=u_j}\neq 0$$

Which completes the construction of the result.

To illustrate this, consider the Example 1 (1-dim Simple Harmonic Oscillator) with constant strength a . Applying the formalism above in (16d) to the term from (11d) yields

$$\left(\frac{(-i\hbar\partial_x)^2}{2m}+ax^2\right)\Psi_{\text{configuration space}}(x,t)=i\hbar\partial_t\Psi_{\text{configuration space}}(x,t)$$

$$F_{t\to\omega}\left(\Psi_{\text{configuration space}}(x,t)=\int_{-\infty}^{\infty}\frac{e^{\frac{ixp}{2\hbar}}}{\sqrt{4\pi\hbar}}L^{-1}_{\left(\substack{(\bar{x};\bar{p})\\ \to(x;p)}\right)}\left[\breve{\Psi}_0(\bar{x},\bar{p},t=0)e^{\frac{-i}{\hbar}t\left(\frac{(-i\hbar\partial_x)^2}{2m}+ax^2\right)\left(\substack{(x)\\ \mapsto(i\hbar\bar{p}+\alpha x)}\right)\left(\substack{(-i\hbar\partial_x)\\ \mapsto(-i\hbar\bar{x}+\gamma p)}\right)}\right]dp\right) \tag{16e}$$

$$\Psi_{\substack{\text{configuration space}\\ \omega\ \text{eigenvalue}}}(x,\omega)=\int_{-\infty}^{\infty}\frac{e^{\frac{ixp}{2\hbar}}}{\sqrt{4\pi\hbar}}L^{-1}_{\left(\substack{(\bar{x};\bar{p})\\ \to(x;p)}\right)}\left[\sqrt{2\pi}\breve{\Psi}_0(\bar{x},\bar{p},t=0)\delta\left(\omega-\frac{1}{\hbar}\left(\frac{\left(-i\hbar\bar{x}+\frac{p}{2}\right)^2}{2m}+a\left(i\hbar\bar{p}+\frac{x}{2}\right)^2\right)\right)\right]dp$$

$$\left(\frac{(-i\hbar\partial_x)^2}{2m}+ax^2-\hbar\omega\right)\Psi_{\substack{\text{configuration space}\\ \omega\ \text{eigenvalue}}}(x,\omega)=0$$

Just to point out, if the potential term of the SHO above ax^2 is replaced with a potential term of arbitrary functional profile as $V(x)$ then (16e) generalizes to

$$\left(\frac{(-i\hbar\partial_x)^2}{2m}+V(x)\right)\Psi_{\text{configuration space}}(x,t)=i\hbar\partial_t\Psi_{\text{configuration space}}(x,t)$$

$$F_{t\to\omega}\left(\Psi_{\text{configuration space}}(x,t)=\int_{-\infty}^{\infty}\frac{e^{\frac{ixp}{2\hbar}}}{\sqrt{4\pi\hbar}}L^{-1}_{\left(\substack{(\bar{x};\bar{p})\\ \to(x;p)}\right)}\left[\breve{\Psi}_0(\bar{x},\bar{p},t=0)e^{\frac{-i}{\hbar}t\left(\frac{(-i\hbar\partial_x)^2}{2m}+V(x)\right)\left(\substack{(x)\\ \mapsto(i\hbar\bar{p}+\alpha x)}\right)\left(\substack{(-i\hbar\partial_x)\\ \mapsto(-i\hbar\bar{x}+\gamma p)}\right)}\right]dp\right) \tag{16f}$$

$$\Psi_{\substack{\text{configuration space}\\ \omega\ \text{eigenvalue}}}(x,\omega)=\int_{-\infty}^{\infty}\frac{e^{\frac{ixp}{2\hbar}}}{\sqrt{4\pi\hbar}}L^{-1}_{\left(\substack{(\bar{x};\bar{p})\\ \to(x;p)}\right)}\left[\sqrt{2\pi}\breve{\Psi}_0(\bar{x},\bar{p},t=0)\delta\left(\omega-\frac{1}{\hbar}\left(\frac{\left(-i\hbar\bar{x}+\frac{p}{2}\right)^2}{2m}+V\left(i\hbar\bar{p}+\frac{x}{2}\right)\right)\right)\right]dp$$

$$\left(\frac{(-i\hbar\partial_x)^2}{2m}+V(x)-\hbar\omega\right)\Psi_{\substack{\text{configuration space}\\ \omega\ \text{eigenvalue}}}(x,\omega)=0$$

It is worth re-emphasizing that the result (16d) of course is immediately applicable to non-seperable Hamiltonians in general, such as Example 3 (Schrödinger Molecular Hamiltonian with Pairwise Coulomb Interaction), yielding the solufion of the radial Molecular Schrodinger Eigenvalue Equation and many others.

Example 7. Exact Quadrature Solution of General Class of Variable-Coefficient Differential Equations

In view of the arbitrary functional form of the Hamiltonian and its connection with a Lagrangian [ref. 11, 12, 13] for the given problem,

$$H\left(x_1,....,x_n;p_1,....,p_n;t\right)=\sum_{j=1}^{n} p_j\dot{x}_j - L\left(x_1,....,x_n;\dot{x}_1,....,\dot{x}_n;t\right)$$
$$p_j=\partial_{\dot{x}_j} L\left(x_1,....,x_n;\dot{x}_1,....,\dot{x}_n;t\right) \tag{17a}$$

Consider now the equations of motion for the case of an external force $F(x_1,....,x_n;\dot{x}_1,....,\dot{x}_n;t)$ (viz the Extended Hamilton Principle) given as

$$\left(-\partial_t\right)\partial_{\dot{x}_j} L\left(x_1,....,x_n;\dot{x}_1,....,\dot{x}_n;t\right)+\partial_{x_j} L\left(x_1,....,x_n;\dot{x}_1,....,\dot{x}_n;t\right)=F\left(x_1,....,x_n;\dot{x}_1,....,\dot{x}_n;t\right) \tag{17b}$$

It follows that (17b) extends naturally to accommodate higher-order derivatives for several variables, though the notation becomes cumbersome [refs 12,13]. As such we will below only construct the results for the case of one dependent variable x as

$$\sum_{j=1}^{n}\left(-D_t\right)^j \partial_{\left(D_t^j x\right)} L\left(x;\dot{x};\ddot{x};\dddot{x};....;D_t^n x;t\right)+\partial_{x_j} L\left(x;\dot{x};\ddot{x};\dddot{x};....;D_t^n x;t\right)=F\left(x;\dot{x};\ddot{x};\dddot{x};....;D_t^n x;t\right)$$

for example $n=4$ yields (17c)

$$\left(D_t^4\partial_{D_t^4 x}-D_t^3\partial_{D_t^3 x}+D_t^2\partial_{D_t^2 x}-\left(D_t\right)\partial_{D_t x}+\partial_x\right)L\left(x;\dot{x};\ddot{x};\dddot{x};\left(\tfrac{dx}{dt}\right)^4;t\right)=F\left(x;\dot{x};\ddot{x};\dddot{x};\left(\tfrac{dx}{dt}\right)^4;t\right)$$

In turn, these order-n derivative Lagrangians are reducible to first-order derivative systems by way of variable substitutions[ref. 12,13]. In this note, the properties in (17b) and (17c) are pointed out explicitly, as they form the essential basis for applying to HOA scheme to any system that can be put in the Hamiltonian form: from higher-order Lagrangian systems of discrete particles to continuous fields[refs. 11,12,13,14.15]: they all admit to HOA solution as they are all ultimately transformable to the Hamiltonian formulation of dynamics. So, without loss of generality, we consider the case of (17b) in one dependent variable x

$$\left(-D_t\right)\partial_{\dot{x}} L\left(x;\dot{x};t\right)+\partial_x L\left(x;\dot{x};t\right)=F\left(x;\dot{x};t\right) \tag{17d}$$

Now to illustrate HOA in the context of this Hamiltonian-Lagrangian connection, take the Lagrangian $L\left(x;\dot{x};t\right)=\dfrac{m\left(\dot{x}^2\right)}{2}-a\left(t\right)x^2$ and no external term $F\left(x;\dot{x};t\right)=0$ from the generalized Hamilton Principle, yielding from (17d)

$$(-D_t)\partial_{\dot{x}}\left(\frac{m(\dot{x}^2)}{2}-a(t)x^2\right)+\partial_x\left(\frac{m(\dot{x}^2)}{2}-a(t)x^2\right)=0$$
$$\frac{m\ddot{x}}{2}+a(t)x=0 \tag{17e}$$

which is the equation of motion for the classical Simple Harmonic Oscillator Lagrangian with arbitrary time-dependent strength $a(t)$ therein. By way of (17a), the Hamiltonian that corresponds to the system in (17e) is $H=\dfrac{p^2}{2m}+a(t)x^2$: this is clearly the Hamiltonian for Example 1 herein: the quantum version of this system. Following Example 1 , the QPSR of this system is given by

$$\left(\frac{\left(-i\hbar\partial_x+\frac{p}{2}\right)^2}{2m}+a(t)\left(i\hbar\partial_p+\frac{x}{2}\right)^2\right)\Psi(x;p;t)=i\hbar\partial_t\Psi(x;p;t)$$

$$\left(\frac{\left(-i\hbar\overline{x}+\frac{p}{2}\right)^2}{2m}+a(t)\left(i\hbar\overline{p}+\frac{x}{2}\right)^2\right)\breve{\Psi}(\overline{x};\overline{p};t)=i\hbar\partial_t\breve{\Psi}(\overline{x};\overline{p};t)$$

$$\Psi(x;p;t)=L^{-1}_{((\overline{x})\to(x))}\left[L^{-1}_{((\overline{p})\to(p))}\left[e^{\frac{-i\left(t\frac{\left(-i\hbar\overline{x}+\frac{p}{2}\right)^2}{2m}+\int_0^t a(u)du\left(i\hbar\overline{p}+\frac{x}{2}\right)^2\right)}{\hbar}}\times\breve{\Psi}_0(\overline{x};\overline{p};t=0)\right]\right] \tag{11d}$$

With the configuration space wavefunction via QPSR wavefunction tor this system given as

$$\Psi_{\text{configuration space}}(x,t)=\int_{-\infty}^{\infty}\frac{e^{\frac{ixp}{2\hbar}}}{\sqrt{4\pi\hbar}}\left[\Psi_0(x,p,t=0)\underbrace{*}_{x,p}\sqrt{\frac{-m}{2\int_0^t a(u)du}}\frac{e^{i\left(\frac{p^2+2mx^2\int_0^t a(u)du}{4\hbar t\int_0^t a(u)du}\right)}}{2\pi\hbar t}\right]dp \tag{11e}$$

Now it follows by way of Ehrenfest's Theorem that the expectation value of the position[also momentum] operator follows its classical counterpart's equations of motion in any framework(e.g., Hamiltonian,Lagrangian, etc]. Following the Recap herein, applying the configuration position operator $\hat{x} \equiv x$ to the system wavefunction (11e) to yield the expectation value of the position operator as

$$\langle x(t)\rangle = \int_{\text{Configuration Space}} \left(\Psi^*_{\text{configuration space}}(x,t)(x)\Psi_{\text{configuration space}}(x,t)\right)dx$$

$$\Psi_{\text{configuration space}}(x,t) = \int_{-\infty}^{\infty} \frac{e^{\frac{ixp}{2\hbar}}}{\sqrt{4\pi\hbar}} \left[\Psi_0(x,p,t=0) \underbrace{*}_{x,p} \sqrt{\frac{-m}{2\int_0^t a(u)du}} \frac{e^{i\left(\frac{p^2+2mx^2\int_0^t a(u)du}{4\hbar t\int_0^t a(u)du}\right)}}{2\pi\hbar t} \right] dp \tag{17f}$$

So

$$\langle x(t)\rangle = \int_{\text{Configuration Space}} \left|\Psi_{\text{configuration space}}(x,t)\right|^2 x dx$$

$$\left|\Psi_{\text{configuration space}}(x,t)\right|^2 = \left| \int_{-\infty}^{\infty} \frac{e^{\frac{ixp}{2\hbar}}}{\sqrt{4\pi\hbar}} \left[\Psi_0(x,p,t=0) \underbrace{*}_{x,p} \sqrt{\frac{-m}{2\int_0^t a(u)du}} \frac{e^{i\left(\frac{p^2+2mx^2\int_0^t a(u)du}{4\hbar t\int_0^t a(u)du}\right)}}{2\pi\hbar t} \right] dp \right|^2 \tag{17g}$$

The upshot of equation (17g) is that $\langle x(t)\rangle$ follows the same equation of motion (17e). Therefore

$$\frac{m\langle \ddot{x}(t)\rangle}{2} + a(t)\langle x(t)\rangle = 0$$

$$\langle x(t)\rangle = \int_{\text{Configuration Space}} \left| \int_{-\infty}^{\infty} \frac{e^{\frac{ixp}{2\hbar}}}{\sqrt{4\pi\hbar}} \left[\Psi_0(x,p,t=0) \underbrace{*}_{x,p} \sqrt{\frac{-m}{2\int_0^t a(u)du}} \frac{e^{i\left(\frac{p^2+2mx^2\int_0^t a(u)du}{4\hbar t\int_0^t a(u)du}\right)}}{2\pi\hbar t} \right] dp \right|^2 x dx \tag{17h}$$

Thus Example 7 above provides an exact analytical quadrature solution to the given related classical equations of motion as well. By way of the Hamiltonian-Lagrangian correspondence, though the variables are different, (17h) provides an exact analytical quadrature solution of the purely mathematical problem of the exact analytical quadrature solution to the second-order linear Ordinary Differential Equation with arbitrary variable coefficient in it's canonical homogeneous form

$$\ddot{y}(t)+b(t)y(t)=0$$

a problem which should require no introduction.

Finally, consider now the Lagrangian $L(x;\dot{x};t)=\frac{5(\dot{x}^2)}{4}+\frac{x^{\frac{5}{2}}}{t^{\frac{1}{2}}}$ and no external term $F(x;\dot{x};t)=0$ from the generalized Hamilton Principle, yielding from (17d)

$$(-D_t)\partial_{\dot{x}}\left(\frac{5(\dot{x}^2)}{4}+\frac{x^{\frac{5}{2}}}{t^{\frac{1}{2}}}\right)+\partial_x\left(\frac{5(\dot{x}^2)}{4}+\frac{x^{\frac{5}{2}}}{t^{\frac{1}{2}}}\right)=0$$

$$\ddot{x}=\frac{x^{\frac{3}{2}}}{t^{\frac{1}{2}}} \tag{17i}$$

following HOA yields

$$\left(\frac{\left(-i\hbar\partial_x+\frac{p}{2}\right)^2}{2(5/2)}-\frac{1}{t^{\frac{1}{2}}}\left(i\hbar\partial_p+\frac{x}{2}\right)^{\frac{5}{2}}\right)\Psi(x;p;t)=i\hbar\partial_t\Psi(x;p;t)\ ,\ni m=\tfrac{5}{2}$$

$$\left(\frac{\left(-i\hbar\overline{x}+\frac{p}{2}\right)^2}{5}-\frac{1}{t^{\frac{1}{2}}}\left(i\hbar\overline{p}+\frac{x}{2}\right)^{\frac{5}{2}}\right)\breve{\Psi}(\overline{x};\overline{p};t)=i\hbar\partial_t\breve{\Psi}(\overline{x};\overline{p};t) \tag{17j}$$

$$\Psi(x;p;t)=L^{-1}_{((\overline{x})\to(x))}\left[L^{-1}_{((\overline{p})\to(p))}\left[\breve{\Psi}_0(\overline{x};\overline{p};t=0)e^{\frac{-i\left(t\frac{\left(-i\hbar\overline{x}+\frac{p}{2}\right)^2}{5}-2t^{\frac{1}{2}}\left(i\hbar\overline{p}+\frac{x}{2}\right)^{\frac{5}{2}}\right)}{\hbar}}\right]\right]$$

By way of the prescription of (17f) through (17h), we arrive at

$$\langle x(t)\rangle = \int_{\text{Configuration Space}} \left|\Psi_{\text{configuration space}}(x,t)\right|^2 x dx$$

$$\left|\Psi_{\text{configuration space}}(x,t)\right|^2 = \left| \int_{-\infty}^{\infty} \frac{e^{\frac{ixp}{2\hbar}}}{\sqrt{4\pi\hbar}} L^{-1}_{\left(\substack{(\bar{x};\bar{p}) \\ \to(x;p)}\right)} \left[\breve{\Psi}_0(\bar{x},\bar{p},t=0) e^{\frac{-i\left(t\frac{\left(-i\hbar\bar{x}+\frac{p}{2}\right)^2}{5} - 2t^{\frac{1}{2}}\left(i\hbar\bar{p}+\frac{x}{2}\right)^{\frac{5}{2}} \right)}{\hbar}} \right] dp \right|^2 \tag{17k}$$

$$\langle \ddot{x}(t)\rangle = \frac{\langle x(t)\rangle^{\frac{3}{2}}}{t^{\frac{1}{2}}}$$

Hence (17k) is an exact quadrature solution of the Thomas-Fermi Equation.

Notwithstanding their intrinsic interest, the few Examples 1-7 herein these Remarks do not even begin to touch the immense vista of problems that may be solved exactly via HOA methods. Indeed, the HOA may be modified to accommodate auxiliary time conditions besides Initial-Value Problems. Moreover, the Integral and Differential Operators in the HOA may be generalized to Discrete Sum/Finite Difference Operators and sundry other extensions. Since classical and quantum dynamical systems admit to Hamiltonian/Lagrangian formulation, then indeed the various other formulations[e.g., Density Functional Theory, Path Integral Formalism, Stochastic Formalisms,etc], as well as derived abstracted mathematical problems[e.g., ref 14,15] may all benefit by HOA treatments. Quoting from [ref. 6] , ".....By definition, solution in one representation implies simultaneous solution in all representations[the Physics is the same regardless of 'pictures']. ....Hence, Quantum Dynamics is now reduced to exact quadratures, as are all the associated purely mathematical problems that are abstracted from the physical formalism." MT21:42.

References

[1] Torres-Vega, G. & Frederick, J. H. (1993). A quantum-mechanical representation in phase space. *Journal of Chemical Physics*, **98**(4), 3103-20.

[2] Li, Q. S. & Lu, J. (2000). Rigorous solutions of diatomic molecule oscillator with empirical potential function in phase space. *Journal of Chemical Physics*, **113**(11), 4565-4571.

[3] Hu, X. G. & Li, Q. S. (1999). Morse oscillator in a quantum phase-space representation: rigorous solutions. Journal of Physics A: *Mathematical and General*, **32**(1), 139-146.

[4] Li, Q. S. & Lu, J. (2001). One-dimensional hydrogen atom in quantum phase-space representation: rigorous solutions. *Chemical Physics Letters*, **336**(1,2), 118-122.

[5] Li, Q. S., Wei, G. M. & Lu, L. Q. (2004). Relationship between the Wigner function and the probability density function in quantum phase space representation. Physical Review A: *Atomic, Molecular, and Optical Physics*, **70**(2), 022105/1-022105/5.

[6] Simpao, V. A. (2004). *Electronic Journal of Theoretical Physics*, **1**, 10-16.

[7] Simpao, V. A. (2008). Toward Chemical Applications of Heaviside Operational Ansatz: Exact Solution of Radial Schrodinger Equation for Nonrelativistic N-particle System with Pairwise 1/rij Radial Potential in Quantum Phase Space [published MAY *Journal of Mathematical Chemistry*.

[8] Simpao, V. A. (2006). Electronic *Journal of Theoretical Physics*, **3**, No. 10, 239-247

[9] Moss, R. E. (1973). Advanced Molecular *Quantum Mechanics*, (Wiley and Sons, NY).

[10] Lindgren, I. (1989). in Lecture Notes in Chemistry, vol 52: Many-Body Methods in Quantum Chemistry, ed. U. Kaldor (*Springer, Berlin*), 293-306.

[11] Goldstein, H., Poole, C. P. & Jr., Safko, J. L. (2000). Classical Mechanics 3rd Edition, *Pearson*.

[12] Whittaker, E. T. (1944). Analytical Dynamics 4th Edition, *Dover New York*.

[13] Weinstock, R. (1974). *Calculus of Variations, Dover*, New York.

[14] de Gossen, M. & Luef, F. (2008). *Lett Math Phys*, **85**, 173-183.

[15] Cosmos. Zachos, David B. Fairlie, & Thomas L. (2005). Curtright, *Quantum Mechanics in Phase Space*, (World Scientific, Singapore.

In: Mathematical Chemistry
Editor: W.I. Hong, pp. 451-456
ISBN: 978-1-60876-894-3

Chapter 7

ON THE ENUMERATION OF VARIOUS NETWORKS

Michael J. Bucknum and Eduardo A. Castro *
Hexagonite, Oceanside, CA, US & Consejo de Investigaciones Científicas y Técnicas (CONICET), Buenos Aires, Argentina
INIFTA, Theoretical Chemistry Division, CONICET-La Plata, Buenos Aires, Argentina

I. Introduction

There is an infinite series of 3-,4-connected, crystalline patterns that emerge from the elaboration of 1,4-cyclohexadiene-like, hydrocarbon building block generating fragments, onto the template of each and every possible tessellation. Probably, the archetypal crystalline pattern in this series is the so-called closoglitter lattice described below. Like all the infinite set of glitter series patterns, as just described, closoglitter is built from elaborating properly a quasi-closest packed tessellation thus with 1,4-cyclohexadiene-like moieties. And thus any arbitrary 2D tessellation can be elaborated like this, with 1,4-cyclohexadiene-like hydrocarbon fragments to produce novel 3D patterns thus. Furthermore it is true, that each and every one of these three-dimensional (3D) patterns, generated from two-dimensional (2D) tessellations, should have a set of "n" π-π* crystal orbitals (CO's), in a frontier manifold that determines the electronic properties of the lattice. Here "n" is the rotational order of the space group that the glitter series pattern lies in. And these π-π* CO's are able to be visualized at some symmetry point of the reciprocal space of the lattice, and are composed of pure p atomic orbitals in their symmetry adapted combinations, and they are beautiful to sketch on paper as they represent various shadings of the innumerable 2D tessellations thus.

II. Structure of the Archetypal Closoglitter Lattice

There is a set of orthohexagonal crystallographic coordinates for a glitter series pattern we call closoglitter in Table 1, and it is comprised of a expanded closest packed layering, as

*E-mail address: eacast@gmail.com (Corresponding author)

thus a quasi-closest packed layering, containing cyclopropane-like hydrocarbon substructures, linked through intervening 1,4-cyclohexadiene-like hydrocarbon building block rings which lie normal to the cyclopropane-like substructures. Like the other members of this series called glitter, isoglitter & neoglitter which are to be described separately; closoglitter is entirely generated from 1,4-cyclohexadiene-like rings. It would perhaps be instructive to generate a computerized drawing of this lattice, together with a band structure, DOS & perhaps an crystal orbital (CO) analysis of the 3 π & 3 π^* crystal orbitals at the symmetry point in the Brillouin Zone (BZ) in which σ-π separation of the crystal orbitals applies.

III. On the Topology of the Glitter Series Networks

Now we have completed the topology of the closoglitter C allotrope and we have enumerated the following topology thus: the net is a 3-,4-connected pattern in a stoichiometric ratio of 1-to-1. The Wells point symbol for the network appears to be $(6^3)(3.6^5)$ = AB, and the corresponding Schlaefli symbol is (5.666667, 3.5) = (n, p), while the form index of the closoglitter network is:

$$1 = n/p = (5.6666667)/(3.5) = 1.61904....... \sim \phi \quad (1)$$

In contrast, the homologous glitter series network called neoglitter, possesses the Wells point symbol of $(6^3)(4.6^5)$ = AB, and the corresponding Schlaefli symbol is (5.7777778, 3.5) = (n, p), while the form index of the neoglitter network is:

$$1 = n/p = (5.7777778)/(3.5) = (\phi)\cdot((2)\cdot(10^{e\cdot\phi/1000} - 1)) \quad (2)$$

And it is apparent to us now that A. F. Wells discovered a entirely profound area of mathematics as well as crystallography and artistic expression, and we are very fortunate in life for being introduced to this pregnant work.

IV. Enumeration of Crystalline Networks in 3-Dimensions

Trigohexagonite is described in the Hexagonite webpages (1,2). This structure-type is rare as probably the very first quaternary stoichiometry network that has been discovered. The network was elaborated partially by one of us (MJB). Our derivation of the trigohexagonite network is perhaps semi-empirical in Nature, where one starts with certain structurally sound principles to work from and attempts to elaborate a sterically plausible lattice thus. It is a semi-empirical enterprise in that one has to use creativity or, say, artistic inspiration in some sense, to complete certain aspects of the crystalline elaboration. Perhaps it is a kind of witchcraft.

The first network we have created called glitter, was generated entirely from an elaboration involving the structural motif of the 1,4-cyclohexadiene molecular fragment, and the strategy there was to produce a network that is a counterpoint to graphene, which is built from benzene-like tilings, and the diamond polytypic series which are built upon a cyclohexane-like generating motif. The result of this line of investigation was to create the spiro quantum chemistry edifice (3), and this outcome was entirely serendipitous and unanticipated by one of us (MJB). This author noticed that glitter can alternatively be generated from the hydrocarbon molecular fragment called 1,3,5,7-tetramethylene-cyclooctane, a molecule first created at DuPont CRD in about 1959.

From this we came to learn that there are a family of crystalline patterns that can be generated from molecular fragments that comprise a homologous hydrocarbon series that we call the exocyclic polyenes. The first member of this series is 1,3-dimethylene-cyclobutane, and this generating fragment molecule can be used to elaborate, in a seamless approach like the elaboration of glitter from 1,3,5,7-tetramethylene-cyclooctane-like fragments, a network that we'll call by a name later and which A. F. Wells derived himself. This network exhibits spiroconjugation effects like glitter but is perhaps a semi-conductor after all.

The next member of the homologous series of exocyclic polyenes is 1,3,5-trimethylene-cyclohexane, and this generating molecular fragment can be used as a starting point to elaborate the trigohexagonite pattern, but the elaboration is not seamless after all, and one uses structural and steric intuition, or perhaps witchcraft, to complete the structural elaboration. In this connection we were able to begin by creating a part of the unit cell that is comprised of a stacking of 1,3,5-trimethylene- cyclohexane substructures, and then we proceeded on to elaborate perhaps other parts of the lattice by inspired structural intuition that became suggested in the very moment one of us (MJB). He would say that it reduced at some point to artisitic-steric reasoning of an intuitive Nature. Then we made a drawing of a hexagonal grid, and plotted the points in the hexagonal unit cell approximately within the grid, and with structural intuition we were able to guess the fractional hexagonal coordinates of 7 of the 10 atoms in the unit cell by sheer inspection.

Then there is 1,3,5,7-tetramethylene-cyclooctane hydrocarbon generating fragment, which seamlessly gives glitter. And then there is 1,3,5,7,9-pentamethylene- cyclodecane, which appears to have approximately a 5-fold axis, and thus it will not be the basis for generating a crystalline pattern. We also made a model of this molecular fragment and it is not sterically regular in any way so as to lend itself to elaboration. Then there is 1,3,5,7,9,11-hexamethylene-cyclododecane which is the basis for generating the network we have called hexagonite (4), and this elaboration is perhaps seamless or semi-seamless in Nature. In this instance, it is advantageous to start with the alternative motif called barralene and to elaborate seamlessly from this generating fragment, rather than the cumbersome 1,3,5,7,9,11-hexamethylene- cyclododecane generating fragment.

And next is the 1,3,5,7,9,11,13-heptamethylene-cyclo[14]ane generating molecular fragment, which is a substructure with perhaps an approximate 7-fold axis, and thus it will not lend itself to the elaboration of a crystalline pattern thus. And then there is 1,3,5,7,9,11,13,15-octamethylene-cyclo[16]ane, which curiously we were able to elaborate into some sort of emerging crystalline pattern which must possess 4-fold rotational symmetry, but perhaps it may exhibit a certain pseudo-8-fold crystalline symmetry somehow.

And then there is the glitter series of networks (5), this includes now glitter, isoglitter, neoglitter and closoglitter, and these structures were arrived at seamlessly by the application of symmetry principles, together with valence principles and including ordinary elementary bonding concepts. Our approach was to create, in the instance of what I have called isoglitter, 1-dimensional substructures of the generating fragment molecule 1,4-cyclohexadiene (which is the common thread joining this series of patterns) as a so-called polyquinoid substructure, and the strategy emerged that these polyquinoid substructures could be crosslinked to others like them to generate an orthorhombic (Pmmm) pattern. We have described this isoglitter lattice together with the glitter lattice in a grey literature PTO publication in August, 1993 and this document is deposited at the Hexagonite webpages (1).

Table 1. Orthohexagonal Coordinates of Closoglitter[1].

Atom #	X	Y	Z
1	0.0000	0.8718	0.0000
2	0.7550	-0.4359	0.0000
3	-0.7550	-0.4359	0.0000
4	0.0000	1.450	1.395
5	0.0000	2.780	1.395
6	0.0000	3.360	0.0000
7	-0.7550	4.665	0.0000
8	0.7550	4.665	0.0000
9	1.255	4.954	1.395
10	2.407	5.619	1.395
11	-1.255	4.665	1.395
12	-2.407	5.619	1.395
(a-axis)	7.325	0.0000	0.0000
(b-axis)	-3.663	6.344	0.0000
(c-axis)	0.0000	0.0000	2.790

[1] Aside from the set of orthohexagonal coordinates given in Table 1 for closoglitter, the other characteristics of this pattern are that it is in space group P6/mmm, there are 12 atoms in the unit cell. You must convert the orthohexagonal coordinates into fractional hexagonal coordinates to do a picture and band structure and DOS, etc. And the way you do this is by using the orthogonal-hexagonal transformation as:

xh = xo + yh(sin 30)

yh = yo/(cos 30)

zh = zo/c

where xh, yh and zh are the fractional hexagonal coordinates you want, and xo, yo and zo are the orthogonal coordinates I have provided you with above. And remember that hat all fractional hexagonal coordinates are positive, but in orthohexagonal systems a set of orthohexagonal coordinates can have some values which are not positive.

Later in 2003 one of us (MJB) discovered a large generating fragment molecule whose name I do not know, but which is comprised of a central cyclobutane ring, and at each vertex of this ring is elaborated a orthogonal 1,4-cyclohexadiene-like substructure. And this beast can be used to effectively create a 3D pattern, which I now call neoglitter, and this pattern as a C allotrope, apparently displays hyperconjugation effects in 3D. And this may result in the delocalization of the 2 π electrons in the lowest-lying π crystal orbital (or perhaps the HOCO π band) of this neoglitter structure, entirely through 3D. This is accomplished in conjunction with interactions of the appropriate symmetry with certain σ levels of the saturated C atoms in the lattice, entirely in analogy with the discrete molecular phenomena described by Hoffmann et al. earlier on. There is another member of this series, which is called closoglitter, and which is a structure analogous to neoglitter but with a central cyclopropane-like moiety in an hexagonal pattern (P6/mmm). The orthohexagonal coordinates are thus given in Table 1 below.

Then there is super[3]prismane and the super[n]prismane series, this super[3]prismane is a lattice that is semi-seamlessly generated using prismane as a molecular generating fragment molecule. It is semi-seamlessly generated, and it is of hexagonal-trigonal symmetry (perhaps

P3m1 or P6/mmm or some higher symmetry member within these supergroup families). Superprismane is a homlogue to super[4]cubane, where super[4]cubane is rectangular-cubic, and it is simple to figure out the crystalline parameters for this, as has been described and deposited at the Hexagonite webpages (1).

Then there is the so-called inorganic series of networks and these are characterized by not being Platonic or Archimedean networks of 3- or 4- coordination exclusively, or hybrid Catalan or Wellsean patterns of 3-,4-coordination (where specifically for the organic series of networks the 3-connected vertices are reasonably trigonal planar and the 4-connected vertices are reasonably tetrahedral). And thus such inorganic networks are comprised of 3- or 4-coordination, or 3-,4-coordination with vertices of higher coordination. Except that there are evidently some inorganic series networks comprised of trigonal planar & square planar coordination that comprise such inorganic crystalline patterns like the Archimedean prototype Waserite pattern discovered by Waser et al. in about 1951 at Rice University and details are described at the Hexagonite webpages (1).

Thus with Moravia (6), and this derivation was seamless reasoning in this case, as we took the octahedral-cuboctahedral space filling of Andreini and produced the Moravia pattern in this way, see (1). Kentuckia (7) was derived from Perovskite by deletion of a corresponding pair of 6-connected, octahedrally coordinated, face-centered vertices, but we saw this pattern in my Mind's Eye separately somehow and made a drawing instantaneously in Milledgeville, see (1). The jubilite pattern was created with Ms. His-cheng Shen, by a complicated and artistically inspired methodology involving trial and error in Olney in 2000-2001 (8). And later it was discovered that indeed it is structurally in correspondence with Waserite & Fluorite, each as end members of a related series of patterns, see (1). And New Yorkite (9) came to one of us (MJB) in Milledgeville instantaneously, see (1) as did the tessellation called lithographite (the name was from Lithonia, east of Atlanta in Georgia) (10). These networks have emerged from a program of intuitive construction principles that emerge from a chemical topology scheme proposed by the British crystallographer A. F. Wells who was also educated at Cambridge University.

In this regard, it is of interest to investigate closoglitter as the prototypical glitter series network, whose coordinates are attached here in Table 1, that may be interesting from the point of view of novel π–π* conjugation principles. This program of work, which began with the discovery of the glitter net, confirmed important characteristics of the phenomenon of spiroconjugation in its band structure, while the related nets of closoglitter & neoglitter crystallinity, apparently exhibit novel characteristics of what is termed hyperconjugation in organic chemistry, in 1-, 2- and 3-dimensions, which it is hoped will be investigated as well.

And there are indeed innumerable such glitter series networks that can be generated from the intuitive construction principles of A. F. Wells, in connection with the versatile hydrocarbon building block fragment 1,4-cyclohexadiene and other hydrocarbon building block fragments like the [n]prismane series. It is known to be the case that A. F. Wells discovered & published perhaps 100's of novel, non-glitter series networks himself in his lifetime, and he also was an accomplished experimentalist and teacher.

References

[1] http://sites.google.com/site/mjbucknum

[2] Bucknum, M. J., Wen, B. & Castro, E. A. (2007). *Global J. Mol. Sci.* **2**(2) 37-44.

[3] Bucknum, M. J. & Castro, E. A. (2004). The Spiro Quantum Chemistry, Nova Science Publishers, New York, (in press), M. J. Bucknum, E. A. Castro, *J. Math. Chem.* **36**(4), 381-408.

[4] Bucknum, M. J. & Castro, E. A. (2006). *J. Chem. Theory & Comput.,* **2**(3), 775-781.

[5] Bucknum, M. J. & Castro, E. A. (2005). *J. Mol. Model.*, **12**, 111-124.

[6] Bucknum, M. J. & Castro, E. A. (2005). *Central Eur. J. Chem.*, **3**(1), 169-173.

[7] Bucknum, M. J. & Castro, E. A. (2006). *Russ. J. Gen. Chem.*, **76**(2), 265-266. Published in Russian language in *Zhurnal Obshchei Khimii,* **76**(2), 280-281.

[8] Bucknum, M. J. & Castro, E. A. (2005). *Int. J. Mol. Sci.,* **6,** 177-187.

[9] Bucknum, M. J. & Castro, E. A. (2008). *Zhurnal Obshchei Khimi,* **78**(1), 38-39, *Russ. J. Gen. Chem.,* **78**(1), 35-36.

[10] Bucknum, M. J. & Castro, E. A. (2006). *MATCH Commun. Math. Comput. Chem.,* **55**(1), 57-64.

In: Mathematical Chemistry
Editor: W.I. Hong, pp. 457-461

ISBN: 978-1-60876-894-3

Chapter 8

CHEMICAL PHYSICS OF PHONONS AND SUPERCONDUCTIVITY: A HEURISTIC APPROACH

Michael J. Bucknum and Eduardo A. Castro[*]
INIFTA, Divisiόn Química Teόrica, Universidad Nacional de La Plata,
Buenos Aires, Argentina

Abstract

Several potential heuristic angles are explored with regard to the parametrization of the superconducting transition temperature, T(c), and its relationship to the chemical physics of bonding. Fruitful angles explored in this paper include the relationship of the gas of Cooper pairs in a superconductor to a van der Waals equation of state for such a fluid and how it may be controlled and exploited through the consideration of such variables as pressure, P, and volume, V, of the fluid of Cooper pairs resident in a superconductor. Other angles explored, in an attempt to get a heuristic handle on the superconducting transition temperature, T(c), for superconducting compositions across the Periodic Table, include the introduction of a Morse's anharmonic chemical bond potential for the Debeye frequency of the phonons as it emerges from the central result for T(c) in superconductors from the Bardeen-Cooper-Schreifer (BCS) formulation of superconductivity. Yet another angle explored in this heuristic reasoning, is the use of a Badger's relationship between chemical bond force constant, k, and equilibrium internuclear distance, $r_{(e)}$, to try to gain some understanding of the nature of the electron-phonon coupling mechanism in superconductors. A relationship between $r_{(e)}$ and $D_{(e)}$ emerges as a result of this line of thought, given as $r_{(e)} \cdot D_{(e)}$ = "critical temperature constant", from which examination of spectroscopic data across the Periodic Table may yield a compositional solution of the appropriate chemical bonding, with the assumption of the proper electron-phonon coupling constant, C, and density of states, N(0), that ultimately may yield more desirable superconductivity transition temperatures.

[*]E-mail address: eacast@gmail.com. (Corresponding author)

I. Superconductivity and the Gas Laws

Considering our Equation (1) from the microscopic version of the BCS theory[1]:

$$k \cdot T(c) = 1.14 \cdot h\,\omega_{(D)} \cdot \exp(-1/\lambda) \quad (1)$$

one can multiply both sides by N_A, the Avogadro constant, to yield a corresponding macroscopic version of this Equation (1), as Equation (2), our macroscopic BCS theory result.

$$N_A \cdot k \cdot T(c) = N_A \cdot 1.14 \cdot h\omega_{(D)} \cdot \exp(-1/\lambda) \quad (2)$$

Using the identities $N_A \cdot k \cdot T(c) = R \cdot T$, and $P \cdot V = R \cdot T$, for an ideal gas, we have Equation (3):

$$P \cdot V = N_A \cdot 1.14 \cdot h\omega_{(D)} \cdot \exp(-1/\lambda \quad (3)$$

Assuming the Bose gas of Cooper pairs inside the superconducting lattice behaves as an ideal gas, then there seemingly is nothing we can do with this macroscopic BCS theory expression at all. P and V would thus be the thermodynamic values of the Bose gas of Cooper pairs as their pressure and volume they occupy, for a mole of Cooper pairs constrained within a mole of superconducting material.

But what if the Bose gas of Cooper pairs is non-ideal, what if it obeys a van der Waals equation of state for a gas, as Equation (4) suggests:

$$(P + (a^2/V^2)) \cdot (V - b) = NA \cdot 1.14 \cdot h\omega_{(D)} \cdot \exp(-1/\lambda) \quad (4)$$

Therefore, if one could measure the vdW parameters, "a" the attractive force between Cooper pairs, and "b" the finite volume occupied by Cooper pairs, one would have a means of controlling the superconductivity mechanism somewhat and possibly affecting the superconducting temperature onset, T(c) through a control of these thermodynamic variables.

II. Superconductivity and the Electron-Phonon Coupling Parameter, λ

An article by the Mao-Hemley group[2] at the Geophysical Laboratory in Washington, D.C., describes measurements of conductivity and superconductivity in B allotropes under pressure. They cite an expression for the electron-phonon coupling of superconductivity, the parameter λ in Equation (1) above. The expression is for s-p metals & superconductors, and is given as Equation (5):

$$\lambda = ((N(0) \cdot C) / \langle\omega^2\rangle)) \quad (5)$$

The parameter "C" is supposed to be a numerical constant, so we don't worry about that. N(0) is the DOS(E_f) = density of states at the Fermi energy, E_f. And, importantly, $\langle\omega^2\rangle = (k/\mu)$ is the averaged phonon frequency squared.

So we see that the phonon frequencies and ultimately the phonon width at the Fermi energy, play an important role twice in the BCS theory result given in Equation (1) here. In the term, $h\omega_{(D)}$, in Equation (1), it seems that the higher the phonon frequencies attained in the BCS mechanism, the higher will be the transition temperature, T(c). But in the electron-phonon coupling parameter, because of the inverse exponential-type function it enters as in Equation (1), it seems that the higher the phonon frequencies contributing to the BCS mechanism, the lower will be the transition temperature, T(c). It is as if the two effects in the BCS theory result of Equation (1) are indeed opposed to each other.

The BCS theory doesn't seem to give a monotonic-type, useful predictive result. It contains parameters that work at cross-purposes to each other. We think, also, judging from the paper we have cited[3], where we note there is yet another expression for the superconducting transition temperature T(c), that indeed there may be a plethora of such expressions for T(c). Indeed, it seems that there is no concensus in the physics community over what the proper physical expression should be for the behavior of T(c) in superconductors. Our earlier result for substituting into (1) the classical Morse potential, U(r), to yield a semiclassical BCS-type theory with maximum contributions, in a continuum of states, from anharmonicity and anharmonic vibrations of the lattice, would seem to work to increase the transition temperature, T(c), were it not for the term in the electron-phonon coupling, given by $\exp(-1/\lambda)$, which suggests such anharmonicity corrections would serve to suppress the superconducting transition temperature, T(c).

Finally, there is the term N(0) which is $DOS(E_f)$. Clearly the bigger the $DOS(E_f)$ the bigger will be the contribution to increasing the transition temperature of superconductivity, T(c). In this instance, and only in this instance, does it seem that a band structure can play a role. And the flatter the bands are at the Fermi energy, the more so will N(0) contribute to a higher transition temperature, T(c).

This seems to be all we can gather from Equation (1) and the other references we have studied thus far. It is odd that the BCS theory has this competition built into it from contributions of the lattice vibrations. It is a conundrum indeed.

III. Superconductivity and the Morse Potential & Badger's Rule

In treating anharmonicity in the BCS expression for superconductivity, one can start[2] with Equation (6), which is the ordinary result from the BCS theory, in which one approximates phonons in a harmonic potential:

$$k \cdot T(c) = 1.14 \cdot h\,\omega_{(D)} \cdot \exp(-1/\lambda) \tag{6}$$

And thus one can use an anharmonic Morse's potential[4], $U(r) = D_{(e)}(1 - \exp^{(a(r-re)2})$, where the anharmonicity parameter is given by, $a = \sqrt{(k/2D_{(e)})}$, to solve the Schroedinger equation for the diatomic molecular potential, and thus to replace the phonon energy term in (6), $h\ \omega(D)$, in order to get an analogous BCS expression for anharmonic phonons as in Equation (7):

$$k \cdot T(c) = 1.14 \cdot (h\,\omega_{(D)} - (h\omega_{(D)})^2/4D_{(e)}) \cdot \exp(-1/\lambda) \tag{7}$$

In (7) we have thus introduced the effect of the bond strength $D_{(e)}$ into the BCS theory of superconductivity.

And from the point of view of Badger's relation[5], a correlation between bond force constant, k, and bond distance, $r_{(e)}$, developed in the 1930's to relate together various spectroscopic information across the Periodic Table as, $k = a(r_{(e)} - b)^{-3}$, where k and $r_{(e)}$ have their usual definitions, and "a" and "b" are empirical fitting constants of the appropriate dimensions, which is the Badger's relation, and one can look again to the BCS expression, this time to Equation (8) for the electron-phonon coupling constant[6]:

$$\lambda = ((N(0) \cdot C)/\langle\omega^2\rangle) \tag{8}$$

and one can replace the term in ω by recognizing the relationship $\omega = \sqrt{(k/\mu)} = \sqrt{(a(r_{(e)} - b)^{-3}/\mu)}$ where μ is the reduced mass of the chemical bond of interest in the electron-phonon coupling phenomena.

Therefore, upon squaring the latter expression for the phonon frequency, in terms of $\omega^2 = (k/\mu) = (a(r_{(e)} - b)^{-3}/\mu)$, we get the following result for the electron-phonon coupling parameter from the BCS theory:

$$\lambda = ((N(0) \cdot C) \cdot (a(r_{(e)} - b)^3 . \mu)) \tag{9}$$

where the electron-phonon coupling constant is in terms of the variables in N(0), the density of states at the Fermi level, the Badger chemical bond parameters a and b, the equilibrium internuclear distance of interest $r_{(e)}$, the reduced mass, μ, of the bond of interest and the electron-phonon coupling constant C.

If we thus introduce all of the changes made above into Equation (6), we would have a BCS expression that introduces both the phonon bond dissociation energy, $D_{(e)}$, and the phonon equilibrium internuclear distance, $r_{(e)}$, explicitly into the overall BCS picture of superconductivity as in Equation (10):

$$k \cdot T(c) = 1.14 \cdot (h\,\omega_{(D)} - (h\omega_{(D)})2/4D(e)) \cdot \exp\text{-}(1/((N(0) \cdot C) \cdot (a(r_{(e)} - b)3 \cdot \mu)) \tag{10}$$

Thus one has effectively transformed the BCS expression, in terms of phonon frequencies, $\omega_{(D)}$, into an equivalent, viable expression in terms of phonon properties of dissociation energy, $D_{(e)}$ and equilibrium bond length, $r_{(e)}$, that are in fact more intuitive quantities to deal with and that have been determined spectroscopically quite accurately across the Periodic Table.

Equation (10), and the effects of chemical bond length, $r_{(e)}$, and chemical bond strength $D_{(e)}$, can thus be simply interpreted for the chemical bond that has given rise to the phonon of interest in the given superconductivity mechanism, by making some obvious assumptions. Thus by assuming that the given maximum values for the associated density of states of the material at the Fermi energy, N(0), and simultaneously a maximum in the electron-phonon coupling constant C hold for the purposes of testing out the effect of these chemical bond parameters on the overall critical temperature, $T_{(c)}$, of various superconducting compositions, one would analyze Equation (10) to find that as $r_{(e)}$ gets larger, to a power of 3, the corresponding critical transition temperature gets larger, $T_{(c)}$. By an analogous algebraic analysis, one can see that as $D_{(e)}$ gets larger, similarly $T_{(c)}$ gets larger as well. So one would like to maximize both these chemical bond parameters, $r_{(e)}$ and $D_{(e)}$, in order to maximize the critical superconducting temperature in a given composition. The approach one would take, in such a case, would be of course to scan the spectroscopic data across the Periodic Table, and see which diatomic chemical bond potentials maximize the product function given in Equation (11) below:

$$r_{(e)} \, . \, D_{(e)} = \text{"critical temperature constant"} \tag{11}$$

It seems to be the case, in this connection, that such expressions as Equation (10) and Equation (11) lead to a more chemically intuitive understanding of the phonons involved in the superconductivity mechanism. One can thus use (10) and (11) to get a more predictive handle on the superconductivity transition temperature, a central result emerging out of the BCS theory, using chemically intuitive concepts such as the spectroscopic chemical bond parameters given by $r_{(e)}$, and $D_{(e)}$.

References

[1] Bardeen, J., Cooper, L. N. & Schreiffer, J. R. (1957). *Physical Review*, **108**, 1175.
[2] Eremets, M. I., Struzhkin, V. V., Mao, H. K. & Hemley, R. J. (2001). *Science*, **293**, 272.
[3] Ibid [2]
[4] Pauling, L. (1960). The Nature of the Chemical Bond, 3rd edition, Cornell University Press, Ithaca, NY, 594.
[5] Ibid [4], p.231
[6] Ibid [2]

In: Mathematical Chemistry
Editor: W.I. Hong, pp. 463-467

ISBN: 978-1-60876-894-3

Chapter 9

An Intensity-Enhanced 2D Graphical Representation of DNA and RNA Sequence

Teik-Cheng Lim
School of Science and Technology, SIM University (UniSIM), Republic of Singapore

Abstract

The 2D graphical representation of genetic sequence as proposed by Gates, Nandy and Leong & Morgenthaler have been enhanced by ensuring the number of overlaps is denoted by different line intensities. With the availability of this additional information, distribution of weak and strong H-bonds can be obtained from the intensity-enhanced Gates representation, while pyrimidine and purine distribution can be extracted based on intensity-enhanced Nandy representation. The intensity-enhanced Leong-Morgenthaler representation allows the distribution of keto and amino groups to be plotted. These intensity-enhanced graphical representations are herein illustrated for the first exon of human beta-globin.

Keywords: Graphical representation; genetic sequence; intensity-enhanced.

1. Introduction

While it is obvious that 2D graphical representation of DNA and RNA enable convenient visual inspection of data and therefore aid the discrepancy recognition of almost-similar genetic code sequence, these representations also avail numerical characterization to be performed on the sequences. Comparison of DNA sequence has been useful for understanding the homology of molecules [1] and drafting phylogenetic tree of various species [2]. Although DNA consists of only 4 types of bases: adenine (A), cytocine (C), guanine (G) and thymine (T), the very long nature of DNA chain justifies the development and application of 2D graphical representation. A 2D graphical representation of DNA by Jeffrey [3] denotes four points of a Cartesian coordinate (-1,-1), (-1,1), (1,-1) and (1,1) as A, C, G and T respectively. Starting from the origin, a straight line is plotted towards one of the four corners depending on the base type and stops halfway. The next line representing the next base starts from where the previous line stops. This diagram has been extended to proteins whereby each of the 20 amino acids is represented by its position on the "magic circle" [4]. Extension of the Jeffrey diagram into 3D representation was performed on DNA [5] and protein [6] within a tetrahedron confined by $-1 \le i \le 1$ where $i = x, y, z$. Instead of the straight line halfway to the corner, Gates [7] introduced a system whereby a one unit line is plotted in the (-x), (+y), (+x) and (-y) for G, T, C and A bases respectively.

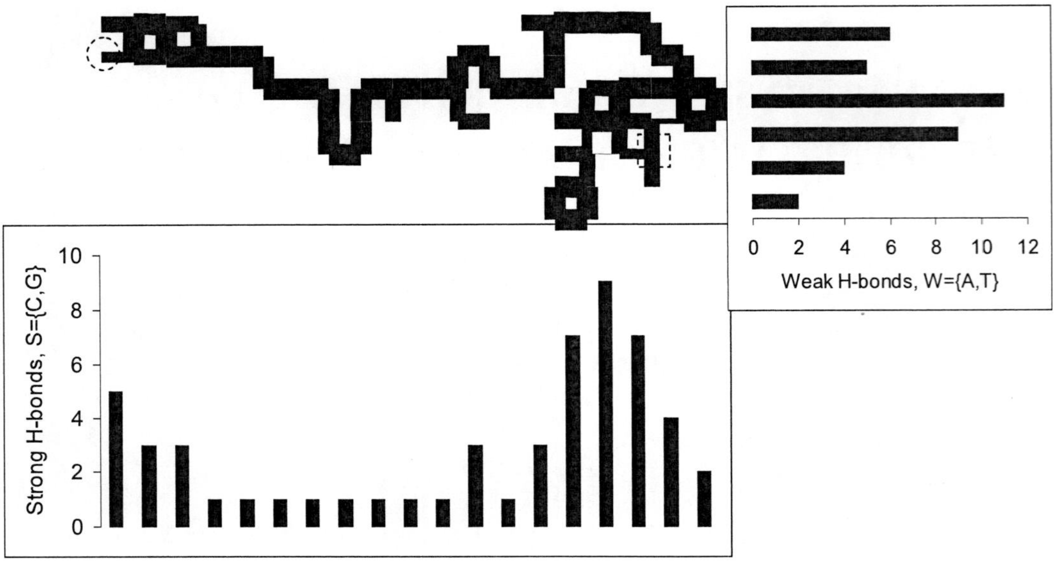

Figure 1. Intensity-enhanced Gates representation for the first exon of human beta-globin.

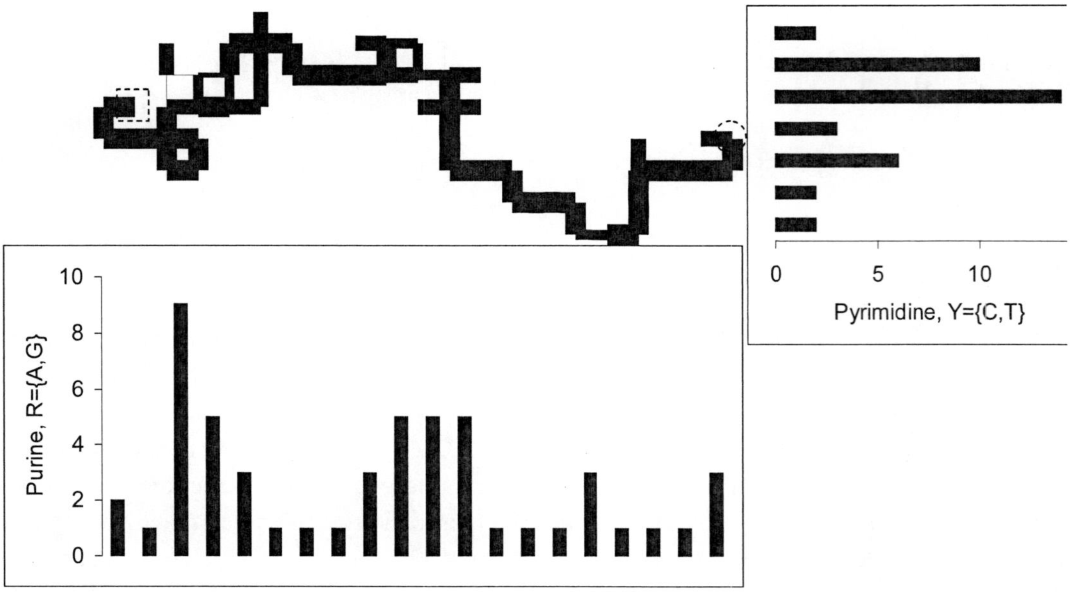

Figure 2. Intensity-enhanced Nandy representation for the first exon of human beta-globin.

Alternative versions were proposed by Nandy [8,9], and Leong and Morgenthaler [10] whereby the same coordinate displacement refer to A, C, G and T for the former and C, T, A and G for the latter. These three schemes are complete in that they pair up all different possible bases for pointing in the opposing direction. Other combinations will lead to mirror images to at least one of the abovementioned schemes [7-10]. Other variations to these schemes were proposed to decrease the graphical representations' level of degeneracy [11-17]. In this chapter, we impose lines of various boldness to the abovementioned schemes in order to reduce the level of degeneracy, as well as to enable the extraction of other information such as weak H-bonds W={A,T}, strong H-bonds S={C,G}, purine R={A,G}, pyrimidine Y={C,T}, amino group M={A,C} and keto group K={G,T}.

2. Intensity-Enhanced Graphical Representation

The intensity-enhanced graphical representation was inspired in part by the 4-color map representation of DNA [18]. In this paper, lines of different boldness are imposed on the graphical representations of Gates, Nandy and Leong-Morgenthaler instead of the square spiral. To appreciate the benefit of bolded lines, we note that the sequences GG, GGA, GGAA, GAGG, GGAG and GAGGAG appear similar in the Nandy scheme, and thereby lead to information loss. In the intensity-enhanced enhancement, each line is accorded a line intensity and a change in boldness is imposed each time the line is overlapped. By imposing line intensities on the graphical representation schemes of Gates, Nandy and Leong-Morgenthaler, one is able to easily see the locations of these graphical representations that have high intensities. We herein define the intensity as the number of overlaps. This additional dimension in line intensities retains the viewing convenience and simplicity of the original 2-D representations, while storing additional data in order to reduce the degeneracy in conventional black and white representations. Theoretically, any line intensity schemes can be used. However, a logical scheme would show the intensity of line boldness being corresponded to the number of overlaps.

3. Example

In this chapter, we impose line intensities on the first exon of human beta-globin (see Appendix). Figures 1, 2 and 3 show the intensity-enhanced versions of Gates, Nandy and Leong-Morgenthaler graphical representations respectively, with the five line intensities. Since the Gates scheme places A and T in the vertical direction, then summation of all vertical lines, including their intensities, along the horizontal direction gives the distribution of weak H-bonds, as shown in Figure 1. The other two types of bases – C and G – in the horizontal direction allows the distribution of strong H-bonds to be plotted by summing all horizontal lines (with their intensities) along the vertical direction. Performing similar summation on the intensity-enhanced Nandy scheme gives the pyrimidine and purine distributions, as depicted in Figure 2, while distributions of the keto and amino groups can be similarly extracted from the intensity-enhanced Leong-Morgenthaler scheme (see Figure3). The interest of extracting the information on weak and strong H-bonds, pyrimidine, purine, keto and amino groups has recently been reflected in recent efforts [19-21]. Due to the obstructing view of the x and y axes, they are being removed. The start point is denoted by a square, which, to a certain extent, resembles the origin. To further enhance the intensity-enhanced graphical representations, a circle is placed at the final point due to its resemblance to the full stop.

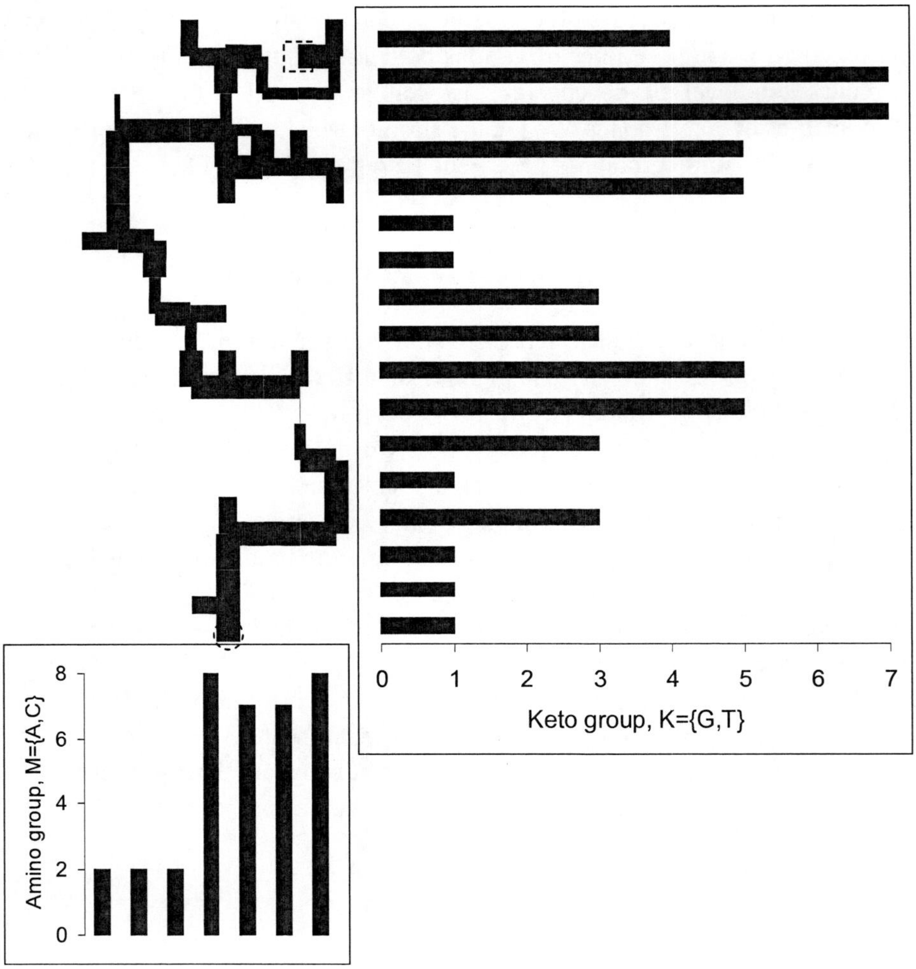

Figure 3. Intensity-enhanced Leong-Morgenthaler representation for the first exon of human beta-globin.

4. Conclusions

Although 2D graphical representations are not as detailed as 3D representations, the attractiveness of the former is the ease of usage – the user only needs to look at one view instead of three. The limitations encountered by the 2D graphical representation schemes of Gates, Nandy and Leong-Morgenthaler are significantly overcome by incorporation of various line boldness to indicate the number of overlap. While the original graphical representations allow genetic code to be compared by viewing the shape of the graphical representation, the intensity-enhanced ones allow greater depth of comparison to be made.

For example, the sequences GAGTG, GTCTG and GTGAG appear to be 100% similar based on the original Nandy representation, while the corresponding intensity-enhanced one show higher intensity in three different location lines. In addition to allowing ease of viewing due to the line intensities, the stored data containing the number of overlaps enable summation of vertical and horizontal lines along the horizontal and vertical directions respectively in order to ploy the distribution of weak and strong H-bonds from intensity-enhanced Gates representation, pyrimidine and purine based on intensity-enhanced Nandy representation, and keto and amino groups based on intensity-enhanced Leong-Morgenthaler representation.

Appendix

The 92 nucleotide bases in the first exon of human beta-globin:

ATGGTGCACCTGACTCCTGAGGAGAAGTCTGCCGTTACTGCCCTGTGGGGCA AGGTGAACGTGGATGAAGTTGGTGGTGAGGCCCTGGGCAG

References

[1] Needleman, S. B. & Wunsch, C. D. (1970). *J. Mol. Biol.*, **48**, 443.
[2] Liao, B. Shan, X., Zhu, W. & Li, R. (2006). *Chem. Phys. Lett.*, **422**, 282.
[3] Jeffrey, H. I. (1990). *Nucleic Acid Res.*, **18**, 2163.
[4] Randic, M., Butina, D. & Zupan, J. (2006). *Chem. Phys. Lett.*, **419**, 528.
[5] Zupan, J. & Randic, M. (2005). *J. Chem. Inf. Model.*, **45**, 309.
[6] Randic, M., Zupan, J. & Balaban, A. T. (2004). *Chem. Phys. Lett.*, **397**, 247.
[7] Gates, M. A. (1986). *J. Theor. Biol.*, **119**, 319.
[8] Nandy, A. (1994). *Curr. Sci.*, **66**, 304.
[9] Nandy, A. (1996). *Curr. Sci.*, **70**, 611.
[10] Leong, P. M. & Morgenthaler, S. (1995). *Comput. Applic. Biosci.*, ***11***, 503.
[11] Bai, F. & Wang, T. (2005). *Chem. Phys. Lett.*, **413**, 458.
[12] Liao, B., Tan, M. & Ding, K. (2005). *Chem. Phys. Lett.*, **414**, 296.
[13] Liao, B., Xiang, X. & Zhu, W. (2006). *J. Comput. Chem.*, **27**, 1196.
[14] Liu, Y., Guo, X., Xu, J., Pan, L. & Wang, S. (2002). *J. Chem. Inf. Comput. Sci.*, **42**, 529.
[15] Guo, X., Randic, M. & Basak, S. C. (2001). *Chem. Phys. Lett.*, **350**, 106.
[16] Randic, M., Vracko, M., Nandy, A. & Basak, S. C. (2000). *J. Chem. Inf. Comput. Sci.* **40**, 1235.
[17] Liao, B. & Ding, K. (2006). *Theor. Computer Sci.*, **358**, 56.
[18] Randic, M., Lers, N., Plavsic, D., Basak, S. C. & Balaban, A. T. (2005). *Chem. Phys. Lett.*, **407**, 205.
[19] Song, J. & Tang, H. (2005). *J. Biochem. Biophys. Methods,* **63**, 228.
[20] Li, C. & Hu, J. (2006). *J. Biochem. Mol. Biol.*, **39**, 292.
[21] Yao, Y. H., Nan, X. Y. & Wang, T. M. (2006). *J. Mol. Struct., Theochem*, **764**, 101.

In: Mathematical Chemistry
Editor: W.I. Hong, pp. 469-507

ISBN 978-1-60876-894-3

Chapter 10

LAGUERRE FUNCTIONS WITH APPLICATIONS FROM MATHEMATICAL CHEMISTRY

Damian Trif*

Department of Applied Mathematics,
Babes-Bolyai University of Cluj-Napoca, Romania

Abstract

Some physically interesting equations are naturally set on a semi-infinite interval and it would be desirable to have accurate and efficient numerical methods for such problems. The usual Laguerre polynomials have very poor resolution properties when compared with other types of orthogonal polynomials. However, we may obtain uniformly convergent Galerkin approximations using scaled Laguerre functions in regular unbounded domains, and construct stable and efficient numerical algorithms for their practical implementations. We introduce and illustrate some algorithmic ideas with several simple problems from mathematical chemistry.

1. Introduction

Some physically interesting equations, e.g. the Schrödinger equation, are naturally set on a semi-infinite interval and it would be desirable to have accurate and efficient numerical methods for such problems.

A boundary value problem on $(0, \infty)$ may be tackled in three ways:

a) truncate the unbounded interval to a bounded one,

b) transform it into a problem over a finite interval by mapping of the independent variable,

c) approximate by classical orthogonal systems on unbounded domains.

The domain truncation is a viable idea only for problems with rapidly (exponential) decaying solutions or when accurate non-reflecting or exact boundary conditions are available at the truncated boundary. The main disadvantage of method b) is that the transformed

*E-mail address: dtrif@math.ubbcluj.ro

equation is usually very complicated. A better approach (we think) is the approximation by classical orthogonal systems on semi-infinite domains.

Most of the previous numerical results used Laguerre *polynomials* which have very poor resolution properties when compared with other types of orthogonal polynomials. This is the main factor which has limited the interests and developments on using Laguerre polynomials for problems in unbounded domains. However, we may obtain uniformly convergent Galerkin approximations using Laguerre *functions* in regular unbounded domains, and construct stable and efficient numerical algorithms for their practical implementations, see [7] for more details.

We remark that it is very important to choose a natural procedure to discretize the continuous problem. For example, let us consider the spectral test problem

$$x\frac{dy}{dx} = \lambda y, \; y(0) = 0, \; x \in (0, \infty).$$

The exact polynomial eigenfunctions are $y_k(x) = x^k$ and the corresponding eigenvalues are $\lambda_k = k$, for $k = 1, 2, ...$

Firstly we use the *pseudospectral* method based on Laguerre functions from the DMS package [15]. By setting

$$z(x) = e^{-\frac{x}{2}} y(x)$$

the problem becomes

$$x\frac{dz}{dx} + \frac{x}{2}z = \lambda z, \; z(0) = 0.$$

The MATLAB commands

```
n=64;[x,D]=lagdif(n,1,1);
A=diag(x)*D+diag(x)/2;B=eye(n);
A(1,:)=zeros(1,n);A(1,1)=1;B(1,1)=1/500;
L=sort(eig(A,B));
```

discretize the equation and impose the boundary condition, giving us the numerical eigenvalues represented in figure 1 as complex numbers. The first small eigenvalues approximate very well the exact corresponding eigenvalues but the others are spurious due to the high sensitivity of the full differentiation matrix D to rounding errors.
If we use the Laguerre *spectral* method implemented in our package LaguerreEig, the MATLAB commands

```
n=64;X=mult(n);D=deriv(n);
A=X*D+X/2;B=eye(n);
A(1,:)=ones(1,n);B(1,:)=1/500;
L=sort(eig(A,B))
```

give the *exact* eigenvalues $\lambda_k = k$ for $k = 1, 2, ..., 63$. Consequently, we *must* calculate the numerical eigenvalues and eigenfunctions of Schrödinger problems, for example, using this Laguerre *spectral* method.

The chapter is organized as follows. In section 2 we describe a spectral approximation procedure based on the matrix formulation of the main operators – the differentiation and the multiplication by the independent variable. In section 3 we present the MATLAB package LaguerreEig which uses these ideas to approximate decaying functions on a

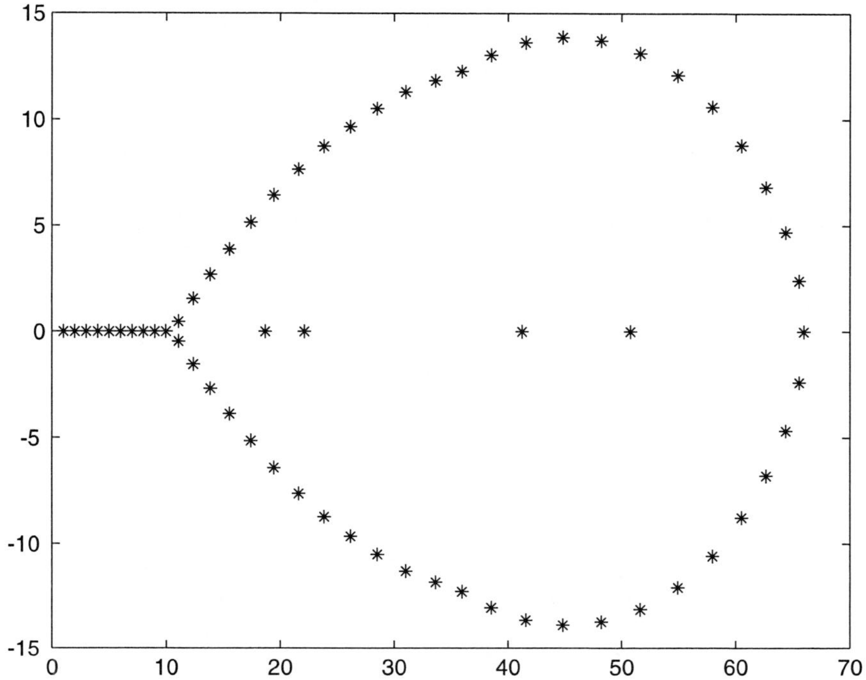

Figure 1. Numerical eigenvalues calculated by DMS.

semi-infinite interval. Section 4 contains some interesting applications of this approach from mathematical chemistry. Some concluding remarks are given in the final section.

2. Theoretical Results

Basic Properties

Let us recall, following [9], some basic properties of the Laguerre polynomials $L_k(x)$, defined by

$$L_k(x) = \frac{1}{k!}e^x \frac{d^k}{dx^k}\left(x^k e^{-x}\right), \quad x \in (0, \infty), \ k = 0, 1, 2, ...$$

They are the eigenfunctions of the singular Sturm-Liouville problem

$$\frac{d}{dx}\left(xe^{-x}\frac{d}{dx}u(x)\right) + \lambda e^{-x}u(x) = 0, \ x \in (0, \infty)$$

corresponding to the eigenvalues $\lambda_k = k$, $k = 0, 1, 2, ...$and scaled to verify $L_k(0) = 1$. The Laguerre polynomials fulfill the recurrence relations

$$(k+1)L_{k+1}(x) = (2k + 1 - x)L_k(x) - kL_{k-1}(x), \ k = 1, 2, ...$$
$$L_k(x) = \frac{d}{dx}L_k(x) - \frac{d}{dx}L_{k+1}(x), \ k = 0, 1, 2, ...$$

Let $\omega(x) = e^{-x}$ be the weight function of the space $L^2_\omega(0,\infty)$, which is a Hilbert space with the following inner product and norm

$$(u,v)_\omega = \int_0^\infty u(x)v(x)\omega(x)dx, \ \|u\|_\omega = (u,u)_\omega^{1/2}.$$

The set of Laguerre polynomials form an orthonormal basis for $L^2_\omega(0,\infty)$, i.e.,

$$\int_0^\infty L_k(x)L_m(x)\omega(x)dx = \delta_{k,m} = \begin{cases} 0, & k \neq m \\ 1, & k = m \end{cases}$$

and any function $u \in L^2_\omega(0,\infty)$ can be expanded as

$$u(x) = \sum_{k=0}^\infty \widehat{u}_k L_k(x), \ \widehat{u}_k = \int_0^\infty u(x)L_k(x)\omega(x)dx.$$

In order to analyze the approximation properties of the Laguerre polynomials, let us define the generalized weight function $\omega_\alpha(x) = x^\alpha e^{-x}$. The orthogonal projection

$$\pi_N : L^2_\omega(0,\infty) \to P_N,$$

where P_N is the set of all algebraic polynomials of degree $\leq N$, is defined by

$$(u - \pi_N u, v_N)_\omega = 0, \ \forall v_N \in P_N$$

and it is clear that

$$\pi_N u(x) = \sum_{k=0}^N \widehat{u}_k L_k(x).$$

Let us define the weighted space

$$B^m(0,\infty) = \left\{ u | \frac{d^k}{dx^k} u \in L^2_{\omega_k}(0,\infty), 0 \leq k \leq m \right\}.$$

The fundamental Laguerre approximation result is

THEOREM 1. *For any $m \geq 0$ and $u \in B^m(0,\infty)$,*

$$\left\| \frac{d^k}{dx^k}(\pi_N u - u) \right\|_{\omega_k} \lesssim N^{(k-m)/2} \left\| \frac{d^m}{dx^m} u \right\|_{\omega_m}, \ 0 \leq k \leq m.$$

Here $A \lesssim B$ means that there exists a generic positive constant c such that $A \leq cB$.

We now recall the Laguerre-Gauss type quadrature. Let $\{x_j\}_{j=0}^N$ be the zeros of $L_{N+1}(x)$ and

$$w_j = \frac{1}{(N+1)^2} \frac{x_j}{[L_N(x_j)]^2}, \ 0 \leq j \leq N$$

the weights. Then,

$$\int_0^\infty p(x)e^{-x}dx = \sum_{j=0}^N p(x_j)w_j, \ \forall p \in P_{2N+1}.$$

If $f(x)$ decreases or increases at a rate that is slower than exponential, then

$$\int_0^\infty f(x)e^{-x}dx \simeq \sum_{j=0}^{N} f(x_j)w_j$$

with an error

$$E_N = \frac{[(N+1)!]^2}{[2(N+1)]!} f^{((2N+2)}(\xi)$$

where $\xi > 0$.

If we define the interpolation operator

$$I_N : C[0,\infty) \to P_N,\ (I_N u)(x_j) = u(x_j),\ j = 0, i, ..., N$$

we have the approximation result:

THEOREM 2. *Assuming* $u \in C[0,\infty)$, $u \in B^m(0,\infty)$ *and* $\frac{d}{dx}u \in B^{m-1}(0,\infty)$ *with* $m \geq 1$ *then*

$$\|I_N u - u\|_\omega \lesssim N^{(1-m)/2}\left(\left\|\frac{d^m}{dx^m}u\right\|_{\omega_{m-1}} + (\ln N)^{1/2}\left\|\frac{d^m}{dx^m}u\right\|_{\omega_m}\right).$$

We remark that Laguerre polynomials are only suitable for the approximations of functions with algebraic or exponential growth at infinity. Moreover, although the error estimated in Theorem 1 is of spectral type, it only implies meaningful pointwise approximation for a short interval, due to the exponential decay weight in the norm. For problems with some decay properties at infinity, it is more appropriate to use the Laguerre functions

$$\mathcal{L}_k(x) = e^{-x/2}L_k(x),\ x \in (0,\infty),\ k = 0,1,2,...$$

It is clear that the Laguerre functions are orthonormal,

$$\int_0^\infty \mathcal{L}_k(x)\mathcal{L}_m(x)dx = \delta_{k,m},\ \forall k,m = 0,1,2,...$$

and form a complete system for $L^2(0,\infty)$. We have the following recurrence relations

$$(k+1)\mathcal{L}_{k+1}(x) = (2k+1-x)\mathcal{L}_k(x) - k\mathcal{L}_{k-1}(x),$$
$$\mathcal{L}_0(x) = e^{-x/2},\ \mathcal{L}_1(x) = (1-x)e^{-x/2}$$

and

$$\frac{d}{dx}\mathcal{L}_k(x) = \frac{d}{dx}\mathcal{L}_{k-1}(x) - \frac{\mathcal{L}_k(x)+\mathcal{L}_{k-1}(x)}{2}.$$

It is straightforward to obtain the Laguerre function approximation, because for any $u \in L^2(0,\infty)$ we have $ue^{x/2} \in L^2_\omega(0,\infty)$. Let us denote $\mathbf{w}_k(x) = x^k$,

$$\mathcal{P}_N = \left\{v|v = e^{-x/2}p \text{ with } p \in P_N\right\},$$
$$\boldsymbol{\pi}_N : L^2(0,\infty) \to \mathcal{P}_N,\ \boldsymbol{\pi}_N u = e^{-x/2}\pi_N(ue^{x/2}),$$
$$\frac{\mathbf{d}}{\mathbf{d}x}u = \frac{d}{dx}u + \frac{1}{2}u,\ \frac{\mathbf{d}^2}{\mathbf{d}x^2} = \frac{\mathbf{d}}{\mathbf{d}x}\left(\frac{\mathbf{d}}{\mathbf{d}x}u\right),...$$
$$\mathbf{B}^m(0,\infty) = \left\{u|\frac{\mathbf{d}^k}{\mathbf{d}x^k}u \in L^2_{\mathbf{w}_k}(0,\infty),0 \leq k \leq m\right\}.$$

We have the following approximation result:

THEOREM 3. *For any $m \geq 0$ and $u \in B^m(0,\infty)$,*

$$\left\| \frac{\mathrm{d}^k}{\mathrm{d}x^k} (\pi_N u - u) \right\|_{\mathbf{w}_k} \lesssim N^{(k-m)/2} \left\| \frac{\mathrm{d}^m}{\mathrm{d}x^m} u \right\|_{\mathbf{w}_m}.$$

We also have an extension of Theorem 2 for the interpolation operator

$$\mathbf{I}_N : C[0,\infty) \to \mathbf{P}_N, \ (\mathbf{I}_N u)(x) = e^{-x/2} I_N \left(u e^{x/2} \right)$$

by observing that

$$(\mathbf{I}_N u)(x_j) = u(x_j), \ j = 0, .., N.$$

THEOREM 4. *Assuming $u \in C[0,\infty)$, $u \in \mathbf{B}^m(0,\infty)$ and $\frac{\mathrm{d}}{\mathrm{d}x} u \in \mathbf{B}^{m-1}(0,\infty)$ with $m \geq 1$ then*

$$\| \mathbf{I}_N u - u \| \lesssim N^{(1-m)/2} \left(\left\| \frac{\mathrm{d}^m}{\mathrm{d}x^m} u \right\|_{\mathbf{w}_{m-1}} + (\ln N)^{1/2} \left\| \frac{\mathrm{d}^m}{\mathrm{d}x^m} u \right\|_{\mathbf{w}_m} \right).$$

Matrix Formulation of the Main Operators

The Laguerre spectral method to approximate a function u on a semi-infinite interval is to expand u into a series of the form

$$u(x) = \sum_{k=0}^{\infty} c_k \mathcal{L}_k(x)$$

where $\mathcal{L}_k(x)$ are the Laguerre functions. We search for an approximation of u as

$$u_N(x) = \sum_{k=0}^{N-1} c_k \mathcal{L}_k(x)$$

and we say that the vector $(c_0, ..., c_{N-1})^T$ discretizes the function $u(x)$ in the *spectral space*. For a suitable set of nodes $x = (x_1, ..., x_N)^T$ we say that the vector

$$(u(x_1), ..., u(x_N))^T$$

discretizes the function $u(x)$ in the *physical space*. The spectral discretization is useful for linear operators while the physical discretization works well for nonlinear operators. For the practical implementation, we define the vector c and the matrix t by

$$c = (c_0, c_1, c_2, ...)^T, \ t^T(x) = (\mathcal{L}_0(x), \mathcal{L}_1(x), \mathcal{L}_2(x), ...)$$

so that $u(x) = t^T(x)\, c$. Here the upperscript T means the transpose of the corresponding vector or matrix.

Let us find now a matrix X for which $xu(x) = t^T(x)(Xc)$, i.e. *if the coefficients of $u(x)$ are c, then the coefficients of $xu(x)$ are Xc.* By using the properties of Laguerre polynomials we have

$$(1 + 2k - x)L_k(x) - kL_{k-1}(x) - (k+1)L_{k+1}(x) = 0$$

with the same recurrence for the functions $\mathcal{L}_k$ and then

$$x\mathcal{L}_k(x) = (2k+1)\mathcal{L}_k(x) - k\mathcal{L}_{k-1}(x) - (k+1)\mathcal{L}_{k+1}(x)$$

where $\mathcal{L}_0(x) = e^{-x/2}$. Consequently,

$$\begin{aligned} x \cdot \mathcal{L}_0 &= \mathcal{L}_0 - \mathcal{L}_1 \\ x \cdot \mathcal{L}_1 &= 3\mathcal{L}_1 - \mathcal{L}_0 - 2\mathcal{L}_2 \\ x \cdot \mathcal{L}_2 &= 5\mathcal{L}_2 - 2\mathcal{L}_1 - 3\mathcal{L}_3 \\ &\ldots\ldots\ldots\ldots\ldots \end{aligned}$$

and the non-zero elements of X are

$$X_{ii} = 2i-1,\ X_{i,i+1} = X_{i+1,i} = -i, \quad i = 1, 2, \ldots$$

Of course, in practical calculation we must truncate this infinite matrix to have a finite order N, so that we obtain only an approximation of the exact result.

Then, in general,

$$x^m u(x) = t^T(x)(X^m c)$$

and

$$f(x)u(x) = t^T(x)(f(X)c)$$

for analytical functions f, i.e.

$$f(x) = \sum_{k=0}^{\infty} f_k \frac{x^k}{k!}.$$

Moreover,

$$\frac{u(x)}{x^m} = t^T(x)\left(X^{-m}c\right)$$

if the l.h.s. has no singularity at the origin. Of course, X^m is a banded matrix for small m but, generally, the matrix version `funm(X)` of the scalar function $f(x)$ or $X^{-m} = [inv(X)]^m$ are no longer sparse matrices.

A special case appears if we have to multiply $f(x)u(x)$ when f is given by its physical values at the grid x, not by a formula. If the coefficients of $u(x)$ are c, then $t^T * c$ are the physical values of $u(x)$ at the grid x, $diag(f(x)) * t^T * c$ are the physical values of $f(x)u(x)$ and, finally, $\left(t^T\right)^{-1} * diag(f(x)) * t^T * c$ are the coefficients of $f(x)u(x)$. If we denote by

$$F = \left(t^T\right)^{-1} * diag(f(x)) * t^T$$

then *the coefficients of* $f(x)u(x)$ *are approximated by* $F * c$.

Similarly, let us find the differentiation matrix D giving $\dfrac{du}{dx} = t^T(x)(Dc)$, i.e. *if the coefficients of* $u(x)$ *are* c, *then the coefficients of the derivative* $\frac{du}{dx}$ *are* Dc. The derivatives of L_k satisfy $L_k' = L_{k-1}' - L_{k-1}$ so that

$$\mathcal{L}_k' = \left[e^{-x/2}L_k\right]' = \mathcal{L}_{k-1}' - \frac{\mathcal{L}_k + \mathcal{L}_{k-1}}{2}.$$

We have

$$\mathcal{L}_0' = -\frac{1}{2}\mathcal{L}_0 \tag{1}$$
$$\mathcal{L}_1' = -\mathcal{L}_0 - \frac{1}{2}\mathcal{L}_1$$
$$\mathcal{L}_2' = -\mathcal{L}_0 - \mathcal{L}_1 - \frac{1}{2}\mathcal{L}_2$$
$$\ldots\ldots\ldots\ldots\ldots$$

so that the non-zero elements of D are

$$D_{ii} = -\frac{1}{2}, \quad D_{i,j} = -1, \ i = 1, 2, \ldots \ \text{ and } i < j.$$

and we truncate this matrix to the order N. Moreover, we have

$$\frac{d^m u}{dx^m} = t^T(x)\,(D^m c), \ m = 2, 3, \ldots$$

3. Implementation

LaguerreEig

The MATLAB package `LaguerreEig`, which could be freely obtained from [12], contains the files `pd.m`, `x2t.m`, `mult.m` and `deriv.m`

a) The function `pd.m` (physical domain),

```
function [x,w]=pd(n)
% From:   Edward Neuman
% http://www.math.siu.edu/matlab/tutorial5.pdf, p.27
d=-(1:n-1);f=1:2:2*n-1;J=diag(d,-1)+diag(f)+diag(d,1);
[u,v]=eig(J);[x,j]=sort(diag(v));w=(u(1,:).^2);w=w(j);
```

calculates the nodes (column) and weights (row) for the Gauss-Laguerre quadrature, i.e.

$$I \equiv \int_0^\infty e^{-x} f(x)dx \approx \sum_{k=1}^{N} f(x_k)w_k = \texttt{w*f(x)} \tag{2}$$

For example, let us calculate the Euler Gamma function

$$\Gamma(n+1) = \int_0^\infty e^{-x}x^n dx$$

for $n = 6$, with the exact value $\Gamma(7) = 6! = 720$. The code is

```
N=16;[x,w]=pd(N);f=x.^6;I=w*f
```

with the result `I=7.200000000000005e+002`

b) The function `x2t.m`

```
function t=x2t(n,x)
x=x(:)';
if n>=1 t(1,:)=ones(1,length(x)).*exp(-x/2);end;
```

```
if n>=2 t(2,:)=t(1,:)-x.*exp(-x/2);end;
if n>=3 for i=3:n
t(i,:)=(2*i-3-x)/(i-1).*t(i-1,:)-(i-2)/(i-1)*t(i-2,:);
end;end;
```

calculates the values of the Laguerre functions $\mathcal{L}_k(x)$, $k = 0, 1, ..., N-1$ at the points x, *which can be arbitrary nodes*, not only the Gauss-Laguerre nodes given by `pd.m`.

Consequently, if

$$f(x) = c_0\mathcal{L}_0(x) + c_1\mathcal{L}_1(x) + ... + c_{N-1}\mathcal{L}_{N-1}(x)$$

and $c^T = (c_0, ..., c_{N-1})$, we have $f(x) = t^T * c$ for $x \in [0, \infty)$.

Of course, we can use a Clenshaw-type algorithm to calculate the physical values of f, but for problems where this transformation from spectral to physical discretization is performed many times, we prefer to use the conversion matrix t. This matrix t is calculated only once at the beginning of the code and next, the matrix by vector product $t^T * c$ is faster than the Clenshaw procedure.

The code `x2t.m` can also be used to transform the physical representation $f(x)$ of f at the grid x into the spectral representation $c = \left(t^T\right)^{-1} * f(x)$ for moderate values of N. Again, we can calculate the coefficients c from their expressions

$$c_k = \int_0^\infty f(x)\mathcal{L}_k(x)dx = \int_0^\infty e^{-x}\left(e^x * f(x)\right)\mathcal{L}_k(x)dx, \; k = 0, 1, ..., N-1$$

that is

$$c = t * diag\,(w) * diag(e^x) * f(x).$$

In fact,

$$t * diag\,(w) * diag(e^x) \approx \left(t^T\right)^{-1}$$

with an error less than 4.75×10^{-13} for $N = 128$ and, moreover, this expression involves only matrix by vector products. But for moderate values of N (i.e. for $N \leq 256$) there are no important advantages, while for bigger N we lose the accuracy or we even get overflow from the calculation of e^x.

Consequently, the code `x2t.m` based on the calculation of the matrix t will be used for transformations between the spectral representation c and the physical representation $f(x)$ of the function f at the Gauss-Laguerre nodes x

$$f(x) = t^T * c, \; c = \left(t^T\right)^{-1} * f(x).$$

c) The files `deriv.m` and `mult.m` calculate the matrices D respectively X.

```
function D=deriv(n)
D=zeros(n);
for i=1:n, D(i,i)=-1/2;for j=i+1:n,D(i,j)=-1;end
end

function X=mult(n)
X=zeros(n);
for i=1:n-1,X(i,i)=2*i-1;X(i,i+1)=-i;X(i+1,i)=-i;end
```

```
X(n,n)=2*n-1;
```

Let us consider, as an example, the following problem

$$-\frac{d^2u}{dx^2}+u=[16\sin(4x)+8\cos(4x)]\,e^{-x},\ x\in(0,\infty)$$
$$u(0)=0,\ \lim_{x\to\infty}u(x)=0$$

with the exact solution $u(x)=\sin(4x)e^{-x}$. The matrix form of the problem is

$$\left(-D^2+I\right)c=b$$

where $c=(c_0,...,c_{N-1})^T$ are the coefficients of the solution (column vector),

$$u(x)=\sum_{k=0}^{N-1}c_k\mathcal{L}_k(x)$$

while $b=(b_0,...,b_{N-1})^T$ are the coefficients of the r.h.s. The boundary condition

$$u(0)=\sum_{k=0}^{N-1}c_k\mathcal{L}_k(0)=\sum_{k=0}^{N-1}c_k=0$$

is implemented by replacing the last row of the matrix $A=-D^2+I$ by ones and the last entry of b by zero (the tau method). The MATLAB code is

```
N=80; [x,w]=pd(N);t=x2t(N,x);D=deriv(N);
A=-D^2+eye(N);
B=t'\((16*sin(4*x)+8*cos(4*x)).*exp(-x));
A(N,:)=ones(1,N);B(N)=0;% boundary condition u(0)=0
c=A\B;
u=t'*c;% the solution u at the points x
```

and the calculated solution approximates the exact solution at the same points x with a maximum error less than 10^{-3}.

Scaling

If we consider a problem whose solution decays at infinity, i.e. the solution is negligible outside an effective interval, the collocation points which fall outside this interval are useless. The performance of Laguerre spectral method can be significantly enhanced by making a coordinate transform $y=\beta x$, so that the extreme collocation points are close to the endpoints of the effective interval.

Let us illustrate this idea by numerically verifying that

$$\int_0^\infty e^{-\frac{x^2}{2}}dx=\sqrt{\frac{\pi}{2}}.$$

In order to apply the Laguerre-Gauss quadrature, we transform the above integral to

$$I=\int_0^\infty e^{-x}e^{x-\frac{x^2}{2}}dx.$$

The MATLAB code

```
N=16;[x,w]=pd(N);f=exp(x-x.^2/2);I=w*f;
```

gives us `I=1.253294859437643`. The exact value is

$$I = \sqrt{\pi/2} = 1.25331413731550$$

so the error is about 2×10^{-5}.

Now, let us transform the integral to

$$I = \int_0^\infty e^{-\beta x} e^{\beta x - \frac{x^2}{2}} dx = \frac{1}{\beta}\int_0^\infty e^{-y} e^{y - \frac{y^2}{2\beta^2}} dy$$

by using the coordinate transform $y = \beta x$. Again, the MATLAB code with b$\equiv \beta = 6$

```
N=16;b=6;[y,w]=pd(N);f=exp(y-y.^2/2/b^2);I=w*f/b;
```

gives us `I=1.253314137315454` with an error about 5×10^{-14}.

In this case, for $N = 16$ we have the collocation nodes $y_k \in (0, 52)$ but the integrand $e^{-y^2/2}$ becomes negligible for $y > 9$. Therefore, $x_{\max} = y_{\max}/\beta = 9$ leads us to $\beta = y_{\max}/9 \simeq 6$.

Generally speaking, if we have to calculate

$$I = \int_0^\infty e^{-\beta x} f(x) dx$$

then, by using the coordinate transform $y = \beta x$, we obtain

$$I = \frac{1}{\beta}\int_0^\infty e^{-y} f(\frac{y}{\beta}) dy = \sum_{k=0}^{N-1} \frac{w_k}{\beta} f(\frac{y_k}{\beta})$$

where `[y,w]=pd(N)`. Finally, we can write

$$\int_0^\infty e^{-\beta x} f(x) dx \simeq \sum_{k=0}^{N-1} \mathbf{w}_k f(\mathbf{x}_k)$$

where

$$\mathbf{w}_k = \frac{w_k}{\beta}, \ \mathbf{x}_k = \frac{y_k}{\beta}. \tag{3}$$

As another example, let us consider again the previous problem

$$-\frac{d^2u}{dx^2} + u = [16\sin(4x) + 8\cos(4x)]\, e^{-x}, \ x \in (0,\infty)$$
$$u(0) = 0, \ \lim_{x\to\infty} u(x) = 0$$

with the exact solution $u(x) = \sin(4x)e^{-x}$. For $N = 80$ the maximum error was less than 10^{-3}. But if we search for the approximate solution as

$$u(x) = \sum_{k=0}^{N-1} c_k \mathcal{L}_k(\beta x) = \sum_{k=0}^{N-1} c_k e^{-\frac{\beta x}{2}} L_k(\beta x)$$

the coordinate transform $y = \beta x$ leads us to the problem

$$-\beta^2 \frac{d^2 v}{dy^2} + v = \left[16\sin(\frac{4y}{\beta}) + 8\cos(\frac{4y}{\beta})\right] e^{-\frac{y}{\beta}},\ y \in (0,\infty)$$
$$v(0) = 0,\ \lim_{y\to\infty} v(y) = 0$$

where $v(y) = u(y/\beta)$. We can now use the basic scheme

```
N=80;b=10; [y,w]=pd(N);t=x2t(N,y);D=deriv(N);
A=-b^2*D^2+eye(N);
B=t'\((16*sin(4*y/b)+8*cos(4*y/b)).*exp(-y/b));
A(N,:)=ones(1,N);B(N)=0;c=A\B;
uex=sin(4*y/b).*exp(-y/b);
u=t'*c;% the solution u at the points x=y/b
figure(1);semilogy(y/b,abs(u-uex),'.');
```

and we obtain a maximum error at the points $x = y/\beta$ less than 2.6×10^{-9}, see figure 2. Here, for $N = 80$, we have $y \in (0, 300)$ and then $x = y/\beta \in (0, 30)$. But the exact

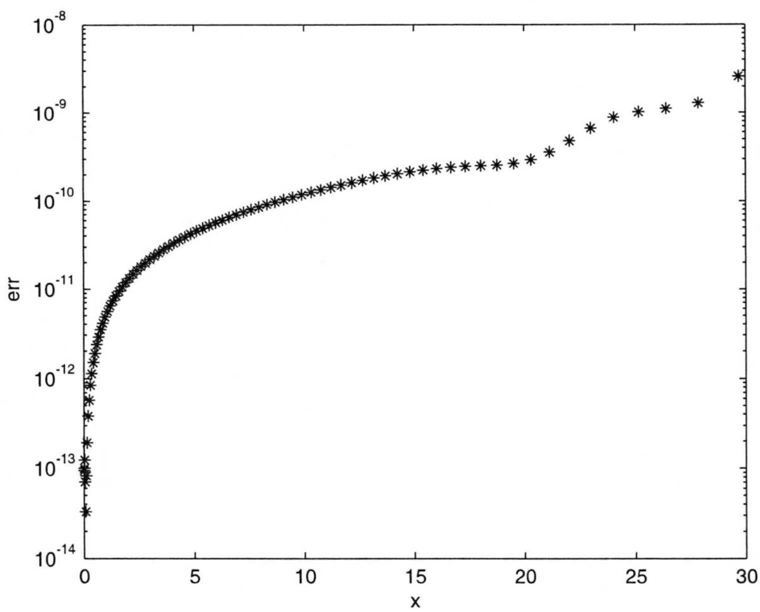

Figure 2. Errors for the modified Laguerre functions.

solution $u(x) = \sin(4x)e^{-x}$ as well as the r.h.s.

$$[16\sin(4x) + 8\cos(4x)]\, e^{-x}$$

become negligible for $x > 30$.

In fact, starting with

```
[y,w]=pd(N);D=deriv(N);Y=mult(N);
```

we must replace y by $x = y/\beta$, D by βD and Y by $X = Y/\beta$.

The modified Laguerre polynomials $L_k^{(\beta)}(x) \equiv L_k(\beta x)$ or functions $\mathcal{L}_k^{(\beta)}(x) \equiv \mathcal{L}_k(\beta x)$ have many advantages for the numerical integration of initial value problems for ordinary differential equations, cf. [3], [2]. The algorithms based on these modified polynomials or functions are easy to implement, especially for nonlinear problems and provide the global convergence in the norm of certain weighted Sobolev space (for modified Laguerre polynomials) or in the space $L^2(0,\infty)$ (for modified Laguerre functions). In particular, the numerical solutions possess the spectral accuracy – the convergence order is proportional to the smoothness of the exact solution. Therefore, these methods are more suitable for long-time integration.

Precisely, let $\omega_\beta(x) = e^{-\beta x}$, $\beta > 0$ and consider the weighted space $L^2_{\omega_\beta}(0,\infty)$ with the inner product and norm

$$(u,v)_{\omega_\beta} = \int_0^\infty \omega_\beta(x)u(x)v(x)dx, \ \|u\|_{\omega_\beta} = (u,u)^{1/2}_{\omega_\beta}.$$

The set of modified Laguerre polynomials $L_k^{(\beta)}(x) \equiv L_k(\beta x)$ is a complete orthogonal system for $L^2_{\omega_\beta}(0,\infty)$, i.e.

$$\left(L_k^{(\beta)}, L_m^{(\beta)}\right)_{\omega_\beta} = \frac{1}{\beta}\delta_{k,m}$$

$$u(x) = \sum_{k=0}^{\infty} \widehat{u}_k L_k^{(\beta)}(x), \ \widehat{u}_k = \beta\left(u, L_k^{(\beta)}\right)_{\omega_\beta}, \ \forall u \in L^2_{\omega_\beta}(0,\infty).$$

Let N be any positive integer and $P_N(0,\infty)$ the set of all algebraic polynomials of degree at most N. For any $p \in P_{2N+1}(0,\infty)$ we have

$$\int_0^\infty \omega_\beta(x)p(x)dx \simeq \sum_{k=0}^{N} p\left(x^N_{\beta,k}\right) w^N_{\beta,k}$$

where $x^N_{\beta,k}, w^N_{\beta,k}$ can be calculated from (3).

Let us consider now the model problem

$$\begin{aligned} \frac{d}{dx}u(x) &= f(u(x),x), \, x > 0 \\ u(0) &= u_0 \end{aligned} \tag{4}$$

where we assume that there exists a real number γ so that

$$(f(z_1,x) - f(z_2,x))(z_1 - z_2) \le \gamma |z_1 - z_2|^2, \ \forall z_1, z_2 \in \mathbb{R}.$$

For example, if f fulfills the Lipschitz condition with the Lipschitz constant L

$$|f(z_1,x) - f(z_2,x)| \le L|z_1 - z_2|, \forall z_1, z_2 \in \mathbb{R},$$

the above condition is verified with $\gamma = L$.

We approximate $U(x)$ by $u^N(x) \in P_{N+1}(0,\infty)$ so that

$$\frac{d}{dx}u^N(x^N_{\beta,k}) = f\left(u^N(x^N_{\beta,k}), x^N_{\beta,k}\right), 0 \le k \le N \qquad (5)$$
$$u^N(0) = u_0.$$

We have a final result, see [2],

$$\left\|u - u^N\right\|^2_{\omega_\beta} = \left(1+\beta^{-2}\right)\left(1 + \frac{\beta^2}{\beta - 2\gamma} + \frac{1}{(\beta-2\gamma)^2}\right)\mathcal{O}\left((\beta N)^{1-r}\ln N\right), \; r > 1$$

that is (5) has a global convergence and spectral accuracy in $L^2_{\omega_\beta}(0,\infty)$ if $\beta > 2\gamma$. So, a suitable choice of β may improve the numerical accuracy.

The above error is measured in the norm of the space $L^2_{\omega_\beta}(0,\infty)$ and a small error in this weighted norm does not imply a small one in the maximum norm. Moreover, if the exact solution decays fast enough as $x \to \infty$ the above evaluation is not the most appropriate. Another integration process for the same model problem (4) uses modified Laguerre functions

$$\mathcal{L}^{(\beta)}_k(x) = \mathcal{L}_k(\beta x) = e^{-\frac{\beta x}{2}} L^{(\beta)}_k(x) = e^{-\frac{\beta x}{2}} L_k(\beta x)$$

and gives global convergence and spectral accuracy in usual $L^2(0,\infty)$ norm.

Precisely, consider now the space $L^2(0,\infty)$ with the inner product and norm

$$(u,v) = \int_0^\infty u(x)v(x)dx, \;\; \|u\| = (u,u)^{1/2}.$$

The set of modified Laguerre functions $\mathcal{L}^{(\beta)}_k(x)$ is a complete orthogonal system for $L^2(0,\infty)$, i.e.

$$\left(\mathcal{L}^{(\beta)}_k, \mathcal{L}^{(\beta)}_m\right) = \frac{1}{\beta}\delta_{k,m}$$

$$u(x) = \sum_{k=0}^\infty \widehat{u}_k \mathcal{L}^{(\beta)}_k(x), \; \widehat{u}_k = \beta\left(u, \mathcal{L}^{(\beta)}_k\right), \; \forall u \in L^2(0,\infty).$$

Let us consider the same model problem

$$\frac{d}{dx}u(x) = f(u(x), x), x > 0$$
$$u(0) = u_0$$

where we now assume that there exists a real number γ_0 so that

$$\left(f(z_1,x) - f(z_2,x)\right)(z_1 - z_2) \le -\gamma_0\left|z_1 - z_2\right|^2, \; \forall z_1, z_2 \in \mathbb{R}.$$

We approximate $u(x)$ by

$$u^N(x) \in span\left\{\mathcal{L}^{(\beta)}_0, \mathcal{L}^{(\beta)}_1, ..., \mathcal{L}^{(\beta)}_{N+1}\right\}$$

so that

$$\frac{d}{dx}u^N(x^N_{\beta,k}) = f\left(u^N(x^N_{\beta,k}), x^N_{\beta,k}\right), 0 \le k \le N \qquad (6)$$
$$u^N(0) = u_0.$$

We have a final result, see [2] for more details,

$$\left\|u - u^N\right\|^2 = \frac{c\left(1+\beta^{-1/2}+\beta^{1/2}\right)}{\min\left(\beta, 2\gamma_0\right)}\mathcal{O}\left(\beta N\right)^{1/2-r/2}\left(\ln N\right)^{1/2}, \; r > 1$$
$$\left|u(\infty) - u^N(\infty)\right| = \frac{c\left(1+\beta^{-1/2}+\beta^{1/2}\right)}{\sqrt{\min\left(\beta, 2\gamma_0\right)}}\mathcal{O}\left(\left(\beta N\right)^{1/2-r/2}\left(\ln N\right)^{1/2}\right),$$

that is (6) has a global convergence and spectral accuracy in $L^2(0,\infty)$. So, the point-wise numerical error decays rapidly as the mode N increases and for any fixed N the norm $\left\|u - u^N\right\|$ is bounded, thus $u(t) - u^N(t) \to 0$ as $t \to \infty$.

4. Applications

4.1. The Arrhenius Integral

The relationship between the rate a reaction proceeds and its temperature is determined by the Arrhenius Equation $\kappa(\tau) = Ae^{-\frac{E}{R\tau}}$ where k is a temperature-dependent rate coefficient, A is a constant, E is the activation energy and R is the universal gas constant. Therefore, the rate of the reaction at a certain instant t is

$$\frac{d\varphi}{dt} = \kappa(\tau)f(\varphi)$$

where φ is the reacted fraction at the absolute temperature τ and $f(\varphi)$ is the reaction model. Supposing that the temperature variation is conducted at a constant rate $\beta = \frac{d\tau}{dt}$, the previous equation becomes

$$\frac{d\varphi}{d\tau} = \frac{A}{\beta}e^{-\frac{E}{R\tau}}f(\varphi).$$

Rearranging and integrating the above equation for temperature τ varying from 0 to T and the fraction of conversion φ varying from 0 to α, we obtain

$$\int_0^\alpha \frac{d\varphi}{f(\varphi)} = \frac{A}{\beta}\int_0^T e^{-\frac{E}{R\tau}}d\tau = \frac{AE}{\beta R}\int_x^\infty \frac{e^{-z}}{z^2}dz$$

where $x = \frac{E}{RT}$, $z = \frac{E}{R\tau}$. Let us denote by

$$I(x) = \int_x^\infty \frac{e^{-z}}{z^2}dz$$

the integral of the Arrhenius equation, which does not have an exact analytical solution. The lack of accuracy of usual analytical approximations can cause problems in the estimation of the kinetic parameters (due to the very large value of the factor $\frac{AE}{\beta R}$). So, we must obtain more accurate values.

Integration by parts leads us to

$$I(x) = \frac{e^{-x}}{x} - \int_x^\infty \frac{e^{-z}}{z} dz$$

and by changing the variable $z = x + y$ we obtain

$$I(x) = \frac{e^{-x}}{x} \left(1 - x \int_0^\infty \frac{e^{-y}}{x+y} dy \right).$$

The above integral can be numerically calculated at any $x > 0$ by using `LaguerreEig`. The code is the following

```
function [Iarrh,Imaple]=arrhenius(x,n)
[y,w]=pd(n);
f=1./(x+y);
Iarrh=exp(-x)/x*(1-x*(w*f));
xx=vpa(x,50);
Imaple=exp(-xx)/xx-maple('Ei',1,xx);
```

and figure 3 shows the relative errors for the numerical values obtained by the package `LaguerreEig` with $N = 16$ or for those obtained by Capela & al. [1]. The reference values are those calculated by MAPLE using the function `Ei` (Exponential integral).

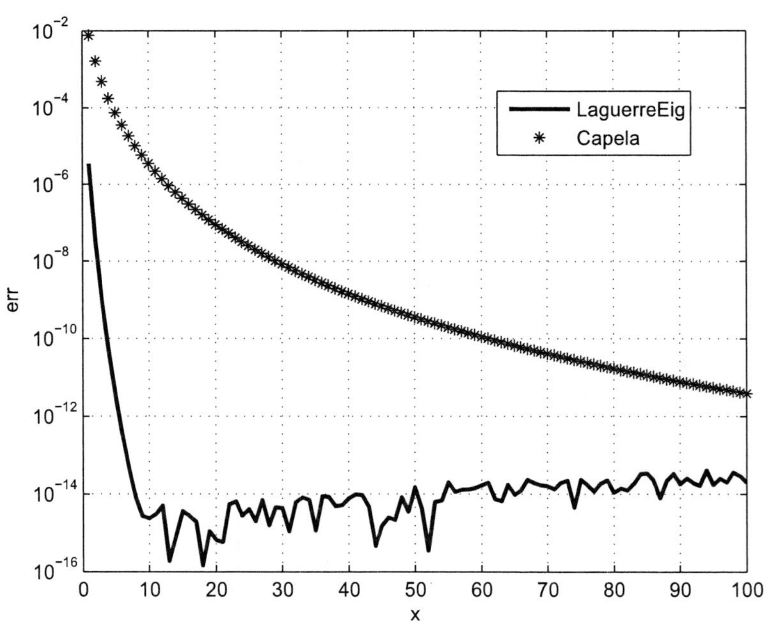

Figure 3. Errors for the Arrhenius integral.

In fact, we have

$$I(x) = \frac{e^{-x}}{x}\left(1 - x\sum_{k=1}^{N}\frac{w_k}{x + y_k}\right) \tag{7}$$

where the nodes y_k and the weights w_k are given by `[y,w]=pd(N)`. For $N = 4$ for example, the above analytical approximation coincides (with a 10^{-14} accuracy) with the corresponding approximation from [1]

$$I(x) \simeq \frac{\exp(-x)}{x}\frac{x^3 + 14x^2 + 46x + 24}{x^4 + 16x^3 + 72x^2 + 96x + 24}.$$

For $N = 64$ the relative errors are less than 10^{-14} for $x = 2, 3, ..., 100$ and less than 10^{-12} for $x = 1$. The table 1 shows the nodes and the weights for $N = 4$ in (7).

Table 1. Nodes and weights for the Gauss-Laguerre quadrature for $N = 4$

y	w
y_1=0.322547689619392	w_1=0.603154104341633
y_2=1.745761101158346	w_2=0.357418692437800
y_3=4.536620296921129	w_3=0.038887908515005
y_4=9.395070912301133	w_4=0.000539294705561

4.2. Singular Linear Problems

There are many problems, for example spherically symmetric problems in quantum mechanics, in which the coefficients of derivatives of different orders degenerate. The package `LaguerreEig` gives reasonably good numerical results even for such problems.

Let us consider an example from [14],

$$\begin{aligned} -\Delta U(x,\theta) + \mu U(x,\theta) &= f(x,\theta),\ \mu > 0,\ \text{in}\ \Omega = (0,\infty)\times(0,2\pi), \\ U(x,\theta + 2\pi) &= U(x,\theta),\ \text{in}\ \Omega \\ |U(0,\theta)| < \infty,\ &\lim_{x\to\infty} xU(x,\theta) = 0,\ \theta \in (0,2\pi) \end{aligned}$$

where

$$\Delta U(x,\theta) = \frac{\partial^2 U}{\partial x^2} + \frac{1}{x}\frac{\partial U}{\partial x} + \frac{1}{x^2}\frac{\partial^2 U}{\partial \theta^2}.$$

We consider only the symmetric case in which $U = U(x)$ and $f = f(x)$ are independent of θ. The above problem is reduced to

$$\begin{aligned} -\frac{d^2}{dx^2}U(x) - \frac{1}{x}\frac{d}{dx}U(x) + \mu U(x) &= f(x),\ x \in (0,\infty) \\ |U(0)| < \infty,\ \lim_{x\to\infty} xU(x) &= 0. \end{aligned}$$

For numerical simulations in LaguerreEig we take a particular problem with the exact solution $U(x) = e^{-\gamma x} \sin(hx)$, i.e.,

$$f(x) = e^{-\gamma x} \left[\sin(hx) \left(-\gamma^2 + h^2 + \mu + \frac{\gamma}{x} \right) + \cos(hx) \left(2\gamma h - \frac{h}{x} \right) \right]$$

and we choose $\gamma = 1$, $h = 4$, $\mu = 5$ and $N = 96$. The exact solution as well as the r.h.s. f become negligible for $x > 80$ so that we perform a scaling with the parameter $\beta = x(end)/80$ where $x(end)$ is the last node x_N. Using the matrices D and X (the differentiation matrix and the product with the independent variable matrix respectively), the code singular.m which numerically solves this problem is

```
N=96;g=1;h=4;m=5;
[x,w]=pd(N);t=x2t(N,x);D=deriv(N);X=mult(N);b=x(end)/80
A=-b^2*D^2-b^2*inv(X)*D+m*eye(N);
B=t'\(exp(-g*x/b).*(sin(h*x/b).*(-g^2+h^2+m+b*g./x)+...
    cos(h*x/b).*(2*g*h-b*h./x)));
A(N,:)=ones(1,N);B(N)=0;% implements U(0)=0
c=A\B;% the coefficients of the solution
uex=sin(h*x/b).*exp(-g*x/b);
ind=find(x/b<80);
ucalc=t'*c;% the values of the solution at the nodes x/b
semilogy(x(ind)/b,abs(ucalc(ind)-uex(ind)),'+');
```

We run the code twice, once with the scaling factor b=1 and next with b=x(end)/80 i.e. b=4.5 for $N = 96$. The errors are represented in figure 4. We remark the important

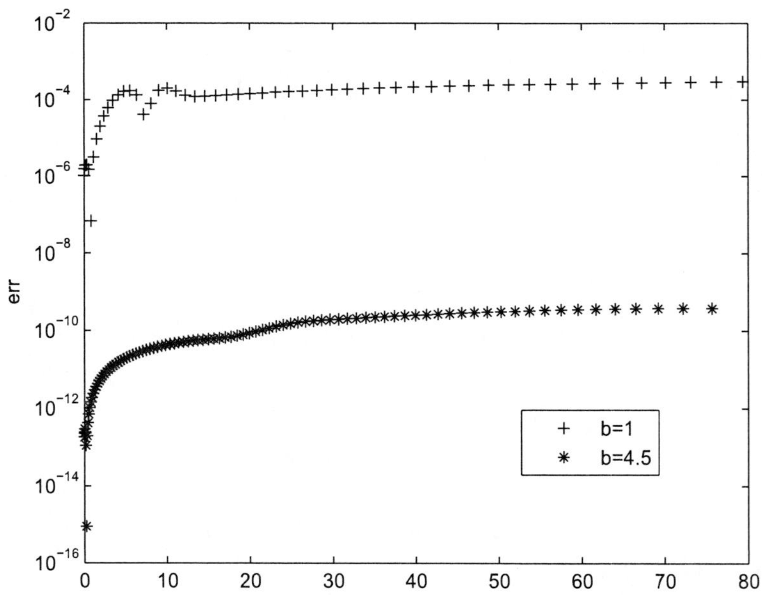

Figure 4. The errors for the singular problem.

improvement of the accuracy by using the scaling. Figure 5 shows the convergence rates for $\beta = 1$ and $\beta = x(end)/80$ and exhibits the spectral accuracy of the method.

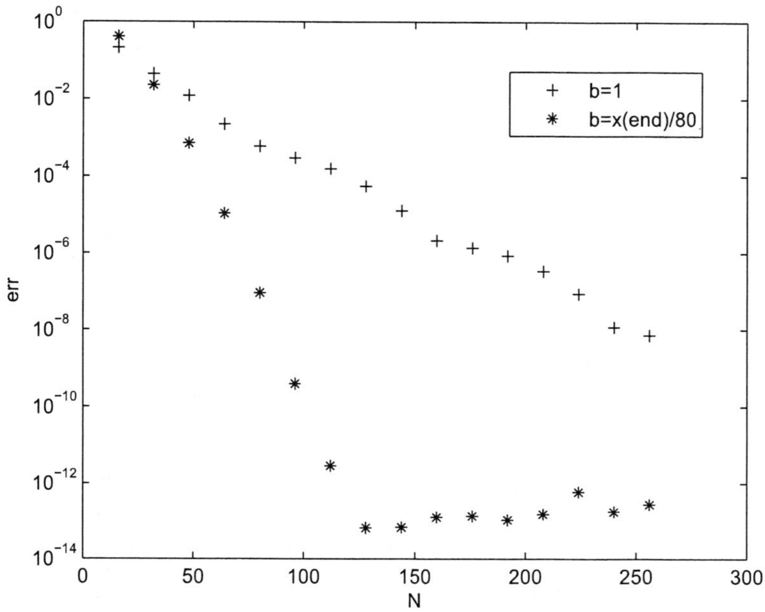

Figure 5. Convergence rates for the singular problem.

4.3. Eigenvectors and Eigenvalues

The Schrödinger Equation

The one-dimensional Schrödinger equation

$$-\frac{\hbar}{2m}y''(x) + (V(x) - E)\,y(x) = 0$$

is the fundamental equation of quantum mechanics. Here m is the mass of the particle and $\hbar$ is Planck's constant over 2π, $V(x)$ is a potential function describing a potential field, an eigenvalue E is an energy level and its corresponding eigenfunction y is the wave function of a particle. We consider here the Schrödinger equation in natural units

$$u''(x) = (V(x) - E)\,u(x).$$

The package `LaguerreEig` mainly calculates a finite number of numerical eigenvalues and eigenfunctions of Schrödinger problems

$$-\frac{d^2u}{dx^2} + V(x)u = Eu, \quad x \in (0, \infty) \tag{8}$$
$$u(0) = 0, \; u(\infty) = 0.$$

By discretizing, we get

$$\left[-D^2+V\left(X\right)\right]c=\lambda c$$

and the boundary condition

$$u(0)=c_0+c_1+...+c_{N-1}=0.$$

If we define the matrix

$$A=-D^2+V(X)$$

the vector c satisfies the generalized eigenvalues/eigenvectors problem $\mathcal{A}c=\mathcal{B}\lambda c$ where

$$\begin{gathered}\mathcal{A}(1:N-1,1:N)=A(1:N-1,1:N)\\ \mathcal{A}(N,1:N)=ones(1,N)\\ \mathcal{B}\left(1:N-1,1:N-1\right)=I_{N-1}\\ \mathcal{B}\left(N,1:N\right)=ones(1,N)/\lambda^*.\end{gathered}$$

(the tau-method). Here the other elements of $\mathcal{B}$ are zero and λ^* is a large numerical value. So, $\mathcal{B}$ is a non-singular matrix, every eigenvector c corresponding to an eigenvalue $\lambda\neq\lambda^*$ satisfies the boundary condition while λ^* itself becomes a known spurious eigenvalue.

For example, let us consider the following singular problem, the *hydrogen atom equation*,

$$\begin{gathered}-u''+\left(\frac{2}{x^2}-\frac{1}{x}\right)u=Eu,\ x\in(0,\infty)\\ u(0)=u(\infty)=0\end{gathered}$$

which has a Coulomb type potential and it is a particular case of the radial Schrödinger equation,

$$u''=\left(\frac{l(l+1)}{x^2}-\frac{Z}{x}+V(x)-E\right)u,\ x\in(0,\infty).$$

Here x is the distance from a spherically symmetric nucleus, Z and l are constants and the term $\frac{l(l+1)}{x^2}$ is the centrifugal component of the effective potential. This potential is singular at the origin and we construct its matrix approximation using X^{-1}. The code `hydrogen.m` is

```
n=200;[x,w]=pd(n);t=x2t(n,x);D=deriv(n);X=mult(n);
Xm1=inv(X);V=2*Xm1^2-Xm1;% the potential
A=-D^2+V;A(n,:)=ones(1,n);
B=zeros(n);B(n,:)=ones(1,n)/(1.e10);
for i=1:n-1 B(i,i)=1;end;
[v,l]=eig(A,B);L=diag(l);[L,ind]=sort(L);psi=v(:,ind);
```

The numerical computation using `hydrogen` gives the results in table 2, where the first column contains the first ten calculated eigenvalues and the second column contains the exact eigenvalues $E_k=-1/(2k+4)^2$, $k=0,1,...$ Figure 6 shows the eigenfunction corresponding to λ_4, produced by the command

```
plot(x,t'*psi(:,5));
```

Table 2. Calculated and exact eigenvalues for the hydrogen equation

LaguerreEig	Exact
-0.06250000000001	-0.06250000000000
-0.02777777777784	-0.02777777777778
-0.01562500000005	-0.01562500000000
-0.00999999999994	-0.01000000000000
-0.00694444444434	-0.00694444444444
-0.00510204081622	-0.00510204081633
-0.00390624999989	-0.00390625000000
-0.00308641975303	-0.00308641975309
-0.00249999999983	-0.00250000000000
-0.00206611560005	-0.00206611570248

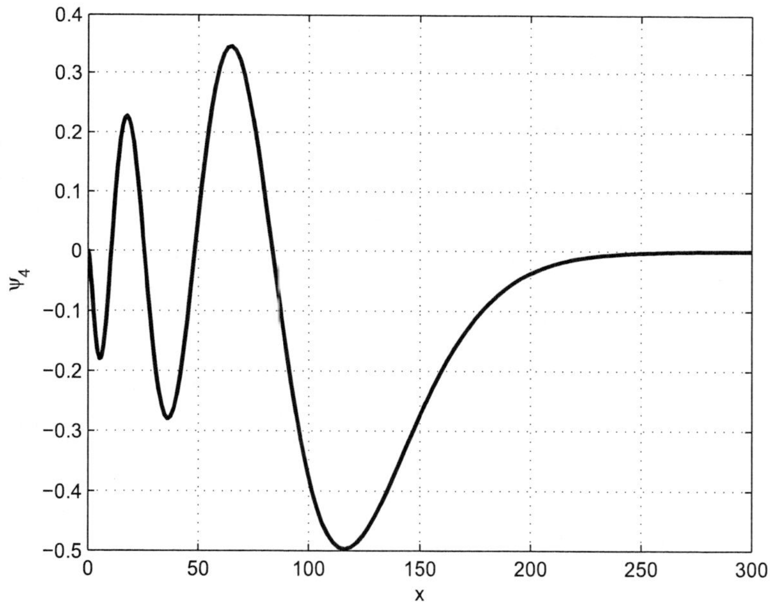

Figure 6. The 5th eigenfunction for the hydrogen equation.

We can use the scaling $y = \beta x$ and the equation for $v(y) = u(y/\beta)$ becomes

$$-\beta^2 v'' + \left(\frac{2\beta^2}{y^2} - \frac{\beta}{y}\right) v = Ev,\ y \in (0, \infty)$$

or

$$-v'' + \left(\frac{2}{y^2} - \frac{1}{\beta y}\right) v = \frac{E}{\beta^2} v,\ y \in (0, \infty)\,.$$

With $\beta = 0.1$ the errors for the first 10 eigenvalues become less than 1.25×10^{-14}.

Another test problem from the package *MATSLISE* [5] has a chemical model potential,

$$u''(x) = \left(\frac{2}{x^2} + \frac{-1 + 5e^{-2x}}{x} - E\right) u(x), \; x \in (0, \infty)$$
$$u(0) = u(\infty) = 0$$

with the first 5 eigenvalues calculated by *MATSLISE* and *LaguerreEig* in table 3.

Table 3. First 5 eigenvalues for the chemical model potential

k	MATSLISE	LaguerreEig
0	-0.061681846633	-0.061681846633353
1	-0.027498099943	-0.027498099943024
2	-0.015501561691	-0.015501561691089
3	-0.009935496851	-0.009935496850881
4	-0.006906701382	-0.006906701382088

The code `chemical.m` is the same as `hydrogen.m` with the line corresponding to V changed into

```
Xm1=inv(X);V=2*Xm1^2+Xm1*(5*expm(-2*X)-eye(n));
```

4.3.1. The Fokker-Planck Equation

Many physical, chemical, biological, and economical systems can be described by the Fokker–Planck equation with various coefficients. The one-dimensional Fokker-Planck equation is of the form

$$\frac{\partial P(x,t)}{\partial t} = \frac{\partial}{\partial x}\left(A(x)P(x,t) + \frac{\partial}{\partial x}\left(B(x,t)P(x,t)\right)\right),$$

where $A(x)$ and $B(x)$ are the drift and diffusion coefficients, respectively. This equation has a stationary solution

$$P_0(x) = \frac{1}{B(x)} e^{-\int_0^x \frac{A(s)}{B(s)} ds}.$$

If we set $P(x,t) = P_0(x)g(x,t)$, then the equation for $g(x,t)$ is

$$\frac{\partial g(x,t)}{\partial t} = -A(x)\frac{\partial g(x,t)}{\partial x} + B(x)\frac{\partial^2 g(x,t)}{\partial x^2} \equiv -Lg(x,t).$$

This equation admits a solution in terms of the eigenvalues λ_k and eigenfunctions $\phi_k(x)$ of the form

$$P(x,t) = P_0(x) \sum_{k=0}^{\infty} a_k e^{-\lambda_k t} \phi_k(x)$$

where the Fokker-Planck eigenvalue equation is given by

$$L\phi_k(x) = \lambda_k \phi_k(x).$$

The coefficients a_k are calculated from the initial condition $P(x,0)$. In principle, the Fokker–Planck eigenvalue problem can be solved directly, but, however, it is more convenient to work computationally with a Schrödinger equation.

If we transform the independent variable x to a new variable y,

$$y = \int^x \frac{1}{\sqrt{B(s)}} ds$$

and we define

$$\psi_k(y) = \sqrt{\widetilde{P}_0(x(y))}\phi_k(x(y)),$$

where $\widetilde{P}_0(y) = \sqrt{B(x(y))}P_0(x(y))$, then the Fokker-Planck eigenvalue equation becomes a Schrödinger equation

$$-\frac{d^2\psi_k(y)}{dy^2} + V(y)\psi_k(y) = \lambda_k \psi_k(y)$$

where the effective potential function is

$$V(y) = \frac{1}{4}W^2(y) - \frac{1}{2}\frac{dW(y)}{dy}$$

and

$$W(y) = \frac{A}{\sqrt{B}} + \frac{B'}{2\sqrt{B}}.$$

See [10] for more details.

As an application of `LaguerreEig`, let us consider a Lorentz gas model, which arises in the study of electron relaxation in gases,

$$\frac{\partial P(x,t)}{\partial t} = \frac{\partial}{\partial x}\left[\left(2x^2 - 3\right)P(x,t)\right] + \frac{\partial^2}{\partial x^2}\left[xP(x,t)\right],$$

with the eigenvalue equation

$$-x\frac{d^2\phi_k}{dx^2} + \left(2x^2 - 3\right)\frac{d\phi_k}{dx} = \lambda_k \phi_k, \; x \in (0, \infty).$$

The effective potential for the corresponding Schrödinger equation is

$$V(y) = \frac{y^6}{64} - y^2 + \frac{15}{4y^2},$$

which is bounded below but is has singularities both at 0 and ∞. The corresponding code is

```
n=128;b=50;
[x,w]=pd(n);t=x2t(n,x);D=deriv(n);X=mult(n);Xm1=inv(X);
```

```
A=-b^2*D^2+(X^6/b^6)/64-X^2/b^2+b^2*15/4*Xm1^2;
A(n,:)=ones(1,n);B=zeros(n);B(n,:)=ones(1,n)/(1.e10);
for i=1:n-1, B(i,i)=1;end;
[v,l]=eig(A,B);L=diag(l);[L,ind]=sort(L);
psi=v(:,ind);% the eigenfunctions
plot(x,t'*psi(:,1))% plot the first eigenfunction
```

and some calculated eigenvalues (compared with those of Shizgal [10]) are given in table 4.

Table 4. Eigenvalues for the Lorentz gas model

k	Shizgal	LaguerreEig
0	0	0.00000000018888
1	4.68340	4.683395161769944
2	10.11252	10.112518798079810
3	16.42968	16.429684162299736
6	40.05238	40.052378550211714
10	80.44794	80.447936685495009
15	142.4461	142.4461211461990
20	215.1631	215.1631277757547
30	387.623	387.6226878180584
40	590.867	590.8675571908099

The ground state eigenfunction ψ_0 is the equilibrium distribution with zero eigenvalue. The ψ_0, ψ_1, ψ_2, ψ_3 and ψ_6 eigenfunctions are represented in figure 7.

4.4. Nonlinear Steady Problems

4.4.1. Thomas-Fermi Equation

Let us consider the Thomas-Fermi nonlinear ordinary differential equation in semi-infinite interval

$$\sqrt{x}\frac{d^2u}{dx^2} = u^{3/2}(x), \quad x \in (0,\infty)$$
$$u(0) = 1, \quad \lim_{x\to\infty} u(x) = 0.$$

This equation appears in the problem of determining the effective nuclear charge in heavy atoms.

In order to avoid complex values of $u^{3/2}(x)$ due to numerical errors that make u negative (even if it is small) for some large values of x, we use the package `LaguerreEig` for the equivalent problem

$$x\left(\frac{d^2u}{dx^2}\right)^2 = u^3(x), \; u(0) = 1$$

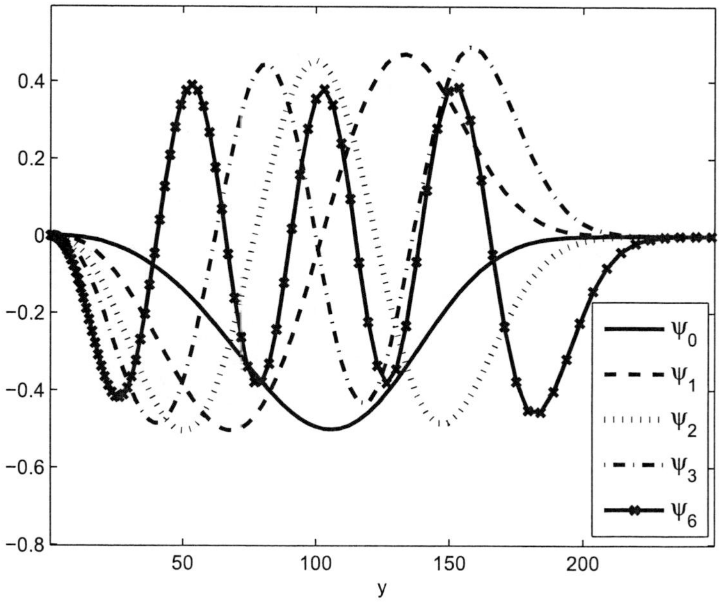

Figure 7. Some Lorentz-Fokker-Planck eigenfunctions.

and, moreover, with the scaling $y = \beta x$, $z(y) = u(y/\beta)$, we are led to

$$\frac{y}{\beta}\left(\beta^2 \frac{d^2 z}{dy^2}\right)^2 - z^3 = 0.$$

The nonlinear part can be treated more conveniently in the physical space, even if the solution z is represented in the spectral space of the coefficients.

The main code which solves the above nonlinear algebraic system for the entries of Z is

```
function [Z,sol,up0]=thomas_fermi(N,b)
[y,w]=pd(N);t=x2t(N,y);D=deriv(N);D2=D^2;Y=mult(N);
Zinit=t'\exp(-y/b);% first approximation of the solution
options=optimset('Display','iter','TolX',1.e-12,...
    'TolFun',1.e-12,'MaxFunEvals',100000);
[Z,fval] = fsolve(@fermi,Zinit,options);sol=t'*Z;
%[Z,fval] = fsolve(@fermi1,Zinit,options);sol=t'*Z;
yy=[0;b*0.5;b*1;b*2;b*5];tyy=x2t(N,yy);res=tyy'*Z;
display([yy/b,res]);up0=b*sum(D*Z)% the results
function F=fermi(Z)
F=(y/b).*(t'*(b^2*D2*Z)).^2-((t'*Z).^3);F(N)=sum(Z)-1;
end;
function F=fermi1(Z)
F=sqrtm(Y)*D2*Z-t'\((t'*Z/b).^(3/2));F(N)=sum(Z)-1;
end;end
```

where the column vector Z contains the coefficients of the unknown function

$$z(y) = \sum_{k=0}^{N-1} Z_{k+1}\psi_k(y).$$

The results for $N = 64$ and $b = 9$ are given in table 5, compared with those of Parand [6].

Table 5. The solution of the Thomas-Fermi equation

x	Parand	LaguerreEig
0	1	0.999999999999999
0.5	0.606700008	0.606982920409492
1	0.424333179	0.424003188256783
2	0.242678587	0.243012730321167
5	0.078277758	0.078871443835748

Of course, the problem can be discretized without squaring,

$$\sqrt{\frac{y}{\beta}}\beta^2\frac{d^2z}{dy^2} - z^{3/2} = 0,\ z(0) = 1$$

using the function `function F=fermi1(Z)` in `fsolve` and we obtain almost the same solution.

The initial slope $u'(0)$ plays a major role in determining the energy of a neutral atom in the Thomas–Fermi approximation,

$$E = \frac{6}{7}\left(\frac{4\pi}{3}\right)^{\frac{2}{3}} \mathbf{Z}^{\frac{7}{3}}u'(0)$$

where $\mathbf{Z}$ is the nuclear charge. The initial slope $u'(0)$ calculated in [6] is $u'(0) = -1.5880702966$ while `LaguerreEig` gives $u'(0) = -1.551898886790704$.

We remark that the behavior of the exact solution is $u(x) \sim O\left(x^{-3}\right)$ for $x \to \infty$ so that it does not match perfectly with the behavior of the approximation which is $O\left(e^{-x/2}\right)$. This leads to a slow convergence as N increases. For example, for $N = 64$, the coefficients Z decay as in figure 8, with the last entry $Z_{64} = 0.000206341958803$.

4.4.2. The Gross-Pitaevskii Equation

We consider now the time-independent Gross–Pitaevskii equation which describes the dilute Bose-condensed atoms in harmonic trap at zero temperature, of dimensionless form [4]

$$\left[-\frac{d^2}{dx^2} + \frac{x^2}{4} + \alpha\frac{\Phi^2(x)}{x^2} - \beta\right]\Phi(x) = 0,\ x \in (0, \infty)$$

$$\int_0^\infty \Phi^2(x)dx = 1,\ \Phi(0) = 0,\ \Phi(x) \sim Ce^{-\frac{x^2}{4}+\left(\beta-\frac{1}{2}\right)\ln x} \text{ for } x \to \infty.$$

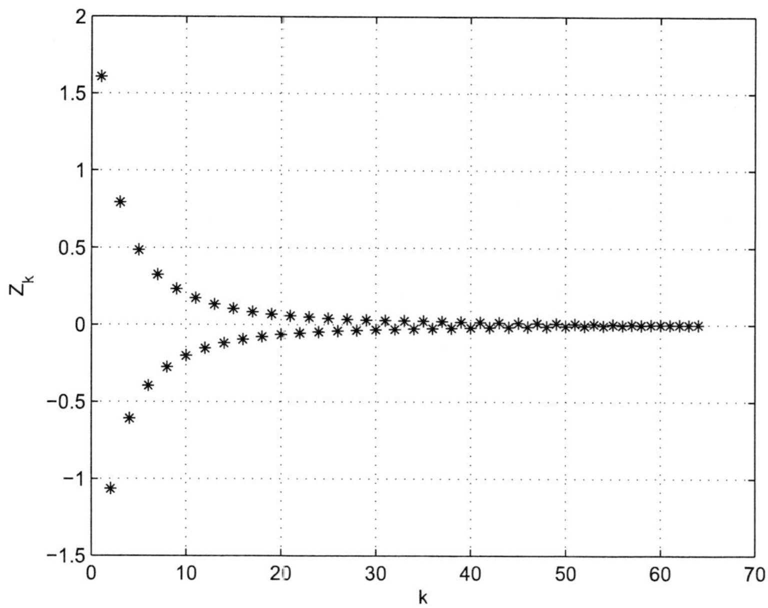

Figure 8. The coefficients of the Thomas-Fermi solution for $N = 64, b = 9$.

We solve this nonlinear problem with `LaguerreEig` and we obtain the eigenvalues β and the corresponding wavefunctions $\Phi(x)$ for different values of the nonlinear coefficient α.

Using the operators from `LaguerreEig`, the discrete form of the problem is a nonlinear system $F(U) = 0$ for the coefficients of the solution $(U_1, ..., U_N)^T$ **and** for the eigenvalues $\beta \equiv U_{N+1}$ which reads

$$F(1:N) = t'(-b^2D^2 + X^2/b^2/4 - U(N+1)I_N)U(1:N) + \\ + \alpha b^2 (t'U(1:N))\hat{\ }3/x^2$$

$$F(N) = \sum_{k=1}^{N} U(k) \text{ (the boundary condition } u(0) = 0)$$

$$F(N+1) = \sum_{k=1}^{N} \frac{w_k}{b} e^{x_k + \log((t'U(1:N))\hat{\ }2)_k} - 1 \text{ (the norm condition).}$$

Here b is the scaling parameter. Given `N` , `alpha` and the starting vector `uinit`, we need $U(1:N)$ the coefficients of the eigenfunction, $U(N+1)$ the corresponding eigenvalue β and the derivative of the eigenfunction at $x = 0$,

$$\Phi'(0) = b \sum_{k=1}^{N} (DU(1:N))_k.$$

The MATLAB code is `gross_pitaevskii.m`,

```
function [U,phip0]=gross_pitaevskii(N,alpha)
```

```
[x,w]=pd(N);t=x2t(N,x);D=deriv(N);D2=D^2;X=mult(N);
X2=X^2;b=1;
uinit(1:N,1)=4*t'\exp(-(x/b).^2/2);% the starting vector
uinit(N+1,1)=2;% the value of beta
options=optimset('Display','iter','TolX',1.e-12,...
    'TolFun',1.e-12,'MaxFunEvals',100000);
U = fsolve(@(U) myfun(U,alpha),uinit,options);
if U(1)<0, U(1:N)=-U(1:N);end;
sol=t'*U(1:N,1);beta=U(N+1),phip0=sum(b*D*U(1:N,1))
ind=find(x/b<10);
plot(x(ind)/b,b*sol(ind)./x(ind));grid;
function F=myfun(U,alpha)
F=zeros(size(U));
F(1:end-1,1)=t'*(-b^2*D2*U(1:end-1)+X2/4/b^2*U(1:end-1)-...
    U(end)*U(1:end-1))+alpha*b^2*(((t'*U(1:end-1)).^3)./...
    (x.^2));
F(end-1,1)=ones(1,length(U)-1)*U(1:end-1);
F(end,1)=w/b*(exp(x+log((t'*U(1:end-1)).^2)))-1;
end;end;
```

For example, in the attractive nonlinearity case $\alpha = -1.5839$, we take as starting vector $u_{init} = 4e^{-\frac{1}{2}\left(\frac{x}{b}\right)^2}$, $\beta_{init} = 2$ and we obtain for $N = 32$ the results in table 6.

Table 6. Gross-Pitaevskii initial slope of an eigenfunction and the corresponding eigenvalue

N	b	$\Phi'(0)$	β
32	1	1.626444514	0.60164313
64	1	1.628416135	0.60007897
96	1	1.628415819	0.60007813
128	1	1.628415812	0.60007813
32	10	1.628415813	0.60007813
Hua		1.62844	0.60006

The last row contains the corresponding results from Hua [4]. Now, due to the faster decay of the solution at infinity, the convergence is faster than for the Thomas-Fermi example and the scaling is very efficient. For example, the coefficients U decay as in figure 9.

For the repulsive nonlinearity case $\alpha = 10$ we can take the same starting vector

$$u_{init} = 4e^{-\frac{1}{2}\left(\frac{x}{b}\right)^2}$$

with $N = 32$, $b = 10$ but with $\beta_{init} = 2, 4, 6, 9$. We obtain the ground state wavefunction and the first three excited states wavefunctions in figure 10, represented for convenience as $\phi(x)/x$.

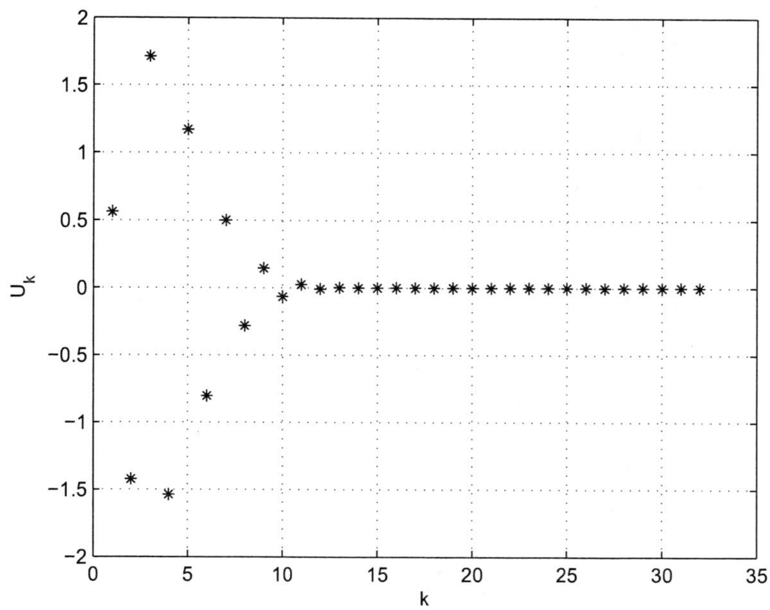

Figure 9. The coeficients of the Gross-Pitaevskii solution for $N = 32, b = 10$.

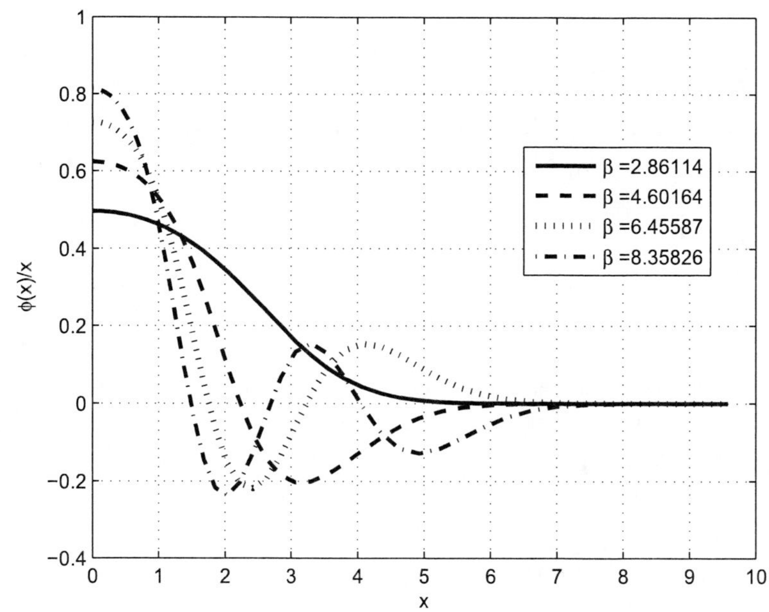

Figure 10. Four eigenfunctions and the corresponding eigenvalues for the Gross-Pitiaevskii problem.

4.5. Nonlinear Evolution Problems

We consider here, as a very known example, the Korteweg–de Vries equation on semi-infinite interval

$$\frac{\partial u}{\partial t}+u\frac{\partial u}{\partial x}+\frac{\partial^3 u}{\partial x^3}=0,\ x\in(0,\infty)\,,\ t\in(0,T]$$

with the exact (soliton) solution

$$u(x,t)=12\kappa^2\text{sech}^2\left(\kappa x-4\kappa^3 t-x_0\right)$$

so we have the boundary and initial conditions

$$u(0,t)=12\kappa^2\text{sech}^2\left(-4\kappa^3 t-x_0\right),\ u(\infty,t)=0,$$
$$u(x,0)=12\kappa^2\text{sech}^2\left(\kappa x-x_0\right).$$

We take as in [8] (where the theory of this example is given) $\kappa=0.3$, $x_0=4$ and we use `LaguerreEig`. This time the form of the solution is

$$u(x,t)=\sum_{k=0}^{N-1}c_k(t)\mathcal{L}_k(x).$$

The linear part of the KdV equation is discretized using the coefficients $c_k(t)$ while for the nonlinear part we pass to the physical space, we perform the multiplication $u\frac{\partial u}{\partial x}$ and, finally, we pass back to the spectral space using the conversion matrix generated by the code `T=x2t(N,x)`. Precisely, in the physical space the problem becomes

$$T'\frac{dc}{dt}=-\left(T'c\right).*\left(T'\left(Dc\right)\right)-T'D^3c$$

where c is the column vector of the coefficients c_k, T' means the transpose of T and $.*$ means the multiplication element by element of the two vectors. In the spectral space we have

$$\frac{dc}{dt}=-T'^{-1}\left[\left(T'c\right).*\left(T'\left(Dc\right)\right)\right]-D^3c$$
$$c(0)=T'^{-1}\left[12\kappa^2\text{sech}^2\left(\kappa x-x_0\right)\right]$$

where x is the vector of the Gauss-Laguerre nodes generated by `[x,w]=pd(N)`. Here, the differential system for the unknown functions $c_k(t)$, $k=0,...,N-1$ contains N equations. We keep only the first $N-1$ differential equations and the last one is replaced by the boundary condition

$$0=\sum_{k=0}^{N-1}c_k(t)\mathcal{L}_k(0)-12\kappa^2\text{sech}^2\left(-4\kappa^3 t-x_0\right).$$

So, we obtain in fact a differential-algebraic system which can be solved by a stiff solver of MATLAB (e.g. `ode23t`). The general form of such a system is

$$M\frac{dc}{dt}=F\left(c_0,...,c_{N-1}\right),$$
$$c(0)=f$$

and in our case the mass matrix M is

$$M = \begin{pmatrix} 1 & 0 & \cdots & 0 \\ 0 & \ddots & & \\ \vdots & & 1 & \\ 0 & & & 0 \end{pmatrix}.$$

The code kdv.m with scaling is

```
function [t,c]=kdv(N,b)
k=0.3;x0=4;M=eye(N);M(N,N)=0;
[x,w]=pd(N);T=x2t(N,x);D=deriv(N);
c0=T'\(12*k^2*sech(k*x/b-x0/b).^2);% the initial condition
options=odeset('Mass',M,'RelTol',1.e-8,'AbsTol',1.e-8);
[t,c]=ode23t(@f,[0,40],c0,options);
figure(1);semilogy(x/b,abs(12*k^2*sech(k*x/b-4*k^3*t(end)...
    -x0/b).^2-T'*c(end,:)'));
figure(2);plot(x/b,T'*c(end,:)');

function dcdt=f(t,c)
dcdt=-T'\((T'*c).*(T'*(b*D*c)))-b^3*D*(D*(D*c));
% the boundary condition
dcdt(N)=sum(c)-12*k^2*sech(-x0/b-4*k^3*t).^2;
end;end
```

Here the input parameters are the dimension of the discretization N and the scaling factor b, while the output parameters are the column vector t of the time levels where the solution was calculated and each row j of the matrix c contains the calculated solution $(c_0, ..., c_{N-1})$ at the corresponding time level t_j. The calculated values are the coefficients of the solution $u(x,t)$ but we can also calculate the physical values `T'*c'` as a matrix whose columns j contain the physical values of the solution $u(x,t_j)$ at the Gauss-Laguerre nodes. In the above example, the integration interval was $[0,40]$ and we have the numerical error at $t = 40$ in figure 11 obtained with the command `[t,c]=kdv(64,6);`

4.6. Dynamical Systems

4.6.1. The Brusselator

Let us consider, as the first example of dynamical systems, one of the simplest but fundamental models which displays biological and chemical oscillations, the diffusion-free Brusselator model

$$\begin{gathered} u_1' = B - (A+1)u_1 + u_1^2 u_2 \\ u_2' = Au_1 - u_1^2 u_2 \\ u_1(0) = u_1^0,\ u_2(0) = u_2^0 \end{gathered}$$

where u_1 and u_2 represent concentrations of two substances and $A > 0$, $B > 0$ are constants. The only critical point of the system is $(u_1^*, u_2^*) = (B, A/B)$ and for the numerical experiment with `LaguerreEig` we take $A = 3$, $B = 1$ and the initial conditions

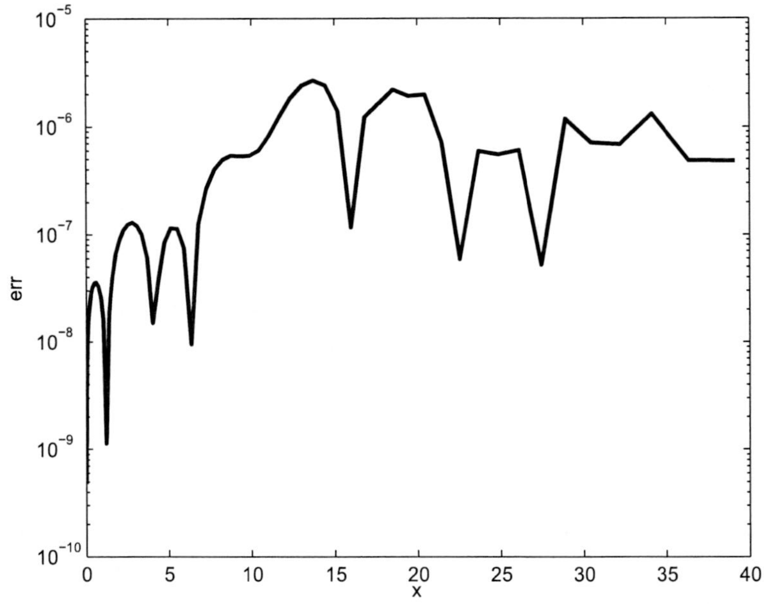

Figure 11. The error for the kdv problem at $t = 40$.

$u_1(0) = 1.5$, $u_2(0) = 3$. As it is well known, for $A > 1 + B^2$ this system exhibits limit cycles.

First at all, we must prepare the system for numerical simulation with `LaguerreEig`. The scaling $y = bt$ leads us to the problem

$$\begin{gathered} bv_1' = B - (A+1)v_1 + v_1^2 v_2 \\ bv_2' = Av_1 - v_1^2 v_2 \\ v_1(0) = u_1^0,\ v_2(0) = u_2^0 \end{gathered}$$

where $u(t) = u\left(y/b\right) = v(y)$.

If the dependent variables v_1 and v_2 oscillate when $y \to \infty$ (the stable limit cycle case), `LaguerreEig` cannot approximate them directly by using the Laguerre functions $\mathcal{L}_k(y) = e^{-y/2}L_k(y)$, $k = 0, 1, 2, ...$ We perform a change of variables

$$z_1(y) = e^{-y/2}v_1(y),\ z_2(y) = e^{-y/2}v_2(y)$$

so that the new dependent variables z_1 and z_2 can be efficiently approximated by using the Laguerre functions. This change of variables must be used even if v_1 and v_2 do not oscillate but have a non-zero limit when $y \to \infty$ (the stable critical point case). The new system is

$$\begin{gathered} b\left(z_1' + \frac{1}{2}z_1\right) - Be^{-y/2} + (A+1)\,z_1 - e^y z_1^2 z_2 = 0 \\ b\left(z_2' + \frac{1}{2}z_2\right) - Az_1 + e^y z_1^2 z_2 = 0 \\ z_1(0) = u_1^0, z_2(0) = u_2^0 \end{gathered}$$

and the original solution $u(t)$ is given by

$$u(t) = v(y) = e^{y/2} z(y)$$

at the points $t = y/b$, where y are the Gauss-Laguerre nodes.

The discretization of z_1 and z_2 is performed into the spectral space. The discretization of the nonlinear expression $z_1^2 z_2$ begins with the coefficients c_1 and c_2 of z_1 and z_2 respectively, then z_1 and z_2 are passed into the physical space where the physical values of $e^{y_k} z_1^2(y_k) z_2(y_k)$ at the nodes y_k are calculated and finally we pass back into the spectral space representation of this expression. Consequently, the discretized form of the problem is

$$b\left(Dz_1 + \frac{1}{2}z_1\right) - BT'^{-1}e^{-y/2} + (A+1)\,z_1 - T'^{-1}\left[e^y\left(T'z_1\right)^2\left(T'z_2\right)\right] = 0$$

$$b\left(Dz_2 + \frac{1}{2}z_2\right) - Az_1 + T'^{-1}\left[e^y\left(T'z_1\right)^2\left(T'z_2\right)\right] = 0$$

$$\sum_{k=0}^{N-1} c_{1,k} = u_1^0, \quad \sum_{k=0}^{N-1} c_{2,k} = u_2^0$$

where T is the transfer matrix between spectral and physical representations, given by `T=x2t(N,y)`.

Finally, we must remark that the accuracy of `LaguerreEig` depends on N and b, i.e. the numerical errors with bigger N are smaller. But it is not always efficient to use very big N, especially when we must solve nonlinear systems by Newton iterations. On the other hand, the distance between the adjacent interpolation nodes y_k increases rapidly as N increase, especially for the nodes which are located far from the origin. This feature is one of the advantages of the Laguerre-Gauss interpolation approximation, because we can use a moderate N to evaluate the unknown function at large argument. But it is also its shortcoming: if the exact solution oscillates rapidly between two large adjacent interpolation nodes, then we may lose information about the structure of exact solution between those nodes.

To remedy this deficiency, we may use the scheme of `LaguerreEig` repeatedly on small intervals $[y, y+h]$ to move forward in time, like in Runge-Kutta methods implemented in the MATLAB `ode` solvers. This means that we choose fixed small $N = 10$ and $h = 0.1$, for example, and a fixed sufficiently big scaling factor b ($b = 40$) so that the great part of the nodes $y_k/b \in [0, h]$, see [2] for more details. We perform the method successively on $[0, h]$, $[h, 2h]$ and so on, taking the last calculated values of the solutions as initial conditions for the next interval. Of course, we may change N, h and b at every interval according to the behavior of the calculated solutions but this adaptive scheme is not yet implemented.

The MATLAB code `brusselator.m` is the following. Here, the coefficients of the unknown functions z_1 and z_2 of the system are assembled into a $2N$ column vector c so that `c(1:N)` contains the coefficients of z_1 while `c(N+1:2*N)` contains the coefficients of z_2.

```
function [arg,res]=brusselator(N,b)
A=3;B=1;h=0.1; [y,w]=pd(N);t=x2t(N,y);D=deriv(N);
cinit(1)=1.5;cinit(2)=3;arg=[0];res=[1.5,3];
```

```
c=zeros(2*N,1);c(1)=1.5;c(N+1)=3;
options=optimset('Display','off','TolX',1.e-12,...
    'TolFun',1.e-12,'MaxFunEvals',100000);
for contor=1:1000
    [c,fval,u,output] = fsolve(@f,c,options);
    xx=b*(0:0.01:h)';txx=x2t(N,xx);
    sol(:,1)=exp(xx/2).*(txx'*c(1:N));
    sol(:,2)=exp(xx/2).*(txx'*c(N+1:2*N));
    arg=[arg;(contor-1)*h+xx(2:end)/b];
    res=[res;sol(2:end,:)];
    cinit(1)=sol(end,1);cinit(2)=sol(end,2);
    display([contor,output.iterations]);
end
figure(1);plot(res(:,1),res(:,2),'r-');grid;hold on;
figure(2);plot(arg,res(:,1),'b.',arg,res(:,2),'r.');
grid;hold on
function F=f(c)
c1=c(1:N);c2=c(N+1:2*N);E=t'\(exp(y).*((t'*c1).^2).*...
    (t'*c2));
F(1:N)=b*(D*c1+c1/2)-B*t'\exp(-y/2)+(A+1)*c1-E;
F(N)=sum(c1)-cinit(1);
F(N+1:2*N)=b*(D*c2+c2/2)-A*c1+E;
F(2*N)=sum(c2)-cinit(2);
end
end
```

Figure 12 shows the results for $u_1(t)$ for $t \in [0, 100]$ and figure 13 shows the errors between u_1 calculated by `LaguerreEig` and u_1 calculated by the MATLAB integrator `ode45` with `RelTol=AbsTol=1.e-14`. The computing time was about 4 secs. for ode45, about 20 secs. for the stiff solver ode15s and about 32 secs. for our `LaguerreEig` unoptimized code. Of course, we do not claim that `LaguerreEig` is in any way superior, but we consider that it is interesting even for dynamical systems.

If we change the parameters to $A = 1.9$ and $B = 1$, we have no more limit cycles. The solutions calculated by LaguerreEig are given in figure 14.

4.6.2. The Lotka-Volterra Model

The Lotka-Volterra model of oscillating chemical reactions is

$$u_1' = Au_1 - Bu_1u_2$$
$$u_2' = -Cu_2 + Bu_1u_2$$

where u_1, u_2 are the concentrations of two substances and A, B, C are positive constants. After scaling $y = bt$ and dependent variables change $z_1(y) = e^{-y/2}v_1(y)$,

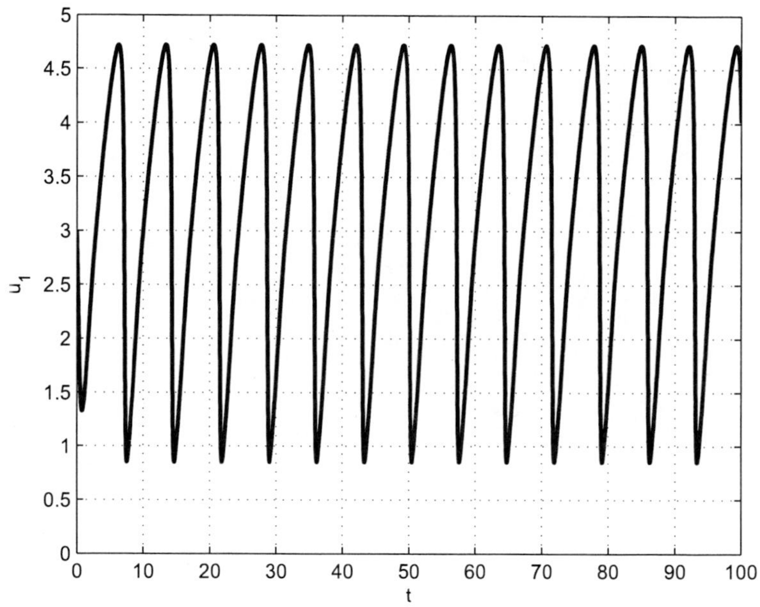

Figure 12. u_1 calculated by LaguerreEig for A=3,B=1.

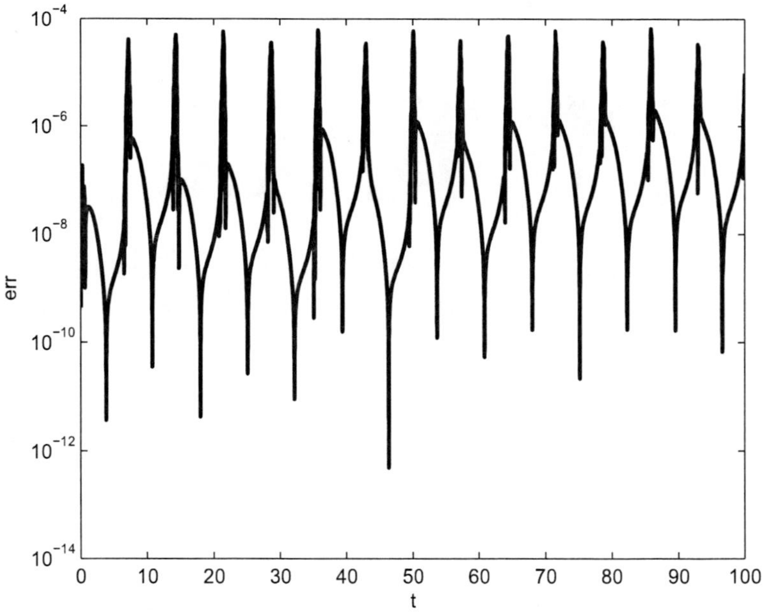

Figure 13. The errors between u_1 calculated by LaguerreEig and u_1 calculated by ode45 for A=3, B=1.

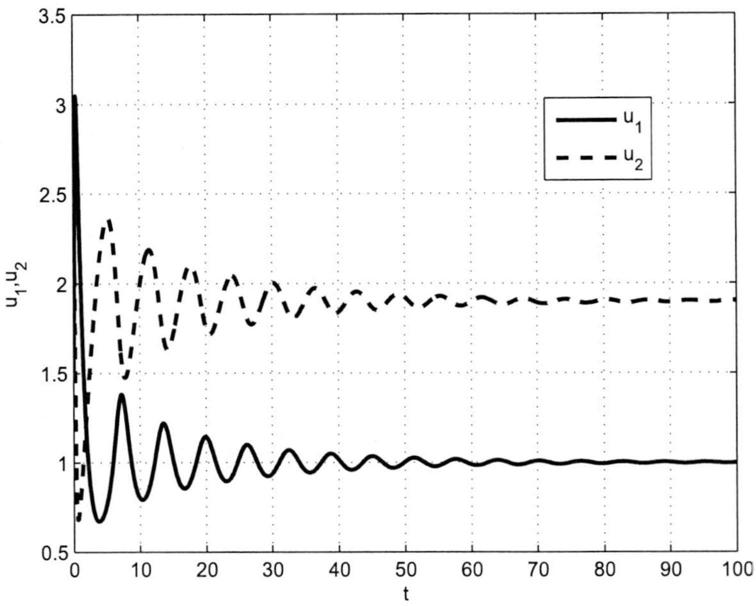

Figure 14. u_1, u_2 calculated by LaguerreEig for A=1.9, B=1.

$z_2(y) = e^{-y/2}v_2(y)$ we obtain the following system

$$b\left(z_1' + \frac{1}{2}z_1\right) - Az_1 + Be^{y/2}z_1z_2 = 0,$$
$$b\left(z_2' + \frac{1}{2}z_2\right) + Cz_2 - Be^{y/2}z_1z_2 = 0.$$

The values of the parameters are $A = 0.1$, $B = 0.2$, $C = 0.05$ and the initial conditions are $u_1(0) = u_2(0) = 0.5$. We choose again $N = 10$, $b = 40$, $h = 0.5$ and we perform 185 steps to complete the first period of the solution.

The code is the same as for the Brusselator, of course with the corresponding changes of the parameters and the structure of the problem

```
function [arg,res]=lotka(N,b)
A=0.1;B=0.2;C=0.05;h=0.5; [y,w]=pd(N);
t=x2t(N,y);D=deriv(N);
cinit(1)=0.5;cinit(2)=0.5;arg=[0];res=[0.5,0.5];
c=zeros(2*N,1);c(1)=0.5;c(N+1)=0.5;
options=optimset('Display','off','TolX',1.e-12,...
'TolFun',1.e-12,'MaxFunEvals',100000);
for contor=1:185
..........
end
..........
```

```
function F=f(c)
c1=c(1:N);c2=c(N+1:2*N);E=t'\((t'*c1).*(t'*c2).*exp(y/2));
F(1:N)=b*(D*c1+c1/2)-A*c1+B*E;
F(N)=sum(c1)-cinit(1);
F(N+1:2*N)=b*(D*c2+c2/2)+C*c2-B*E;
F(2*N)=sum(c2)-cinit(2);
end
end
```

The results are presented in figure 15 and the errors from the results calculated by ode45 are less than 10^{-12}.

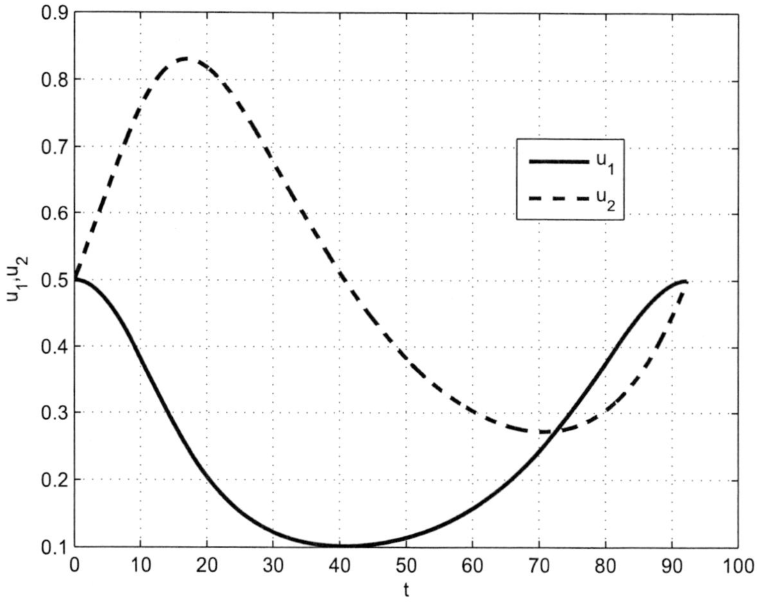

Figure 15. The solutions for the Lotka-Volterra model, the first period.

The Lotka-Volterra system has as invariant $\Lambda = Bu_1 + Bu_2 - C\ln(u_1) - A\ln(u_2)$. For our calculations, we have indeed

$$\max\Lambda - \min\Lambda < 2.06 \times 10^{-12}, \forall t \in [0, 92.3].$$

5. Conclusion

This chapter presents the advantages of the spectral method based on Laguerre functions with scaling for problems on semi-infinite intervals. As it was pointed out by Gottlieb and Orszag, Laguerre polynomials/functions have very poor resolution properties when compared with other types of orthogonal polynomials. Although the Laguerre-spectral methods show a theoretical spectral convergence rate, the actual error decays considerably slower than the Chebyshev or Legendre spectral method for similar problems in finite intervals.

This is one of the main reasons why Laguerre polynomials/functions were rarely used in practice. However, the resolution of Laguerre functions can be greatly improved by using a proper scaling factor.

The main factor responsible for the poor resolution of Laguerre polynomials and Laguerre functions is that usually a significant portion of the Laguerre–Gauss points is located outside of the interested interval. Thus it makes sense to scale the function so that all the effective collocation points are inside of this interval. There exists a sufficiently big scaling factor for ensuring the global spectral convergence, therefore, the presented method is applicable to a large class of problems.

The theoretical results presented in the previous sections are illustrated with various numerical examples which may give confidence in using the Laguerre functions with scaling. Calculations of integrals on half-line, solving problems for singular linear or nonlinear equations, including eigenproblems for Schrödinger-type equations, even the study of the evolution of some dynamical systems or evolutive partial differential equations, are all performed with a simple algebraic transcript of the original problem and with a high accuracy.

All the algorithms used in the chapter are coded into `LaguerreEig`, a MATLAB package. The corresponding complete programs are included in the chapter and also can be obtained from [12]. As demonstrated in this chapter, we believe that properly formulated spectral methods using Laguerre functions are competitive with the classical and recent procedures for problems in unbounded domains.

References

[1] Capela, J. M. V.; Capela, M. V.; Ribeiro, C. A. *J. Math. Chem.* 2009, 45, 769-775.

[2] Guo, B-Y.; Wang, Z-Q. *Comput. Meth. Appl. Mech. Engrg.* 2007, 196, 3726-3741.

[3] Guo, B-Y.; Wang, Z-Q.; Tian, H-J.; Wang, L-L. *Math. Comp.* 2008, 77, 181–199.

[4] Hua, W.; Liu, X.; Ding, P. *J. Math. Chem.* 2006, *40*, 243-255.

[5] Ledoux, V. (2007). Ph.D.Thesis [pdf].
http://users.ugent.be/~vledoux/pdf/doctoraatVLedoux.pdf

[6] Parand, K.; Shahini, M. *Physics Letters A* 2009, 373, 210–213.

[7] Shen, J. *Siam J. Numer. Math.* 2000, 38, 1113-1133.

[8] Shen, J.; Wang, L-L. *DCDS-B* 2006, 6, 1381-1402.

[9] Shen, J.; Wang, L-L. *Commun. Comput. Phys.* 2009, 5, 195-241.

[10] Shizgal, B. D.; Chen, H. *J. Chem. Phys.* 1996, 104, 4137-4150.

[11] Siyyam, H. I. *J. Comp. Anal. Appl.* 2001, 3, 173-182.

[12] Trif, D. (2009), LaguerreEig [zip].
http://www.mathworks.com/matlabcentral/fileexchange/24266.

[13] Vigo-Aguiar, J.; Ramos, H. *J. Math. Chem*. 2006, 40, 71-83.

[14] Wang, Z-Q.; Guo, B-Y. *DCDS-B* 2009, 11, 1019-1038.

[15] Weideman, J. A.; Reddy, S. C. *ACM TOMS 2000*, 26, 465-519.

In: Mathematical Chemistry
Editor: W.I. Hong, pp. 509-535

ISBN: 978-1-60876-894-3

Chapter 11

STABILITY OF CHEMICAL AND PHASE EQUILIBRIUM: ALTERNATIVE FORMS OF EQUATIONS FOR THERMODYNAMIC ANALYSIS

A.M. Toikka*

Saint-Petersburg State University, Department of Chemical Thermodynamics & Kinetics, Universitetsky pr., Saint-Petersburg, Russia

Abstract

The chapter presents different approaches to the analysis of the stability of chemical and phase equilibrium in multicomponent heterogeneous systems. The traditional approach for the analysis of thermodynamic stability is presented. Le Chatelier – Braun principle is considered as alternative approach to the stability criteria. New variants of formulations of this statement are presented. The interconnection between non-conjugated thermodynamic parameters is also discussed on the base of Le Chatelier – Braun principle. Thermodynamic inequalities, both stability conditions and Le Chatelier – Braun principle, are considered for the case of phase transitions in multicomponent systems without chemical reactions. Thermodynamic inequalities are presented for the systems with chemical interactions. The general system of thermodynamic inequalities and chains of inequalities for non-reacting and reacting systems are considered at different external conditions. The possibility is discussed to describe different types of physicochemical processes on the base of the generalized concept of affinity. The illustration is given of the application of thermodynamic inequalities to various physicochemical systems and processes: e.g. the shifting of phase and chemical equilibrium in systems under different external conditions. Various transformations of the stability matrix are presented as variants of mathematical formulations of stability conditions. The possibility is discussed to use the stability criteria for the characterisation of the non-equilibrium process. Inequalities for the second derivatives of thermodynamic potentials give an opportunity to compare the changes of thermodynamic parameters in equilibrium and non-equilibrium processes. Processes in homogeneous and heterogeneous systems are discussed, including the processes accompanied by the generation of new phases. The interpretation is given for so-called jumps in the values of the second derivatives of thermodynamic potentials.

* E-mail address: toikka@yandex.ru. (Corresponding author)

1. Introduction

Stability criteria are of special significance in the thermodynamic theory. Both theoretical and applied studies pay special attention to the thermodynamic equilibrium. The latter is a state that a system can't leave spontaneously. Therefore, here one should take into account the elements of the Second Law meaning related to the establishment of the existence of entropy and absolute temperature. The Second Law statement about the directedness of natural processes is reflected by the fact that the state of equilibrium is the final state, i.e. the result of any process. The direct formulation of the conditions of chemical equilibrium doesn't require, at least explicitly, to introduce the most important element of the Second Law, the rising of entropy. Meanwhile, the formulation of stability criteria enables one to include the essence of the Second Law in full, i.e. including the entropy increase statement in isolated or adiabatic systems, or the positivity of the inner entropy change in an arbitrary system. This reflects the significance of stability criteria in both fundamental and applied sense. The criteria provide us with the capability to include aspects of irreversible thermodynamics into investigations of thermodynamic equilibrium. It may be stated that stability criteria form a bridge between the equilibrium and non-equilibrium theory. Including elements related to irreversible processes into the equilibrium theory gave rise to several implicit and explicit logical contradictions in investigations and interpretations of thermodynamic stability of the equilibrium. It resulted also in the development of alternative mathematical formulations.

In this Chapter, we will consider several mathematical formulations of the conditions related to the concepts of thermodynamic stability. We will limit ourselves to the cases of phase and chemical equilibrium being among most important for applications.

2. Traditional Approach to the Analysis of Thermodynamic Stability

The classical mathematical formulation of the Second Law in the phenomenological thermodynamics determines only the direction of real processes as accompanied by the positive inner entropy change. Thus, possible consequences of the Second Law are qualitative restrictions to real processes (irreversible) and usually get the form of inequalities. Quantitative characterization is given by irreversible thermodynamics, whose essential problem is getting an answer to the question: *How does entropy change quantitatively in the course of a real process*? The solution requires involving new basic concepts and regularities. Unfortunately, these are not as general as the Second Law, which is one of the principle laws of nature. Therefore, in what follows we will limit our considerations to the classical interpretation of the Second Law without touching upon the results obtained on the basis of irreversible thermodynamics. Main attention is paid to the theory of stability constructed according to Gibbs' and relative approaches.

The generally accepted concept of thermodynamic equilibrium and its stability belongs to Gibbs. In his fundamental work [1] he avoids a consecutive treatise on thermodynamics, but, in essence, he postulates conditions of equilibrium and its stability. Such axiomatic construction of the theory is formally associated with analytical mechanics, namely its variation principles. In the full form, conditions of equilibrium are given as

$$(\delta S)_{U,V,m_1,m_2,...m_n} \leq 0 \quad , \tag{2.1}$$

$$(\delta U)_{S,V,m_1,m_2,...m_n} \geq 0 \quad , \tag{2.2}$$

where S is entropy, U – internal energy of the system, V – volume, m_i – the amount (e.g., the number of moles) of the species *i*, n - the number of species. As a result we obtain the following principle of equilibrium in the entropy form:

"For the equilibrium of any isolated system it is necessary and sufficient that in all possible variations of the state of the system which do not alter its energy, the variation of its entropy shall either vanish or be negative" [1].

In both equations (2.1) and (2.2) the symbol δ denotes a virtual displacement; accordingly both δS and δU are infinitesimal values of the first order. It is evident that in this case these equations do not clarify the type of the *U* or *S* extreme (maximum or minimum). Nevertheless the typical errors are attempts to use inequalities (2.1) or (2.2) for the determination of the type of extreme. Such unclearness in the interpretation of the inequality signs < and > in conditions (2.1) and (2.2.) had been mentioned and criticised in books [2] and [3].

Most monographs on thermodynamics present relations expressing principles of equilibrium as having only the equality sign. The inequality sign is considered to be of less importance for the solution of practical problems and of significance only in specific cases. The inequality sign in Eq. (2.1) and (2.2) determines conditions for a sharp extreme that can occur in the case of unidirectional shifts from the equilibrium state. In particular, the full form of the equilibrium principle (equality-inequality) turns useful in studies of chemical equilibrium [2, 4].

The type of the extreme should be determined when taking into account variances of higher orders. I.e. when the function S (or U) is expanded into Taylor series the infinitesimal members of higher orders have to be taken into account. In this case the law of entropy rising gives the sign for the changes of these functions. According to Gibbs the following general equations should be valid:

$$(\Delta S)_{U,V,m_1,m_2,...m_n} \leq 0, \tag{2.3}$$

$$(\Delta U)_{S,V,m_1,m_2,...m_n} \geq 0, \tag{2.4}$$

where symbol Δ signifies an infinitesimal though definite increment of a function. These equations determine the criteria of thermodynamic stability. Similar to mechanics few types of the equilibrium should be distinguish: stable, metastable and neutral. For both stable and metastable equilibrium the < sign stays in equation (2.3) (> in equation (2.4)). In both cases Δ can mean not only an infinitesimal increment, but also a finite difference in U and S. The difference between the stable and metastable states means that for the first one both equation (2.3) and (2.4) are always valid for any finite difference. For a metastable state at least one finite displacement exists that doesn't obey either of conditions expressed by equations (2.3) or (2.4). Therefore the analysis of stability contains in itself the problem formulated by Gibbs himself (when he discussed one of the form of stability conditions – inequality $\Delta U > T\Delta S - P\Delta V + \mu_1\Delta m_1 + \ldots + \mu_n\Delta m_n$): *"it must be possible to assign limits within which it shall hold true of finite differences"* [1]. To solve this problem one needs to know the phase diagram of a system [2], thus classical works on the thermodynamic stability of equilibrium states consider the general case combining both stability and metastability, which means using infinitesimal Δ in equations (2.3) or (2.4).

The equality sign refers to a neutral, or indifferent, equilibrium [1, 5]. All other states differing from the equilibrium ones (stable, metastable, and neutral) are non-equilibrium states, Eq. (2.3) and (2.4) are not valid for them, the states corresponding to the unstable range of the phase diagram.

In analysing stability of phases one can express the stability conditions using variables drawn from the fundamental equations. The condition

$$\Delta T\Delta S - \Delta P\Delta V + \sum_{i=1}^{n} \Delta\mu_i \Delta m_i > 0 \tag{2.5}$$

is the necessary and sufficient stability criterion for a phase with regard to continuous changes of state [1]. In equation (2.5) T – is absolute temperature, P – pressure, μ_i – chemical potential of species *i*. This inequality leads to the condition of convexity for the surface of internal energy

$$\delta^2 U \geq 0 \,, \tag{2.6}$$

that is widely used in the theoretical analysis of stability for specific systems. Equation (2.6) results from equation (2.5) if the function U is expanded into Taylor series and only the infinitesimal members of second order are taken into account:

$$\begin{aligned} &\frac{\partial T}{\partial S}(\delta S)^2 - \frac{\partial P}{\partial V}(\delta V)^2 + \sum_{i=1}^{n}\frac{\partial \mu_i}{\partial m_i}(\delta m_i)^2 + \frac{\partial T}{\partial V}\delta S\delta V - \\ &-\frac{\partial P}{\partial S}\delta S\delta V + \sum_{i=1}^{n}\frac{\partial T}{\partial m_i}\delta S\delta m_i + \sum_{i=1}^{n}\frac{\partial \mu_i}{\partial S}\delta S\delta m_i + \\ &+\sum_{i=1}^{n}\frac{\partial \mu_i}{\partial V}\delta V\delta m_i - \sum_{i=1}^{n}\frac{\partial P}{\partial m_i}\delta V\delta m_i + \sum_{i\neq k}^{n}\sum_{k\neq i}^{n}\frac{\partial \mu_i}{\partial m_k}\delta m_i\delta m_k \geq 0 \end{aligned} \tag{2.7}$$

The $\geq$ sign appeared because infinitesimal members of higher order have been neglected. Such approach is suitable for many tasks. However for some problems, e.g. critical phenomena consideration one should take into account the infinitesimal members of higher orders.

Left side of the equation is a quadratic form of extensive parameters S, V, m_i. Accordingly condition (2.6) may be presented in matrix form as follows:

$$\begin{vmatrix} \frac{\partial T}{\partial S} & \frac{\partial T}{\partial V} & \frac{\partial T}{\partial m_1} & \cdots & \frac{\partial T}{\partial m_n} \\ \frac{\partial(-P)}{\partial S} & \frac{\partial(-P)}{\partial V} & \frac{\partial(-P)}{\partial m_1} & \cdots & \frac{\partial(-P)}{\partial m_n} \\ \frac{\partial \mu_1}{\partial S} & \frac{\partial \mu_1}{\partial V} & \frac{\partial \mu_1}{\partial m_1} & \cdots & \frac{\partial \mu_1}{\partial m_n} \\ \cdots\cdots & \cdots\cdots & \cdots\cdots & \cdots & \cdots\cdots \\ \frac{\partial \mu_n}{\partial S} & \frac{\partial \mu_n}{\partial V} & \frac{\partial \mu_n}{\partial m_1} & \cdots & \frac{\partial \mu_n}{\partial m_n} \end{vmatrix} \geq 0 \qquad (2.8)$$

This is one of the forms of so-called stability matrix. It may be constructed on the base of any thermodynamic function. We will consider some forms of stability matrix in the following sections.

The case of stability of chemical equilibrium should be treated in the similar way. Let us consider the chemical reaction in multicomponent system

$$\sum_i \nu_i R_i = 0, \qquad (2.9)$$

where R_i denotes species i, ν_i is stoichiometric number of species i, which is positive for the products of the reaction and negative for initial substances (reagents). In the run of the reaction the changes of any reacting species must satisfy equation (2.9.) and are presented by the following equation:

$$dm_i = \nu_i d\xi, \qquad (2.10)$$

where ξ is the extent of reaction. Accordingly it follows from (2.2) that in chemical equilibrium

$$\left(\sum_i \nu_i \mu_i\right) \delta\xi \geq 0 \qquad (2.11)$$

where the symbol δ denotes a virtual displacement of chemical equilibrium. The inequality sign in Eq. (2.11) again corresponds to the case of unidirectional shifts from the equilibrium state. Such unidirectional shift means that the reaction is irreversible: only products exist in the state of chemical equilibrium. For reversible reaction the equality sign is valid. The value $\sum_i \nu_i \mu_i$ may be presented as ΔG, Gibbs energy of the reaction. In the following sections we will use also the affinity A [4] which is determined by equation:

$$A = -\sum_i \nu_i \mu_i. \qquad (2.12)$$

and the condition of chemical equilibrium (2.11) may be presented in following form:

$$A\delta\xi \le 0. \tag{2.13}$$

It should be emphasized again that equations like equilibrium conditions, with the inequality sign $\le$ also, don't determine the stability because δ denotes infinitesimal values of the first order. The stability of chemical equilibrium is determined by the derivative of affinity with respect to the extent of reaction ξ [4], e.g. for $T, P = const$:

$$\left(\frac{dA}{d\xi}\right)_{T,P} < 0. \tag{2.14}$$

In this section we will confine ourselves to making these general and basic notes with regard to traditional approach (i.e. Gibbs method) for the conditions of stability in thermodynamics. Gibbs also distinguished few types of stability: in respect to continuous (equation (2.5)) and discontinuous changes of state. The last type corresponds to the case when after disturbance a new phase could appear [1, 5]. The detailed description of Gibbs theory of stability and relative approaches can be found in the cited books [1-5] together with other publications (e.g., [6-15]). Numerous papers concerning applications of stability conditions (e.g., [16-25]) also are mostly based on Gibbs' theory. The development of Gibbs' thermodynamic theory on the axiomatic base had been made by Tisza [6]. Most of other works concerning axiomatic base of thermodynamics usually don't include a detailed consideration of stability conditions (e.g. [26]).

3. Le Chatelier – Braun Principle as an Alternative Approach to the Stability Criteria

Le Chatelier principle is one of the example of a rule when incorrect formulation leads to errors in the results. Commonly, the formulation of the principle is like this:

If a system at equilibrium experiences external impacts, the latter cause counter-action forces to appear in the system that tend to decrease the effect of these impacts.

Such qualitative formulation leads to numerous incorrect results. Accordingly many authors including Ehrenfest [27] and Plank [28] had studied thermodynamic aspects of the principle and obtained mathematical formulations of this statement. In the simplest form the principle could be considered like a stability condition [2]. More important results, which were firstly obtained by Ehrenfest [27], deal with the behavior of systems under the influence of different number and type of external forces. Such results could be considered as both Le Chatelier – Braun principle for complex systems and alternative forms of stability conditions.

Here and in the following sections we will use the general notations for the intensive and extensive thermodynamic parameters in the fundamental equation for the internal energy

$$dU = TdS - PdV + \sum_{j=1}^{n} \mu_j dm_j \cdot \tag{3.1}$$

For intensive parameters

$$T, (-P), \mu_j \equiv X_i \tag{3.2}$$

For extensive parameters

$$S, V, m_j \equiv Y_i \tag{3.3}$$

Then the stability condition (2.7) could be rewritten as follows:

$$\sum_i dX_i dY_i \geq 0 \tag{3.4}$$

where for real physical changes of state we use the symbol d instead of δ (used for virtual displacements). From (3.4) the partial forms for stability conditions follow:

$$\frac{\partial X_i}{\partial Y_i} \geq 0, \quad \frac{\partial Y_i}{\partial X_i} \geq 0 \tag{3.5}$$

The only limitation for these partial derivatives follows from Gibbs equation

$$\sum_{i=1}^{n+2} Y_i dX_i = 0; \tag{3.6}$$

according to this equation not more then n intensive parameters can be constant.

Equations (3.5) directly lead to the conclusions which should be considered as Le Chatelier – Braun principle (the first rule). For example it follows from (3.5):

$$\frac{\partial T}{\partial S} > 0 \cdot \tag{3.7}$$

It means that heating leads to the increase of temperature. The driving force of this process is a difference in temperature of the environment (T_e) and temperature of the system (T): $T_e - T > 0$. After heating this difference (accordingly the driving force of the heat-transfer process) will decrease.

The similar consideration can be made for the chemical equilibrium. In the case of a reaction between the species of a multicomponent system the following equations is valid:

$$T\left(\frac{\partial S}{\partial T}\right)_{P,\sum_i^n \nu_i\mu_i=0} > 0 \tag{3.8}$$

(subscript means constant pressure and chemical equilibrium, $\sum_i^n \nu_i \mu_i = 0$). The temperature rising should be accompanied by the increasing of the entropy and enthalpy: the system absorbs heat. Accordingly, the rise of the temperature leads to the shifting of chemical equilibrium in the direction of the endothermic process.

According to (3.4), in equations (3.5) the same partial derivatives may be written for the different ways of fixing other independent parameters, e.g. $\left(\frac{\partial S}{\partial T}\right)_P$ or $\left(\frac{\partial S}{\partial T}\right)_V$. The difference between the values of these derivatives can be also estimated by the stability criteria. This case is analogous to interpreting the known thermodynamic condition

$$C_P > C_V > 0\,, \tag{3.9}$$

where C_P and C_V are heat capacities at constant pressure and volume respectively. The thermodynamic relationships of this type are referred to as the Le Chatelier – Braun principle by many authors (see, e.g., [2, 5, 27-33]). Now we will consider such thermodynamic differential inequalities (like (3.9)) as a second rule related to Le Chatelier – Braun principle and will analyse the values of second derivatives of thermodynamic potentials (i.e. $\frac{\partial X_i}{\partial Y_i}$ and $\frac{\partial Y_i}{\partial X_i}$) with different parameter fixing.

Ehrenfest [27] proposed the thermodynamic equation

$$\left(\frac{\partial Y_i}{\partial X_i}\right)_{X_k} \geq \left(\frac{\partial Y_i}{\partial X_i}\right)_{Y_k} \geq 0 \tag{3.10}$$

for the mathematical expression of Le Chatelier – Braun principle. This equation means that the value of the change of extensive parameter Y_i (generalized coordinate) after influence of conjugative intensive parameter X_i (generalized force) depends on the way of fixing another parameters. In the case of the derivative $\left(\frac{\partial Y_i}{\partial X_i}\right)_{X_k}$ the generalized force X_k is fixed, in the second case (derivative $\left(\frac{\partial Y_i}{\partial X_i}\right)_{Y_k}$) this force is not fixed and could influence the process.

Accordingly it follows from equation (3.10) that the action of the force X_k reduces the external influence of the force X_i on the generalized coordinate Y_i. It is evident that the chain of inequalities in equation (3.9) directly follows from equation (3.10).

Rusanov and Shultz derived equations that generalized Erhenfest's results. More general mathematical expression of Le Chatelier – Braun principle (a second rule – for the shifting of equilibrium under the action of secondary forces [30]) was presented as the chain of inequalities, e.g.:

$$\left(\frac{\partial Y_i}{\partial X_i}\right)_{Y_1,X_2,...X_{i-1},X_{i+1},...X_{n+2}} \geq \left(\frac{\partial Y_i}{\partial X_i}\right)_{Y_1,Y_2,X_3,...X_{i-1},X_{i+1},...X_{n+2}} \geq ... \tag{3.11}$$
$$\geq \left(\frac{\partial Y_i}{\partial X_i}\right)_{Y_1,Y_2,Y_3,...Y_{i-1},Y_{i+1},...Y_{n+1},X_{n+2}} \geq \left(\frac{\partial Y_i}{\partial X_i}\right)_{Y_1,Y_2,Y_3,...Y_{i-1},Y_{i+1},...Y_{n+1},Y_{n+2}} \geq 0.$$

The analogical chain is valid for inversed derivatives $\dfrac{\partial X_i}{\partial Y_i}$.

These inequalities could be presented in shorter form on the base of the concept of generalized thermodynamic potential Ψ_k [2]

$$\Psi_k = U - \sum_{i=1}^{k} X_i Y_i \tag{3.12}$$

Differentiation of (3.12) (for different values *k*) leads to the derivatives $\dfrac{\partial X_i}{\partial Y_i}$ with different ways of parameter fixing and the following chain of inequalities for second derivatives of Ψ_k is valid [34]:

$$\frac{\partial^2 U}{\partial Y_j^2} = \frac{\partial^2 \Psi_0}{\partial Y_j^2} \geq \frac{\partial^2 \Psi_1}{\partial Y_j^2} \geq \frac{\partial^2 \Psi_2}{\partial Y_j^2} \geq ... \geq \frac{\partial^2 \Psi_n}{\partial Y_j^2} \geq 0, \ (j = n+1), \tag{3.13}$$

where the condition $j = n+1$ adds no limitation because parameter numbering is arbitrary. An analogical result for differences in the second derivatives of thermodynamic potential was obtained by Gibbs also [1].

The third rule, which should be considered also as Le Chatelier – Braun principle concerns the correlation between non-conjugated thermodynamic variables. Let's present the differentials of extensive parameters

$$\left.\begin{aligned} dY_k &= \frac{\partial Y_k}{\partial X_k} dX_k + \frac{\partial Y_k}{\partial X_l} dX_l \\ dY_l &= \frac{\partial Y_l}{\partial X_k} dX_k + \frac{\partial Y_l}{\partial X_l} dX_l \end{aligned}\right\} \tag{3.14}$$

and solve this system of equation in following way:

$$(dX_l)_{Y_l} \frac{\begin{vmatrix} \frac{\partial Y_k}{\partial X_k} & (dY_k)_{Y_l} \\ \frac{\partial Y_l}{\partial X_k} & 0 \end{vmatrix}}{\begin{vmatrix} \frac{\partial Y_k}{\partial X_k} & \frac{\partial Y_k}{\partial X_l} \\ \frac{\partial Y_l}{\partial X_k} & \frac{\partial Y_l}{\partial X_l} \end{vmatrix}} = \frac{-\left(\frac{\partial Y_l}{\partial X_k}\right)(dY_k)_{Y_l}}{\Delta}, \tag{3.15}$$

where determinant in the denominator, Δ is ≥ 0 due to stability conditions. After transformation to the equation (using the first of equations (3.14))

$$\left(\frac{\partial Y_k}{\partial X_l}\right)_{Y_l} = \frac{\Delta}{-\frac{\partial Y_l}{\partial X_k}} = \frac{\Delta}{-\left(\frac{\partial Y_k}{\partial X_l}\right)_{X_k}} \tag{3.16}$$

the following conditions can be obtained ($l = k+1$):

$$\text{if} \left(\frac{\partial Y_k}{\partial X_{k+1}}\right)_{X_k} > 0, \text{ then} \left(\frac{\partial Y_k}{\partial X_{k+1}}\right)_{Y_{k+1}} < 0, \tag{3.17}$$

$$\text{if} \left(\frac{\partial Y_k}{\partial X_{k+1}}\right)_{X_k} < 0, \text{ then} \left(\frac{\partial Y_k}{\partial X_{k+1}}\right)_{Y_{k+1}} > 0, \tag{3.18}$$

which are the third types of rules reflecting Le Chatelier – Braun principle. The thermal expansion coefficient

$$\alpha = \frac{1}{V}\left(\frac{\partial V}{\partial T}\right)_P$$

can be considered as an example of these condition. Equations (3.17) give the following inequalities in this case:

$$\left(\frac{\partial V}{\partial T}\right)_P > 0, \quad \left(\frac{\partial V}{\partial T}\right)_S < 0, \tag{3.19}$$

or

$$\text{if } \alpha = \frac{1}{V}\left(\frac{\partial V}{\partial T}\right)_P \equiv \alpha_P > 0, \text{ then } \alpha_S = \frac{1}{V}\left(\frac{\partial V}{\partial T}\right)_S < 0$$

The negative thermal expansion coefficients α_P are considered as unusual phenomena. Nevertheless such effect is described in literature [35]. This property is of practical importance and there are numerous recent papers describing this phenomena including the synthesis of materials with negative α_P and experimental research (e.g., see [36-38]). It follows from (3.18) that $\alpha_P < 0$ when $\alpha_S > 0$.

4. Thermodynamic Inequalities for the Case of Phase Transitions in Multicomponent Systems without Chemical Reactions

Phase stability is a main object in the research of stability of equilibrium since Gibb's works [1]. These investigations concern conditions of phase existence, some limitations on thermodynamic parameters of homogeneous and heterogeneous systems, formation of new phases (e.g. liquid phase splitting), thermodynamic consideration of critical phenomena, stability limits, etc. Some general and special aspects of phase stability have been studied and discussed in numerous monographs, books and original papers, including classical works of Van der Waals [39], other books (e.g. cited works [2-17, 25] and modern research. Now we will discuss some special problems concerning both phase stability and phase transitions in multicomponent systems without chemical reactions.

The usual condition of the stability of heterogeneous system as a whole requires that all coexisting phases of this system must be stable with respect to continuous changes of state (see e.g. [2, 5]). Such condition is a result of summarizing of thermodynamic inequalities, the necessary stability conditions for every phase of an r-phase multicomponent system:

$$\sum_i dX_i^{(k)} dY_i^{(k)} > 0 \quad (k = 1,2,...r), \tag{4.1}$$

where superscript *(k)* denotes index of a phase of an r-phase system. Summing by all phases gives

$$\sum_i dX_i^{(g)} dY_i^{(g)} > 0 \tag{4.2}$$

where superscript *(g)* denotes that the value belongs to the entire system, a heterogeneous complex of all coexisting phases. The condition (4.2) is necessary for the stability of a heterogeneous system, but the sum could be positive also in the case of the < sign for some of items. The additional insufficiently of this approach is stipulated by the way of derivation of condition (4.2): it is based on the disturbances of internal equilibrium of separate phases. The disturbances of equilibrium between coexisting phases are not taken into account.

Let us consider the problem of the stability of heterogeneous system proceeding from the condition of stability between coexisting phases. For the sake of simplicity the case of isolated system will be considered[1]:

$$\sum_{k=1}^{r} dY_i^{(k)} = 0 \quad (i = 1,2,...n+2) \tag{4.3}$$

Let's introduce the internal parameter ξ_i to characterize the change of an extensive value Y_i under a disturbance, i.e. a deviation from phase equilibrium [2]. For every phase

$$dY_i^{(k)} = a_i^{(k)} d\xi_i, \tag{4.4}$$

where

$$a_i^{(k)} = \frac{dY_i^{(k)}}{d\xi_i} \tag{4.5}$$

is generally a variable depending on the way of choosing of internal parameter ξ_i and on the type (peculiarities) of the disturbance. Let $a_i^{(k)}$ be constant for a small disturbance. Then the stability conditions for every phase will be written in following form:

$$\sum_i dX_i^{(k)} d\left(a_i^{(k)} \xi_i\right) = \sum_i a_i^{(k)} dX_i^{(k)} d\xi_i > 0 \cdot \tag{4.6}$$

The sum of inequalities (4.6) for all phases leads to the necessary stability condition:

[1] In general case the violating of isolation condition for any parameter may be taken into consideration by proper balance equations.

$$\sum_i d\xi_i \left(a_i^{(1)} dX_i^{(1)} + a_i^{(2)} dX_i^{(2)} + \ldots + a_i^{(r)} dX_i^{(r)} \right) > 0 \cdot \tag{4.7}$$

This equation is similar to stability conditions for the chemical equilibrium in the case of several chemical reactions [4, 32] (at $T, P = const$):

$$\sum_\rho \sum_{\rho'} \frac{\partial A_\rho}{\partial \xi_{\rho'}} d\xi_\rho d\xi_{\rho'} = \sum_\rho dA_\rho d\xi_\rho = \sum_\rho d\xi_\rho \left(\nu_{1\rho} d\mu_{1\rho} + \nu_{2\rho} d\mu_{2\rho} + \ldots \right) < 0, \tag{4.8}$$

where ρ and ρ' are numbers of reactions; A_ρ, ξ_ρ (or $\xi_{\rho'}$) – affinity and extent of reaction; $\nu_{i\rho}$ - stoichiometric number of species *i* in reaction ρ; $d\mu_{i\rho}$ – the change of chemical potential of specie *i* caused by the disturbance of chemical equilibrium of reaction ρ. Values $\nu_{i\rho}$ are analogs of $a_i^{(k)}$ and are constant for any variations.

The left side of equation (4.7) may be considered as a second variation of internal energy in an isolated system. In the general case of a non-isolated system it would be a part of the second variation of internal energy due to internal processes in the system, $\left(\partial^2 U\right)_{\text{int}}$. The changes in internal energy depending on environment interactions $\left(\partial^2 U\right)_{ext}$ can be treated as quasistatic ones: they can be any and don't determine the thermodynamic stability of the system. Hence

$$\left(\partial^2 U\right)_{\text{int}} = \partial^2 U - \left(\partial^2 U\right)_{ext} > 0, \tag{4.9}$$

and equation (4.7) determines internal stability of a heterogeneous system relative to the processes disturbing the equilibrium between phases. This type of stability characterizes phase stability of the system.

The internal parameter ξ_i should be taken together with the variables of fundamental equations and additional conditions [2] but it isn't a parameter of the state that characterizes the equilibrium. For example in a two-phase system under additional condition

$$m_i^{(1)} + m_i^{(2)} = const \tag{4.10}$$

the internal parameter could be determined as

$$\xi_i = m_i^{(1)} \tag{4.11}$$

Taking into account condition (4.10) it is possible to set any reliable values of the distribution of the component between phases $\frac{m_i^{(2)}}{m_i^{(1)}}$ both in equilibrium and nonequilibrium processes. Coefficient a_1 is unit, a_2 is equal to -1,

For two-phase system and $T, P = const$ expressions (4.4), (4.7) and (4.11) lead to

$$\sum_{i=1}^{n} dm_i^{(1)} d(\mu_i^{(1)} - \mu_i^{(2)}) > 0, \tag{4.12}$$

where $Y_i = m_i$ $X_i = \mu_i$. This inequality was used for the analysis of the processes of shifting the equilibrium [40].

For the sake of simplicity the multiphase equilibrium will be considered as a case of a three-phase system under $T, P = const$. Again, let $\xi_i = m_i^{(1)}$, $Y_i = m_i$ $X_i = \mu_i$; inequality (4.7) is presented as follows

$$\sum_{i=1}^{n} dm_i^{(1)} d(\mu_i^{(1)} + a_i^{(2)} \mu_i^{(2)} - a_i^{(3)} \mu_i^{(3)}) > 0. \tag{4.13}$$

Here we again suppose that $a_i^{(k)}$ are constant for small disturbances. Coefficient $a_i^{(1)}$ is unit, values

$$a_i^{(2)} = \frac{dm_i^{(2)}}{dm_i^{(1)}}, \quad a_i^{(3)} = \frac{dm_i^{(3)}}{dm_i^{(1)}} \tag{4.14}$$

are parameters of the mass-transfer process between phases and are determined by the nature of the process. The problem of calculation $a_i^{(2)}$ and $a_i^{(3)}$ is similar to the problem of calculation of stochiometric numbers in reaction between isomers

$$\nu_1 R_1 + \nu_2 R_2 + \nu_3 R_3 = 0 \tag{4.15}$$

where all species may be converted from one to another:

$$R_1 \leftrightarrow R_2, \quad R_1 \leftrightarrow R_3, \quad R_2 \leftrightarrow R_3 \tag{4.16}$$

Though stoichiometric equations (4.16) are correct they do not give useful and necessary thermodynamic information for the condition of chemical equilibrium

$$\sum_{i=1}^{3} \nu_i \mu_i = 0 \tag{4.17}$$

where all species are presented.

A three-phase equilibrium may be considered also in other way as a total combination of two-phase equilibria.

5. Thermodynamic Inequalities for the Systems with Chemical Interactions

In this section and the following one we keep notification for intensive and extensive parameters (X_i, Y_i) including amounts (here – moles) and chemical potential of substances which don't participate in chemical reactions. For reacting species numbers of moles and chemical potentials are denoted as usual, m_i and μ_i. We also claim that the system is closed for reacting species. The following equation for any pairs of X_i, Y_i holds:

$$\left(\frac{\partial X_1}{\partial Y_1}\right)_{\sum\limits_i \nu_i\mu_i} = \left(\frac{\partial X_1}{\partial Y_1}\right)_{\xi} + \left(\frac{\partial X_1}{\partial \xi}\right)_{Y_1}\left(\frac{\partial \xi}{\partial Y_1}\right)_{\sum\limits_i \nu_i\mu_i} \tag{5.1}$$

where the subscript $\sum\limits_i \nu_i\mu_i$ means that the affinity ($A = -\sum\limits_i \nu_i\mu_i$) is constant ($\sum\limits_i \nu_i\mu_i = 0$ is for chemical equilibrium), subscript ξ indicates that the reaction doesn't proceed, parameters X_1, Y_1 were chosen because the numbering of these parameters is arbitrary. Due to the condition of reciprocity and stability condition the second term at the right side of (5.1) is negative and the following inequality holds [41]:

$$\left(\frac{\partial X_1}{\partial Y_1}\right)_{\sum\limits_i \nu_i\mu_i} < \left(\frac{\partial X_1}{\partial Y_1}\right)_{\xi} \tag{5.2}$$

For inversed derivatives

$$\left(\frac{\partial Y_1}{\partial X_1}\right)_{\sum\limits_i \nu_i\mu_i} > \left(\frac{\partial Y_1}{\partial X_1}\right)_{\xi} \tag{5.3}$$

or

$$\left(\frac{\partial Y_1}{\partial X_1}\right)_{A} > \left(\frac{\partial Y_1}{\partial X_1}\right)_{\xi}. \tag{5.4}$$

It should be noted that the condition $\xi = const$ means that the composition of the reacting mixture is also constant (the reaction is frozen, catalyst is absent, or an inhibitor presents), The condition $A = const$ means that the composition may change, e.g. if $A = 0$ there is a process of equilibrium shifting and the reaction proceeds (establishment of chemical equilibrium under new conditions). It is also important that the difference between derivatives has a finite value and can be calculated: it is a value of the second member in the right side of equation (5.1), $\left(\frac{\partial X_1}{\partial \xi}\right)_{Y_1}\left(\frac{\partial \xi}{\partial Y_1}\right)_{\sum_i \nu_i \mu_i}$ [2].

Let us consider some example of the application of inequalities (5.2) – (5.4). For entropy and temperature (5.4) gives

$$\left(\frac{\partial S}{\partial T}\right)_A > \left(\frac{\partial S}{\partial T}\right)_\xi > 0 \cdot \tag{5.5}$$

Accordingly for heat capacity the following conditions holds:

$$c_{p,A} > c_{p,\xi} > 0 \text{ or } c_{V,A} > c_{V,\xi} > 0 . \tag{5.6}$$

These conditions are the same as well-known conclusion for chemical equilibrium, when $A = 0$ (see [4]): heat capacity under constant composition should be always smaller than heat capacity during equilibrium transformation.

For isothermal and isoentropic compressibility $\kappa_T = -\frac{1}{V}\left(\frac{\partial V}{\partial P}\right)_T$, $\kappa_S = -\frac{1}{V}\left(\frac{\partial V}{\partial P}\right)_S$ it follows from (5.4) that

$$\kappa_{T,A} > \kappa_{T,\xi} > 0 \text{ and } \kappa_{S,A} > \kappa_{S,\xi} > 0; \tag{5.7}$$

for chemical equilibrium this condition is also presented in the book by Prigogine and Defay [4].

The case of several chemical reactions might be considered in the same way. The rlation between derivatives $\frac{\partial X_k}{\partial Y_k}$ for the cases of different number of chemical reactions is established by the equation

$$\left(\frac{\partial X_k}{\partial Y_k}\right)_{\xi_1,\xi_2,\ldots\xi_{\rho-1},A_\rho,A_{\rho+1},\ldots} = \left(\frac{\partial X_k}{\partial Y_k}\right)_{\xi_1,\xi_2,\ldots\xi_\rho,A_{\rho+1},\ldots} + \left[\left(\frac{\partial X_k}{\partial \xi_\rho}\right)_{Y_k}\left(\frac{\partial \xi_\rho}{\partial Y_k}\right)_{A_\rho}\right]_{\xi_1,\xi_2,\ldots\xi_{\rho-1},A_{\rho+1},\ldots} , \tag{5.8}$$

[2] The same conclusions concerning the *finite* differences between derivatives are valid for other inequalities considered in this chapter (e.g. (3.10), (3.11), (5.9), (5.10), (6.1), (6.2)).

where subscripts indicate constant affinities or constant reaction extents (1, 2, ... $\rho-1$, ρ, $\rho+1$,...) Here symbols X_k, Y_k again denote $S, T, V, (-P)$ and chemical potential of substances which don't participate in chemical reactions. The derivative at the left side of equation (5.8) refers to constant affinity A_ρ (particularly, if $A_\rho = 0$ to the chemical equilibrium of the reaction ρ). The first derivative at the right side correspond to the constant extent of the reaction ρ: this reaction doesn't proceed. Other conditions in the subscripts are the same: reactions $1,2,...\rho-1$ are frozen, reactions $\rho+1, \rho+2....$ may proceed. After some transformation of (5.8) and due to the stability condition (2.14) we will obtain that the second term in the right side of (5.8) is negative and the following inequality holds:

$$\left(\frac{\partial X_k}{\partial Y_k}\right)_{\xi_1,\xi_2,...\xi_{\rho-1},A_\rho,A_{\rho+1},...} < \left(\frac{\partial X_k}{\partial Y_k}\right)_{\xi_1,\xi_2,...\xi_\rho,A_{\rho+1},...} \tag{5.9}$$

In the case of S independent reactions ρ has values $1,2,..., S$ and the following chain of inequalities holds:

$$0 < \left(\frac{\partial X_k}{\partial Y_k}\right)_{A_1,A_2,...,A_s} < \left(\frac{\partial X_k}{\partial Y_k}\right)_{A_1,A_2,...,A_{s-1},\xi_s} < ...$$
$$... < \left(\frac{\partial X_k}{\partial Y_k}\right)_{A_1,\xi_2,\xi_3,...,\xi_s} < \left(\frac{\partial X_k}{\partial Y_k}\right)_{\xi_1,\xi_2,...\xi_s} \tag{5.10}$$

Both inequalities (5.2)-(5.4) and the chain of inequalities (5.10) can be treated as mathematical expressions for rules of the Le Chatelier – Braun principle, such as inequality (3.11) and (3.13).

6. The General System of Thermodynamic Inequalities

The results described in the previous sections show that the general system of thermodynamic inequalities can be constructed. The similar relationships for both phase transitions and chemical equilibria demonstrate the unity of chemical and phase processes from the thermodynamic point of view. Additionally to the inequalities for phase equilibria (section 4) some other relationships for phase transitions can be established. For example, for a two-phase system the following chain of inequalities is valid [25, 40, 41]:

$$\left(\frac{\partial X_1}{\partial Y_1}\right)_{Y_2^{(1)},Y_3^{(1)},...Y_{n+2}^{(1)}} > \left(\frac{\partial X_1}{\partial Y_1}\right)_{(X_2^{(1)}-X_2^{(2)}),Y_3^{(1)},...Y_{n+2}^{(1)}} >$$

$$> ... > \left(\frac{\partial X_1}{\partial Y_1}\right)_{(X_2^{(1)}-X_2^{(2)}),(X_3^{(1)}-X_3^{(2)}),...(X_{n+1}^{(1)}-X_{n+1}^{(2)}),Y_{n+2}^{(1)}} > 0, \tag{6.1}$$

$$Y_2, Y_3,...Y_{n+2} = const,$$

where the symbols without superscripts refer to the entire heterogeneous system. The condition $Y_i^{(1)} = const$ (when $Y_i = const$) indicates the absence of a phase transition for the parameter *i*: entropies of phases, volumes of phases, or amounts of given substances in phases are constant. The condition $X_i^{(1)} - X_i^{(2)} = const$ means the corresponding phase transition and, in particular, when $X_i^{(1)} - X_i^{(2)} = 0$, the equilibrium of the phases with regard to the parameter *i* (partial equilibrium). According to (6.1), from the comparison of the derivatives

$$\left(\frac{\partial X_1}{\partial Y_1}\right)_{Y_2^{(1)},Y_3^{(1)},...Y_{n+2}^{(1)}} > \left(\frac{\partial X_1}{\partial Y_1}\right)_{(X_i^{(1)}-X_i^{(2)})=0,Y_{n+2}^{(1)}} > 0 \tag{6.2}$$

$$(i = 2,3,...n-1)$$

it follows that the derivative $\frac{\partial X_1}{\partial Y_1}$ is always larger in the case when the compositions of the phases are constant (lack of equilibrium) than in the case when the equilibrium between the phases is maintained. The similar results had been discussed also by Prigogine and Defay [4] in the frame of stability problem, moderation theorems and Le Chatelier – Braun principle.

Other correlations of stability with respect to chemical and phase processes are evident due to the generalized concepts of affinity and internal parameter [2]. Though affinity is usually considered as the element of the thermodynamic theory of chemical reactions the application of this value for the describing of phase processes was presented already in the book [4].

The affinity of the phase process in an *r*-phase system for the substance *i* can be introduced as follows

$$A_i = -\sum_{k=1}^{r} a_i^{(k)} \mu_i^{(k)} \tag{6.3}$$

Substituting this formula in the equation under $T, P = const$ leads to the relationship

$$\sum_i d\xi_i \left(a_i^{(1)} dX_i^{(1)} + a_i^{(2)} dX_i^{(2)} + \ldots + a_i^{(r)} dX_i^{(r)}\right) =$$
$$= \sum_{i=1}^{n} d\xi_i \left(a_i^{(1)} d\mu_i^{(1)} + a_i^{(2)} d\mu_i^{(2)} + \ldots + a_i^{(r)} d\mu_i^{(r)}\right) = \qquad (6.4)$$
$$= -\sum_{i=1}^{n} d\xi_i dA_i = -\sum_{i=1}^{n} \sum_{j=1}^{n} \frac{\partial A_i}{\partial \xi_j} d\xi_i d\xi_j > 0$$

This form of the condition of stability of phase equilibrium

$$-\sum_{i=1}^{n} \sum_{j=1}^{n} \frac{\partial A_i}{\partial \xi_j} d\xi_i d\xi_j > 0 \qquad (6.5)$$

coincides with stability conditions for chemical equilibrium in the case of several chemical reactions (see (4.8)):

$$-\sum_{\rho} \sum_{\rho'} \frac{\partial A_\rho}{\partial \xi_{\rho'}} d\xi_\rho d\xi_{\rho'} > 0 \cdot \qquad (6.6)$$

Inequalities (6.5) and (6.6) characterize phase and chemical stability of multiphase multicomponent systems with several chemical reactions under $T, P = const$. The system of these expressions may be treated as a general system of thermodynamic inequalities. This result could be extended also to the case of variable temperature and pressure.

7. Transformations of the Stability Matrix

The matrix form of stability conditions is useful for solving some general and special problems. The most known form of the stability matrix, the left side of equation (2.8), is the matrix of a quadratic form that is a second differential of internal energy (2.7). The general form of the stability matrix should be built with the second derivatives of a generalized thermodynamic potential Ψ_k (equation (3.12)). Such approach gives the opportunity to present stability conditions by the second variations of any thermodynamic potential and demonstrates the equivalence of the use of these potentials (Gibbs and Helmholtz energy, enthalpy, etc.) to estimate stability. Though the equivalence of thermodynamic potentials is a well-known fact, the discussion of the possibility to apply stability conditions presented by different potentials often meet misunderstanding. For example, the following wrong statement can be found in some text-books: *stability condition presented by second variations of Gibbs energy is insufficiency to determine thermal and mechanical stability.* Such error in interpreting stability conditions should be explained by the incomplete presentation of the stability matrix in second derivatives of Gibbs energy. Opposite to internal energy (equations (2.6)-(2.8)) stability conditions in terms of Gibbs energy include two quadratic forms: the first form depends on extensive parameters $m_1, m_2, \ldots m_n$ only, the second form depends on intensive parameters T, P. As a result we get different conditions for the curvature of the surface of Gibbs energy: convexity down at $T, P = const$ and convexity up at $m_1, m_2, \ldots m_n = const$. The last convexity condition determines the thermal and mechanical stability. In general the surface of Gibbs energy is convexo-concave.

The stability conditions in terms of a generalized thermodynamic potential Ψ_k, which is a function of n+2 parameters can be present as follows [2]:

$$\left[\delta^2 \Psi_k (Y_{k+1}, Y_{k+2}, \ldots Y_{n+2})\right]_{X_1, X_2, \ldots X_k} - \\ - \left[\delta^2 \Psi_k (X_1, X_2, \ldots X_k)\right]_{Y_{k+1}, Y_{k+2}, \ldots Y_{n+2}} > 0 \tag{7.1}$$

where the first term at the left side is a quadratic form of extensive parameters, the second one is a quadratic form of intensive parameters. This condition must be valid if the following inequalities for quadratic forms hold:

$$\left[\delta^2 \Psi_k (Y_{k+1}, Y_{k+2}, \ldots Y_{n-1})\right]_{X_1, X_2, \ldots X_k, Y_{n+2}} > 0, \tag{7.2}$$

$$\left[\delta^2 \Psi_k (X_1, X_2, \ldots X_k)\right]_{Y_{k+1}, Y_{k+2}, \ldots Y_{n+2}} < 0; \tag{7.3}$$

the first quadratic form is positive definite, while the second one is negative definite. The condition $Y_{n+2} = const$ in the subscript of (7.2) is added commonly to avoid the variation due to the changes in the total mass of a virtual phase only. In a similar way we get the condition for the matrix of the first quadratic form (7.2), which depends on extensive parameters only:

- *all major minors should be positive.*

For the matrix of the second quadratic form (7.3) which depends on intensive parameters only the following conditions hold:

- *all major minors of even order should be positive,*
- *all major minors of odd order should be negative.*

The usual illustration of the conditions for the matrixes of the quadratic form of intensive or extensive parameters is again the case of Gibbs energy (e.g. see [2]). For the sake of simplicity we will consider a ternary system and molar Gibbs energy g together with mole fractions x_i as composition variables. Due to (7.2) and (7.3) the stability conditions would be presented as sets of inequalities for matrixes:

$$\begin{vmatrix} \dfrac{\partial^2 g}{\partial x_1^2} & \dfrac{\partial^2 g}{\partial x_1 \partial x_2} \\ \dfrac{\partial^2 g}{\partial x_2 \partial x_1} & \dfrac{\partial^2 g}{\partial x_2^2} \end{vmatrix} > 0, \quad \frac{\partial^2 g}{\partial x_1^2} > 0, \quad \frac{\partial^2 g}{\partial x_2^2} > 0 \quad (7.4)$$

$$\begin{vmatrix} \dfrac{\partial^2 g}{\partial T^2} & \dfrac{\partial^2 g}{\partial T \partial P} \\ \dfrac{\partial^2 g}{\partial P \partial T} & \dfrac{\partial^2 g}{\partial P^2} \end{vmatrix} > 0, \quad \frac{\partial^2 g}{\partial T^2} < 0, \quad \frac{\partial^2 g}{\partial P^2} < 0 \quad (7.5)$$

It is evident now what the conditions of thermal and mechanical stability will be in terms of the second deviation of Gibbs energy. These and other considerations of stability matrixes in terms of different thermodynamic potentials are described quite well in numerous thermodynamic works (e.g. [2, 4, 6]). Therefore we restricted our discussion by general remarks. It could be mentioned also that inequalities (7.5) lead both to the conditions for heat capacities

$$C_P > 0 \text{ and } C_V > 0$$

and to condition for compressibility, e.g.

$$\kappa_T > 0$$

Equations (7.5) also give

$$C_P > C_V > 0,$$

i.e. the result related to Le Chatelier – Braun principle.

Since considered stability conditions include only second variations of thermodynamic potentials they can't be applied to the analysis of more complex problems like critical states. In the last case the consideration of the variations of higher order is necessary [2, 4-6]. Such approach leads to important results but the analysis and the way of presentation of final expressions become rather complicated. To avoid these troubles Gorovitz *et al* proposed the method which is based in principle on the transformation of the matrices to the diagonal form with the use of new (transformed) composition variables [42-44]. This approach gives one an opportunity to get clear analytical presentations of the properties of critical states and to present relatively simple graphical illustration for different types of the critical phase's behavior.

The interesting and very important result concerning transformation of the stability matrix was obtained by Gibbs [1]. The reducing of the stability matrix is also due to the replacement of variables. As a result the stability matrix presents a product of derivatives.

While such stability matrix transformation has been clear described already in Gibbs' work [1] it was used relatively rarely and in some special cases only (e.g. [2, 4, 6]). Moreover, since Le Chatelier's edition of the Gibbs' work [45] some mistakes in the presentation of this result are common.

According to Gibbs the determinant (2.8) at $V = const$, i.e.

$$\begin{vmatrix} \frac{\partial^2 U}{\partial S^2} & \frac{\partial^2 U}{\partial m_1 \partial S} & \cdots & \frac{\partial^2 U}{\partial m_n \partial S} \\ \frac{\partial^2 U}{\partial S \partial m_1} & \frac{\partial^2 U}{\partial m_1^2} & \cdots & \frac{\partial^2 U}{\partial m_n \partial m_1} \\ \cdots & \cdots & \cdots & \cdots \\ \frac{\partial^2 U}{\partial S \partial m_n} & \frac{\partial^2 U}{\partial m_1 \partial m_n} & \cdots & \frac{\partial^2 U}{\partial m_n^2} \end{vmatrix} = \left| \frac{\partial^2 U}{\partial Y_i \partial Y_j} \right| \tag{7.6}$$

presents itself a product of partial derivatives

$$\left(\frac{\partial T}{\partial S}\right)_{V,m_1,\ldots m_n} \cdot \left(\frac{\partial \mu_1}{\partial m_1}\right)_{T,V,m_2,\ldots m_n} \cdot \left(\frac{\partial \mu_2}{\partial m_2}\right)_{T,V,\mu_1,m_3\ldots m_n} \times \ldots$$
$$\ldots \times \left(\frac{\partial \mu_n}{\partial m_n}\right)_{T,V,\mu_1,\ldots \mu_{n-1}} = \left| \frac{\partial^2 U}{\partial Y_i \partial Y_j} \right| \tag{7.7}$$

The common errors in the interpretation of this result are related to the mistakes in subscripts in the left side of equation (7.7). The derivation of (7.7) can be made in different ways. The simple ground was proposed in the work [34].

8. The Use of Stability Criteria to Characterise Nonequilibrium Systems and Processes

Regardless of the term "equilibrium", the thermodynamic theory of the stability of equilibrium is based on the concept of virtual displacement (or disturbance). The disturbance, both virtual and real, should be treated as an irreversible process. Thus stability criteria are the "bridge" between equilibrium and nonequilibrium thermodynamics. In this section we present some examples of the use of stability conditions to characterise the non-equilibrium systems and processes.

In previous sections we discussed the results following from condition (6.2), which enable us to estimate the difference between thermodynamic parameters of the system in the state of phase equilibrium and when there is a lack of equilibrium. The inequality (6.2) can be interpreted in a wider way: if the process of establishing the equilibrium between phases relative to the parameters i is slower than the process of changing X_1 and Y_1 then the derivative $\frac{\partial X_1}{\partial Y_1}$ has a larger value than at the instant of establishing the equilibrium. For example, for the partial pressure of the component 1 with regard to the relationship

$$\mu_1 = \mu_1^\circ(T) + RT\ln p_1,$$

where p_1 is a partial pressure of the substance 1, $\mu_1^\circ(T)$ - standard value of chemical potential of substance 1 at given temperature T, we obtain at $T, P = const$

$$\left(\frac{\partial \ln p_1}{\partial m_1}\right)_{m_2^{(1)}, m_3^{(1)},\ldots} > \left(\frac{\partial \ln p_1}{\partial m_1}\right)^{equilibrium} > 0 \tag{8.1}$$

i.e. on introducing a component into the system its partial pressure increases, but then somewhat decreases in the process of establishing the equilibrium with regard to all the components. This result illustrates the possibility to use the criteria of stability of the *equilibrium* to characterise the *non-equilibrium* process.

Analogous relationships can be obtained for other parameters being second derivatives of thermodynamic potentials. For example, at $P = const$ for a two-phase multicomponent system

$$\left(\frac{\partial S}{\partial T}\right)_{P,\mu_i^{(1)}=\mu_i^{(2)}} > \left(\frac{\partial S}{\partial T}\right)_{P,m_i^{(1)}} > 0, \tag{8.2}$$

$$m_i = m_1, m_2 \ldots = const.$$

Since the heat capacity at constant pressure, c_P, can be expressed by the derivatives in (8.2)

$$c_P = T\left(\frac{\partial S}{\partial T}\right)_P,$$

then the statement follows from (8.2): *if under the change of temperature of the system the stable phase equilibrium is not established (or has no time to establish) then the heat capacity turns lower than the equilibrium heat capacity.*

Of practical interest is the next limiting case when the initial state of the system is a single-phase state, i.e.

$$Y_1^{(1)} = Y_1,\ Y_2^{(1)} = Y_2, \ldots Y_{n+2}^{(1)} = Y_{n+2},$$

but the second phase can appear when the state changes. Then it follows from equation (6.2) that, for example, the parameter $\frac{\partial X_1}{\partial Y_1}$ in a metastable homogeneous state has a larger value than in the corresponding stable two-phase state. It was mentioned before in this chapter that the difference between the derivatives in equation (6.2) has a finite value [41]. It follows then that the transition from a homogeneous range to the inhomogeneous one is accompanied by the jump in the value of $\frac{\partial X_1}{\partial Y_1}$. The latter conclusion can be compared to some known results for the configuration effects and heat capacities [4, 46, 47].

The analogous differential thermodynamic inequalities for the systems with chemical reactions also follow from the stability criteria [48].

9. Conclusion

In this chapter we considered some general and special form of thermodynamic inequalities that include stability conditions and also rules related to Le Chatelier – Braun principle. The conditions of phase and chemical equilibrium were discussed and analyzed on common base. That gives the opportunity to construct the general system of thermodynamic inequalities, e.g. with the use of generalized conception of affinity. Le Chatelier – Braun principle in correct mathematical formulation was treated as a part of thermodynamic stability theory. Rules based on this principle can be used for the estimation of phase and chemical equilibrium shifting. Moreover, corresponding inequalities (and chains of inequalities) determine the relative intensity of the thermodynamic parameters changes under different external conditions. A special interest presents the opportunity to use thermodynamic inequalities derived as stability conditions of equilibrium systems for the case of nonequilibrium processes, caused by the disturbance of equilibrium. Thus stability conditions can be considered as a "bridge" between equilibrium and non-equilibrium thermodynamics.

The chapter does not cover all aspects of thermodynamic stability theory (e.g. application to critical phenomena) because we restrict the topic by several alternative forms of equations for thermodynamic analysis. The application of stability conditions to solve practical problems of the chemical and phase equilibrium are described in cited monographs and text-books, also in original papers (e.g. [16-25]).

Acknowledgement

Author is grateful to Dr. Viktor Rakhimov for the useful discussion and remarks. This work was partly supported by Russian Foundation for Basic Research (grant 09-03-00812).

References

[1] Gibbs, JW. *The Collected Works. V. 1. Thermodynamics*, Longmans, Green and Co, New York, London, Toronto, 1931.

[2] Münster, A. *Chemische Thermodynamik* (in German), Academie–Verlag, Berlin, 1969.

[3] Haase, R. *Thermodynamik der Irreversiblen Prozesse* (in German), Dr. Dietrich Steinkopff Verlag, Darmstadt, 1963.

[4] Prigogine, I; Defay, R. *Chemical Thermodynamics*, Longmans, Green and Co., London, New York, Toronto, 1954.

[5] Storonkin, AV. *Termodinamika geterogennyh system (in Russian) (Thermodynamics of Heterogeneous Systems)*, Part 1&2, Publishing house of Leningrad University, Leningrad, 1967.

[6] Tisza, L. *Generalized Thermodynamics*, MIT. Press, Cambridge (Mass.), 1966.

[7] Schottky, W; Ulich, H; Wagner, C. *Thermodynamik*. Springer, Berlin, 1929.

[8] Prausnitz, JM; Lichtenthaler, RN; Azevedo, EG. *Molecular Thermodynamics of Fluid-Phase Equilibria*: 3rd ed., Prentice Hall PTR, Upper Saddle River, New Jersey, 1999.

[9] Tester, JW; Modell, M. *Thermodynamics and its Applications*, 3rd Edition, Prentice Hall PTR, Upper Saddle River, New Jersey, 1997.

[10] Kondepudi, D; Prigogine, I. *Modern Thermodynamics. From Heat Engines to Dissipative Structures*, Wiley & Sons, Chichester, 1999.

[11] Debenedetti, PG. *Metastable Liquids. Concepts and Principles*, Princeton University Press, Princeton, New Jersey, 1996.

[12] O'Connell, JP; Halle, JM. *Thermodynamics. Fundamental for Applications*, Cambridge University Press, 2005.

[13] Elliott, JR; Lira, CT. *Introductory Chemical Engineering Thermodynamics, Prentice-Hall*, 1999.

[14] Guggenheim, EA. *Thermodynamics*, North-Holland Publishing Company, Amsterdam, 1957.

[15] *A Commentary On The Scientific Writings of JW. Gibbs.* V. I and II. Eds. Donnan FG. & Haas A. Yale Univ. Press, New Haven, 1936.

[16] Michelsen, ML. The Isothermal Flash Problem. Part I. Stability, *Fluid Phase Equilibria*, 1982, 9, 1-19.

[17] Michelsen, ML. The Isothermal Flash Problem. Part II. Phase-Split Calculation, *Fluid Phase Equilibria*, 1982, 9, 21-40.

[18] Wasylkiewicz, SK; Sridhar, LN; Doherty, MF; Malone, MF. Global Stability Analysis and Calculation of Liquid-Liquid Equilibrium in Multicomponent Mixtures, *Ind. Eng. Chem. Res.*, 1996, 35, 1395-1408.

[19] Wasylkiewicz, SK; Ung, S. Global phase stability analysis for heterogeneous reactive mixtures and calculation of reactive liquid–liquid and vapor–liquid–liquid equilibria, *Fluid Phase Equilibria*, 2000, 175, 253-272.

[20] Tessier, SR; Brennecke, JF; Stadtherr, MA. Reliable Phase Stability Analysis for *Excess Gibbs Energy Models, Chem Eng. Sci.*, 2000, 55, 1785-1796.

[21] Harding, ST; Floudas, CA. Phase Stability with Cubic Equations of State: A Global Optimization Approach, *AIChE Journal*, 2000, 46, 1422-1440.

[22] Solokhin, MA; Solokhin, AV; Timofeev, VS. Phase-Equilibrium Stability Criterion in Terms of the Eigenvalues of the Hessian Matrix of the Gibbs Potential, *Theoret. Found. Chem. Eng.*, 2002, 36, 444-446.

[23] Nichita, DV; Gomez, S; Luna E. Phase stability analysis with cubic equations of state by using a global optimization method, *Fluid Phase Equilibria*, 2002, 194-197, 411-437.

[24] Cai, J; Prausnitz, JM. Stability criteria for a polydisperse system, *Fluid Phase Equilibria*, 2003, 206, 41-59.

[25] Toikka, AM; Jenkins, JD. Conditions of thermodynamic equilibrium and stability as a basis for the practical calculation of vapour–liquid equilbria, *Chem. Eng. Journal*, 2002, 89, 1-27.

[26] Giles, R. *Mathematical Foundation of Thermodynamics*, Pergamon, Oxford, 1964.

[27] Ehrenfest, P. Le Chatelier-Braun principle and thermodynamic laws of reciprocity (in Russian), *Zh. Rus. Fiz.-khim. Obschestva* (*J. Russian Phys.-Chem. Soc.*), 1909, 41, 347-364.

[28] Plank, M. Das prinzip von Le Chatelier und Braun (in German), *Ann. Phys.*, 1934, 19, 759-768.

[29] Rusanov, AI. *Phasengleichgewichten und Grenzflächenerscheinungen* (in German), Akademie-Verlag, Berlin, 1978.

[30] Rusanov, AI; Shultz, MM. On the short form of Le Chatelier –Braun principle (in Russian), *Vestnik Leningrad Univ., Ser. 4, Phys., chem.*, 1960, No 4, 60-65.

[31] Toikka, AM. Le Chatelier –Braun principle and some problems of the stability theory (in Russian), in book: *Thermodynamics of Heterogeneous Systems and Theory of Surface Phenomena,* Issue 7, Publishing house of Leningrad University, Leningrad, 1985, (31-53).

[32] Glandsdorff, I; Prigogine, I. *Thermodynamics of Structure, Stability and Fluctuations*, Wiley-Interscience, N.Y., 1971.

[33] Kubo, R. *Thermodynamics*, North-Holland Publishing Company, Amsterdam, 1968.

[34] Toikka, AM. Some formulation of Le Chatelier –Braun principle, *Zh. Fiz. Khim. (J. Phys. Chem. USSR*) 1990, 64, 2557-2559.

[35] Wallace, DC. *Thermodynamics of Crystals*, Dover Publications, Inc, Mineola/N.Y., 1998.

[36] Huang, RJ; Xu, W; Xu, XD; Li, LF; Pan, XQ; Evans D. Negative thermal expansion and electrical properties of $Mn_3(Cu_{0.6}Nb_xGe_{0.4-x})N$ (x = 0.05–0.25) compounds, *Mat. Lett.*, 2008, 62, 2381-2384.

[37] Isobe, T; Kato, Y; Mizutani, M; Ota, T; Daimon K. Pressureless sintering of negative thermal expansion $ZrW_2O_8/Zr_2WP_2O_{12}$ composites, *Mat. Lett.*, 2008, 62, 3913-3915.

[38] Perrin, C; Mangelinck, D; Nemouchi, F; Labar, J; Lavoie, C; Bergman, C; Gas, P. Nickel silicides and germanides: Phases formation, kinetics and thermal expansion, *Mat. Sci. Eng.*, B 2008, 154-155, 163-167.

[39] Van der Waals, JD; Konstamm, P. *Lehbruch der Thermostatik* (in German), Verlag Von Johann Ambrosius Barth, Leipzig, 1927.

[40] Toikka, AM. Equilibrium displacement in heterogeneous system and Le Chatelier-Braun principle, *Zh. Fiz. Khim. (Russ. J. Phys. Chem.*), 1994, 68, 1515-1517.

[41] Toikka, AM. Thermodynamic inequalities for complicated equilibrium systems (in Russian), in book: *Thermodynamics of Heterogeneous Systems and Theory of Surface Phenomena,* Issue 9, Publishing house of St.-Petersburg University, St.-Petersburg, 1992, (51-64).

[42] Gorovitz, BI; Sokolov, VA; Markuzin, NP. Critical phenomena in the ternary splitting systems. I. The local structure of phase diagrams in the vicinity of critical points for the case of single degeneration of the stability matrix, *Zh. Fiz. Khim.* (*Russ. J. Phys. Chem.*), 1991, 65, 1109-1114.

[43] Gorovitz, BI; Sokolov, VA; Markuzin, NP. Critical phenomena in the ternary splitting systems. II. The local structure of phase diagrams in the vicinity of critical points for the

case of full degeneration of the stability matrix, *Zh. Fiz. Khim.* (*Russ. J. Phys. Chem.*) 1991, 65, 1115-1118.

[44] Gorovitz, BI; Markuzin, NP; Sokolov, VA. Critical phenomena in the ternary splitting systems. IV. The forms of bimodal and spinodal in the critical point for the case of single degeneration of the stability matrix, *Zh. Fiz. Khim.* (*Russ. J. Phys. Chem.*), 1991, 65, 1197-1199.

[45] Gibbs, J. *Equilibre des systèmes chimiques*. Traduit par Le Chatelier. Paris, 1899.

[46] Filippov, VK. Effect of temperature and pressure on entropy and volume of the closes heterogeneous system (in Russian), in book: *Thermodynamics of Heterogeneous Systems and Theory of Surface Phenomena,* Issue 6, Publishing house of Leningrad University, Leningrad, 1982, (3-75).

[47] Filippov, VK; Shultz MM. On the link between phase composition, temperature and pressure in closed heterogeneous systems (in Russian), in book: *Thermodynamics of Heterogeneous Systems and Theory of Surface Phenomena*, Issue 5, Publishing house of Leningrad University, Leningrad, 1979.

[48] Toikka, AM. Thermodynamic inequalities for the systems with chemical reactions (in Russian), *Vestnik. Leningrad Univ., Ser.4, Phys., chem.,* 1994, No 3, 62-65.

Reviewed by Alexander Slobodov, Professor of Physical Chemistry, Saint-Petersburg Technological Institute (Technical University), Russia.

In: Mathematical Chemistry
Editor: W.I. Hong, pp. 537-559

ISBN 978-1-60876-894-3

Chapter 12

LIMITATIONS IN USING THE SANDERSON'S PRINCIPLE TO ESTIMATE GLOBAL REACTIVITY

Jorge Martínez[*]
Departamento de Química, Campus Santiago San Joaquín
Universidad Técnica Federico Santa María.
Av. Vicuña Mackenna
San Joaquín, Santiago, Chile.

Abstract

The Sanderson's geometric mean equalization principle for electronegativity has allowed expressions for chemical potential and molecular hardness to be derived within the framework of the Conceptual Density Functional Theory for obtaining molecular electronic properties of an entire molecule or supramolecule from the isolated fragments. Along with the determination of electronic properties, in former articles it has been demonstrated that Sanderson's scheme can be very useful as a method for rationalizing chemical reactions when both N (the number of electrons) and $v(\mathbf{r})$ (the external potential) change.The main problem that arises when using this approximation to estimate molecular properties is that the bonding potential and redistribution of the electron density among the fragments are not considered, and structural relaxation due to new specific through-bond and through-space interactions is not allowed within this scheme. Therefore this principle presents an important restriction to be used. In this chapter a suitable example to expose this limitation is mathematically developed by means of a general deduced mathematical expression for a polymeric growing chain according to this principle. This reveals that a task for improving this scheme is pending.

Keywords: Conceptual DFT Electronic Chemical Potential Molecular Hardness Electrophilicity Metallocenes Catalysis Polymerization

1. Summary

It is important to mention that all specific conclusions obtained in this analysis are concerning the polymerization processes so that the under or over–estimations of some

[*]E-mail address: jmartiar@gmail.com

values cannot be generalized to other systems. Most important to remark are the general conclusions related to limitations of additivity schemes to be used on any molecular system.

The main global reactivity indexes coming from the Conceptual DFT: chemical potential μ, molecular hardness η and electrophilicity ω, are analyzed in terms of their convergence behavior when an alkyl chain grows during an ethylene polymerization process catalyzed by a modelled metallocene cationic structure. Expressions that quantify μ and η values of a supra–molecular structure from their molecular fragments are deduced. This comes from Sanderson's electronegativity equalization principle[1, 2, 3]. Meanwhile the expression for obtaining ω is deduced from μ and η. In this chapter it will be demonstrated that the convergence requirement for each of these three global indexes is satisfied as the polymeric chain grows, but the respective asymptotic value is incorrect. This reveals some limitations suggesting that some precautions should be considered before using this principle and some improvements should be implemented on the equalization principle. The importance to determine these global properties is based on the fact that some properties could be connected with the reactivity indexes mentioned above. As an example, it has been demonstrated that the polarizability can be expressed as a function of the molecular hardness[4, 5]. Then as there is a mathematical expression to determine the molecular weights by means of light scattering measurement[6] depending on the polarizability, the latter might be replaced by a function that depends on the molecular hardness in order to obtain approximated values of molecular weights before starting an experimental procedure of synthesis.

2. Theoretical Background and Computational Methods

2.1. Theoretical Background

The Qualitative Density–Functional Theory of Chemical Reactivity also called Conceptual DFT[7, 8] has provided a set of reactivity descriptors based on the fact that in the canonical ensemble, the electronic energy E can be written as a function of N, the total number of electrons and as a functional of $v(\mathbf{r})$, the external potential; then $E \equiv E[N, v(\mathbf{r})]$. Assuming differentiability of E with respect to N and $v(\mathbf{r})$, a series of response functions emerges. The first of these functions is the chemical potential that characterizes the escaping tendency of electrons from the equilibrium system defined as[7]:

$$\mu = \left(\frac{\partial E}{\partial N}\right)_{v(\mathbf{r})}, \tag{1}$$

The second function is the molecular hardness, which measures the resistance to charge transfer defined as[8]:

$$\eta = \frac{1}{2}\left(\frac{\partial^2 E}{\partial N^2}\right)_{v(\mathbf{r})} = \frac{1}{2}\left(\frac{\partial \mu}{\partial N}\right)_{v(\mathbf{r})}. \tag{2}$$

Both μ and η are global properties involved in the reactivity of molecular systems[9]. A three–points finite difference approximation leads to the following working equations for these quantities[7, 9]

$$\mu \approx -\frac{1}{2}(\mathrm{I}+\mathrm{A}) \tag{3}$$

$$\eta \approx \frac{1}{2}(\mathrm{I}-\mathrm{A}), \tag{4}$$

where I and A are the first vertical ionization potential and the first electronic affinity of the neutral molecule, respectively. Further approximation using the Koopmans' theorem[10, 11] from the closed–shell Hartree–Fock theory[12] (I $\approx -\varepsilon_{\mathrm{HOMO}}$ and A $\approx -\varepsilon_{\mathrm{LUMO}}$) allows one to write μ and η in terms of energies of the lower unoccupied molecular orbital and higher occupied molecular orbital, LUMO ($\varepsilon_{\mathrm{LUMO}}$) and HOMO ($\varepsilon_{\mathrm{HOMO}}$), respectively:

$$\mu \approx \frac{1}{2}(\varepsilon_{\mathrm{LUMO}}+\varepsilon_{\mathrm{HOMO}}), \tag{5}$$

$$\eta \approx \frac{1}{2}(\varepsilon_{\mathrm{LUMO}}-\varepsilon_{\mathrm{HOMO}}). \tag{6}$$

Third, following Parr *et al.*[13], the square of the chemical potential of a chemical species divided by two times its chemical hardness, defines its electrophilicity index:

$$\omega = \frac{\mu^2}{2\eta}, \tag{7}$$

that measures the energetic stabilization when the system gets electrons from the surroundings. The Equations (3); (4) or (5); (6) can be used on Equation (7).
In addition, Sanderson[1, 2] established his equalization principle for electronegativity: *The electronegativity of a molecule is given by the geometric mean of the electronegativities of isolated atoms*. Since the chemical potential and the electronegativity are related as $\mu = -\chi$[14], under this proposed principle a general formula to obtain the chemical potential has been proposed. When a number of n_f atoms (or molecular fragments) are put together to form a molecular structure, the chemical potential can be estimated by means of Equation (8):

$$\overline{\mu} = -\left\{\prod_{i=1}^{n_f} |\mu_i|\right\}^{\frac{1}{n_f}}, \tag{8}$$

where μ_i is the chemical potential of the ith–atom (or fragment).
The derivative of Equation (8) with respect to N (the total number of electrons in the system) provides the Equation (9)[15, 16], which corresponds to an estimation of the molecular hardness of the system formed by n_f atoms (or fragments):

$$\overline{\eta} = \frac{\overline{\mu}}{n_f}\sum_{i=1}^{n_f}\frac{\eta_i}{\mu_i}, \tag{9}$$

where $\overline{\mu}$ is given by Equation (8), whereas μ_i and η_i correspond to the chemical potential and molecular hardness of the ith–atom (or fragment), respectively.

Both Equations (8) and (9) can be adapted to the polymerization process catalyzed by metallocenes providing proper working equations that depend on the number of monomeric units.

2.2. Computational Methods

The molecular systems were geometrically optimized according to the Schlegel's algorithm [17] at the DFT level of theory; in particular, the functionals used were the Becke–3 for exchange and Lee–Parr–Yang (B3LYP) for correlation[18, 19, 20, 21]. The 6–31G(d)[22, 23] basis set was used for non–metallic atoms; LANL2DZ[24, 25] pseudo–potentials were used for metallic atoms of Group IV B (Ti, Zr and Hf). All calculations were carried out using the GAUSSIAN 03[26] code.

3. Analysis of Convergence

It is interesting to analyze the convergence of reactivity indexes as the polymeric chain grows. Chemical potential μ, molecular hardness η and electrophilicity ω have been chosen on the one hand because the two former correspond to the first and second order indexes in the canonical ensemble respectively within the framework of Conceptual DFT and the last one because it directly depends on μ and η. The motivation to carry out this study is that some polymeric properties can be related with these reactivity indexes which could help develop new polymers by estimating those experimental properties using these reactivity indexes. But first, it is mandatory to obtain a set of suitable mathematical expressions to be applied in growing polymeric chains based on reactivity indexes of the respective monomer. Once these expressions are obtained, an analysis of convergence on them must be carried out as the polymeric chain grows. In this case, a cationic catalyst is generated by means of MAO (methylalumoxane); such a cationic catalyst interacts with an ethylene molecule as Figure 1 (b) indicates at the beginning of the polymerization process. As a consequence, a new cationic metallocenic species is generated, thus keeping the polymerization reaction until a long alkyl chain is obtained as Figure 2 indicates. Since an alkyl chain can adopt multiple conformations as it grows, it will become more and more complicated to analyze the evolution of the reactivity descriptors. In order to avoid that problem, the most simple situation will be used, meaning that a straight growing alkyl chain will be considered, so that one set of μ, η and ω global reactivity indexes will be expected to be obtained according to that conformation. As each expression has to be created, a method of mathematical proof as mathematical induction will be used in order to check the validity of the deduced expressions.

3.1. Building the Chemical Potential

Defining μ_0 as the obtained chemical potential of cationic species during the activation step of the polymerization $Cp_2MCH_3^+$ and because this cation has not been coordinated yet with any ethylene molecule, the zero subscript on reactivity indexes will be used for this species. In the case of any ethylene molecule C_2H_4, its respective chemical potential is μ_{et}.

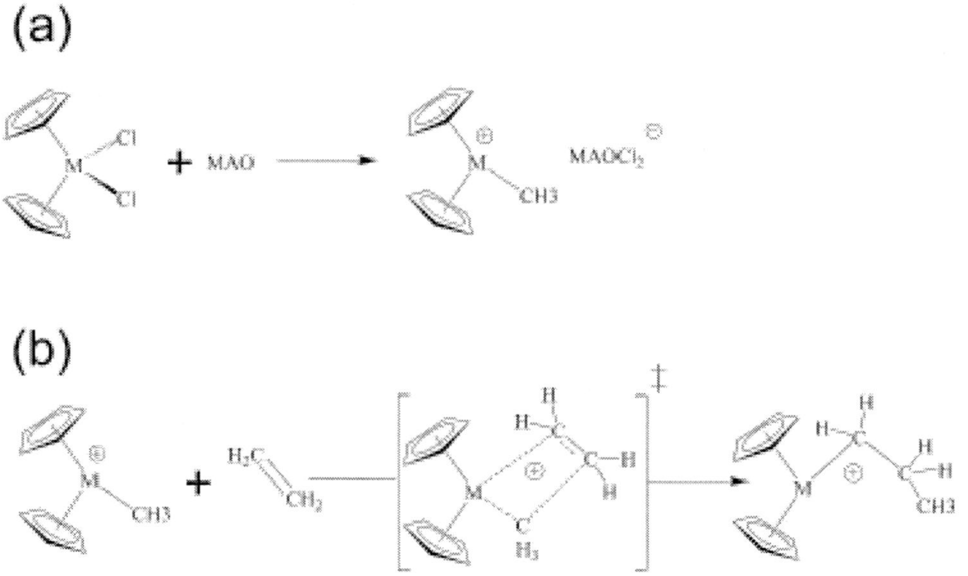

Figure 1. (a) Activation of metallocene catalyst by means of MAO, methylalumoxane; (b) ethylene coordination to start the polymerization process.

Since the polymerization process by coordination implies a successive addition of ethylene molecules, an additive expression for the growing alkyl chain will be written based on the Sanderson's principle.

According to the Sanderson's principle, the very first coordination between one cationic catalytic species and one ethylene molecule will produce the product of the model reaction for the initiation step of this polymerization; this coordination is indicated by Figure 1 (b).

The resulting chemical potential will be labelled as $\bar{\mu}_1$. As new ethylene molecules are incorporated into the cationic species, the alkyl chain will increase its length as depicted by Figure 2. The product of coordination of a second ethylene molecule has two new carbon atoms so that it will have a different chemical potential in comparison with the separated initial cation and ethylene molecule. After n additions of ethylene molecules, a $\bar{\mu}_n$ chemical potential will be obtained. A convergent value is expected for the chemical potential. As the alkyl chain can adopt multiple conformations, it will be considered only a straight conformation as Figure 3 indicates in order to avoid energetic changes given by conformational changes.

In order to assess the possibility to use the additivity scheme given by the Sanderson's principle to obtain the chemical potential of a polymeric chain as mentioned before, a convergence analysis on this reactivity index will be carried out as the polymeric chain grows. In the case of polymerization by coordination, the Equation (8) reduces to a two–term prod-

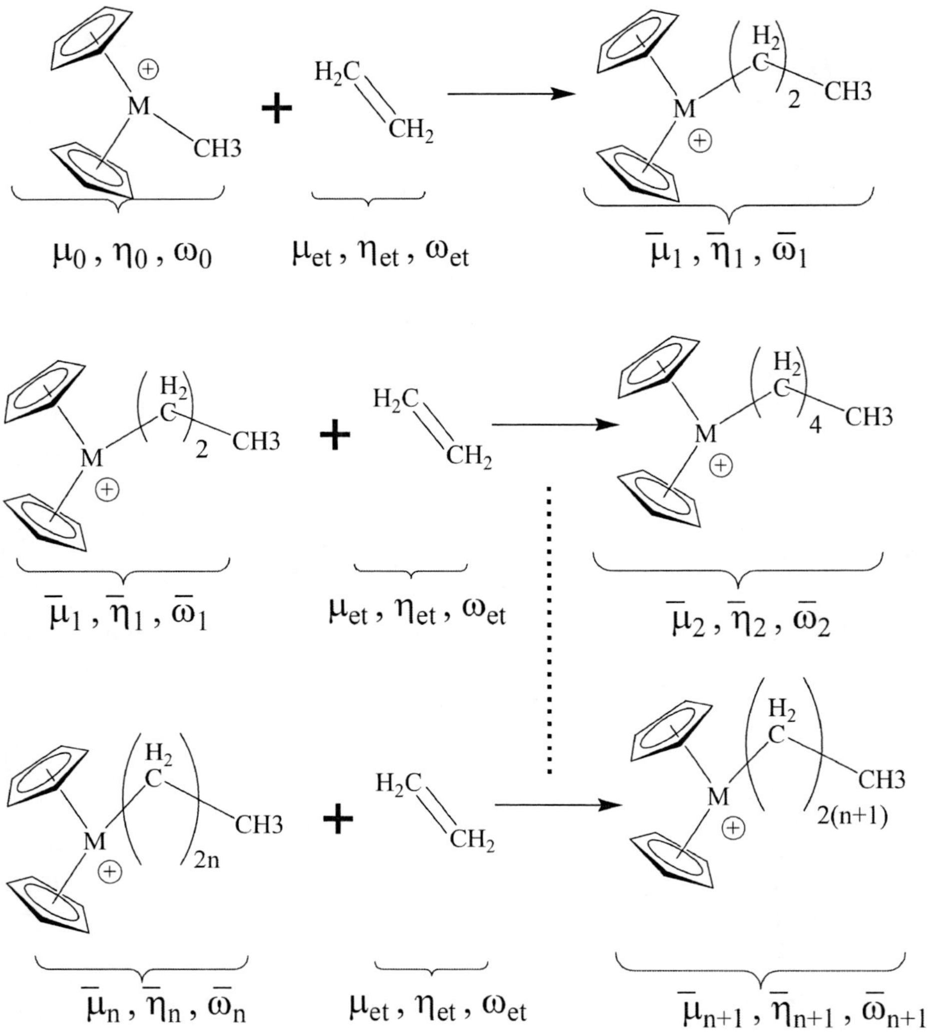

Figure 2. Polymerization process.

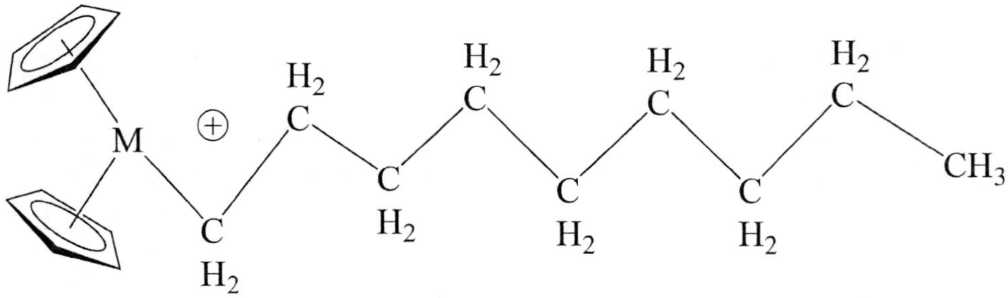

Figure 3. A model of a straight alkyl chain increasing its length.

uct:

$$\begin{aligned}
\overline{\mu}_1 &= -\{|\mu_0| \cdot |\mu_{et}|\}^{\frac{1}{2}} \\
\overline{\mu}_2 &= -\{|\overline{\mu}_1| \cdot |\mu_{et}|\}^{\frac{1}{2}} \\
\overline{\mu}_3 &= -\{|\overline{\mu}_2| \cdot |\mu_{et}|\}^{\frac{1}{2}} \\
\ldots &= \ldots \\
\overline{\mu}_{k+1} &= -\{|\overline{\mu}_k| \cdot |\mu_{et}|\}^{\frac{1}{2}}
\end{aligned}$$

where $k + 1$ is the number of inserted ethylene molecules into the cationic species thus producing the cationic species $Cp_2M(CH_2)_{2(k+1)}CH_3^+$, so that $\overline{\mu}_0 \equiv \mu_0$. It is important to establish this identity because $\overline{\mu}_0$ comes from quantum mechanical calculations or from experimental measurements, but not from the additivity scheme. Replacing the mathematical expression for obtaining $\overline{\mu}_1$ into the expression $\overline{\mu}_2$, the following is obtained:

$$\begin{aligned}
\overline{\mu}_2 &= -\{|\overline{\mu}_1| \cdot |\mu_{et}|\}^{\frac{1}{2}} \\
\overline{\mu}_2 &= -\left\{\left| -(|\mu_0| \cdot |\mu_{et}|)^{\frac{1}{2}} \right| \cdot |\mu_{et}|\right\}^{\frac{1}{2}} \\
\overline{\mu}_2 &= -\left\{\left| (|\mu_0| \cdot |\mu_{et}|)^{\frac{1}{2}} \right| \cdot |\mu_{et}|\right\}^{\frac{1}{2}} \\
\overline{\mu}_2 &= -\left\{\left| (|\mu_0| \cdot |\mu_{et}|)^{\frac{1}{4}} \right| \cdot |\mu_{et}|^{\frac{1}{2}}\right\} \\
\overline{\mu}_2 &= -\left\{\left| \left(|\mu_0|^{\frac{1}{4}} \cdot |\mu_{et}|^{\frac{1}{4}}\right) \right| \cdot |\mu_{et}|^{\frac{1}{2}}\right\} \\
\overline{\mu}_2 &= -\left\{\left||\mu_0|^{\frac{1}{4}} \cdot |\mu_{et}|^{\frac{1}{4}}\right| \cdot |\mu_{et}|^{\frac{1}{2}}\right\} \\
\overline{\mu}_2 &= -\left\{\left||\mu_0|^{\frac{1}{4}}\right| \cdot \left||\mu_{et}|^{\frac{1}{4}}\right| \cdot |\mu_{et}|^{\frac{1}{2}}\right\} \\
\overline{\mu}_2 &= -\left\{|\mu_0|^{\frac{1}{4}} \cdot |\mu_{et}|^{\frac{1}{4}} \cdot |\mu_{et}|^{\frac{1}{2}}\right\} \\
\overline{\mu}_2 &= -\left\{|\mu_0|^{\frac{1}{4}} \cdot |\mu_{et}|^{\frac{1}{2}+\frac{1}{4}}\right\}.
\end{aligned}$$

The same algebraic analysis is developed to obtain $\overline{\mu}_3$:

$$\overline{\mu}_3 = -\left\{|\mu_0|^{\frac{1}{8}} \cdot |\mu_{et}|^{\frac{1}{2}+\frac{1}{4}+\frac{1}{8}}\right\}.$$

This suggests the following expression:

$$\overline{\mu}_n = -\left\{|\mu_0|^{\frac{1}{2^n}} \cdot |\mu_{et}|^{\frac{1}{2}+\frac{1}{2^2}+\frac{1}{2^3}+\cdots\frac{1}{2^n}}\right\}, \tag{10}$$

which must be demonstrated through a proof by induction. If the Equation (10) for obtaining $\overline{\mu}_n$ is a true expression, the following suggested identity must be satisfied:

$$\overline{\mu}_{n+1} = -\left\{|\mu_0|^{\frac{1}{2^{n+1}}} \cdot |\mu_{\rm et}|^{\frac{1}{2}+\frac{1}{2^2}+\frac{1}{2^3}+\cdots\frac{1}{2^{n+1}}}\right\},$$

where $\overline{\mu}_{n+1} \equiv -\{|\overline{\mu}_n| \cdot |\mu_{\rm et}|\}^{\frac{1}{2}}$. By using Equation (8):

$$\overline{\mu}_n = -\left\{|\mu_0|^{\frac{1}{2^n}} \cdot |\mu_{\rm et}|^{\frac{1}{2}+\frac{1}{2^2}+\frac{1}{2^3}+\cdots\frac{1}{2^n}}\right\} /|\ |$$

$$|\overline{\mu}_n| = |\mu_0|^{\frac{1}{2^n}} \cdot |\mu_{\rm et}|^{\frac{1}{2}+\frac{1}{2^2}+\frac{1}{2^3}+\cdots\frac{1}{2^n}} / \cdot |\mu_{\rm et}|$$

$$|\overline{\mu}_n| \cdot |\mu_{\rm et}| = |\mu_0|^{\frac{1}{2^n}} \cdot |\mu_{\rm et}|^{\frac{1}{2}+\frac{1}{2^2}+\frac{1}{2^3}+\cdots\frac{1}{2^n}} \cdot |\mu_{\rm et}|$$

$$|\overline{\mu}_n| \cdot |\mu_{\rm et}| = |\mu_0|^{\frac{1}{2^n}} \cdot |\mu_{\rm et}|^{1+\frac{1}{2}+\frac{1}{2^2}+\frac{1}{2^3}+\cdots\frac{1}{2^n}}$$

$$|\overline{\mu}_n| \cdot |\mu_{\rm et}| = |\mu_0|^{\frac{1}{2^n}} \cdot |\mu_{\rm et}|^{1+\frac{1}{2}+\frac{1}{2^2}+\frac{1}{2^3}+\cdots\frac{1}{2^n}} /(\ )^{\frac{1}{2}}$$

$$|\overline{\mu}_n|^{\frac{1}{2}} \cdot |\mu_{\rm et}|^{\frac{1}{2}} = |\mu_0|^{\frac{1}{2^{n+1}}} \cdot |\mu_{\rm et}|^{\frac{1}{2}+\frac{1}{2^2}+\frac{1}{2^3}+\cdots\frac{1}{2^n}+\frac{1}{2^{n+1}}}$$

$$|\overline{\mu}_n|^{\frac{1}{2}} \cdot |\mu_{\rm et}|^{\frac{1}{2}} = |\mu_0|^{\frac{1}{2^{n+1}}} \cdot |\mu_{\rm et}|^{\frac{1}{2}+\frac{1}{2^2}+\frac{1}{2^3}+\cdots\frac{1}{2^n}+\frac{1}{2^{n+1}}} / \cdot (-1)$$

$$-|\overline{\mu}_n|^{\frac{1}{2}} \cdot |\mu_{\rm et}|^{\frac{1}{2}} = -\left\{|\mu_0|^{\frac{1}{2^{n+1}}} \cdot |\mu_{\rm et}|^{\frac{1}{2}+\frac{1}{2^2}+\frac{1}{2^3}+\cdots\frac{1}{2^n}+\frac{1}{2^{n+1}}}\right\}$$

$$-\{|\overline{\mu}_n| \cdot |\mu_{\rm et}|\}^{\frac{1}{2}} = -\left\{|\mu_0|^{\frac{1}{2^{n+1}}} \cdot |\mu_{\rm et}|^{\frac{1}{2}+\frac{1}{2^2}+\frac{1}{2^3}+\cdots\frac{1}{2^n}+\frac{1}{2^{n+1}}}\right\}.$$

Replacing $\overline{\mu}_{n+1}$ by $-\{|\overline{\mu}_n| \cdot |\mu_{\rm et}|\}^{\frac{1}{2}}$, it has been demonstrated that:

$$\overline{\mu}_{n+1} = -\left\{|\mu_0|^{\frac{1}{2^{n+1}}} \cdot |\mu_{\rm et}|^{\frac{1}{2}+\frac{1}{2^2}+\frac{1}{2^3}+\cdots\frac{1}{2^n}+\frac{1}{2^{n+1}}}\right\}$$

After n additions of ethylene molecules into the initial cationic species $Cp_2MCH_3^+$, a final cationic species is obtained $Cp_2M(CH_2)_{2n}CH_3^+$; its chemical potential will be given by the Equation (10).

The exponent of $|\mu_{\rm et}|$ from the Equation (10) is a geometric convergent series:

$$S = \frac{1}{2} + \frac{1}{2^2} + \frac{1}{2^3} + \cdots \frac{1}{2^{n-1}} + \frac{1}{2^n},$$

so that by multiplying S by 2, the following is obtained:

$$2S = 1 + \frac{1}{2} + \frac{1}{2^2} + \cdots \frac{1}{2^{n-1}},$$

and the substraction $2S - S$ allows one to obtain:

$$S = 1 - \frac{1}{2^n} \equiv 1 - 2^{-n},$$

so that Equation (11) is obtained:

$$\overline{\mu}_n = -|\mu_0|^{2^{-n}} \cdot |\mu_{\text{et}}|^{1-2^{-n}} \tag{11}$$

Now the convergence of this expression will be analyzed.
Successive additions of ethylene into the catalytic cationic should lead into a constant chemical potential value, thus indicating that an intrinsic reactivity property has been reached for the polymeric chain. That can be obtained by taking the limit of Equation (11) when n tends to be infinite:

$$\lim_{n \longrightarrow \infty} \overline{\mu}_n = -|\mu_{\text{et}}|,$$

$$\lim_{n \longrightarrow \infty} \overline{\mu}_n = \mu_{\text{et}}. \tag{12}$$

Although the Equation (11) is convergent, its limit value corresponds to the chemical potential of the ethylene molecule which is a wrong value; after each insertion of one ethylene molecule into the catalyst, the ethylene molecule is turned into an alkyl fragment forming the alkyl chain. The limit value is consistent according to the expected value from a geometric mean given by Equation (8), but it does not make sense physically.

3.2. Building the Molecular Hardness

The molecular hardness is related to the polarizability [27]. When a molecular system is formed by n_f atoms (or fragments), it is expressed as Equation (9) indicates, where $\overline{\mu}$ is given by Equation (8). In the case of the proposed polymeric reaction by coordination, after each insertion of an ethylene molecule, the molecular hardness changes similar to the chemical potential changes:

$$\begin{aligned}
\overline{\eta}_1 &= \frac{\overline{\mu}_1}{2} \cdot \left\{ \frac{\overline{\eta}_0}{\overline{\mu}_0} + \frac{\eta_{\text{et}}}{\mu_{\text{et}}} \right\} \\
\overline{\eta}_2 &= \frac{\overline{\mu}_2}{2} \cdot \left\{ \frac{\overline{\eta}_1}{\overline{\mu}_1} + \frac{\eta_{\text{et}}}{\mu_{\text{et}}} \right\} \\
\overline{\eta}_3 &= \frac{\overline{\mu}_3}{2} \cdot \left\{ \frac{\overline{\eta}_2}{\overline{\mu}_2} + \frac{\eta_{\text{et}}}{\mu_{\text{et}}} \right\} \\
\ldots &= \ldots \\
\overline{\eta}_{k+1} &= \frac{\overline{\mu}_{k+1}}{2} \cdot \left\{ \frac{\overline{\eta}_k}{\overline{\mu}_k} + \frac{\eta_{\text{et}}}{\mu_{\text{et}}} \right\},
\end{aligned}$$

where $k+1$ has the same meaning as explained for obtaining the chemical potential. Figure 2 indicates that the input conditions are given by $\overline{\mu}_0 \equiv \mu_0$ and $\overline{\eta}_0 \equiv \eta_0$, the chemical potential and the molecular hardness of the catalyst species, respectively.

The index $\overline{\mu}_1$ given by Equation (11) is used in Equation (9) to obtain $\overline{\eta}_1$ so that a new expression in terms of μ_0, η_0, μ_{et} and η_{et} is obtained:

$$\begin{aligned}
\overline{\eta}_1 &= \frac{\overline{\mu}_1}{2} \cdot \left\{ \frac{\overline{\eta}_0}{\overline{\mu}_0} + \frac{\eta_{et}}{\mu_{et}} \right\} \\
\overline{\eta}_1 &= \frac{-|\mu_0|^{\frac{1}{2^1}} \cdot |\mu_{et}|^{1-\frac{1}{2^1}}}{2} \cdot \left\{ \frac{\overline{\eta}_0}{\overline{\mu}_0} + \frac{\eta_{et}}{\mu_{et}} \right\} \\
\overline{\eta}_1 &= \frac{-|\mu_0|^{\frac{1}{2}} \cdot |\mu_{et}|^{\frac{1}{2}}}{2} \cdot \left\{ \frac{\overline{\eta}_0}{\overline{\mu}_0} + \frac{\eta_{et}}{\mu_{et}} \right\} \\
\overline{\eta}_1 &= -\frac{1}{2}|\mu_0|^{\frac{1}{2}} \cdot |\mu_{et}|^{\frac{1}{2}} \cdot \left\{ \frac{\overline{\eta}_0}{\overline{\mu}_0} + \frac{\eta_{et}}{\mu_{et}} \right\} .
\end{aligned}$$

Since the chemical potential is always a negative quantity, the identity $\mu \equiv -|\mu|$ is satisfied without restrictions. Replacing μ_0 by $-|\mu_0|$ and μ_{et} by $-|\mu_{et}|$ along with the definitions $\overline{\mu}_0 \equiv \mu_0$ y $\overline{\eta}_0 \equiv \eta_0$ then the following algebraic development is carried out:

$$\begin{aligned}
\overline{\eta}_1 &= -\frac{1}{2}|\mu_0|^{\frac{1}{2}} \cdot |\mu_{et}|^{\frac{1}{2}} \cdot \left\{ \frac{\eta_0}{-|\mu_0|} + \frac{\eta_{et}}{-|\mu_{et}|} \right\} \\
\overline{\eta}_1 &= -\frac{1}{2}|\mu_0|^{\frac{1}{2}} \cdot |\mu_{et}|^{\frac{1}{2}} \cdot (-1) \cdot \left\{ \frac{\eta_0}{|\mu_0|} + \frac{\eta_{et}}{|\mu_{et}|} \right\}
\end{aligned}$$

$$\begin{aligned}
\overline{\eta}_1 &= \frac{1}{2}|\mu_0|^{\frac{1}{2}} \cdot |\mu_{et}|^{\frac{1}{2}} \cdot \left\{ \frac{\eta_0}{|\mu_0|} + \frac{\eta_{et}}{|\mu_{et}|} \right\} \\
\overline{\eta}_1 &= \frac{1}{2} \cdot \left\{ \frac{\eta_0 \cdot |\mu_0|^{\frac{1}{2}} \cdot |\mu_{et}|^{\frac{1}{2}}}{|\mu_0|} + \frac{\eta_{et} \cdot |\mu_0|^{\frac{1}{2}} \cdot |\mu_{et}|^{\frac{1}{2}}}{|\mu_{et}|} \right\} \\
\overline{\eta}_1 &= \frac{1}{2} \cdot \left\{ \frac{\eta_0 \cdot |\mu_{et}|^{\frac{1}{2}}}{|\mu_0|^{\frac{1}{2}}} + \frac{\eta_{et} \cdot |\mu_0|^{\frac{1}{2}}}{|\mu_{et}|^{\frac{1}{2}}} \right\} \\
\overline{\eta}_1 &= \frac{1}{2} \cdot \eta_0 \cdot |\mu_{et}|^{\frac{1}{2}} \cdot |\mu_0|^{-\frac{1}{2}} + \frac{1}{2} \cdot \eta_{et} \cdot |\mu_0|^{\frac{1}{2}} \cdot |\mu_{et}|^{-\frac{1}{2}}
\end{aligned}$$

Similarly the reader can check that an expression for $\overline{\eta}_2$ is obtained by following the same procedure:

$$\overline{\eta}_2 = \frac{1}{2^2} \cdot \eta_0 \cdot |\mu_{et}|^{\frac{1}{2}+\frac{1}{2^2}} \cdot |\mu_0|^{-(\frac{1}{2}+\frac{1}{2^2})} + \left(\frac{1}{2} + \frac{1}{2^2}\right) \cdot \eta_{et} \cdot |\mu_0|^{\frac{1}{2^2}} \cdot |\mu_{et}|^{-\frac{1}{2^2}} .$$

thus suggesting a pattern to be used in building up the molecular hardness for this system that depends on n ethylene molecules used in forming the polymeric alkyl chain. The proposed expression is given by:

$$\overline{\eta}_n = \frac{1}{2^n}\cdot\eta_0\cdot|\mu_{\text{et}}|^{\frac{1}{2}+\frac{1}{2^2}+\cdots\frac{1}{2^n}}\cdot|\mu_0|^{-(\frac{1}{2}+\frac{1}{2^2}+\cdots\frac{1}{2^n})}+\left(\frac{1}{2}+\frac{1}{2^2}+\cdots\frac{1}{2^n}\right)\cdot\eta_{\text{et}}\cdot|\mu_0|^{\frac{1}{2^n}}\cdot|\mu_{\text{et}}|^{-\frac{1}{2^n}}$$

$$\overline{\eta}_n = \frac{1}{2^n}\cdot\eta_0\cdot|\mu_{\text{et}}|^{1-\frac{1}{2^n}}\cdot|\mu_0|^{-(1-\frac{1}{2^n})}+\left(1-\frac{1}{2^n}\right)\cdot\eta_{\text{et}}\cdot|\mu_0|^{\frac{1}{2^n}}\cdot|\mu_{\text{et}}|^{-\frac{1}{2^n}}$$

$$\overline{\eta}_n = 2^{-n}\cdot\eta_0\cdot|\mu_{\text{et}}|^{1-2^{-n}}\cdot|\mu_0|^{-(1-2^{-n})}+(1-2^{-n})\cdot\eta_{\text{et}}\cdot|\mu_0|^{2^{-n}}\cdot|\mu_{\text{et}}|^{-2^{-n}}$$

$$\overline{\eta}_n = 2^{-n}\cdot\eta_0\cdot|\mu_{\text{et}}|^{1-2^{-n}}\cdot|\mu_0|^{2^{-n}-1}+(1-2^{-n})\cdot\eta_{\text{et}}\cdot|\mu_0|^{2^{-n}}\cdot|\mu_{\text{et}}|^{-2^{-n}} \tag{13}$$

which must be demonstrated through a proof by induction too. If Equation (13) to obtain $\overline{\eta}_n$ is a true expression, the following identity must be satisfied:

$$\overline{\eta}_{n+1} = \frac{1}{2^{n+1}}\cdot\eta_0\cdot|\mu_{\text{et}}|^{\frac{1}{2}+\frac{1}{2^2}+\cdots\frac{1}{2^{n+1}}}\cdot|\mu_0|^{-(\frac{1}{2}+\frac{1}{2^2}+\cdots\frac{1}{2^{n+1}})}+ \left(\frac{1}{2}+\frac{1}{2^2}+\cdots\frac{1}{2^{n+1}}\right)\cdot\eta_{\text{et}}\cdot|\mu_0|^{\frac{1}{2^{n+1}}}\cdot|\mu_{\text{et}}|^{-\frac{1}{2^{n+1}}}$$

$$\overline{\eta}_{n+1} = \frac{1}{2^{n+1}}\cdot\eta_0\cdot|\mu_{\text{et}}|^{1-\frac{1}{2^{n+1}}}\cdot|\mu_0|^{-(1-\frac{1}{2^{n+1}})}+ \left(1-\frac{1}{2^{n+1}}\right)\cdot\eta_{\text{et}}\cdot|\mu_0|^{\frac{1}{2^{n+1}}}\cdot|\mu_{\text{et}}|^{-\frac{1}{2^{n+1}}}$$

$$\overline{\eta}_{n+1} = 2^{-(n+1)}\cdot\eta_0\cdot|\mu_{\text{et}}|^{1-2^{-(n+1)}}\cdot|\mu_0|^{-1+2^{-(n+1)})}+ \left[1-2^{-(n+1)}\right]\cdot\eta_{\text{et}}\cdot|\mu_0|^{2^{-(n+1)}}\cdot|\mu_{\text{et}}|^{-2^{-(n+1)}}.$$

The reader should remember that

$$\overline{\eta}_{n+1}\equiv\frac{\overline{\mu}_{n+1}}{2}\cdot\left\{\frac{\overline{\eta}_n}{\overline{\mu}_n}+\frac{\eta_{\text{et}}}{\mu_{\text{et}}}\right\}.$$

Since this is a more complicated algebraic expression than the chemical potential, an expression for $\frac{\overline{\eta}_n}{\overline{\mu}_n}$ will be obtained first. Second, the result of a summation between such a quotient with the quotient $\frac{\eta_{\text{et}}}{\mu_{\text{et}}}$ will be multiplied by $\frac{\overline{\mu}_{n+1}}{2}$. In addition, since the Equation (11) has been demonstrated already, it can be used during this demonstration.
Then $\frac{\overline{\eta}_n}{\overline{\mu}_n}$ must be found:

$$\frac{\overline{\eta}_n}{\overline{\mu}_n} = \frac{\frac{1}{2^n}\cdot\eta_0\cdot|\mu_{\text{et}}|^{\frac{1}{2}+\frac{1}{2^2}+\cdots\frac{1}{2^n}}\cdot|\mu_0|^{-(\frac{1}{2}+\frac{1}{2^2}+\cdots\frac{1}{2^n})}+\left(\frac{1}{2}+\frac{1}{2^2}+\cdots\frac{1}{2^n}\right)\cdot\eta_{\text{et}}\cdot|\mu_0|^{\frac{1}{2^n}}\cdot|\mu_{\text{et}}|^{-\frac{1}{2^n}}}{-|\mu_0|^{2^{-n}}\cdot|\mu_{\text{et}}|^{1-2^{-n}}}$$

$$\frac{\overline{\eta}_n}{\overline{\mu}_n} = \frac{\frac{1}{2^n}\cdot\eta_0\cdot|\mu_{\text{et}}|^{\frac{1}{2}+\frac{1}{2^2}+\cdots\frac{1}{2^n}}\cdot|\mu_0|^{-(\frac{1}{2}+\frac{1}{2^2}+\cdots\frac{1}{2^n})}}{-|\mu_0|^{2^{-n}}\cdot|\mu_{\text{et}}|^{1-2^{-n}}} + \frac{\left(\frac{1}{2}+\frac{1}{2^2}+\cdots\frac{1}{2^n}\right)\cdot\eta_{\text{et}}\cdot|\mu_0|^{\frac{1}{2^n}}\cdot|\mu_{\text{et}}|^{-\frac{1}{2^n}}}{-|\mu_0|^{2^{-n}}\cdot|\mu_{\text{et}}|^{1-2^{-n}}},$$

the convergent series is taken into account too: $\frac{1}{2}+\frac{1}{2^2}+\frac{1}{2^3}+\cdots\frac{1}{2^{n-1}}+\frac{1}{2^n}$. By replacing:

$$\frac{\overline{\eta}_n}{\overline{\mu}_n} = \frac{\frac{1}{2^n}\cdot\eta_0\cdot|\mu_{\text{et}}|^{1-\frac{1}{2^n}}\cdot|\mu_0|^{-(1-\frac{1}{2^n})}}{-|\mu_0|^{2^{-n}}\cdot|\mu_{\text{et}}|^{1-2^{-n}}} + \frac{\left(1-\frac{1}{2^n}\right)\cdot\eta_{\text{et}}\cdot|\mu_0|^{\frac{1}{2^n}}\cdot|\mu_{\text{et}}|^{-\frac{1}{2^n}}}{-|\mu_0|^{2^{-n}}\cdot|\mu_{\text{et}}|^{1-2^{-n}}}$$

$$\frac{\overline{\eta}_n}{\overline{\mu}_n} = \frac{2^{-n}\cdot\eta_0\cdot|\mu_{\text{et}}|^{1-2^{-n}}\cdot|\mu_0|^{-(1-2^{-n})}}{-|\mu_0|^{2^{-n}}\cdot|\mu_{\text{et}}|^{1-2^{-n}}} + \frac{(1-2^{-n})\cdot\eta_{\text{et}}\cdot|\mu_0|^{2^{-n}}\cdot|\mu_{\text{et}}|^{-2^{-n}}}{-|\mu_0|^{2^{-n}}\cdot|\mu_{\text{et}}|^{1-2^{-n}}}$$

$$\begin{aligned}\frac{\overline{\eta}_n}{\overline{\mu}_n} &= -2^{-n}\cdot\eta_0\cdot|\mu_{\text{et}}|^{1-2^{-n}-1+2^{-n}}\cdot|\mu_0|^{-1+2^{-n}-2^{-n}} \\ &\quad -\left(1-2^{-n}\right)\cdot\eta_{\text{et}}\cdot|\mu_0|^{2^{-n}-2^{-n}}\cdot|\mu_{\text{et}}|^{-2^{-n}-1+2^{-n}}\end{aligned}$$

$$\frac{\overline{\eta}_n}{\overline{\mu}_n} = -2^{-n}\cdot\eta_0\cdot|\mu_{\text{et}}|^0\cdot|\mu_0|^{-1}-\left(1-2^{-n}\right)\cdot\eta_{\text{et}}\cdot|\mu_0|^0\cdot|\mu_{\text{et}}|^{-1}.$$

and secondly, the amount $\frac{\eta_{\text{et}}}{\mu_{\text{et}}}$ is added and the demonstration is ready to be continued:

$$\begin{aligned}
\frac{\overline{\eta}_n}{\overline{\mu}_n} &= -2^{-n}\cdot\eta_0\cdot|\mu_0|^{-1}-\left(1-2^{-n}\right)\cdot\eta_{\text{et}}\cdot|\mu_{\text{et}}|^{-1}/+\frac{\eta_{\text{et}}}{\mu_{\text{et}}} \\
\frac{\overline{\eta}_n}{\overline{\mu}_n}+\frac{\eta_{\text{et}}}{\mu_{\text{et}}} &= -2^{-n}\cdot\eta_0\cdot|\mu_0|^{-1}-\left(1-2^{-n}\right)\cdot\eta_{\text{et}}\cdot|\mu_{\text{et}}|^{-1}+\frac{\eta_{\text{et}}}{\mu_{\text{et}}} \\
\frac{\overline{\eta}_n}{\overline{\mu}_n}+\frac{\eta_{\text{et}}}{\mu_{\text{et}}} &= -2^{-n}\cdot\eta_0\cdot|\mu_0|^{-1}-\left(1-2^{-n}\right)\cdot\eta_{\text{et}}\cdot|\mu_{\text{et}}|^{-1}-\frac{\eta_{\text{et}}}{|\mu_{\text{et}}|} \\
\frac{\overline{\eta}_n}{\overline{\mu}_n}+\frac{\eta_{\text{et}}}{\mu_{\text{et}}} &= -2^{-n}\cdot\eta_0\cdot|\mu_0|^{-1}-\left(1-2^{-n}\right)\cdot\eta_{\text{et}}\cdot|\mu_{\text{et}}|^{-1}-\eta_{\text{et}}\cdot|\mu_{\text{et}}|^{-1} \\
\frac{\overline{\eta}_n}{\overline{\mu}_n}+\frac{\eta_{\text{et}}}{\mu_{\text{et}}} &= -2^{-n}\cdot\eta_0\cdot|\mu_0|^{-1}-\left(1-2^{-n}+1\right)\cdot\eta_{\text{et}}\cdot|\mu_{\text{et}}|^{-1} \\
\frac{\overline{\eta}_n}{\overline{\mu}_n}+\frac{\eta_{\text{et}}}{\mu_{\text{et}}} &= -2^{-n}\cdot\eta_0\cdot|\mu_0|^{-1}-\left(2-2^{-n}\right)\cdot\eta_{\text{et}}\cdot|\mu_{\text{et}}|^{-1}.
\end{aligned}$$

Finally, this expression is multiplied by $\frac{\overline{\mu}_{n+1}}{2}$ for finishing the demonstration:

$$\frac{\overline{\eta}_n}{\overline{\mu}_n} + \frac{\eta_{\text{et}}}{\mu_{\text{et}}} = -2^{-n} \cdot \eta_0 \cdot |\mu_0|^{-1} - \left(2 - 2^{-n}\right) \cdot \eta_{\text{et}} \cdot |\mu_{\text{et}}|^{-1} / \cdot \frac{\overline{\mu}_{n+1}}{2}$$

$$\frac{\overline{\mu}_{n+1}}{2} \cdot \left\{\frac{\overline{\eta}_n}{\overline{\mu}_n} + \frac{\eta_{\text{et}}}{\mu_{\text{et}}}\right\} = \frac{\overline{\mu}_{n+1}}{2} \cdot \left\{-2^{-n} \cdot \eta_0 \cdot |\mu_0|^{-1} - \left(2 - 2^{-n}\right) \cdot \eta_{\text{et}} \cdot |\mu_{\text{et}}|^{-1}\right\}$$

$$\frac{\overline{\mu}_{n+1}}{2} \cdot \left\{\frac{\overline{\eta}_n}{\overline{\mu}_n} + \frac{\eta_{\text{et}}}{\mu_{\text{et}}}\right\} = \frac{-|\mu_0|^{2^{-(n+1)}} \cdot |\mu_{\text{et}}|^{1-2^{-(n+1)}}}{2} \cdot \left\{-2^{-n} \cdot \eta_0 \cdot |\mu_0|^{-1} - \left(2 - 2^{-n}\right) \cdot \eta_{\text{et}} \cdot |\mu_{\text{et}}|^{-1}\right\}$$

$$\frac{\overline{\mu}_{n+1}}{2} \cdot \left\{\frac{\overline{\eta}_n}{\overline{\mu}_n} + \frac{\eta_{\text{et}}}{\mu_{\text{et}}}\right\} = \frac{|\mu_0|^{2^{-(n+1)}} \cdot |\mu_{\text{et}}|^{1-2^{-(n+1)}}}{2} \cdot \left\{2^{-n} \cdot \eta_0 \cdot |\mu_0|^{-1} + \left(2 - 2^{-n}\right) \cdot \eta_{\text{et}} \cdot |\mu_{\text{et}}|^{-1}\right\}$$

$$\frac{\overline{\mu}_{n+1}}{2} \cdot \left\{\frac{\overline{\eta}_n}{\overline{\mu}_n} + \frac{\eta_{\text{et}}}{\mu_{\text{et}}}\right\} = \frac{|\mu_0|^{2^{-(n+1)}} \cdot |\mu_{\text{et}}|^{1-2^{-(n+1)}}}{2} \cdot 2^{-n} \cdot \eta_0 \cdot |\mu_0|^{-1} + \frac{|\mu_0|^{2^{-(n+1)}} \cdot |\mu_{\text{et}}|^{1-2^{-(n+1)}}}{2} \cdot \left(2 - 2^{-n}\right) \cdot \eta_{\text{et}} \cdot |\mu_{\text{et}}|^{-1}$$

$$\frac{\overline{\mu}_{n+1}}{2} \cdot \left\{\frac{\overline{\eta}_n}{\overline{\mu}_n} + \frac{\eta_{\text{et}}}{\mu_{\text{et}}}\right\} = |\mu_0|^{2^{-(n+1)}} \cdot |\mu_{\text{et}}|^{1-2^{-(n+1)}} \cdot 2^{-1} \cdot 2^{-n} \cdot \eta_0 \cdot |\mu_0|^{-1} + |\mu_0|^{2^{-(n+1)}} \cdot |\mu_{\text{et}}|^{1-2^{-(n+1)}} \cdot 2^{-1} \cdot \left(2 - 2^{-n}\right) \cdot \eta_{\text{et}} \cdot |\mu_{\text{et}}|^{-1}$$

$$\frac{\overline{\mu}_{n+1}}{2} \cdot \left\{\frac{\overline{\eta}_n}{\overline{\mu}_n} + \frac{\eta_{\text{et}}}{\mu_{\text{et}}}\right\} = |\mu_0|^{2^{-(n+1)}} \cdot |\mu_{\text{et}}|^{1-2^{-(n+1)}} \cdot 2^{-(n+1)} \cdot \eta_0 \cdot |\mu_0|^{-1} + |\mu_0|^{2^{-(n+1)}} \cdot |\mu_{\text{et}}|^{1-2^{-(n+1)}} \cdot \left[1 - 2^{-(n+1)}\right] \cdot \eta_{\text{et}} \cdot |\mu_{\text{et}}|^{-1}$$

$$\frac{\overline{\mu}_{n+1}}{2} \cdot \left\{\frac{\overline{\eta}_n}{\overline{\mu}_n} + \frac{\eta_{\text{et}}}{\mu_{\text{et}}}\right\} = |\mu_0|^{2^{-(n+1)}-1} \cdot |\mu_{\text{et}}|^{1-2^{-(n+1)}} \cdot 2^{-(n+1)} \cdot \eta_0 + |\mu_0|^{2^{-(n+1)}} \cdot |\mu_{\text{et}}|^{-2^{-(n+1)}} \cdot \left[1 - 2^{-(n+1)}\right] \cdot \eta_{\text{et}}$$

$$\frac{\overline{\mu}_{n+1}}{2} \cdot \left\{\frac{\overline{\eta}_n}{\overline{\mu}_n} + \frac{\eta_{\text{et}}}{\mu_{\text{et}}}\right\} = 2^{-(n+1)} \cdot \eta_0 \cdot |\mu_{\text{et}}|^{1-2^{-(n+1)}} \cdot |\mu_0|^{2^{-(n+1)}-1} + \left[1 - 2^{-(n+1)}\right] \cdot \eta_{\text{et}} \cdot |\mu_0|^{2^{-(n+1)}} \cdot |\mu_{\text{et}}|^{-2^{-(n+1)}} .$$

Since $\overline{\eta}_{n+1} \equiv \frac{\overline{\mu}_{n+1}}{2} \cdot \left\{\frac{\overline{\eta}_n}{\overline{\mu}_n} + \frac{\eta_{\text{et}}}{\mu_{\text{et}}}\right\}$, the following expression has been demonstrated:

$$\overline{\eta}_{n+1} = 2^{-(n+1)} \cdot \eta_0 \cdot |\mu_{\text{et}}|^{1-2^{-(n+1)}} \cdot |\mu_0|^{2^{-(n+1)}-1} + \left[1 - 2^{-(n+1)}\right] \cdot \eta_{\text{et}} \cdot |\mu_0|^{2^{-(n+1)}} \cdot |\mu_{\text{et}}|^{-2^{-(n+1)}} .$$

Therefore after n additions of ethylene molecules into the cationic species $Cp_2MCH_3^+$, it is concluded that a hypothetical cationic species $Cp_2M(CH_2)_{2n}CH_3^+$ is obtained and its molecular hardness will be given by Equation (13).

Then a constant value is expected to be obtained when $n \longrightarrow \infty$:

$$\lim_{n \longrightarrow \infty} \overline{\eta}_n = \eta_{\mathrm{et}} \,. \tag{14}$$

Like the chemical potential, an asymptotic value of the molecular hardness is obtained, but it is not physically consistent because it corresponds to the ethylene's molecular hardness.

3.3. Building the Electrophilicity

An additivity scheme for obtaining ω is easier because using the definition given by $\frac{\mu^2}{2\eta}$ should provide an expression of the electrophilicity in this polymeric context, and therefore it is not necessary to carry out a proof by induction. Replacing μ by $\overline{\mu}$ from Equation (11) and η by $\overline{\eta}$ from Equation (13), the algebraic development is carried out as follows:

$$\overline{\omega}_n = \frac{(\overline{\mu}_n)^2}{2 \cdot \overline{\eta}_n}$$

$$\overline{\omega}_n = \frac{\left(-|\mu_0|^{2^{-n}} \cdot |\mu_{\mathrm{et}}|^{1-2^{-n}}\right)^2}{2 \cdot \left\{2^{-n} \cdot \eta_0 \cdot |\mu_{\mathrm{et}}|^{1-2^{-n}} \cdot |\mu_0|^{2^{-n}-1} + (1-2^{-n}) \cdot \eta_{\mathrm{et}} \cdot |\mu_0|^{2^{-n}} \cdot |\mu_{\mathrm{et}}|^{-2^{-n}}\right\}}$$

$$\overline{\omega}_n = \frac{|\mu_0|^{2^{1-n}} \cdot |\mu_{\mathrm{et}}|^{2-2^{1-n}}}{2 \cdot \left\{2^{-n} \cdot \eta_0 \cdot |\mu_{\mathrm{et}}|^{1-2^{-n}} \cdot |\mu_0|^{2^{-n}-1} + (1-2^{-n}) \cdot \eta_{\mathrm{et}} \cdot |\mu_0|^{2^{-n}} \cdot |\mu_{\mathrm{et}}|^{-2^{-n}}\right\}}$$

Both numerator and denominator are multiplied by the following product $|\mu_0|^{1-2^{-n}} \cdot |\mu_{\mathrm{et}}|^{2^{-n}}$ thus giving:

$$\overline{\omega}_n = \frac{|\mu_0|^{1+2^{-n}} \cdot |\mu_{\mathrm{et}}|^{2-2^{-n}}}{2 \cdot \{2^{-n} \cdot \eta_0 \cdot |\mu_{\mathrm{et}}| + (1-2^{-n}) \cdot \eta_{\mathrm{et}} \cdot |\mu_0|\}}$$

$$\overline{\omega}_n = \frac{|\mu_0|^{1+2^{-n}} \cdot |\mu_{\mathrm{et}}|^{2-2^{-n}}}{2^{1-n} \cdot \eta_0 \cdot |\mu_{\mathrm{et}}| + (2-2^{1-n}) \cdot \eta_{\mathrm{et}} \cdot |\mu_0|} \,. \tag{15}$$

After analyzing the convergence of Equation (15), it is concluded that it presents the same error reported in the cases of chemical potential and molecular hardness.

$$\lim_{n \longrightarrow \infty} \overline{\omega}_n = \frac{|\mu_{\mathrm{et}}|^2}{2 \cdot \eta_{\mathrm{et}}}$$

$$\lim_{n \longrightarrow \infty} \overline{\omega}_n = \frac{\mu_{\mathrm{et}}^2}{2 \cdot \eta_{\mathrm{et}}} \,. \tag{16}$$

So far Equations (11), (13) and (15) are mathematically consistent according to the scheme of additivity, but they are physically inconsistent. This problem is given by the scheme of additivity by itself because it does not include the relaxation effects which could neglected in forming little clusters of atoms or molecular aggregates[15, 3, 28] but not in polymerization processes. This failure indicates that the Sanderson's principle has a serious limitation to be used in some fields of Chemistry. However, it is not very clear how much the deviation there is when comparing true values of chemical potential, molecular hardness and electrophilicity for a long alkyl chain; these deviations and a temporary way to correct them will be shown in next section.

4. Deviations in Using the Additivity Schemes in Ethylene Polymerization Catalyzed by Metallocenes.

As demonstrated before, as the alkyl chain grows, an asymptotic behavior on electronic properties has been found for expressions to measure the μ, η and ω which are written in terms of the catalytic species and the monomer global index values along with the number of monomers n. These expressions will be compared with the true values obtained for a long alkyl chain whose values were obtained from quantum mechanical calculations. These will be considered reference values. The true dependence of μ, η and ω values on n, the number of ethylene molecules, are shown by Figure 4.

Quantum mechanical calculations have been performed on alkyl growing chains not including the complex cationic catalysts. This is because the analysis is focused on when the number of ethylene units has been increased so that the influence of a catalysts becomes negligible; whereas the use of the Sanderson scheme has been used on alkyl growing chain interacting with the complex cationic catalysts in order to explicitly use the additivity scheme and check how the catalytic influence on global reactivity indexes becomes more and more negligible. This allows a proper comparison to be established with the pure alkyl growing chain.

4.1. Chemical Potential

The limit value of μ_n estimated from pure quantum chemical calculations is given by Figure 4 (a): $-57.81\,\text{kcal}\cdot\text{mol}^{-1}$, this will be used as a reference value. Then a comparison with respect to the limit values obtained from Equation (11) will be carried out:

$$\overline{\mu}_n = -|\mu_0|^{2^{-n}} \cdot |\mu_{\text{et}}|^{1-2^{-n}}.$$

The respective profile is depicted by Figure 5 (a) as a function of n, the number of ethylene molecules that have been used to form the alkyl chain by means of the metallocene cationic species as a catalyst.

As can be seen, the trend of $\overline{\mu}_n$ is qualitatively the same trend exhibited by μ_n from Figure 4 (a) and although the three different catalytic systems have been considered (based on Ti, Zr and Hf) in this comparison, the trend is only one as expected because the longer is the alkyl chain the more negligible the catalytic species is. However, the limit value of $\overline{\mu}_n$ when $n \longrightarrow \infty$ is $\mu_{\text{et}} = -77.76\,\text{kcal}\cdot\text{mol}^{-1}$, it differs from $-57.81\,\text{kcal}\cdot\text{mol}^{-1}$, thus indicating that the additivity scheme underestimates the chemical potential.

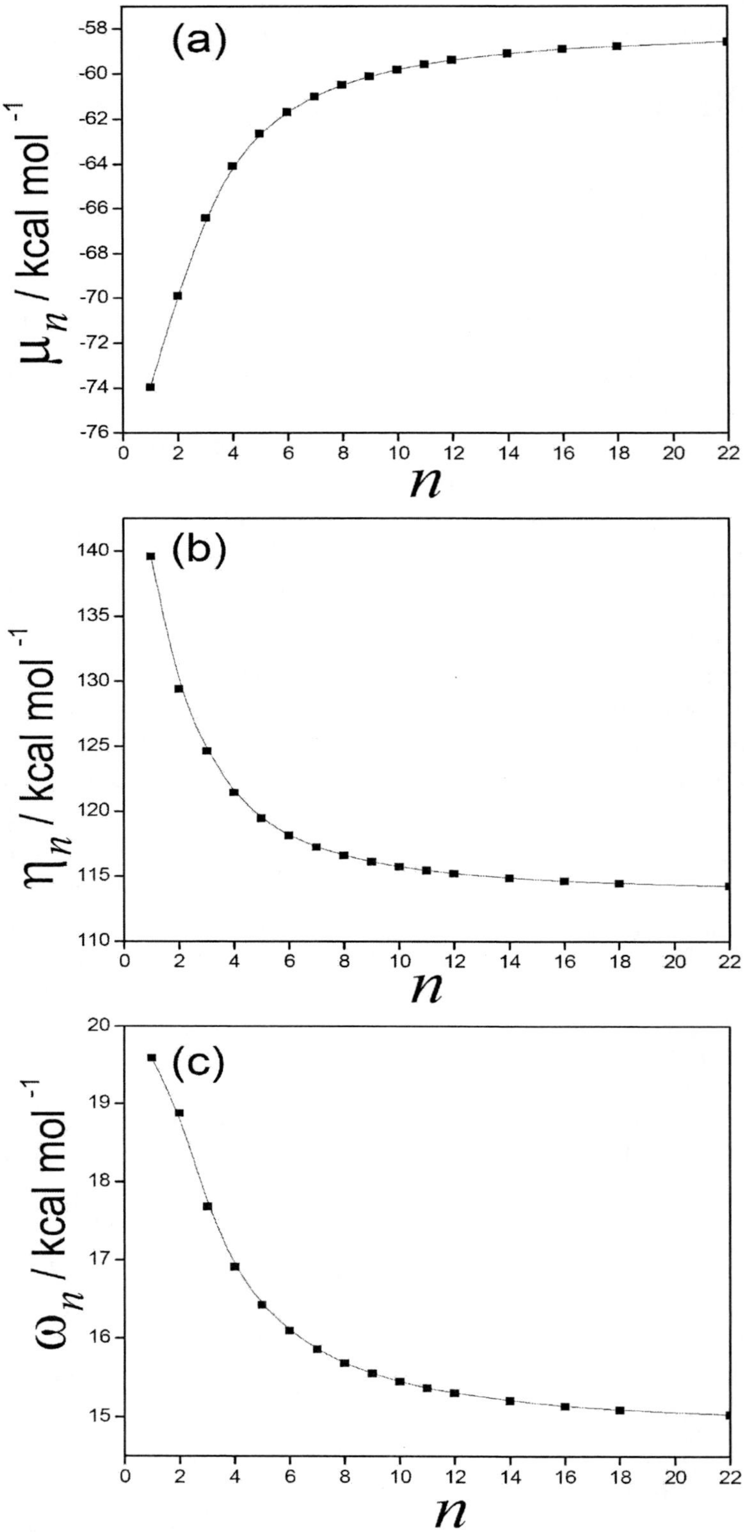

Figure 4. From quantum chemical calculations on a pure alkyl chain (a) chemical potential (b) molecular hardness (c) electrophilicity.

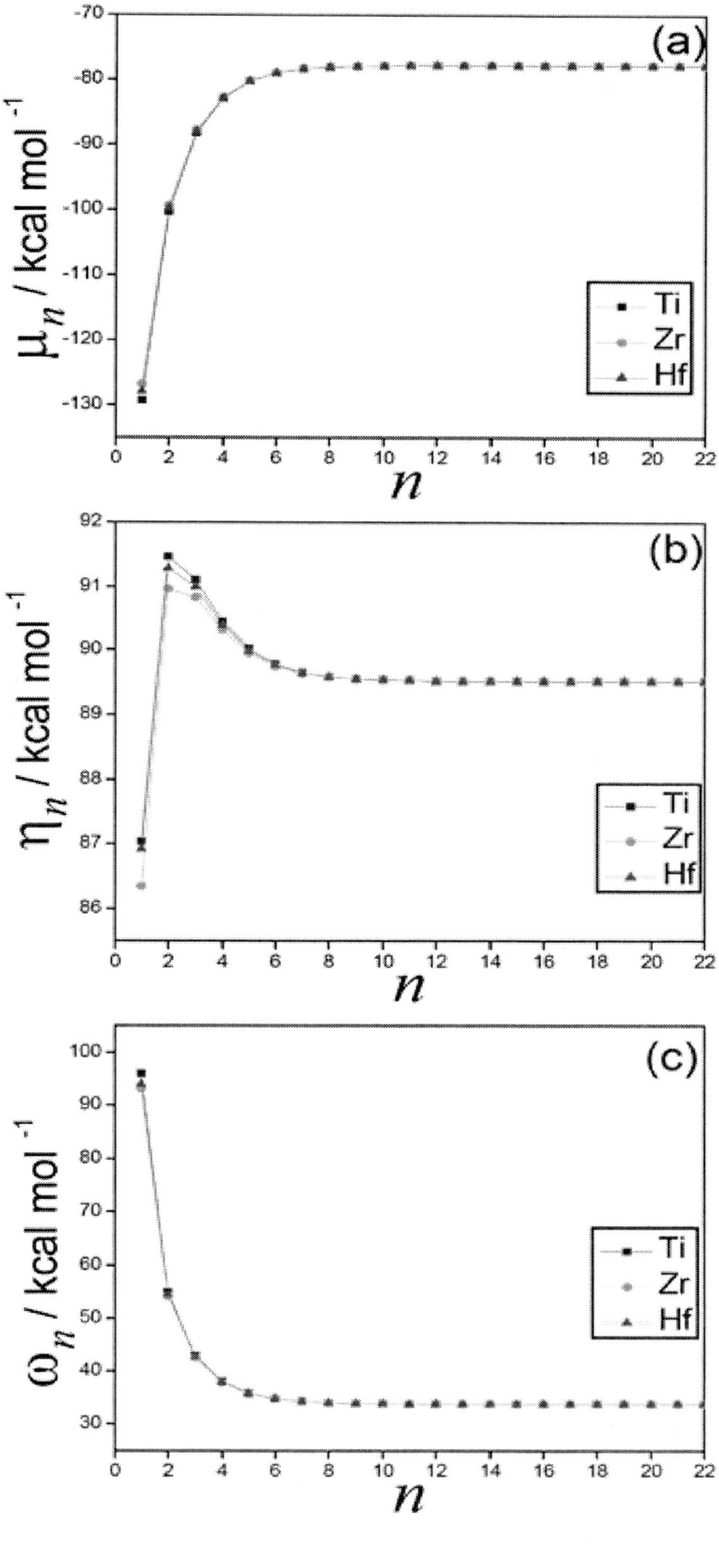

Figure 5. From the Sanderson's scheme (a) chemical potential (b) molecular hardness (c) electrophilicity.

The reader can verify the limit value given by the Equation (11). Chemical potential values for ethylene is $-77.76\,\mathrm{kcal}\cdot\mathrm{mol}^{-1}$ and for the cationic complexes $\mathrm{Cp_2MCH_3^+}$ are : -215.01 kcal·mol^{-1}; -206.79 kcal·mol^{-1} and -210.21 kcal·mol^{-1} with M=Ti,Zr and Hf, respectively.

4.2. Molecular Hardness

The η_n asymptotic value from Figure 4 (b) is: $113.40\,\mathrm{kcal}\cdot\mathrm{mol}^{-1}$ which will be considered the reference value of this global reactivity index in this polymerization process.

As demonstrated, Equation (13):

$$\overline{\eta}_n = 2^{-n}\cdot\eta_0\cdot|\mu_{\mathrm{et}}|^{1-2^{-n}}\cdot|\mu_0|^{2^{-n}-1} + \left(1-2^{-n}\right)\cdot\eta_{\mathrm{et}}\cdot|\mu_0|^{2^{-n}}\cdot|\mu_{\mathrm{et}}|^{-2^{-n}}.$$

provides the value for this global property in a growing alkyl chain on the one hand in terms of μ_0 and η_0 [the catalytic species $\mathrm{Cp_2MCH_3^+}$] and on the other hand, in terms of μ_{et} and η_{et} [the monomer ethylene $\mathrm{C_2H_4}$]. The profile of this global index is given by Figure 5 (b). Although at the beginning a slight increase is observed due to the predominant influence of each metal complex in the system, after the second ethylene insertion, the behavior of $\overline{\eta}_n$ is the same that η_n shows in Figure 4 (b). Again, for the three metallic molecular systems (Ti, Zr and Hf), this global index converges into one single value because, as explained in the case of chemical potential, the electronic properties of the alkyl chain will prevail over the electronic properties of the catalytic system as the alkyl chain increases in size. But, the observed trend from Figure 5 (b) indicates that this asymptotic value of $\overline{\eta}_n$ when $n \longrightarrow \infty$ is $\eta_{\mathrm{et}} = 89.53\,\mathrm{kcal}\cdot\mathrm{mol}^{-1}$ which is different in comparison with $113.40\,\mathrm{kcal}\cdot\mathrm{mol}^{-1}$. This indicates that the additivity scheme underestimates the molecular hardness.

The molecular hardness value for ethylene is $113.40\,\mathrm{kcal}\cdot\mathrm{mol}^{-1}$ and for the cationic complexes $\mathrm{Cp_2MCH_3^+}$ are shown as follows: 41.89 kcal·mol^{-1}; 43.52 kcal·mol^{-1} and 43.78 kcal·mol^{-1} for M=Ti,Zr and Hf, respectively.

4.3. Electrophilicity

The ω_n asymptotic value from Figure 4 (c) is: $14.54\,\mathrm{kcal}\cdot\mathrm{mol}^{-1}$, it will be considered the reference value of this third global reactivity index during the polymerization process.

Like chemical potential and molecular hardness, the electrophilicity in this polymerization process is analyzed through Equation (15). Figure 4 (c) indicates the behavior of this global reactivity index from quantum chemical calculations. This global descriptor depends on μ_0, η_0, μ_{et} and η_{et}. Figure 5 (c) shows a decrease in $\overline{\mu}_n$ until getting the expected asymptotic behavior. The obtained value is 33.77 kcal·mol^{-1}, the ethylene's electrophilicity. Clearly this is not the correct value for the polymeric chain.

After this analysis, amounts to correct deviations on these global descriptors are shown as follows in Figure 6:

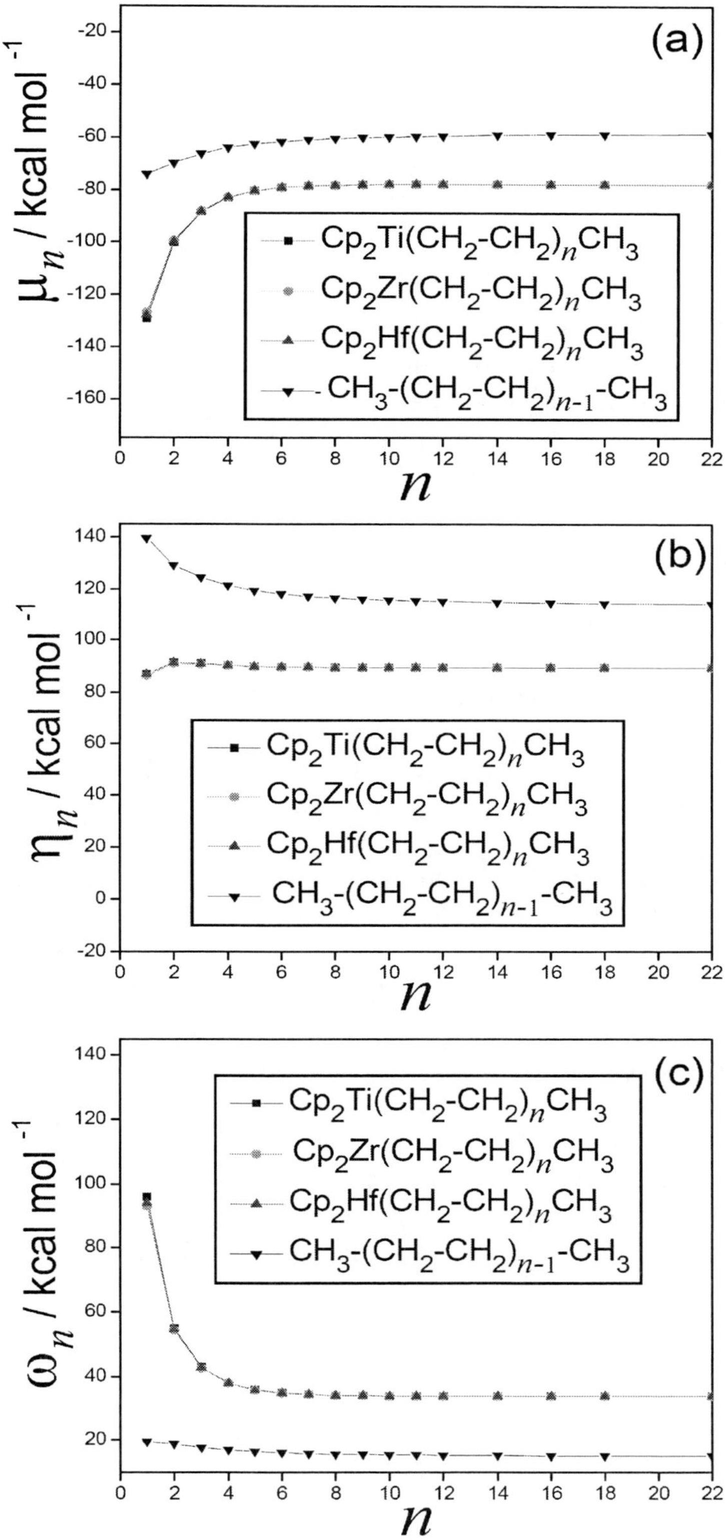

Figure 6. Comparing the Sanderson's scheme and quantum chemical calculations (a) chemical potential (b) molecular hardness (c) electrophilicity.

$$\mu_n = \overline{\mu}_n + \Delta\mu \Longrightarrow \Delta\mu = 19.95\,\text{kcal}\cdot\text{mol}^{-1}\,,$$

$$\eta_n = \overline{\eta}_n + \Delta\eta \Longrightarrow \Delta\eta = 23.87\,\text{kcal}\cdot\text{mol}^{-1}\,,$$

$$\omega_n = \overline{\omega}_n - \Delta\omega \Longrightarrow \Delta\omega = 19.23\,\text{kcal}\cdot\text{mol}^{-1}\,.$$

These amounts $\{\Delta\mu, \Delta\eta, \Delta\omega\}$ include the relaxation effects which are absent within the Sanderson's scheme. These relaxation effects are related to the formation and destruction of bonds; conformational rearrangements have not been considered because the whole analysis was done with a straight alkyl chain (extended conformation) as Figure 3 indicates. The μ and η values are underestimated when using the Sanderson's additivity scheme, whereas the electrophilicity is overestimated. These statements can be better understood for η when analyzing the polymeric chain; when its length increases, a saturated hydrocarbon chain is obtained which cannot have any π electrons in comparison with ethylene. Therefore it is expected that in this particular process the polymeric chain has a higher molecular hardness than ethylene.

Although these equations for estimating the chemical potential (11); molecular hardness (13) and electrophilicity (15) have proven to be insufficient to reproduce the respective global reactivity index value from quantum mechanical calculations at the same level of theory, they should be tested in many other polymeric processes by following the same procedure exposed here in order to find out whether a pattern to systematically correct this limitation can be established or not so generating a set of $\{\Delta\mu, \Delta\eta, \Delta\omega\}$ values to add relaxation effects. The reader should note that the main goal is not to find the exact experimental value from these equations but to rationally reproduce the respective value of reactivity from minimum information: just taking into account the cationic catalyst and the monomer at the same level of theory.

The lack of physical support observed in Equations (8) and (9) could be a reason for the failure of Equations (11), (13) and (15). If the true amount of electrons is taken into account from each fragment then a more rationalized expression should be obtained. According to that proposition, when the proportion of electrons given by each fragment is considered, the following expression is obtained for the particular case of two fragments, 1 and 2; they are interacting to form one molecular structure having a chemical potential estimated by the following expression:

$$\overline{\mu} = -|\mu_1|^{\frac{N_1}{N}} \cdot |\mu_2|^{\frac{N_2}{N}}\,,$$

where $N = N_1 + N_2$, where N_1 is the number of electrons of fragment 1 and N_2, the number of electrons of fragment 2; so that an alternative expression for Equation (8) for the case of n_f fragments is obtained:

$$\overline{\mu} = -\left\{\prod_{i=1}^{n_f} |\mu_i|^{\frac{N_i}{N}}\right\}\,, \tag{17}$$

where $N = \sum_i^{n_f} N_i$. The reader can easily verify that Equation (8) is a specific case of Equation (17) when $N_1 = N_2 = \ldots = N_{n_f}$ and then one would argue this is the desired expression because it includes a generalized geometric mean.

Another alternative expression, based on an ordinary arithmetic mean, is given as follows in the case of two fragments:

$$\overline{\mu} = -\left\{ \frac{N_1}{N} \cdot \mu_1 + \frac{N_2}{N} \cdot \mu_2 \right\} .$$

whose generalized expression can be written as follows:

$$\overline{\mu} = -\sum_i^{n_f} \frac{N_i}{N} \cdot \mu_i \tag{18}$$

As a consequence, Equations (17) and (18) take into account more physical support than Equation (8) does; however, these considerations are not relevant as these expressions are analyzed in the context of a polymerization process. This will be demonstrated as follows. From the proposed Equation (17), the following expression is obtained by following the same procedure to obtain Equation (11):

$$\overline{\mu}_n = -|\mu_0|^{\frac{N_0}{N_0 + nN_{\text{et}}}} \cdot |\mu_{\text{et}}|^{\frac{nN_{\text{et}}}{N_0 + nN_{\text{et}}}} , \tag{19}$$

where N_0 is the number of electrons provided by the cationic catalyst and N_{et} is the number of electrons of one ethylene molecule, therefore $N_0 + nN_{\text{et}}$ is the total number of electrons of the system after n ethylene additions. The limit when $n \longrightarrow \infty$ gives the electronic potential of ethylene again:

$$\lim_{n \longrightarrow \infty} \overline{\mu}_n = -|\mu_{\text{et}}| = \mu_{\text{et}} , \tag{20}$$

The same situation occurs when Equation (18) is written in a general way depending on n inserted monomers to build the alkyl chain:

$$\overline{\mu}_n = -\left\{ \frac{N_0}{N_0 + nN_{\text{et}}} |\mu_0| + n \frac{N_{\text{et}}}{N_0 + nN_{\text{et}}} |\mu_{\text{et}}| \right\} , \tag{21}$$

An analysis of convergence when $n \longrightarrow \infty$ indicates this limit exists and it corresponds to the exactly same situation. This suggests that additivity schemes should be revisited in order to solve this problem. Expressions for η and ω are algebraically much more complicated and they have not been included in this analysis, however they exhibit the same problem. The development of new additivity schemes is a pending task.

5. Conclusions

Since expressions (11), (13) and (15) are convergent, they can be corrected by adding on each a constant amount that provides relaxation effects that the Sanderson's scheme does not include. Such an amount should be characteristic for each polymeric chain. However, this is a temporary solution that should be replaced by a robust model that can reach the

same reactivity values that are obtained from quantum mechanical calculations in polymeric chains. This chapter invites the reader to search for a better expression that includes physical support to establish a proper additivity scheme for intensive properties as the chemical potential, molecular hardness and electrophilicity indexes.
Because of the purely mathematical nature of expressions (8) and (9), expressions (11), (13) and (15) are not restricted to be used in polymerization processes but also in any other process where a molecular structure A reacts with a molecular structure B following a growth pattern A–B; A–B–B; A–B–B–B and so on.

References

[1] R. T. Sanderson. *Science*, **114**: 670, 1951.

[2] R. G. Parr and L. J. Bartolotti. *J. Am. Chem. Soc.*, **104**: 3801, 1982.

[3] S. Gutiérrez-Oliva, P. Jaque, and A. Toro-Labbé. *J. Phys. Chem. A*, **104**: 8955, 2000.

[4] Y. Simón-Manso and P. Fuentealba. *J. Phys. Chem. A*, **102**: 2029, 1998.

[5] J. Garza and J. Robles. *Phys. Rev. A*, **47**: 2680, 1993.

[6] Javier Areizaga, M. Milagros Cortázar, José M. Elorza, and Juan J. Iruin. *Polímeros*. Editorial Síntesis, Madrid, 2002.

[7] R.G. Parr and W. Yang. *Density–Functional Theory of Atoms and Molecules*. Oxford Univ. Press, New York, 1989.

[8] R.G. Pearson. *J. Chem. Educ.*, **64**: 561, 1987.

[9] P. Geerlings, F. De Proft, and W. Langenaeker. *Chem. Rev.*, **103**: 1793, 2003.

[10] A. Szabo and N.S. Ostlund. *Modern Quantum Chemistry: Introduction to Advanced Electronic Structure Theory*. Dover Publications, INC., New York, 1996.

[11] T.A. Koopmans. *Physica*, **1**: 104, 1933.

[12] P. Politzer and F. Abu-Awwad. *Theor. Chem. Acc.*, **99**: 83, 1998.

[13] R.G. Parr, L. v. Szentpály, and S. Liu. *J. Am. Chem. Soc.*, **121**: 1922, 1999.

[14] R.S.J. Mülliken. *Chem. Phys.*, **2**: 782, 1934.

[15] P. Jaque and A. Toro-Labbé. *J. Phys. Chem. A*, **104**: 995, 2000.

[16] G. I. Cárdenas-Jirón, A. Toro-Labbé, and M. Solà. *J. Mol. Struct. (THEOCHEM)*, **390**: 79, 1997.

[17] H.B. Schlegel. *J. Comp. Chem.*, **3**: 214, 1982.

[18] A.D. Becke. *J. Chem. Phys.*, **98**: 5648, 1993.

[19] C. Lee, W. Yang, and R.G. Parr. *Phys. Rev. B*, **37**: 785, 1988.

[20] B. Miehlich, A. Savin, H. Stoll, and H. Preuss. *Chem. Phys. Lett.*, **157**: 200, 1989.

[21] S.H. Vosko, L. Wilk, and M. Nussair. *Can. J. Phys.*, **58**: 1200, 1980.

[22] P.C. Hariharan and J.A. Pople. *Theor. Chim. Acta*, **28**: 213, 1973.

[23] M.J. Frisch, J.A. Pople, and J.S. Binkley. *J. Chem. Phys.*, **80**: 3265, 1984.

[24] P.J. Hay and W.R. Wadt. *J. Chem. Phys.*, **82**: 270, 1985.

[25] W.R. Wadt and P.J. Hay. *J. Chem. Phys.*, **82**: 284, 1985.

[26] M.J. Frisch, G.W. Trucks, H.B. Schlegel, G.E. Scuseria, M.A. Robb, J.R. Cheeseman, J.A. Montgomery Jr., T. Vreven, K.N. Kudin, J.C. Burant, J.M. Millam, S.S. Iyengar, J. Tomasi, V. Barone, B. Mennucci, M. Cossi, G. Scalmani, N. Rega, G.A. Petersson, H. Nakatsuji, M. Hada, M. Ehara, K. Toyota, R. Fukuda, J. Hasegawa, M. Ishida, T. Nakajima, Y. Honda, O. Kitao, H. Nakai, M. Klene, X. Li, J.E. Knox, H.P. Hratchian, J.B. Cross, V. Bakken, C. Adamo, J. Jaramillo, R. Gomperts, R.E. Stratmann, O. Yazyev, A.J. Austin, R. Cammi, C. Pomelli, J.W. Ochterski, P.Y. Ayala, K. Morokuma, G.A. Voth, P. Salvador, J.J. Dannenberg, V.G. Zakrzewski, S. Dapprich, A.D. Daniels, M.C. Strain, O. Farkas, D.K. Malick, A.D. Rabuck, K. Raghavachari, J.B. Foresman, J.V. Ortiz, Q. Cui, A.G. Baboul, S. Clifford, J. Cioslowski, B.B. Stefanov, G. Liu, A. Liashenko, P. Piskorz, I. Komaromi, R.L. Martin, D.J. Fox, T. Keith, M.A. Al-Laham, C.Y. Peng, A. Nanayakkara, M. Challacombe, P.M.W. Gill, B. Johnson, W. Chen, M.W. Wong, C. Gonzalez, and J.A. Pople. Gaussian 03, Revision B.04. Gaussian, Inc., Pittsburgh, PA, 2003.

[27] S. Hati and D. Datta. *J. Phys. Chem.*, **98**: 10451, 1994.

[28] P. Jaque and A. Toro-Labbé. *J. Chem. Phys.*, **117**: 3208, 2002.

In: Mathematical Chemistry
Editor: W.I. Hong, pp. 561-605

ISBN 978-1-60876-894-3

Chapter 13

A GENERAL METHOD FOR CENTRAL POTENTIALS IN QUANTUM MECHANICS

Amlan K. Roy*
Division of Chemical Sciences, Indian Institute of Science Education and Research (IISER), Block FC, Sector III, Salt Lake, India.

Abstract

This article focuses on a recently developed generalized pseudospectral method for accurate, efficient treatment of certain central potentials of interest in various branches in quantum mechanics, usually having singularity. Essentially this allows *optimal, nonuniform* spatial discretization of the pertinent single-particle Schrödinger equation satisfying Dirichlet boundary condition leading to standard diagonalization of symmetric matrices. Its validity and feasibility have been demonstrated for a wide range of important potentials such as Hulthén, Yukawa, generalized spiked harmonic oscillators, Hellmann, Coulomb potentials without/with various perturbations (for instance, linear and quadratic) etc. Although initially designed for *singular* potentials, this has also been remarkably successful for various other cases such as power-law, logarithmic, harmonic potentials containing higher order perturbations, 3D rational potentials as well as confinement studies. Furthermore, a large number of low-, moderately high-, high-lying multiply excited Rydberg states such as singly, doubly excited He as well as triply excited *hollow* $2l2l'2l''(n \geq 2)$ and $3l3l'3l''$ *doubly-hollow* resonances in many-electron atoms have been treated by this approach within a Hohenberg-Kohn-Sham density functional theory with great success. This offers very high-quality results for *both ground and higher lying* states for arbitrary values of potential parameters (covering both *weak and strong* coupling) with equal ease and efficacy. In all cases, excellent agreement with literature results are observed; in many cases this surpasses the accuracy of all other existing results while in other occasions our results are comparable to the best ones available in literature. This provides a simple general efficient route towards the understanding of a number of singular and other potentials of relevance in quantum mechanics including multiply excited Rydberg series of many-electron systems. This helps predicting many new states for the first time.

*E-mail address: akroy@iiserkol.ac.in, akroy@chem.ucla.edu

1. Background and Motivation

Study of singular potentials in quantum mechanics is almost as old as quantum mechanics itself. Many important areas in physics and chemistry, such as atomic, molecular, solid-state, nuclear and particle physics, field theory, astrophysics, etc., frequently demand quantum mechanical solutions where the governing Hamiltonian contains certain central potential (typically having a *singularity*). Often this also includes an extra external perturbation term characterizing the physical system under investigation. *Exact* analytical solution of the respective Schrödinger equation could be obtained only for a handful of idealized, model situations, such as the harmonic oscillator or Coulomb potential, which are unfortunately quite inadequate for majority of our realistic problems. Thus, leaving aside a very few privileged cases, for almost all practical purposes, recourse must be taken to approximation methods. Consequently, an impressive amount of approximate analytical as well as numerical methods have been proposed over the decades, employing a variety of attractive elegant techniques for their studies. In general, singular potentials pose more spectacular difficulties and challenges than the regular ones. Therefore, development of a *general* method which can offer accurate reliable results on such potentials has constituted one of the most fruitful and active areas of research for long time, and still this continues to grow in time.

Because of the difficulties concerning physical interpretation of attractive singular potentials, scientific community concluded that no significance could be attached to any singular potential as regards singularity at the center of force. Also, the mathematical difficulties in tackling these potentials made these more formidable. Historically, probably the first important observation was made for well-known Coulomb potentials in a relativistic case. Plesset [1], while investigating the Dirac equation for an electron in a Coulomb field of a fictitious nucleus of charge $\alpha Z > 137$, surprisingly found that the essential distinction between attractive and repulsive potential was lost; more precisely all potentials tend to display characteristic features of an attractive potential near the origin. They both produce large and small components of wave function behaving in that region like a power of r times a factor,

$$\exp\left[\pm i \int^{r} V(r)\, dr\right],$$

which is in sharp contrast with the non-relativistic case, where $V(r)$ rather than $\sqrt{V(r)}$ appears in the exponential. The Klein-Gordon equation also shows same behavior. Non-uniqueness of the solution was also noted by Case [2], who resolved this dilemma by specifying one bound-state energy, and determining the rest of the bound-state spectrum by imposing orthogonality on wave functions.

In another stimulating paper [3] in this direction, the authors argued that physical interactions in real-world problems were more likely to be highly singular rather than regular (non-singular), and naturally their studies were more relevant than their counterparts. They also showed that singular potentials display Regge behavior in more simple terms than the regular potentials. Finding effective potentials for field-theoretical interactions by means of Bethe-Salpeter or quasi-potential equations has given further impetus to the subject. At the same time, in a pioneering work [4], a correlation between renormalizability attributes of a field theory and nature of the effective potential was established. The effective potential for super-normalizable, renormalizable and non-renormalizable theories were found to be regular, transitional and singular respectively. Further work on peratization approximation in the context of field theoretical studies of weak interactions [5, 6] has generated much interest in the study of singular potentials. In the elementary particle scattering, short-range interactions between such particles are described by repulsive singular potentials. In molecular physics as well, long-range and short-range part of the inter-atomic and inter-molecular forces are represented by various singular potentials. The long-range part includes purely electrostatic forces between polar molecules or induction forces between a polar and non-polar molecule or dispersion forces between two non-polar molecules; corresponding potential is attractive singular when extended to the origin. Whereas the short-range force develops due to an overlap of their electron clouds as atoms or molecules approach each other at shorter distances. This force is typically represented phenomenologically by a repulsive singular (or sometimes non-singular) potential, such as a Lennard-Jones potential.

The preceding examples illustrate some of the broad application areas where singular potentials serve as mathematical models for certain concepts (underlying interaction forces). Time is ripe now to mention a few words about some of the specific potentials. The literature is vast and here we restrict ourselves to only a few selective ones. One of the deceptively simple one-dimensional potentials $V(x) = -e^2/|x|$ has been studied by many workers, [7, 8, 9, 10, 11, 12, 13, 14, 15, 16, 17, 18] mainly because of following reasons: (a) exact solvability (b) unfortunate formal resemblance to its three-dimensional counterpart, the H atom, which has brought it the name "one-dimensional H atom problem" (c) diverse physical applications such as in the exciton in high-temperature superconductivity, semiconductors, polymers, 1D electron gas at the helium surface and Wigner crystal, etc. Being a function of $|x|$, this is not analytic (e.g., $x=0$ is not just a pole) and note that the independent variable spans the whole x axis including origin. Acceptable solutions must satisfy the wave equation over entire range of x. Its exact solution first appeared in 1959 [7], where two unusual features were observed for this 1D system. First, the discrete bound-state spectrum was found to be degenerate and ground state corresponded to infinite binding energy. A later work [11] employing momentum representation ascribed these rather peculiar features to the hidden $O(2)$ symmetry, which was criticized later. Thereafter, This system has been studied by a wide range of mathematical methods, such as generalized Laplace transform, Fourier transform, quantum phase space representation, momentum space representation, etc. For the past several decades, this apparently "simple" system, initially considered a pedagogical problem, has thrown some intriguing controversy among the researchers, such as the degeneracy in ground and excited states, hidden symmetry connection, etc., some of which still remain unresolved. In contrast to the 1D H atom potential, the other 1D Coulomb potential $V(x) = -Ze^2/x$ has been explored relatively less [19, 20].

Quantum mechanical solution of the celebrated 3D Coulomb problem, the H atom, was published as early as in 1926 by Schrödinger in a series of papers by solving an eigenvalue equation for energy of this system. Quantum mechanical description of H atom via a central Coulomb potential holds a unique distinction of one of the very few realistic physical systems to offer separable and exactly solvable solutions within *both* non-relativistic and relativistic picture [21]. Innumerable works have been done in the following years to obtain valuable insights into this simplest atomic system in 2,3 and N dimensions, from both mathematical and physical perspectives [22, 23, 24, 25, 26, 27, 28]. Although many textbooks on elementary quantum mechanics present H atom as a closed case, this prototypical system continues to offer many new and interesting features, such as its dynamic nature, or its evolution with time under the influence of a strong electromagnetic field or its behavior in a reference frame of arbitrary dimensions. It is now well-known that most of the 3D results actually have an N-dimensional counterpart, including Runge-Lenz vector, symmetry properties, Clebsch-Gordan coefficients, etc. Symmetry in these higher dimensional systems manifests into the separation of corresponding time-independent Schrödinger equation into a radial and angular part in hyperspherical coordinates, in distinct analogy to the spherical coordinates in 3D and polar coordinates in 2D. In other words, the angular solutions emerge as eigenfunctions of a generalized angular momentum operator, usually referred to as *hyperspherical harmonics*. Another interesting feature of the N-dimensional Coulomb problem is that its solution is connected to the D ($= 2N - 2$) dimensional harmonic oscillator potential [28]. In other words, there exists a transformation which will turn the radial equation for a generalized N-dimensional Coulomb potential into that of a D-dimensional harmonic oscillator. This link consists of a map $r^2 = \rho$ and the two are related in following manner,

$$R_{nl}(N;\rho) = \Lambda \Phi_{n',L}(D;r). \tag{1}$$

Here, the constant Λ arises due to the respective eigenfunctions being normalized in different dimensions and is given by,

$$\Lambda = \left(\frac{1}{2[n + \frac{1}{2}(N-3)]^{N+1}} \right)^{1/2}. \tag{2}$$

This equation holds only for certain values of N and D, and there are also certain relations between the respective quantum numbers that must be fulfilled, such as,

$$D = 2N - 2, \quad n' = 2(n-1), \quad L = 2l \tag{3}$$

The singular potentials of $1/r^n$, with $n \geq 2$ are of significant current interest. The $n = 2$ potential has relevance in three-body problem in nuclear physics, as well as point-dipole interactions in molecular physics [29, 30]. Historically, one of the first important difficult cases in dealing with highly singular potentials was encountered in the quantum mechanical study of a strongly attractive $1/r^2$ term in the Hamiltonian [2]. This potential shows numerous fascinating features rich in physics and mathematics. For example, being uniquely and interestingly placed in the borderline of so-called regular and singular potentials, this defines a *transition* point in non-relativistic quantum mechanics [31]. The $1/r^2$ interaction (in addition to the two-dimensional delta function), is also shown to exhibit the phenomenon of anomalous symmetry breaking, wherein a symmetry present in the system at a classical level is broken by introducing quantization in to the picture [32]. Case $n=3$ is used to describe the tensor force between nucleons in nuclear physics. Also in the perturbation theory of nuclear interactions [33], proper renormalization of this potential constitutes an important step. Interaction of an atom with a flat wall, at short distances, is governed by an attractive van der Waals potential proportional to $-1/r^3$, while at larger distances, by the highly retarded Casimir-Polder potential proportional to $-1/r^4$ [34]. $n=4$ potential also describes the interaction between a charge and an induced dipole [35].

Strictly speaking, inverse fourth-power potential is the *only true* singular potential (in the context of non-relativistic quantum mechanics, these are having a singularity at least as great as the inverse square at the center of force [31]) and one of the very few besides the afore-mentioned Coulomb (r^{-1}) as well as harmonic (r^2), Morse and Pöschl-Teller potentials, etc., to offer *exact* analytical solutions (in terms of Mathieu functions) [31, 36]. $n = 5$ corresponds to a perturbation correction to the tensor force in the nuclear potential [33]. Both $n = 6$ and 7 are connected to the London and Casimir-Polder type van der Waals forces [37]. Scattering of atoms by a conducting sphere is represented by a $-1/r^6$ potential for small distances and a $-1/r^7$ potential for large distances [38]. Inter-atomic and intermolecular forces at short distances (as strongly repulsive due to overlap of the electron clouds) are usually represented by singular potentials such as Lennard-Jones (12,6) which has a $\sim 1/r^{12}$ behavior.

The layout of the chapter is as follows. Section II gives an overview of the distinguishing characteristics of singular and regular potentials. Necessary details of the current methodology is summarized in Section III. Section IV makes a discussion on results of some central potentials (both singular and non-singular) using this method with relevance reference to the literature. Finally we end with a few concluding remarks in Section V.

2. Regular vs. Singular

This article exclusively deals with the non-relativistic quantum mechanical situation, while making only casual glances on its relativistic counterpart in some occasions. Since a majority of the potentials considered in this work are singular, it would be useful for our future discussion, to differentiate these from regular potentials. The motion of a single non-relativistic particle in presence of a spherically symmetrical potential $V(|\mathbf{r}|) \equiv V(r)$ is governed by the following time-independent Schrödinger equation in 3D space (henceforth, $\hbar = m = 1$ is assumed, unless otherwise mentioned);

$$-\frac{1}{2}\nabla^2\psi(\mathbf{r}) + [V(\mathbf{r}) - E]\psi(\mathbf{r}) = 0 \tag{4}$$

Depending on the boundary conditions imposed on such a wave function at large distances, three different types of physical situations can be associated with this equation, namely, bound, scattering and resonant-state problems. Alternatively, a solution can be prescribed at the boundary conditions at the origin. The wave function can be resolved into a sum of products of an r-dependent term and an angular term. In case of scattering, we thus have the familiar partial-wave expansion,

$$\psi(\mathbf{r}) = \sqrt{\frac{k}{\pi}}\,\frac{1}{4\pi kr}\sum_{l=0}^{\infty}(2l+1)\, i^l P_l(\cos\theta)\, u_l(r). \tag{5}$$

Where k is related to energy E by the expression $k = \sqrt{\frac{2mE}{\hbar^2}}$ and $P_l(\cos\theta)$ signifies the Legendre polynomial. For bound states of angular momentum l, only one appropriately normalized term would appear from this summation. The corresponding radial wave function $u_l(r)$ then satisfies a differential equation of the following form,

$$\left[\frac{1}{2}\frac{d^2}{dr^2} + k^2 - \frac{l(l+1)}{r^2}\right] u_l(k,r) = V(r)u_l(k,r) \tag{6}$$

Most of the studies of singular potentials eventually leads to the understanding this radial equation and its solutions.

Following [31], a potential $V(r)$ is defined as *regular* at $r=0$ if,

$$\lim_{r\to 0} r^2 V(r) = 0 \tag{7}$$

and *singular* at $r=0$ if,

$$\lim_{r\to 0} r^2 V(r) = \pm\infty \tag{8}$$

If the limiting values in (6) and (7) are finite, $V(r)$ is termed as *transition* potential. A singular potential is classified as repulsive or attractive according to whether the limiting value is, respectively, $+\infty$ or $-\infty$. The principal difference between a singular and non-singular potential lies in the fact that in the latter case, solutions of the Schrödinger equation subject to quadratic integrability condition forms a complete orthonormal set, while in the former case, solutions are too numerous and hence, over-complete.

In relativistic domain, however, the scenario is much different from the above non-relativistic situation. Much weaker infinities in the potential are "highly singular", so that within a relativistic wave equation, even a Coulomb potential is highly singular [2]. Relativistic motion leads to a different singularity criterion; *viz.*, in the vicinity of $r = 0$, these potentials give an infinite value for the limit,

$$\lim_{r\to 0} r|V(r)|, \tag{9}$$

in contrast to Eq. (8) of the non-relativistic case. The regular potentials are defined by a vanishing value of this limit, while a finite, non-zero value characterizes a transition potential, which includes all those potentials exhibiting Coulomb-like behavior at $r=0$. Finally, in contrast to the non-relativistic situation, all singular potentials behave like attractively singular in the equations of motion, in the relativistic regime.

While the repulsive singular potentials offer no outstanding problems regarding their physical interpretation, corresponding attractive case causes serious concern. Physical solutions can be determined uniquely for the former, while it is not so for the latter. From a classical point of view, a particle moving in an attractive singular potential gives rise to well-defined scattering only if certain conditions are fulfilled (such as the impact parameter exceeds certain threshold critical value). For any other motions including bounded motions, the particle falls to the origin with an infinite velocity. Furthermore, both bound or scattering trajectories are ill-defined unless trajectory tangents are matched and also energy as well as angular momentum conservation is maintained at the center of force. The scattering problem for an attractive singular potential is also not resolved in quantum picture as in the classical case, and is never defined without extraneous physical assumptions (an arbitrary phase parameter remains to be assigned). In the bound-state scenario, the attractive singular potentials offer a non-unique spectrum consisting of an infinite number of bound states with no lower bounds on the energy. For pure power-law potentials, within non-relativistic quantum mechanics, transition from regular to singular behavior begins to occur at r^{-2}, in much the same way as this happens with the Coulomb potential in the relativistic domain.

3. The Generalized Pseudospectral Method

In this article, we are concerned with the accurate bound-state solutions within non-relativistic quantum mechanics. As already mentioned, the radial Schrödinger equation for a singular potential can be solved *exactly* only in one case, inverse fourth power, where wave functions are obtained in the form of a modified Mathieu functions of generally complex arguments. Thus, approximation methods must be invoked in all other cases. A large number of approximate analytic, semi-analytic, numerical methods have been suggested over the past several decades. These are appropriately discussed in Section IV. In this section we briefly summarize the essential features of our method used in this current work in order to solve the radial eigenvalue problem. For a more detailed account on this method, see [39, 40, 41, 42, 43, 44, 45, 46, 47, 48, 49, 50, 51, 52] and the references therein.

The desired time-independent radial Schrödinger equation for a single particle in a non-relativistic case can be written as,

$$\hat{H}(r)\,\phi(r) = \varepsilon\,\phi(r). \tag{10}$$

The Hamiltonian operator includes usual kinetic and potential energy operators (symbols have their usual meanings),

$$\hat{H}(r) = -\frac{1}{2}\,\frac{d^2}{dr^2} + v(r), \tag{11}$$

with

$$v(r) = V(r) + \frac{\ell(\ell+1)}{2r^2} \tag{12}$$

and $V(r)$ is the potential in question. Generally speaking, finite-difference spatial discretization schemes often require a large number of grid points to achieve good accuracy presumably because of the fact that majority of these methods employ a uniform mesh (non-uniform schemes are used in a few occasions as well, e.g., in [53]). The generalized pseudospectral (GPS) method, however, can give *non-uniform* and optimal spatial discretization accurately, allowing one to work with a denser mesh at shorter r regions and a coarser mesh at larger r. Additionally, the GPS method is computationally orders of magnitude faster than the finite-difference or finite-element methods.

The principal feature of this scheme lies in approximating a function $f(x)$ defined in the interval $x \in [-1, 1]$ by a polynomial $f_N(x)$ of order N

$$f(x) \cong f_N(x) = \sum_{j=0}^{N} f(x_j)\,g_j(x), \tag{13}$$

such that the approximation is *exact* at *collocation points* x_j, i.e.,

$$f_N(x_j) = f(x_j). \tag{14}$$

In what follows we employ the Legendre pseudospectral method using $x_0 = -1$, $x_N = 1$, where $x_j (j = 1, \ldots, N-1)$ are obtainable from the roots of first derivatives of Legendre polynomial $P_N(x)$ with respect to x, i.e.,

$$P'_N(x_j) = 0. \tag{15}$$

$g_j(x)$ in Eq. (13), called cardinal functions, are given by the following expression,

$$g_j(x) = -\frac{1}{N(N+1)P_N(x_j)} \; \frac{(1-x^2)\, P'_N(x)}{x - x_j}. \tag{16}$$

They have the unique property $g_j(x_{j'}) = \delta_{j'j}$. Now the semi-infinite domain $r \in [0, \infty]$ is mapped onto the finite domain $x \in [-1, 1]$ by a transformation $r = r(x)$. One can make use of the following algebraic nonlinear mapping,

$$r = r(x) = L \; \frac{1+x}{1-x+\alpha}, \tag{17}$$

where L and $\alpha = 2L/r_{max}$ may be termed as the mapping parameters. Furthermore, introducing a relation of the form,

$$\psi(r(x)) = \sqrt{r'(x)} f(x) \tag{18}$$

in conjunction with a symmetrization procedure [54, 55], eventually leads to a Hamiltonian in the following transformed form as below,

$$\hat{H}(x) = -\frac{1}{2} \frac{1}{r'(x)} \frac{d^2}{dx^2} \frac{1}{r'(x)} + v(r(x)) + v_m(x), \tag{19}$$

where $v_m(x)$ is given by,

$$v_m(x) = \frac{3(r'')^2 - 2r'''r'}{8(r')^4}. \tag{20}$$

The advantage is that this leads to a *symmetric* matrix eigenvalue problem which can be readily solved to give accurate eigenvalues and eigenfunctions. For the particular transformation used in this work, $v_m(x) = 0$. This discretization then leads to following set of coupled equations,

$$\sum_{j=0}^{N} \left[-\frac{1}{2} D^{(2)}_{j'j} + \delta_{j'j}\, v(r(x_j)) + \delta_{j'j}\, v_m(r(x_j)) \right] A_j = E A_{j'}, \quad j = 1, \ldots, N-1, \tag{21}$$

where

$$A_j = \left[r'(x_j) \right]^{1/2} \psi(r(x_j)) \; [P_N(x_j)]^{-1}. \tag{22}$$

and the symmetrized second derivative of the cardinal function, $D^{(2)}_{j'j}$ is given by,

$$D^{(2)}_{j'j} = \left[r'(x_{j'}) \right]^{-1} d^{(2)}_{j'j} \left[r'(x_j) \right]^{-1}, \tag{23}$$

with

$$\begin{aligned} d^{(2)}_{j',j} &= \frac{1}{r'(x)} \frac{(N+1)(N+2)}{6(1-x_j)^2} \frac{1}{r'(x)}, \quad j = j', \\ &= \frac{1}{r'(x_{j'})} \frac{1}{(x_j - x_{j'})^2} \frac{1}{r'(x_j)}, \quad j \neq j'. \end{aligned} \tag{24}$$

It is worth mentioning here that GPS method offers both simplicity of direct finite-difference and/or finite-element method, as well as the fast convergence of finite basis set method. In this sense, one can have the cake and eat it too! It has been rigorously proved [56, 57] that the method guarantees an *exponential* (or alternatively termed, *infinite-order*) convergence for a problem of a *smooth* or *infinitely differentiable* solution (which is often the case) provided that the orthogonal functions of a common singular Sturm-Liouville problem are used. Here the 'exponential convergence' implies that the error of approximate solution decreases asymptotically faster than the algebraic decay of any order. Moreover a GPS scheme having (N+1) grid points is usually equivalent in accuracy to a corresponding basis-set expansion method involving N basis functions. Many other details pertaining to this approach could be found in the references [39, 40, 41, 42, 43, 44, 45, 46, 47, 48, 49, 50, 51, 52].

Finally, a series of test calculations were done with a variety of potentials in the literature for which exact/near-exact solutions have been reported, in order to optimize its performance with respect to the mapping parameters. In this way, the following parameter set ($r_{max}=200$, $\alpha=25$, $N=300$) has been consistently used throughout this work (unless otherwise mentioned), which seemed to be quite satisfactory for the purpose at hand.

4. Results and Discussion

This section presents a discussion on the results obtained by using GPS. As will be evident soon, this has been very successfully applied to a large variety of physical/chemical problems including static and dynamic situations. Here we select a cross-section of the most compelling and testing cases, while the rest of these could be found by inquisitive reader in the references [39, 40, 41, 42, 43, 44, 45, 46, 47, 48, 49, 50, 51, 52]. However, before we move on to the various central potentials representing different physical systems, first we give a brief snapshot of this approach for one particular potential. For this illustrative purpose, we have chosen the 3D quartic oscillator as a prototypical case. Quartic oscillators have found many notable applications. For example, two- and three-dimensional anharmonic oscillators have drawn much interest in far infra-red and microwave regions. Four-membered ring molecules (such as trimethylene oxide, cyclobutanone etc.) are known to have out-of-plane vibrational modes which are predominantly quartic, five-membered ring compounds also have ring puckering modes with a significant quartic contribution to the potential. Therefore the pure quartic oscillator, the mixed quartic-harmonic oscillators and in general, the anharmonic oscillators have been subject to intense study in both quantum field theory and chemical physics for a long period of time and the interest still continues to grow [58, 59, 60, 61, 62, 63, 64]. General analytical solutions of the quartic oscillator are unknown; although some special cases [63] are reported for only certain states of a 1D quartic oscillator where solutions could be found analytically (e.g., the modified quartic oscillator where the potential depends on $|x|$). Thus there has been considerable interest to study the bound-state spectra of these physically as well as chemically important systems using a wide variety of mathematical methods, such as WKB method, phase-integral approach etc.

Each energy level in a 3D isotropic harmonic oscillator is $\frac{1}{2}(\nu+1)(\nu+2)$-fold degenerate. This degeneracy is associated with the two angular momentum quantum numbers l, m.

Table 1. Comparison of pure 3D quartic oscillator energies for some high-lying states; $\nu = 48, 49$ and $l = 0 - 9$. Taken from ref. [46].

ν	l	GPS [46]	Variational [58]	Analytical [59]	Finite-difference [65]	Asymptotic shooting [66]
48	0	250.183358697	250.183351	250.183369	250.183359	250.1833586971
49	1	256.916238928	256.916220	256.916238	256.916239	256.9162389286
48	2	250.096690608	250.096679	250.096671	250.096691	250.0966906080
48	8	249.144812457	249.144801	249.1452		249.1448124575
49	9	255.664161642	255.664146	255.66480		255.6641616427

Each lth level is $(2l + 1)$-fold degenerate, corresponding to the $(2l + 1)$ linearly independent states with $m = l, l-1, \cdots, 0, \cdots, -l$. Upon introduction of a radial perturbation to a harmonic oscillator (such as a pure 3D quartic oscillator) the degeneracy present in a harmonic oscillator is partly removed. For any given ν, levels with different values of l are split, but the m-type degeneracy of each level remains. In other words, l remains a good quantum number; each lth level being $(2l + 1)$-fold degenerate. Only an angular perturbation removes the degeneracy of these m levels. Table I compares some specimen results for odd- and even-parity high-lying excited states of the 3D pure quartic oscillator. Vibrational quantum numbers correspond to $\nu = 48, 49$, and angular momentum quantum number is varied from $l = 0 - 9$. Note that these two quantum numbers must match in parity. The present results are quoted from [46]. It is worth mentioning that while numerous works have been presented for low-lying states (especially the ground states), similar successful attempts for accurate treatment of high-lying states such as those concerned here, have been dramatically less, because of the difficulties faced. Our results are compared with some of the carefully selected best literature values, *viz.*, (a) linear variational calculations involving diagonalization of large order matrices (800×800) [58], (b) approximate analytical formulas derived from a scaled oscillator approach [59] (c) finite-difference numerical result [65] and the (d) asymptotic shooting method [66]. Clearly, amongst these literature values, [66] appears to be the most accurate, where solutions are expressed in terms of finite polynomials, requiring straightforward determination (integration) of zeros of such polynomials. As evident, our present results for all these states match exactly up to the 9th decimal place with those of [66], which manifestly demonstrates the power and efficacy of this method for higher excitations. Similar kind of tests have been made for other cases such as Morse oscillator, an anharmonic oscillator with a quartic perturbation [43], charged harmonic oscillator [42], Hulthén potential [45], harmonic potential including an inverse quartic and sextic perturbation as well as a Coulomb potential with a linear and quadratic coupling [46], etc., which offer conditionally exact solution for certain states. In all such cases, near-exact solutions were obtained from GPS method. Now we proceed for the discussion on individual potentials.

4.1. Spiked Harmonic Oscillator

A class of singular potentials defined by the following Hamiltonian,

$$\mathrm{H} = \mathrm{p}^2 + \mathrm{r}^2 + \lambda \mathrm{r}^{-\alpha} \equiv \mathrm{H}_0 + \lambda \mathrm{r}^{-\alpha}, \quad \mathrm{r} \in [0, \infty], \quad \alpha > 0 \tag{25}$$

where $p = -i\ \partial/\partial r$, has been termed as the spiked harmonic oscillator (SHO). H_0 stands for the simple harmonic oscillator Hamiltonian, whereas the coupling parameter λ and positive constant α determine *strength* of the perturbative potential and *type* of singularity at the origin respectively. Ever since the fascinating work of [67] on its ground-state energies in the mid 70s, an enormous amount of works have been published for its studies [68, 69, 70, 71, 72, 65, 73, 74, 75, 76, 77, 78, 79, 80, 81, 82, 83, 84, 85, 86, 87, 88, 89, 90, 91, 92, 93, 94, 95, 96, 97, 98, 99, 53, 100, 101, 102, 103, 104, 105, 106, 107, 42, 108, 109, 52]. This is not only due to its widespread applications in atomic, molecular, nuclear and particle physics but also because of multitude of inherent interesting properties from mathematical physics point of view. This gives rise to an interesting situation recognized long times ago, i.e., no dominance of either of two terms in the interaction potential for extreme values of λ. In other words, one never deals with *small* perturbations [75, 79]. For all $\lambda \to 0$, $\lambda r^{-\alpha}$ adds an infinite repulsive barrier near the origin. On the other hand, in the limit of $\lambda \to \infty$, one can not ignore the r^2 term. In fact, the harmonic term can never be neglected, for it is needed for the existence of definite ground states [2, 110]. Thus the potential resembles a wide valley extending to ∞ [76]. Another distinctive feature related to this potential is that they exhibit the so-called *Klauder phenomena*, *viz.*, once the perturbation is turned on, complete turn-off is impossible; permanent irreversible vestigial effects of the interaction persists [68, 69, 70, 72]. In particular, a sufficiently singular potential can not be smoothly turned off ($\lambda \to 0$) in the Hamiltonian ($H = H_0 + \lambda V$) to restore the free Hamiltonian H_0. It has been shown that [67] for an SHO, the familiar Rayleigh-Schrödinger perturbation diverges according to the relation $n \geq \frac{1}{\alpha - 2}$, where n is the order of perturbation term. Accordingly, the first-order perturbation correction diverges for $\alpha \geq 2$, second order for $\alpha \geq 5/2$, etc. This potential also exhibits the *super-singularity* phenomenon [71] in the region of $\alpha \geq 5/2$, i.e., every matrix element of the potential becomes infinite. They have been discussed in the interesting context of degeneracy in one-dimensional systems [92]. In another work [107], it was found that, the perturbation theory in λ has an ultraviolet divergence for a range of α, which causes the perturbation series to be ordered by λ^Z; Z being a fraction less than unity.

At this stage it would be convenient for our purpose to discuss a simpler special case of $\alpha = 1$, the so-called *charged harmonic oscillator*, before we proceed for the general stronger spikes ($\alpha \neq 1$). This is a non-supersingular spiked oscillator characterized by a perturbative term of the Coulomb form λ/r, and has been studied in considerable detail [80]. It is possible to identify three distinct regions depending on the value of *effective* coupling constant, *viz.*, (a) *Coulomb* region corresponding to large negative values of λ (b) *strong-coupling* region corresponding to large positive values of λ (c) *weak-coupling* region having small (positive or negative) values of λ. For (a), (b) approximate Rayleigh-Ritz perurbative expansions for ground (and some excited) state energies were developed by means of a combination of hypervirial relation and Hellmann-Feynman theorem. In the Coulomb case, this is given as,

$$E(\lambda) = \lambda^2 \left[-\frac{1}{4} + \frac{12}{\lambda^4} - \frac{1032}{\lambda^8} + \frac{348864}{\lambda^{12}} - \frac{211519200}{\lambda^{16}} + \frac{188054861568}{\lambda^{20}} + \cdots \right] \quad (26)$$

Table 2. Some elementary solutions (in a.u.) of the charged harmonic oscillator ($\alpha = 1$). Ground state energies are given for different values of λ. Results from GPS method are taken from [42] and "Exact" results are quoted from [80].

λ	E(GPS)	E(Exact)	λ	E(GPS)	E(Exact)
0	1.49999999999	1.5	2	2.49999999999	2.5
$\sqrt{20}$	3.50000000000	3.5	$(30+6\sqrt{17})^{1/2}$	4.49999999999	4.5
$(70+6\sqrt{57})^{1/2}$	5.49999999999	5.5	14.450001026966	6.49999999999	6.5

whereas for the strong-coupling case this becomes ($\mu = (2/\lambda)^{1/3}$),

$$E = 3\mu^{-2} + \sqrt{3} + \frac{7\mu^2}{36} + \frac{37\mu^4}{432\sqrt{3}} + \frac{2573\mu^6}{139968} + \frac{168233\mu^8}{2239488\sqrt{27}} + \cdots \tag{27}$$

It is also known that there could be an indirect, somewhat involved path to connect the strong-coupling regime ($\lambda \to \infty$) with the small-coupling regime, as well as to connect Coulomb regime ($\lambda \to -\infty$) with the small-coupling regime. However, as yet no direct link has been found to connect the $+\infty$ and $-\infty$ regions. Another very important characteristic of the charged oscillator is that this offers an infinite set of *elementary solutions* [80] for certain selected values of positive coupling constant only. These solutions are typically of the form of a polynomial multiplied by a Gaussian and are possible for both ground as well excited states. Some of these, taken from [42], are displayed in Table II. Note that "Exact" results have been divided by a 2 factor for consistency with the literature. We see excellent agreement of current method with "Exact" results for all values of λ. Ground states for general λ (both positive and negative) have been reported in [80, 42], whereas first three excited states corresponding to $l = 0, 1, 2, 3$ have been studied in [42]. This concludes our discussion on charged oscillator; more detail could be found in [80, 42].

Let us now return back to SHO. From a variational analysis [71], the eigenvalues of SHO for $2 \leq \alpha \leq 3$ were given by asymptotic series to first order for positive λ. But for $\alpha \geq 3$, the ground-state eigenvalues are given by,

$$E_0(\lambda) = 3 + k\lambda^{\nu} + O(\lambda^{\nu}) \tag{28}$$

whereas for $\alpha = 3$, these are obtained as,

$$E_0(\lambda) = 3 + k'\lambda \log(\lambda) + O(\lambda) \tag{29}$$

where k, k' are to be determined variationally. Later, a modified Rayleigh-Schrödinger series was put forth [67] by exploiting the standard WKB approximation for lowest few orders. This success encouraged other workers to further develop a special perturbation theory, known as *singular perturbation theory* to obtain first few terms of perturbed λ-expansion for different values of α. To this end, asymptotic series for ground-state eigenvalues of the SHO Hamiltonian are explicitly written as,

$$\begin{aligned}
E_0(\alpha \geq 4, \lambda) &= 3 + \frac{4\nu^{2\nu}\Gamma(1-\nu)}{\sqrt{\pi}\Gamma(1+\nu)}\lambda^{\nu} + O(\lambda^{2\nu}) \\
E_0(3 < \alpha < 4, \lambda) &= 3 + \frac{4\nu^{2\nu}\Gamma(1-\nu)}{\sqrt{\pi}\Gamma(1+\nu)}\lambda^{\nu} - \frac{4\nu\Gamma(\frac{3-\frac{1}{\nu}}{2})}{(1-\nu)\sqrt{\pi}}\lambda + O(\lambda^{2\nu}) \\
E_0(\alpha = 3, \lambda) &= 3 - \frac{4}{\sqrt{\pi}}\lambda\log(\lambda) - \frac{10c}{\sqrt{\pi}}\lambda + O(\lambda^2\log^2(\lambda)) \\
E_0(5/2 < \alpha < 3, \lambda) &= 3 + \frac{4\nu^{2\nu}\Gamma(1-\nu)}{\sqrt{\pi}\Gamma(1+\nu)}\lambda^{\nu} + \frac{2\Gamma(\frac{3-\alpha}{2})}{\sqrt{\pi}}\lambda + O(\lambda^{2\nu})
\end{aligned} \quad (30)$$

Here $\nu = 1/(\alpha - 2)$ and $c = 0.5772156649$ is the Euler's constant. It is worth noting that $\alpha = 5/2$ bears a close analogy with the low-density expansion of energy of a many-body boson system at zero temperature [111], *viz.*,

$$E/N = (2\pi\hbar^2/m)\rho a[1 + C_1(\rho a^3)^{1/2} + C_2\rho a^3 \ln(\rho a^3) + C_3\rho a^3 + \cdots] \quad (31)$$

In another development [75], ground-state energy was obtained by making use of a functional space spanned by the solution of Schrödinger equation for a linear harmonic oscillator within a variational framework, followed by standard diagonalization of symmetric matrices. A strong coupling expansion ($\alpha \geq 2$) for positive λ was derived for approximate ground-state energy. For example, for $\lambda = 5/2$, this reads as,

$$E(\lambda) = \frac{9}{5}\left(\frac{5}{9}\right)^{4/9} \lambda^{4/9} + \left(\frac{9}{2}\right)^{1/2} + \cdots \quad (32)$$

For same α, the fourth-order large coupling perturbation expression is,

$$E(\lambda) = \frac{9}{5}\left(\frac{5\lambda}{4}\right)^{4/9} + \left(\frac{9}{2}\right)^{1/2} + \frac{77}{288}\left(\frac{4}{5\lambda}\right)^{4/9} - \frac{1967}{27648}\left(\frac{2}{9}\right)^{1/2}\left(\frac{4}{5\lambda}\right)^{8/9} + \cdots \quad (33)$$

The weak coupling expression using perturbation theory up to second order as well as the strong coupling expansions up to 10th order (in algebraic form) were obtained for the non-singular SHO ($\alpha < 5/2$) through a re-summation technique [76]. In the above work, also an attempt was made to find a path connecting ground-state energy in the weak coupling regime with that of the strong coupling regime. Similar weak-coupling expansions for ground-state energies were given for the special case of $\alpha = 2$ in [84]. Ground-state energy of the particular case $\alpha = 4$ was studied by using a non-orthonormal basis set satisfying the correct boundary conditions [87]. Through a logarithmic perturbation theory [108], closed-form expressions for energy and wave function correction terms were obtained for ground states.

In another analytical method, namely, pseudo-perturbative shifted-l expansion technique (PSLET) [96], it was possible to obtain both eigenvalues and eigenfunctions in one batch. This uses $1/\bar{l}$ as a perturbation expansion parameter, where $\bar{l} = l - \beta$ (l is a quantum number and β is a suitable shift introduced to avoid the trivial case $l = 0$). This has been used to study bound states of D-dimensional spiked harmonic oscillator spectra [97] as well, taking into account the effect of inter-dimensional degeneracies arising out of the isomerism between angular momentum and dimensionality of the central force Schrödinger equation. Other approximate variational methods have also been attempted [78]. These states have been studied by a modified WKB approximation [93]. Upper and lower bounds of ground as well as excited states have been developed [94] by means of envelope theory through some convenient smooth transformation. A rigorous variational method suitable for the complete set of discrete energy eigenvalues has been established [103], which also remains valid for arbitrary angular momentum quantum number, in general N-dimensional case. This was done by expressing the SHO Hamiltonian as a perturbation of the singular Gol'dman and Krivchenkov Hamiltonian H_0. Further it was proved that zeroth order eigenfunctions generated by H_0 form a suitable singularity-adapted basis for the appropriate Hilbert space of the full problem. Ground-state bounds have been investigated by means of potential envelopes [89]. An eigenvalue moment method has been proposed for accurate estimation of ground states of singular potentials [91].

The SHO problem was numerically solved through finite-difference scheme [65, 53], a Lanczos grid method [83]. In the latter, Hamiltonian was discretized on a grid with a 10th order finite-difference formula for kinetic energy and a starting function $r^8e^{-r^2}$. Applications were made for s states for several values of α, λ. Accurate numerical solutions were found [90] by modifying the usual analytic continuation method of [112, 66], which was initially applicable to potentials whose solutions do not have essential singularities. This was possible by introducing a family of non-singular potentials which depend on a parameter R_c and which approach the exact potential as R_c tends to zero. A non-perturbative but completely convergent algorithm [88], and formally identical to the Lanczos method was proposed for ground states. An accurate algorithm was presented where a discretized symmetric expression was derived for the transformed Schrödinger equation [79], which could be efficiently solved by standard routines for tridiagonal matrices efficiently. This involved a coordinate transformation either of the form $r = Kx/(1-x)$ (parametrized Euler transformation) or $1 + Kr = e^{Kx}$ [53], where the adjustable parameter K needs to be chosen judiciously.

In spite of all the above attempts, a general method which can offer accurate reliable results for both these potential parameters for arbitrary states (covering both ground as well as various excited states, especially the higher ones), have been very rare. Thus, for example, physically meaningful and high accuracy results are obtainable by only a few of the methods discussed above. Moreover, some of these methods provide high-quality results for certain type of parameters, while performing rather poorly for other sets. An enormous amount of work has been reported for ground states; excited states have received much less attention presumably due to the inherent difficulties encountered with them. Excepting a very few rare studies (for example, [100, 42]), these have almost exclusively dealt with eigenvalue determination; nature of eigenfunctions has been explored only in some rare occasions. Last, but not the least, some of these methods are often fraught with rather tedious, cumbersome mathematical complexities. The GPS method provides a simple general and easily affordable scheme where all these above mentioned discomfitures are either completely removed or partly alleviated by invoking an *optimal, non-uniform* effective spatial grid. General applicability of the approach is amply demonstrated in Tables III, IV for some representative parameter sets of the potential for arbitrary states. Literature results are

Table 3. Comparison of ground-state energies E (in a.u.) of the SHO with $\alpha = 4$ and 6 for selected values of λ. See [42] for details.

λ	Energy ($\alpha = 4$)		Energy ($\alpha = 6$)	
	GPS[42]	Literature	GPS[42]	Literature
0.001	1.53438158545	1.53438158545[a], 1.534385[b]	1.63992791296	1.63992791296[a]
1	2.24708899168	2.24708899168[a], 2.24709[b,c] 3.3033112560[d]	2.32996998478	2.32996998478[a],2.329970[c] 3.00160451[d]
1000	10.6847312660	10.6847312660[a], 10.68473[b], 10.684731265[d]	6.35930853290	

[a]Ref. [90]. [b]Ref. [87]. [c]Ref. [78]. [d]Ref. [91].

quoted, for comparison, wherever possible. Clearly very accurate results are obtained for a large range of coupling parameters for low as well as high states. More detailed results on energies, expectation values, radial densities through the GPS method are available in [42].

Now we discuss a slightly modified Hamiltonian from that defined in Eq. (25), *viz.*,

$$H = -\frac{d^2}{dr^2} + ar^2 + \frac{b}{r^4} + \frac{c}{r^6}, \quad a, c > 0, r \in [0, \infty] \tag{34}$$

which is a non-trivial generalization of the SHO Hamiltonian and has also been studied in considerable detail in the literature. For example, in [79], a large-order strong coupling expansion, in terms of some *ad hoc* expansion parameter was proposed for large anharmonicity constants corresponding to either of b,c large or both of them large. For small anharmonicity constants, lowest order correction to the perturbed energies were also obtained.

A very important feature of this central singular potential is that it provides conditionally exact solutions for certain values of the potential parameters. From very early stage of the development of quantum mechanics, there is continual interest in solving Schrödinger equation exactly. However, as we know, exact solutions could be found for very few potentials. For systems with one degree of freedom, supersymmetric quantum mechanics together with shape invariance has been found to be one of the most successful techniques for understanding the exact solvability. Of late, due attention has been paid to different class of potentials which are quasi-exactly solvable (QES) and conditionally exact solvable (CES). In the former case, only a finite number of eigenstates can be found exactly. A more interesting case is offered by the CES potential, which is intermediate between the exact solvable and QES potential. For, exact eigenvalues are obtainable only when potential parameters satisfy certain conditions. The wave function ansatz technique is one very common procedure used for this purpose, which is purely mathematical and usually fails to resolve the physical reason for conditional solvability. However, this technique is not straightforward for arbitrary states. Recently, super-potential ansatz technique has also been proposed. Bound states in this potential for $c = 0$ [73] and $c \neq 0$ [74] have been investigated nearly two decades ago. A simplified ansatz for eigenfunctions has offered exact ground-state energy in the following closed form [77],

$$\phi_0(r) = N_0\, r^{(3+b/\sqrt{c})/2}\, e^{-\frac{1}{2}(\sqrt{a}r^2 + \sqrt{c}r^{-2})} \tag{35}$$

Table 4. Comparison of energies of $\ell \neq 0$ states (in a.u.) of the SHO with $\alpha = 4$ (top) and 6 (bottom) for selected λs. Un-superscripted GPS results are quoted from [42].

ℓ	$\lambda = 0.001$	$\lambda = 0.1$	$\lambda = 10$
5	6.50002020182(6.50002020182)[1]	6.50201821626(6.50201821626)[1]	6.68566506197
20	21.5000012507	21.5001250765	21.5124915772
50	51.5000001999	51.5000200019	51.5020000374
5	6.50000577192	6.50057643602	6.55258902874
20	21.5000000675	21.5000067609	21.5006759799
50	51.5000000040	51.5000004123	51.5000412410

[1] Ref. [106].

with the ground-state energy given as,

$$E_0 = \sqrt{a}(4 + b/\sqrt{c}) \tag{36}$$

However, for this to be satisfied, the potential parameters must be related as follows:

$$(2\sqrt{c} + b)^2 = c[(2l + 1)^2 + 8\sqrt{ac}] \tag{37}$$

Later, a similar approach was employed for first excited states [81]. The harmonic potential with an inverse quartic and sextic anharmonicity was solved numerically [85] for some lowest states using B-spline basis sets. Continuing along the same line, four sets of solutions were obtained, including one constraint equation for each set [86]; furthermore it was found that the analytical expression for energy agrees with numerical result for any one among the ground, first and second excited states, depending on the particular constraint condition. Conditional exact solutions were studied by [101, 102] in the light of supersymmetric quantum mechanics and shape invariance. A Hill determinant approach was attempted [82].

In parallel to above works, recently there has been considerable interest in studying the so-called generalized spiked harmonic oscillator(GSHO), defined as,

$$H = -\frac{d^2}{dr^2} + v(r); \quad v(r) = r^2 + \frac{A}{r^2} + \frac{\lambda}{r^\alpha}, \ A \geq 0 \tag{38}$$

Here λ, α are two real parameters and obviously SHO is a special case of GSHO with $A=0$. Both variation and perturbation methods were employed for their studies [95, 98, 99, 100, 105, 106]. Using the exact Gol'dman-Krivchenkov wave functions (which constitute an orthonormal basis for the Hilbert space $L^2(0, \infty)$), compact closed form expressions have been derived [95, 100] for the required singular-potential integrals (or matrix elements) of the form $\langle m|x^{-\alpha}|n\rangle$. Variational bounds [98] have been examined on the light of above as well. A first-order perturbation series was developed [100] for wave functions in terms of generalized hyper-geometric functions. The modified perturbation theory of [67] was extended to obtain perturbation expansions for the GSHO problem [105]. These expressions remain valid for small values of coupling $\lambda > 0$ and corroborate the results of those obtained for SHO [67]. In [106], using a perturbation expansion up to 3rd order, lower and upper bounds of the GSHO ground states for small coupling parameter λ, were estimated. Weak-coupling perturbation expansions for ground states were derived in [99].

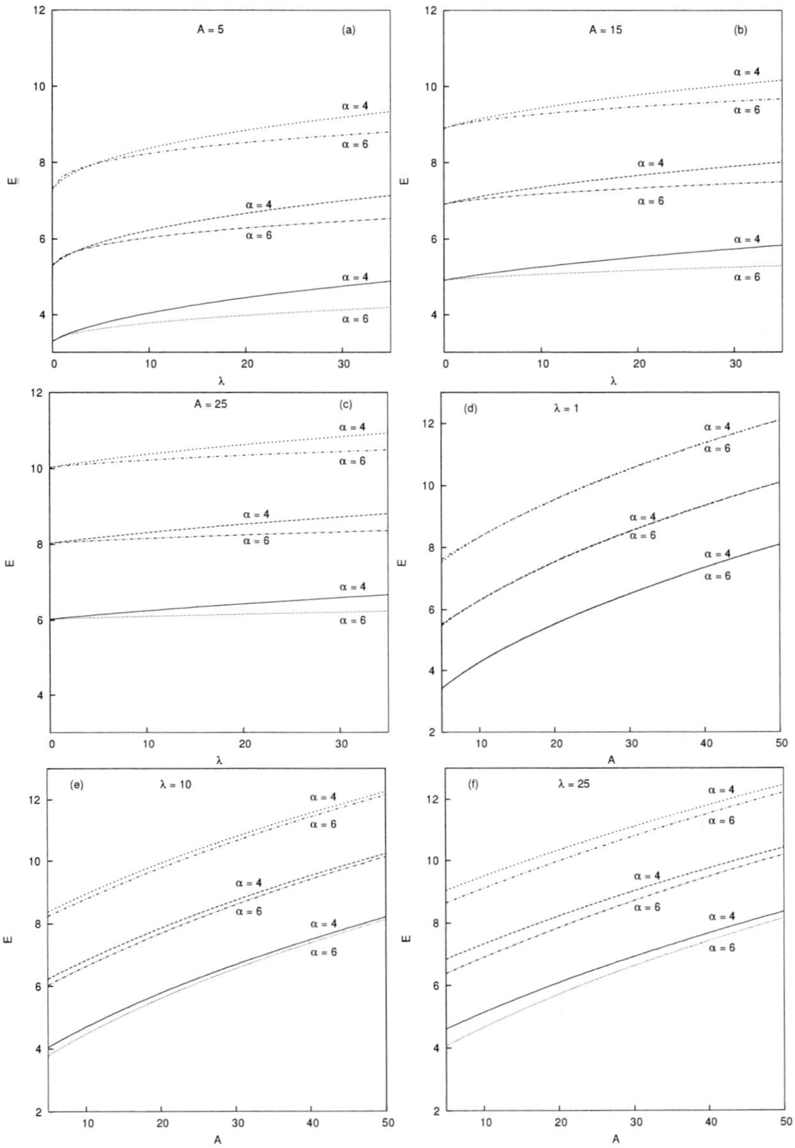

Figure 1. Energy variations of a GSHO with respect to λ (top) and A (bottom) respectively for the first three states belonging to $\ell = 0$ having $\alpha = 4$ and 6. (a) $A = 5$, (b) $A = 15$, (c) $A = 25$; (d) $\lambda = 1$, (e) $\lambda = 10$, (f) $\lambda = 25$. Adopted from [52].

However, to the best of my knowledge, no *direct* results were provided in the above investigations. By means of GPS method, very accurate ground as well excited states, expectation values, radial densities have been reported only lately [52]. A wide range of interaction was considered (both weak and strong coupling) for arbitrary values of n, l. Energy variations with respect to the two parameters λ, A are displayed at the top and bottom panels in Fig. 1. The first three eigenstates of $l = 0$ are studied for $\alpha = 4, 6$ respectively with λ varied from 0–35. As seen clearly, energy varies rather slowly with respect to λ (monotonic increase with increase in λ) for both $\alpha = 4, 6$ (smaller α produces larger effect). In the neighborhood of zero λ, for all values of A, both α give very similar energies. As λ

increases, appreciable separation shows up. An increase in A reduces the slope of λ vs. E plot, eventually becoming flat in (c), signifying even slower variation of E with λ. The bottom panel illustrates A vs E for three λ values (1,10,25) in (d)–(f). The same three states of both $\alpha = 4, 6$ make three separate families. For a fixed λ, energy increases rather promptly relative to the top panel. Interestingly, $\alpha = 4, 6$ plots seem to be virtually identical for $\lambda = 1$ for all three states. Finally separation between $\alpha = 4, 6$ increases gradually with λ as one passes from (d)–(f). For further discussion on these results, consult [52].

4.2. Screened Coulomb Potentials

The screened Coulomb potentials, defined as,

$$V(r) = -\frac{Z}{r} \sum_{k=0}^{\infty} V_k (\lambda r)^k \tag{39}$$

have found significant importance in many areas in physics, chemistry. This can be used to approximate the potential experienced by an electron in an atom where other electrons screen the nuclear charge. In the form of Debye-Hückel potential, it describes the shielding effect in plasmas. An enormous amount of work has been done for understanding many fascinating features these potentials exhibit. In the context of atomic systems, Z is identified as atomic number while the screening constant λ bears different significance in different branches. In what follows, we are concerned with the two simple representatives of the screened Coulomb potential. First one, the Hulthén potential [113, 114], given as,

$$V(r) = -\frac{Z\delta e^{-\delta r}}{1 - e^{-\delta r}} \tag{40}$$

is one of the most important short-range potentials. It has relevance in nuclear and particle physics [115, 116], atomic physics [117, 118], solid-state physics [119, 120], chemical physics [121], etc. This is also a special case of Eckart potential. Other one, the Yukawa potential [122], as given below,

$$V(r) = -\frac{Ze^{-\lambda r}}{r} \tag{41}$$

has also found numerous applications in various branches in physics, chemistry, etc.

They show many similarities, e.g., for small r, they display Coulomb-like behavior,

$$V(r) \to -\frac{Z}{r} \quad \text{for } r \to 0 \tag{42}$$

whereas they decay monotonically exponentially to zero for large r,

$$V(r) \to O(e^{-\gamma r}) \quad \text{for } r \to \infty \tag{43}$$

Use of a scaling transformation $r \to r/Z$ leads to the following well-known relations,

$$\begin{aligned} E(Z, \gamma) &= Z^2 E\left(1, \frac{\gamma}{Z}\right) \\ \psi(Z, \gamma, r) &= Z^{3/2} \psi\left(1, \frac{\gamma}{Z}, Zr\right) \end{aligned} \tag{44}$$

Table 5. Calculated negative eigenvalues E (a.u.) of various states of Hulthén potential as function of δ. Numbers in the parentheses denote δ_c values taken from [124]. Asterisks denote exact analytical values. Taken from [45].

State	δ	−Energy	
		GPS [45]	Literature
$1s(2.0)$	1.97	0.00011249999999	0.0001125*
$2p(0.377)$	0.35	0.00379309814702	0.00379309814702[1]
$4d(0.098)$	0.075	0.00383453307692	0.00383453307692[1]
$8k(0.021)$	0.02	0.0002027526409	
$17s(0.007)$	0.005	0.0001332288062	0.0001332288062*
$4f(0.086)$	0.08	0.00135376897143	
$6h(0.038)$	0.005	0.01147020315553	
$10m(0.013)$	0.01	0.0009003110142	

[1]Ref. [125].

Here $\gamma = \delta$ or λ. Therefore it suffices to consider only the case of $Z = 1$, and develop energy and eigenfunctions as a function of screening parameter δ. A distinctive feature (in contrast to the Coulomb case) is that the number of bound states is *limited* because of the presence of screening parameters. In other words, bound states exist only for certain values of screening parameter below a threshold limit. For Yukawa potential, this value has been estimated quite accurately as 1.19061227±0.00000004 a.u. [123]. The former also has an additional property that it offers *exact* analytical solutions for $l = 0$, not for higher partial waves.

Great many attempts have been made to calculate the bound-state energy, eigenfunction of these potentials, as well their scattering properties. $l \neq 0$ states of Hulthén potential, correct to any order of δ, have been obtained by using an extended version of the analytic perturbation theory [126]. Using a non-perturbative approach, an extension of Ecker-Weizel approximation was used to obtain analytic closed-form solutions for eigenvalues and eigenfunctions for arbitrary angular momenta [118]. Latter authors used non-rigorous but intuitive physical arguments to determine the unknown constants involved in this approximation. To this end, the energy was obtained as,

$$E_{nl} = -\frac{1}{2}(1/n - n\delta/2)^2 + \frac{1}{8}\delta^2 l(l+1)B[4/n^2\delta + 2 - l(l+1)B/n^2] \tag{45}$$

where

$$B = (e-1)^{-1}[(e-1)^2/e - 1] = 0.0501438. \tag{46}$$

A strong-coupling series was suggested for bound state energies of both these potentials [127] within the WKB approximation, with special emphasis on behavior of energies in the neighborhood of critical region. Computation of higher-order perturbation theory and summation methods for divergent perturbation series was also addressed [128]. Later, a path-integral formalism [129] was put forth for s states, where the exact energy spectrum and normalized s-state eigenfunctions were obtained from poles of the Green function and their residues. A shifted $1/N$ expansion technique [130, 131] was also used for these potentials. Further, an algebraic perturbation method based on Lie algebra of the group $SO(2,1)$

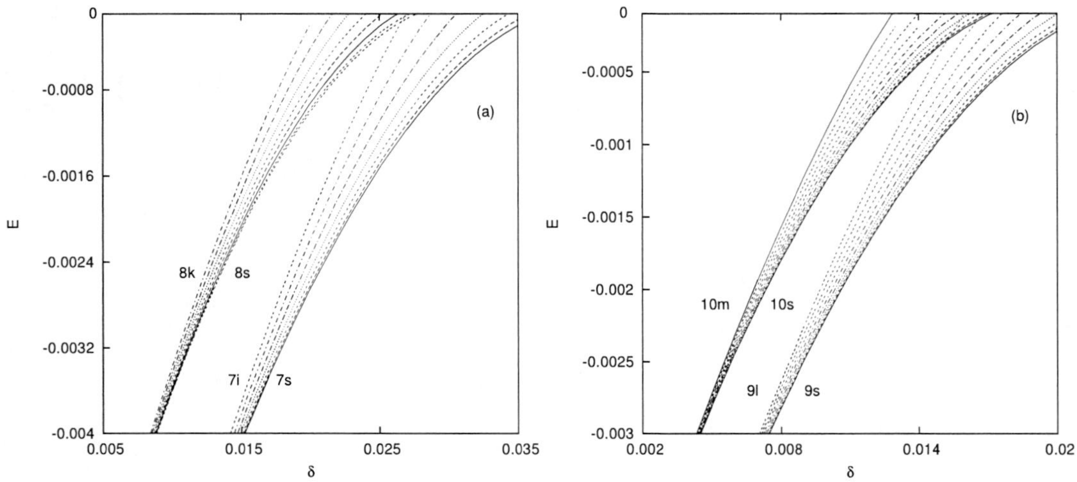

Figure 2. Energy eigenvalues (a.u.) of the Hulthén potential for (a) $n = 7, 8$ and (b) $n = 9, 10$ levels respectively as a function of δ in the vicinity of zero energy. Taken from [45].

which is known to be the dynamical group for a number of spherically symmetric potentials, was developed for s states by [132] and for arbitrary states by [133]. A one-parameter variational calculation has also been reported for energies, oscillator strengths [124]. These were found to be quite satisfactory in low-screening region, but proved inadequate for higher δ. The concept of kinetic potentials was used to construct a global geometrical approximation theory for the energy spectra of these potentials [134]. A very accurate generalized variational method [125] was developed for both these potentials. This utilized trial functions from a linear combination of independent functions. For Hulthén potential, the basis functions used assume the following form,

$$\phi_k = A_k r^k e^{-\beta r}(1 - e^{-\delta r}) \tag{47}$$

where $k = -1, 0, 1, 2, \cdots$ for s states and $k = 0, 1, 2, \cdots$ for $l \neq 0$ states. β is a variational parameter determined by minimizing the energy for a given state and basis size. The constant $A_k = (2\beta + \delta)^{k+1}/\sqrt{(2k+2)!}$ is included to prevent numerical overflow. For Yukawa potential, the basis takes the following form,

$$\psi_k = B_k r^k e^{-\beta r/2}, \quad k = 0, 1, 2, \cdots \tag{48}$$

where B_k, β are normalization constant and variational parameter respectively. An improved variational scheme was proposed [135], wherein a set of variational parameters was introduced into the trial wave function to form a family of independent functions. This potential has also been dealt using supersymmetric quantum mechanics and first-order perturbation theory [136]. Based on the small and large r behavior, threshold and asymptotic properties of energy and eigenfunctions of these potentials were examined critically in [137].

First-order perturbative calculations using Hulthén potential as unperturbed potential, was employed for s states of Yukawa potential [117]. Later, motivated by this, using solutions to a Hulthén-like effective potential as variational trial functions, nonzero l states of

Table 6. Comparison of the negative eigenvalues (a. u.) of Yukawa potential as function of λ. Numbers in the parentheses denote λ_c values. See [45] for details.

State	λ	−Energy		State	λ	−Energy	
		GPS [45]	Literature			GPS [45]	Literature
$4d(0.0581)$	0.01	0.02222779248980	0.02222779248980[1]	$4d$	0.055	0.00049188376726	
$9s(0.016)$	0.01	0.0005858247612	0.000585[2]	$10s$	0.005	0.0015083559307	
$9p(0.015)$	0.01	0.0005665076261	0.000565[2]	$10p$	0.005	0.0015009235029	
$9l(0.0094)$	0.005	0.0021291265596	0.00213[2]	$10l$	0.005	0.0012296811835	
$10m$	0.005	0.0011557947569		$17s$	0.001	0.000919120394	

[1]Ref. [125]. [2]Ref. [146].

Yukawa potential were obtained [138] with reasonable accuracy. It was soon realized that these trial functions provide better variational energies, wave functions with fewer parameters than frequently used hydrogenic or Slater-type functions. Detailed variational results were presented for lowest 45 eigenstates [139]; also probabilities for spontaneous emission in dipole approximation was studied as a function of screening length, for transitions between six lowest states. A combined Padé approximant ([6,6], [6,7]) to the perturbation series for Yukawa potential was presented [140]. A shifted $1/N$ expansion [141, 142, 143] as well as an improved expansion was put forth [144]. These offered convergence, better than usual $1/N$ expansion. Through a combined perturbation theory and continued fractions-Padé approximations at large order [145], very accurate bound states were obtained. A linear combination of atomic orbitals (LCAO) calculation was performed for ground states as function of λ within a variational framework (using Slater basis functions) [123].

Accurate numerical methods have been developed for these potentials. All the 45 eigenstates $1s$ through $n=9, l=8$ were numerically investigated for wider λ for Yukawa potential nearly four decades ago with reasonably good accuracy [146]. Another of them [147] consists in solving the Dirichlet problem in a box with radius n by a Ritz method. The convergence to the eigenfunctions in the norm of Hilbert space $L_2(0, n)$ was proved to be guaranteed.

It may be mentioned that although there are decent number of high-quality results available for these interactions in the literature in *weak*-coupling regions as well as *lower* states, there is a scarcity of such results for *stronger* coupling and for *higher* states. In a GPS study [45], these potentials were calculated very accurately for arbitrary field strengths with special emphasis on these issues. Sample Hulthén potential results from GPS method are collected in Tables V. Numbers in the parentheses denote respective critical screening constants. Literature values are quoted for comparison, wherever possible. As discussed in [45], either for higher states or high screening constants, larger R is needed, whereas eigenvalues are apparently less sensitive with respect to the number of grid points N. In Table V, the s states are presented in the strong δ region. Exact analytical results (denoted by asterisks), available for $l = 0$ states of Hulthén potential are given as,

$$E_n^{exact} = -\frac{\delta^2}{8n^2}\left[\frac{2}{\delta} - n^2\right] \tag{49}$$

with $n^2 < 2/\delta$. Our calculated values for s states completely coincide with exact analytical

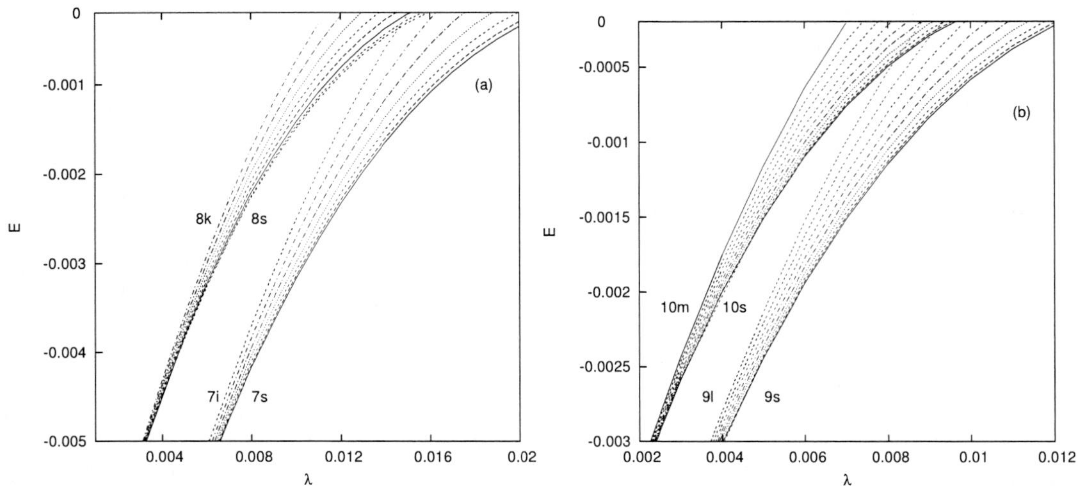

Figure 3. Energy eigenvalues (a.u.) of the Yukawa potential for (a) $n = 7, 8$ and (b) $n = 9, 10$ (right) levels respectively as a function of λ in the vicinity of zero energy. Adopted from [45].

results for all states (up to $n = 17$) covering a whole range of interaction. The δ_c for $l = 0$ state is given exactly by the simple relation $\delta_c = 2/n^2$, whereas for $l \neq 0$ this is approximated by the analytical expression [127],

$$\delta_c = 1/[n\sqrt{2} + 0.1645l + 0.0983l/n]^2 \tag{50}$$

which offers good agreement with numerically determined values [124]. Excellent agreement with literature results has been observed for all the states. As seen, a uniform accuracy is achieved for all states encompassing a large range of interaction, unlike some previous calculations which encountered difficulties in the strong-coupling region. For weaker coupling, present results are superior to all other results except the variational work of [125]. However, in the stronger region, GPS results are superior. We have enlarged the coupling region from all other previous works and current results are so far the most accurate ones in the neighborhood of critical δ. Additionally, Figure 2 depicts the variation of energies with respect to δ for all states belonging to $n = 7, 8$ (a) and $n = 9, 10$ (b) in the neighborhood of zero energy. For small n, there is good resemblance of energy orderings with those of Coulomb potentials; however, this scenario changes with an increase in n. In that case, significant deviation is noticed as well as complex level crossing observed in the vicinity of zero energy, which makes their accurate determination quite difficult. This is more dramatic for latter (e.g., $9k$, $9l$ mixing heavily with $10s$, $10p$, $10d$, $10f$ at around $\delta = 0.015 - 0.017$). Besides, for a given n, separation between states with different l increases with δ. Further discussion on results including radial densities, expectation values etc., could be found in [45].

Let us now turn our focus on the Yukawa potential. Similar results as those Hulthén potential, are presented in Table VI and Fig. 3. No analytical expressions are available for critical screening constants in this case; numerically determined λ_cs, available from [146], are quoted, wherever possible. As in the previous case, here also very good agreement is observed for all states with the best theoretical results. Once again, a large range of interaction (both weak, strong) as well as very high states are considered. As in Hulthén potential, results of [125] are more accurate than ours in the small λ region, but for stronger couplings, present results are superior. Some of the higher states have not been calculated by any method other than those of [146, 137]; our results significantly improve those. Considering all these, present results appear to be the most accurate and reliable for all the states (except $1s$, $2s$) in regions close to δ_c. As n, l increase, accurate calculation of these states become progressively difficult; for $7s - 9l$ states only two attempts [146, 141] are known other than the present one. In Fig. 3, dependence of energy orderings on λ is shown vividly in the neighborhood of $E=0$ for $n=7, 8$ (a) and $n=9, 10$ (b). Very similar conclusions as in Fig. 2 can be drawn: (i) as n increases energy ordering tends to be more complex with the possibility of level crossing becoming higher (ii) energy splitting between states with different l increases with an increase in λ for a particular n. No attempts are known for any of these states with $n > 9$ other than the present one and these may constitute a useful reference for future studies. As an illustration, also some very high-lying states ($17s$) are included in this table. Finally, we note that eigenvalues of Coulomb, Hulthén and Yukawa potentials are known to satisfy the following relation,

$$E_n^{\text{coulomb}} \leq E_{n,\ell}^{\text{Hulthen}}(\delta) \leq E_{n,\ell}^{\text{Yukawa}}(\lambda) \tag{51}$$

This has been found to be true for all the states considered here. For many other interesting features on these potentials, see [45].

4.3. Power-law and Logarithmic Potentials

The power-law and logarithmic potentials have found wide-spread applications in particle physics [148, 149, 150, 151]. The orbital structure of logarithmic potential has been studied in the context of self-consistent modeling of triaxial systems (such as elliptical galaxies), bars in the centers of galaxy discs [152], global dynamics [153], etc. The Coulomb plus power-law potential also serves as a non-relativistic model for the principal part of a quark-quark interaction [154]. The potential is given by $V(r) = A\ \text{sgn}(\nu) r^\nu$, where $r = ||r||$, $A > 0$ and $\nu \neq 0$. For dimension $N = 1$, $\nu > -1$ and for higher dimensions $N \geq 2$, $\nu > -2$. For $\nu = 0$, we have $V(r) = A \ln(r)$, with $A > 0$. It is possible to include logarithmic potential as a limiting case of the power potentials if in place of the potential family $f(r) = \text{sgn}(\nu) r^\nu$, we use $V(r, \nu) = (r^\nu - 1)/\nu$, whose limit as $\nu \to 0$ is $V(r, 0) = \ln(r)$. Eigenvalues of the power-law potential E_{nl}^N can be labeled by two quantum numbers; the total angular momentum $l = 0, 1, 2, \cdots$, and a 'radial' quantum number $n = 1, 2, 3, \cdots$, which represents 1 plus number of nodes in the radial part of wave function. The eigenvalues in $N \geq 2$ spatial dimensions has degeneracy 1 for $l = 0$ and for $l > 0$, the same is given by,

$$\Lambda(N, l) = \frac{(2l + N - 2)(l + N - 3)!}{l!(N - 2)!}, \quad N \geq 2, l > 0 \tag{52}$$

The well-known hydrogenic atom and harmonic oscillator constitute the two exactly solvable cases in N dimensions corresponding to $\nu = -1, 2$ respectively,

$$\begin{aligned} E_{nl}^N(-1) &= -[2(n+l+N/2-3/2)]^{-2}, \quad N \geq 2 \\ E_{nl}^N(2) &= 4n+2l+N-4, \qquad N \geq 2 \end{aligned} \tag{53}$$

Also analytical solutions for linear potential in 1D, as well as s states in 3D could be found from the zeros of Airy function [155].

Numerous attempts [156, 157, 155, 158, 159, 160, 161, 162, 163, 164, 165] have been made in the past years to examine many interesting properties of their solutions utilizing an array of methodologies. For example, ground and excited energy levels of the generalized 1D anharmonic oscillator characterized by the potential $V(x) = x^2 + \lambda x^{2m}$, $m = 2, 3$ were calculated non-perturbatively [156] by a Hill determinant method. A WKB approximation [155] was proposed. It was proved [157] that the eigenvalues $E_n = E_{n0}^1$ of power-law potentials in 1D increase with n at a higher rate for a greater ν. However for any ν, this increase never attains n^2. In general, the dependence of E_{nl}^N on the coupling parameter A may be established by elementary scaling arguments by replacing r by σr. Then one finds that,

$$E_{nl}^N(A) = A^{2/(\nu+2)} E_{nl}^N(1) \tag{54}$$

Thus without any loss of any generality, one can limit further discussion on the case of unit coupling $A = 1$. Lower and upper analytic bounds for ground states were developed for power law potentials [158]. The shifted $1/N$ expansion technique, with some of its variants (such as a modified or large-order expansion) [159, 160, 161] has been quite successful. Dependence of eigenvalues on the power parameter ν has been studied by spectral geometrical arguments [162]. Bounds of eigenvalues for polynomial potentials in N dimension have also been studied semi-classically [163, 164]. A variational method [165] was developed for eigenvalues and eigenfunctions; it provided good results for small ν, but suffered in the higher ν regions.

GPS method [43] has produced very accurate results for both these potentials for states with arbitrary quantum numbers n, l. A detailed comparison with literature results has been made in [43], from which it is abundantly clear that the present scheme offers results which are considerably better than the existing results available. Moreover, as in the previous occasions, here also we obtain both low as well as higher states with equal ease and accuracy. As an illustration, Table VII compares our calculated eigenvalues for two potentials $V(r) = -2^{1.7} r^{-0.2}$ and $V(r) = 2^{7/2} r$, which have been examined earlier quite extensively by other workers. Selected states with $n \leq 4$ and $l \leq 3$ are reported here. In the former case, accurate results are not available for comparison and the present results match closely to all of the results available, while for for the latter case, GPS results are in excellent agreement with the accurate values of [161]. For a more complete discussion, see [43].

4.4. Rational Potentials

Rational, also termed as non-polynomial oscillator (NPO) potentials, defined by,

$$V(r) = r^2 + \frac{\lambda r^2}{1+gr^2}; \quad g > 0, \ \lambda \in (-\infty, \infty) \tag{55}$$

Table 7. Comparison of eigenvalues of the two power-law potentials $V(r)=-2^{1.7}r^{-0.2}$ (left) and $V(r)=2^{7/2}r$ (right) for several n, l values. See [43] for details.

Energy		n	ℓ	Energy	
GPS [43]	Literature			GPS [43]	Literature
−2.68602822	−2.68601[1], −2.6859[2], −2.686[3]	0	0	9.352429641	9.352429643[4], 9.3524296418[5]
−2.04431800	−2.04658[1], −2.0440[2], −2.044[3]	2	0	22.08223931	22.08223931[4], 22.0822393124[5]
−1.81414352		4	0	31.77653434	
−2.02906490	−2.02906[1], −2.0291[2], −2.029[3]	0	3	20.20370253	
−1.90486674	−1.90491[1], −1.9049[2], −1.905[3]	1	3	25.32846149	
−1.73987512		3	3	34.38846804	

[1] Ref. [160]. [2] Ref. [165]. [3] Numerical results, from [160]. [4] Ref. [161]. [5] Exact values, as quoted in [161].

has been one of the most important model systems in quantum mechanics. The relevance in nonlinear Lagrangian field theory [166], laser theory (as the reduction of Fokker-Planck equation of a single-mode laser under suitable conditions) and nonlinear optics [167] was pointed out long times ago. The Schrödinger equation with such an interaction Lagrangian is analogous to a zero-dimensional field theory with a nonlinear Lagrangian in elementary particle physics [156, 168]. Also the 3D analogue was found to produce a sequence of energy levels which is identical to that occurring in the shell model of nucleus [169]. It may be noted that for either of the following situations, $\lambda=0$ or $\lambda=g=0$ or $g<<\lambda$ or large g, the solution behaves as the harmonic oscillator.

The potential in 1D has generated considerable interest among the theoreticians as evidenced by numerous works employing a wide range of methodologies such as variational method, perturbation theory, semi-numerical as well as purely numerical methods. It is possible to obtain *exact* eigenvalues and eigenfunctions of ground and higher states provided the potential parameters obey certain specific relations between them. In [170], existence of a class of solutions (in terms of terminating polynomials or Sturmians of Schrödinger equation with potential $x^2 - \lambda/\{g(1+gx^2)\}, -\infty < x < \infty$) was found out when certain algebraic relations between g, λ are satisfied. In another attempt [171], eigenfunctions were expressed as definite integrals whereas eigenvalues by means of a limiting procedure. Exact even- and odd-parity solutions in the form of products of exponentials and polynomial of x have been investigated [172]. For small $\lambda'(=\lambda/g)$, the eigenvalues were given by,

$$E_n = 2n + 1 + \frac{1}{2}\lambda' - \lambda'(\sqrt{\pi}\, 2^n n!)^{-1} I_n \tag{56}$$

where

$$I_n = \int_0^\infty \exp(-x^2) H_n^2(x)(1-gx^2)/(1+gx^2)dx \tag{57}$$

with $n = 0, 1, 2, \cdots$, and $H_n(x)$ is a Hermite polynomial of order n. Similarly, the first

Table 8. Comparison of some lowest ($n_r = 0$) eigenvalues E (times 2 in a.u.) of 3D NPO for several g and λ with literature data for $\ell = 0, 1, 2$. Exact analytical values are referred by asterisks [180, 176]. GPS results are quoted from [51].

ℓ	g	λ	E: GPS	E: Ref.
0	0.1	−0.46	2.4000000000000	2.400000000000[1],2.4*
0	1	−10	−3.000000000000	−3.000000000000[1],−3*
2	1	−18	−6.999999999999	−7.000000000000[1],−7*
1	0.1	−0.5	3.9999999999999	4.000116[2],4*
1	0.01	−0.041	4.9000000000000	4.899974[2],4.9*
2	10	−1440	−133.000000000000	−133.000000000000[1], −133*

[1] Variational calculation [180]. [2] Shifted 1/N expansion result [181].

four eigenvalues (for large g) were given as,

$$
\begin{aligned}
E_0 &= 1+\lambda'\left(1-\sqrt{\pi}\,g^{-1/2}+\frac{5}{2}g^{-1}\right)\\
E_1 &= 3+\lambda'\left(1-\frac{3}{2}g^{-1}+2\sqrt{\pi}\,g^{-3/2}\right)\\
E_2 &= 5+\lambda'\left(1-\frac{1}{2}\sqrt{\pi}\,g^{-1/2}+\frac{9}{4}g^{-1}\right)\\
E_3 &= 7+\lambda'\left(1-\frac{3}{2}g^{-1}+\frac{3}{2}\sqrt{\pi}\,g^{-3/2}\right)
\end{aligned} \tag{58}
$$

It is possible to supersymmetrize the non-polynomial interaction and in that case, one may find as many as exact analytical solutions (corresponding to ground states of different supersymmetric quantum mechanical system) one wishes [173]. Existence of conditionally exact solutions for 1D NPO has been studied by other authors as well [174, 175, 176]. Possibility of an infinite set of exact solutions of both odd- and even-parity, which could be expressed in terms of a product of exponential and polynomial functions of x^2 for specific relations between λ, g, has been explored in [177, 178, 179] as well.

The literature for 1D NPO is vast; only some of the most important ones are cited here chronologically. A non-perturbative method, in conjunction with a Ritz variational (with a Hermite polynomial basis) method was developed for ground and first two excited states with reasonable success [182]. Using perturbation theory, asymptotic expansions were derived [183] for energies and eigenfunctions in the range of small g and large λ. A combined variational and perturbation theory with properly scaled harmonic oscillator functions as basis set was used as well [184]. A Hill-determinant method [185] was proposed. Ground and first three excited states of the interaction potential were obtained by forming a [6,6] Padé approximant to the energy perturbation series through a hypervirial relation [186]. A perturbed ladder operator method [187] was applied to the resolution of perturbed harmonic oscillator wave equation for cases when the perturbation is expandable in a convergent series of Hermite polynomials. Some other methods are: (i) variety of finite-difference approaches of different flavor [188, 189, 190] with differing accuracy (ii) quasi-polynomial

Table 9. First two ($n_r = 0, 1$) eigenvalues of 3D NPO corresponding to $l = 0 - 3$ for select g, λ along with literature data. Taken from [51].

				Energy	
n_r	ℓ	g	λ	GPS [51]	Literature
0	0	0.1	0.1	3.120081864016	3.1200[1]
1				7.231009980656	7.2312[1]
0	2			7.2439618404219	7.2439618404138<E<7.2439618404260[2], 7.2439618404189[3],7.243927[4],7.244[1]
1				11.317997742355	11.258[1]
0	1	100	100	5.993438790399	– <E<6.389[2], 5.993438873366[3],5.993439[5],5.993565[4],5.9936[1]
1				9.993516159965	9.993516[5],9.994694[4], 9.9946[1]
0	3			9.997153638476	9.9969<E<10.0113[2], 9.997153638602[3],9.997145[5],9.9972[1]
1				13.99715862578	

[1] Ref. [169]. [2] Ref. [200]. [3] Ref. [180]. [4] Ref. [181]. [5] Ref. [193].

solutions with the help of first Heun confluent equation or spheroidal Heun equation [191] (iii) an algebraic perturbation theory, based upon the SO(2,1) dynamical group and a tilting transformation, found to be quite successful for eigenfunctions, eigenvalues in the small g region [192] (iv) supersymmetric as well ordinary WKB method [176] (v) a mixed-continued fraction algorithm [193] (vi) an analytic continuation procedure [194] using a Taylor series, which produces very accurate energies and wave functions (vii) perturbation theory with mixed hypervirial and Hellmann-Feynmann theory [195] (viii) a composite of modified Hill-determinant as used in [172], incorporating an operator method and a vector recurrence [196], gives quite accurate solutions (ix) variational bounds via Rayleigh-Ritz theorem [197] (x) a quadrature discretization technique [198] (xi) purely numerical approach [199], etc. While most of these focus on 1D case, some (such as [193]) deal with both 1D plus 3D and/or N dimensions.

In parallel to the works in 1D, great deal of attention has been paid to investigate 3D NPO eigenvalues, eigenfunctions in the past several decades, although the amount of work is visibly and surprisingly much less compared to the 1D counterpart. Through a supersymmetry-inspired factorization method [201], it was possible to obtain exact algebraic-type solutions under suitable constraints on potential parameters. Right choice of potential parameter leads to compact analytic expressions for exact eigenenergies, eigenfunctions as well the constraint relations. Exact solutions of the NPO in 2D, 3D have been studied in [202] also. Quasi-polynomial solutions in N dimension was suggested by [203] through the Heun confluent equation. Some of the successful works in 3D are quoted here. Through a shifted $1/N$ expansion [169, 181], results of 3D NPO were obtained for 9 sets of n_r, l values for $n = 0 - 4$. An eigenvalue moment method [200] has been quite promising in providing accurate estimates of energy bounds for the 3D NPO. A combined hypervirial and Padé approximation [204] has provided very accurate results on this potential. A unified variational treatment [180] based on the Gol'dman and Krivchenkov Hamiltonian offered very accurate bounds. However, quite unfortunately, while many accurate results for 1D NPO (for example, eigenvalues accurate up to 10 decimal place were obtained in a number of works such as [190, 192, 194, 195, 196, 197, 198]) are available, similar results for *general* states of a 3D NPO for *arbitrary* sets of potential parameters have been obtained in

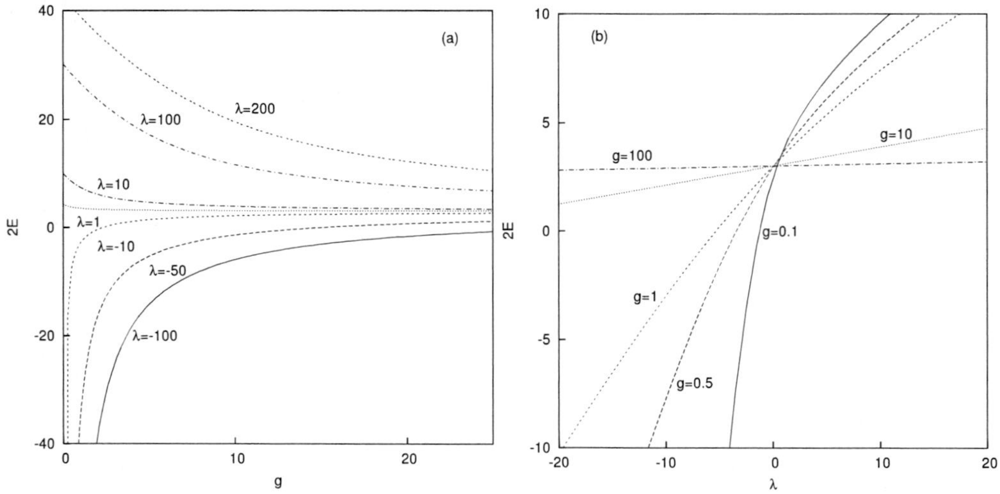

Figure 4. Ground-state energy versus (a) g for fixed λ (b) λ for fixed g, for a 3D NPO.

very few of the works mentioned above. To the best our knowledge, such accurate results could be obtained by only three methods [200, 204, 180]. Even then, leaving aside [204], the other two works deal chiefly with bounds and not provide *direct* eigenvalues. Thus there is a genuine lack of good-quality results for 3D NPO. Moreover, with the rare exception of [169], virtually all these above mentioned works have focused mainly on positive λ, even though it was known for long time that equally well-behaved solutions could be obtained from negative λ provided $g > 0$. Negative λ case has been critically examined in detail for 1D in [196, 197]. In the following paragraph, we will discuss the performance of GPS method.

Table VIII gives a few eigenvalues corresponding to certain levels of 3D NPO for some particular values of g, λ, which offer exact analytical solution (denoted by asterisk). These are available for $\lambda < 0$ and presented for lowest ($n_r = 0$) states of $l = 0 - 2$. Literature results are quoted wherever possible. Some of them were reported long times ago using a shifted 1/N expansion [181]. In all cases, 12-13 place decimal accuracy is easily achieved by the GPS method [51]. These results clearly outperform the best results available so far. Also calculations with $\lambda = 0$ for low as well as high values lead to expected harmonic oscillator solutions promptly (up to 13th place of decimal). This is not presented here and can be found in [51].

Next, Table IX compares the lowest two eigenstates ($n_r = 0, 1$) of $l = 0 - 3$ with the best existing literature data for a sufficiently broad range of interaction (g, λ varied from 0.1–100). For small g, λ, the estimate of bounds obtained in eigenvalue moment method [200] are usually good, but they deteriorate quite badly for larger values of parameters. Significantly improved bounds have been published lately [180]. It is quite clear that our method provides the most accurate direct estimates for all these states. Figure 4 vividly displays the variation of ground-state energy against g for fixed values of λ in (a) and λ for fixed values of g in (b). Similar plots are obtained for higher states as well. For fixed λ, energy steadily decreases as g increases; eventually approaching those of the harmonic oscillator asymptotically, for large g, as expected. With larger λ, this behavior is observed with a greater magnitude. Negative λ shows a correspondingly opposite trend. In (b), changes against λ are seen to be more prominent for smaller g; once again eigenvalues approaching harmonic oscillator energies with an increase in g. Many other features regarding higher states (as high as up to $l = 20$) have been discussed, at length, in [51]. These high-lying states are inherently diffuse and consequently extend over a larger spatial region; hence a larger r_{max} value is needed in order to incorporate these long-range contributions. To our knowledge, only two sets of results are available in the literature; one in the form of lower and upper bounds [200], other is from finite-difference calculations [204]. Our results [51] practically coincide with those from latter in all cases.

In [51], a very detailed analysis was presented for energy variations with respect to interaction parameters (g, λ). For this, 45 sets of (g, λ) were chosen; g values sampled as 0.1, 1, 10, 100, 1000 while λ as -100, -10, -1, -0.1, 0.1, 1, 10, 100, 1000. All 9 states $(1s - 1g)$ of $n = 0 - 4$ corresponding to 9 (n_r, l) pairs of (+)ve λ were studied in [169, 193], while $(-)$ve λ has been studied rarely [169]. Our study extended to 30 states belonging to $n = 0 - 9$. It may be noted that in this context, the spectroscopic notation is more appropriate, i.e., levels are labeled as $n_r + 1$ and l values. So $n_r = 2, l = 1$ signifies a $3p$ level and so on. Interestingly it was found that for $\lambda > 0$, the ordering given as,

$$1s < 1p < 2s < 1d < 2p < 1f < 3s < 2d < 1g < 3p < 2f < 1h < \cdots$$

was obeyed by 23 sets out of 25 sets, excepting (1,100), (10,1000). First occurrence of violation of above ordering was observed between the levels $(1f, 3s)$. Thereafter this ordering is violated in many occasions. It was also numerically verified that as g increases in (1000,0.1) set, NPO eigenenergies approach quantum harmonic oscillator values gradually, and in the limit of $\lambda \to 0$, all levels belonging to a particular n tend to be degenerate. Similar observations hold true for (1000,-0.1) except that, now the eigenvalues approach harmonic oscillator values from below. Out of 20 (g, λ) pairs, with $(-)$ve λ, the ordering given below,

$$1s < 1p < 1d < 2s < 1f < 2p < 1g < 2d < 3s < 1h < 2f < 3p < \cdots$$

was satisfied by 17 sets, while (0.1,-10), (0.1,-100), (1,-100) sets violated this ordering. First instance of such violation occurs for $(2s, 1f)$ and thereafter it happens for several adjacent pairs of states. Finally, it was also noted that, the "usual" or most commonly observed ordering in the cases of $\lambda < 0$ and $\lambda > 0$ for a fixed n follow a mirror-image relationship. For instance, if $n = 9$, the ordering for $\lambda > 0$ is $5p < 4f < 3h < 2j < 1l$ whereas for $\lambda < 0$ this is reversed, i.e., $1l < 2j < 3h < 4f < 5p$.

At this stage, a few words regarding degeneracy issues in 3D NPO are in order. It is well-known that all the (n_r, l) states belonging to a particular n in a 3D harmonic oscillator satisfying $n = 2n_r + l$ are degenerate. For example, there are 3 degenerate states corresponding to (n_r, l) pairs (2,0), (1,2) and (0,4), for $n = 4$. For $\lambda \neq 0$, such degeneracies in a 3D NPO vanishes and these are conveniently analyzed through their respective level spacings, $\Delta E = E_{n_r,l} - E_{n_r',l'}$. First 4 $(n \leq 4)$ such splittings of positive λ were investigated in [169, 193] and more recently in [51], where very similar qualitative features were observed. However, such studies for $\lambda \neq 0$ have been made only recently [51] through GPS method; variation of 12 such splittings possible between certain adjacent levels of $n = 2 - 7$ were considered with respect to the interaction parameters in potential. The first 2 splittings

$E_{1,0} - E_{0,2}$, $E_{1,1} - E_{0,3}$ are related to $n = 2, 3$, whereas the last 3 of them $E_{1,5} - E_{0,7}$, $E_{2,3} - E_{1,5}$ and $E_{3,1} - E_{2,3}$ correspond to $n=7$. Changes in these splittings with respect to λ were followed by varying latter from -0.1 to -100 keeping g fixed at 0.1. The same for g were monitored by changing latter from 1 to 1000 for fixed $\lambda = -100$. All ΔEs tend to increase as $|\lambda|$ increases (eventually approaching a constant value in the limit of $\lambda \rightarrow \pm\infty$) and decrease as g increases. Furthermore, the splittings tend to vanish in the limit of large g for $|\lambda|$. For more detailed discussion on energies, as well as other quantities, see [51].

4.5. Application to Atomic Rydberg and Hollow Resonances

In this subsection, we briefly mention one recent application of GPS method for atomic excited states, with special reference to singly, doubly excited Rydberg resonances in He and triply excited hollow resonances in three electron atoms. For this, the GPS method is used to solve the radial Kohn-Sham (KS) equation within a density functional framework. This approach has been very successfully applied to a broad range of important physical processes in atomic excitations such as multiply excited states, valence as well as core excitations, high-lying Rydberg states, negative atoms, etc. [39, 44, 47, 49]. Dynamical situations have also been treated by this method quite well [40, 41].

Triply excited atomic lithium containing all three electrons in $n \geq 2$ shells leaving the K shell completely empty, constitutes an interesting multi-excited atomic problem. This is a prototypical case of a highly correlated, three-electron system under the influence of a nucleus and thus typifies the well-known four-body Coulombic problem (an ideal system for examining delicate inter-electronic correlation). Since one-step photo-generation of such a state requires coherent excitation of all three available electrons, they pose significantly more difficulty to be produced from ground state by single-photon absorption or electron impact excitations. Besides, their close proximity to more than one thresholds as well presence of an infinite number of open channels offer considerable challenges to both experimentalists and theoreticians. A vast majority of these hollow states are auto-ionizing and have found important practical applications in high-temperature plasma diagnostic by means of high-resolution X-ray spectroscopy.

Development of third-generation, extreme-UV synchrotron radiation as well as availability of several powerful, sophisticated quantum mechanical methodologies, have inspired an overwhelming amount of work in the last three decades towards characterization of these states. Ever since the first electron-He scattering experiment [205], and $2l2l'2l''$ states in Li and highly charged ions in beam-foil experiments [206], many subsequent attempts have been made which helped identify many bound states such as $2p^3\ ^4S^o$ besides some auto-ionizing ones. However, the lowest $2s^22p\ ^2P^o$ resonance in Li was observed in a photo-absorption spectroscopy [207] through a dual laser plasma technique. Thereafter, various higher resonances were found and tentatively classified in a wide range of 140–165 eV. An enormous amount of experimental works have appeared in the literature lately for high-precision determination of these resonance positions, widths, lifetimes, etc (see, for example, [44, 47], for other experimental reference on the subject).

Parallel to the experimental developments, an impressive amount of theoretical works have been reported in literature over the past years, with wide-varying range of complexity, capability, accuracy. However, due to the problems mentioned earlier, accurate and

Table 10. Comparison of singly excited triplet state energies of He (a.u.). Numbers in parentheses in column 3 denote absolute percentage errors with respect to literature data. Taken from [39].

State	−E(X-only)	$-E$(XC)	−E(Ref. [219])
$1s2s\ ^3S$	2.17420 (2.17425[1])	2.17545 (0.0101)	2.17523
$1s3s\ ^3S$	2.06793	2.06890 (0.0102)	2.06869
$1s4s\ ^3S$	2.03606	2.03671 (0.0098)	2.03651
$1s5s\ ^3S$	2.02242	2.02264 (0.0010)	2.02262
$1s7s\ ^3S$	2.01107	2.01115 (0.0010)	2.01113
$1s9s\ ^3S$	2.00658	2.00660 (0.0000)	2.00660
$1s11s\ ^3S$	2.00431	2.00431 (0.0000)	2.00431
$1s13s\ ^3S$	2.00310	2.00310 (0.0000)	2.00310
$1s15s\ ^3S$	2.00231	2.00231 (0.0000)	2.00231
$1s16s\ ^3S$	2.00203	2.00203	

[1] HF result, [220].

dependable characterization of these states has remained a formidable challenge, from a theoretical standpoint. Despite all these, several works are available in the literature. Some of the most successful methods are: (i) truncated diagonalization method [208] (ii) 1/Z expansion method [209] (iii) many-body perturbation theory [210] (iv) state-specific theory [211] (v) configuration interaction [212] (vi) joint saddle point and complex coordinate rotation [213] (vii) a space partition and stabilization procedure [214] (viii) several variants of R-matrix theory [215, 216] (ix) a hyperspherical coordinate approach [217, 218], etc.

Density functional theory (DFT) [221, 222, 223, 224] has emerged as one of the most powerful and successful tools for electronic structure calculation of atoms, molecules, solids in the past four decades. While for ground states its success was conspicuous, it was not so for excited states until only recently. In the last few years, a DFT-based formalism has been proposed for such resonances [39, 44, 47, 49]. This exploited a local non-variational work-function-based exchange potential [225], found to be much more advantageous computationally compared to the non-local Hartree-Fock potential. Earlier, it was demonstrated to be quite successful for *general* atomic excited states [226, 227, 228]. Some of the applications included: singly, doubly, triply excited states, low- and moderately high-lying states, valence and core excitations as well as auto-ionizing and satellite states, etc. However, these all used a Numerov-type finite-difference scheme for discretization of the spatial coordinates and solution of the relevant radial KS equation, which is given as (in atomic units),

$$\left[-\frac{1}{2}\nabla^2 + v_{es}(\mathbf{r}) + v_{xc}(\mathbf{r})\right]\psi_i(\mathbf{r}) = \varepsilon_i\psi_i(\mathbf{r}) \tag{59}$$

where the three terms in left-hand side relate to kinetic, electrostatic and exchange-correlation (XC) contributions. Here $v_{es}(\mathbf{r})$ contains the nuclear-attraction and classical internuclear Coulomb repulsion as,

$$v_{es}(\mathbf{r}) = -\frac{Z}{r} + \int \frac{\rho(\mathbf{r}')}{|\mathbf{r}-\mathbf{r}'|}\mathrm{d}\mathbf{r}' \tag{60}$$

where Z is the nuclear charge. The exchange potential is obtained through a physical interpretation, as the work required to move an electron against an electric field $\mathcal{E}_x(\mathbf{r})$ arising out of its own Fermi-hole charge distribution, $\rho_x(\mathbf{r}, \mathbf{r}')$, and given by a line integral [225],

$$v_x(\mathbf{r}) = -\int_{\infty}^{r} \boldsymbol{\mathcal{E}}_x(\mathbf{r}) \cdot \mathrm{dl}. \tag{61}$$

where

$$\boldsymbol{\mathcal{E}}_x(\mathbf{r}) = \int \frac{\rho_x(\mathbf{r}, \mathbf{r}')(\mathbf{r} - \mathbf{r}')}{|\mathbf{r} - \mathbf{r}'|^3} \, \mathrm{d}\mathbf{r}. \tag{62}$$

For well-defined potentials, work done must be path-independent (irrotational), which is satisfied for spherically symmetric systems such as those studied here. Exchange potential now can be calculated accurately as the Fermi hole is known exactly in terms of single-particle orbitals. Now, within the central-field approximation, $\psi_i(\mathbf{r}) = R_{nl}(r) \; Y_{lm}(\Omega)$. Finally, employing a suitable correlation functional (here we employ one of the most widely used Lee-Yang-Parr potential, [229]), one finally obtains a self-consistent set of orbitals, which gives the electron density as,

$$\rho(\mathbf{r}) = \sum_i |\psi_i(\mathbf{r})|^2. \tag{63}$$

However, due to the presence of Coulomb singularity at origin as well as the long-range nature of Coulomb potential, usual FD methods require a larger number of grid points for accurate calculation even for ground states. Hence it would be even more difficult to capture the long tails for excited states, especially the higher ones such as Rydberg series, we are concerned here. In order to circumvent this problem, GPS method was invoked for accurate efficient solution of the KS equation in this context. In what follows we briefly mention the results on some selected states such as the doubly excited Rydberg states in He and hollow resonances in atomic Li, while [39, 44, 47] could be referred for more complete discussion.

Table X gives a comparison of our calculated GPS results [39] for 1sns ($n = 2 - 16$) ^{3}S states of He with the best available results in literature. Columns 2 and 3 report energies obtained from without and with correlation included. For $n = 2$, the X-only result is only 0.0023% above the HF value [220], indicating excellent accuracy in the present work. For all other states (up to $n = 15$), current energies in both columns 2,3 match very closely with the much elaborate complex rotation calculation [219]. Interestingly, for high-lying Rydberg states ($n > 10$), energies in 3 columns (X only, XC, and reference) are seen to be essentially identical, consistent with the fact that for Rydberg states, asymptotic long-range Coulomb potential (that arises *solely* because of the exchange potential) remains the dominant factor for their electronic structure determination; electron correlation plays very little role. With decrease in n, discrepancies in these 3 columns become noticeable as correlation now plays an increasingly important role. For low n, XC state energies have fallen slightly below the literature values, most probably due to an overestimation caused by the correlation functional used. As n goes to higher values, energy spacings decrease, and present work reproduces this phenomenon very well. It is worth mentioning that for high-lying states, many commonly used quantum mechanical methods encounter a cumbersome problem of self-consistent convergence, chiefly due to the inaccuracies in potential and densities. However, through GPS we succeeded in getting converged results for all states as reported.

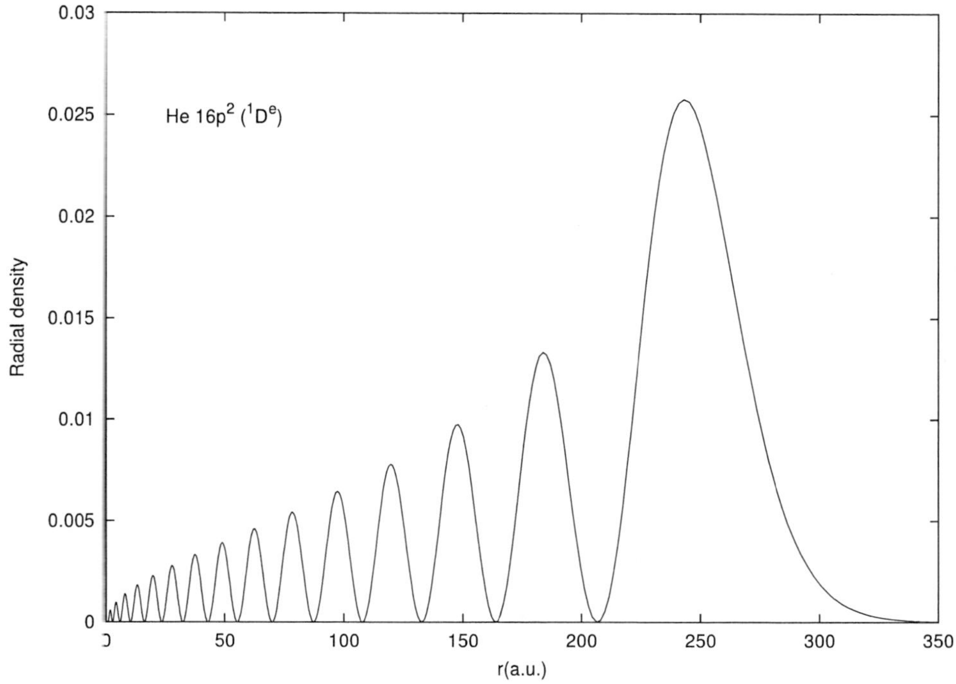

Figure 5. Radial density of He $16p^2\ ^1D^e$ state in a.u. Adopted from [39].

Such results were produced for even higher states by this approach in a straightforward manner (not presented here). Finally even though the method is non-variational, anomaly in energy orderings has not been observed. Figure 5 displays the calculated radial density for $16p^2\ ^1D^e$ state of He. As expected, there are 15 maxima (first peak can be seen after magnification).

Next, some selected hollow states of Li, obtained from GPS method, are reported in Table XI, along with some literature data, in terms of non-relativistic term energies as well as excitation energies. In the literature, usually, only the latter is reported. In a DFT study on low-lying singly excited states of open-shell atoms [228], it was observed that excitation energies from X-only and numerical HF calculations show good agreement with each other. However, excitation energies did not show significant improvements upon inclusion of two correlation energy functionals (one relatively simple local Wigner and the other considerably involved non-local LYP), although excited-state energies were improved significantly. In this table, they both are presented. The latter is estimated with respect to the accurate ground-state energy of Li, -7.47805953 a.u. [231], calculated from a full core plus correlation via a multi-configuration interaction wave function, in order to maintain consistency with literature. The present DFT calculation yields the same as -7.4782839 a.u. For convenience, an independent particle model classification [208, 230] is adopted; thus the six core Li^+ n=2 intra-shell doubly excited states, *viz.*, $2s^2\ ^1S^e$, $2s2p\ ^3P^o$, $2p^2\ ^3P^e$, $2p^2\ ^1D^e$, $2s2p\ ^1P^o$ and $2p^2\ ^1S^e$ are denoted by A,B,C,D,E and F respectively. This table compares the even-parity $\langle A, nd\rangle$ and $\langle D, ns\rangle$ $^2D^e$ Li hollow states with literature data, arising from electronic configurations $2s^2nd$ and $2p^2ns$ having n up to 24,25 respectively. For the first series, to the best of our knowledge, no experimental results are available, whereas for latter,

Table 11. Comparison of state (in a.u.) and excitation energies (in eV) of $2s^2nd$ and $2p^2ns\ ^2D^e$ resonances of Li. 1 a.u.=27.2076544 eV. See [44] for details.

n	$\langle$A,nd$\rangle\ ^2D^e$				$\langle$D,ns$\rangle\ ^2D^e$			
	−E(a.u.)		Excitation energy(eV)		−E(a.u.)		Excitation energy(eV)	
	GPS	Ref.	GPS	Ref.	GPS	Ref.	GPS	Ref.
3	1.9461	1.9614[2]	150.512	149.826[1],150.095[2]	1.8458	1.8069[2]	153.241	152.364[1],154.299[2]
7	1.8995	1.9084[2]	151.780	151.069[1],151.537[2]	1.7618	1.7607[2]	155.526	154.895[1],155.556[2]
9	1.8951	1.9042[2]	151.899	151.185[1],151.652[2]	1.7564	1.7571[2]	155.673	155.026[1],155.654[2]
10	1.8936	1.9029[2]	151.940	151.217[1],151.687[2]	1.7549		155.714	155.065[1]
11	1.8924	1.9020[2]	151.973	151.241[1],151.712[2]	1.7538		155.744	155.102[1]
12	1.8914		152.000	151.260[1]	1.7530		155.765	155.122[1]
16	1.8893		152.057	151.302[1]	1.7512		155.814	155.169[1]
20	1.8884		152.082	151.321[1]	1.7504		155.836	155.191[1]
22	1.8882		152.087	151.327[1]	1.7501		155.844	155.197[1]
24	1.8880		152.094		1.7499		155.850	155.202[1]
25					1.7498		155.853	155.205[1]

[1]Ref. [216]. [2]Ref. [230].

only the lowest state is detected experimentally at 144.77 eV in photo-electron spectroscopy [232]. DFT excitation energy matches excellently with the experimental value (only 0.043 eV lower with a deviation of 0.03%). The term energies are slightly underestimated in all cases with respect to the truncated diagonalization result [230], whereas for latter, some overestimation is noticed for n=2,4–7, which could occur either due to (a) non-variational nature of the exchange potential employed and/or (b) inadequacy of LYP correlation energy functional. Higher resonances (up to n=22,25) for the two series have been theoretically investigated through the R-matrix method [216]; DFT excitation energies show discrepancies in the range of 0.46–0.50% and 0.04–0.58% for them. Finally Fig. 6 depicts radial densities for some selective hollow states of Li, which show the characteristic shell structures through superposition of radial densities.

In [44], DFT calculations have also been reported for twelve $2l2l'nl''$ (n≥2) triply excited hollow resonance series of Li, *viz.*, $2s^2ns\ ^2S^e$, $2s^2np\ ^2P^o$, $2s^2nd\ ^2D^e$, $2p^2ns\ ^2D^e, ^4P^e$, $2s2pns\ ^4P^o$, $2s2pnp\ ^4D^e$, $2p^2np\ ^2F^o, ^4D^o$, $2p^2nd\ ^2G^e$, $^4F^e$ and $2s2pnd\ ^4F^o$, covering a total of about 270 low-, moderately high- and high-lying states. Except for one state, the discrepancy in state energies remained well within 0.98%, whereas GPS excitation energies deviated by 0.02–0.58% with respect to the literature results. As a further extension, additional calculations were also performed for 37 $3l3l'nl''$ (n≥3) doubly hollow resonances *viz.*, $3l3l'nl''(3 \leq n \leq 6)$ ($^2S^e$, $^2P^o$, $^2D^e$, $^2F^o$, $^4S^o$, $^4P^{e,o}$, $^4D^{e,o}$ and $^4F^o$) (seven are n=3 intra-shell type) of Li with both K, L shells empty (up to n=6) in the photon energy range 175.63–180.51 eV, having different symmetries and multiplicities. In this occasion also, DFT calculation shows quite good agreement with recent literature data. In this case, however, literature data is far more sparse compared to the hollow resonances. The most distinctive feature of these are: (a) they are weaker by about an order of magnitude, compared to their hollow cousins (b) broad and having much larger widths [233]. The major difficulty in handling such hollow resonances at higher photon energies arises mainly due to a very rapid increase in the density of triply and other lower excited states of same symmetry, as well as of large number of available open channels, leading to very strong, complicated correlation

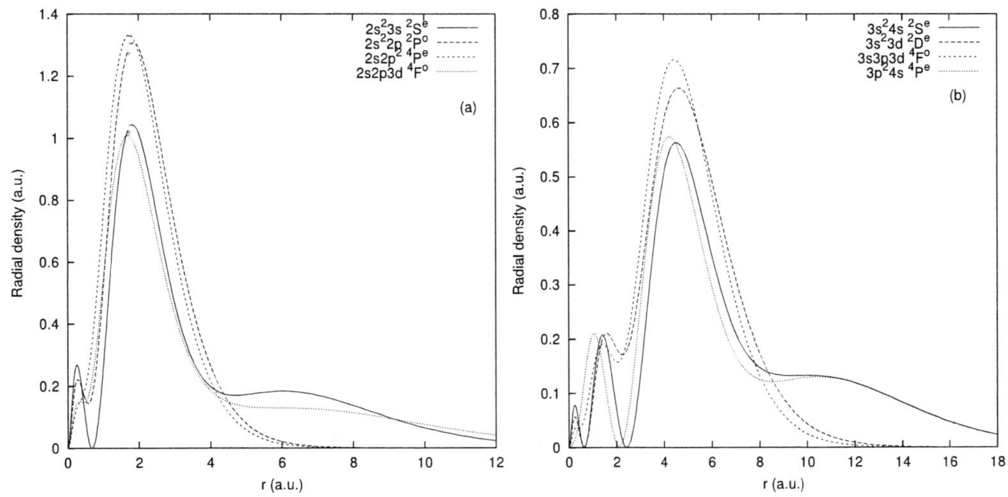

Figure 6. The radial densities (a.u.) of Li for (a) $2s^23s\ ^2S^e$, $2s^22p\ ^2P^o$, $2s2p^2\ ^4P^e$, $2s2p3d\ ^4F^o$ and (b) $3s^24s\ ^2S^e$, $3s^23d\ ^2D^e$, $3s3p3d\ ^4F^o$, $3p^24s\ ^4P^e$ states respectively. Taken from [44].

effects. Nevertheless, some attempts have been made to study these states. Some of the prominent theoretical works include: (a) complex scaling method having correlated basis functions constructed from B splines [234] (b) state specific theory [235], etc. For more detailed discussion as well as available experimental and theoretical works on these, see [44]. Hollow resonances of Li-isoelectronic series have also been studied successfully by this method in [47], where 8 $2l2l'nl''$ ($3 \leq n \leq 6$) hollow resonance series, namely, $2s^2ns\ ^2S^e$, $2s^2np\ ^2P^o$, $2s^2nd\ ^2D^e$, $2p^2ns\ ^2D^e$, $2s2pns\ ^4P^o$, $2s2pnp\ ^4D^e$, $2p^2np\ ^4D^o$, of all the 7 positive ions from Z=4–10 were reported.

5. Concluding Remarks

Quantum mechanics has nowadays spread applications widely in almost every imaginable area in contemporary science and technology, including physics, chemistry, materials science, nanoscience, etc. The focus of this chapter, central potentials, especially those *singular* or near-singular, take the centrestage in atomic, molecular, optical physics and chemistry. However, any exactly solvable quantum system, more so for the singular potentials, are very scarce. Innumerable attempts have been made to solve the respective Schrödinger equation almost ever since the early inception of quantum mechanics. While variational and perturbative approaches remain the most commonly employed, many other attractive, elegant formalisms are available these days, which offer physically meaningful and quite accurate results for various quantum mechanical systems of interest. An enormous number of analytical, semi-numerical, purely numerical techniques have also been developed over the decades for this purpose. Nevertheless, as discussed above, there is still a great need for better approaches. Because, for all the physical systems covered in this work (and probably also true for many other situations not touched upon here) many of these methods would be satisfactory for certain ranges of potential parameters and less successful for other sets.

Moreover, a vast majority of these methods work well for lower states; outstanding difficulties and challenges are encountered for higher states. Additionally extraction of radial density, as well as other expectation values are not straightforward task. Very few methods would satisfy *all* these criteria. In essence, there is a great need for a methodology which can satisfy all these criteria.

Here we have presented an account of the development of GPS method in the context of central (both singular and non-singular) potentials in quantum mechanics. Motivation, background, need for such a method as well its details have been discussed at some length. Although initially designed for Coulombic singular systems, its success for other singular systems as well as for other non-singular central potentials was realized promptly. The formalism is quite *general*, in the sense that it delivers uniformly accurate results for *lower and higher* states of a *broad* spectrum of potentials (describing a variety of physical systems) covering a *wide* range of interaction/coupling. Its usefulness and applicability was demonstrated for some specific potentials in Section IV, such as spiked oscillator, Hulthén, Yukawa, power-law, logarithmic, non-polynomial oscillator, and lastly, some Rydberg and hollow resonances in atomic systems, etc. It has been successfully applied to certain other systems as well, which have not been mentioned at all (such as Hellmann potential or Coulomb potential perturbed by a linear and quadratic coupling) and these could be found in refs [39, 40, 41, 42, 43, 44, 45, 46, 47, 48, 49, 50, 51, 52]. Eigenvalues, eigenfunctions, radial densities, spatial expectation values are obtained in a simple manner. A comparison with literature data reveals that in most cases, results obtained are quite competitive to the best ones or surpasses the accuracy of existing best results available. In almost all cases, it helps to estimate many new states for the first time, which could constitute useful references for future investigations. Finally, it is hoped that this method will continue to remain as a valuable tool for many other physical systems in the future.

Acknowledgments

I express my sincere gratitude and thanks to Prof. F. Columbus, for giving this opportunity to present my work in this exciting area in this edition. I am very much thankful to Nova Publishers, NY, USA, for their generous support and extending deadline for submission. Valuable discussion with Prof. P. Panigrahi is gratefully acknowledged. It is a pleasure to thank my colleagues at IISER, Kolkata, for providing a friendly working environment.

References

[1] M. S. Plesset. *Phys. Rev.*, **41**:278, 1932.

[2] K. M. Case. *Phys. Rev.*, **80**:797, 1950.

[3] E. Predazzi and T. Reggae. *Nuovo Cimento*, **24**:518, 1962.

[4] A. Bastai, L. Bertocchi, S. Fubini, G. Furlan, and M. Tonin. *Nuovo Cimento*, **30**:1512, 1963.

[5] G. Feinberg and A. Pais. *Phys. Rev.*, **131**:2724, 1963.

[6] G. Feinberg and A. Pais. *Phys. Rev.*, **133**:B477, 1964.

[7] R. Loudon. *Am. J. Phys.*, **27**:649, 1959.

[8] R. J. Elliott and R. Loudon. *J. Phys. Chem. Solids*, **15**:196, 1960.

[9] M. W. Cole and M. H. Cohen. *Phys. Rev. Lett.*, **23**:1238, 1969.

[10] M. Andrews. *Am. J. Phys.*, **44**:1064, 1976.

[11] L. S. Davtyan, G. S. Pogosyan, A. N. Sissakian, and V. M. Ter-Antonyan. *J. Phys. A*, **20**:2765, 1987.

[12] H. N. Núñez Yépez, C. A. Vargas, and A. L. Salas Brito. *Eur. J. Phys.*, **8**:189, 1987.

[13] W. Fisher, H. Leschke, and P. Müller. *J. Math. Phys.*, **36**:2313, 1995.

[14] P. Kurasov. *J. Phys. A*, **29**:1767, 1996.

[15] W. Fisher, H. Leschke, and P. Müller. *J. Phys. A*, **30**:5579, 1997.

[16] A. N. Gordeyev and S. C. Chhajlany. *J. Phys. A*, **30**:6893, 1997.

[17] Qian-Shu Li and J. Li. *Chem. Phys. Lett.*, **336**:118, 2001.

[18] G. Palma and U. Raff. *Can. J. Phys.*, **84**:787, 2006.

[19] J. A. Reyes and M. del Castillo-Mussot. *J. Phys. A*, **32**:2017, 1999.

[20] Y. Ran, L. Xue, S. Hu, and Ru-Keng Su. *J. Phys. A*, **33**:9265, 2000.

[21] C. Semay, R. Ceuleneer, and B. Silvestre-Brac. *J. Math. Phys.*, **34**:2215, 1993.

[22] M. M. Nieto. *Am. J. Phys.*, **47**:1067, 1979.

[23] S. Cavalli, V. Aquilanti, and C. Coletti. *J. Chem. Phys.*, **85**:1362, 1986.

[24] D. A. Morales. *Int. J. Quant. Chem.*, **57**:7, 1986.

[25] J. Avery. *Int. J. Quant. Chem*, **41**:673, 1992.

[26] V. Aquilanti and J. Avery. *Chem. Phys. Lett.*, **267**:1, 1997.

[27] V. Aquilanti, S. Cavalli, and C. Coletti. *Chem. Phys.*, **214**:1, 1997.

[28] S. Nouri. *Phys. Rev. A.*, **60**:1702, 1999.

[29] P. F. Bedaque, H.-W. Hammer, and U. von Kolck. *Nucl. Phys. A*, **646**:444, 1999.

[30] P. F. Bedaque, H.-W. Hammer, and U. von Kolck. *Phys. Rev. Lett.*, **82**:463, 1999.

[31] W. M. Frank, D. J. Land, and R. M. Spector. *Rev. Mod. Phys.*, **43**:36, 1971.

[32] S. A. Coon and B. R. Holstein. *Am. J. Phys.*, **70**:513, 2002.

[33] U. von Kolck. *Prog. Part. Nucl. Phys.*, **43**:409, 1999.

[34] H. B. G. Casimir and D. Polder. *Phys. Rev.*, **73**:360, 1948.

[35] E. Vogt and H. Wannier. *Phys. Rev.*, **95**:1190, 1954.

[36] R. G. Newton. *Scattering Theory of Waves and Particles*. McGraw-Hill, New York, 1982.

[37] L. D. Landau and E. M. Lifshitz. *Quantum Mechanics*. Pergamon Press, London, 1965.

[38] F. Arnecke, H. Friedrich, and J. Madroñero. *Phys. Rev. A*, **75**:042903, 2007.

[39] A. K. Roy and S. I. Chu. *Phys. Rev. A*, **65**:052508, 2002.

[40] A. K. Roy and S. I. Chu. *Phys. Rev. A*, **65**:043402, 2002.

[41] A. K. Roy and S. I. Chu. *J. Phys. B*, **35**:2075, 2002.

[42] A. K. Roy. *Phys. Lett. A*, **321**:231, 2004.

[43] A. K. Roy. *J. Phys. G*, **30**:269, 2004.

[44] A. K. Roy. *J. Phys. B*, **37**:4369, 2004.

[45] A. K. Roy. *Pramana-J. Phys.*, **38**:2189, 2005.

[46] A. K. Roy. *Int. J. Quant. Chem.*, **104**:861, 2005.

[47] A. K. Roy. *J. Phys. B.*, **38**:1591, 2005.

[48] K. D. Sen and A. K. Roy. *Phys. Lett. A.*, **357**:112, 2006.

[49] A. K. Roy and A. F. Jalbout. *Chem. Phys. Lett.*, **445**:355, 2007.

[50] A. K. Roy, A. F. Jalbout, and E. I. Proynov. *J. Math. Chem.*, **44**:260, 2008.

[51] A. K. Roy, A. F. Jalbout, and E. I. Proynov. *Int. J. Quant. Chem.*, **108**:827, 2008.

[52] A. K. Roy and A. F. Jalbout. *J. Mol. Struct: Theochem*, **853**:27, 2008.

[53] J. P. Killingbeck, G. Jolicard, and A. Grosjean. *J. Phys. A*, **34**:L367, 2001.

[54] G. Yao and S. I. Chu. *Chem. Phys. Lett.*, **204**:381, 1993.

[55] J. Wang, S. I. Chu, and C. Laughlin. *Phys. Rev. A*, **50**:3028, 1994.

[56] D. Gottlieb, M. Yousuff, and S. A. Orszag. In R. G. Voigt, D. Gottlieb, and M. Y. Hussaini, editors, *Spectral Methods for Partial Differential Equations*, Philadelphia, 1984. SIAM.

[57] C. Canuto, M. Y. Hussaini, A. Quarteroni, and T. A. Zang. *Spectral Methods in Fluid Dynamics*. Springer, Berlin, 1988.

[58] S. Bell, R. Davidson, and P. A. Warsop. *J. Phys. B*, **3**:123, 1970.

[59] P. M. Mathews, M. Seetharaman, and S. Raghavan. *J. Phys. B*, **15**:103, 1982.

[60] P. E. Shanley. *Ann. Phys.*, **186**:292, 1988.

[61] P. E. Shanley. *Ann. Phys.*, **186**:325, 1988.

[62] M. Lakshmanan and P. Kaliappan. *Phys. Rev. A*, **49**:3296, 1994.

[63] L. Skála, J. Dvořák, and V. Kapsa. *Int. J. Theor. Phys.*, **36**:2953, 1997.

[64] E. Delabaere and F. Pham. *Ann. Phys.*, **261**:180, 1997.

[65] J. P. Killingbeck. *J. Phys. B*, **15**:829, 1982.

[66] A. Holubec, A. D. Stauffer, P. Acacia, and J. A. Stauffer. *J. Phys. A*, **23**:4081, 1990.

[67] E. M. Harrell. *Ann. Phys.*, **105**:379, 1977.

[68] J. R. Klauder. *Phys. Lett.*, **47**:523, 1973.

[69] B. DeFacio and C. L. Hammer. *J. Math. Phys.*, **15**:1071, 1974.

[70] H. Ezawa, J. R. Klauder, and L. A. Shepp. *J. Math. Phys.*, **16**:783, 1975.

[71] L. C. Detwiler and J. R. Klauder. *Phys. Rev. D*, **11**:1436, 1975.

[72] J. R. Klauder. *Science*, **199**:735, 1978.

[73] M. Znojil. *J. Math. Phys.*, **23**:30, 1989.

[74] M. Znojil. *J. Math. Phys.*, **31**:108, 1990.

[75] V. C. Aguilera-Navarro, G. A. Estévez, and R. Guardiola. *J. Math. Phys.*, **31**:99, 1990.

[76] V. C. Aguilera-Navarro and R. Guardiola. *J. Math. Phys.*, **32**:2135, 1991.

[77] R. S. Kaushal. *Ann. Phys. (NY)*, **206**:90, 1991.

[78] F. M. Fernández. *Phys. Lett. A*, **160**:511, 1991.

[79] R. Guardiola and J. Ros. *J. Phys. A*, **25**:1351, 1992.

[80] V. C. Aguilera-Navarro, F. M. Fernández, R. Guardiola, and J. Ros. *J. Math. Phys.*, **25**:6379, 1992.

[81] R. S. Kaushal and D. Parashar. *Phys. Lett. A*, **170**:335, 1992.

[82] M. Znojil. *Phys. Lett. A*, **169**:415, 1992.

[83] W. Solano-Torres, G. A. Estévez and F. M. Fernández, and G. C. Groenenboom. *J. Phys. A*, **25**:3427, 1992.

[84] E. S. Estévez-Bretón and G. A. Estévez-Bretón. *J. Math. Phys.*, **34**:437, 1993.

[85] M. Landtman. *Phys. Lett. A*, **175**:147, 1993.

[86] Y. P. Varshni. *Phys. Lett. A*, **183**:9, 1993.

[87] V. C. Aguilera-Navarro and A. L. Coelho and N. Ullah. *Phys. Rev. A*, **49**:1477, 1994.

[88] H. G. Miller. *J. Math. Phys.*, **35**:2229, 1994.

[89] R. L. Hall and N. Saad. *Can. J. Phys.*, **73**:493, 1995.

[90] E. Buendía and F. J. Gálvez and A. Puertas. *J. Phys. A*, **28**:6731, 1995.

[91] C. R. Handy. *Phys. Lett. A*, **216**:15, 1996.

[92] K. Bhattacharyya and R. K. Pathak. *Int. J. Quant. Chem.*, **59**:219, 1996.

[93] J. Trost and H. Friedrich. *Phys. Lett. A*, **228**:127, 1997.

[94] R. L. Hall and N. Saad. *J. Phys. A*, **31**:963, 1998.

[95] R. L. Hall, N. Saad, and A. von Keviczky. *J. Math. Phys.*, **39**:6345, 1998.

[96] O. Mustafa and M. Odeh. *J. Phys. B*, **32**:3055, 1999.

[97] O. Mustafa and M. Odeh. *J. Phys. A*, **33**:5207, 2000.

[98] R. L. Hall and N. Saad. *J. Phys. A*, **33**:569, 2000.

[99] R. L. Hall and N. Saad. *J. Phys. A*, **33**:5531, 2000.

[100] R. L. Hall, N. Saad, and A. von Keviczky. *J. Phys. A*, **34**:1169, 2001.

[101] B. Chakrabarti and T. K. Das. *J. Phys. A*, **35**:4701, 2002.

[102] B. Chakrabarti and T. K. Das. *Mod. Phys. Lett. A*, **17**:1367, 2002.

[103] R. L. Hall, N. Saad, and A. B. von Keviczky. *J. Math. Phys.*, **43**:94, 2002.

[104] R. Guardiola and F. M. Fernández. *Eur. Phys. J. D*, **21**:19, 2002.

[105] N. Saad, R. L. Hall, and A. B. von Keviczky. *J. Math. Phys.*, **44**:5021, 2003.

[106] N. Saad, R. L. Hall, and A. B. von Keviczky. *J. Phys. A*, **36**:487, 2003.

[107] J. K. Bhattacharjee and S. Bhattacharyya. *J. Phys. A*, **36**:L223, 2003.

[108] S. K. Bandyopadhyay and K. Bhattacharyya. *Int. J. Quant. Chem.*, **103**:19, 2005.

[109] S. K. Bandyopadhyay and K. Bhattacharyya. *Int. J. Quant. Chem.*, **106**:390, 2005.

[110] R. M. Spector. *J. Math. Phys.*, **5**:1185, 1964.

[111] G. A. Baker, Jr. *Rev. Mod. Phys.*, **43**:479, 1971.

[112] A. Holubec and A. D. Stauffer. *J. Phys. A*, **18**:2141, 1985.

[113] L. Hulthén. *Ark. Mat. Astron. Fys. A*, **28**:5, 1942.

[114] L. Hulthén. *Ark. Mat. Astron. Fys. B*, **29**:1, 1942.

[115] B. Durand and L. Durand. *Phys. Rev. D*, **23**:1092, 1981.

[116] W. van Dijk. *Phys. Rev. C*, **40**:1437, 1989.

[117] C. S. Lam and Y. P. Varshni. *Phys. Rev. A*, **4**:1875, 1971.

[118] R. Dutt and U. Mukherji. *Phys. Lett.*, **90A**:395, 1982.

[119] C. S. Lam and Y. P. Varshni. *Phys. Rev. A*, **6**:1391, 1972.

[120] C. S. Lam and Y. P. Varshni. *Phys. Status Solidi*, **B29**:103, 1978.

[121] J. A. Olson and D. A. Micha. *J. Chem. Phys.*, **68**:4352, 1978.

[122] H. Yukawa. *Proc. Phys. Math. Soc. Jpn.*, **17**:48, 1935.

[123] O. A. Gomes, H. Chacham, and J. R. Mohallem. *Phys. Rev. A*, **50**:228, 1994.

[124] Y. P. Varshni. *Phys. Rev. A*, **41**:4682, 1990.

[125] C. Stubbins. *Phys. Rev. A*, **48**:220, 1993.

[126] C. S. Lai and W. C. Lin. *Phys. Lett.*, **78A**:335, 1980.

[127] S. H. Patil. *J. Phys. A*, **17**:575, 1984.

[128] V. S. Popov and V. M. Weinberg. *Phys. Lett.*, **107A**:371, 1985.

[129] J. M. Cai, P. Y. Cai, and A. Inomata. *Phys. Rev. A*, **34**:4621, 1986.

[130] A. Z. Tang and F. T. Chan. *Phys. Rev. A*, **35**:911, 1987.

[131] B. Roy and R. Roychoudhury. *J. Phys. A*, **20**:3051, 1987.

[132] B. I. Dunlap and L. Armstrong, Jr. *Phys. Rev. A*, **6**:1370, 1972.

[133] P. Matthys and H. De Meyer. *Phys. Rev. A*, **38**:1168, 1988.

[134] R. L. Hall. *J. Phys. A*, **25**:1373, 1992.

[135] V. Fessatidis, J. D. Mancini, Q. Haider, Y. Zhou, and L. Greco. *Phys. Lett. A*, **242**:74, 1998.

[136] B. Gönül and O. Özer and Y. Cançelic and M. Koçak. *Phys. Lett. A*, **275**:238, 2000.

[137] S. H. Patil. *J. Phys. A*, **34**:3153, 2001.

[138] R. L. Greene and C. Aldrich. *Phys. Rev. A*, **14**:2363, 1976.

[139] K. M. Roussel and R. F. O'Connell. *Phys. Rev. A*, **9**:52, 1974.

[140] C. S. Lai. *Phys. Rev. A*, **23**:455, 1981.

[141] T. Imbo, A. Pagnamenta, and U. Sukhatme. *Phys. Lett.*, **105A**:183, 1984.

[142] G. Moreno and A. Zepeda. *J. Phys. B*, **17**:21, 1984.

[143] A. Chatterjee. *J. Phys. A*, **18**:1193, 1985.

[144] H. Christiansen and L. N. Epele and H. Fanchiotti and C. A. Garcia Canal. *Phys. Rev. A*, **40**:1760, 1989.

[145] E. R. Vrscay. *Phys. Rev. A*, **33**:1433, 1986.

[146] F. J. Rogers, H. C. Graboske, Jr., and D. J. Harwood. *Phys. Rev. A*, **1**:1577, 1970.

[147] M. A. Núñez. *Phys. Rev. A*, **47**:3620, 1993.

[148] C. Quigg and J. L. Rosner. *Phys. Lett. B*, **71**:153, 1977.

[149] S. N.Jena and D. P. Rath. *Phys. Rev. D*, **34**:196, 1986.

[150] H. Akcay and H. Ciftci. *J. Phys. G*, **22**:455, 1996.

[151] S. N. Jena, P. Panda, and T. C. Tripathy. *J. Phys. G*, **27**:227, 2001.

[152] M. Schwarszchild. *Astrophys. J.*, **232**:236, 1979.

[153] C. Stoica and A. Font. *J. Phys. A*, **36**:7693, 2003.

[154] H. Ciftci, R. L. Hall, and Q. D. Katatbeh. *J. Phys. A*, **36**:7001, 2003.

[155] C. Quigg and J. L. Rosner. *Phys. Rep.*, **56**:206, 1979.

[156] S. N. Biswas, K. Datt, R. P. Saxena, P. K. Srivastava, and V. S. Varma. *J. Math. Phys.*, **14**:1190, 1973.

[157] M. M. Nieto and L. M. Simons. *Am. J. Phys.*, **47**:634, 1979.

[158] R. E. Crandall and M. H. Reno. *J. Math. Phys.*, **23**:64, 1982.

[159] U. Sukhatme and T. Imbo. *Phys. Rev. D*, **28**:418, 1983.

[160] T. Imbo, A. Pagnamenta, and U. Sukhatme. *Phys. Rev. D*, **29**:1669, 1984.

[161] S. A. Maluendes, F. M. Fernández, A. M. Mesón, and E. A. Castro. *Phys. Rev. D*, **34**:1835, 1986.

[162] R. L. Hall. *Phys. Rev. A*, **39**:5500, 1989.

[163] R. L. Hall and N. Saad. *J. Math. Phys.*, **38**:4909, 1997.

[164] R. L. Hall and N. Saad. *J. Phys. A*, **36**:7173, 2003.

[165] H. Çiftci and E. Ateşer and H. Koru. *J. Phys. A*, **36**:3821, 2003.

[166] H. Risken and H. D. Vollmer. *Z. Phys. D*, **201**:323, 1967.

[167] H. Haken. Laser Theory. In S. Flügge and L. Genzel, editors, *Encyclopedia of Physics, vol XXV/2c*, Berlin, 1970. Springer-Verlag.

[168] S. N. Biswas, K. Datt, R. P. Saxena, P. K. Srivastava, and V. S. Varma. *Phys. Rev. D*, **14**:3617, 1978.

[169] Y. P. Varshni. *Phys. Rev. A*, **36**:3009, 1987.

[170] R. R. Whitehead, A. Watt, G. P. Flessas, and M. A. Nagarajan. *J. Phys. A*, **15**:1217, 1982.

[171] G. P. Flessas. *J. Phys. A*, **15**:L97, 1982.

[172] R. N. Chaudhuri and B. Mukherjee. *J. Phys. A*, **16**:4031, 1983.

[173] P. Roy and R. Roychoudhury. *Phys. Lett. A*, **122**:275, 1987.

[174] G. P. Flessas. *Phys. Lett.*, **83A**:121, 1981.

[175] V. S. Varma. *J. Phys. A*, **14**:L489, 1981.

[176] P. Roy and R. Roychoudhury Y. P. Varshni. *J. Phys. A*, **21**:1589, 1988.

[177] M. H. Blecher and P. G. L. Leach. *J. Phys. A*, **20**:5923, 1987.

[178] J. A. C. Gallas. *J. Phys. A*, **21**:3393, 1988.

[179] G. V. Berghe and H. E. De Meyer. *J. Phys. A*, **22**:1705, 1989.

[180] N. Saad, R. L. Hall, and H. Ciftci. *J. Phys. A*, **39**:7745, 2006.

[181] B. Roy, R. Roychoudhury, and P. Roy. *J. Phys. A*, **21**:1579, 1988.

[182] A. K. Mitra. *J. Math. Phys.*, **19**:2018, 1978.

[183] R. S. Kaushal. *J. Phys. A*, **12**:L253, 1979.

[184] N. Bessis and G. Bessis. *J. Math. Phys.*, **21**:2780, 1980.

[185] A. Hautot. *J. Comp. Phys.*, **39**:72, 1981.

[186] C. S. Lai and H. E. Lin. *J. Phys. A*, **15**:1495, 1982.

[187] N. Bessis, G. Bessis, and G. Hadinger. *J. Phys. A*, **16**:497, 1983.

[188] V. Fack and G. V. Berghe. *J. Phys. A*, **18**:3355, 1985.

[189] M. R. M. Witwit. *J. Phys. A*, **25**:503, 1992.

[190] M. R. M. Witwit. *J. Comp. Appl. Math. Phys.*, **69**:331, 1996.

[191] G. Marcilhacy and R. Pons. *J. Phys. A*, **18**:2441, 1985.

[192] V. Fack, H. De Meyer, and G. V. Berghe. *J. Math. Phys.*, **27**:1340, 1986.

[193] H. Scherrer, H. Risken, and T. Leiber. *Phys. Rev. A*, **38**:3949, 1988.

[194] R. J. W. Hodgson. *J. Phys. A*, **21**:1563, 1988.

[195] M. R. M. Witwit and J. P. Killingbeck. *Can. J. Phys.*, **70**:1261, 1992.

[196] R. K. Agrawal and V. S. Varma. *Phys. Rev. A*, **48**:1921, 1993.

[197] C. Stubbins and M. Cornstein. *Phys. Lett. A*, **202**:34, 1995.

[198] H. Chen and B. D. Shizgal. *J. Math. Chem.*, **24**:321, 1998.

[199] H. Ishikawa. *J. Phys. A*, **35**:4453, 2002.

[200] C. R. Handy, H. Hayes, D. V. Stephens, J. Joshua, and S. Summerour. *J. Phys. A*, **26**:2635, 1993.

[201] R. Adhikari, R. Dutt, and Y. P. Varshni. *J. Math. Phys.*, **32**:447, 1991.

[202] S. K. Bose and N. Varma. *Phys. Lett. A*, **141**:141, 1989.

[203] R. Pons and G. Marcilhacy. *Phys. Lett. A*, **152**:235, 1991.

[204] M. R. M. Witwit. *J. Phys. A*, **24**:5291, 1991.

[205] C. E. Kuyatt, J. A. Simpson, and S. R. Mielczrek. *Phys. Rev.*, **138**:A385, 1965.

[206] R. Bruch, G. Paul, J. Andrä, and L. Lipsky. *Phys. Rev. A*, **12**:1808, 1975.

[207] L. M. Kiernan, E. T. Kennedy, J-P. Mosnier, J. T. Costelli, and B. F. Sonntag. *Phys. Rev. Lett.*, **72**:2359, 1994.

[208] M. J. Conneely and L. Lipsky. *Phys. Rev. A*, **61**:032506, 2000.

[209] U. I. Safronova and V. S. Senashenko. *J. Phys. B*, **11**:2623, 1978.

[210] R. L. Simons, H. P. Kelly, and R. Bruch. *Phys. Rev. A*, **19**:682, 1979.

[211] C. A. Nicolaides and N. A. Piangos. *J. Phys. B*, **34**:99, 2001.

[212] N. Vacek and J. E. Hansen. *J. Phys. B*, **25**:883, 1992.

[213] Y. Zhang and K. T. Chung. *Phys. Rev. A*, **58**:1098, 1998.

[214] H. Bachau. *J. Phys. B*, **29**:4365, 1996.

[215] K. Berrington and S. Nakazaki. *J. Phys. B*, **31**:313, 1998.

[216] H. L. Zhou, S. T. Manson, P. Faucher, and L. Vo Ky. *Phys. Rev. A*, **62**:012707, 2000.

[217] T. Morishita and C. D. Lin. *Phys. Rev. A*, **59**:1835, 1999.

[218] T. Morishita and C. D. Lin. *Phys. Rev. A*, **67**:022511, 2003.

[219] A. Bürgers, D. Wintgen, and J-M. Rost. *J. Phys. B*, **28**:3163, 1995.

[220] C. F. Fischer. *The Hartree-Fock Method for Atoms*. John Wiley, New York, 1977.

[221] R. G. Parr and W. Yang. *Density Functional Theory of Atoms and Molecules*. Oxford University Press, New York, 1989.

[222] D. Joulbert (Eds). *Density Functionals: Theory and Applications*. Springer, Berlin, 1998.

[223] J. F. Dobson, G. Vignale, and M. P. Das (Eds). *Density Functional Theory: Recent Progress and New Directions*. Plenum, New York, 1998.

[224] W. Koch and M. C. Holthausen. *A Chemist's guide to Density Functional Theory*. John Wiley, New York, 2001.

[225] M. Harbola and V. Sahni. *Phys. Rev. Lett.*, **62**:489, 1989.

[226] A. K. Roy, R. Singh, and B. M. Deb. *J. Phys. B*, **30**:4763, 1997.

[227] A. K. Roy, R. Singh, and B. M. Deb. *Int. J. Quant. Chem.*, **65**:317, 1997.

[228] R. Singh, A. K. Roy, and B. M. Deb. *Chem. Phys. Lett.*, **296**:530, 1998.

[229] C. Lee, Y. Wang, and R. G. Parr. *Phys. Rev. B*, **37**:785, 1988.

[230] M. J. Conneeely and L. Lipsky. *At. Data Nucl. Data Tables*, **82**:115, 2002.

[231] K. T. Chung. *Phys. Rev. A*, **44**:5421, 1991.

[232] D. Cubaynes, S. Diehl, L. Journel, B. Rouvellou, J.-M. Bizau, S. Al Moussalami, F. J. Wuilleumier, N. Berrah, L. Vo Ky, P. Faucher, A. Hibbert, C. Blancard, E. T. Kennedy, T. J. Morgan, J. Bozek, and A. S. Schlachter. *Phys. Rev. Lett.*, 77:2194, 1996.

[233] Y. Azuma, F. Koike, J. W. Cooper, T. Nagata, G. Kutluk, E. Shigemasa, R. Wehlitz, and I. A. Sellin. *Phys. Rev. Lett.*, **79**:2419, 1997.

[234] L. B. Madsen, P. Schlagheck, and P. Lambropoulos. *Phys. Rev. A*, **62**:062719, 2000.

[235] N. A. Piangos and C. A. Nicolaides. *Phys. Rev. A*, **67**:052501, 2003.

In: Mathematical Chemistry
Editor: W.I. Hong, pp. 607-612

ISBN 978-1-60876-894-3

Chapter 14

A Review of Reduction Methods for Multiple Time Scale Stochastic Reaction Networks

***Chang Hyeong Lee*[1,*] *and Roger Lui*[2,†]**
[1]School of Mechanical and Advanced Materials Engineering,
Ulsan National Institute of Science and Technology(UNIST),
Ulsan, South Korea
[2]Department of Mathematical Sciences, Worcester Polytechnic Institute,
Worcester, MA, US

Abstract

Simulations of complex stochastic reaction networks are often performed by Monte Carlo methods such as the Gillespie algorithm. In case that a stochastic reaction network contains certain fast reactions, simulations of the fast reactions by Monte Carlo methods are computationally inefficient and expensive, and researches have focused on reduction methods for fast reactions to reduce the computational intensity of the simulations. In this article, we first review reduction methods for multiple time scale stochastic reaction networks and then discuss possible future works on such methods.

1. Recent Results on Reduction Methods

When a chemical or biological system is characterized by the interactions of a small number of molecules, the stochastic fluctuations, or noise, inherent in molecular interactions may affect the dynamics of the system [1]. To understand such effects, researchers have modeled each molecular species as a discrete random entity. In the discrete stochastic modeling the reaction events are modeled by a continuous-time Markov jump process and the number of each individual species is a random variable. Let $n_i(t)$ be the number of molecules of species i at time t and let $\mathbf{n}(t) = (n_1(t), n_2(t), \ldots, n_s(t))$ be a random vector with probability density $p(\mathbf{n}, t)$. For a reaction system with r reactions the governing

*E-mail address: chlee@unist.ac.kr (Corresponding author)
†E-mail address: rlui@wpi.edu

equation, called chemical master equation (CME), is

$$\frac{dp(\mathbf{n},t)}{dt} = \sum_{k=1}^{r} [a_k(\mathbf{n}-V_k)p(\mathbf{n}-V_k,t) - a_k(\mathbf{n})p(\mathbf{n},t)] \tag{1}$$

where a_k is the propensity function of the kth reaction and V_k denotes the kth column of the stoichiometry matrix V [2].

The master equation can be written equivalently as a Kolmogorov equation in matrix form by labeling all accessible states of $\mathbf{n}$ and computing the transition rates between them. Let $\mathbf{m}$ denote the k possible states of $\mathbf{n}$. Then the evolution of the system is described by

$$\frac{d\mathbf{p}(t)}{dt} = K\,\mathbf{p}(t) \tag{2}$$

where $\mathbf{p}(t)$ is a k-vector whose i^{th} component is Prob$\{\mathbf{n}(t) = \mathbf{m}^i\}$, K is a matrix of transition rates between the states defined, for $i \neq j$, by

$$K_{ij} = \begin{cases} a_\ell(\mathbf{n}) & \text{if } \mathbf{m}^i = \mathbf{m}^j + V_\ell \quad \text{for some } \ell = 1,\ldots,r \\ 0 & \text{otherwise} \end{cases}$$

and $K_{ii} = -\sum_{j\neq i} K_{ji}$.

Equation (1) is often impossible to solve. Also, the solution of (2), namely $\mathbf{p}(t) = e^{Kt}\mathbf{p}(0)$, is difficult or impossible to compute if K is large or if it is infinite-dimensional when there are infinitely many states. Therefore, researchers rely on simulating the solutions of (1) by Monte Carlo methods such as the Gillespie stochastic simulation algorithm (SSA) [3].

If a reaction network has multiple time scale that includes both fast and slow reactions, then simulations of the fast reactions by Monte Carlo methods are often intensive because the SSA simulates every reaction event that occurs in the network. There have been two different approaches for reducing the computational load for simulation of such a network. One is time-leaping method such as the τ-leaping method, and the other is the reduction of fast kinetics, after partition of all reactions into fast and slow ones, under the so-call quasi-steady-state assumption (QSSA) or quasi-equilibrium assumption (QEA). In this article, we focus on the latter. Readers who are interested in the time-leaping method may refer to [4] and references therein.

One of the first analytical works in the reduction of fast kinetics was due to Janssen who proposed a method for adiabatic elimination of fast variables in stochastic description of chemical reaction networks [5, 6]. Starting with a master equation description of the network, Janssen showed in a few examples that intermediate chemical species are eliminated after reactions are partitioned into fast and slow ones, and then he derived a reduced CME that shows a coarse description of the stochastic dynamics of networks. Haseltine and Rawling, and Rao and Arkin were the first to make important contributions in implementation of efficient SSA for multiple time scale reaction networks [7, 8]. Haseltine and Rawling used the extents, the number of reactions occurred in a given time interval, as variables, and derived a CME in terms of the extents $\boldsymbol{x}$ by using the relation $\mathbf{n}(t) = \mathbf{n}(0) + V\boldsymbol{x}(t)$ [7]. After separating the m-dimensional variable $\boldsymbol{x}$ into l-dimensional fast variable $\boldsymbol{z}$ and $(m-l)$-dimensional slow variable $\boldsymbol{y}$ and summing over $\boldsymbol{y}$, they derived the CME for probability of $\boldsymbol{z}$

$$\frac{dP(\boldsymbol{z};t)}{dt} = \sum_{k=1}^{l} b_k(\boldsymbol{z}-\delta_k^z)P(\boldsymbol{z}-\delta_k^z;t) - b_k(\boldsymbol{z})P(\boldsymbol{z};t)$$

where $b_k(\boldsymbol{z}) = \sum_{\boldsymbol{y}} a_k(\boldsymbol{y},\boldsymbol{z})P(\boldsymbol{y}|\boldsymbol{z};t)$, a_k is the propensity of kth fast reaction and δ_k^z is the kth column of the $l \times l$ unit matrix δ^z. Under the assumption that the volume Ω of the system is of sufficient magnitude, Haseltine and Rawlings approximated the above CME for $\boldsymbol{z}$ by a stochastic differential equation (SDE) and implemented a modified Gillespie algorithm, by putting the SDE as a constraint on the probability $P(\boldsymbol{y};t)$ of the slow variable $\boldsymbol{y}$. They showed that it accurately simulates the mean and standard deviation of each species in two examples such as an intracellular viral infection model and also simulation speed of the algorithm is much faster than that of the exact Gillespie algorithm. While Haseltine and Rawlings used the extents as variables, Rao and Arkin utilized the number of species as variables [8]. After dividing the set of species $\boldsymbol{x}$ into primary species $\boldsymbol{y}$ and intermediate species $\boldsymbol{z}$, they postulated two conditions on the conditional variable $\boldsymbol{z}|\boldsymbol{y}$ that $\boldsymbol{z}|\boldsymbol{y}$ is Markovian and it is in the quasi-steady-state on the slow time scale. Under the two assumptions, the authors successfully estimated the steady-state probability $P(\boldsymbol{z}|\boldsymbol{y})$ and used it to approximate the probability $P(\boldsymbol{y};t)$ of the primary species $\boldsymbol{y}$. They implemented a modified version of the Gillespie algorithm and, to show the accuracy and the efficiency of their algorithm, they illustrated simulation results of three examples such as an enzyme-substrate model. In the enzyme-substrate model, Rao and Arkin proposed a stochastic version of Michaelis-Menten approximation. When the molecular number of substrate is larger than that of enzyme initially, the numerical results by the approximation compared well with those of the exact Gillespie algorithm, but as they mentioned in their paper, when the molecular number of enzyme is larger than that of substrate , result by their approximation method diverges significantly and the method cannot be applied. [1]

Since the works of Haseltine and Rawlings, and Rao and Arkins, reduction methods for implementation of more efficient SSA for multiple time scale reaction networks have been proposed by other researchers. Salis and Kaznessis proposed a "Next Reaction" hybrid method which partitions multiple time scale reaction networks dynamically and solves fast reactions using an SDE approximation [9]. Their numerical results compared well to exact stochastic simulation of the full system. Following Haseltine and Rawlings's way of separation of variables into fast and slow ones, Goutsias proposed an approximation method of fast kinetics [10]. Defining the degree of advancement (DA) or the extent as variables and using a Taylor series expansion of the slow propensity functions, he showed in a rigorous way that the propensity functions of fast reactions can be approximated by conditional means of the DAs of fast reactions given DAs of slow reactions. More specifically, he obtained an approximate master equation on a coarse time scale:

$$\frac{dP(\boldsymbol{z}_s,t)}{dt} = \sum_{m\in\mathcal{M}_s} a_m^{(t)}(\boldsymbol{z}_s-\boldsymbol{e}_m)P(\boldsymbol{z}_s-\boldsymbol{e}_m,t) - a_m^{(t)}(\boldsymbol{z}_s)P(\boldsymbol{z}_s,t)$$

[1] A reduction method was presented which works well in both cases of the enzyme-substrate model [16].

where $a_m^{(t)}(\boldsymbol{z}_s) = \sum_{\boldsymbol{z}_f} a_m(\boldsymbol{z}_s, \boldsymbol{z}_f)P(\boldsymbol{z}_f|\boldsymbol{z}_s, t), m \in \mathcal{M}_f$, the variable $\boldsymbol{z}_f$ and $\boldsymbol{z}_s$ denote the DAs of fast and slow reactions, respectively, $\mathcal{M}_f$ and $\mathcal{M}_s$ denote the sets of fast and slow reactions, respectively, a_m is the propensity of the mth-reaction, and $\boldsymbol{e}_m$ is a unit vector. Moreover, Goutsias showed that under the linear dependence of propensity of slow reactions on fast DAs, the function $a_m^{(t)}$ is computed as $a_m^{(t)}(\boldsymbol{z}_s) = a_m(\boldsymbol{z}_s, \boldsymbol{\mu}(\boldsymbol{z}_s, t))$ where $\boldsymbol{\mu}(\boldsymbol{z}_s, t)$ is the vector of $\mu_m(\boldsymbol{z}_s, t)$ that is the conditional expectation of DAs of the mth fast reaction given slow DAs $\boldsymbol{z}_s$. Cao *et al.* proposed a slow time scale stochastic simulation algorithm [11]. Under the assumption that reaction channels can be partitioned into fast and slow ones, they first defined fast and slow species according to whether the species are affected by occurrence of fast reactions or not, and defined virtual fast processes after switching off all slow reactions. Utilizing first and second moments of steady-state distribution of the fast processes, they successfully approximated propensity functions of slow reactions in a system with stable fast processes. Their numerical results for a system with one independent fast species were compared well with the results by the exact Gillespie algorithm. As the authors mentioned, there may be difficulties in applying their method to systems with more than one independent fast variable, since it is difficult to find first and second moments at steady-state of fast processes in such systems. E *et al.* proposed a nested stochastic simulation algorithm that is a modified SSA with inner loops for simulation of fast reactions and outer loops for simulation of slow reactions [12]. Hellander and Lötstedt proposed a hybrid method that uses deterministic description for fast kinetics and stochastic description for slow kinetics [13].

Another approach on the reduction method has been made on the Kolmogorov system $d\boldsymbol{p}/dt = K\boldsymbol{p}$. In this approach, unlike the Monte-Carlo method such as SSA, approximation of the exact solution of the Kolmogorov system has been sought. Munsky and Khammash proposed the finite state projection (FSP) method which approximates the exponential solution of the Kolmogorov system [14]. Using the FSP method and singular perturbation theory, Peleš *et al.* presented a model reduction method which approximates master equations to first order of accuracy for certain linear reaction systems [15]. Recently, for multiple time scale reaction networks whose fast subsystem has unique equilibrium probability, Lee and Lui presented a rigorous reduction method that leads to a system of reduced equations whose solutions can successively approximate the exact probability to any degree of accuracy[16].

2. Future Work

A promising area of research on the reduction method will be the development of an adaptive hybrid method. The reduction methods described in the previous section are based on a priori partition of reactions into fast and slow ones. However, in many real complex reaction networks, reactions that occur rapidly in the initial time period can slow down in a long time scale or vice versa, and thus it is not practically simple to partition reactions a priori into fast and slow ones. In this case, most of the previous reduction methods may not be applied and one may need to have a new method which is automatically adaptive to change of regimes of fast and slow kinetics. Developing such an adaptive and accurate hybrid method for simulation of complex networks will be a promising and challenging future work in this area.

Another work to be done in the future will be on the reduction methods on the Kolmogorov system of infinite dimension. The reduction methods developed so far on the Kolmogorov systems can be used mainly for reaction networks which have bounded state space. When the state space of a reaction network is unbounded, one has to consider all transitions between infinitely many possible states to describe the system dynamics. In other words, one has to deal with infinite dimensional Kolmogorov system $d\boldsymbol{p}/dt = K\boldsymbol{p}$ with infinite dimensional transition matrix K and probability vector $\boldsymbol{p}$. It is mathematically difficult to develop a reduction method for Kolmogorov system of such networks because very little is known about ODE systems of infinite dimension. This problem is important and should be solved because many reaction networks, especially open networks, have unbounded state space.

Another possible work will be implementation of stochastic simulation algorithm for systems that has fast subsystem with non-unique equilibrium probabilities. In most of the previous reduction methods, it is assumed that the fast subsystems in a multiple time scale network have unique equilibrium probabilities. However, in many real chemical systems, this assumption is not satisfied. For example, if a fast reaction is a branching reaction, which is very common in chemical reaction networks, the equilibrium probability of the fast subsystem is not uniquely determined. In this case, the previous reduction methods may not be applied. Observing this fact, Lee and Lui presented a reduction method for Kolmogorov system of multiple time scale closed reaction networks whose fast subsystems have non-unique equilibrium probabilities [17]. They showed that if the system is a closed linear network, a reduced system under the QSSA has the form of an ODE system with nonhomogeneous terms regardless of its network topology. However, as mentioned in the paper, reduction methods on the Kolmogorov system may not be practical when one deals with a large reaction network since identification of all states and transitions in the network requires tremendous effort. For such networks, stochastic simulation algorithms should be utilized. The difficulty in utilizing SSA for such systems is that the propensity of the reduced master equation may not be found easily since the propensity under the QSSA is expressed in terms of equilibrium probability of the fast subsystem that is not uniquely determined. It is still an open problem how to obtain a reduced chemical master equation for such networks and to implement a modified form of the SSA from the master equation.

Acknowledgement

C.H. Lee acknowledges that this work was supported by the 2009 Research Fund of the UNIST(Ulsan National Institute of Science and Technology).

References

[1] D. A. McQuarrie. Stochastic approach to chemical kinetics. *J. Appl. Prob.*, **4**:413-478 (1967)

[2] D. T. Gillespie, A rigorous derivation of the chemical master equation. *Physica A* **188**:404-425 (1992)

[3] D. T. Gillespie, Exact simulation of coupled chemical reactions. *J. Phys. Chem.* **81**:2340-2361 (1977)

[4] D. T. Gillespie, Stochastic simulation of chemical kinetics. *Annu. Rev. Phys. Chem.*, **58**:35-55 (2007).

[5] J. A. M. Janssen. The elimination of fast variables in complex chemical reactions. II. Mesoscopic level (reducible case) *J. Stat. Phys.*, **57**:171-186 (1989).

[6] J. A. M. Janssen. The elimination of fast variables in complex chemical reactions. III. Mesoscopic level (irreducible case). *J. Stat. Phys.*, **57**:187-198 (1989).

[7] Eric L. Haseltine and James B. Rawlings. Approximate simulation of coupled fast and slow reactions for stochastic chemical kinetics. *J. Chem. Phys.*, **117**(15):6959 (2002).

[8] C. V. Rao and A. P. Arkin. Stochastic chemical kinetics and the quasi-steady state assumption: Application to the Gillespie algorithm. *J. Chem. Phys.*, **118**:4999 (2003).

[9] H. Salis and Y. Kaznessis. Accurate hybrid stochastic simulation of a system of coupled chemical or biochemical reactions. *J. Chem. Phys.*, **122**:054103 (2005).

[10] J. Goutsias. Quasiequilibrium approximation of fast reaction kinetics in stochastic biochemical systems. *J. Chem. Phys.*, **122**:184102 (2005).

[11] Y. Cao, D. T. Gillespie, and L. R. Petzold. The slow-scale stochastic simulation algorithm. *J. Chem. Phys.*, **122**:014116 (2005).

[12] E. Weinan, Di Liu, and Eric Vanden-Eijnden. Nested stochastic simulation algorithm for chemical kinetic systems with disparate rates. *J. Chem. Phys.*, **123**:194107 (2005).

[13] A. Hellander and P. Lötstedt. Hybrid method for the chemical master equation. *J. Comput. Phys.*, **227**:100-122 (2007).

[14] B. Munsky and M. Khammash, The finite state projection algorithm for the solution of the chemical master equation. *J. Chem. Phys.*, **124**:044104 (2006).

[15] S. Peleš, B. Munsky, and M. Khammash. Reduction and solution of the chemical master equation using time scale separation and finite state projection. *J. Chem. Phys.*, **125**:204104 (2006).

[16] C. H. Lee and R. Lui. A reduction method for multiple time scale stochastic reaction networks. *J. Math. Chem.*, **46**:1292-1321 (2009).

[17] C. H. Lee and R. Lui. A reduction method for multiple time scale stochastic reaction networks with non-unique equilibrium probability. *J. Math. Chem.*, **47**:750-770 (2010).

INDEX

A

B

C

D

E

F

G

H

O

P

T

U

V

W